Y0-BSD-790

Chemical Synthesis

NATO ASI Series

Advanced Science Institutes Series

A Series presenting the results of activities sponsored by the NATO Science Committee, which aims at the dissemination of advanced scientific and technological knowledge, with a view to strengthening links between scientific communities.

The Series is published by an international board of publishers in conjunction with the NATO Scientific Affairs Division

A Life Sciences **B Physics**	Plenum Publishing Corporation London and New York
C Mathematical and Physical Sciences **D Behavioural and Social Sciences** **E Applied Sciences**	Kluwer Academic Publishers Dordrecht, Boston and London
F Computer and Systems Sciences **G Ecological Sciences** **H Cell Biology** **I Global Environmental Change**	Springer-Verlag Berlin, Heidelberg, New York, London, Paris and Tokyo

PARTNERSHIP SUB-SERIES

1. Disarmament Technologies	Kluwer Academic Publishers
2. Environment	Springer-Verlag / Kluwer Academic Publishers
3. High Technology	Kluwer Academic Publishers
4. Science and Technology Policy	Kluwer Academic Publishers
5. Computer Networking	Kluwer Academic Publishers

The Partnership Sub-Series incorporates activities undertaken in collaboration with NATO's Cooperation Partners, the countries of the CIS and Central and Eastern Europe, in Priority Areas of concern to those countries.

NATO-PCO-DATA BASE

The electronic index to the NATO ASI Series provides full bibliographical references (with keywords and/or abstracts) to more than 50000 contributions from international scientists published in all sections of the NATO ASI Series.
Access to the NATO-PCO-DATA BASE is possible in two ways:

– via online FILE 128 (NATO-PCO-DATA BASE) hosted by ESRIN, Via Galileo Galilei, I-00044 Frascati, Italy.

– via CD-ROM "NATO-PCO-DATA BASE" with user-friendly retrieval software in English, French and German (© WTV GmbH and DATAWARE Technologies Inc. 1989).

The CD-ROM can be ordered through any member of the Board of Publishers or through NATO-PCO, Overijse, Belgium.

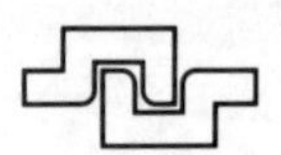

Series E: Applied Sciences - Vol. 320

Chemical Synthesis

Gnosis to Prognosis

edited by

Chryssostomos Chatgilialoglu
Consiglio Nazionale delle Ricerche,
I.Co.C.E.A.,
Bologna, Italy

and

Victor Snieckus
Department of Chemistry,
University of Waterloo,
Waterloo, Ontario, Canada

Kluwer Academic Publishers

Dordrecht / Boston / London

Published in cooperation with NATO Scientific Affairs Division

Proceedings of the NATO Advanced Study Institute on
Chemical Synthesis: Gnosis to Prognosis
Ravello, Italy
May 8–19, 1994

A C.I.P. Catalogue record for this book is available from the Library of Congress.

ISBN 0-7923-4041-8

Published by Kluwer Academic Publishers,
P.O. Box 17, 3300 AA Dordrecht, The Netherlands.

Kluwer Academic Publishers incorporates the publishing programmes of
D. Reidel, Martinus Nijhoff, Dr W. Junk and MTP Press.

Sold and distributed in the U.S.A. and Canada
by Kluwer Academic Publishers,
101 Philip Drive, Norwell, MA 02061, U.S.A.

In all other countries, sold and distributed
by Kluwer Academic Publishers Group,
P.O. Box 322, 3300 AH Dordrecht, The Netherlands.

Printed on acid-free paper

Printed in the Netherlands

Table of Contents

PREFACE

Chemical Synthesis: Gnosis to Prognosis

(Χημική Σύνθεση: από τη Γνώση στη Πρόγνωση)

"....other things being equal, that field has the most merit which contributes most heavily to, and illuminates most brightly, its neighbouring scientific disciplines[1]

One hundred scientists, a blend of students, industrialists, and academics from twenty countries gathered to circumscribe, understand, and elaborate this topic in the magical setting of Ravello, Italy. The mandate of this workshop? To survey existing knowledge, assess current work, and discuss the future directions of chemical synthesis as it impinges on three exciting interdisciplinary themes of science in the 1990's: bioactive molecules, man-made chemical materials, and molecular recognition. This tempting but inexact menu summoned diverse students and scientists who wished to seriously reflect upon, dissect, and eject ideas and own experiences into open debate on this topic, which is at a crossroad in internal evolution and impact on the life and material sciences.

The group arrived from many directions and in various forms of transportation, matters soon forgotten, when it found itself in the village which nurtured Wagner's inspiration and set to work immediately to ponder the question which has received extensive thought, prediction, and caveat from illustrious chemists over a period of time [2], two of which, to the delight of all, in presence among the Lectures.

During the intense ten days of activities, the three themes were addressed in a deliberate potpourri fashion which, from the outset, provided the desired response: enthusiastic, unbridled, and open-ended questions and discussion. It continued outside of the Chapel, at breakfast, over lunch and dinner, and was buoyantly evident in the garden coffee breaks hosted by the congenial staff and management.

Lectures on chemical synthesis at its fundamental levels, indicative of the vibrant health of the science, ranged from specific modern methods (Barton, Cainelli, Snieckus), to asymmetric catalysis (Brunner, Kagan), to target-oriented strategies (Hanessian). The disappearing biology-organic synthesis boundaries were clearly evident from contemplation of biomimetic catalysis (Breslow), design of enzyme mimics (Sanders) and inhibitors (Bartlett), catalytic antibodies (Hilvert), and a tour de force "why" topic (Eschenmoser). Molecular recognition and function delivered a range of stimulating salvos on receptors, self-assembly and -organizaton, and sequence-specific modification of nucleic acids (Hélène, Reinhoudt, Stoddart, Vögtle). The material science theme saw the interdisciplinary bridges shrink further with presentations par excellence on macroscopic non-covalent bond interactions (Lehn, Whitesides) and molecular electronic devices (Wrighton[3]). The sophisticated synthetic chemistry ongoing in the polymer arena was masterfully illuminated from perspectives of radical (Chatgilialoglu) and transition metal catalyzed (Ciardelli and Waymouth) reaction perspectives.

Three Panel Moderators posed provocative questions concerning the value of total synthesis in the 21st Century (Heathcock), the relationship between synthesis and the life sciences (Arigoni), and the construction of giant molecular modules (Michl). These sessions, unharassed by time limitations and suit-and-tie formality (and once assisted by rain), became playing fields for delving deeper into the three themes but never losing sight of prognosis. As hopefully evident from the taped transcripts, the multilogue ranged from serious to mirthful, all resolved to place quality substance in front of the audience and the readers of the workshop proceedings.

For this event, we warmly thank the Lecturers, the Panel Coordinators and Members, the Discussion Leaders, and, most of all, the participants. The Lecturers covered a broad range of subjects in excellent fashion, with careful attention to the diverse experience of the audience, including students. They orchestrated their talks with a view to establish interconnecting links among speakers in related areas, overcoming internal chemtalk barriers. The Discussion Leaders set the atmosphere by initiating and encouraging relaxed inquiry. The Panel Coordinators took full control of their difficult tasks, prepared an agenda, and catalyzed broad-ranging response; the Panel Members provided food for thought and unhesitently went into unknown territories for the same end result. Finally, all participants, by their questions, interaction, suggestions, and overall attentiveness made the our tasks easy and happy ones.

In summary, the Ravello NATO ASI was a very special workshop which we remember as an intense but collegial scientific discussion, a forum for interaction among individuals from broadly different cultures and languages, and the reinforcement or initiation of cameraderie which continues. We hope that this volume reflects all of these aspects, provides current status reports in diversified fields, and, most significantly, acts as a benchmark for the discipline of organic synthesis as it surveys and reaches towards its central contributions in the 21st Century.

Behind (usually in front) of all named conference organizers are the truly responsible individuals. From this viewpoint, we wish to thank most warmly Carla Ferreri and Barb Weber who, although distanced geographically, brought it to fruition. Another assumed task, the operation of projection, was carried out with great attention to detail by Massimo Capobianco, Thanasis Gimisis and Marco Lucarini. The Organizing Committee (H.J. Bestmann, A. Eschenmoser, C.H. Heathcock, and J-M. Lehn) not only reinforced our plans for this unusual program but also vigorously partook in discussion and panel deliberation. Our special thanks are due to NATO, Ente Provinciale del Turismo di Salerno and Area della Ricerca di Bologna (CNR) who kindly provided the financial support.

Chryssostomos Chatgilialoglu　　　　Victor Snieckus

1, Weinberg, A.M. *Minerva* **1963**, *2*, 159.

2 Inter alia, "Synthesis must always be carried out by plan, and the synthetic frontier can be defined only in terms of the degree to which realistic planning is possible, utilizing all of the intellectual and physical tools available." [Woodward, R.B. In *Perspectives in Organic Chemistry*; Todd, A.R., Ed.; Interscience: New York, 1956]. "What the synthetic target provides is stimulation and direction. Challenged by the necessity of moving towards a given structure, the synthesis often leads to problems that are complementary to those chosen in open-ended preparative research on reactivity." [Eschenmoser, A.; Wintner, C.E. *Science (Washington, D.C.)* **1977**, *196*, 1410]. We have travelled far since 1828 and the interest attached to 'total synthesis' has disappeared." [Robinson, R. *J. Chem. Soc.* **1936**, 1079]; "This statement was not really true in 1936. It will still not be true in a hundred years." G. Stork, Prix Roussel Lecture, 8 juin, 1978. "The best industrial synthesis may attain average yields of over 90% for a 30-40 step synthesis. It is clear, therefore, that the synthesis of natural products remains an objective of great scientific value and of great social and economic significance." [Barton, D.R.H. *Pure Appl. Chem.* **1977**, *49*, 1241]. "... it will still be the chemists skilled in synthesis who will succeed in preparing the most interesting targets and exploring the most challenging themes ..." [Seebach, D. *Angew. Chem. Int. Ed. Engl.* **1990**, *29*, 1320].

3. Wrighton, M. article not published.

DEDICATION

Dear Professor Prelog,

We were overjoyed to read your concurrence to be our Honorary Chairman but then downcast when we received your hesitant second letter. As the opening day approached, we realized and then accepted that we would not delight in the benefit of your broad knowledge, deep insight, and personal warmth and humor in a workshop at which all of these characteristics would have found expression.

> "... we believed that it was not worthwhile to compete..., but rather, given sufficient trust, that we should communicate our results to each other. Unnecessary duplication, priority arguments, and disappointments are thus avoided, and one also learns much from one's partners."[1]

Although to quote you this briefly is to do injustice, to quote is to convey to you and to the readers that the spirit of these words, in context of interaction and discussion, prevailed at Ravello. That spirit, we hope, is also transmitted in the essays and panel discussions of this volume where you will find themes which reflect your multifaceted achievements and interests in organic chemistry.

We did miss you. Because through your insight, the self-assembly *vs* self-recognition definition argument would have been clarified; through your wisdom, the uncertain impact of combinatorial chemistry would have received prognosis; through your humor, the repartee on the term "chemzymes" would have, we are confident, taken another twist.

We know you are well because we saw you at Linde Oberstrass near the ETH recently. And so we offer this volume, dear colleague, with the sincere wish that you will, without multimedia distraction, delight in the ongoing changes in organic synthesis for whose foundations you have been a major architect.

Chryssostomos Chatgilialoglu — Victor Snieckus

1. Prelog, V. *My 132 Semesters of Chemistry Studies*; Seeman, J.J., Ed.; American Chemical Society: Washington, D.C., 1991; p 37.

ASYMMETRIC SYNTHESIS :

I - FUNDAMENTALS AND RECENT ADVANCES
II - SOME ASPECTS OF ASYMMETRIC CATALYSIS WITH TRANSITION - METAL COMPLEXES

H. B. KAGAN
Institut Universitaire de France,
Institut de Chimie Moléculaire d'Orsay
Laboratoire de Synthèse Asymétrique
Université Paris-Sud, 91405 Orsay, France

ABSTRACT: The main approaches for the synthesis of enantiomerically enriched compounds are presented. Stoichiometric asymmetric synthesis is exemplified by some routes to chiral sulfoxides or chiral ferrocenes with planar chirality. Catalytic reactions using enantiomerically impure auxiliaries are described. Nonlinear effects, eg departure to proportionality between the ee's of product and chiral auxiliary are often observed and are discussed with a simple kinetic model.

I. FUNDAMENTALS AND RECENT ADVANCES

I.1. Introduction

Asymmetric synthesis has emerged as a major preparative method, widely used in organic chemistry and in the total synthesis of natural products[1], and which is also of interest for industrial chemistry[2]. The importance of enantiomerically pure compounds is connected with the applications in pharmaceutical industries, since very often the biological activity is strongly linked to the absolute configuration. In this article the historical developments of asymmetric synthesis will be briefly presented, as well as the main methods to prepare enantiomerically enriched compounds. Then recent asymmetric synthesis of two classes of compounds will be discussed : i) *Sulfoxides*, chiral at sulfur ii) *Ferrocenes with planar chirality*. The last part of the article will be devoted to *asymmetric catalysis* with transition-metal complexes. The cases of *asymmetric oxidation of sulfides* to sulfoxides and *nonlinear effects in asymmetric catalysis* will be mainly considered.

I.2. Historical background

The first discussions concerning the conditions for creating optically active compounds in the laboratory may be traced to Pasteur and Le Bel. The first mention of the expression "asymmetric synthesis" can be found in the work of E. Fischer in 1894 concerning his stereochemical studies on sugars[3,4]. He observed the formation of unequal amounts of cyanohydrins in the Kiliani reaction applied to aldehydic sugars. In 1904, Marckwald reported an asymmetric synthesis of 2-methyl-butanoic acid by decarboxylation of 2-ethyl 2-methyl-malonic acid in the presence of an alkaloid[5]. The importance and the mechanism of this reaction were later the subject of much debate. However, this paper remains significant because Marckwald defined asymmetric synthesis, as follows: "It is a reaction giving optically active products from symmetrical

C. Chatgilialoglu and V. Snieckus (eds.), Chemical Synthesis, 1–23.

compounds using optically auxiliaries but excluding any separation by analytical process". Even before 1950, one finds many reports on asymmetric synthesis and the name "asymmetric induction" was already defined as the phenomena which produces the stereochemical control. Amongst the main contributors in this area, one may quote A. M. Mc Kenzie, P. D. Ritchie, E. Erlenmeyer, G. Vavon, J. A. Mills, W. Doering, R. H. Baker, and H. Mosher. During the 1950 - 1960 period, conformational analysis began to be used, mainly by V. Prelog or D. Cram, for the interpretation of asymmetric induction in combination with steric effects. The next decade (1960-1970) saw the increase in reported asymmetric syntheses with a slow improvement in product ee's. It was also the beginning of asymmetric catalysis. Finally in the period 1970-1990 many useful asymmetric syntheses have been discovered, with ee's often higher than 90-95%, especially in asymmetric catalysis. The main results obtained before 1970 were reviewed in the book of Morrison and Mosher in 1971[6]. These authors defined asymmetric synthesis as "A reaction in which one achiral unit in an ensemble of molecules (substrate) is transformed into a chiral unit in such a way that stereoisomeric products (enantio or diastereo) are formed in unequal amounts". This definition allows to consider all the processes where stereoselectivity is involved, but for preparative work the definition of Marckwald remains valid. Modern asymmetric synthesis has been reviewed in many books or articles[7].

I.3. Main methods to prepare enantiomerically enriched compounds

I.3.1. Asymmetric synthesis is performed on an achiral substrate with the help of a chiral auxiliary which is temporarily bound to the substrate or which is external to it (in the reagent or the catalyst). The first case is named *diastereoselective asymmetric synthesis*, and the second one *enantioselective asymmetric synthesis* to define the product being formed as a mixture of diastereomers and a mixture of enantiomers, respectively[12]. The two processes are indicated in Figure 1.

Diastereoselective asymmetric synthesis

$$A \xrightarrow[Z^*]{i} Z^*\text{-}A \xrightarrow[B]{ii} Z^*\text{-}A^*\text{-}B \xrightarrow{iii} A^*\text{-}B + Z^*$$

Enantioselective asymmetric synthesis

$$A + B\text{-}Z^* \xrightarrow{iv} A^*\text{-}B + Z^*$$

$$A + B \xrightarrow[Cat\text{--}Z^*]{iv} A^*\text{-}B$$

A : achiral precursor (eg ketone)
B : reagent
Z* : chiral auxiliary
A* - B : chiral product (eg alcohol)
Z* - A* - B : chiral diastereomer (with high de)

i) Bonding of the chiral auxiliary
ii) Diastereoselective reaction
iii) Removal of the chiral auxiliary
iv) Enantioselective reaction

Principles of asymmetric synthesis

Figure 1.

Some achiral precursors in enantioselective asymmetric synthesis are indicated in Figure 2. In order to be used in diastereoselective asymmetric syntheses they need to bear a functional group able to give a temporary linkage to the chiral auxiliary. The asymmetric synthesis is a process of *dessymmetrization of an achiral precursor* (removal of all the elements of symmetry reflection, eg S_n elements).

I. No chiral subunits

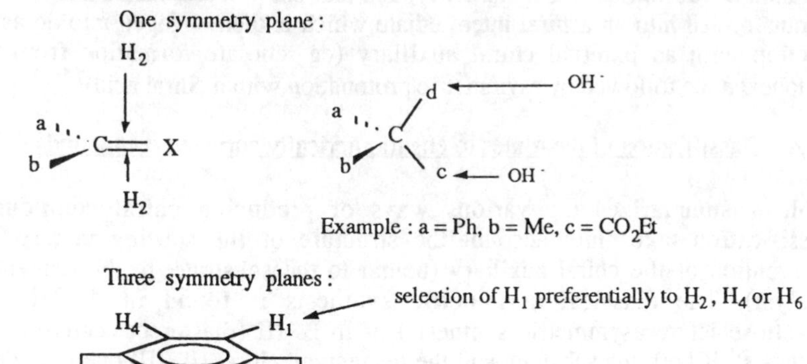

II. Pairs of chiral subunits of opposite absolute configuration

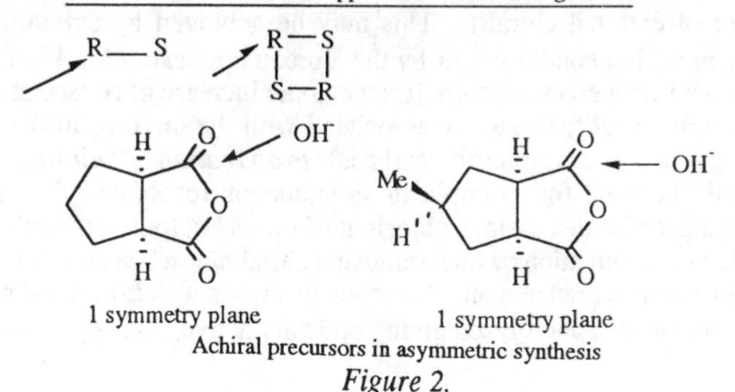

Achiral precursors in asymmetric synthesis

Figure 2.

I.3.2. Resolution

Resolution of a racemic mixture was discovered by Pasteur in the last century. It remains an useful method to prepare enantiomerically pure compounds, although the yield in the desired enantiomer cannot exceed 50%[13]. It is realized by the reaction of stoichiometric amounts of a chiral auxiliary which will produce a 1: 1 mixture of diastereomers, generally easy to separate. Removal of the chiral auxiliary generates the desired enantiomer. A special case of resolution is one in which the racemic compound crystallizes as a conglomerate. Here, a chiral seed can propagate the production of crystals of the same absolute configuration (spontaneous resolution)[13]. In the very rare case where an *in situ* racemization of the unwanted enantiomer may be combined with the above crystallization, one may, in principle, realize full transformation (*asymmetric transformation*) of the racemic starting material into one enantiomer. Asymmetric transformation of a racemic mixture into one enantiomeric product is also a possibility when a chiral auxiliary (external or internal) controls the stereoselectivity of a reaction on a racemic mixture in combination with a fast *in situ* racemization of the starting material. *Deracemization* is a closely related one-pot process, where a racemic mixture is transformed into an achiral intermediate which is then engaged into an asymmetric reaction with an external chiral auxiliary (eg. enolate formation from 1-methyl cyclohexanone followed by asymmetric protonation with a chiral acid)[14].

I. 3.4. Classification of the routes to enantiomerically enriched compounds

Table 1 summarizes the various ways of producing chiral compounds. The classification takes into account the structure of the starting material and the intervention of the chiral auxiliary (bound to the substrate, to the reagent, or to a catalyst). The classical asymmetric synthesis is found in A. III (Table 1, (enantioselective asymmetric synthesis) or in B. III (diastereoselective asymmetric synthesis). If both the substrate and the reagent are chiral (B. III) one may deal with a *double asymmetric induction*[15]. "Reagent control" could be very useful for the stereoselective introduction of asymmetric centers in a chiral molecule, thanks to the external chirality. A chiral precursor (from the chiral pool) is sometimes transformed into a more complex chiral molecule (the chiron approach)[17]; it is referred to as "building block" in entry B. Entries C and D make the distinction between a racemic or a partially resolved mixture, because, in the later case, it is possible to amplify the ee without the help of external chirality. This may be achieved by crystallization, by chromatography in achiral conditions, or by the Horeau duplication[16]. *Amplification of chirality* is not a well-defined expression. It refers to an increase of ee (see entries C and D), *multiplication of chirality* is usually associated with the increase in the amount of material with high ee (eg. asymmetric catalysis), *propagation of chirality* is also ill-defined and could be used for example in spontaneous resolution, for asymmetric catalysis or to characterize discussions on origin of optical activity on earth. *Transfers of chirality* are the transformations which remove a chiral unit while creating a new one. This is known for many rearrangements. It occurs also when a chiral crystal with achiral components (eg. alkenes) is photolyzed giving optically active [2+2] cycloadducts[18].

I. 3.5. Examples of high stereocontrol in the early asymmetric syntheses

The first enantioselective asymmetric synthesis with ee in the range of 90% was the asymmetric hydroboration of alkenes described in 1961 by H. C. Brown et al[19]. In this classic study (Figure 3), (R)-(-)2 butanol of 90% ee was isolated using commercial (+)-α-pinene of about 90% ee. More recently, the authors prepared enantiopure α-pinene and were able to produce 2-butanol with 98% ee. Asymmetric hydroboration became an established and an important class of asymmetric syntheses with very high ee's.

The first diastereoselective asymmetric synthesis with ee > 98% seems to be the asymmetric synthesis of aspartic acid of Horeau et al in 1968[20]. The chiral auxiliary was a chiral β-aminoalcohol.

TABLE 1. Preparation of chiral compounds

	Process / Starting Material	I No chemical reaction	II Chemistry without external chiral auxiliary	III Chemistry with external chiral auxiliary
A	Achiral precursor	Crystallization —> Chiral crystals (eg. glycine)		- Stoichiometric asymmetric synthesis - Catalytic asymmetric synthesis
B	Chiral precursor or achiral precursor bound to a chiral auxiliary		- Transfer of chirality (in solution or solid state) - Building block - Creation of chiral units	- Building block - Creation of chiral units
C	Racemic mixture	Resolution : - Crystallization (spontaneous resolution) - Chiral chromatography	Crystallization (spontaneous resolution) coupled with an *in situ* fast racemization	- Separation via diastereomers - Kinetic resolution with or without creation of new chiral units - Asymmetric Transformation - Deracemization
D	Partially resolved mixture	Resolution : - Crystallization - Chiral or achiral chromatography	Amplification of ee (eg. Horeau duplication)	- Separation via diastereomers - Kinetic resolution with or without creation of new chiral units - Asymmetric transformation - Deracemization

(R)-(-) 90% ee

Asymmetric hydroboration of alkenes (ref. 19)

Figure 3.

The three steps of this simple and efficient process are indicated in Figure 4, the actual conformations are not known. The two phenyl groups allow the smooth release of the product by a double hydrogenolysis, which unfortunately destroy the chiral auxiliary.

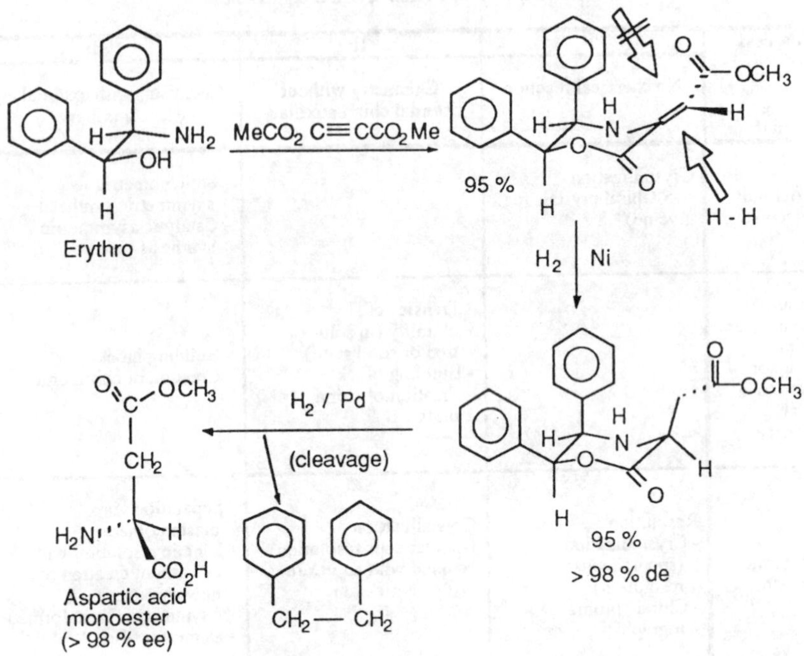

Asymmetric synthesis of aspartic acid (ref. 20)

Figure 4.

I. 3.6. Modern asymmetric synthesis

Many chiral auxiliaries are commercially available, allowing the execution of various types of asymmetric synthesis (ee > 90%) with recovery of the chiral auxiliary. This area has been extensively reviewed[7] and will not be discussed here because lack of place. As recent examples we will develop the case of the asymmetric synthesis of *sulfoxides* and of *ferrocenes with planar chirality. Asymmetric catalysis* will be considered in part II of this chapter.

I. 4. Asymmetric synthesis of sulfoxides

I. 4.1. Interest of chiral sulfoxide

Some chiral sulfoxides have biological activity associated with a given configuration at sulfur. Chiral sulfoxides may also be useful in material science (eg ferroelectric liquid crystals). However the main interest of these compounds is related to their usefulness as chiral auxiliaries in asymmetric synthesis[21-25]. The sulfinyl moiety increases the acidity of the α-hydrogens, allowing for facile formation of carbanions which can undergo asymmetric addition on aldehydes or imines. Several classes of vinyl sulfoxides are suitable substrates for Diels-Alder reactions and Michael additions. Allylic sulfoxides can rearrange into allylic alcohols or can be transformed into anions giving highly asymmetric 1,4 - additions to conjugated ketones. Acyclic β-keto sulfoxides are stereoselectively reduced into syn or anti- β-hydroxysulfoxides depending on the experimental conditions. These compounds are excellent starting materials in many total syntheses of natural products. The sulfinyl fragment is easily removed at a desired stage by reductive or oxidative methods, or by β-elimination.

I. 4.2. Main preparations of chiral sulfoxides

There are several ways to produce chiral sulfoxides. The resolution of a racemic mixture is seldom used. The main approach is the *Andersen method* which is based on the stereoselective transformation of a sulfinate deriving from a cheap alcohol, usually (-)-menthol[26] (Figure 5).

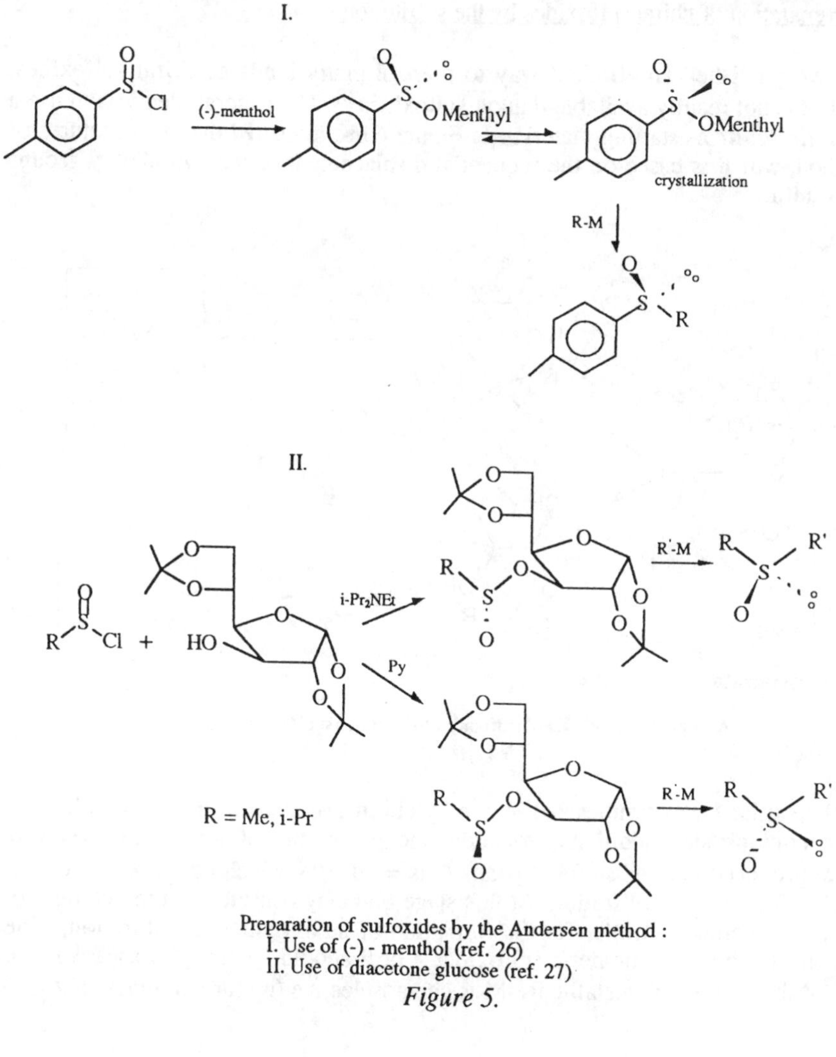

I.

(-)-menthol

O Menthyl

OMenthyl

crystallization

R-M

R

II.

i-Pr$_2$NEt

Py

R'-M

R = Me, i-Pr

Preparation of sulfoxides by the Andersen method :
I. Use of (-) - menthol (ref. 26)
II. Use of diacetone glucose (ref. 27)
Figure 5.

The preparation of menthyl sulfinate from menthol and R-S(O)Cl provides a mixture of epimers at sulfur, which has to be separated. In suitable cases (such as menthyl p-tolylsulfinate) one can combine a slow crystallization with an *in situ* epimerization catalyzed by HCl, giving an excellent yield of diastereoisomerically pure sulfinate[25]. Various types of organometallic reagents afford p-tolyl-sulfoxides (~100% ee) with a clean inversion of stereochemistry at sulfur. This method is especially convenient when diastereomeric sulfinates are easily separated by crystallization. The preparation of dialkyl sulfoxides is difficult by this method. A useful modification, recently described by Alcudia et al, involves the diastereoselective preparation of a methyl or n-propylsulfinate of diacetone glucose (Figure 5)[27]. Asymmetric oxidation of sulfides is a very direct route which, until recently, was unsatisfactory. In Section II, the asymmetric oxidation of sulfides using hydroperoxides in the presence of chiral titanium complexes is developed.

I. 4.3. Preparation of chiral sulfoxides by the sulfite route

In 1991 we published an efficient way to prepare many kinds of chiral sulfoxides, especially the not readily available dialkyl sulfoxides[28]. The general idea was to use a *chiral cyclic sulfite* as starting material. In Figure 6 are indicated the main features of this method, which is based on the sequential displacement of the two alkoxy groups bound to sulfur.

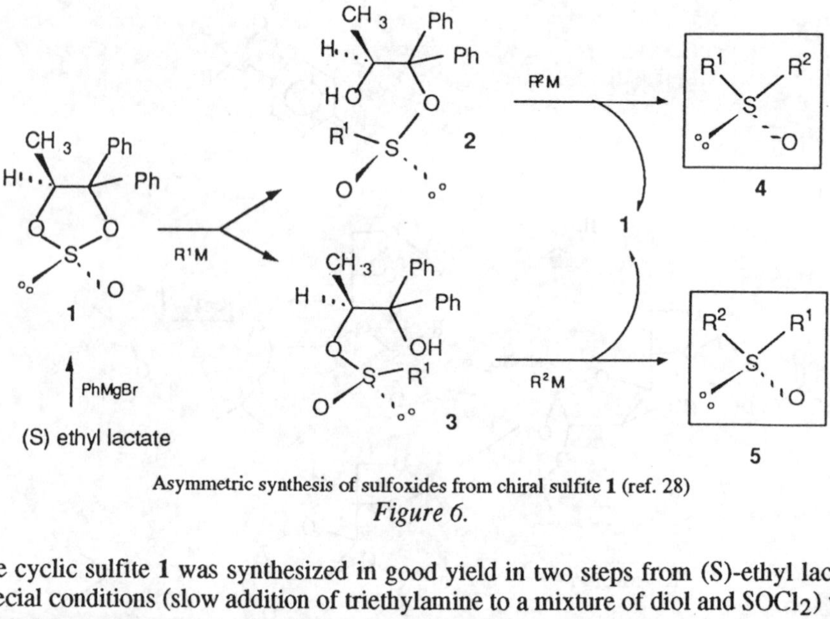

Asymmetric synthesis of sulfoxides from chiral sulfite **1** (ref. 28)

Figure 6.

The cyclic sulfite **1** was synthesized in good yield in two steps from (S)-ethyl lactate. Special conditions (slow addition of triethylamine to a mixture of diol and $SOCl_2$) were needed to produce the trans-sulfite **1** (trans / cis = 90 : 10) which could be isolated in 70% yield after one crystallization. At this stage one may consider that the chiral diol has acted as a chiral controller for the formation of an asymmetric sulfur atom. The conversion of **1** into enantiomeric sulfoxides **4** or **5** should be straightforward if the addition of the first organometallic R^1-M is chemoselective (no further attack of **2** or **3** by R^1M which will give symmetrical sulfoxide R^1-S(O)-R^1) and regioselective (preferred formation of sulfinate **2** or **3**) as well as stereoselective (inversion of stereochemistry at each step). The last condition is the easiest to assume since there are many examples in sulfur chemistry of nucleophilic substitutions ocurring with inversion. The chemoselectivity should be high with t-alkyl Grignard reagents since the analogous reaction with acylclic sulfites has been reported to give sulfinates, whreas the use of primary alkyl Grignards results in the formation of symmetrical sulfoxides. We found that cyclic sulfite **1** reacts with Grignard reagents such as t-BuMgCl or MesitylMgBr in THF at -78°C, leading to a large excess of the crystalline **2** (**2/3** = 90:10) from which diastereomerically pure **2** (R^1=t-Bu or R^1=Mes) could be isolated by crystallization in 70% yield. The sulfinates **2** are then treated by 2 equivalents of R^2M. The yield of this step (*cf* the Andersen method) is almost quantitative, giving sulfoxides **4** with ee ~ 100% (measured by ^1H nmr in the presence of a chiral shift reagent, by polarimetry or by chiral hplc). Some known chiral sulfoxides were obtained (eg t-Bu-S(O)-Ph or Mes-S(O)-Ph) as well as many new compounds. Taking into account the known absolute configuration of known sulfoxides **4**, one concludes that starting sulfite **1** has *trans* stereochemistry ((R) configuration at sulfur) if one assumes inversion of stereochemistry at each step. This has been confirmed by the X-rays crystal structures of **1** and **2** (R=t-Bu). It was unexpected and synthetically interesting to discover that Grignard reagents R^1M (with R^1 = Me or n-octyl) are also able to react with cyclic sulfites. Moreover, the sulfinates have structure **3**, isomeric with the cases where R^1=t-Bu or Mes. The regioselectivity of the attack is excellent (R^1= Me, **3** /**2** = 90 : 10, R^1 = n-octyl, **3** / **2** = 95:5). Permutation of the order of attack of the two organometallics allowed us to prepare the two enantiomers of methyl n-octyl sulfoxide (100%ee) from (S) - **1**. Some of the sulfoxides (100% ee) prepared by the sulfite route are listed in Figure 7. Various aspects of this method, including a mechanistic discussion, may be found in ref. 28.

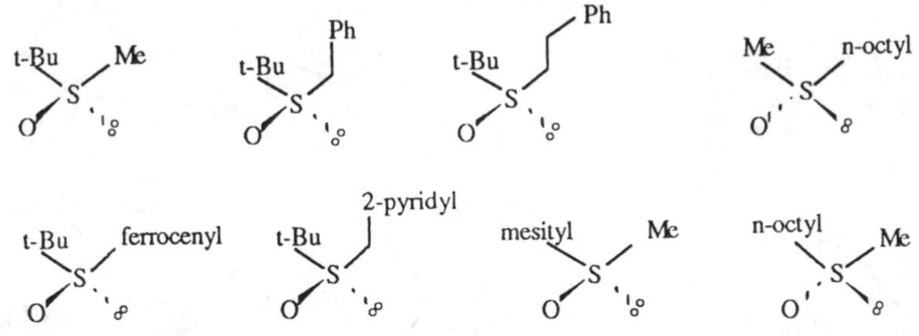

Some enantiomerically pure sulfoxides prepared from sulfite (S)-1 (ref 28)
R^1 and R^2 of Figure 6 are on the left and on the right respectively on the drawings
Figure 7.

I. 5 . Asymmetric synthesis of ferrocenes with planar chirality

I. 5.1. Interest of chiral ferrocenes

Ferrocenes become chiral upon attachment of a chiral unit to the ferrocene system or by intrinsic chirality (planar chirality) arising from the substitution pattern of ferrocenes as in **6-8**.

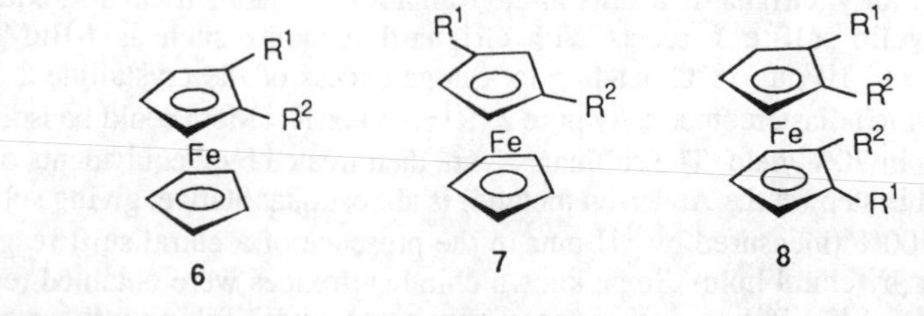

Such compounds are of interest as building blocks to prepare chiral auxiliaries, chiral ligands, biologically active compounds or materials.

I. 5.2. Preparations of chiral ferrocenes with planar chirality

The resolution of a racemic mixture by chemical or biochemical methods is the main route to ferrocenes with planar chirality. In pioneering work in 1970, Ugi et al resolved the ferrocenyl amine **9** (Figure 8) and found that the amino group directs the ortho-deprotonation with a good diastereoselectivity[29]. By this method many compounds were later prepared, including chiral ligands.

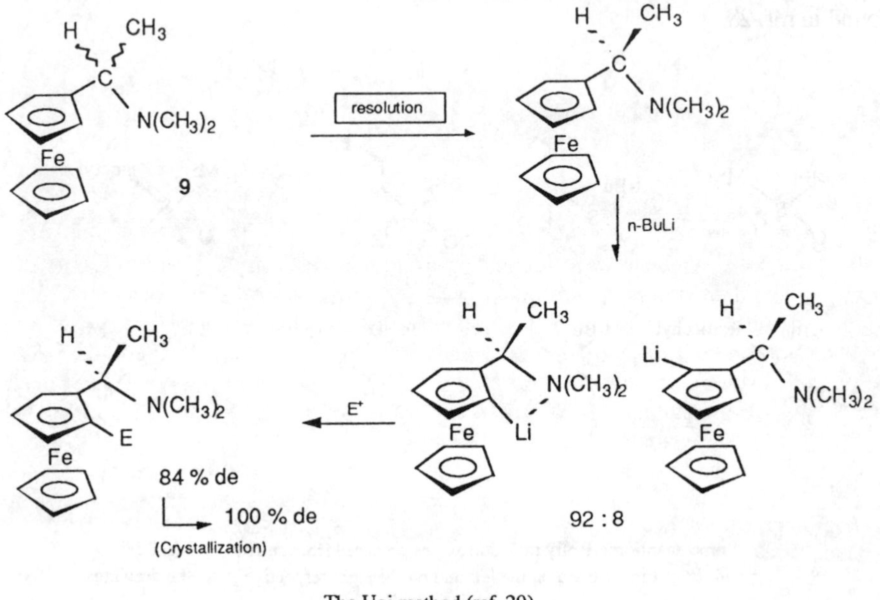

The Ugi method (ref. 29)

Figure 8.

We wished to explore the use of asymmetric synthesis for the preparation of ferrocenes bearing planar chirality. The main approaches are summarized in Figure 9. We attempted the route I, taking R=CH_2NMe_2 and using the combination sec-Buli / sparteine as the chiral base. Unfortunately a racemic ferrocenyl lithium compound was formed, as established by some electrophilic quenchings. We recently successfully developed route II *(vide infra)*.

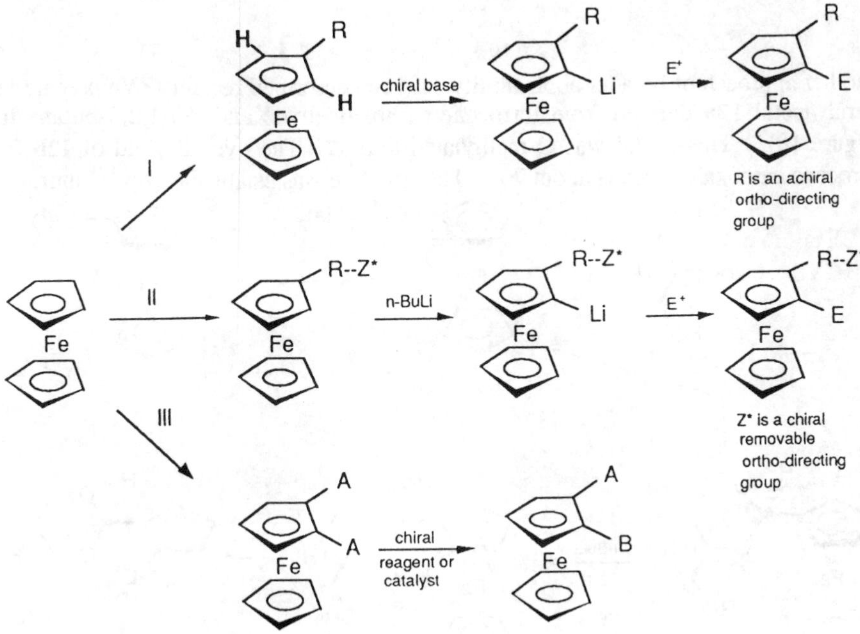

Some possibilities of asymmetric synthesis of ferrocenes with planar chirality

Figure 9.

I. 5.3. Ferrocenyl sulfoxides

The asymmetric synthesis of ferrocenyl sulfoxides is very easy to perform by the sulfite route or by the Andersen method[30]. The subsequent ortho deprotonation can be achieved with an almost perfect diastereoselectivity by the proper choice of the experimental conditions. The sulfite route allows the preparation of (S)-ferrocenyl t-butyl sulfoxide **10** (100% ee) in good yield. The assignement of (S) configuration was based on the mechanism of reaction (inversion of stereochemistry) and has been confirmed by X-ray analysis of **11**; the favored conformation forces oxygen to face one of the ortho sites and then orientes the lithiation. In order to be general the above asymmetric synthesis should be able to give access to ferrocenes with planar chirality and devoid of the sulfinyl group, we are working in this direction.

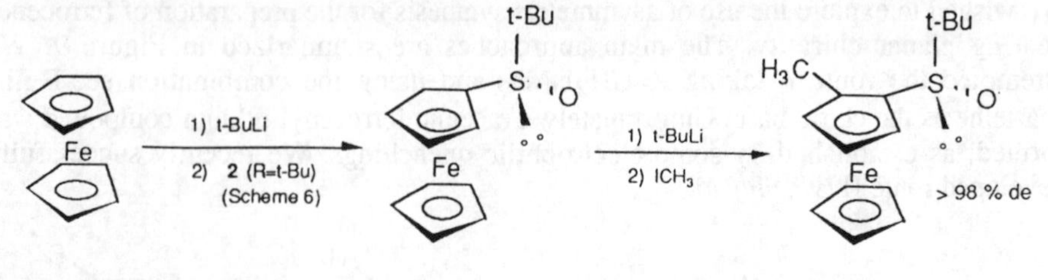

t-Bu t-Bu H$_3$C S O S O Fe 1) t-BuLi 2) 2 (R=t-Bu) (Scheme 6) Fe 1) t-BuLi 2) ICH$_3$ Fe > 98 % de

10 11

Another approach of broader applicability has been described recently. We prepared the chiral acetal **12a** derived from ferrocene carboxaldehyde and (S)-1,2,5-butane triol (Figure 10)[31] This acetal was O-methylated into **12b**.The overall yield of **12b** from ferrocene carboxaldehyde is about 70%. The structure was established by ^{1}H nmr.

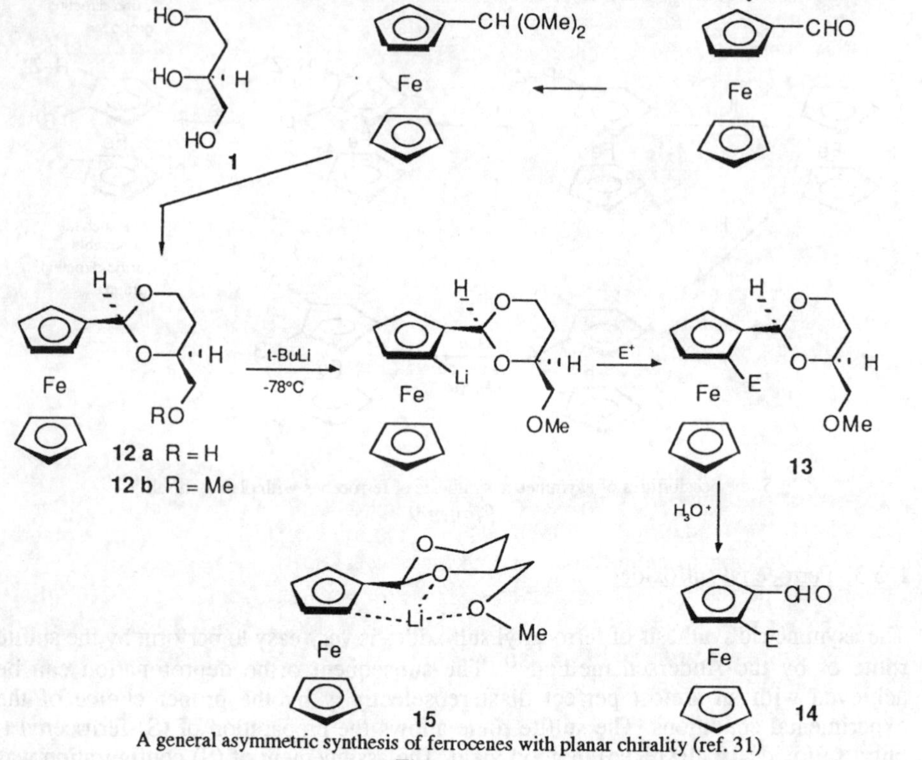

HO HO H HO 1 CH (OMe)$_2$ Fe CHO Fe

H O O H Fe RO 12 a R = H 12 b R = Me t-BuLi -78°C H O Li O H Fe OMe E$^+$ H O E O H Fe OMe 13 H$_3$O$^+$

O O Li O Me Fe 15 CHO E Fe 14

A general asymmetric synthesis of ferrocenes with planar chirality (ref. 31)

Figure 10.

We found that a clean ortho-deprotonation by t-Buli occurred at -78°C. The diastereoselectivity is almost complete since electrophilic quenching by various electrophiles (eg $ClSiMe_3$, $ClCO_2Me$, $ClPPh_2$) shows only one diastereomer in the crude product (^{1}H nmr). This has been confirmed by further transformation into aldehydes **14**. Some of these aldehydes (E=$SiMe_2$ or CO_2Me) have been already prepared. Chemical correlations indicated that the ee is approximately 100% (specific rotations), and that the ortho deprotonation occurred on the proS carbon atom, presumably through an organolithio compound where the methoxy group participates as depicted in **15**.

Chiral ferrocene carboxaldehydes **14** are now easily available in very high ee's, with predictable absolute configurations and are key compounds to reach a large range of ferrocenes with planar chirality.

II. ASYMMETRIC CATALYSIS WITH TRANSITION-METAL COMPLEXES

II.1. Introduction

Asymmetric catalysis is the most economical way to prepare chiral compounds if one takes into consideration the amount of chiral auxiliary which is used. Biochemical reactions are very efficient in many asymmetric syntheses; however, we will consider only the chemical approaches here. The 1970s saw the appearance of reports of asymmetric syntheses with ee's in the range of 80-90% derived through the use of organic or organometallic chiral catalysts. Transition-metal complexes bearing chiral ligands were especially successful for carrying out many asymmetric reactions. Hydrogenation of alkenes catalyzed by rhodium / diphosphine complexes[32-35] or codimerization of olefines catalyzed by nickel-monophosphine[36] complexes were the first examples where an appreciable amount of stereoselectivity could be achieved. There is now a great number of asymmetric catalytic reactions with ee's higher than 95%; and several reviews summarize this growing field[37-40]. Only two points will be discussed in this article: the *asymmetric oxidation of sulfoxides* and the *nonlinear effects* in asymmetric catalysis.

II. 2. Asymmetric oxidation of sulfides

In Part I, Section 4.2 the preparation of chiral sulfoxides was summarized. Asymmetric oxidation of sulfides by chemical means gave for a long time ee's lower than 10%. It was only ten years ago that new methods of oxidation of sulfides emerged, mainly the use of chiral oxaziridines[41] or of hydroperoxides in the presence of some chiral titanium complexes[42,43]. The Sharpless reagent for asymmetric epoxidation of allylic alcohols by hydroperoxides is a 1:1 combination of $Ti(OiPr)_4$ and diethyl tartrate (DET)[44]. Traces of water in the titanium complex catalyst are detrimental to high ee's; it was subsequently found that employment of molecular sieves protects the catalyst and restores the high enantioselectivity[45]. We discovered by sirendipity that addition of one equivalent of water deactivates the titanium complex for epoxidation of allylic alcohols, giving a new complex which efficiently mediates the asymmetric oxidation of sulfides into sulfoxides[42,46]. An optimization of the process led to use the combination $Ti(OiPr)_4$ / (R,R)-DET / H_2O=1:2:1, the oxidant being t-BuOOH. The reaction was run at -20°C in dichloromethane. The reagent was prepared at room temperature by first mixing $Ti(OiPr)_4$ and DET and then adding the water. It was hypothesized that water produced a partial hydrolysis of the titanium / DET complex into a bimetallic species with a μ-oxo Ti-O-Ti system. By this procedure many sulfides have been oxidized, some typical results are listed in Figure 11.

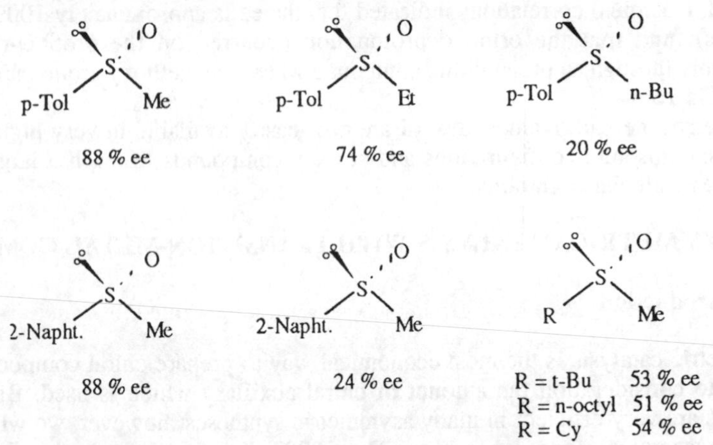

p-Tol S Me 88 % ee
p-Tol S Et 74 % ee
p-Tol S n-Bu 20 % ee

2-Napht. S Me 88 % ee
2-Napht. S Me 24 % ee
R S Me
R = t-Bu 53 % ee
R = n-octyl 51 % ee
R = Cy 54 % ee

Asymmetric synthesis of sulfoxides by oxidation of sulfides with the combination $Ti(OiPr)_4$ / (R,R)-DET / H_2O / t-BuOOH = 1 : 2 : 1 : 1 (refs. 42, 46)

Figure 11.

It is apparent that the highest ee's were obtained when the sulfur was flanked by an aryl group and a small alkyl group. This led to a simple model for prediction of absolute configuration (Figure 12). There are no exception to this rule, which may correspond to the oxidation via a binuclear titanium complex as represented in Figure 12 with diethyl tartrate acting as a tridentate unit.

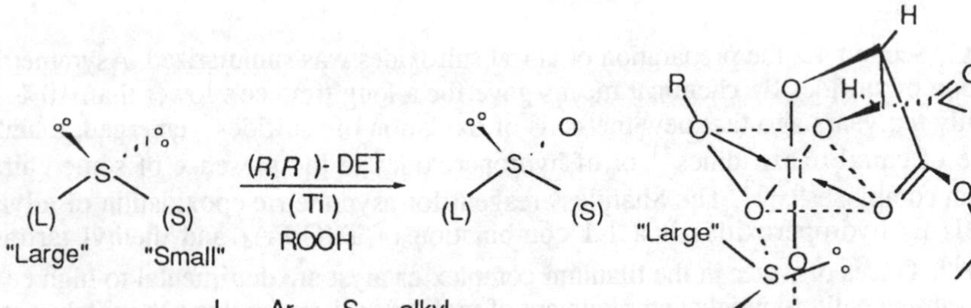

(L) "Large" (S) "Small"
(R,R)- DET (Ti) ROOH
(L) (S)
L = Ar S = alkyl
L = t-Bu S = n-alkyl
L = C≡C S = Me
"Large" "Small" [Ti

Prediction of the absolute configuration in asymmetric sulfoxidation (refs 46, 47)

Figure 12.

The efficiency of the enantioselective oxidation was significantly improved by replacing t-BuOOH by cumene hydroperoxide (CHP), eg using CHP, methyl tolyl sulfoxide and methyl n-octyl sulfoxide were formed in 90% ee and 80% ee (instead of

88% and 53% respectively with t-BuOOH). The reaction could be run catalytically (up to 0.15 equiv. of Ti complex).

This method has been widely applied, eg see refs[48-50]. We recently extended this method to the synthesis of *ferrocenyl sulfoxides*[51]. Initial erratic results obliged us to carefully define the experimental protocol. The isolation procedure (with purification of sulfoxides by flash-chromatography) will be discussed in the next paragraph (II.3.). Another important parameter is the preparation of the chiral titanium complex. We found that the temperature and time of premixing $Ti(OiPr)_4$ and DET before water addition and the ageing time are crucial for high enantioselectivity in the oxidation. It seems that many types of active titanium species can be formed, of which only one or a few number are stereoselective for the oxidation of a given sulfide. It is also beneficial to use an excess of hydroperoxide (2 equiv.) in order to increase the reaction rate (quite low for ferrocenyl sulfoxides). Some asymmetric syntheses of almost enantiopure ferrocenyl sulfoxides are now possible by this procedure (eg phenyl or p-tolyl ferrocenyl sulfoxide). The reagent was prepared by adding $Ti(OiPr)_4$ to DET at 27°C in CH_2Cl_2 maintaining the temperature for 3 min before the addition of water. After 20 min at 27°C the temperature is decreased to -23°C, before the addition of the sulfoxide and cumene hydroperoxide. This applies, with some modification, to the oxidation of aryl methyl sulfoxides, providing ee's in the range of 98-99%. It is hoped that the "tuning" of the titanium complex by a control of its preparation will allow to significantly increase the enantioselectivity of the oxidation of other classes of sulfides. The next goal will be to work under truly catalytic conditions (<5% equiv.) without the loss of enantioselectivity. Highly active catalysts (<1% equiv.) were recently described but enantioselectivity are at the moment lower than 90%[52].

II. 3. Enantiomeric fractionation by flash chromatography

Part of the irreproducibility of the above asymmetric oxidation of sulfides was due to the use of chiral hplc for the measurement of ee's (which can be performed on very small amounts of material). We found that the various fractions of sulfoxides collected by flash chromatography (used to remove sulfide, trace of sulfone and cumyl alcohol) have different ee's[53]. It is surprising that "achiral chromatography" involving silica gel and ethyl acetate is able to differentiate enantiomers of a solute. In fact enantiomeric fractionation by achiral chromatography has been already described by several authors[54-56]. These authors proposed to consider auto-association of the solute in the mobile and (or) the stationnary phase as the key factor leading to enantiomeric fractionation. A theoretical model has recently confirmed the validity of these hypotheses[57]. We found many cases of partially resolved sulfoxides for which the first fractions were highly enriched (till 99% ee), for two examples see Figure 13.

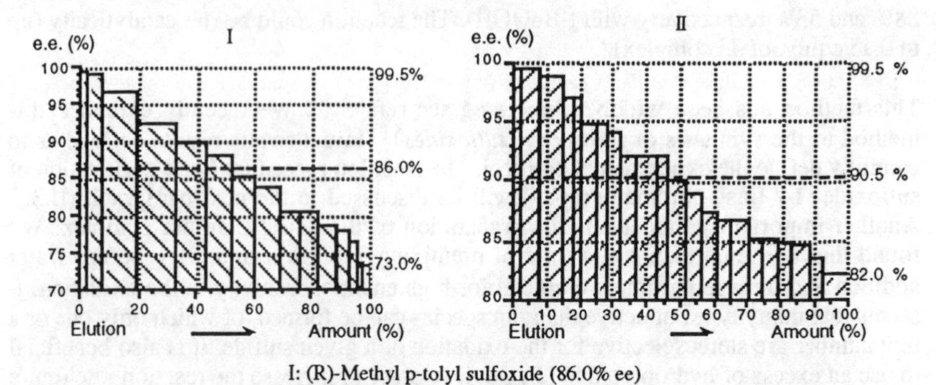

I: (R)-Methyl p-tolyl sulfoxide (86.0% ee)
II: (R)-Methyl ferrocenyl sulfoxide (90.5% ee)

Enantiomeric fractionation by flash-chromatography (ref.53)

Figure 13

When such a fractionation occurs in flash chromatography, it is important to mix together all the fractions before making an ee measurement.

The autoassociations which are responsible for the ee fractionation are for example, of the type $(\text{--R--R})_n$, $(\text{--S--S})_n$ or $(\text{--R--S--})_n$. The two first one are enantiomeric and have identical chromatographic behavior. The third agglomerate of racemic composition is a diastereomeric entity of the first two, and does not necessarily have the same properties. This can play a role not only in the chromatographic behavior but also in various processes unable to differentiate betwen enantiomers. Horeau showed that sometimes optical purity and enantiomeric excess are not linearly correlated because of association through hydrogen bonds[58]. Nmr spectra may show some difference between enantiomerically pure and partially resolved compounds. The general problem of diastereomeric interactions in a mixture of enantiomers has been discussed by Horeau and Guetté for physical phenomena[59], while Wynberg and Feringa pointed out their significancy in stereoselective transformations[60]. In the next section, we aim to show that the enantiomeric excess of a chiral ligand for asymmetric catalysis may give unexpected phenomena called *nonlinear effects*.

II. 4. Nonlinear effects in asymmetric catalysis

There are many examples in the literature where partially resolved chiral auxiliaries or chiral ligands were used for asymmetric synthesis. A correction was usually made by dividing the ee (%) of the product (EE_{prod}) by the ee (%) of the auxiliary (ee_{aux}) and multiplying by 100. The value thus obtained (EE_0) is the ee (%) predicted for the product prepared with enantiomerically pure chiral auxiliary or ligand. This calculation corresponds to equation [1] where all the enantiomeric excesses are taken with absolute values ≤1.

$$EE_{prod} = EE_0 \, ee_{aux} \qquad [1]$$

In 1986, in collaboration with Professor Agami, we criticized the foundations of this classical treatment[61]. Why must one exercise great caution in using equation [1]? The reason is easy to understand in the case of asymmetric catalysis by a complex carrying two chiral ligands. The M L_RL_R and M L_SL_S catalysts are mirror-images and will behave exactly the same, except for the absolute configuration of the product. In contrast, the "meso" ML_RL_S complex is a diastereomer of the previous two and will have a different reactivity, producing a racemic product. If the meso complex is unreactive and abundant, it will store some ligand of racemic composition, thereby increasing the enantiomeric excess of the remaining ligand which will be in the homochiral complexes (ML_RL_R or ML_SL_S). One then predicts a positive nonlinear effect ((+)-NLE), as shown by the curve 3 in Figure 14. The straight line 1 corresponds to equation [1] while curve 2 is a case of a negative nonlinear effect. A (-)-NLE is expected if the meso complex is abundant and more reactive than the homochiral complex.

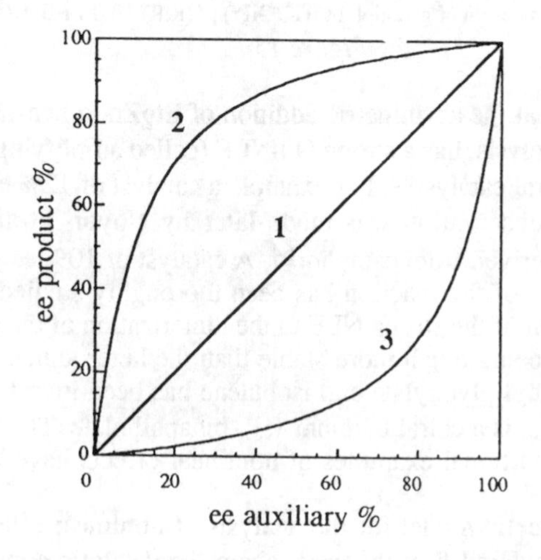

Positive nonlinear effects (curve 3) or negative nonlinear effect (curve 2) . Line 1 represents the proportionality between ee_{aux} and $ee_{product}$. It is arbitrarily assumed here that enantiopure chiral auxiliary gives 100% ee in the asymmetric synthesis (ref. 61).

Figure 14

In 1986, we discovered the first three cases of nonlinear effects in asymmetric synthesis: the Sharpless epoxidation of geraniol ((+)-NLE), the asymmetric oxidation of p-tolyl methyl sulfide by our titanium reagent ((-)-NLE), and the proline catalyzed asymmetric aldolization of a triketone ((-)-NLE). The mechanism of the last reaction was studied by Agami et al and found to be second-order with respect to proline[62].
In Figure 15, the NLE concerning the asymmetric epoxidation of geraniol is reproduced.

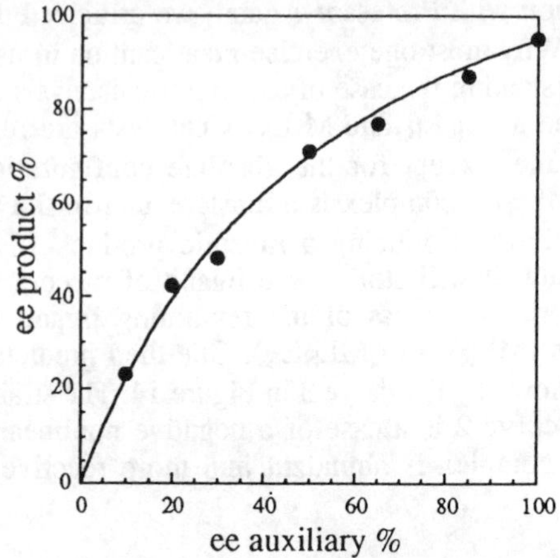

(+)-NLE in asymmetric epoxidation of geraniol by Ti((OiPr)$_4$ / (R,R)-DET / t-BuOOH = 1:1:2 (ref.61)

Figure 15

In 1988, Oguni found that the asymmetric addition of Et_2Zn to benzaldehyde, yielding PhCH(OH)Et after hydrolysis, has a strong (+)-NLE (called amplifying effect) with a β-aminoalcohol as the chiral catalyst[63]. For example a catalyst of 12% ee gives a product of 78% ee. A similar observation was made later by Noyori et al, who used a β-aminoalcohol (DAIB) derived from camphor[64]. A catalyst of 10% ee gave a product of 82% ee. The mechanism of the reaction has been thoroughly studied by Noyori et al. They attributed the origin of the strong NLE to the dimerization of the zinc derivative of DAIB, the meso dimer being much more stable than the heterodimer. The asymmetric ene reaction between ethyl glyoxylate and isobutene has been investigated by Mikami and Nakai[65]. The catalyst is a chiral titanium 1,1' -binaphtholate. These authors found a high (+)-NLE. Many additional examples of nonlinear effects have been reported, eg see refs 66-68.

We studied a simple kinetic model for the analysis of nonlinear effects and for some predictions[69]. We considered first the case where the catalytic complexes are of the type ML_2 (Figure 16). We called x, y, z, the relative amounts of the three complexes assumed to be in fast equlibrium compared to the reaction rates. The apparent rate constants were named $k_{RR} = k_{SS}$ and k_{RS}. The two enantiomers of the product are generated by three chanels. We retained the hypothesis that the chiral auxiliary L (of enantiomeric excess ee_{aux}) was entirely bound to the metal. Under these conditions, it is possible to calculate the ee of the product (EE_{prod}), which is given by equation [2] :

$$EE_{prod} = EEo\ ee_{aux} \frac{1+\beta}{1+g\beta} \qquad [2]$$

$$\beta = z / (x+y) \qquad g = K_{RS} / K_R$$

$$L_R + L_S + M \rightleftharpoons \rightleftharpoons \rightleftharpoons ML_RL_R + ML_SL_S + ML_RL_S$$

ML_RL_R (x)	ML_SL_S (y)	ML_RL_S (z)
$k_R \downarrow$	$k_S \downarrow$	$k_{RS} \downarrow$
Products E_o	Products E_o	Products
[R]>[S]	[R]<[S]	[R]=[S]

$$\underset{x}{ML_RL_R} + \underset{y}{ML_SL_S} \rightleftharpoons \rightleftharpoons 2\,\underset{z}{ML_RL_S} \qquad K = z/xy$$

Figure 16

The parameters in equation [2] are defined in Figure 16. If g = 1 (same rate constants for the meso complex and the homochiral complexes) or β=0 (no meso complex), the fraction becomes equal to 1 and equation [2] is transformed into equation [1]. In all the other cases there will be a NLE. In order to compute and to draw the curves $EE_{prod} = f(ee_{aux})$ it is necessary to evaluate β as a function of ee_{aux} and of K. This is possible and gives an analytical solution which will be not reported here. The special case where there is statistical distribution of ligands L_R and L_S into the three complexes corresponds to K=4. The curves for various values of g are plotted in Figure 17. It is apparent that the smaller the reactivity g of the meso complex, the larger the amplification effect ((+)-NLE). Correspondingly, the higher the g, the larger the (-)-NLE. For values of K much higher than 4, indicating a large amount of the meso complex, the positive NLE increases, giving a very sharp amplification.

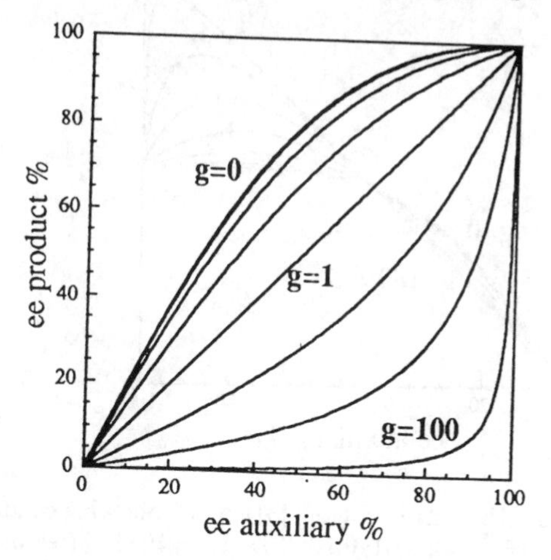

Computed-drawn curves for ML_2 catalysts in the following condition : K=4 (statistical distribution of ligand) g=0; 0.01; 0.10; 0.33; 1; 3; 10; 100 (ref. 69)

Figure 17.

Similar calculations were performed with ML_3 and ML_4 complexes. The degree of complexity increases with the number of ligands, resulting in diversity of curve shapes. For example, the ML_3 system generates four complexes: $ML_RL_RL_R$, $ML_SL_SL_S$, $ML_RL_RL_S$ and $ML_SL_SL_R$. The same hypotheses and parameters as for the ML_2 system were taken, ie a fast equilibrium between the four complexes (K is the equilibrium constant between homochiral and heterochiral complexes), β (relative amounts of heterochiral and homochiral complexes) and g (ratio of apparent first-order rate constants of heterochiral and homochiral complexes). The calculation is difficult, but simplification occurs by selecting the case where there is a statistical distribution of ligands L_R and L_S between the four complexes. One finds that the enantiomeric excess of the product is given by equation [3].

$$EE_{prod} = EE_0\, ee_{aux} \frac{3 + 3g\, EE'_0 / EE_0 + ee^2_{aux} (1 - 3g EE'_0 / EE_0)}{1 + 3g + 3ee^2_{aux} (1-g)} \qquad [3]$$

In this equation one finds EE_0 and EE'_0 which are the ee's given by the enantiomerically pure homochiral and heterochiral complexes respectively. Here the heterochiral complex is not a meso structure and will also provide a chiral product. Numerical simulations show interesting features. If $|EE_0| > |EE'_0|$ the curves drawn for various g values have shapes similar to curves of ML_2 system (Figure 18). When $|EE'_0| > |EE_0|$ one finds examples where EE_{prod} is *improved if ee_{aux} is lower than 100%* (Figure 18).

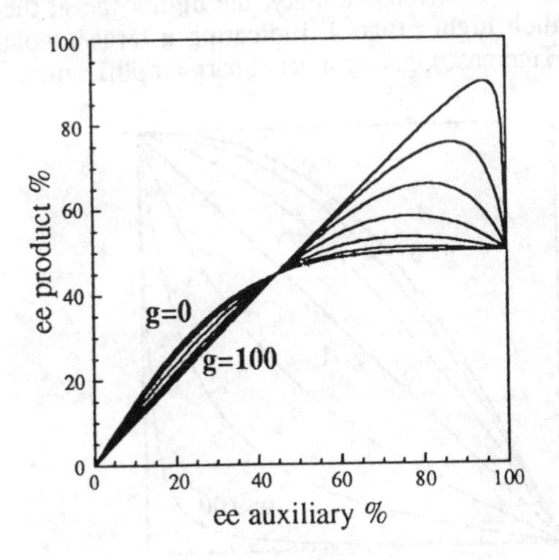

Computed-drawn curves for ML_3 catalysts (equ. [3]) in the following condition : EE_0 = 50%, EE_0' = 100% ; g=0; 0.01; 0.10; 0.33; 1; 3; 10; 100 (ref. 69)

Figure 18

This is an unconventional way to optimize the efficiency of an asymmetric synthesis. This prediction has not yet been verified experimentally; however, it finds analogy in the optimal biological activity which is not always related to 100% ee for the drug or the compound (eg some insect pheromones). This arises from a multi-sites biological

response. Calculations with ML_4, or more generally, ML_n systems has been also carried out. The curves $EE_{prod} = f\ (ee_{aux})$ could be drawn. With ML_4 system it is possible to get multi-shapes curves.
In conclusion, enantiomeric pure ligands of opposite absolute configuration will give similar catalytic systems apart from the absolute configuration. When the ligand is a mixture of enantiomers additional diastereomeric complexes usually occur, inside or outside the catalytic cycle, which may generate a nonlinear effect. The above discussion has been focused on ML_n species. This is a symbolical representation, what is important is the number of ligands in a molecular entity (ML_2 is equivalent to $(ML)_2$ from this point of view). The model developed here is very simplified, for example it assumes that there is no modification of the catalyst by interaction with the product, as sometimes observed[70]. A nonlinear effect makes extrapolation of the product ee based on only one experiment a dangerous undertaking for ligands which are not 100 % ee. The presence of a NLE is also informative for mechanistic discussion. A strong (+)-NLE is a process of "amplification" of the ee which can give rise to prebiotic models of propagation of optical activity on earth (combined to an autocatalytic process).

II. 5. Conclusions

Asymmetric catalysis with transition-metal complexes is becoming a powerful method for the preparation of enantiomerically enriched compounds with very high ee's. It stimulates the understanding of many catalytic systems and the preparation of new mono and polydentate chiral ligands. Efficient asymmetric catalyses are also possible with chiral organic catalysts (amines, amino acids etc.) but examples are more limited. Nonlinear effects are very informative of diastereomeric species, even those arising by formation of agglomerates. We did not discuss here the many applications of enzymatic reactions to asymmetric synthesis. A new area is now of growing interest, that of catalytic antibodies, which is expected to generate tailor-made asymmetric catalysts for a given reaction.

REFERENCES

1. a) Morrison, J. D. (ed) (1982-1989), *Asymmetric Synthesis*, **1-5**, Academic Press, New York.
   b) Seebach, D. (1990) Organic synthesis-where now?, *Angew. Chem. Int. Ed.* **29**, 1320-1367.
   c) Noyori, R. (1994) *Asymmetric Catalysis in Organic Synthesis*, J. Wiley, New York.
2. a) Kagan H. B. (1988) Asymmetric catalysis in organic synthesis with industrial perspectives, *Bull. Soc. Chim. Fr.*, 846-853.
   b) Crosby, J. (1991) Synthesis of optically active compounds: a large scale perspective, *Tetrahedron* **47**, 4789-4846.
   c) Nugent, W. A., Rajan Babu, T. V. and Burk, M. J. (1993) Beyond Nature"s chiral pool: enantioselective catalysis in industry, *Science* **259**, 479-483.
3. Fischer, E. (1894) *Ber. Dtsch. Chem. Ges.* **27**, 3231.
4. Lichtenhaler, F. W. (1992) Emil Fischer's proofs of the configuration of sugars: a centenial tribute *Angew. Chem. Int. Ed. Engl.* **32**, 1541.
5. Marckwald, W. (1904), *Ber. Dtsch. Chem. Ges.* **37**, 1368.
6. Morrison J. D. and Mosher H. S. (1971) *Asymmetric Organic Reactions*, Prentice Hall, Englewoods Cliffs, NJ.
7. For example see refs. 1-5, 8-11
8. Izumi, Y.and Tai, A.(1977) *Stereo-differentiating reactions*, Kodansha and Academic Press, New York.
9. Kagan, H. B. and Fiaud, J. C. (1978) New approaches in asymmetric synthesis, *Topics in Stereochem.* **10**, 175-285.

10. Eliel, E. L. and Otsuka, S. (1982) *Asymmetric reactions and processes in chemistry*, ACS Symposium Series, Washington D.C.
11. Enantioselective synthesis (1992) *Chem. Rev.* **92**, 739-1140
12. Izumi , Y. (1971) Methods of asymmetric synthesis-Enantioselective catalytic hydrogenation, *Angew. Chem. Int. Ed. Engl.* **10**, 871-881.
13. Jacques J. and Collet A. (1981) *Enantiomers, Racemates and Resolutions*, Wiley Interscience, J. Wiley , New York.
14. Duhamel, L. and Plaquevent, J. C. (1978) Deracemization by enantioselective protonation. A new method for the enantiomeric enrichment of α-amino acids, *J. Am. Chem. Soc.* **100**, 7415-7416.
15. Horeau A., Kagan, H. B. and Vigneron, J. P. (1968) Synthèses asymétriques par double induction, *Bull. Soc. Chim. Fr.* , 3795-3797.
16. Masamune, S., Choy, W., Petersen J.S. and Sita, L.R. (1985) Double asymmetric synthesis and a new strategy for stereochemical control in organic synthesis, *Angew. Chem. Int. Ed. Engl.* **24**, 1-30.
17. Hanessian S. (1983) *Total synthesis of natural products : the "Chiron approach"*, Pergamon Press, New York.
18. Green, B. S., Lahav, M. and Rabinovitch, (1979) Asymmetric synthesis via reaction in chiral crystals, *Acc. Chem. Res* **12**, 191-197.
19. Brown H. C., and Zweifel, G. (1961) Hydroboration as a convenient procedure for the asymmetric synthesis of alcohols of high optical purity, *J. Am. Chem. Soc.* **83**, 486-487.
20. Vigneron, J. P., Kagan, H. B. and Horeau A. (1968) Synthèse asymétrique de l'acide aspartique optiquement pur, *Tetrahedron Lett.* 5681-5683.
21. Solladié, G. (1981) Asymmetric synthesis using nucleophilic reagents concerning a chiral sulfoxide group, *Synthesis*, 185-196.
22. Barbachyn, M. R. and Johnson, C. R. (1984) Optical activation and utilization of compounds containing chiral sulfur centers, in Morrison, J. D. (ed.) *Asymmetric Synthesis*, **4**, pp. 227-261, Academic Press, New york.
23. Mikolajczyk, M. and Drabowicz, J. (1982), *Top. Stereochem.* **13**, 333.
24. Posner, G. H. (1988) Asymmetric synthesis using α-sulfinyl carbanions and β-unsaturated sulfoxides, in Patai, S., Rappoport, Z. and Stirling, C. J. M. (eds), *The Chemistry of Sulphones and Sulphoxides*, Chapter 16, pp. 823-848.
25. Drabowicz, J., Kielbasinski, P. and Mikolajczyk, M., (1988) in *The Chemistry of Sulphones and Sulphoxides*, Patai, S. Rappoport, Z. and Stirling, C. J. M. (eds), Chapter 8, pp. 233-253.
26. Andersen, K. K. (1988) Stereochemistry, conformation and chiroptical properties of sulfoxides, in Patai, S., Rappoport, Z., Stirling, and C. J. M. (eds), *The Chemistry of Sulphones and Sulphoxides*, Chapter 3, pp. 55-92.
27. Fernandez, I., Khiar, N., Llera, J. M. and Alcudia, F. (1992) Asymmetric synthesis of alkane- and arenesulfinates of diacetone-D-glucose: an improved and general route to both enantiomerically pure sulfoxides, *J. Org. Chem..* **57**, 6789-6796.
28. Rebière, F.; Samuel, O.; Ricard, L. and Kagan, H. B. (1991) A general route to enantiomerically pure sulfoxides from a chiral sulfite, *J. Org. Chem.* **56**, 5991-5999.
29. Marquarding, D., Klusacek, H.;,Gokel, G., Hoffmann, P. and Ugi, I.(1970) Correlation of central and planar chirality in ferrocene derivatives , *J. Am. Chem. Soc.* **92**, 5389-5393.
30. Rebière, F., Riant, O.;, Ricard, L., Kagan, H. B. (1993) Asymmetric synthesis and highly diastereoselective ortho-lithiation of ferrocenyl sulfoxides. Application to the synthesis of ferrocenyl derivatives with planar chirality., *Angew. Chem. Int. Ed. Eng.* **32**, 568-570.
31. Riant, O.; Samuel, O. and Kagan, H. B. (1993) A general asymmetric synthesis of ferrocenes with planar chirality, *J . Am. Chem. Soc.* **115**, 5835-5836.
32. Kagan H. B. and Dang, T. P. (1972) Asymmetric catalysis with transition metal complexes I. A catalytic system of rhodium (I) with a new chiral diphosphine, *J. Am. Chem. Soc* . **94**, 6429.
33. Vineyard, B. D., Knowles, W. S., Sabacky, M. J., Bachman, G. L. and Weinkauff, D. J. (1977) Asymmetric hydrogenation. Rhodium chiral biphosphine catalyst, *J. Am. Chem. Soc.* **99**, 5946-5952.
34. Achiwa, K. (1976) Asymmetric hydrogenation with new chiral functionalized biphosphine- rhodium complexes, *J. Am. Chem. Soc.* **98**, 8265-8266.
35. Fryzuk, M. D. and Bosnich, B. (1977) Asymmetric synthesis of optically active amino acids by catalytic hydrogenation, *J. Am. Chem. Soc.* **99**, 6262-6267.
36. Bogdanovic, B., Henc, B., Tösler, A.; Meister, B., Pauling, H. and Willke, G. (1973) Asymmetric synthesis with the aid of homogeneous transition-metal catalysts, *Angew. Chem. Int. Ed. Engl.* **12**, 954-964.
37. Ojima, I. (1993) *Catalytic Asymmetric Synthesis*, VCH, New York.
38. Bosnich, B. (1986) *Asymmetric Catalysis* , NATO ASI Series, Nijhoff, Dordrecht, .
39. Morrison, J. D. (1985) Asymmetric Catalysis, *Asymmetric Synthesis*, **5**, Academic Press, New York.
40. Kagan, H. B. (1982) Asymmetric synthesis using organometallic catalysts, in Wilkinson, G. (ed), *Comprehensive Organometallic Chemistry* **8**, pp.483-498
41. Davis, F. A., Mc Caulley, J. P. Jr and Harakal, M. E. (1984) Chiral sulfamides: optically active 2-sulfamyloxaziridines. High enantioselectivity in the asymmetric oxidation of sulfides to sulfoxides, *J. Org. Chem..* **49**, 1465-1467.

42. Pitchen, P., Dunach, E., Deshmukh, M. N. and Kagan, H. B. (1984) An efficient asymmetric oxidation of sulfides to sulfoxides, *J. Am. Chem. Soc.* **106**, 8188-8193.
43. Di Furia, F., Modena, G. and Seraglia, R. (1984) Synthesis of chiral sulfoxides by metal-catalyzed oxidation with t-butyl hydroperoxide, *Synthesis*, 325-326.
44. Katsuki, T. and Sharpless, K. B. (1980) The first practical method for asymmetric epoxidation, *J. Am. Chem. Soc.* **102**, 5974-5976.
45. Hanson, R. M. and Sharpless, K. B. (1986) Procedure for the catalytic asymmetric epoxidation of allylic alcohols in the presence of molecular sieves, *J. Org. Chem.* **51**, 1922-1925.
46. Kagan, H. B. and Rebière, F. (1990) Some routes to chiral sulfoxides of very high enantiomeric excesses, *Synlett*, 643-650.
47. Zhao, S., Samuel, O. and Kagan, H. B. (1987) Asymmetric oxidation of sulfides mediated by chiral titanium complexes: mechanistic and synthetic aspects, *Tetrahedron* **43**, 5135-5144.
48. Beckwith, A. L. J. and Boate D. R. (1986) Stereochemistry for intramolecular homolytic substitution at the sulphur atom of a chiral sulphoxide, *J. Chem. Soc. Chem. Commun.* 189-190.
49. Davis, R., Kern, J. R., Kurtz, L. J. and Pfister, J. R. (1988) Enantioselective synthesis of dihydropyridine sulfone, *J. Am. Chem. Soc.* **110**, 7873-1874.
50. Tanaka, J., Higa, T., Bernardinelli, G. and Jefford, C. W. (1989) Sulfur-containing polybromoindoles from the red alga Laurencia Brongniartii, *Tetrahedron* **45**, 7301-7310.
51. Diter, P., Samuel, O., Taudien, S. and Kagan, H. B. (1994) Highly enantioselective oxidation of ferrocenyl sulfides, *Tetrahedron:Asymmetry* **5**, 549-552.
52. Kagan H. B. (1993), Aymmetric oxidation of sulfides, in Ojima I. (ed.), *Catalytic Asymmetric Synthesis*, VCH, New York , pp.202 - 226.
53. Diter, P., Taudien ,S.; Samuel, O. and Kagan, H. B. (1994) Enantiomeric enrichment of sulfoxides by preparative flash chromatography on an achiral phase, *J. Org. Chem.* **59**, 370-373.
54. Cundy, K. C. and Crooks, P. A. (1983) Unexpected phenomenon in the HPLC analysis of racemic ^{13}C labelled nicotine. Separation of enantiomers in a totally achiral system, *J. Chromatogr.* **281**, 17-33.
55. Charles, R. and Gil -Av, E. (1984) Self-amplification of optical activity by chromatography on an achiral adsorbent, *J. Chromatogr.* **298**, 516-520.
56. Tsai, W. L., Hermann, K., Hug, E., Rohde, B. and Dreiding, A. (1985) Enantiomer-differentiation induced by an enantiomeric excess during chromatography with achiral phases, *Helv. Chim. Acta* **68**, 2238-2243.
57. Jung, M. and Schurig, V. (1992) Computer simulation of three scenarios for the separation of non-racemic mixtures by chromatography on achiral stationary phases, *J. Chromatogr.*, 161-166.
58. Horeau, A. (1969) Interaction d'énantiomères en solution: influence sur le pouvoir rotatoire, pureté optique et pureté énantiomérique, *Tetrahedron Lett.* 3121-3124.
59. Horeau, A. and Guetté, J. P. (1974) Intéractions diastéréoisomères d'antipodes en phase liquide, *Tetrahedron* **30**, 1923-1931.
60. Wynberg, H. and Feringa, B. (1976) Enantiomeric recognition and interactions, *Tetrahedron* **32**, 2831-2834.
61. Puchot, C., Samuel, O., Dunach, E., Zhao, S., Agami, C. and Kagan, H. B. (1986) Nonlinear effects in asymmetric synthesis. Examples in asymmetric oxidation and in asymmetric aldolization reactions, *J. Am. Chem. Soc.* **108**, 2353-2357.
62. Agami, C. (1988) Mechanism of the proline-catalyzed enantioselective aldol reaction, recent advances, *Bull. Soc. Chim. Fr.*, 499-507 and references quoted therein.
63. Oguni, N., Matsuda, Y. and Kaneko, T. (1988) Asymmetric amplifying phenomena in enantioselective addition of diethylzinc to benzaldehyde, *J. Am. Chem. Soc.* **110**, 7877-7878.
64. Noyori, R. and Kitamura, M. (1991) Enantioselective addition of organometallic reagents to carbonyl compounds: chirality, transfer, multiplication and amplification, *Angew. Chem. Int. Ed. Engl.* **30**, 49-69.
65. Mikami, K., Terada, M., Narisawa, S. and Nakai, T. (1992) Asymmetric catalysis for carbonyl-ene reactions, *Synlett*, , 255-265.
66. Tanaka, K., Matsui, J., Kawabata, Y., Suzuki, H. and Watanabe, A. (1991) Chiral amplification i n the synthesis of (R)-muscone by conjugate addition of chiral alkoxydimethylcuprate to (E)-cyclopentadec-2-one, *J. Chem. Soc. Chem. Commun* , 1631-1634.
67. Zhou, Q. L. and Pfaltz A. (1994) Chiral mercaptoaryl-oxazolines as ligands in enantioselective copper-catalyzed 1,4-additions of Grignard reagents to enones, *Tetrahedron* **50**, 4467-4478.
68. De Vries, A. H. M., Jansen, J. P. G. A. and Feringa, B. L. (1994) Enantioselective conjugate addition of diethylzinc to chalcones catalyzed by chiral Ni(II) aminoalcohols complexes, *Tetrahedron* **50**, 4479-4491
69. Zhao, S. H., Samuel, O., Rainford, D., Guillaneux, D. and Kagan, H. B. (1994) Nonlinear effects in asymmetric catalysis, *J. Am. Chem. Soc.*, **116**, 9430-9439.
70. Albert, A. H. and Wynberg, H. (1990) Enantioselective autoinduction in the aldol condensation of ethyl acetate and benzaldehyde: selective precipitation of an optically inactive Li-O aggregate, *J. Chem. Soc. Chem. Commun.*, 453-454.

Addition of Enolates and Metalloalkyls to Imines. Stereospecific Synthesis of β-Lactams, Amines, Aziridines and Aminols

G.CAINELLI*, M.PANUNZIO, D.GIACOMINI,

Dipartimento di Chimica "G.Ciamician" and C.F.S.M.-CNR
via Selmi, 2 I-40126 Bologna

G.MARTELLI, G.SPUNTA AND E.BANDINI

I.Co.C.E.A. Area di Ricerca
via P. Gobetti, 101 I-40129 Bologna

1. Introduction

The chemistry of the azomethine group with enolates represents a heteroatom variant of the well known aldol condensation. However, there are some differences between the two reactions. The most evident one is the much lower reactivity displayed by the nitrogen containing double bond in comparison with the isoelectronic carbonyl group. This fact could be explained by considering that the nitrogen anion, formed in the addition of the enolate to the azomethine group, is more basic by several orders of magnitude compared to the enolate itself whereas the opposite is true in the case of the aldol condensation. As a consequence, if we consider the free anions, the equilibrium of the addition of an enolate to an azomethine group will be shifted to the side of the starting materials while, in the case of the aldol condensation, the equilibrium lies on the opposite side. However because the first step of the addition is the coordination of the imine nitrogen to the metal

C. Chatgilialoglu and V. Snieckus (eds.), Chemical Synthesis, 25–60.

cation of the enolate, the ease of the addition depends on the Lewis acid properties of the enolate cation. For instance, lithium and zinc enolates add better than sodium or potassium enolates. The reversibility of the addition chiefly depends on the nitrogen-metal bond strength of the adduct as well as on the reaction conditions. Thus lithium and, at least under certain circumstances, zinc enolates, lead to irreversible additions while sodium and potassium enolates add reversibly. In the latter case, the addition can be brought to completion by coupling it with an irreversible reaction that removes the adduct during its formation. The nitrogen substituent plays an important role in the addition. Electron withdrawing groups, e.g. aryl or silyl groups, reducing the basicity of the nitrogen atom in the adduct, favour the addition whereas alkyl groups do the opposite. Thus lithium enolates of esters easily add to *N*-aryl or *N*-silylimines but fail to react with *N*-alkylimines which, in contrast, react with the corresponding more Lewis acidic zinc enolates.

2. Synthesis of β-lactams

An interesting aspect of the imine chemistry is represented by the addition of ester enolates to an imine.[1] In this case, the first formed β-amidoester can undergo addition-elimination to give the corresponding β-lactam (Scheme 1). This reaction, reported first in 1943 by Gilman and Speer[2] using Reformasky reagents, has become one of the most important methods for the preparation of β-lactams.

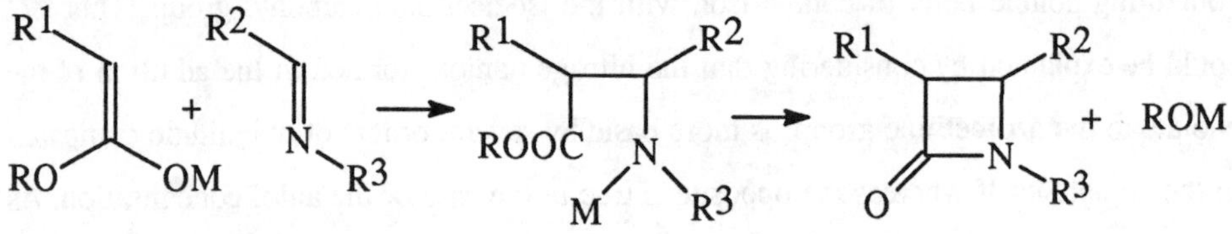

R1 R2 + RO OM N R3 → R1 R2 ROOC N M R3 → R1 R2 O N R3 + ROM

M = Li, Na, K, Al, Zn etc.

Scheme 1

In this essay, we discuss studies concerning the synthesis of non classical β-lactam antibiotics using the ester enolate imine condensation route. As starting imines, we have particularly studied *N*-trimethylsilylimines (Scheme 2).

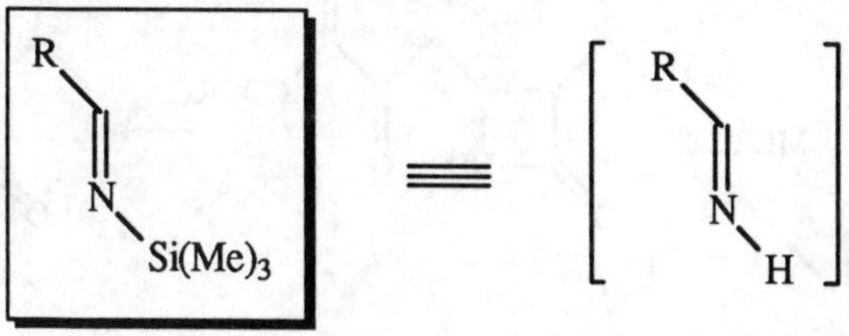

Scheme 2

The *N*-silylimines are monomeric compounds that are stable under anhydrous conditions. Since the metal-nitrogen bond is easily hydrolyzed, the silylimines can be considered a protected, stabilised form of the corresponding elusive imines of ammonia which are known to be very unstable, readily polymerising to triazines and other products.

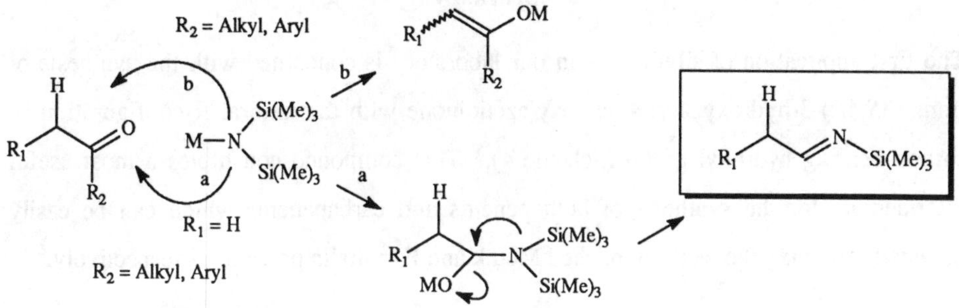

Scheme 3

The most practical and high yielding method to prepare silylimines involves the addition-elimination reaction of lithium hexamethyldisilylamide to an aromatic or aliphatic aldehyde.[3] Enolizable ketones fail to produce silylimines because the strongly basic amide attacks α-hydrogens quantitatively affording the corresponding enolates. On the other hand, enolizable aldehydes, being less sterically demanding, add hexamethyldisilylamide faster than they are deprotonated thus affording the corresponding silylimines in high yield. We have been able to prepare in this way, for the first time, a number of aliphatic silylimines including those substituted in the α or β position by silyloxy and alkoxy groups (Scheme 3).[4]

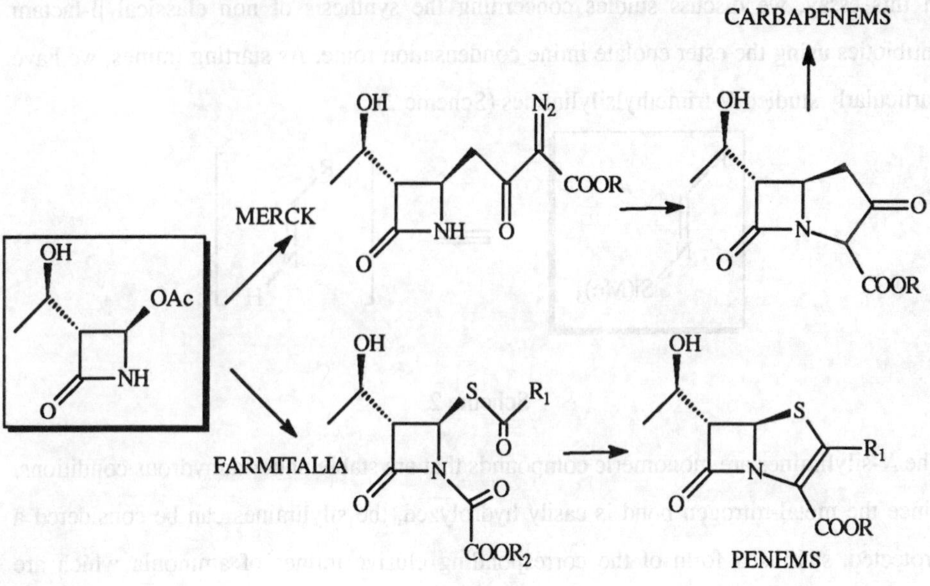

CARBAPENEMS

MERCK

FARMITALIA

PENEMS

Scheme 4

The first application of silylimines in our laboratory is concerned with the synthesis of trans (3*S*,4*R*)-3-hydroxyethyl-4-acetoxy-azetidinone with the natural *R* configuration of carbon bearing hydroxyl group (Scheme 4).[5] This compound constitutes a most useful intermediate for the synthesis of both penems and carbapenems which can be easily obtained following the well established Merck and Farmitalia procedures respectively.[6]

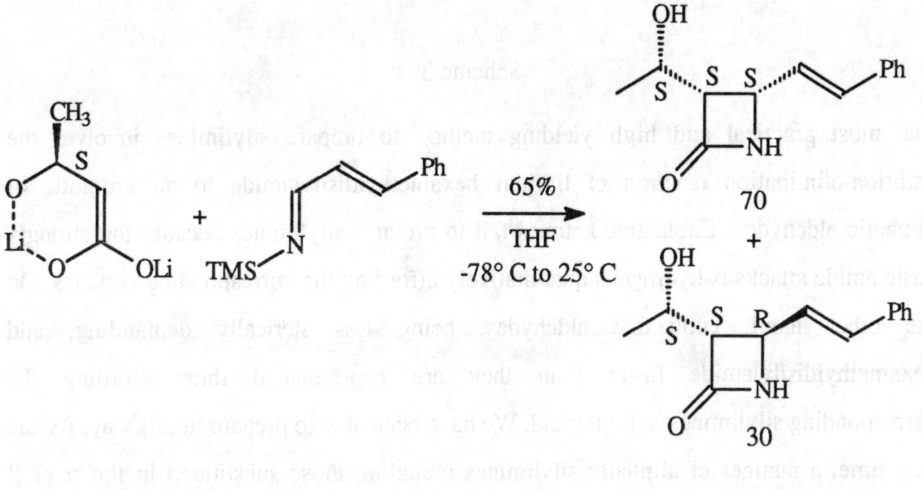

65%

THF

-78° C to 25° C

70

30

Scheme 5

The key step in our synthesis is the condensation between *N*-trimethylsilylimine of cinnamic aldehyde and lithium enolate of ethyl 3(*S*)-hydroxy butanoate in THF at -78°. The reaction is very clean, affording in good yields exclusively two *N*-unsubstituted *cis* and *trans* azetidinones in a 7:3 ratio. The silicon-nitrogen bond is cleaved during the aqueous work up. This constitutes an interesting advantage over the use of *p*-methoxyphenyl or benzyl substituted imines which requires subsequent deprotection. The reaction proceeds with excellent diastereofacial selectivity. The stereocenter with *S* configuration originally present in the nucleophilic partner induces, in a 1,2-*lk* manner,[7] with a diastereoselectivity of more than 94%, the correct natural *S* configuration at C_3 of both azetidinones. In contrast, the simple diastereoselection that controls the stereochemistry of the carbon atom C_4 and therefore the *cis-trans* configuration of β-lactam ring, is less stringent since the reaction gives rise to a mixture of *cis-trans* azetidinones in a ratio of 7:3. The simple diastereoselection in these systems is highly dependent upon many factors which include solvent polarity and structure of the imine side chain.

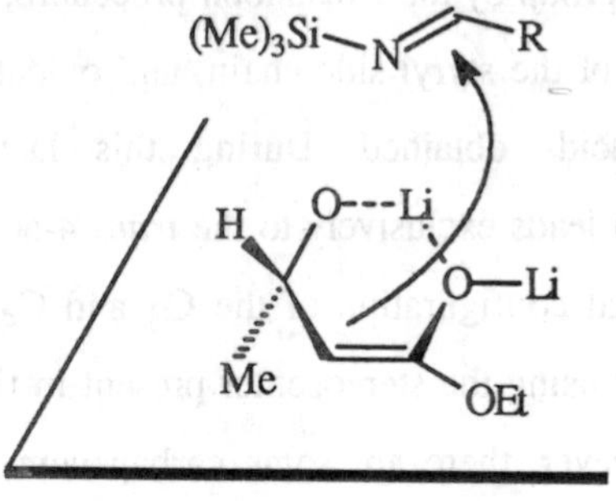

Scheme 6

The high 1,2-*lk* induction on C_3 can be explained by assuming a cyclic structure of the enolate as result of the coordinating effect of one of the lithium cations so that a preferred attack of the electrophilic imine from the less hindered face of the diastereotopic plane takes place (Scheme 6).

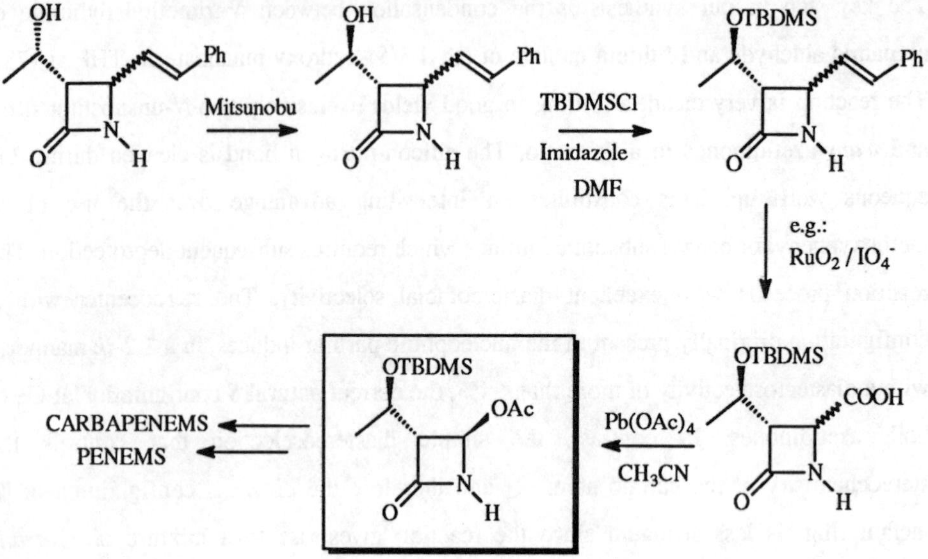

Scheme 7

The two diastereomeric β-lactams obtained are then elaborated to *trans* (3*S*,4*R*)-3-hydroxyethyl-4-acetoxy-azetidinone by a simple process that includes inversion of configuration of hydroxyl group by the Mitsunobu procedure, in order to get the natural *R* configuration, cleavage of the styryl side chain, and oxidative decarboxylation of the mixture of carboxylic acids obtained. During this last process, a quantitative equilibration occurs which leads exclusively to the *trans* 4-acetoxy azetidinone (Scheme 7). In this way, the natural configuration of the C_3 and C_4 stereogenic centers of β-lactam ring are introduced using the stereocenter present in the nucleophilic component of the condensation. However, there are some carbapenem antibiotics which bear no asymmetric center in the C_3 side chain. For instance, the carbapenem PS-5 has a simple ethyl group attached to C_3 while monobactams bear an amido group in the same position. For an enantioselective synthesis of these compounds, we have incorporated the asymmetry in the electrophilic component of the condensation starting from enantiomerically pure silylimine of lactic or mandelic aldehydes.

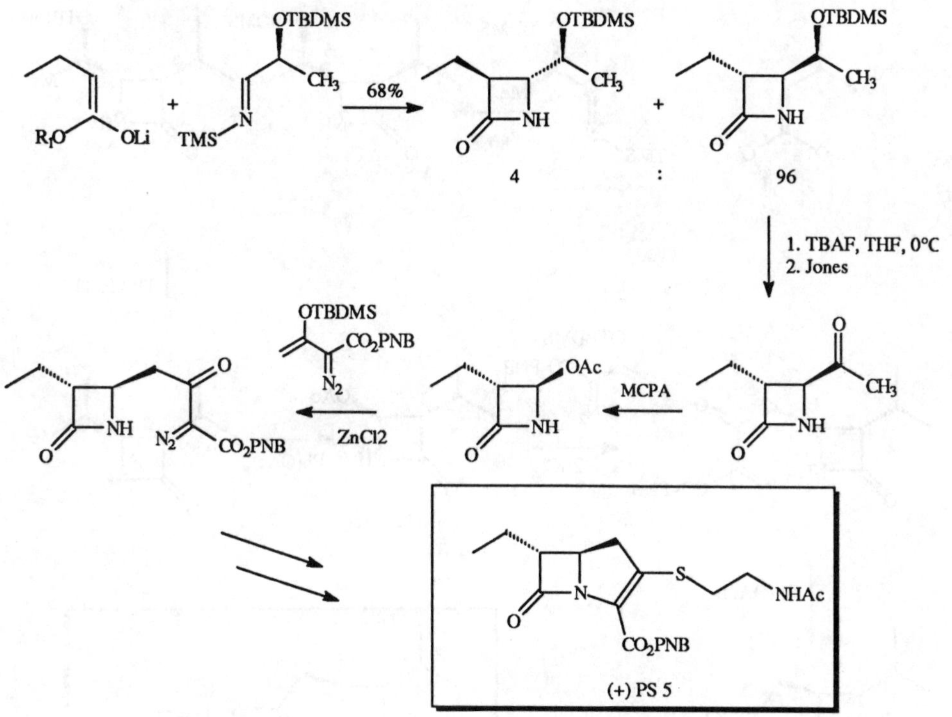

Scheme 8

The synthesis of PS-5 was achieved by treatment of silylimine of *O*-protected (*S*) lactic aldehyde with the lithium enolate of *t*-butyl butanoate.[8] The reaction is highly diastereoselective affording almost completely the *trans* azetidinone with the natural configuration at C_3. This azetidinone was converted, by sequential Jones and Baeyer-Villiger oxidation of hydroxyethyl side chain to the 4-acetoxy derivative that represents a most useful chiral building block for the synthesis of final carbapenem PS-5 *via* the Merck procedure (Scheme 8).

A higher yielding and shorter procedure was developed for the synthesis of PS-6.[9]

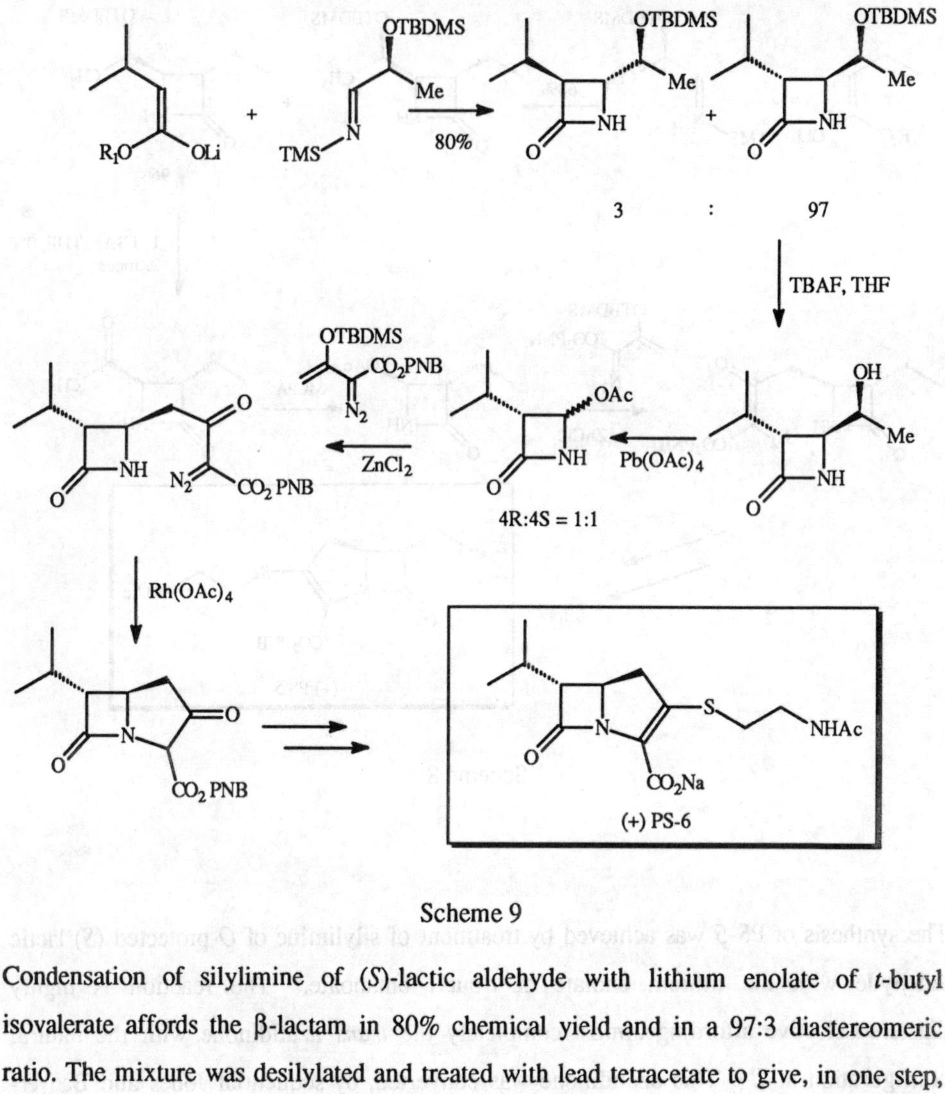

Scheme 9

Condensation of silylimine of (*S*)-lactic aldehyde with lithium enolate of *t*-butyl isovalerate affords the β-lactam in 80% chemical yield and in a 97:3 diastereomeric ratio. The mixture was desilylated and treated with lead tetracetate to give, in one step, through a radical fragmentation reaction, the 4-acetoxy derivative as a 1:1 4(*R*):4(*S*) epimeric mixture. The lack of stereospecificity is not easy to rationalize expecially if one considers that the analogous lead tetraacetate induced oxidative decarboxylation is completely *trans* stereoselective. Both reactions should have the same radical intermediate. However, this lack of stereospecificity is not important for the success of the synthesis since the mixture of diastereoisomers exclusively affords the *trans* 4-substituted azetidinone by the subsequent Merck procedure (Scheme 9).

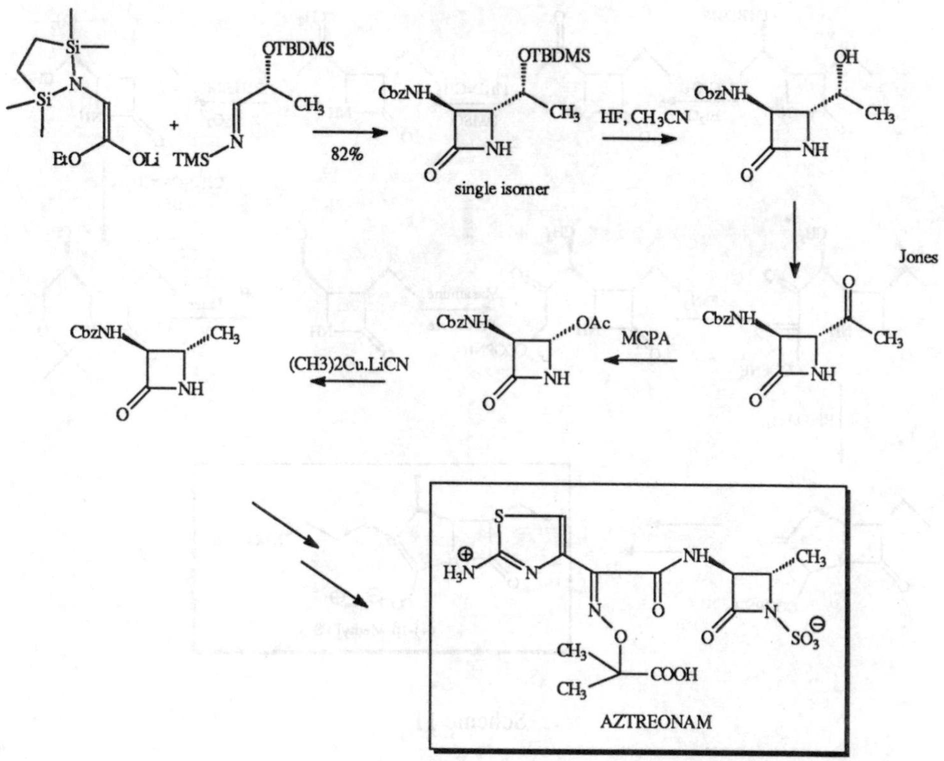

Scheme 10

In general, the monobactams are prepared by cyclisation of suitably modified amino acids such as serine or threonine.[10] An alternative, versatile preparation of Aztreonam and related 4-substituted monobactams is illustrated by our approach to optically active azetidinones *via* condensation of the lactic silylimine with the STABASE *N*-protected form of glycine ethyl ester.[11] Once again a very high *trans*-diastereoselectivity of 98-99% has been achieved. In this case, in order to obtain the natural C_3-*R* configuration and taking into account the 1,2-*lk* induction typical of the reaction, the (*R*)-lactic aldehyde must be used. Protection of the amino group as the carbobenzoxy derivative and elaboration of the C_4 side chain following the already described sequence, affords, in good overall yields, the 4-acetoxy-azetidinone. Finally the C_4 methyl group of Aztreonam is introduced *via* cuprate chemistry (Scheme 10).

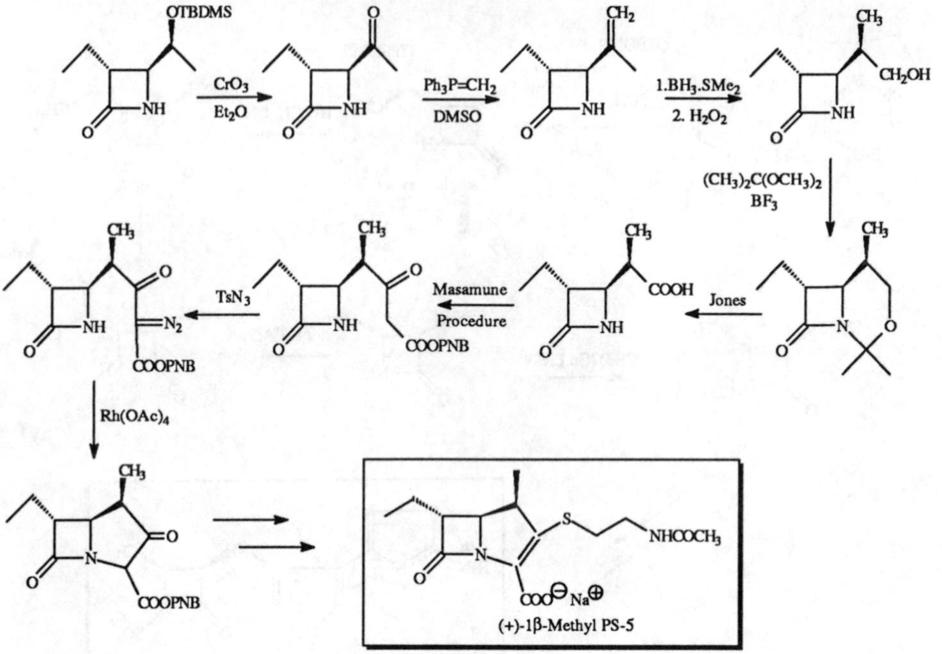

Scheme 11

Using chiral silylimines, the azetidinone synthesis can also be applied for the enantioselective preparation of 1β-methyl carbapenems.[12] As an illustration of this methodology, we prepared, for the first time, 1β-methyl PS-5 starting from (3*S*,4*S*)-3-ethyl-4-hydroxyethyl-azetidinone, a key intermediate in the synthesis of PS-5. The synthesis involves oxidation to a methyl ketone, Wittig methylenation, and hydroboration of the resulting methylene derivative to give, almost exclusively, a primary alcohol with the correct *R* configuration of the methyl bearing stereocenter. This alcohol is then converted into the final target following already known procedures (Scheme 11).

In all these reactions involving silylimines of lactic and mandelic aldehydes, the diastereoface selection, at least under the conditions described, is very high (92-100%) and the simple diastereoselection is total. In fact, only *trans*-azetidinones have been obtained and traces of the corresponding *cis* derivatives had never been detected.

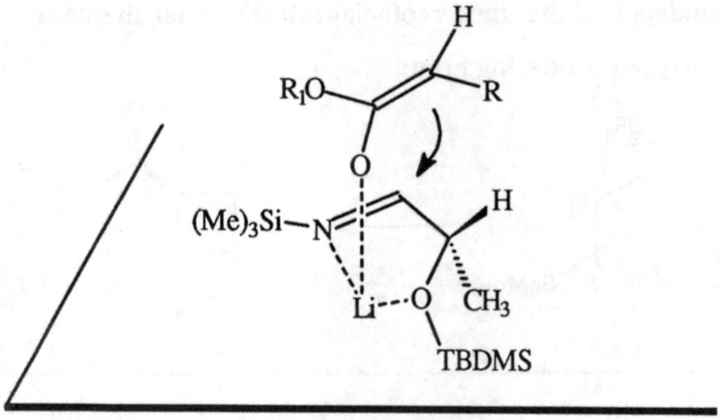

Scheme 12

The observed very high 1,2-*lk* induction of the stereocenter present in the side chain of the imine upon the C_4 stereocenter of the azetidinone may be rationalized *via* a Cram cyclic model by assuming coplanarity between oxygen and nitrogen atom of the imine due to chelation of lithium cation so that the nucleophile attacks from the less hindered diastereotopic face.

The facial diastereoselectivity of these reactions is, however, strongly influenced by the nature of cations and by structure and electronic properties of *O*-protecting groups of the imine. With lithium enolates, the *syn* diastereofacial selection is predominant (73-100%). With the relatively small triethylsilyloxygroup, the *syn* selectivity is complete whereas, by increasing the size of the protecting group, the amount of the *anti* isomer increases. With the large triisopropylsilyloxy group, the *syn:anti* ratio becomes 73:27 and the *t*-butyldiphenylsyliloxy group, gives intermediate results. Sodium enolates appear to be more *anti* diastereoselective especially with the large triisopropyl- and *t*-butyldiphenyl-silyloxy groups. In these cases, a complete reversal of the diastereofacial selection is observed. The *syn*/*anti* ratio changes from 73:27 to 17:83. In contrast, alkoxy groups, e.g. benzyloxy and even the relatively large *t*-butoxy group, afford *syn* isomers exclusively even in the case of the sodium enolate.

Table 1: Dependence of the diastereofacial selectivity on the nature of the enolate cation and the oxygen protecting group

R, tBuO, OM, + , OR_1, N, $Si(Me)_3$ → R, OR_1, NH, O, *Syn* + R, OR_1, NH, O, *Anti*

R	R_1	M = Li *Syn-Anti*	Yield %	M = Na *Syn-Anti*	Yield, %
nBu	Et_3Si	>98 : <2	40	89 : 11	35
nBu	$tBuMe_2Si$	96 : 4	63	70 : 30	38
iPr	$tBuMe_2Si$	98: 2	65	10 : 90	60
Si N— Si	$tBuMe_2Si$	>98 : <2	85	40 : 60	73
nBu	Ph_2tBuSi	90 : 10	53	7 : 93	35
nBu	iPr_3Si	73 : 27	60	17 : 83	51
nBu	Bn	>98 : <2	42	>98 : <2	10
nBu	tBu	>98 : <2	25	>98 : <2	20

In order to razionalize these results, one can assume a competition between a cyclic chelated Cram model, which should lead to the *syn* isomer, and an open chain Felkin-Ahn model leading to preponderant formation of the *anti* isomer. The increased *anti* diastereofacial selection observed in the case of less coordinating sodium enolates points in this direction. The very high *syn* diastereoselectivity invariably shown by the alkoxy derivatives, even in the case of the sodium enolates, could originate from the higher basicity of alkoxy groups in comparison with silyloxy groups, consistent with the operation of the chelated model.[13]

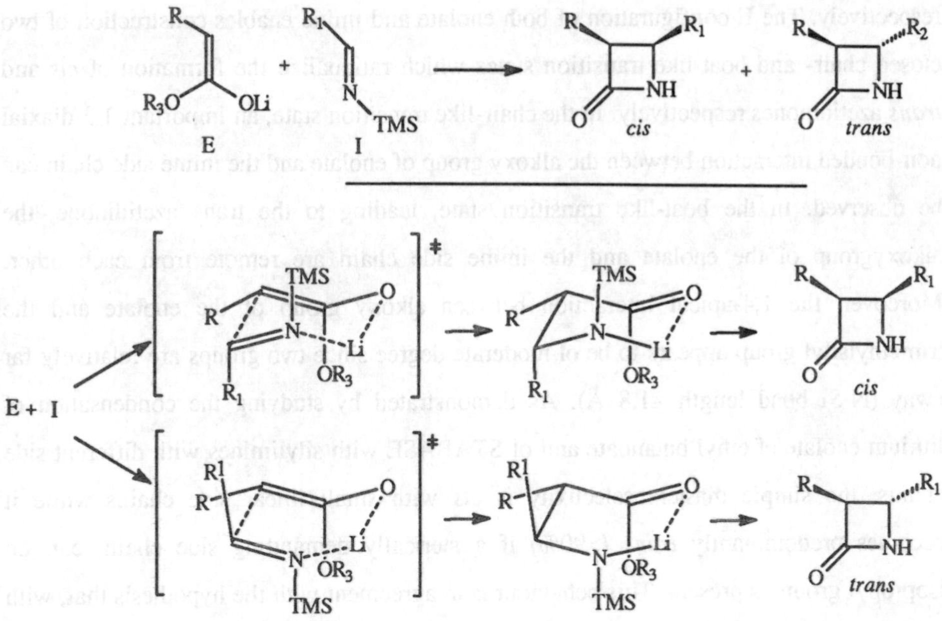

Scheme 13

As we have already seen, the simple diastereoselection of ester enolate condensations with silylimines of lactic and mandelic aldehydes invariably leads to the exclusive formation of *trans* azetidinones. The formation of the β-lactam ring proceeds in two steps: addition of enolate to imine followed by cyclisation. With lithium enolates, under the reported conditions, both addition and cyclisation are irreversible. The first step is the activation of imine by coordination of nitrogen to the metal atom of enolate to form an electrostatic complex. The coordination brings the two reactants in close proximity and this enhances the reaction rate. The next step is the nucleophilic attack of the enolate on the electrophilic carbon of the imine. During this C-C bond formation, the stereochemistry of final azetidinones is determined. The addition is thus likely to proceed through a rigid, closed, chair- or boat-like transition state as first proposed by Zimmerman and Traxel for the aldol condensation (Scheme 13). As well known, the structures of these transition states are dependent on the configuration (E or Z) of both enolate and imine. Ample literature[14] suggests that the configuration of most imines is E. Moreover, the E enolate configuration derived by deprotonations of esters using lithium amides or STABASE, has been demonstrated by Ireland[15] and Overman[16] respectively. The E configuration of both enolate and imine enables construction of two closed chair- and boat-like transition states which rationalize the formation of *cis* and *trans* azetidinones respectively. In the chair-like transition state, an important 1,3-diaxial non-bonded interaction between the alkoxy group of enolate and the imine side chain can be observed. In the boat-like transition state, leading to the trans azetidinone, the alkoxygroup of the enolate and the imine side chain are remote from each other. Moreover, the 1,4-apical interaction between alkoxy group of the enolate and the trimethylsilyl group appears to be of moderate degree since two groups are relatively far away (N-Si bond length ≅1.8 Å). As demonstrated by studying the condensation of lithium enolate of ethyl butanoate and of STABASE with silylimines with different side chains, the simple diastereoselectivity is *cis* with small linear side chains while it becomes predominantly *trans* (>90%) if a sterically demanding side chain, e.g. an isopropyl group, is present. This behaviour is in agreement with the hypothesis that, with large imine side chains, the non bonded 1,3-interaction in the chair-like transition state becomes too large so that the boat-like transition state corresponds to the lowest energy

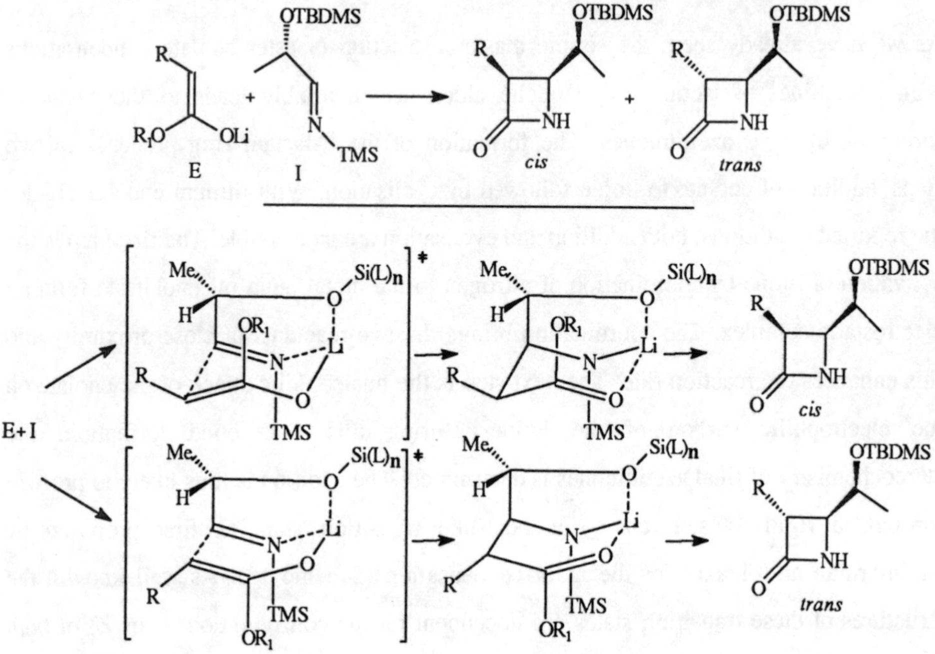

Scheme 14

and leads to the preponderant formation of the *trans* isomer.

Previously it has been shown that the reaction of α-silyloxyimines with enolates affords exclusively *trans*-azetidinones. In this case, where a methyl group of the isopropyl side chain has been substituted by a bulky silyloxy group, the energy difference between boat-like and chair-like transition state becomes even larger so that the *trans* azetidinone is the only isomer formed (Scheme 14).

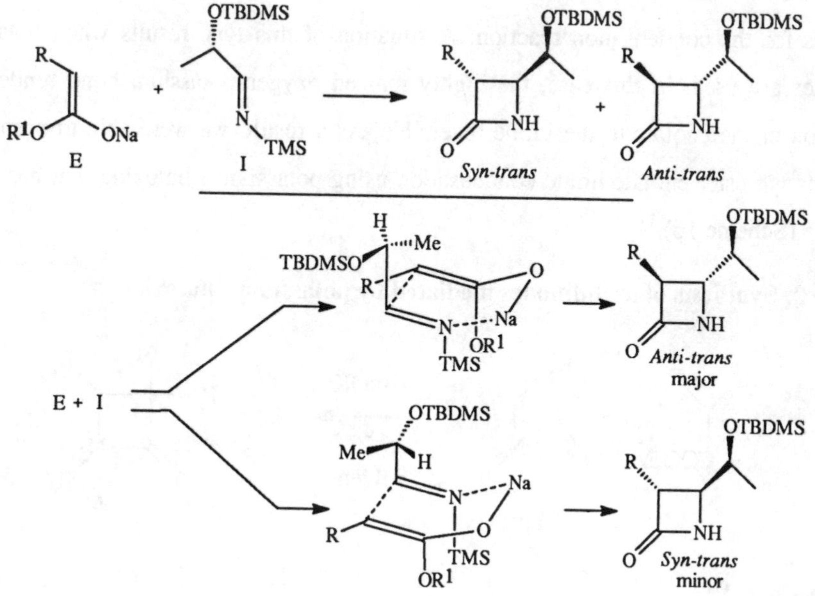

Scheme 15

It appears that the diastereofacial selectivity may be correlated with the nature of the enolate cation and the *O*-protecting group. The sodium cation, known to be a weaker Lewis acid, cannot efficiently coordinate the silyloxy group of imine especially if it is large, so that facial selectivity follows, at least in part, a Felkin-Ahn model leading to preponderant formation of the *anti* isomer. However, the more basic imine nitrogen atom still coordinates the sodium cation so that the simple distereoselectivity still proceeds via a closed boat-like Zimmerman-Traxel transition state leading exclusively to the formation of *trans*-azetidinones (Scheme 15).

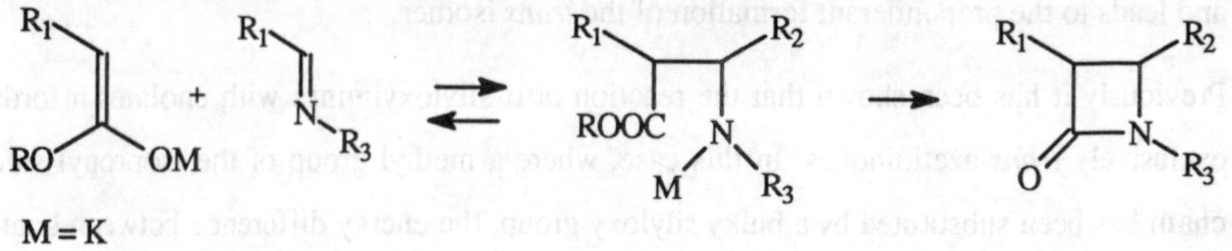

Scheme 16

If the ester enolate reaction involves reversible addition while the cyclisation remains irreversible, it should be possible to use weaker bases than the generally used lithium amides for the condensation reaction. A situation of this type results when potassium enolates are used. In this case, the highly ionized oxygen-potassium bond renders the addition of the enolate to the imine reversible. As a result, we were able to prepare β-lactams *via* ester enolate imine condensation using potassium *t*-butoxide that has a pK_a of 16.5 (Scheme 16).[17]

Table 2: Synthesis of azetidinones mediated by potassium *t*-butoxide

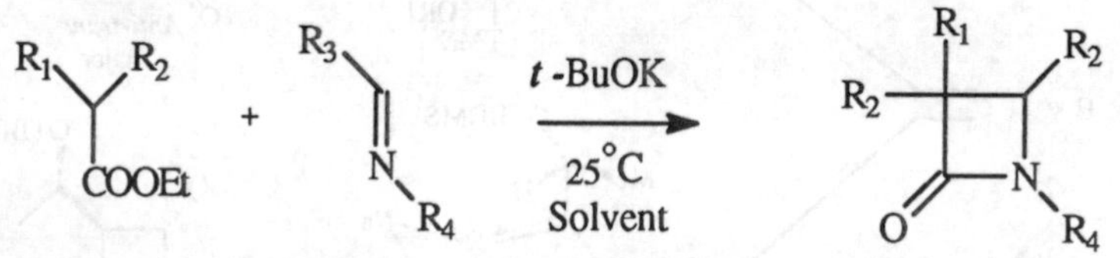

Entry	R_1	R_2	R_3	R_4	Solvent	Yield, %
1	Me	Me	Ph	PMP	DMF	88
2	Me	OTBDMS	Ph	PMP	DMF	60
3	OEt	OEt	Ph	PMP	DMF	88
4	Me	OTBDMS	2-Furyl	PMP	DMF	69
5	OEt	OEt	2-Furyl	PMP	DMF	40
6	Me	Me	Ph	$SiMe_3$	DMF	22
7	Me	Me	H	*i* Bu	DMF	20
8	OEt	OEt	*m*-MeO-Ph	PMP	THF	50
9	OEt	OEt	*o*-MeO-Ph	PMP	THF	32
10	OEt	OEt	*p*-Cl-Ph	PMP	THF	75

The condensation is carried out by stirring a solution of ester and imine at room temperature followed by addition of a solution of potassium *t*-butoxide in THF or DMF. Best results are obtained using 3 equivalents of ester and base per equivalent of imine. The excess of the two components will shift the equilibrium to the right thus increasing the concentration of adduct. As expected, only disubstituted esters give clean reactions. Monosubstituted ones, as for instance *t*-butyl *n*-butanoate, afford complex mixtures. The yields are strictly dependent also upon the imine moiety. *N*-arylimines work quite well while *N*-silylimine of benzaldehyde gives rise to the β-lactam only in 20% yield together with 40% of the amide arising from intermolecular attack of β-amido ester anion on the excess of ester present (Table 2).

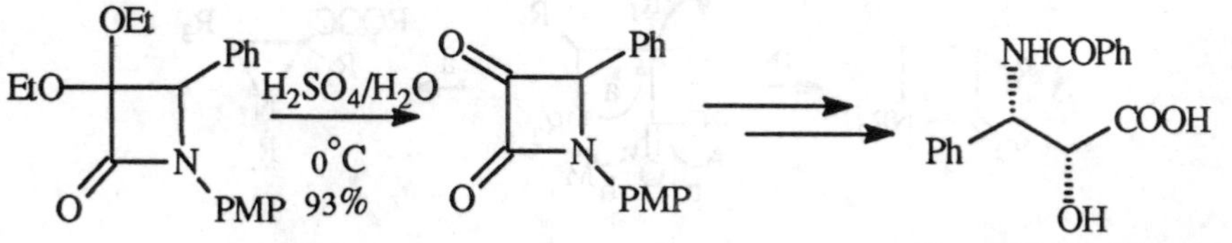

OEt
EtO
Ph
H_2SO_4/H_2O
0°C
O N PMP 93%
O Ph N O PMP
NHCOPh
COOH
Ph
OH
Taxol C-13 side chai

Scheme 17

An interesting application of this methodology concerns the potassium t-butoxide induced preparation of the 3,3'-diethoxy-4-phenyl-azetidinone which represents an interesting intermediate for the synthesis of the taxol C_{13} side chain (Scheme 17). For this purpose, the 3,3'-diethoxy-azetidinone was prepared in 90% yield by treatment of the ethyl ester of diethoxyacetic acid with the *p*-methoxyphenylimine of benzaldehyde. This compounds is then hydrolyzed with sulfuric acid to give the azetidin-2,3-dione in 93% yield. The effectiveness of this compound as a synthon of the taxol side chain has already been demostrated.[18]

3. Synthesis of Aziridines

As discussed above, an ester enolate adds to an imine to give a β-amido ester which then cyclizes to a β-lactam *via* an addition-elimination reaction. The addition can be reversible or irreversible depending on the cation of the enolate while the cyclization is, under the usual reaction conditions, irreversible. A special case is represented by enolates of α-halo substituted esters. Here, the initially formed aldol may, in principle, undergo reaction by two different irreversible pathways: addition-elimination to give azetidinones or nucleophilic displacement of the halogen atom to afford aziridines (Scheme 18).

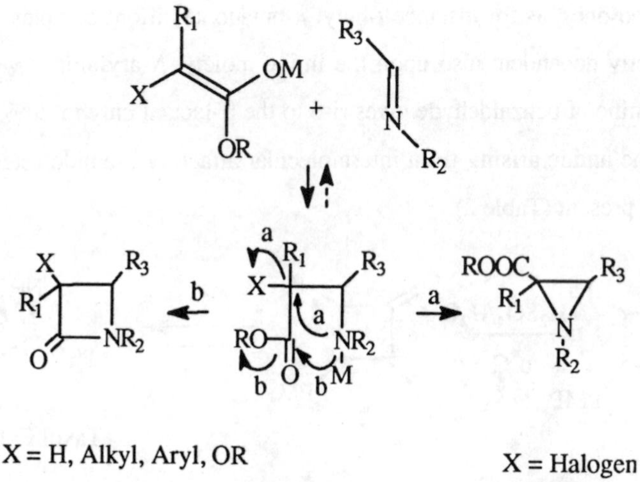

X = H, Alkyl, Aryl, OR

X = Halogen

Scheme 18

The importance of aziridines is well documented.[19] Their applications to the synthesis of α-amino acids, diamines and β-lactams make aziridines interesting synthetic targets. Aziridines have been obtained by treatment of *N*-trimethylsilylimines with lithium enolates of α-haloesters in THF at -30° C. Traces of azetidinones are not detected (Table 3).[20] Enolizable as well as non-enolizable *N*-trimethylsilylimines may be used. The choice of the leaving group is critical. Thus iodide as well as sulfonates, e.g. mesylates and tosylates, give only traces of the expected aziridines. The low yields may be ascribed to the instability of the ester enolates at the reaction temperature which cannot be maintened lower than -30° C. In fact, *t*-butyl ester enolates, known to be more stable than the corresponding ethyl esters, give better yields (entries 1 and 2, Table 3). The failure of the enolates bearing iodide or sulfonate in the α-position is again ascribed to their instability under the reaction conditions.

Table 3: Synthesis of aziridines from α-haloester enolates and silylimines

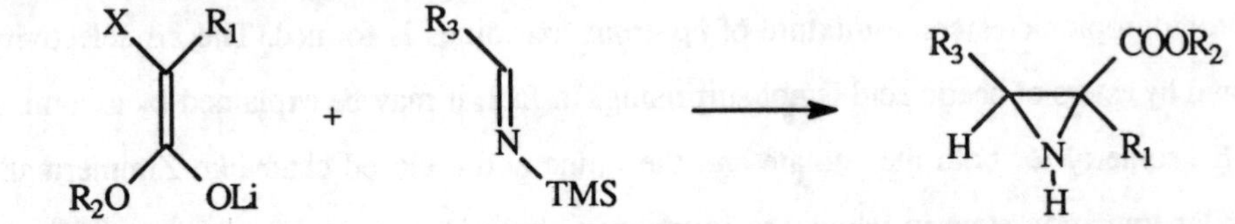

Entry	X	R^1	R^2	R^3	Yield, %	R^3:$COOR^2$ *cis:trans*
1	Cl	H	Et	n-Hexyl	22	>98:<2
2	Cl	H	*t*-Bu	n-Hexyl	52	>98:<2
3	Br	H	*t*-Bu	n-Hexyl	40	>98:<2
4	I	H	*t*-Bu	n-Hexyl	traces	n d
5	OTs	H	*t*-Bu	n-Hexyl	traces	n d
6	OMs	H	*t*-Bu	n-Hexyl	traces	n d
7	Cl	H	*t*-Bu	Phenyl	60	>98:<2
8	Br	H	*t*-Bu	n-Heptyl	40	>98:<2
9	Br	H	*t*-Bu	n-Octyl	40	>98:<2
10	Br	H	*t*-Bu	Cyclohexyl	10	>98:<2
11	Br	Me	*t*-Bu	n-Hexyl	60	40:60

n d = not determined

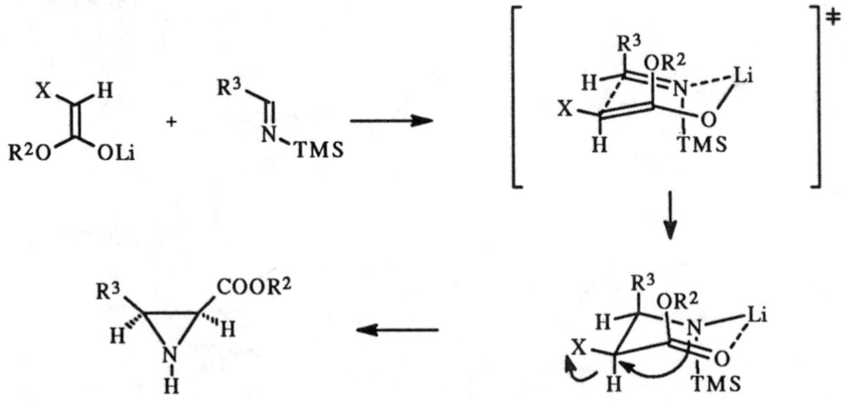

Scheme 19

The reaction is *cis*-stereospecific when esters of acetic acid are used while, in the case of α-bromopropionic esters, a mixture of *cis-trans* aziridines is formed. The *cis*-selectivity shown by esters of acetic acid is not surprising. In fact, it may be explained by assuming an E geometry for both the enolate and the imine and a closed chair-like Zimmerman-Traxler transition state in which the imine side chain is in an axial position while the halogen atom is in an equatorial location. The subsequent nucleophilic displacement of the halogen atom in the resulting intermediate leads directly to the formation of the *cis*-aziridine (Scheme 19).

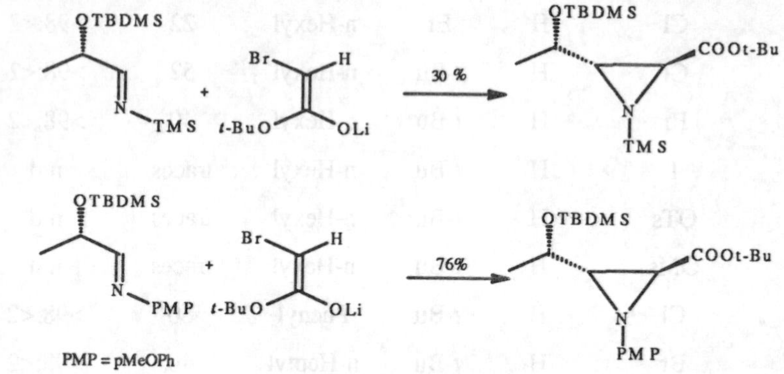

OTBDMS; N–TMS; + Br, H, t-BuO, OLi; 30 %; OTBDMS, COOt-Bu, N, TMS

OTBDMS; N–PMP; + Br, H, t-BuO, OLi; 76%; OTBDMS, COOt-Bu, N, PMP

PMP = pMeOPh

Scheme 20

We have also prepared aziridines starting from silylimines and *p*-methoxyphenylimines of (*S*)-lactic aldehyde protected as their *t*-butyldimethylsilyl ethers. Only *syn*-trans isomers with the shown configuration were isolated. These compounds represent

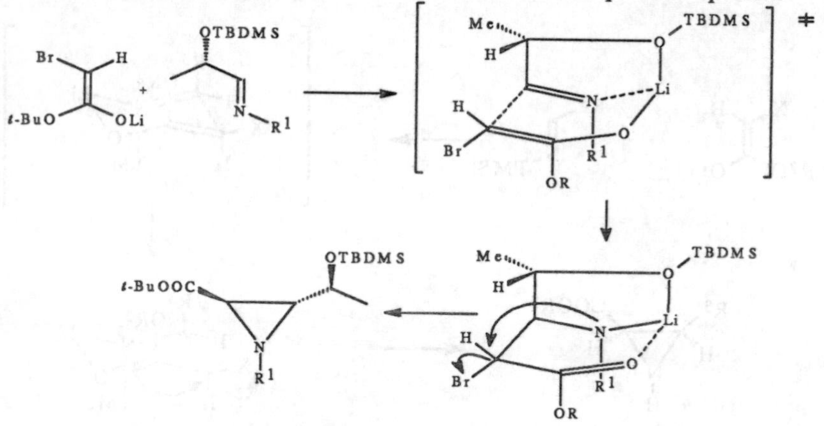

Br, H, t-BuO, OLi; + OTBDMS, N, R1; Me, TBDMS, H, O, N, Li, H, O, Br, R1, OR; ‡

t-BuOOC, OTBDMS, N, R1; Me, TBDMS, H, O, Li, N, H, O, Br, R1, OR

Scheme 21

interesting, enantiomerically pure, potential starting materials for the preparation of substituted α- and β-amino acids.

The complete diastereoselectivity, induced and simple, shown by this reaction is not unexpected if one considers the corresponding use of these imines in the preparation of β-lactams.[8] Here, in fact, the formation of the β-amido esters occurs via a closed boat-like transition state that appears to be more stable than the corresponding chair-like one. The aldol intermediate then undergoes cyclization *via* a nucleophilic substitution to give the final aziridine product (Scheme 21).

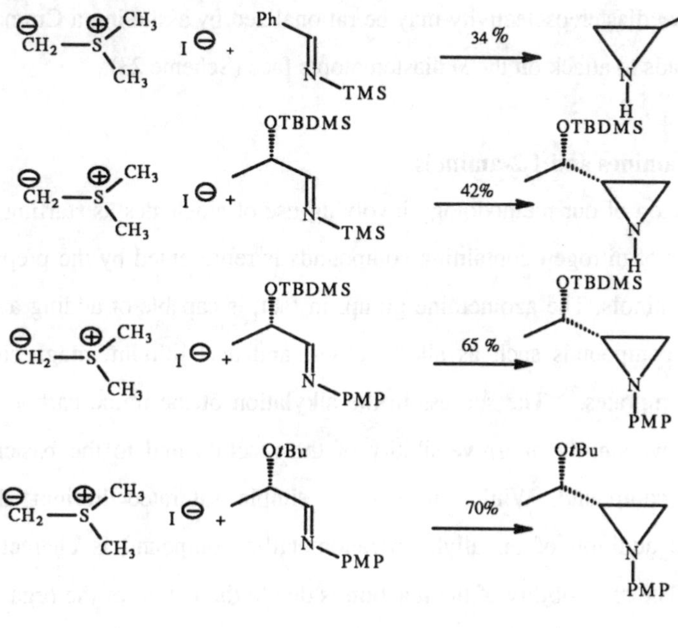

Scheme 22

Sulfur ylides represent another family of nucleophiles structurally coupled with a leaving group. It is therefore to be expected that they will react with imines to give aziridines via an addition-substitution sequence. In fact, the sulfur ylid obtained by metallation of trimethylsulfonium iodide with *n*-butyl lithium reacts at -30° C in THF with both silylimines and *p*-methoxyphenyl imines to give the corresponding aziridines. Again a single isomer is obtained in the case of (*S*)-lactaldehyde imines (Scheme 22).

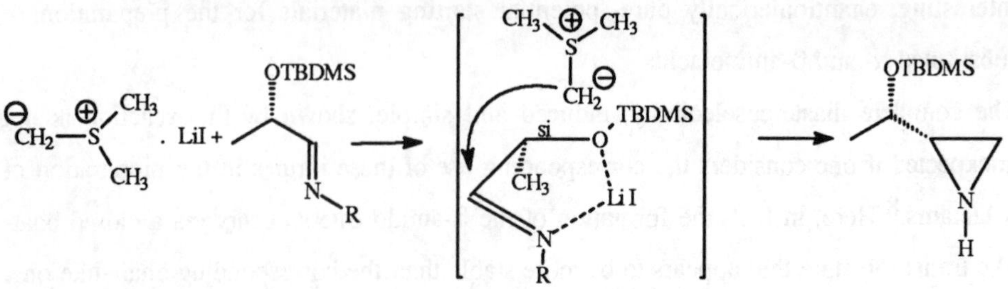

Scheme 23

The observed *syn* diastereoselectivity may be rationalized by assuming a Cram chelation model which leads to attack on the *si* diastereotopic face (Scheme 23).

4. Synthesis of amines and 1,2-aminols

A logical extension of our methodology involving use of aldehydes as starting materials for the synthesis of nitrogen containing compounds is represented by the preparation of amines and α-aminols. The azomethine group, in fact, is capable of adding a variety of organometallic compounds such as alkyl-, allyl-, and aryl lithium, magnesium, zinc, boron, tin, and cuprates.[21] The success in the alkylation of the imine carbon is strictly related to the reversibility or irreversibility of the reaction and to the basicity of the organometallic compound. While, in general, simple saturated lithium alkyls add irreversibly, the addition of an allylic organometallic compound is characterized by reversibility.[22] The reversibility of the reaction is due to the nature of the organometallic compounds and the steric hindrance of the *N*-substituent. A crucial role is played by the basicity of the organometallic reagent. With enolizable imines, a competitive deprotonation of acidic α-imine hydrogens may take place. Even relatively small differences in the basicity of the reagent are important. Thus, for instance, alkyl magnesium reagents are prone to deprotonation whereas allyl and benzyl ones generally add to imines.

Recently, we have studied the addition of organometallic compounds to aluminum-imines[23] and silylimines[24] which are easily prepared from nitriles and aldehydes by treatment with diisobutyl aluminum hydride (DIBAH) and lithium hexamethyl disilyl amide (LHMDSA) respectively. The advantage of these imines is their ease of hydrolysis to directly afford primary amines.

Initially, we have considered the *N*-aluminum imines which may be obtained by DIBAH reduction of nitriles at -78° C in THF. Treatment of the aluminum imines thus obtained with 2 equivalents of alkyl or allyl lithium and benzyl and allyl magnesium chloride at room temperature for 48 hours leads, after hydrolysis, to the expected primary amines (Table 4). The reaction appears to be quite general. In fact, aromatic as well as enolizable aliphatic aluminum imines react with a variety of organometallic reagents to afford good to excellent yields of primary amines after hydrolysis. The stoichiometry of the

Table 4: Synthesis of primary amines from nitriles via aluminum imines

$R_2CH{=}N{-}Al(iBu)_2 + R_1{-}M \rightarrow R_2CH{=}N^{\ominus}{-}Al(iBu)_2R_1\ M^{\oplus} + R_1{-}M \rightarrow R_2R_1CH{-}N(M){-}Al^{\ominus}(iBu)_2R_1\ M^{\oplus} \rightarrow R_2R_1CH{-}NH_2$

Entry	R^2	R^1	M	Yield, %
1	Phenyl	Benzyl	MgCl	46
2	Phenyl	Allyl	MgCl	98
3	Phenyl	n-Butyl	Li	92
4	Phenyl	sec-Butyl	Li	80
5	Phenyl	ter-Butyl	Li	65
6	2-Furyl	Allyl	MgCl	72
7	2-Furyl	n-Butyl	Li	80
8	2-Thienyl	n-Butyl	Li	65
9	2-Thienyl	Allyl	MgCl	98
10	p-Methoxy-phenyl	n-Butyl	Li	70
11	n-Butyl	Allyl	MgCl	96
12	n-Octyl	Allyl	MgCl	89
13	5-(Tritylamino)-pentyl	Allyl	MgCl	50

reaction, requiring two equivalents of the organometallic compound, suggests that the first attack of the nucleophile takes place on the aluminum leading to an ate-complex which then reacts with a second equivalent of the nucleophile to give the final addition product.

The corresponding aryl [25] and alkyl silylimines[26] undergo reaction in the same way but much faster requiring only one equivalent of the organometallic compound (Table 5). With silylimines, the reaction is complete in less that one hour at -78°C. The yields of amines are somewhat lower than those obtained with aluminum imines. In fact, the trimethylsilylimines behave in part as silylating reagents toward organometallic compounds.

Table 5: Synthesis of primary amines from aldehydes via silylimines

$R_1CH=N-Si(CH_3)_3 + R_2-M \longrightarrow R_1CH(R_2)N(M)Si(CH_3)_3 \longrightarrow R_1CH(R_2)NH_2$

Entry	R_1	R_2	M	Yield, %
1	Phenyl	Allyl	MgBr	97
2	p-MeOPh	Phenyl	MgBr	56
3	Phenyl	n-Butyl	Li	53
4	n-Heptyl	Allyl	MgBr	60
5	n-Heptyl	n-Butyl	Li	40

The difference in the reactivity of silylimines in comparison with the corresponding aluminum imines is due to different Lewis acid properties of the two metals. The aluminum imine undergoes reaction first with one molecule of the organometallic reagent to give a negative charged ate-complex while the less Lewis acidic silicon atom is incapable of such reactivity. The neutral trisubstituted silicon and the negative charged trisubstituted aluminum substituent have a completely different influence on the energy of the LUMO orbital of the azomethine group. Thus, the key orbital interaction which controls the addition of a nucleophile to an imine is the HOMO of the nucleophile and the LUMO of the imine.[27]

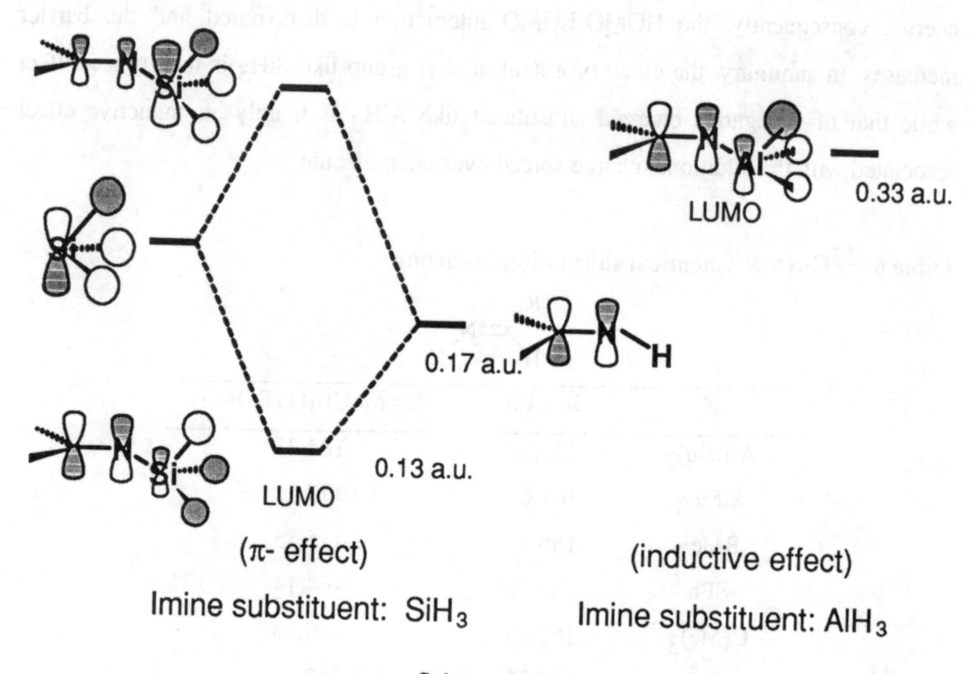

Scheme 24

The stabilizing effect of this interaction depends on the energy gap between the HOMO and the LUMO levels (Scheme 24). Thus, if the energy of the LUMO is decreased, the interaction becomes more stabilizing and the barrier is reduced. Consider now the effect of a substituent in the imine moiety on the energy of the LUMO. Specifically, we consider the replacement of the imine hydrogen with a neutral substituent like SiH_3 and with a formally charged substituent like AlH_3. We can rationalize the effect of a silyl group in terms of the orbital interaction diagram which illustrates the effect of the interaction of the LUMO of the imine fragment with the p^* MO of the silicon hydride fragment. This orbital is higher in energy than the imine LUMO. Consequently, when the two orbitals interact to give the orbitals of the substituted imine, the imine LUMO becomes lower in energy, the LUMO-HOMO interaction is favoured, and the barrier decreases. On the other hand, the effect of a charged substituent is mainly due to the additional negative charge which is spread over the whole molecule, as shown by *ab initio* computations.[27] Because of this negative charge, the full set of MOs increases in energy: consequently, the HOMO-LUMO interaction is disfavoured and the barrier increases. In summary, the effect of a neutral silyl group like SiH_3 is mainly a π-effect while that of a negative charged substituent like AlH_3 is mainly an inductive effect associated with the additional charge spread over the molecule.

Table 6: ^{13}C-NMR Chemical shift of imine carbon

R(H)C=N–X

X	R = Ph	R = MeCH(OTBDMS)
$Al(iBu)_2$	173.77	184.35
$SiMe_3$	167.61	177.11
BMe_2	166.15	175.15
SPh	159.38	164.14
$C(Me)_3$	153.85	160.33
OH	147.00	152.20
OMe	147.22	151.56
NMe_2	136.33	138.70

The nature of the nitrogen substituent has a significant influence on the electron density distribution of the imine function. Table 6 shows the influence of nitrogen substituents on the ^{13}C chemical shift of the imine carbon.[28] Benzal imines exhibit lower chemical shifts than lactic aldimines, a manifestation of the partial positive charge at the benzal carbon atom resulting from delocalization. Another feature is the observed deshielding effect as a function of the greater metallic character at the nitrogen substituent. The order of deshielding follows aluminum imines > silylimines. If the degree of deshielding may be considered a measure of the electrophilic character of the imine carbon, the aluminum imines should be the most reactive toward the attack of a nucleophile. However, we know that, with aluminum imines, the first attack occurs on the aluminum atom to give an ate complex which then undergoes addition of the second equivalent of the nucleophile. In contrast, silylimines add directly to the imine carbon. So they are, in fact, the most reactive imines among those reported in Table 6.

A particularly promising synthetic extension of these methodologies for the preparation of primary amines comprises the use of metalloimines bearing α-oxygen substitution. These imines are potentially interesting starting materials for the synthesis of various amino alcohol. We have therefore decided to study the addition of organometallic compounds to the aluminum- and silylimines of the lactic and mandelic aldehydes which have been protected as *t*-butyldimethylsilyl ethers.

Table 7: Synthesis of 1,2 aminols from nitriles *via* aluminum imines

OTBDMS OTBDMS OTBDMS

2 R_1—Li + R_2 ... N—Al(iBu)$_2$ ⟶ R_2 ... R_1 NH_2 + R_2 ... R_1 NH_2

Syn *Anti*

Entry	R_2	R_1	*Syn : Anti*	Yield, %
1	Phenyl	Allyl	91 : 9	43
2	Phenyl	n-Butyl	91 : 9	48
3	Phenyl	sec-Butyl	67 : 33	35
4	Phenyl	ter-Butyl	11 : 89	27
5	Phenyl	pentyl	86 : 14	40
6	Phenyl	hexyl	89 : 11	35
7	Phenyl	methyl	75 : 25	29
8	Methyl	allyl	72 : 28	47
9	Methyl	n-Butyl	76 : 24	57
10	Methyl	ter-Butyl	45 : 55	60

The aluminum imines are readily available by the addition of one equivalent of DIBAH to commercially available mandelonitrile and lactonitrile protected as *t*-butyldimethysilyl ethers in pentane at -78° C. The use of pentane as solvent is important in order to get acceptable yields of aluminum imines; lower yields in THF may be partially ascribed to a competitive reductive cleavage of the *t*-butyldimethylsilyl group. *In situ* treatment of the obtained aluminum imines with two equivalents of the organometallic compound at room temperature for 24 hours affords a mixture of *syn* and *anti* aminols. Lithium alkyls and allyl magnesium chloride have been used as nucleophiles.

The results (Table 7) show that nearly all types of lithium alkyls may be employed. Analysis of the diastereomeric ratio clearly shows a predominance of the *syn* isomer in almost all cases. However, an interesting trend is the increasing amount of *anti* isomer formed on going from primary to secondary and tertiary butyllithium. A complete reversal of the diastereofacial selectivity does, in fact, occur. For *n*-BuLi, the *syn-anti* ratio is 91:9; for *t*-BuLi is 12:88. [23]

Table 8: Synthesis of 1,2 aminols from lactic or mandelic aldehydes via silylimines

OTBDMS, R, N, TMS + R_1—M → OTBDMS, R_1, R, NH_2 (*Syn*) + OTBDMS, R_1, R, NH_2 (*Anti*)

Entry	R	R_1	*Syn* : *Anti*	Yield, %
1	Methyl	AllylMgBr	25 : 75	50
2	Methyl	AllylMgCl	4 : 96	66
3	Methyl	BenzylMgCl	>98 : <2	50
4	Methyl	CyclohexylmethylLi	80 : 20	70
5	Methyl	n-ButylLi	98 : 2	46
6	Methyl	sec-ButylLi	57 : 43	39
7	Phenyl	BenzylMgCl	>98 : <2	66
8	Phenyl	AllylMgCl	18 : 82	65

As expected, α-aminols can also be prepared starting from silylimines of α-silyloxy aldehydes and lithium alkyls and allyl- or benzylmagnesium halides. The reaction proceeds very fast even at -78° C and requires only one equivalent of the organometallic reagent. The diastereomeric ratio shows a pronounced *syn* selectivity that can reach 98-100% as illustrated for benzyl magnesium chloride. Again, *sec*-butyllithium gives a larger amout of *anti*-adduct in comparison with *n*-butyl lithium. Most interestingly, contrary to the case of the corresponding aluminum derivatives, allylmetals invariably lead to the predominant formation of the *anti* isomer in 75-95% yield.[24]

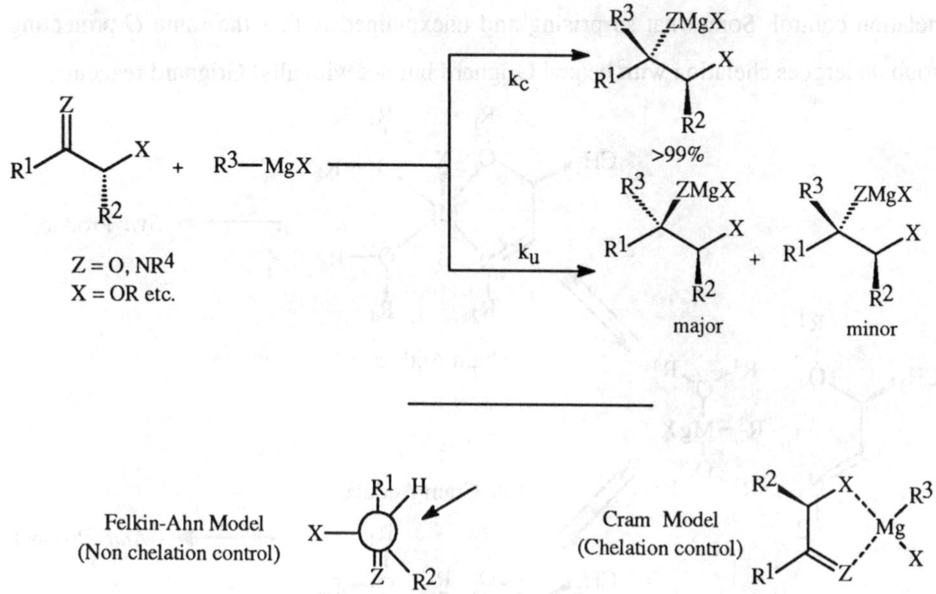

Z = O, NR4
X = OR etc.

R^3—MgX

k_c >99%

k_u major minor

Felkin-Ahn Model
(Non chelation control)

Cram Model
(Chelation control)

Scheme 25

To explain the high *syn* diastereofacial selectivity generally observed in the addition of an organometallic reagent to α-heterosubstituted carbonyl compounds and imines, Cram[29] proposed a cyclic model in which an unsaturated bond and a α-heterofunction are held in an approximately coplanar arrangement by simultaneous chelation with the metal; the nucleophile then adds from the side of the smaller of the two remaining substituents. It has been found by House[30] and Ashby,[31] at least for the case of Grignard reagents, that the reaction is first order in the organometallic compound. This result strongly suggests that the alkyl group is transferred to the sp^2 carbon from the substrate-organometallic complex.

To explain the *syn/anti* diastereomeric ratio observed, a partitioning of the reaction path *via* a chelated transition state, which exclusively gives the *syn* isomer, and an unchelated one of comparable activation energy that affords, probably via a Felkin-Ahn model, a preponderant amount of the *anti* diastereoisomer is generally assumed. This partition has been named by Reetz chelation and non chelation control (Scheme 25).[32]

According to this interpretation, the very high *syn* diatereofacial selectivity observed in the addition of benzyl magnesium chloride to silylimines is due to chelation control whereas the high *anti* selectivity, seen with allyl magnesium halides, is the result of non chelation control. Somewhat surprising and unexplained is that the same *O*-protecting group undergoes chelation with benzyl Grignard but not with allyl Grignard reagents.

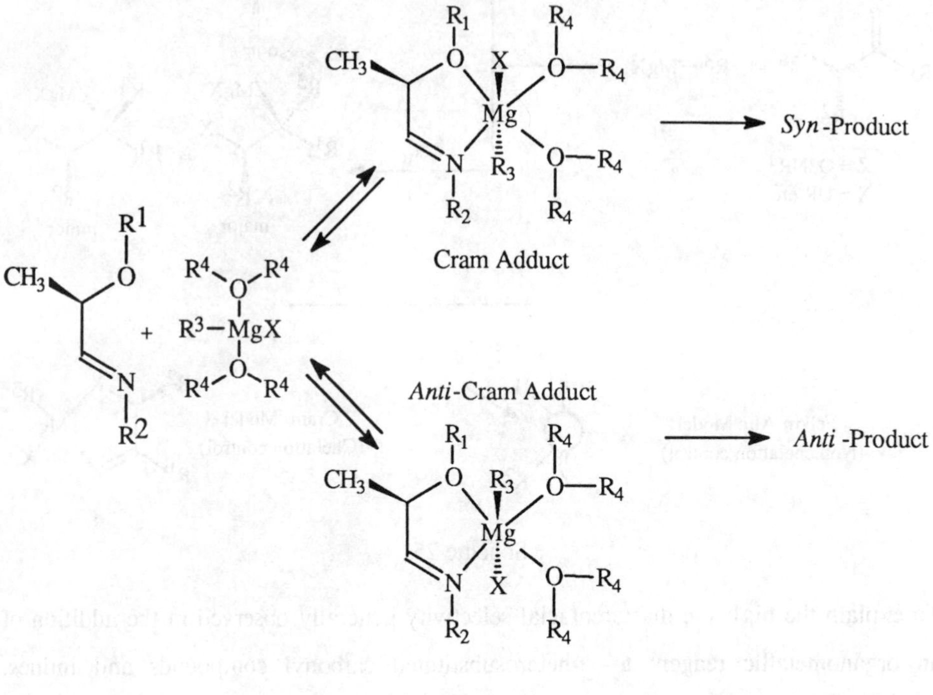

Cram Adduct → *Syn*-Product

Anti-Cram Adduct → *Anti*-Product

Scheme 26

The greater basicity and, presumably, the higher chelation ability of the azomethine compared to the carbonyl, allows the alternate but tentative explanation that both *syn* and *anti* isomers are formed by a chelation control reaction. If we assume that the attacking Grignard reagent in a desaggregating solvent such as THF is monomeric and double solvated and, furthermore, that the reaction proceeds via a rapid formation of a complex between imine and organomagnesium reagent followed by rate determining alkyl transfer, the addition could be formulated to take place by the formation of two chelated complexes (Scheme 26). Hexacoordinated complexes of this kind have several diastereoisomers. We are interested here, however, in the two complexes in which the alkyl group is in an axial position and *cis* or *trans* related with the methyl group of the lactic imine moiety. Diastereoisomers with the alkyl group in an equatorial position apparently do not lead to the final product. The Cram-adduct leads to the *syn* product while the *anti*-Cram adduct should afford the *anti* product. If the Curtin-Hammett principle is operative and the activation energies of the two alkyl transfer processes leading to *syn* and *anti* isomers are reasonably assumed to be similar, the final *syn-anti* ratio will essentially depend upon the relative stability of the two chelated complexes. This rationalization of the stereochemical results requires the assumption that the benzyl Grignard and allyl Grignard additions favor Cram and *anti*-Cram adducts respectively. The increase in the amount of *anti* adduct by increasing the size of the alkyl group, as observed with silyl and aluminum imines, may be rationalized by assuming an equilibration between Cram and *anti*-Cram adducts which favors the latter.[33] This explanation requires amplification: allyl magnesium chloride is *anti* stereoselective with silyl imines and *syn* stereoselective with aluminumimines. It is known that the addition of allylmetals to imines is reversible.[21] The equilibrium is dependent on the nitrogen substituent and the reaction conditions. In the case of allylmagnesium chloride addition to the ate-complex of aluminum imines, the negative charge in the α-position of the imine nitrogen and the reaction conditions (room temperature for 24 hours) favour equilibration and therefore the formation of more stable *syn* adduct. However, a very high *anti* diastereofacial selectivity has been observed using organocopper-boron trifluoride (Table 9).[34]

Treatment of silylimines of *O*-protected mandelic or lactic aldehydes with two equivalents of organocopper reagent, prepared from the corresponding Grignard, leads in moderate to good yields to the α-aminols with high *anti*-diastereoselectivity. Thus, a remarkable metal-tuning has been observed. In fact, if the copper reagent is generated from a lithium alkyl, a *syn* stereoselectivity is observed (entries 6 and 7, Table 9).

Table 9: Synthesis of 1,2 aminols from silylimines and organocopper reagents

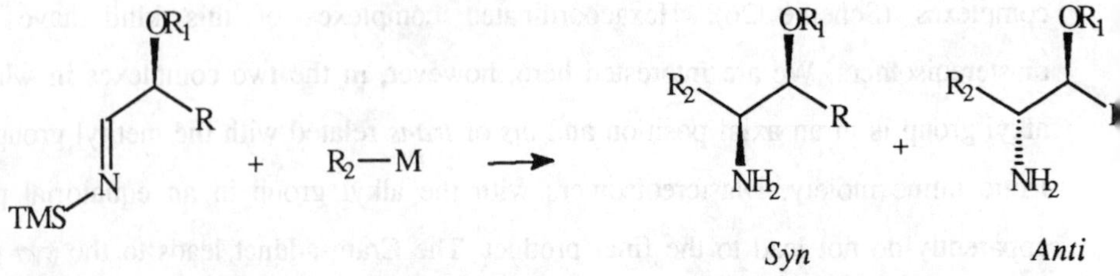

Entry	R	R^1	R^2	*Anti* : *Syn*	Yield %
1	Methyl	TBDMS	n-ButylCu.MgBrI.BF_3	>98 : <2	84
2	Methyl	ter-Butyl	n-ButylCu.MgBrI.BF_3	>98 : <2	40
3	Methyl	Trityl	n-ButylCu.MgBrI.BF_3	92 : 8	86
4	Methyl	TBDMS	AllylCu.LiI.BF_3	91 : 9	54
5	Methyl	TBDMS	AllylCu.MgClI.BF_3	>98 : <2	63
6	Methyl	TBDMS	CyclopropylCu.LiI.BF_3	10 : 90	51
7	Methyl	TBDMS	CyclopropylCu.MgBrI.BF_3	95 : 5	40
8	Methyl	TBDMS	CyclohexylCH_2Cu.MgBrI.BF_3	>98 : <2	70
9	Phenyl	TBDMS	n-ButylCu.MgBrI.BF_3	>98 : <2	24
10	Phenyl	ter-Butyl	n-ButylCu.MgBrI.BF_3	97 : 3	37
11	Phenyl	ter-Butyl	MethylCu.MgBrI.BF_3	>98 : <2	30
12	Phenyl	TBDMS	AllylCu.MgClI.BF_3	81 : 19	67

The very high *anti* diastereoselectivity in the addition of Grignard derived reagents may be rationalized in terms of an open chain Conforth dipolar model. Boron trifluoride is a good Lewis acid capable of complexing oxygen and nitrogen functions but incapable of forming chelated complexes. In our case, two molecules of Lewis acid coordinate both nitrogen and oxygen atoms of the imine resulting in the formation of a complex with a rigid antiperiplanar conformation due to electrostatic repulsion. The attack of the organocopper reagent occurs from the less hindered diastereotopic face, probably *via* an open chain transition state (Scheme 27).

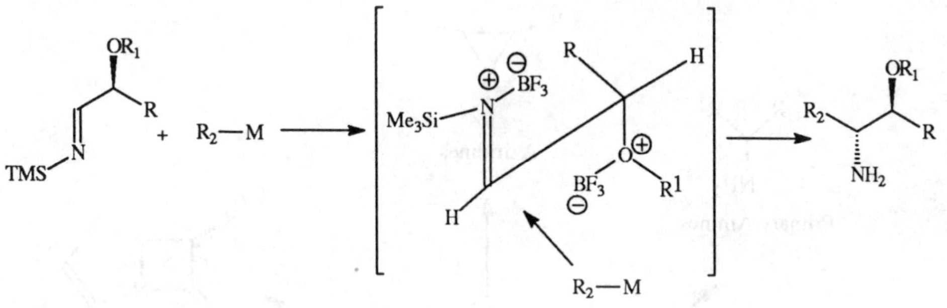

Scheme 27

5. Conclusion

In conclusion, we have prepared, for the first time, chiral *N*-aluminum- and *N*-trimethylsilyl imines starting from aldehydes and nitriles. These compounds represent interesting starting materials for the preparation of a variety of biologically active nitrogen containing compounds.

Through the ester enolate-imine route, we have been able to prepare with high diastereo- and enantio-selectivity carbapenems such as PS-5 and PS-6 as well as 1β-methyl carbapenems and monobactams. A promising application is represented by the stereospecific synthesis of chiral aziridines, potential starting materials for the preparation of α- and β-amino acids. Metallo imines also undergo reaction with a number of organometallic compounds to give primary amines and chiral 1,2-aminols.

We studied the stereochemical outcome of all these reactions and its dependence on the nature of the cations and the structure of the starting imines involved and we have proposed a tentative rationalization of the stereochemical results in order to overcome the limits of the classical models usually employed.

Work is in progress in our laboratory in order to further develop the use of metallo imines in the synthesis of complex molecules and better understand the stereochemistry of the nucleophilic addition to these species.

Aziridines

Primary Amines

Carbapenems
1β-Methyl-carbapenems
Penems

Syn α-Aminols

Monobactams

Anti α-Aminols

References and footnotes

[1] Hart, D.J.; Ha, D.C.; *Chem Rev.* **1989**, 89, 1447; van Koten, G.; van der Steen, F.H.; *Tetrahedron* **1991**, 47, 7503; Georg, G.I.; "Studies in Natural Product Chemistry" Rahman, A-ur (Ed.) Vol. 2 Elsevier Science, Amsterdam.

[2] Gilman, H.; Speeter, M.; *J.Am.Chem.Soc.* **1943**, 65, 2255.

[3] Kruger, C.; Rochow, E.G.; Wannagat, U.; *Chem.Ber.* **1963**, 96, 2132.

[4] Cainelli, G.; Giacomini, D.; Panunzio, M.; Martelli, G.; Spunta, G.; *Tetrahedron Lett.* **1987**, 28, 5369.

[5] Cainelli, G.; Contento, M.; Giacomini, D.; Panunzio, M.; *Tetrahedron Lett.* **1985**, 26, 937.

[6] Shih, D.H.; Baker, F.; Cama, L.; Christensen, B.G.; *Heterocycles* **1984**, 21,29. Perrone, E.; Franceschi, G.; "Synthesis of Penems" in *Recent Progress in the Chemical Synthesis of Antibiotics* G.Lukacs and M.Ohno Eds. Springer Verlag 1990 p 613.

[7] Seebach, D.; Prelog, V.; *Angew. Chem. Int. Ed. Ingl.* **1982**, 21, 654.

[8] Cainelli, G.; Panunzio, M.; Giacomini, D.; Martelli, G.; Spunta, G.; *J.Am.Chem.Soc.* **1988**, 110, 6879.

[9] Andreoli, P.; Cainelli, G.; Panunzio, M.; Bandini, E.; Martelli, G.; Spunta, G.; *J.Org.Chem.* **1991**, 56, 5984.

[10] Thomas, R.C.; "Synthetic Aspects of Monocyclic β-Lactam Antibiotics" in *Recent Progress in the Chemical Synthesis of Antibiotics* G.Lukacs and M.Ohno Eds. Springer Verlag 1990 p 533.

[11] Andreoli, P.; Billi, L.; Cainelli, G.; Panunzio, M.; Bandini, E.; Martelli, G.; Spunta, G.; *Tetrahedron* **1991**, 47, 9061.

[12] Bandini, E.; Cainelli, G.; Giacomini, D.; Martelli, G.; Panunzio, M.; Spunta, G.; *Bioorg. Med. Chem. Lett.* **1993**, 3, 2347.

[13] Blake, J.F.; Jorgensen, W.L.; *J.Org.Chem.* **1991**, 56, 6052.

[14] Hart, D.J.; Lee, C.S.; Pirkle, W.H.; Hyol; M.H.; Tsipouras, A.; *J.Am.Chem.Soc.* **1986**, 108, 6054 and references cited therein.

[15] Ireland, R.E.; Mueller, R.H.; Willard, A.K.; *J.Am.Chem.Soc.* **1976**, 98, 2868.

[16] Overman, L.E.; Osawa, T.; *J.Am.Chem.Soc.* **1985**, 107, 1698.

[17] Reeve, W.; Erikson, C.M.; Aluotto, P.F.; *Can.J.Chem.* **1979**, 57, 2747.

[18] Brieva, R.; Crich, J.Z.; Sih, C.; *J.Org.Chem.* **1993**, 58, 1068.

[19] Padwa, A.; Woolhouse, A.D.; "Aziridines, Azirines and Fused Ring Derivatives" in *Comprehensive Heterocyclic Chemisty,* Lwowski, W. (Ed.) Pergamon Press **1984** Vol. 7.

[20] Cainelli, G.; Panunzio, M.; Giacomini, D.; *Tetrahedron Lett.* **1991**, 32, 121.

[21] Courtois, G.; Miginiac, L.; *J.Organometal.Chem.* **1974**, 69, 1;Yamamoto, Y.; Asao, N.; *Chem.Rev.* **1993**, 93, 2207.

[22] Miginiac, L.; Mauze, B.; *Bull.Soc.Chim.Fr.* **1968**, 3832.

[23] Cainelli, G.; Panunzio, M.; Contento, M.; Giacomini, D.; Mezzina, E.; Giovagnioli, D.; *Tetrahedron* **1993**, 49, 3809.

[24] Cainelli, G.; Giacomini, D.; Mezzina, E.; Panunzio, M.; Zarantonello, P.; *Tetrahedron Lett.* **1991**, 32, 2967.

[25] Hart, D.; Kanai, K.; Thomas, D.G.; Yang, T.K.; *J.Org.Chem.* **1983**, 48, 289.

[26] Cainelli, G.; Giacomini, D.; Panunzio, M.; unpublished results.

[27] Bernardi, F.; Bongini, A.; Cainelli, G.; Robb, M.A.; Suzzi Valli, G.; *J.Org.Chem.* **1993**, 58, 750.

[28] Bongini, A.; Giacomini, D.; Panunzio, M.; Suzzi Valli, G.; Zarantonello, P.; *Spectrochimica Acta Part A* **1995,** 000.

[29] Cram, D.J.; Kopecky, K.R.; *J.Am.Chem.Soc.* **1959**, 81, 2748.

[30] House, H.O.; Oliver, J.E.; *J.Org.Chem.* **1968**, 33, 929.

[31] Laemmle, J.; Ashby, E.C.; Neumann, H.M.; *J.Am.Chem.Soc.* **1971**, 93, 5120.

[32] Reetz, M.T.; *Angew. Chem. Int. Ed. Engl.* **1984**, 23, 556.

[33] Work is in progress in order to confirm our hypothesis by means of theoretical calculations.

[34] Cainelli, G.; Giacomini, D.; Panunzio, M.; Zarantonello, P.; *Tetrahedron Lett.* **1992**, 33, 7783.

TARGET-DRIVEN ORGANIC SYNTHESIS: REFLECTIONS ON THE PAST, PROSPECTS FOR THE FUTURE

Stephen Hanessian
Université de Montréal
Department of Chemistry, C.P. 6128, Centre-ville, Montréal, Canada, H3C 3J7

1. Abstract

Organic synthesis is undergoing a dramatic change with regard to the choice and relevance of target molecules. This is instigated by a conscious effort on the part of the synthetic chemist to bridge the gap between biology and chemistry and to gain better understanding of interactions between organic molecules with their biological counterparts. Natural and unnatural products are being actively pursued on several fronts with the aid of sophisticated tools such as X-ray crystallography, high field NMR spectroscopy and computer-assisted methods. This article discusses aspects of target-driven synthesis with highlights of research activities in the author's laboratory over the past ten years.

The levels of achievement in the total synthesis of natural products and related molecules over the past five decades have been a yardstick by which to follow the evolution of the creativity, ingenuity and courage of synthetic organic chemists. Indeed, a cursory glance at the structures of representative molecules synthesized to date reflects the heights to which this subdiscipline has been elevated [1]. Suffice it to marvel at the achievements in the fifties and sixties, when techniques such as I.R., and U.V., coupled with well-taught laboratory skills (crystallization, distillation and limited chromatography) were the handful of available tools to synthetic chemists. On a relative scale of assessing elegance and ingenuity, such syntheses could be regarded as "classics" in the annals of organic synthesis [2]. Subsequent achievements must be compared accordingly, particularly since we are not necessarily so far ahead in the time it takes to complete a total synthesis of a "medium-sized" molecule in spite of the advent of sophisticated instrumentation and powerful methodology. These remarks in no way diminish the recent impressive accomplishments of synthetic organic chemists in the area of natural products. The fact is, that whereas total synthesis was within the grasp of

C. Chatgilialoglu and V. Snieckus (eds.), Chemical Synthesis, 61–90.

relatively few practitioners three of four decades ago, it is now within reach of all those who wish to climb such mountains, and are well equipped to do it.

The total synthesis of natural products has had enormous appeal to the community of chemists in general over the years. In spite of recently evoked opinions on the subject [3], it is the contention of this author that a judiciously chosen target molecule for synthesis remains an intellectually and practically most rewarding challenge. Since young coworkers are invariably involved, the exercise of designing synthesis strategies and rendering them to practice in the laboratory is an invaluable training ground for future independent scientists. The question to ask therefore is not whether or not to be involved in natural product synthesis, but rather, what molecule to synthesize and why? Figure 1 illustrates different facets of targeted syntheses and why we do what we do from the germination of an idea to its realization.

WHY WE DO WHAT WE DO IN TOTAL SYNTHESIS

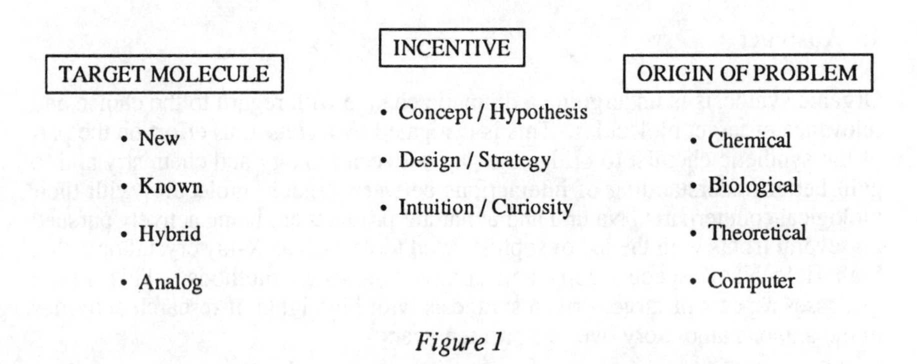

Figure 1

Thus, the impetus to embark upon a synthetic endeavor may come from knowledge acquired through a biological phenomenon, a chemical observation, information provided by computer technology, or other sources. In each case, there will be different incentives for designing and synthesizing a target molecule. A problem arising from the originally observed biological activity of a given molecule may bring forward a concept or a hypothesis which must be proved or disproved through the availability of analogous molecules or a cleverly conceived structure. Other reasons such as curiosity, intellectual satisfaction and a desire to deepen our understanding of phenomena at the molecular level may also prevail, leading to the need to synthesize a given target molecule.

There are a number of valid reasons to synthesize a *known* natural product, beyond the realm of student training. Accessibility through total synthesis is often the only means to provide enough material for pharmacological evaluation, and to eventually produce high volume quantities through chemical manufacture in the plant. In fact, total synthesis using efficient and economically viable processes may in some cases surpass the capacity of a fermentation source for a natural product. A number of antibiotics which, with few exceptions are enantiomerically pure molecules, are presently produced *via* total synthesis and sold in tonnage quantities [4].

A well designed strategy toward a biologically relevant natural product may provide valuable intermediates for a host of non-chemical studies, such as the evaluation of toxicity and resistance, understanding pharmacological behavior, and elucidating metabolic and biosynthetic pathways to mention a few. One of the more important aspects of natural product synthesis utilizing versatile strategies is the possibility to generate analogs that are not directly available form natural sources. Since nature is primarily concerned with the production of single enantiomers, chemical synthesis can, through man-made design or through its "flaws", give rise to enantiomers or diastereomers. These in turn could be invaluable tools for understanding the mode of action and other subtle effects of the natural product itself at the molecular level. How could we have known that (*R*)-penicillamine obtained by synthesis had mutagenic properties, while the natural (*S*)-isomer was an antiarthritic agent, were if not for the ability of synthetic chemists to produce the unnatural isomer? [5]. The lack of predictions in the relationship between stereochemical features in a pair of enantiomers for example, and their pharmacological effects has necessitated the institution of strict guidelines for the drug industry [6]. Thus, a racemic form of a drug candidate must undergo rigorous biological evaluation of *both* of its enantiomers before approval can be given for subsequent clinical studies.

The total synthesis of natural products and their structural variants by chemical means may indirectly expand our understanding of structure-activity relationships of the parent molecule. For example, the biological activity and therapeutic profile of many antibiotics of natural origin have been improved by chemical modification from a knowledge of their modes of action, inactivation by endogenous enzymes and other molecular events.

One can therefore justifiably defend the cause of total synthesis, provided that the criteria for relevance are satisfied, and the need for combining innovation with practicality is realistically achieved. Broadening our vision of total synthesis as a research endeavor, we can envisage several additional scenarios, objectives, goals and other facets yet unknown to keep us committed to excellence. Figure 2 summarizes the ever-increasing ramifications of total synthesis of natural and unnatural molecules.

Great advances on the biological front have opened up enormous possibilities for synthetic work, particularly of medicinally important "small" molecules. The design and synthesis of enzyme inhibitors, antagonists, and agonists for example, has involved a great deal of ingenuity in stereocontrolled organic synthesis.

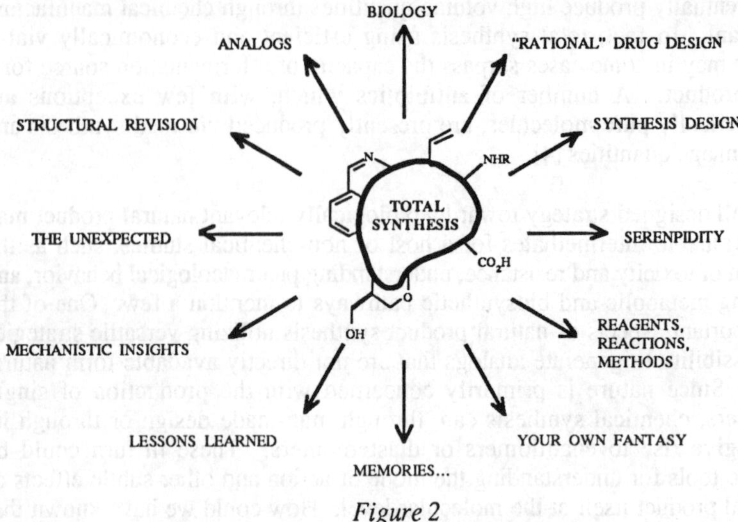

Figure 2

Target molecules may be related to natural products or they may be totally "unnatural", arising from a knowledge of the three dimensional X-ray structure of an enzyme's active site for example. Extensive structure-activity data on a series of compounds in combination with X-ray crystallography, molecular modeling, and computational techniques may suggest a "lead compound" for synthesis [7]. Indeed, it is through this type of total synthesis that molecules exhibiting nanomolar levels of biological activity *in vitro* and *in vivo* have been attained. The emphasis on total synthesis has therefore shifted in part from real natural products such as those offered by fermentation, etc. to man-made molecules based on biological or physico-chemical parameters.

Thus, the synthetic chemist of the third millennium has compelling reasons to continue to explore the long and arduous routes to target molecules of choice. Such molecules may be new natural products, analogs of known structures, hybrid structures, or totally novel entities.

In the following pages, I shall summarize various aspects of target-driven syntheses using representative examples from our own laboratory, and following some of the guidelines shown in Figures 1 and 2. For the purposes of practicality and brevity, I shall focus on a number of synthetic endeavors, discussing our rationales and highlighting any relevant aspects to organic, bio-organic and medicinal chemistry. By target-driven synthesis, I mean the pursuit of an objective by the most appropriate method already available or discovered during the process. Thus, the emphasis is on the attainment of the target structure, rather than on the search of molecules that can be synthesized by a given method.

2. Synthesis Design of Clinically Relevant Molecules Using the Chiron Approach

2.1. PENEM FCE-22101

Woodward's original synthesis of the penems [8], provided an impetus for the design of related structures in which various features of penams and carbapenems were included in hybrid structures. Farmitalia's FCE 22101 [9] and related compounds from the Schering-Plough group [10] in the early 1980's, paved the way for a rapid expansion of synthetic work in this area. Following a visit to the Farmitalia laboratories in Milan in 1981, we were given the opportunity to investigate a number of novel synthetic approaches to their lead structure in my own laboratory. The challenges were clear and the mandate was obvious. We were to design a totally novel route to the intended penem with the hope of providing the process group with procedures capable of being scaled-up to kg quantities. Having the assurance that the entire fate of the project did not rest solely on my shoulders, since an active effort was on-going in Milan as well, we embarked on this ambitious project in the microenvironment of an academic research laboratory. Little did we know, that major research efforts had also been committed to related structures in other parts of the world in a number of pharmaceutical laboratories. Our first objective was to devise a synthesis of a presently commercially available building block for carbapenem and penem synthesis. Our disconnective analysis for penem FCE 22101 is shown in Scheme 1, where L-threonine emerged as a chiral non-racemic template for the construction of the 4-acyloxyazetidinone [11], which was our first objective [12]. The route we developed using L-threonine was in fact amenable to scale-up and led to the production of the first batches of the desired azetidinone for further elaboration to the penem using an ingenious carbonyl coupling strategy developed in the Farmitalia group [13], and also reported by the Schering group [10]. We had also pursued our second objective which was to complete the synthesis of the intended penem itself. This was accomplished by exploiting nitro-olefin chemistry through the use of the Gomper reagent [14], and the chemoselective manipulation of reactive intermediates. Although conceived and executed in Montreal, the total synthesis of penem FCE 22101 is a credit to the technical skills and dedication of three Farmitalia chemists who spent sabbatical periods in my laboratory.

In another project, we applied free-radical allylation methodology to prepare 6-β-allyl penicillin sulfones (Scheme 1) which exhibited interesting β-lactamase inhibitory activity [15].

2.2. THIENAMYCIN

The successful completion of the penem FCE 22101 synthesis provided additional interest and impetus in our laboratory to explore other aspects of nitro-olefin chemistry in conjunction with the total synthesis of thienamycin [16]. Work at the Merck laboratories had demonstrated several elegant syntheses of thienamycin and related carbapenems [17]. Their ingenious application of rhodium-catalyzed diazo-insertion reactions for the assembly of the carbapenem skeleton with minimal protective groups is noteworthy. The literature is abound with methods dealing with the synthesis of azetidinone intermediates [18] directed at the total synthesis of such carbapenems which have remarkable antibacterial activities.

Our synthesis of thienamycin (Scheme 1) relied on the incorporation of a nitro-olefinic appendage on an azetidinone precursor and a kinetically controlled intramolecular Michael addition to construct the carbapenam nucleus [16]. Subsequent manipulation would lead eventually to the desired thienamycin. Although the synthesis is not economically viable, it did provide us with the opportunity to apply novel methodology in the construction of the somewhat delicately functionalized carbapenem skeleton. Although the immediate precursor of the target molecule was prepared as planned, the removal of the protective O-t-butyldimethylsilyl group required a major effort. Only after extensive experimentation with various variations of fluoride-catalyzed reactions, it was possible to achieve partial desilylation, which required the recycling of the unreacted material. The much maligned resistance of protective groups to cleavage in the penultimate stages of multi-step syntheses can be a frustrating experience. In spite of such obstacles, there is great satisfaction in devising synthetic strategies for the synthesis of clinically important molecules such as

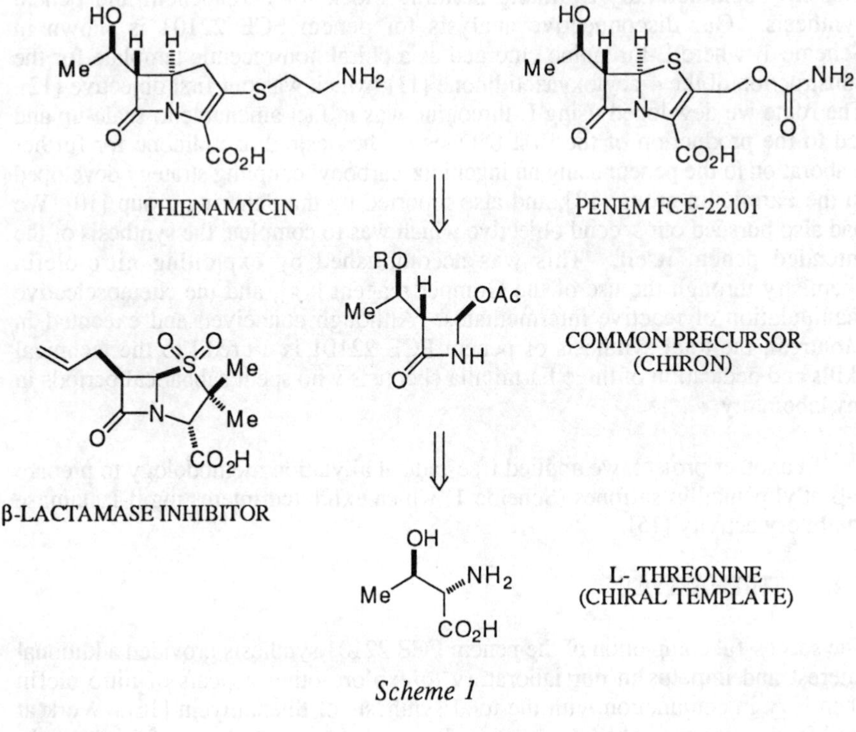

THIENAMYCIN

PENEM FCE-22101

COMMON PRECURSOR
(CHIRON)

β-LACTAMASE INHIBITOR

L- THREONINE
(CHIRAL TEMPLATE)

Scheme 1

thienamycin, to test out the validity of the design, and to complete the task of preparing sufficient quantities for biological testing.

2.3. DIHYDROMEVINOLIN

The quest for effective hypocholesterolemic agents during the past decade or so has culminated with the discovery of mevinolin (Mevacor), a member of the mevinic acid group of fungal metabolites [19]. Because of their important pharmacologic activity and their interesting structures, a large number of groups have contributed to methodology aimed at their total synthesis [20]. Dihydromevinolin exhibits similar biological properties to mevinolin, but it is produced only in small quantities during the fermentation process. Two total syntheses using different approaches were already reported [21] when we embarked on our synthesis. Our disconnective analysis, shown in Scheme 2, illustrates another example of a design element in natural product synthesis utilizing the Chiron approach [22,23]. Thus, L-glutamic acid was chosen as a chiral template which would eventually be incorporated in ring A of the octahydronaphthalene

DIHYDROMEVINOLIN

L-Glutamic acid

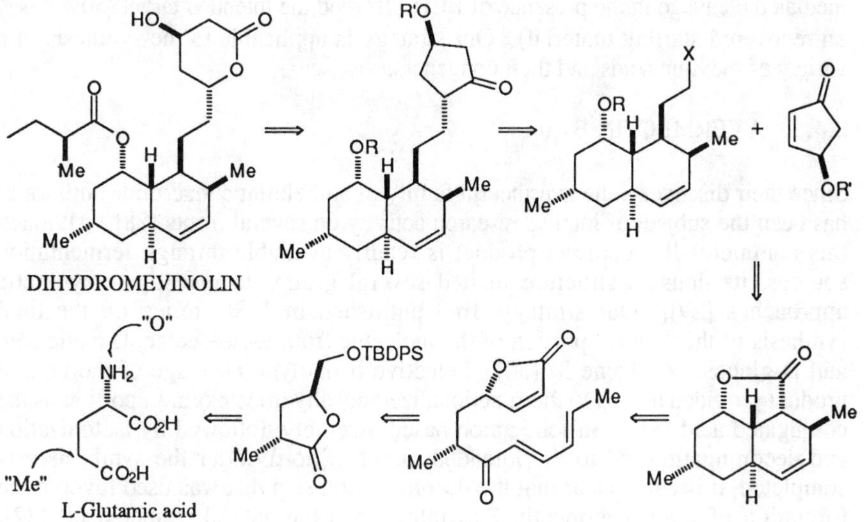

Scheme 2

unit. While the amino acid motif is far removed, at least visually, from the intended substructure, its lactone "equivalent", easily obtained in a few steps, is not. Thus it was our previous experience in manipulating butyrolactone and butenolide motifs [24], that led us to design a strategy in which a butenolide having an appropriate dienic appendage could serve as a suitable template for an intramolecular Diels-Alder reaction en route to the octahydronaphthalene intermediate. The connecting two carbon appendage to the lactone ring was introduced *via* a nitroalkane mediated Michael addition [25] to an enantiomerically enriched cyclopentenone derivative, readily available *via* a chemo-enzymatic route [26]. The next challenge was the oxidative expansion of the cyclopentanone ring to the desired lactone. Although the pattern of substitution was suitable for such an oxidation, it was clear that a Baeyer-Villiger reaction would not be chemoselective due to the presence of the double bond in ring B. This problem was already considered before embarking on such a strategy and a series of model oxidative ring expansions were effected with rewarding results. Bis-(trimethylsilyl)peroxide, in conjunction with $BF_3{\bullet}Et_2O$, which had been shown to effect Baeyer-Villiger oxidations of simple ketones without affecting olefinic bonds was in fact the reagent of choice [27], and proved to be admirably adaptable to the critical oxidation in our advanced intermediate. Having crossed several functional, stereochemical and chemoselective hurdles, we were once again faced with the choice (and removal) of a protective group. What may prove to be highly compatible through several difficult chemical transformations, may ultimately exhibit its own signs of battle fatigue. Such was the case when we attempted to remove a benzyl ether from the lactone portion of dihydromevinolin. Again, because of the presence of the double bond, hydrogenolysis was out of the question. After many attempts with a variety of reagents, the classical Lewis acid mediated cleavage in the presence of BCl_3 afforded the intended target (50% based on recovered starting material). Our strategy is applicable to the synthesis of a variety of mevinic acids and their congeners.

2.4. AVERMECTIN B_{1A}

Since their discovery, the avermectin family of anthelmintic macrolide antibiotics has been the subject of intense research activity on several fronts [28]. Although this commercially lucrative product is readily available through fermentation sources, its unusual structure incited several groups to investigate synthetic approaches [29]. Our strategy, first published in 1986, relied on the total synthesis of the "upper" portion of the molecule from L-isoleucine, L-malic acid and D-glucose (Scheme 3) [30]. Selective ozonolytic cleavage of the natural product provided access to the functionalized hexahydrobenzofuran portion as the conjugated acid [31]. Sulfone anion based assembly, followed by lactonization and deconjugation led to the intended target. Shortly after the synthesis was completed, it became clear that the deconjugation step that was used favored the formation of 2-*epi* avermectin B_{1a} rather than the natural isomer itself [32], although a minor portion of the latter was isolated in the initial deconjugation experiments. A solution to this dilemma was soon found by using imidazole in refluxing benzene as an acid-base system in the deconjugation reaction, where an equilibrium mixture was reached containing the desired isomer as a major product [33]. A major impasse was thus cleared and a lesson learned for all.

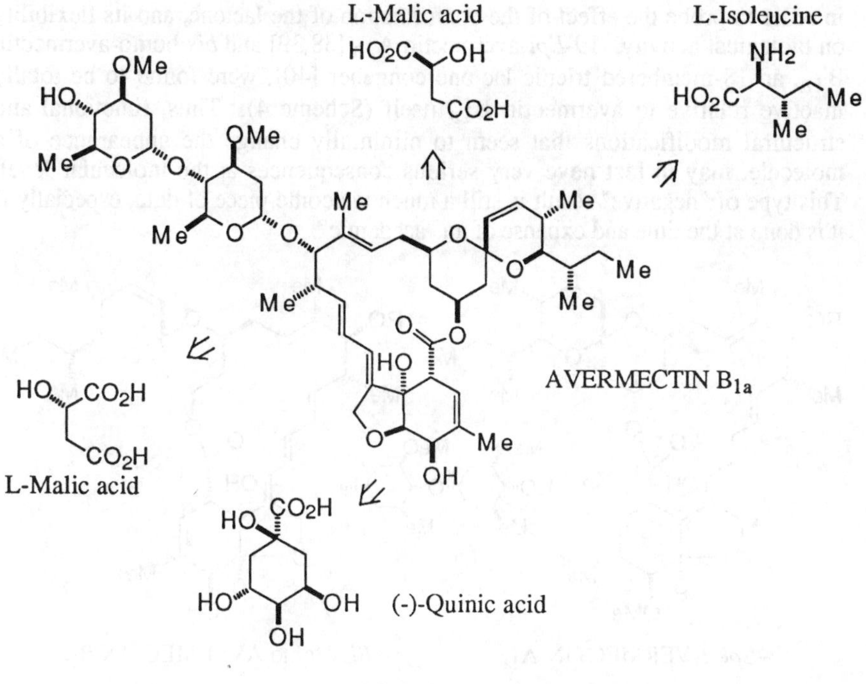

L-Malic acid

L-Isoleucine

AVERMECTIN B_{1a}

L-Malic acid

(-)-Quinic acid

Scheme 3

The *total* synthesis of avermectin B_{1a} aglycone was accomplished using the previously discussed strategy for securing the C_{11}–C_{28} "upper" segment, and the enantiomerically pure hexahydrobenzofuran subunit which was synthesized from (–)-quinic acid [34]. Attachment of the disaccharide portion was done using a glycoside synthesis protocol previously developed in our laboratory [35] and successfully used by Woodward [36] and others [37].

Although none of the presently available syntheses of the avermectins will ever replace the fermentation process for their production, much innovation has come forth as a result of these synthetic studies. For example, our own studies on the selective cleavage of specific double bonds by controlled ozonolysis [31] gave access to valuable substructures of the natural product for a variety of biological and analytical tests. The fascinating effect of imidazole in the deconjugation of the α,β-unsaturated (conjugated) lactone, has been utilized by some investigators in "closing out" their syntheses [29].

Intrigued by the macrolide structure, we constructed a CPK model of avermectin B_{1a} aglycone, only to discover, perhaps not surprisingly, that the macrolactone ring is quite rigid, and that there is practically no "hole" as one might be led to believe from schematic drawings. We then prepared two semi-synthetic analogs from the natural product by systematic chemical manipulation, in order to probe the effect of the conformation of the lactone, and its flexibility on biological activity. 19-*Epi*-avermectin A_{1a} [38,39] and *bis*-homo-avermectin B_{1a}, an 18-membered trienic lactone congener [40], were found to be totally inactive relative to avermectin B_{1a} itself (Scheme 4). Thus, functional and structural modifications that seem to minimally change the appearance of a molecule, may in fact have very serious consequences at the molecular level. This type of "negative" result is still a much welcome piece of data, especially if it is done at the time and expense of an "academic".

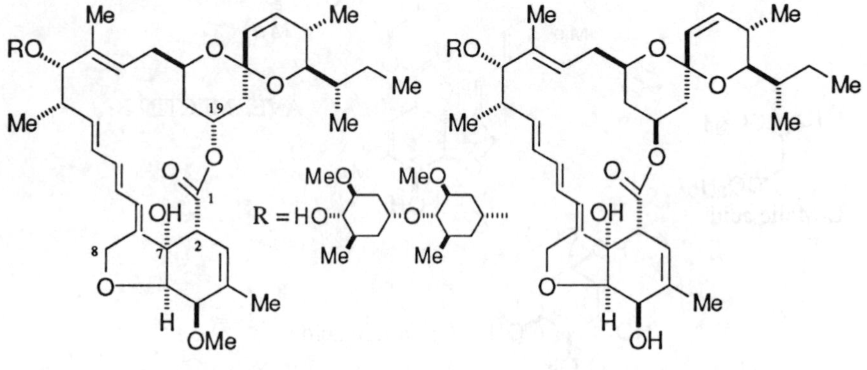

19–*Epi*–AVERMECTIN A_{1a} *Bis*-Homo AVERMECTIN B_{1a}

Scheme 4

2.5. TETRAHYDROLIPSTATIN

Lipstatin and its tetrahydro derivative are representative members of a class of potent inhibitors of pancreatic lipase [41]. Phase II clinical studies have demonstrated the potential for tetrahydrolipstatin as a marketable product against obesity and as a cholesterol lowering agent.

Strategies directed towards the total synthesis of tetrahydrolipstatin are particularly relevant since the molecule will most likely be produced *via* chemical synthesis. To date there are several reports on the total synthesis, each representing different approaches and variable levels of efficiency with regard to number of steps, stereochemical control and practicality [42]. Our total synthesis of tetrahydrolipstatin capitalizes on a combination of internal asymmetric inductions with a high degree of stereocontrol (Scheme 5) [43]. The readily available (*R*)-3-benzyloxytetradecanal was prepared from L-malic acid in order to obtain an enantiomerically pure standard. Alternatively, more practical routes relied on asymmetric allylboronation or on resolution and recycling the wrong enantiomer, followed by benzylation and oxidative cleavage of the double bond. A series of chain extensions involved 1,3-asymmetric induction by the benzyloxy group, and 1,2-asymmetric induction by a β-alkoxyester as previously shown by Fräter [44].

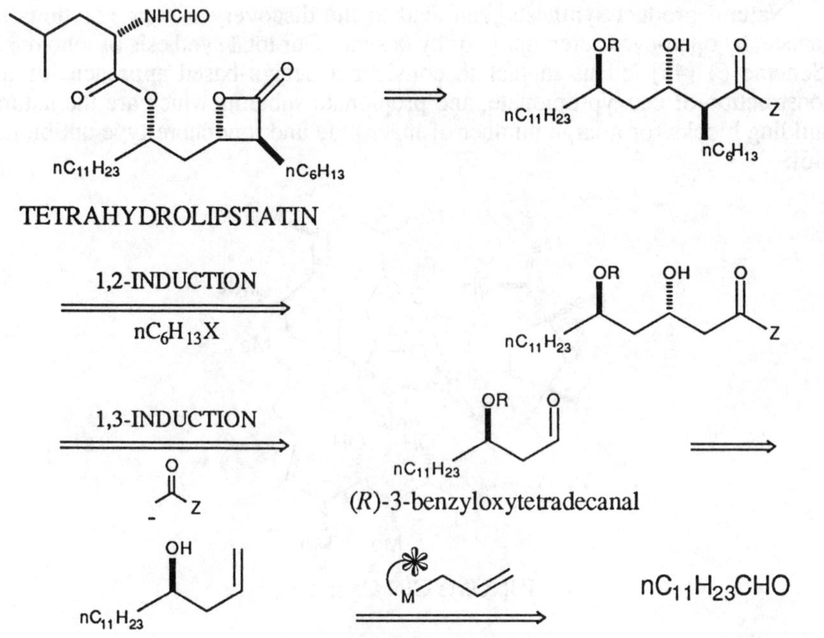

TETRAHYDROLIPSTATIN

1,2-INDUCTION

$nC_6H_{13}X$

1,3-INDUCTION

(*R*)-3-benzyloxytetradecanal

$nC_{11}H_{23}CHO$

Scheme 5

Ultimately, after appropriate functional adjustments, tetrahydrolipstatin was obtained in 38% overall yield from the β-benzyloxyaldehyde in seven steps. The route starting from lauraldehyde *via* allylation and resolution, or *via* asymmetric allylboronation (84% ee at -78°C) comprises 10 steps and 26% overall yield. Although the above outlined synthesis is short in number of steps and quite efficient, it is not clear if it meets the cost effective conditions required for large scale production of tetrahydrolipstatin, particularly because of the low temperature required for some reactions.

3. The Discovery of New Synthetic Methods through Natural Product Synthesis

As previously explained, target-driven total synthesis implies that the prime objective is to reach the intended target by the most expedient, practical, and hopefully innovative method. Achieving such an objective without heavily "borrowing" from already tested synthetic methods may be a very tall order. It is most rewarding indeed when the target molecule is synthesized using totally novel key reactions, in high overall yield, in optimal enantiomeric purity, and in relatively few steps.

Natural product synthesis, can lead to the discovery of new reactions by chance, by open-eyed serendipity, or by design. Our total synthesis of ionomycin (Scheme 6) [45] led us in fact to consider a design-based approach for the construction of deoxypropionate, and propionate subunits which are the natural building blocks for a large number of macrolide and ionophore-type antibiotics [46].

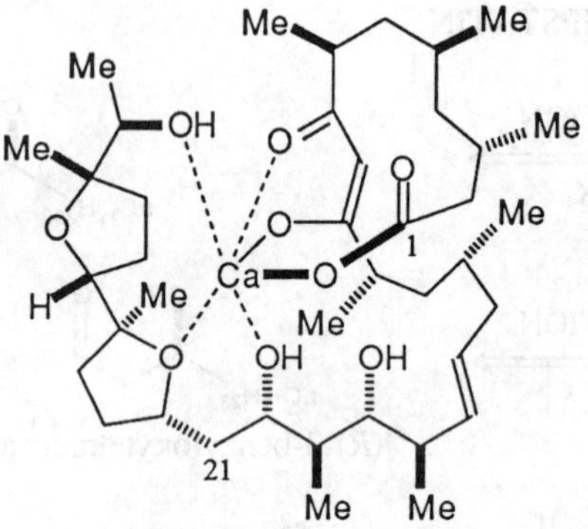

IONOMYCIN Ca salt

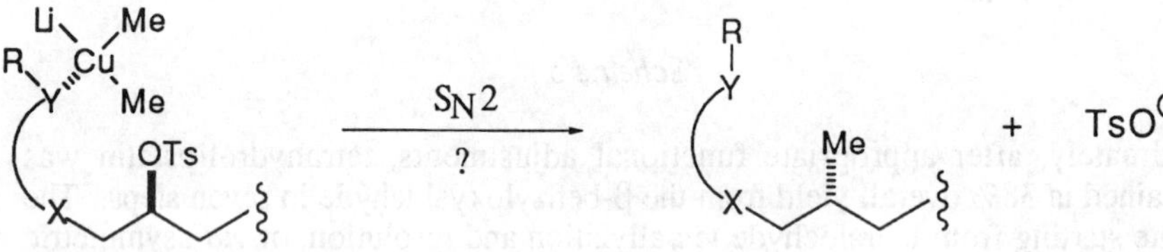

Scheme 6

Our initial approach for the construction of the C_1–C_{21} acyclic portion of ionomycin had relied on a method developed in our laboratory for the construction of propionate subunits with any substitution pattern. This involved the stereocontrolled manipulation of 4-(*S*)-hydroxymethyl butenolide frameworks (derived from L-glutamic acid), their extension, replication and repetition of the sequence (Scheme 7) [24,47]. While this versatile strategy was suitable for carbon chains containing vicinal methyl-hydroxy-methyl substitution patterns, it was not very practical for deoxypropionate motifs (methyl-methylene-methyl) such as one finds in the C_1–C_{14} portion of ionomycin. Clearly, the deoxygenation step would lengthen the sequence leading to such subunits from the butenolide approach. Faced with this problem, it occurred to us that a direct C-methylation of a suitably functionalized carbon chain, as in an S_N2 substitution reaction should perhaps be considered. Capitalizing on the proximity of a phenylthio ether to the displaceable group, we were pleased that *direct* substitution could be achieved with lithium dimethyl cuprate with complete inversion of configuration and virtually no elimination [48]. This discovery was extended to other dialkyl cuprates in S_N2-like displacement reactions of secondary tosyloxy groups utilizing a methylthiomethyl, a methoxymethyl, or a phenylthio ether group as coordination/chelation sites for the reagent with subsequent inversion. As might be expected, there was a preferred distance between the hetero-atom and the tosyloxy group for maximum efficiency in displacement (Scheme 8).

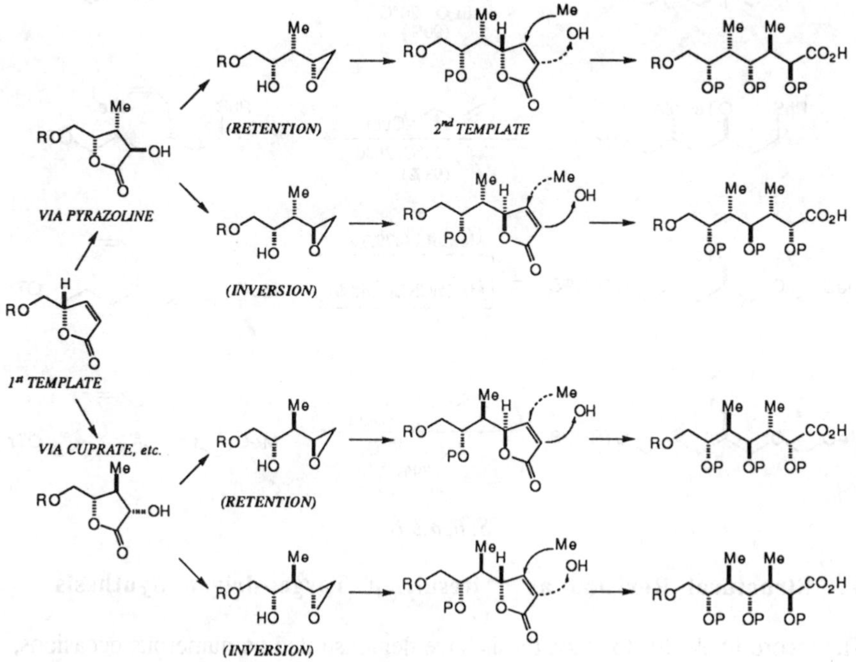

Scheme 7

Although the total synthesis of ionomycin [45] was subsequently completed using well precedented methods, the lasting value of the project will probably rest on the discovery of the S_N2-like displacements of secondary tosyloxy groups with dialkylcuprates with little if any elimination. Such design-based discoveries of new methods or extensions of existing methods, are even more pleasing when they are the result of a desire to optimize the steps of a target-driven synthesis. An aldol strategy has been used for the total synthesis of ionomycin by Evans and coworkers [49].

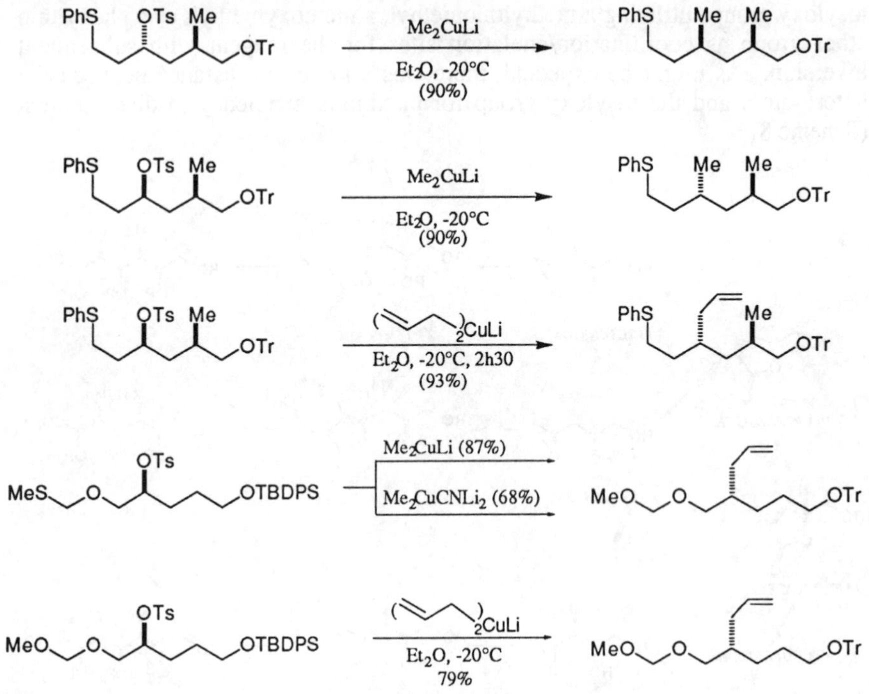

Scheme 8

4. Structural Revision as a Result of Target-driven Synthesis

The record books for total synthesis have demonstrated on numerous occasions, the revision of previously assigned structures as a result of comparing the physical data for synthetic and naturally-derived samples. The most commonly encountered revisions of original structures are reassignments of absolute configurations (wrong enantiomer), structural changes, and configurational changes at certain stereogenic carbon atoms.

Our interest in the polyoxin group of antibiotics [50] was focused on the nature of polyoximic acid [51], a deceptively simple azetidine carboxylic acid derivative which had eluded the efforts of synthetic chemists [52] since the original publication of the structure. Only recently was a synthesis reported for *trans*-polyoximic acid [53], the presumed component of the natural product based on low field NMR studies [51]. A total synthesis of this acid in our laboratory [54,55], and comparison of NMR spectra with a sample kindly provided by Professor Isono showed that they were different. The sample of the naturally-derived amino acid was suitable for X-ray analysis (even after 20 years of isolation and storage!), and it proved to be *cis*-polyoximic acid [54]. A new synthesis of this isomer was undertaken and confirmed its structural as well as stereochemical identity (Scheme 9) [55]. The stereochemical revision of polyoximic acid could have important consequences on its mode of biosynthesis, particularly with regard to the formation of the *cis*, rather than a *trans*-double bond, implying the involvement of a *syn*-periplanar enzymatic elimination of a *pro-R* hydrogen from C-4 of isoleucine. It is also of interest that the synthesis provides enantiomerically pure material. Isolation of polyoxyimic acid from polyoxin by hydrolysis results in complete racemization [51].

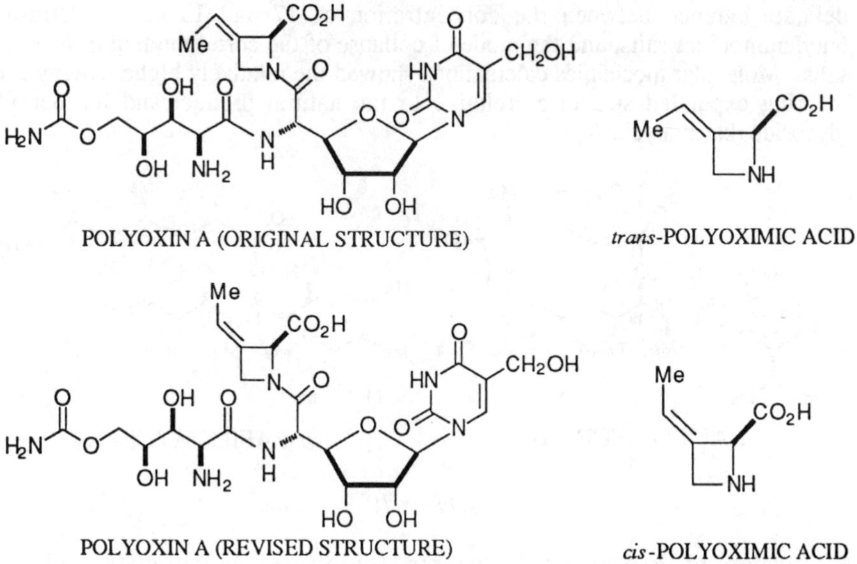

POLYOXIN A (ORIGINAL STRUCTURE)

trans-POLYOXIMIC ACID

POLYOXIN A (REVISED STRUCTURE)

cis-POLYOXIMIC ACID

Scheme 9

5. Serendipity and the Unexpected during Total Synthesis

One of the most pleasant aspects of total synthesis involves the observation of serendipitous and often unexpected chemical reactions. Indeed the annals of synthesis have reported innumerable examples of such events. Sometimes failed reactions or astute observations of reactivity have resulted in the formulation of rules (e.g. Woodward-Hoffmann, Baldwin, etc.). We recount here a recent serendipitous reaction experienced during our studies on the total synthesis and chemical modification of bafilomycin A_1 [56], an unusual 16-membered macrolide antibiotic [57]. In an attempt to obtain a branched derivative, we explored the reaction of 7-keto bafilomycin A_2-TMS ether with methyl copper. We were intrigued by the formation of an isomeric product without the inclusion of a C-methyl group. When the same reaction was applied to 7,21-di-TMS bafilomycin A_2, a new product was formed in high yield. Deprotection of the silyl group led to a crystalline product which proved to be the 18-membered isomer as evidenced by single-crystal X-ray diffraction analysis [58] (Scheme 10).

Thus, an intriguing ring-expansion reaction had taken place, presumably through formation of an alkoxy copper species, attack on the lactone carbonyl group and stereoelectronically favored collapse of an orthoester salt. Interestingly, treatment of the newly formed *iso*-bafilomycin A_2 with fluoride ion in THF, slowly gave the original 16-membered parent structure through a tetra-n-butyl ammonium alkoxide mediated ring contraction. When a dilute solution of fluoride ion in methylene chloride was used, no ring-contraction took place, thus showing a delicate balance between the concentration of 17 and 15 alkoxy tetra-n-butylammonium salts, and the mode of collapse of the corresponding orthoester salts. Molecular mechanics calculations showed the relatively higher energy for the ring expanded structures relative to the natural product and its methyl glycoside (bafilomycin A_2).

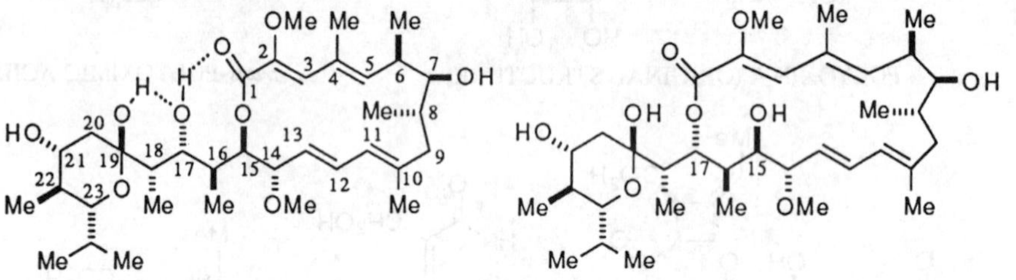

BAFILOMYCIN A_1 *iso*-BAFILOMYCIN A_1

Scheme 10

6. Design of Prototypical Enzyme Inhibitors, Antagonists and the Like

6.1. QUANTAMYCIN

Significant advances in the biological sciences have greatly helped our understanding of processes that take place at the molecular level. Inhibition of protein biosynthesis in bacteria at the ribosomal level by interfering with specific subunits is well documented [59]. In the early eighties, we were presented with a challenging opportunity to synthesize a totally unnatural product which was essentially generated on a computer screen, in conjunction with a hypothetical model for preferred ribosomal binding [60]. Thus was born quantamycin [61], half lincomycin-half peptidyl t-RNA substructure. This model theoretically combines features present in lincomycin and the terminal unit of peptidyl t-RNA represented as f-Met 5'-adenylic acid methyl ester. In its energy minimized conformation, it shows a HOMO associated with the oxygen atom of the amide carbonyl. Such a conformation ($\Delta E = 0.0$ kcal/mol) shows a structure that nicely converges on the f-Met A structure ($\Delta E = 8.4$ kcal/mol), and is much lower in energy. The virtual convergence of presumed critical sites (carbonyl, OH, SMe, amide NH, etc.) was also of interest (Scheme 11).

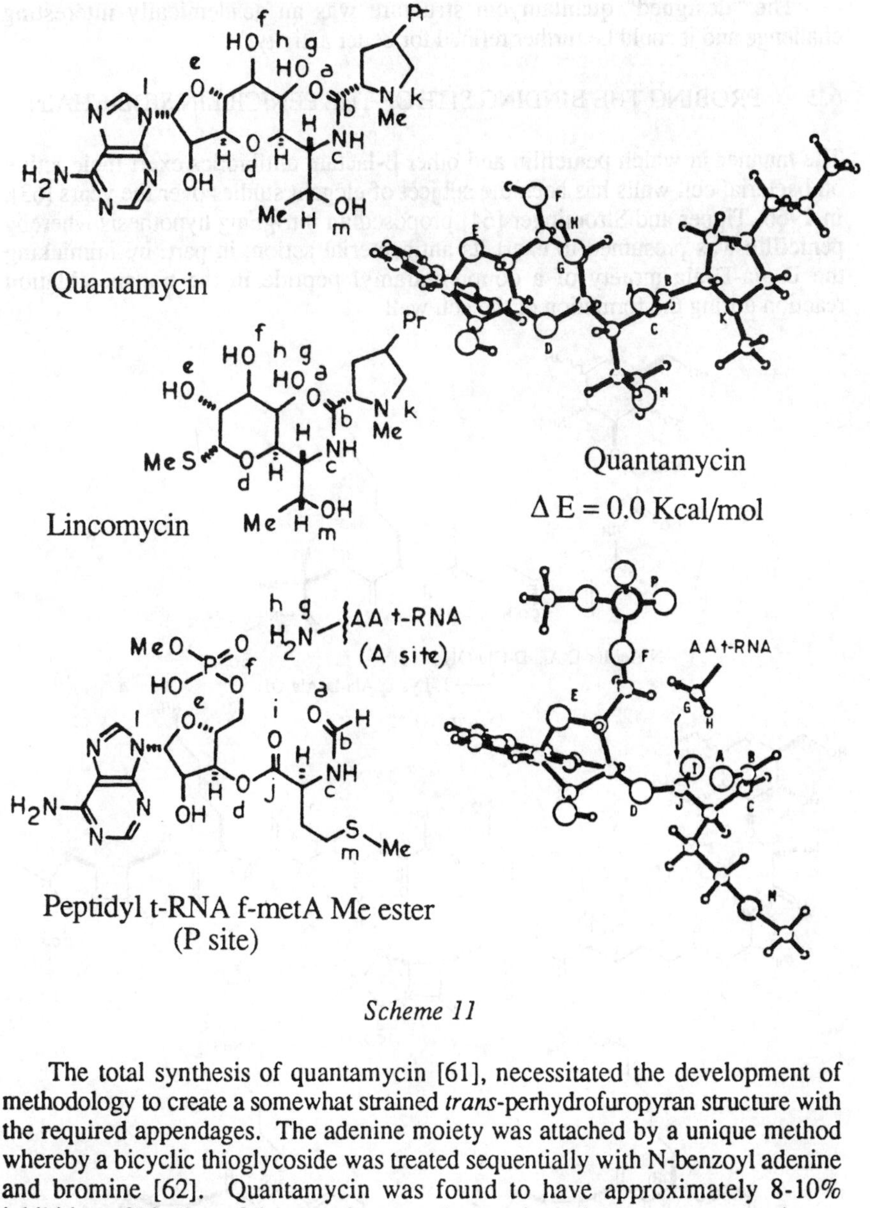

Quantamycin

Lincomycin

Quantamycin

$\Delta E = 0.0$ Kcal/mol

Peptidyl t-RNA f-metA Me ester
(P site)

AA t-RNA

Scheme 11

The total synthesis of quantamycin [61], necessitated the development of methodology to create a somewhat strained *trans*-perhydrofuropyran structure with the required appendages. The adenine moiety was attached by a unique method whereby a bicyclic thioglycoside was treated sequentially with N-benzoyl adenine and bromine [62]. Quantamycin was found to have approximately 8-10% inhibition of binding of lincomycin or erythromycin at the same concentration to ribosomes from *Streptomyces*. However, no *in vitro* antibacterial activity was found, which may be an intrinsic problem, or a pharmacodynamic effect due to improper partitioning.

The "designed" quantamycin structure was an academically interesting challenge and it could be further refined for better activity.

6.2. PROBING THE BINDING SITE OF THE PENICILLIN SIDE-CHAIN

The manner in which penicillin and other β-lactam antibiotics exert their action on bacterial cell walls has been the subject of elegant studies over the years [63]. In 1966, Tipper and Strominger [64] proposed an intriguing hypothesis whereby penicillin was presumed to exert its antibacterial action, in part, by mimicking the D-ala-D-ala moiety of a donor muramyl peptide in the transpeptidation reaction during the formation of the cell wall.

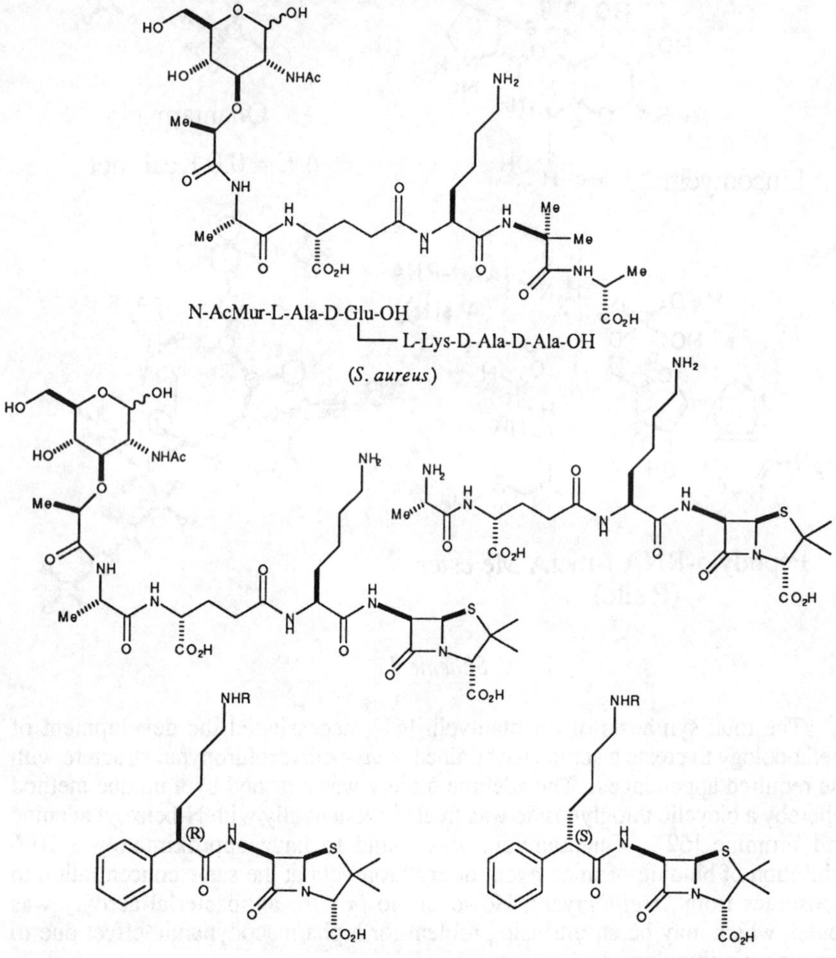

N-AcMur-L-Ala-D-Glu-OH
L-Lys-D-Ala-D-Ala-OH

(*S. aureus*)

Scheme 12

Intrigued by this hypothesis and based on other extensive SAR studies, we reasoned that the replacement of the D-ala-D-ala portion in such muramylpeptides by a 6-aminopenicillanic acid as the terminal dipeptide mimic, might lead to a potentially novel entity and hopefully validate the Tipper-Strominger hypothesis [64] (Scheme 12). Disappointingly, this structure was found to be devoid of antibacterial activity [65]. It then occurred to us that the aromatic group in penicillin G which was used as the standard, was important perhaps as a hydrophobic entity. This led to the design and synthesis of a hybrid structure in which a lysine mimic was incorporated in Pen G, thus simulating a "basic" side-chain environment for possible interaction with penicillin binding proteins as in the natural substrate, while maintaining the "benzyl" portion. Both (*R*)- and (*S*)-isomers of the α-(4-aminobutyl) Pen G were synthesized by stereocontrolled processes [66]. Although antibacterial activity was once again disappointing, there was an interesting stereochemical effect on the binding of these isomeric analogs to PBP's from *E.coli* and *S.aureus*. It is possible that an unfavorable distortion of the aromatic ring compared to Pen G itself is an important factor in these hybrid structures, coupled with other subtle effects that have yet to be understood concerning the nature, orientation and stereochemistry of the side-chain in their natural environment.

6.3. PEPTIDOMIMETIC ANALOG OF MURAMYLPEPTIDE (MPD)

The muramyl peptides have been the focus of intense research efforts over the years in view of their non-specific immunostimulant activity [67]. A large

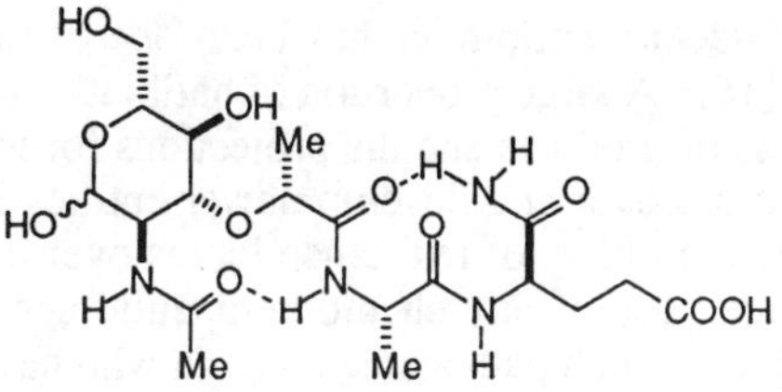

N-ACETYLMURAMYL-L-Ala-D-iGln (MDP)

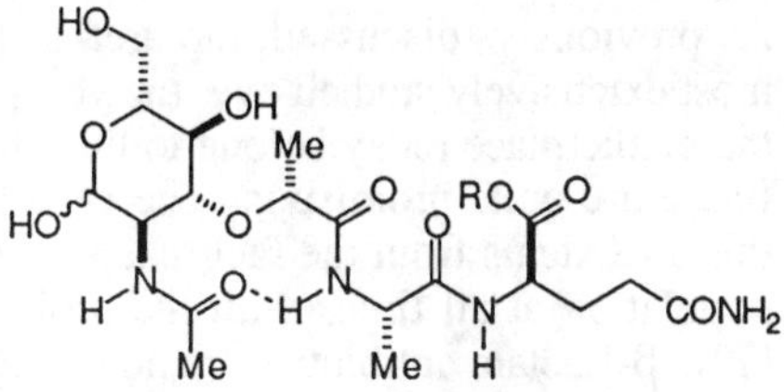

R = n-Bu, MURABUTIDE
R = Me, MURAMETHIDE

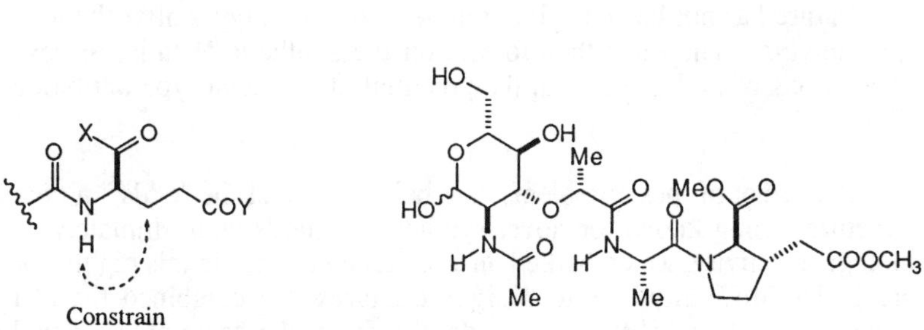

Constrain

Scheme 13

number of MDP analogs have been synthesized in the quest to improve their activity and to dissociate various types of activity and side-effects. Acyclic peptidic analogs [68] and carbocyclic analogs [69] have also been reported. Based on these studies as well as a suggestion from ^{1}H NMR of the existence of an S-shaped conformation consisting of two adjacent β-turns involving H-bonding [70], we envisaged a conformationally constrained analog in which the first β-turn was unperturbed, while the shape of the peptidic portion was altered in a constrained form. This led us to explore ways in which D-pyroglutamic acid could be chemically modified and incorporated into the MDP molecule replacing the D-i-Gln moiety [71] (Scheme 13). The presence of the methyl ester in the peptidomimetic structure made it somewhat similar to muramethide which contains an ester group rather than an amide on the α-carboxylic acid of D-glutamic acid. The intended mimetic structure did not show any release of IL-1α and IL-1β from macrophages up to 200 μg/mL. Much remains to be studied and understood regarding the conformation and shape of the peptide chain in these MDP-type molecules, since, for example, changing the substituents of the carboxylic acid group (ester or amide) has profound effects in dissociating immunostimulating activity from pyrogenicity. Thus, a commitment to systematically study the role and conformation of the peptidic segment of MDP, perhaps through the design of other peptidomimetics, may be a worthwhile research endeavor.

6.4. PROTOTYPES FOR STRUCTURALLY NOVEL β-LACTAM ANTIBIOTICS

As previously discussed, the area of β-lactam antibiotics has been one of the most extensively studied over the years [18]. A large proportion of antibiotics on the marketplace today belong to this class of products and the projections for the future are quite promising. The need to discover and develop newer entities in this area stems from the fact that a serious problem of resistance looms over the population at all times with the prolonged use of any chemotherapeutic agent [72]. β-Lactam antibiotics are no strangers to such pathogenic bacteria who have deployed themselves with potent defense mechanisms with the evolution of β-lactamases through their intricate genetic systems.

Nature has not been as giving in new structural types after the long awaited thienamycin. The onus therefore is on the synthetic chemist to develop new generations of molecules with the potential of β-lactam-type antibiotic activity [73].

Cognizant of these problems, and being aware of the "odds" of creating new structures using known or novel synthetic methods with ultimately interesting biological activity, we embarked on a research program in this highly competitive area. The difficulty was to design structures that combined the elements of novelty and feasibility, independently from the technology developed in pharmaceutical research laboratories with much larger personnel and enormous financial resources. In spite of such odds, we were able to design and synthesize a number of prototypical carbacephems, and carbapenems which combined structural and functional features in such a way so as to provide novel and unique structures. The structures of these novel compounds and their evaluation as antibacterial agents and/or inhibitors of β-lactamases are under study (Scheme 14). Regardless of the eventual outcome of biological tests, it was highly satisfying to be able to carry out multi-step synthetic transformations with delicately balanced functionality present in β-lactam-type motifs, and to reach the target structures with less than a handful of young coworkers.

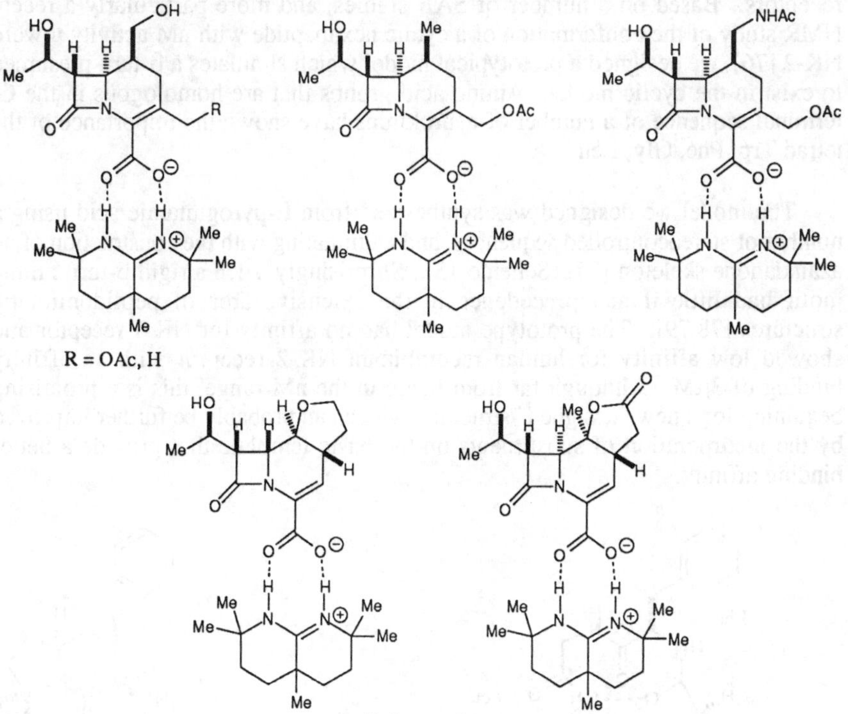

R = OAc, H

Scheme 14

What the future will bring in the area of β-lactam antibiotics may depend on the activities of new structural entities of the types such as those shown in Scheme 14, or others derived from nature or by synthesis. A better understanding of the three-dimensional structures of some essential PBP's and β-lactamases will undoubtedly lead to clearer rationales for the design of new generation β-lactam antibiotics and inhibitors [63].

6.5. A PROTOTYPE MODEL FOR NK-2 ANTAGONIST

The tachykinins and related neurokinins are a series of peptides that are generally associated with physiological events leading to pain and inflammation [75]. These substances have affinities for certain receptors such as NK-1, NK-2 and NK-3 and manifest their activities as mediators of neurotransmission or neuromodulation in the central and peripheral nervous systems.

Because of the importance of the area, many research groups have carried out extensive investigations aimed at developing selective antagonists to these receptors. Based on a number of SAR studies, and more particularly a recent NMR study of the conformation of a cyclic hexapeptide with nM activity toward NK-2 [76], we designed a prototypical model which simulates a β-turn presumed to exist in the cyclic model. Amino acid groups that are homologous in the C-terminal sequence of a number of neurokinins have shown the importance of the tetrad Trp, Phe, Gly, Leu.

The model we designed was synthesized from L-pyroglutamic acid using a number of stereocontrolled sequences, and culminating with the construction of an azaindanone skeleton [77] (Scheme 15). Surprisingly, such a rigid β-turn mimic motif had little if any precedence in the extensive area of peptidomimetic structures [78,79]. The prototype model had no affinity for NK-1 receptor and showed low affinity for human recombinant NK-2 receptor with an affinity binding of 3μM. Although far from being in the nM range, this is a promising beginning for a new "designed" structure, which can probably be further improved by the incorporation of substituents on the basic template that provide a better binding affinity.

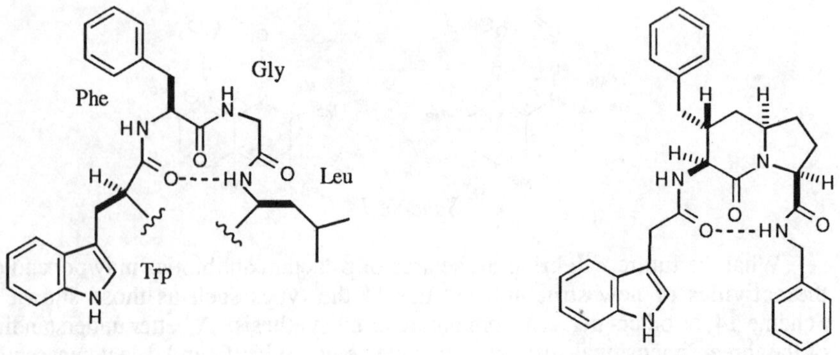

Phe Gly Leu Trp

Partial structure of cyclic hexapeptide

Prototype model antagonist

Scheme 15

6.6. DESIGN AND SYNTHESIS OF A PROTOTYPICAL NON PEPTIDIC INHIBITOR OF THE ENZYME RENIN

The search for a non-peptidic inhibitor of the enzyme renin has been the subject of intensive investigations on several fronts in the quest for an antihypertensive agent [80]. Phenomenal progress has in fact been made in biology, in X-ray crystallography and in synthesis, culminating with inhibitor molecules with activities in the nM range [81]. Based on the knowledge of the action of renin on the Leu-Val scissile bond in angiotensinogen, two types of inhibitor templates have been devised, each with their own optimized functionalities (Scheme 16). It was our intention to probe the hydrophobic site of the P3 subunit in angiotensinogen and to devise a "small" molecule potential inhibitor, based on extensive SAR studies done at the Ciba laboratories [82]. The proposed structure [82] comprised a Type II motif with the familiar L-*threo* amino alcohol unit, and two C-methyl groups which were chosen for synthetic convenience in the prototypical model. This structure was elaborated using L-mandelic acid as a chiral template through a series of functionalizations in which C-C, C-O and C-N bonds were introduced with excellent stereochemical control [83]. Several of the intermediates were crystalline solids which were amenable to X-ray crystal structure elucidation. The prototypical model as the free amine (but not as the Boc derivative), exhibited weak but promising activity (IC_{50} = 37μM) against the enzyme. It is clear that this level of activity is far too weak compared to the presently known front-runners in this highly competitive area. It is, however, one of the smallest synthetic structures to show activity, and it is possible that further refinements will lead to stronger inhibition of the enzyme.

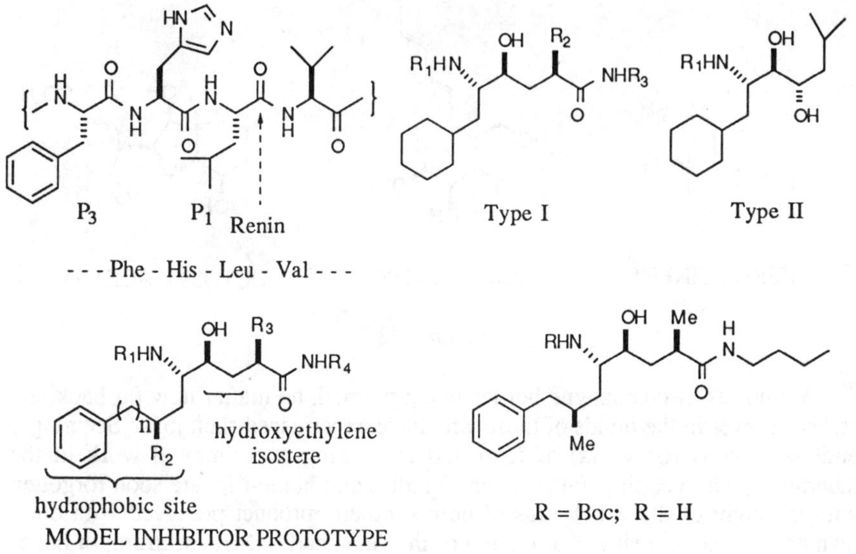

P_3 P_1 Renin

--- Phe - His - Leu - Val ---

Type I

Type II

hydroxyethylene isostere

hydrophobic site

MODEL INHIBITOR PROTOTYPE

R = Boc; R = H

Scheme 16

6.7. CHERISHED MEMORIES, LESSONS LEARNED, PET PEEVES AND FANTASIES THROUGH SYNTHESIS

In the preceding pages, I have highlighted in capsule format some of the challenging projects that have interested us in the total synthesis of molecules within a specific area of medicinal significance. As previously explained, it is the choice of target molecule that was the prime motive for the studies, and in each instance, emphasis was put on the attainment of the goal, rather than on the exploration of methodology and its adaptation to whichever target suited it best.

Our activities in the area of total synthesis have been extended to other types of target molecules, where we have capitalized on "in-house" expertise and methodology, while being open-minded about its merits and limitations. Some cherished memories of finished syntheses in recent years are illustrated in structure format [84] (Scheme 17).

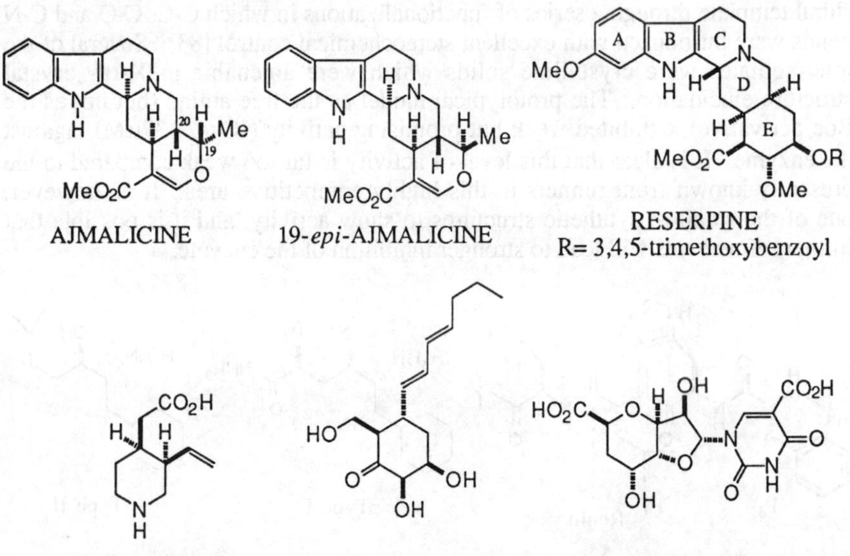

AJMALICINE 19-*epi*-AJMALICINE RESERPINE
R= 3,4,5-trimethoxybenzoyl

MEROQUINENE PALYTANTIN OCTOSYL ACID A

Scheme 17

Memories of champagne bottles being popped, no matter how far back, are etched forever in the minds of those who have experienced such joys. Seemingly endless months (or years) of long and frustrating experimental work in the laboratory, while coping with graduate studies and home-life, are soon forgotten with the sight of those few mgs of pure synthetic product produced with one's own hands. How great is that feeling of the "matched" NMR spectrum, or better yet, the mixed melting point and other physical constants that spell the familiar term... "and the product was found to be identical to a sample of the natural product in all respects". The smile (or big grin) on the "boss's" face, the handshake, the congratulatory words, and above all the walking-on-air feeling are memorable events that have lasting effects.

Indeed, while the young collaborator may well be on his or her way to a nice job soon after the party is over, the quest for the conquest of the next target molecule continues at the hands of another coworker within the group. Thus, the excitement (and frustrations) never really end for the senior investigator, for no sooner the paper appears in print, it becomes part of the past and new frontiers must be opened, crossed and conquered.

In the process, we have learned a great deal from failed reactions, particularly by probing into mechanistic reasons and technical details. Turning failure into success especially through logic has been a most fulfilling experience and one which makes "believers" out of young and impressionable collaborators. The power of observation and paying attention to detail regarding experimental conditions will sooner or later lead to open-eyed serendipity and eventually to a successful outcome. The practice of total synthesis in conjunction with our appreciation for mechanistic rationales, and the design-based process of attaining a realistic goal constitute a superb training ground for a career in research.

Achievements in total synthesis have reached impressively high levels and one must approach the prospects of a new synthesis with new criteria. In the absence of important pharmacological activity, or some anticipated "function", total synthesis is becoming more and more difficult to justify nowadays, at least for funding agencies. Expectations from the community of synthetic organic chemists are very high in view of the truly complicated structures that have been made today. How then can one combine the elements of relevance, methodology, practicality and innovation in the choice of a structurally "impressive" target molecule? There is no simple answer to this question and it is asked only to accentuate the plight of the synthetic organic chemist today. There are in fact many other questions... How can we do state-of-the-art synthesis in conjunction with a biologically important problem, while maintaining interest and progress on all fronts and for all concerned?... Can we finish the synthesis before the theory on the presumed mode of action changes or the interest wanes?... Where can we find such research problems or collaborations without infringement or selling one's chemical soul to confidentiality agreements and patent rights? Should the role of the academic be primarily one of providing the best possible training to young collaborators?... These and other questions are omnipresent in the mind of most if not all synthetic organic chemists today and answers are not easy or general [85].

In the final analysis, it is a state of mind, a love of the métier, a fascination with molecules, an insatiable appetite to uncover new frontiers, personal ambitions and other motives that provide a balance of responsibilities allowing us to function in the best professional way we can, even in the face of adversity.

Returning to the laboratory, it is clear that there are *many* important unsolved problems in organic synthesis in general that can be addressed. In spite of enormous advances in methodology and in instrumental techniques, we are still at the mercy of protective groups for a great many transformations. They are blessings in the beginning of sequences and may become curses toward the end. We are so limited in ways to assemble segments of molecules en route to relatively large target molecules. Reactivity becomes a problem, particularly when anionic chemistry is utilized, and the presence of hetero-atoms certainly contributes to the problem. Reactivity is also a concern in many other instances, especially when chemoselective transformations are required, and when the contributions of steric, electronic and related effects are difficult to predict.

Although biotechnology has made tremendous progress in recent years, it cannot solve the synthetic chemist's problems by itself. By its very nature, the discipline of organic synthesis is based on logic, and logic must eventually prevail even in the face of persistent difficulties. The creativity and inventiveness of the synthetic organic chemist is his/her strongest asset. Past and present experience has shown that given the opportunity, the synthetic organic chemist can excel in other sub-disciplines as well. Thus, the biological and material sciences are well within the grasp of synthetic chemistry and the coming years will see an ever increasing involvement of synthetic organic chemists in these areas with great rewards for mankind.

7. Acknowledgements

The author is grateful for financial support from NSERCC (operating grant, medicinal chemistry chair program), le Ministère de l'Éducation du Québec, and a number of industrial research laboratories worldwide. Special thanks go to the dozens of dedicated coworkers who have participated in the projects discussed in this article.

8. References

1. For some authoritative monographs, see Corey, E.J. and Cheng, X.M. (1989) *The Logic of Chemical Synthesis*, Wiley, New York, N.Y.; (1973-1992) *The Total Synthesis of Natural Products*, ApSimon, J. ed., Wiley, New York, N.Y., vol. 1-8; (1988-1994) *Studies in Natural Product Synthesis*, Atta-ur-Rahman ed. Elsevier, Amsterdam Press, vol. 1-13; (1984-1993) *Strategies and Tactics in Organic Synthesis*, Lindberg, T. Ed., Academic Press, Orlando, Fla., vol. 1-3.
2. Todd, A.R. (1956) *Perspectives in Organic Synthesis*, Ed. Interscience, New York, N.Y.
3. For some incisive and provocative views on the state of the art in organic synthesis, see Seebach, D. (1990) *Angew. Chem. Int. Ed. Engl.* **29**, 1320.
4. Collins, A.N., Sheldrake, G.N., Crosby, J., eds (1992) *Chirality in Industry,* J. Wiley & Sons, New York, N.Y.
5. Sheldon R.A. (1993), *Chirotechnology. Industrial Synthesis of Optically Active Compounds,* M. Dekker, Inc., New York, NY.
6. See for example, Ariens, E.J. (1986) *Med. Res. Rev.* **6**, 451; (1987) **7**, 367; (1988) **8**, 309; Gross, M. (1989) *Ann. Rep. Med. Chem.* **25**, 323; Brossi, A. (1994) *Med. Chem. Rev.* **14**, 665; (1994)*Chem. Eng. News* , Sept. 19, p. 38.

7. Gund, P., Halgren, T.A., Smith, G.M. (1987), *Ann. Rep. Med. Chem.* **22**, 275; Richards, W.G. ed. (1989) *Computer Aided Molecular Design*, VCH Publishers Inc. New York, N.Y.
8. Woodward, R.B. (1977) *Recent Advances in Chemistry of β-Lactam Antibiotics*, Elks, J. ed., Chemical Society London; spec. Publ. *Chem. Soc.*, **28**, 167; Woodward, R.B. (1978) *J. Am. Chem. Soc.* **101**, 6296. Ernest, I., Gosteli, J., Greengrass, C.W., Holick, W., Jackman, D.E., Pfandler, H. R., Woodward, R.B. (1978) *J. Am. Chem. Soc.* **100**, 8214; Pfandler, H.R., Gosteli, J., Woodward, R.B. (1980) *J. Am. Chem. Soc.* **102**, 2039; Ernest, I. (1982) *Chemistry and Biology of the β-Lactam Antibiotics*, Morin, R.B., Gorman, M., eds.; Academic Press, New York, N.Y., p. 315.
9. Franceschi, G., Foglio, M., Alpegiani, M. Battistini, C., Bedeschi, A., Perrone, E., Zarini, F.; Arcamone, F., Della Bruna, C., SanFillipo, A. (1983) *J. Antibiot.* **35**, 938 and references therein; Franceschi, G.; Alpegiani, M., Battistini, C., Bedeschi, A., Perrone, E., Zarini, F. (1987) *Pure Appl. Chem.* **59**, 467; (1991) *Penem Antibiotics* Mitsuhashi, S., Franceschi, G. eds., Japan Scientific Societies, Press, Tokyo; Springer Verlag, New York, N.Y.
10. Afonso, A., Hon, F., Weinstein, J., Ganguly, A.K., McPhail, A. T. (1982) *J.Am. Chem. Soc.* **104**, 6138; Gala, D., Chiu, J.S., Ganguly, A.K., Girijavallabhan, V.M., Jaret, R.S., Jenkins, J.K., McCombie, S.W., Nyce, P.L., Rosenhouse, S., Steinman, M. (1992)*Tetrahedron* **48**, 1175.
11. Hanessian, S., Bedeschi, A., Battistini, C., Mongelli, N. (1985) *J. Am. Chem. Soc.* **107**, 1438; (1985) *Lect. Heterocycl. Chem.* **8**, 43; and refereces cited therein.
12. 4-Acetoxy-3-[(*R*)-1'-hydroxyethyl]-1'-*O*-[*tert*-butyldimethylsilyl-2-azetidinone is commercially available, Kaneka America Corp., New York, N.Y.; Chackalamannil, Naomi F., Kirkup, M., Afonso, A., Ganguly, A.K. (1988) *J. Org. Chem.* **53**, 450; Shiozaki, M.; Ishida, N., Hiraoka, T., Maruyama, H. (1981) *Tetrahedron Lett.* **22**, 5205; Shimohigashi, Y., Waki, M., Izumiya, N. (1979) *Bull. Chem. Soc. Jpn.* **52**, 949; See also: Yanagisawa, H., Ando, A., Shiozaki, M., Hiraoka, T. (1983) *Tetrahedron Lett.* **24**, 1037.
13. Battistini, C., Scarafile, C., Foglio, M., Franceschi, G. (1984), *Tetrahedron Lett.* **25**, 2395; Perrone, E., Alpegiani, M., Bedeschi, A., Giudici, F., Franceschi, G. (1984) *Tetrahedron Lett.*. **25**, 2399.
14. Gomper, R., Schaeffer, H. (1967) *Chem. Ber.* **100**, 591.
15. Hanessian, S., Alpegiani, M. (1986) *Tetrahedron Lett.*, **27**, 4857; (1989) *Tetrahedron*, **45**, 941.
16. Hanessian, S., Desilets, D., Bennani, Y. (1990) *J. Org. Chem.* **55**, 3098.
17. See Saltzmann, T.N., Ratcliffe, R.W., Christensen, B.G., Bouffard, F.A. (1980) *J. Am. Chem. Soc.* **102**, 6161; Mellilo, D.G., Cretovich, R.J., Ryan, K,M., Sletzinger, M. (1986) *J. Org. Chem.* **51**, 1498; Karady, S., Amato, J., Reamer, R.A., Weinstock, L.M. (1981) *J. Am. Chem. Soc.* **103**, 6765.
18. For a recent monograph, see Gunda, G. ed. (1992) *The Organic Chemistry of β-Lactams* VCH publishers, New York, N.Y.; see also Georg, G.I. (1989) *Studies in Natural Product Chemistry*, Atta-ur-Rahman, ed. Elsevier Science, Amsterdam, vol. 4, p. 431; Nagahara, T., Kametani, T. (1987) *Heterocycles* **25**, 729; Kametani, T., Fukumoto, K., Ihara, M. (1982) *Heterocycles* **17**, 463; Ratcliffe, R.W., Albers-Schönberg G (1982) *Chemistry and Biology of β-Lactam Antibiotics*, Morin, R.B., Gorman, M., eds., Academic Press, New York, N.Y. vol. 2, p. 227.
19. Endo, A. (1979) *J. Antibiot.* **32** 854; Alberts, A.W., Chen, J., Kuron, G., Hunt, V., Huff, J., Hoffman, C.H., Rothrock, J., Lopez, M., Joshua, H., Harris, E., Patchett, A.A., Monaghan, R., Currie, S., Stapley, E., Albers-Schönberg, G., Hensens, O., Hirschfield, J., Hoogsteen, K., Liesch (1980) *J. Proc. Nat. Acad. Sci. U.S.A.* **77**, 3957.
20. For a review, see, Rosen, T., Heathcock, C.H. (1986) *Tetrahedron* **42**, 4209.
21. Falck, J.R., Yang, Y.L. (1984) *Tetrahedron Lett.* **25**, 3563; Hecker, S.J., Heathcock, C.H. (1986) *J. Am. Chem. Soc.* **108**, 4586.
22. Hanessian, S., Roy, P.J., Petrini, M., Hodges, P.J., Di Fabio, R., Garganico, G. (1990) *J. Org. Chem.* **55**, 5766.
23. Hanessian, S. (1983) *The Total Synthesis of Natural Products: The Chiron Approach* , Pergamon Press, New York, NY.
24. Hanessian, S. (1989)*Aldrichimica Acta* **22**, 1; Hanessian, S., Murray, P.J. (1987) *Tetrahedron* **43**, 5072; Hanessian, S., Murray, P.J., Sahoo, S.P. (1985) *Tetrahedron Lett.* **26**, 5627.
25. Rosini, G., Ballini, R., Sorrenti, P. (1983) *Synthesis*, 1014; Rosini, G., Marotta, E., Ballini, R., Petrini, M. (1986) *Synthesis*, 1024.
26. Suzuki, M., Yanagisawa, A., Noyori, R. (1985) *J. Am. Chem. Soc.* **107**, 3348; (1983) *Tetrahedron Lett.* **24**, 1187.

27. Matsubara, S., Takai, K., Nozaki, H. (1983) *Bull. Chem. Soc. Jpn.* **56**, 2029; Suzuki, M.; Takada, H.; Noyori, R. (1982) *J. Org. Chem.* **47**, 902.
28. For leading references, see (1989) *Ivermectin and Abamectin*, Campbell, W.C., ed., Springer-Verlag, New York, N.Y.; Blizzard, T., Fisher, M.H., Mrozik, H., Shih, T.L. (1990) *Recent Progress in the Chemical Synthesis of Antibiotics*, Lukacs, G., Ohno, M. eds. Springer-Verlag, New York, N.Y., Chapter 3, p. 65; Fisher, M.H., Mrozik, H. (1985) *Macrolide Antibiotics*, Omura, S., ed., Academic Press, Orlando, Fla., p. 553; Fisher, M.H. (1985) *Recent Advances in the Chemistry of Insect Control*, Janes N.F. ed., Royal Society of Chemistry, London, p. 53.
29. For the total synthesis of avermectin A_{1a}, see Danishefsky, S.J., Armistead, D.M. Winnott, F.E., Selnick, H.G., Hungate, R. (1989) *J. Am. Chem. Soc.* **111**, 2967; for avermectin B_{1a}, see Ford, M.J., Knight, J.G., Ley, S.V., Vile, S. (1990) *Synlett* 331; for avermectin B_{1a} aglycone, see White, J.D., Bolton, G.L. (1990) *J. Am. Chem. Soc.* **112**, 1626; see also Perron, F., Albizati, K.F. (1989) *Chem. Rev.* **89**, 1617; Crimmins, M.T., Hollis, W.J., Jr., O'Mahoney, R. (1988) *Studies in Natural Product Chemistry*, Atta-ur-Rahman, ed., Elsevier Amsterdam, vol. 13, part A, p. 435; Vaillancourt, V., Pratt, N.E., Perron, F., Albizati, K.F. (1992) *The Total Synthesis of Natural Products*, ApSimon, J., ed., Wiley, New York, N.Y. **8**, 533.
30. Hanessian, S., Ugolini, A., Hodges, P.J., Dubé, D., André, C. (1987) *Pure Appl. Chem.* **59**, 299.
31. Hanessian, S., Ugolini, A., Hodges, P.J., Dubé, D. (1986) *Tetrahedron Lett.* **27**, 2699.
32. For other observations concerning conjugation and deconjugation in the avermectin series, see ref. 30; see also Pivnichny, J.V., Shim, J.-S.K., Simmerman, L.A. (1983) *J. Pharm. Sci.* **72**, 1447; Fraser-Reid, B., Wolleb, H., Faghih, R., Barchi, J., Jr. (1987) *J. Am. Chem. Soc.* **109**, 933.
33. Hanessian, S., Dubé, D., Hodges, P.J., André, C. (1987) *J. Am. Chem. Soc.* **109**, 7063.
34. Dubé, D. (1988) Ph.D. thesis, Université de Montréal.
35. Hanessian, S., Bacquet, C., Lehong, N. (1980) *Carbohydr. Res.* **80**, C17.
36. Woodward, R.B., et al. (1981) *J. Am. Chem. Soc.* **103**, 3215.
37. Wuts, P.G.M., Bigelow, S.S. (1983) *J. Org. Chem.* **48**, 3489; Blizzard, T.A., Margiatto, G.M., Mrozik, H., Fisher, M.H. (1993) *J. Org. Chem.* **58**, 3201.
38. Hanessian, S., Chemla, P. (1991) *Tetrahedron Lett.* **32**, 2719.
39. Blizzard, T., Bostrom, L., Margiatto, G., Mrozik, H., Fisher, M.H. (1991) *Tetrahedron Lett.* **32**, 2723.
40. Hanessian, S., Chemla, P., Tu, Y. (1993) *Tetrahedron Lett.* **34**, 1407.
41. Westley, J. W. (1977) *Adv. Appl. Microbiol.* **22**, 177.
42. Barbier, P.R., Schneider, F., Widmer, U. (1987) *Helv. Chim. Acta* **70**, 196; Barbier, P.R., Schneider, F. (1988) *J. Org. Chem.* **53**, 1218; (1987) *Helv. Chim. Acta* **70**, 196; Pons, J.-M., Kocienski, P. (1989) *Tetrahedron Lett.* **30**, 1853; (1994) *Synthesis* 1294; Fleming, I., Lawrence, N.J. (1990) *Tetrahedron Lett.* **31**, 3645; Chadha, N.K., Batcho, A.D., Tang, P.C., Courtney, L.F., Cook, C.M., Wovkulich, P.M., Uskokovic, M.R. (1991) *J. Org. Chem.* **56**, 4714; Case-Green, S.C., Davies, S.G., Hedgecock, C.J.P. (1991) *Synlett* 781.
43. Hanessian, S., Tehim, A., Chen, P. (1993) *J.Org. Chem.*, **58**, 7768.
44. Fräter, G., Müller, V., Günther, W. (1984) *Tetrahedron* **40**, 1269; see also Seebach, D., Aebi, J., Wasmuth, D. (1984) *Org. Syn.* **63**, 109.
45. Hanessian, S. Cooke, N.G., Dehoff, B., Sakito, Y. (1990) *J. Am. Chem. Soc.* **112**, 5277.
46. Weibel, E.K., Hadvary, P., Hochuli, E., Kupfer, E., Lengsfeld, H. (1987) *J. Antibiot.* **40**, 1081; Hochuli, E., Kupfer, E., Maurer, R., Meister, W., Mercadal, Y., Schmidt, K. (1987) *J. Antibiot.* **40**, 1086; Kitahara, M., Asawo, M., Naganawa, H., Maeda, K., Homada, M., Aoyagi, T., Umezawa, H., Itaka, Y., Nakamura, H. (1987) *J. Antibiot.* **40**, 1647.
47. Hanessian, S., Murray, P.J. (1986) *Can. J. Chem.* **64**, 2231 ; Hanessian, S., Murray, P.J. (1987) *Tetrahedron* **43**, 5055; (1987) *J. Org. Chem.* **52**, 1171.
48. Hanessian, S., Thavonekham, B., Dehoff, B. (1989) *J.Org. Chem.* **54**, 5831.
49. Evans, D.A., Dow, R.L., Shih, T.L., Takacs, J.M., Zahler, R. (1990) *J. Am. Chem. Soc.* **112**, 5291.
50. Suzuki, S., Isono, K., Nagatsu, J., Mizutani, T., Kawashina, K., Mizuno, T. (1965) *J. Antibiotics* **A18**, 131; Isono, K., Nagatsu, J., Kawashima, Y., Suzuki, S. (1965) *Agr. Biol. Chem., (Tokyo)* **29**, 848; Suzuki, S., Isono, K., Nagatsu, K., Kawashima, Y., Yamagata, K., Sasaki, K., Hashimoto, K. (1966) *Agr. Biol. Chem. (Tokyo)* **30**, 817; Isono, K., Kobinata, K., Suzuki, S. (1968) *Agr. Biol. Chem.* **32**, 792.
51. Isono, K., Asahi, K., Suzuki, S. (1969) *J. Am. Chem. Soc.* **91**, 7490 and references cited therein.
52. For previous approaches to synthesis, see Baumann, H., Duthaler, R.O. (1988) *Helv. Chim. Acta* **71** 1025, 1035.

53. For a recent total synthesis of racemic *trans*-polyoximic acid, see Emmer, G. (1992) *Tetrahedron* **48**, 7165.
54. Hanessian, S., Fu, J.-M., Chiara, J.L., Di Fabio, R. (1993) *Tetrahedron Lett.* **34**, 4157.
55. Hanessian, S., Fu, J.-M., Tu, Y., Isono, K. (1993) *Tetrahedron Lett.* **34**, 4153.
56. For an aldol route to the assembly of bafilomycin A_1, see Evans, D.A., Calter, M.A. (1993) *Tetrahedron Lett.* **34**, 6871; for the synthesis of the C_{13}-C_{25} segment of bafilomycin A_1, see Roush, W.R., Bannister, T.T. (1992) *Tetrahedron Lett.* **33**, 3587; Roush, W.R., Bannister, T.D., Wendt, M.D. (1993) *Tetrahedron Lett.* **34**, 8387; Paterson, I.; Bower, S.; McLoed, M.D. (1995) *Tetrahedron Lett.* **36**, 175; see also, Hanessian, S., Gai, Y., Meng, Q., Olivier, E., Wang, W., unpublished results.
57. Warner, G., Hagenmaier, H., Drautz, H., Baumgartner, A., Zähner, H. (1984) *J. Antibiotics* **37**, 110.
58. Hanessian, S., Meng, Q. (1994) *Tetrahedron Lett.* **35**, 5393.
59. Gale, E.F., Cundliffe, E., Reynolds, P.E., Richmond, M.H., Waring, M.J. (1972) *The Molecular Basis of Antibiotic Action,* J. Wiley & Sons, New York, N.Y., p. 278.
60. Cheney, B.V. (1974) *J. Med. Chem.* **17**, 590.
61. Hanessian, S., Sato, K., Liak, T.J., Danh, N., Dixit, D. Cheney, B.V. (1984) *J. Am. Chem. Soc.* **106**, 6114.
62. Hanessian, S., Liak, T.J., Dixit, D. (1981) *Pure and Appl. Chem.* **53**, 129; (1981) *Carbohydr. Res.* **88**, C14.
63. For reviews, see Georgopapadakou, N.H. (1988) *Antimicrobial Agents Ann.* **3**, 409; (1983) *Ann. Rep. Med. Chem.* **18**, ; Spratt, B. (1983) *J. Gen. Microbiol.* **129**, 1247; Neuhaus, F.C., Georgopapadakou, N.H. eds. (1992) *Emerging Targets in Antibacterial and Antifungal Chemotherapy* , Sutcliffe, J.; Georgopapadakou, N.H. eds., Chapman and Hall, New York, N.Y., p. 205.
64. Tipper, D.J., Strominger, J.L. (1965) *Proc. Natl. Acad. Sci. U.S.A.* **54**, 1133; Strominger, J.L., Blumberg, P.M., Suginaka, H., Umbreit, J., Wickus (1971) *Proc. R. Soc. Lond. B.* **79**, 369.
65. For a related study see, Strynadka, N.C.J., Adachi, H., Jensen, S.E., Johns, K., Seilecki, A., Betzel, C., Sutoh, K., James, M.N.G. (1992) *Nature* **79**, 369.
66. Hanessian, S., Couture, C. (1993) *Bioorg. Med. Chem. Lett.* **11**, 2323
67. Barton, D.H.R., Dalko, P., Gero, S.D., Quiclet-Sire, B., Stütz, P. (1989) *J. Org. Chem.* **54**, 3764.
68. Hönig, H., Danklmaier, J. (1990) *Liebigs Ann. Chem.* 145 and references cited therein; Takeno, H., Okada, S., Yonishi, S., Hommi, K., Nakaguchi, O., Kitaura, Y., Hashimoto, M. (1984) *Chem. Pharm. Bull.* **32**, 2932; see also Ivanov, V.T., Andronova, T.M., Bezrukov, M.V., Rar, V.A., Makarov, E.A., Lozmin, S.A., Astapova, M.V., Barkova, T.I., Nesmeynov, V.A. (1987) *Pure and Appl. Chem.* **59**, 317.
69. Barton, D.H.R., Dalko, P., Gero, S.D., Quiclet-Sire, B., Stütz, P. (1989) *J. Org. Chem.* **54**, 3764; Becker, B.; Thiem, J. (1994) *Tetrahedron Asymmetry* **5**, 2339.
70. Sizun, P., Perley, B., Level, M., Lefrancier, P., Fermandjian, S. (1988) *Tetrahedron* **44**, 991; Fermandjian, S., Perley, B., Level, M., Lefrancier, P. (1987) *Carbohydr. Res.* **162**, 23; Chapman, B.E., Batley, M., Redmond, J.W. (1982) *Aust. J. Chem.* **35**, 489; McFarlane, E.F., Martinic, C. (1983) *Aust. J. Chem.* **36**, 1087.
71. Hanessian, S., Ratovelomanana, V., Couture, C.A. (1991) *Synlett* **4**, 222.
72. Cohen, M.L. (1992) *Science* **257**, 1050; Neu, H.C. (1992), *Science* **257**, 1064.
73. See for example, (1992) *Recent Advances in the Chemistry of Anti-Infective Agents,* Bentley, P.H., Ponsford, R., eds., Royal Society of Chemistry Special Publication, No. 119; T. Graham House, Science Park, Cambridge, U.K.
74. Hanessian, S., Rozema, M. (1994) *Bioorg. Med. Chem. Lett.* **4**, 2279; Hanessian, S.; Reddy. B. (1994) *Bioorg. Med. Chem. Lett.* **4**, 2285.
75. For selected reviews, see Reese, D.C. (1993) *Ann. Rep. Med. Chem.* **28**, 59; Maggi, A.-C.; Quartara, L.; Giuliani, S., Patacchini, R. (1993) *Drugs of the Future* **18**, 155; Watling, K.J., Guard, S. (1992) *Neurotransmissions* **26**, 43.
76. Malikayil, J.A., Harbeson, S.L. (1992) *Int. J. Peptide Protein Res.* **39**, 497.
77. Hanessian, S., Ronan, B., Laoui, A. (1994) *Bioorg. Med. Chem. Lett.* **4**, 1397.
78. For recent reviews on peptidomimetics see, Gante, J. (1994) *Angew. Chem. Int. Ed. Engl.* **33**, 1699; Giannis, A., Kolter, T. (1993) *Angew. Chem. Int. Ed. Engl.* **32**, 1244; Kahn, M. (1993) *Synlett*, 821; Olson, G.L. et al. (1993) *J. Med. Chem.* **36**, 3039; Hirschmann, R. (1991) *Angew. Chem. Int. Ed. Engl.* **30**, 1278; Hölzemann, G. (1991) *Kontakte*, 3; Ball, J.B., Alewood, P.F. (1990) *J. Mol. Recogn.* **3**, 55; Morgan, B.A., Gainor, J.A. (1989) *Ann. Rep. Med. Chem.* **24**, 243; Kahn, M., Wilke, S. Chen, B., Fujita, K., Lee, Y.-H., Johnson, M.E. (1993) *Current Drugs,* 1157.
79. See for example Nagai, U., Sato, K. (1985) *Tetrahedron Lett.* **26**, 647; Moeller, K.D., Hanau, C.E., d'Avignon, A. (1994) *Tetrahedron Lett.* **35**, 825.

80. Waeber, B., Nussberger, J., Brunner, H. (1986) *Handbook of Hypertension*, vol. 8: *Pathophysiology of Hypertension-Regulatory Mechanisms*, Zanchetti, A., Tarazi, R.C. eds., Elsevier Science Publishers, B.V. Amsterdam.
81. For recent reviews, see, Ocain, T.D., Abou-Gharbia, M. (1991) *Drugs of the Future* **16**, 37; Greenlee, W.J. (1990) *Med. Chem. Rev.* **10**, 173; (1987) *Pharmaceutical Res.* **4**, 364 Antonaccio, M.J., Wright, J.J. (1987) *Progr. Drug. Res.* **31**, 161; Plattner, J.J., Kleinert, H.D. (1987) *Ann. Rep. Med. Chem.* **22**, 63.
82. Dr. J. Maibaum, Ciba Pharmaceuticals, personal communication. Details on these studies will be published elsewhere.
83. Hanessian, S., Sadagopan, R. (1994) *Bioorg. Med. Chem. Lett.* **4**, 1697.
84. For octosyl acid A, see Hanessian, S., Kloss, J. (1986) *J. Am. Chem. Soc.* **108**, 2758; Meroquinene, S. Hanessian, S., Faucher, A.-M. and Léger, S. (1990) *Tetrahedron*, **46**, 231; Palytantin, Hanessian, S., Sakito, Y., Dhanoa and Baptistella, L. (1989)*Tetrahedron* **45**, 6623; Ajmalicine, 19-epi-ajmalicine, Hanessian, S., Faucher, A.-M. (1991) *J.Org. Chem.* **56**, 2947; Reserpine, Hanessian, S., Pan, J., Carnell, A., Bouchard, H., Lesage, L. unpublished results.
85. For a round-table discussion on the subject, see p. 589-607 in this monograph.

OPTICALLY ACTIVE TRANSITION METAL COMPOUNDS CONTAINING CHIRAL TRANSITION METAL ATOMS

HENRI BRUNNER
Universität Regensburg
Institut für Anorganische Chemie
D-93040 Regensburg (Germany)

The asymmetric carbon atom C(a,b,c,d) (Scheme 1, top) dominates stereochemistry since 120 years. Other main group elements, surrounded by four different substituents, have been obtained in optically active form, e.g. silicon in the compound

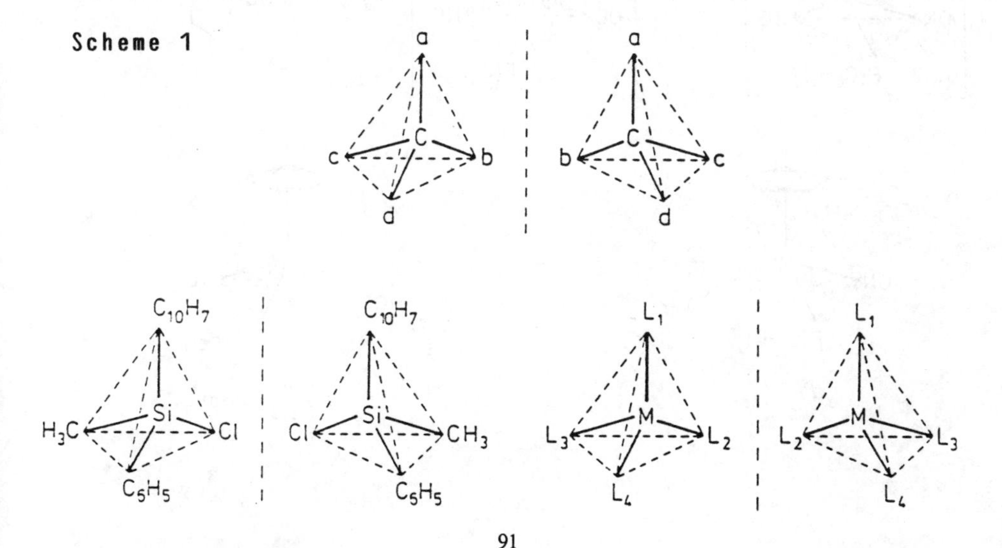

C. Chatgilialoglu and V. Snieckus (eds.), Chemical Synthesis, 91–111.

Si(α-Np,Ph,Me,Cl) (Scheme 1, bottom left), a field which has been pioneered by L.H. Sommer and R.J.P. Corriu [1]. Resolution of compounds containing four different substituents was extended to other main group elements, e.g. N, P, S [2].

In the transition metal series there are the famous optically active octahedral metal tris-chelate complexes, e.g. $[Co(en)_3]X_3$ [3]. The first optically active octahedral compound was resolved by A. Werner in 1911 [4]. Until recently, however, there were no optically active compounds, in which a transition metal M is surrounded by four different ligands L_1 - L_4 (Scheme 1, bottom right) [5].

The first transition metal complex containing four different substituents to be resolved was the manganese complex shown in Scheme 2, top [6,7]. In the cation of this complex, the Mn atom is surrounded by a cyclopentadienyl, a carbonyl, a nitrosyl, and a triphenylphosphine ligand. This complex can be prepared from the commercially available compound $(C_5H_5)Mn(CO)_3$ in two steps yielding $[(C_5H_5)Mn(CO)(NO)PPh_3]PF_6$ as a racemate [8].

Scheme 2

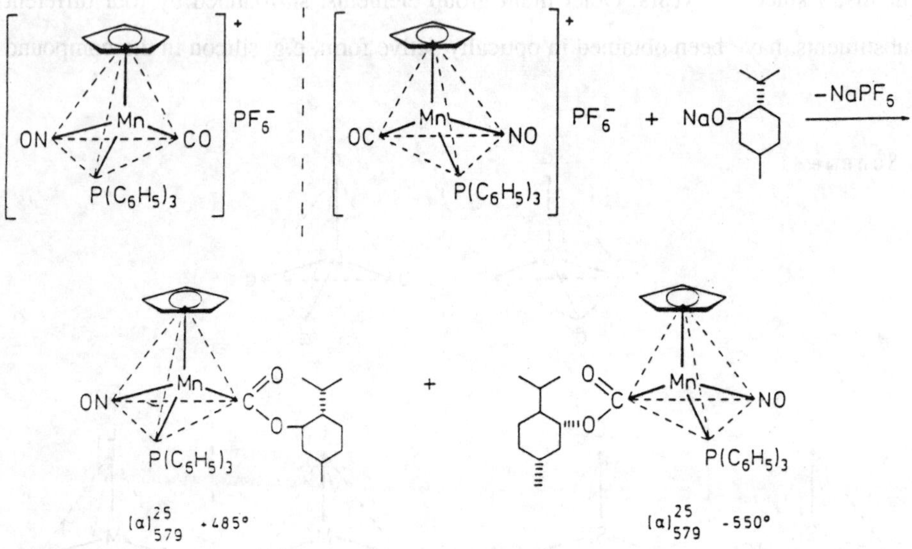

To resolve the racemate, it is reacted with the Na derivative of the natural optically active alcohol (1R,3R,4S)-menthol. $NaPF_6$ is eliminated and the mentholate ion adds to the carbon atom of the carbonyl ligand in the cationic starting material. Neutral ester-like addition products are formed, which contain the same asymmetric centers of natural menthol and which differ only in the configuration of the manganese atom (Scheme 2, bottom). The two diastereomers can be separated on the basis of solubility differences. The $(+)_{579}$-rotating diastereomer is soluble in petroleum ether, whereas the $(-)_{579}$-rotating diastereomer is insoluble in this solvent [6].

To complete the resolution, the optically active auxiliary has to be removed from the separated diastereomers. This can be done by passing a stream of gaseous HCl through a benzene solution of $(+)_{579}$- and $(-)_{579}$-$(C_5H_5)Mn(NO)(PPh_3)COO$menthyl. In this acid cleavage, the C-Omenthyl bond is broken. Menthol is eliminated, transforming the ester group into the carbonyl group. The counter ion to balance the positive charge on the complex first is Cl^-. In a metathetical reaction Cl^- is replaced by PF_6^- [7].

Scheme 3

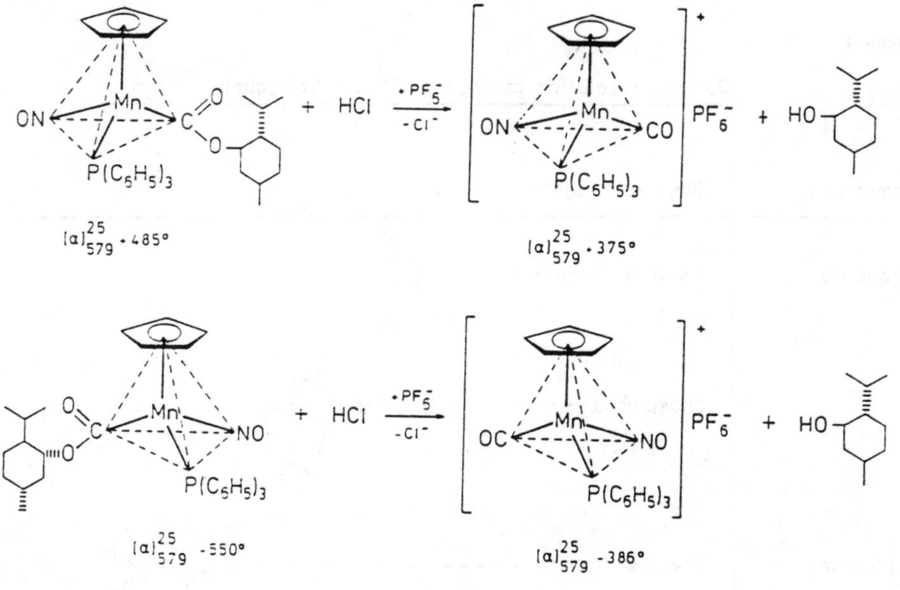

ON, Mn, C, O, O, $P(C_6H_5)_3$ + HCl $\xrightarrow[-Cl^-]{+PF_6^-}$ [ON, Mn, CO, $P(C_6H_5)_3$]$^+$ PF_6^- + HO

$[\alpha]^{25}_{579}$ +485° $[\alpha]^{25}_{579}$ +375°

O, C, O, Mn, NO, $P(C_6H_5)_3$ + HCl $\xrightarrow[-Cl^-]{+PF_6^-}$ [OC, Mn, NO, $P(C_6H_5)_3$]$^+$ PF_6^- + HO

$[\alpha]^{25}_{579}$ -550° $[\alpha]^{25}_{579}$ -386°

Starting with the $(+)_{579}$-diastereomer, the $(+)_{579}$-PF_6 salt (Scheme 3, top) is formed, whilst the $(-)_{579}$-diastereomer is converted into the $(-)_{579}$-PF_6 salt (Scheme 3, bottom) [7]. The diastereomers shown on the left side of Scheme 3 contain a total of four asymmetric centers each, the asymmetric Mn atom and three asymmetric carbon atoms in the menthyl part of the molecule. After elimination of menthol the asymmetric manganese atom is the only source of optical activity in the resolved complexes $[(C_5H_5)Mn(CO)(NO)PPh_3]PF_6$. The optical rotations of the two mirror image isomers have opposite signs and the same magnitude within the limits of error. Schemes 2 and 3 describe the first resolution of a transition metal complex with four different substituents [6,7], having so-called three-legged piano stool geometry.

Scheme 4 shows some optically active resolving agents used in the resolution of organometallic transition metal compounds of tetrahedral, octahedral, and square pyramidal geometry. Schemes 2 and 3 demonstrate the application of the menthoxide ion. The aminophosphine shown will be used in an example discussed later on. The pyridine imine chelate ligand has been the chiral auxiliary for the resolution of octahedral compounds [9,10], not described in detail here, and for the resolution of square pyramidal compounds to be discussed next [11].

Scheme 4

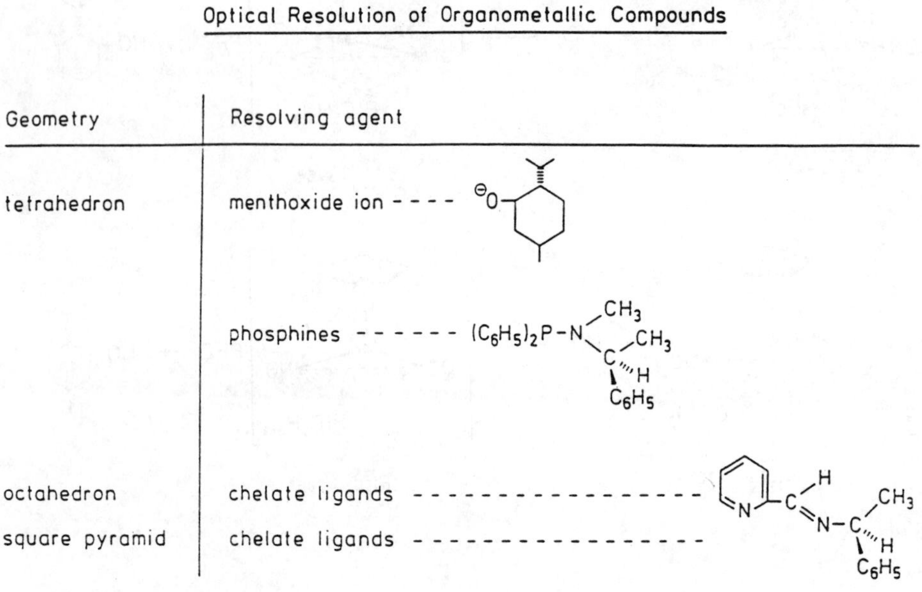

Optical Resolution of Organometallic Compounds

Geometry	Resolving agent
tetrahedron	menthoxide ion ---- $^{\ominus}O$–
	phosphines ------ $(C_6H_5)_2P-N(CH_3)-C(H)(CH_3)(C_6H_5)$
octahedron	chelate ligands ---------------- $C_5H_4N-C(H)=N-C(H)(CH_3)(C_6H_5)$
square pyramid	chelate ligands ----------------

Permutation of four different ligands L_1-L_4 around the positions of a tetrahedron generates two isomers, image and mirror image. A square pyramidal complex in the most general case contains five different ligands L_1-L_5 (Scheme 5, top). Permutation of five different ligands around the positions of a square pyramid results in 30 different isomers, subdividing into 15 pairs of enantiomers [11]. It is interesting to note that the isomer situations of a square pyramid and an octahedron are identical because the phantom ligand opposite to L_1 in the square pyramid of Scheme 5, top is equivalent to a ligand L_6 of an octahedron.

Scheme 5

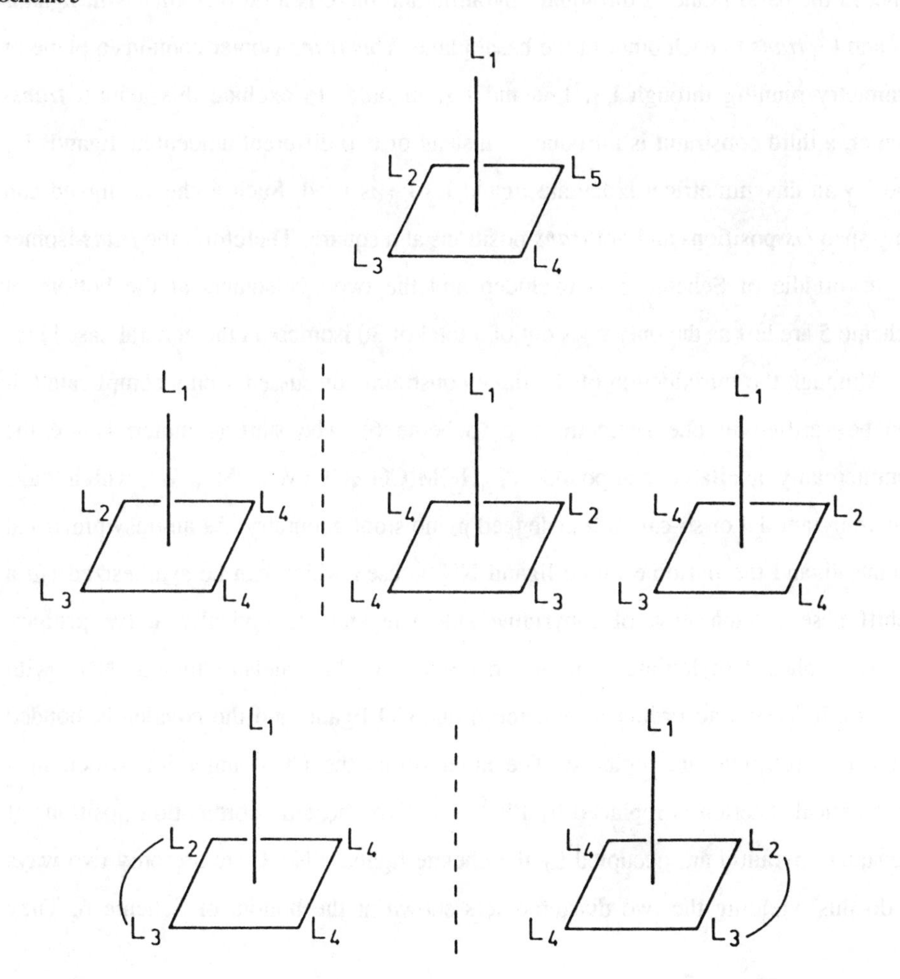

In order to reduce the number of 30 isomers possible in the most general case of a square pyramid with five different ligands, a number of constraints have to be made. Scheme 5 shows how the introduction of three constraints reduces the number of 30 isomers in the most general case to only two, image and mirror image, shown in the bottom line of Scheme 5 [11].

The first constraint is to choose L_1 such that it always stays on top of the pyramid. The second constraint is to make two ligands at the basal plane identical: $L_4 = L_5$. These two constraints reduce the number of 30 isomers to only three, shown in the middle of Scheme 5. There is a pair of enantiomers with ligands L_2 and L_3 *cis* to each other in the basal plane of the square pyramid and there is a third isomer with ligands L_2 and L_3 *trans* to each other in the basal plane. This *trans*-isomer contains a plane of symmetry running through L_1, L_2, and L_3. In order to exclude this achiral *trans*-isomer, a third constraint is introduced. Instead of two different unidentate ligands L_2 and L_3 an unsymmetrical bidentate ligand L_2-L_3 is used. Such a chelate ligand can only span *cis*-positions and not *trans*-positions at a square. Therefore, the *trans*-isomer in the middle of Scheme 5 is excluded and the two *cis*-isomers at the bottom of Scheme 5 are left as the only ones out of a total of 30 isomers in the general case [11].

Although the introduction of the three constraints discussed seems complicated, it can be verified in one synthetic step (Scheme 6). The starting materials are the commercially available compounds $(C_5H_5)M(CO)_3Cl$ (M = Mo, W) which have square pyramidal or so-called four-legged piano stool geometry. As an unsymmetrical chelate ligand the pyridine imine ligand NN' is used which can be synthesized *via* a Schiff base condensation of 2-pyridine aldehyde and the optically active primary amine 1-phenylethylamine. In the reaction of the chelate ligand NN' with $(C_5H_5)M(CO)_3Cl$ according to Scheme 6 one CO ligand and the covalently bonded chlorine substituent are replaced. The latter forms the Cl^- counter ion which in a metathetical reaction is replaced by PF_6^-. The two vacated coordination positions at the square pyramid are occupied by the chelate ligand NN'. There are only two ways to do this, yielding the two diastereomers shown at the bottom of Scheme 6. They contain the same chiral auxiliary NN' and they differ only in the configuration at the metal atom [11].

Scheme 6

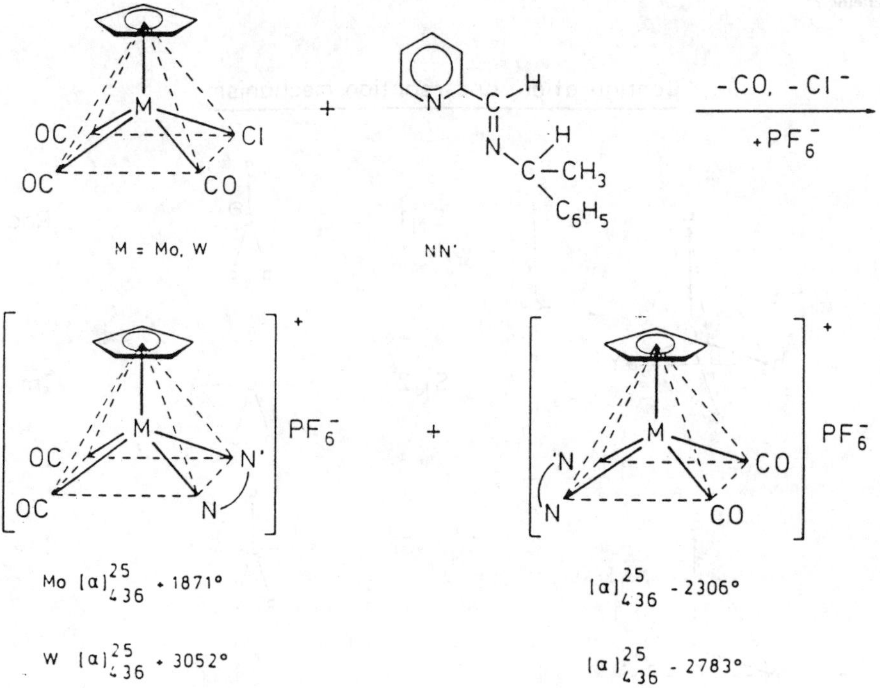

M = Mo, W

NN'

$-CO, -Cl^-$ / $+PF_6^-$

Mo $[\alpha]^{25}_{436}$ +1871°

W $[\alpha]^{25}_{436}$ +3052°

$[\alpha]^{25}_{436}$ -2306°

$[\alpha]^{25}_{436}$ -2783°

The two diastereomers could be separated for Mo and W into the $(+)_{436}$- and $(-)_{436}$-rotating diastereomers on the basis of solubility differences. They are examples of the concept developed in Scheme 5: (i) one of the ligands, in this case the cyclopentadienyl ligand, always stays on top of the pyramid, (ii) two ligands at the basal square are identical, (iii) the remaining two positions are occupied by an unsymmetrical chelate ligand NN'. Then, out of a total of 30 possible isomers only two are left, image and mirror image [11].

In the elucidation of reaction mechanisms in organic chemistry, stereochemical investigations played an extremely important role. In their classical studies in the 1930's, Ingold and Hughes showed that in nucleophilic substitution reactions either racemization or inversion takes place at the asymmetric carbon atom, depending on whether the reaction occurs by an S_N1 mechanism via a planar carbenium ion or by an S_N2 mechanism via a Walden inversion (Scheme 7, top).

Scheme 7

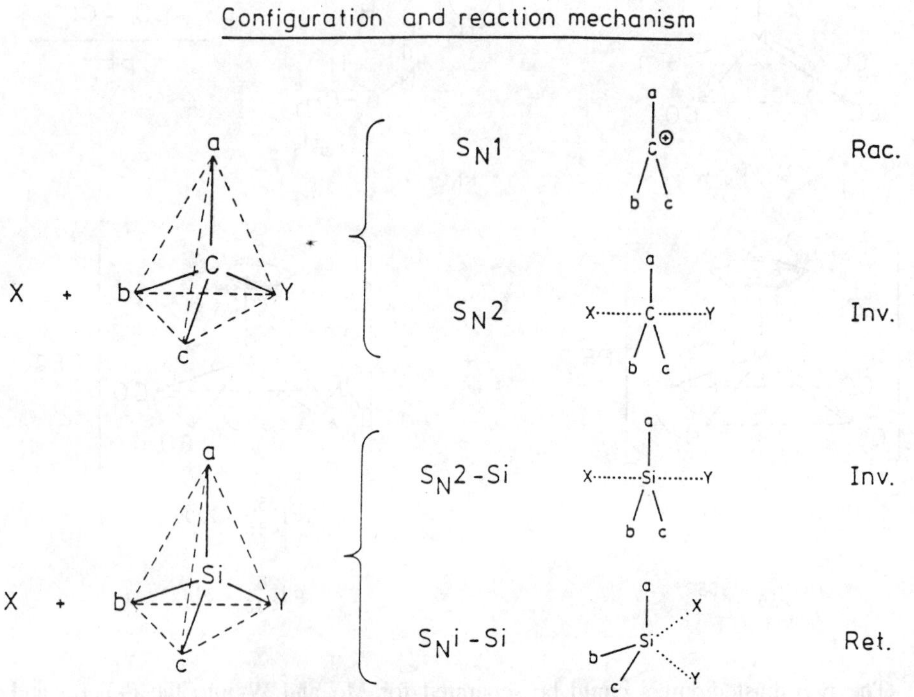

Configuration and reaction mechanism

X + b–C(a)(c)–Y: S_N1 → Rac.; S_N2 → Inv.

X + b–Si(a)(c)–Y: S_N2-Si → Inv.; S_Ni-Si → Ret.

The same nucleophilic substitution at the asymmetric Si atom takes place by an inversion mechanism called S_N2-Si, similar to the organic S_N2 reaction and a retention mechanism called S_Ni-Si, rare in organic chemistry (Scheme 7, bottom) [1]. Thus, C and Si already behave differently. With the help of the new optically active transition metal compounds the stereochemical course of reactions of organometallic compounds can be studied. In Scheme 8 an overview shows that retention reactions, inversion reactions and racemization reactions of various types have been observed [12].

Scheme 8

Stereochemistry of Reactions of Organometallic Compounds

1. Retention reactions
   a) Reactions on the ligands
   b) Dissociation reactions via chiral intermediates
2. Inversion reactions
   a) Two ligands exchange their roles
   b) Walden inversion
3. Racemisation reactions
   a) Successive role exchange of ligands
   b) Successive Walden inversions
   c) Dissociation reactions via planar intermediates
   d) Intramolecular pseudo rotation

The first representative example will be discussed in Scheme 9, demonstrating the value of optically active labeled organometallic compounds for the establishment of the steric course of reactions at transition metal centers. Scheme 9 shows an Fe compound containing four different substituents, a C_5H_5, a CO, a PPh_3 and a COOmenthyl ligand. For the derivatives of (-)-menthol two diastereomers with respect to the Fe configuration are possible, one of them exhibiting $(+)_{546}$- and the other $(-)_{546}$-rotation [13-15]. They can be separated on the basis of solubility differences similar to the Mn complexes shown in Schemes 2 and 3. In Scheme 9 only the $(+)_{546}$-diastereomer is depicted.

Its reaction with methyllithium results in the elimination of lithium mentholate and an acetyl substituent is formed at the Fe atom (Scheme 9, top). Starting with the $(+)_{546}$-menthyl ester, the acetyl complex of $(-)_{546}$-rotation is obtained and the $(-)_{546}$-menthyl ester gives the $(+)_{546}$-acetyl complex [13-15]. The optical rotations of starting material and product have opposite signs and the CD spectra of (+)-menthyl ester and (-)-acetyl derivative are almost mirror images (Scheme 9, bottom) [14].

These chiroptical data suggest an image/mirror image relationship between starting material and product, which was ultimately confirmed by the X-ray determination of the absolute configuration at the Fe atom in the starting material and the reaction product [16-18]. Thus, attack of methyllithium definitely does not occur at the ester group of the starting material as expected but at the carbonyl group, transforming it into the new functional group, the acetyl substituent. The mentholate ion dissociates

Scheme 9

$[\alpha]^{25}_{546}$ +72° $[\alpha]^{25}_{546}$ −285°

$[\alpha]^{25}_{546}$ −120° $[\alpha]^{25}_{546}$ +287°

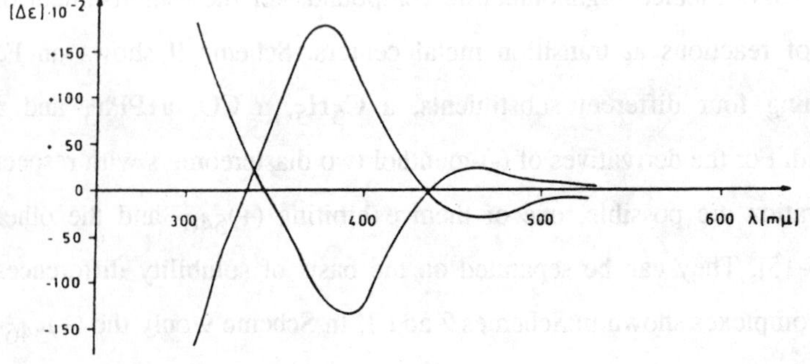

from the ester group, the former functional group, converting it into the carbonyl group of the product. Thus, in the reaction of Scheme 9 two substituents of the Fe complex change their roles, the carbonyl group and the functional group. This results in an inversion of metal configuration, although none of the bonds starting from the Fe atom is cleaved.

The compounds (+)- and (-)-C_5H_5)Fe(CO)(PPh_3)COMe (Scheme 10) have

Scheme 10

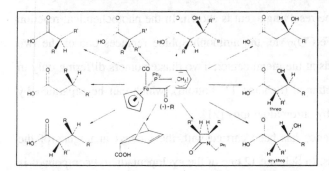

become commercial optically active auxiliaries which allow the synthesis of a variety of interesting compounds via enolate chemistry of the acetyl substituent. The chirality at the Fe atom controls the formation of new asymmetric C atoms in α-, β- and partly γ-position in alkylation reactions, aldol reactions, Michael reactions etc. in high enantioselectivity [19,20].

In Scheme 11 the formation of diastereomeric Fe complexes using an optically active aminophosphine as the resolving agent is shown. In the photochemical reaction with the prochiral $(C_5H_5)Fe(CO)_2Me$, the aminophosphine replaces one of the two enantiotopic carbonyl ligands at the metal center. Two diastereomers differing only in the Fe configuration are obtained (Scheme 11, bottom), which can be separated by fractional crystallization or by chromatography [21].

The same reaction sequence has been carried with the indenyl analogues of the cyclopentadienyl compounds. In Scheme 12 one of the cyclopentadienyl compounds is confronted with its indenyl analogue. For both compounds the absolute configuration at the Fe atom has been determined by X-ray crystallography [22]. The two ORTEPs

Scheme 11

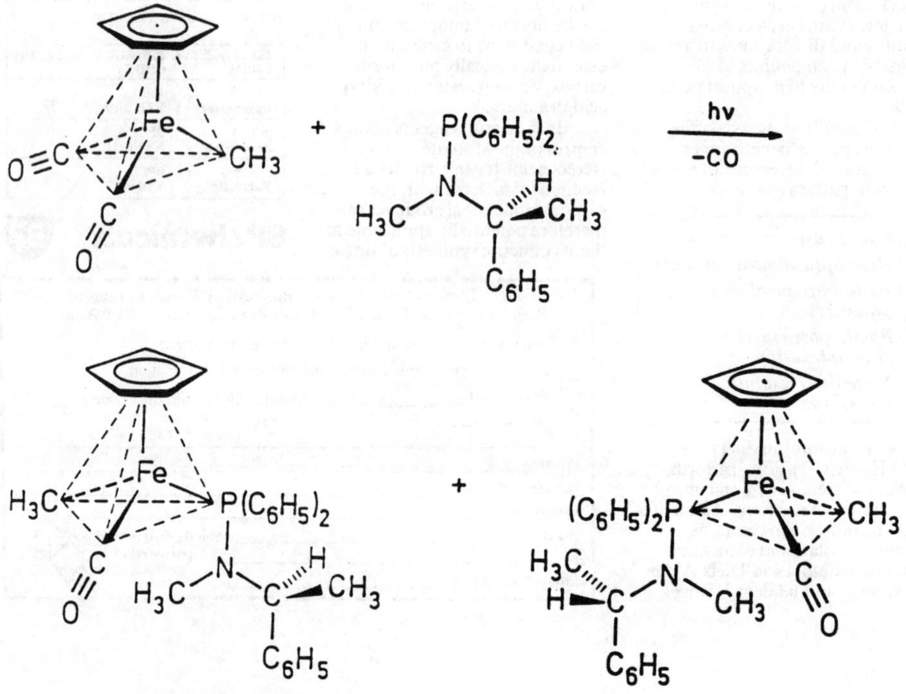

at the bottom of Scheme 12 reveal a surprising similarity of the two compounds in particular as far as the conformation of the amino phosphine ligand is concerned, which has a lot of degrees of freedom.

The cyclopentadienyl and the indenyl derivatives shown in Scheme 12 are configurationally stable in solution at room temperature. However, a change of the configuration at the Fe atom occurs on heating these solutions. The epimerization can be followed by polarimetry and also by ^{1}H nmr spectroscopy. It is strictly of first order. The half life of the cyclopentadienyl compound in C_6D_6 solution at 70 °C is 70 minutes, whereas the half life of the indenyl analogue in C_6D_6 solution at 50 °C is 37 minutes [22]. Thus, the rate difference amounts to a factor of 10-20. This small rate difference is due to a somewhat increased steric hindrance in the indenyl compound. No specific "indenyl" effect including a ring slippage is operating.

Scheme 12

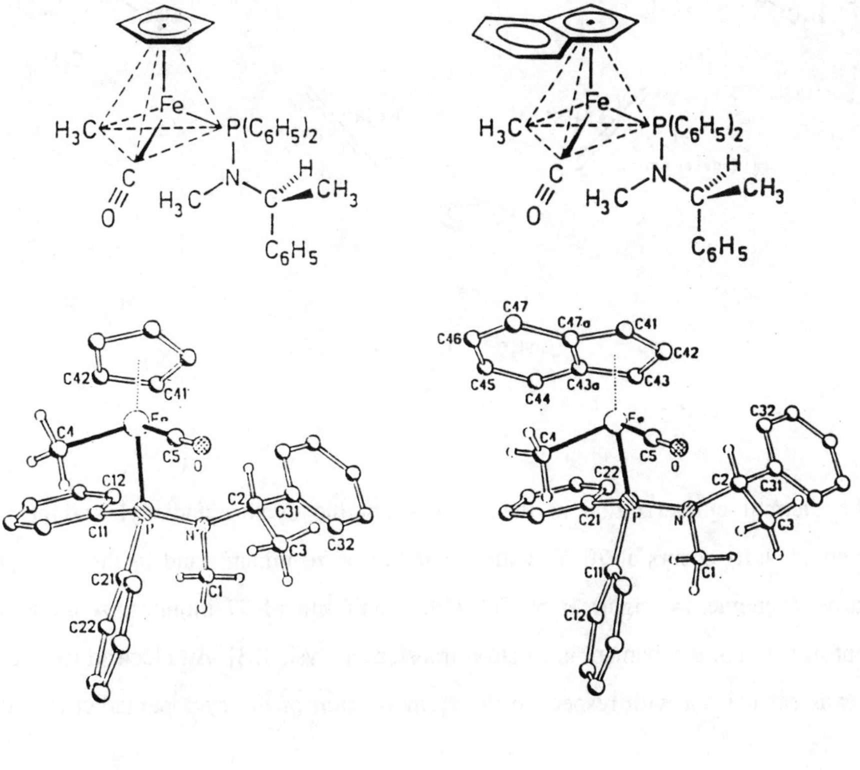

Added phosphine does not influence the 1st order kinetics. However, in the presence of a large excess of a phosphine, different from that bound in the complex, the epimerization is accompanied by phosphine exchange. Thus, the epimerization at the chiral Fe atom occurs *via* a rate determining phosphine dissociation forming a planar intermediate $(C_5H_5)Fe(CO)Me$, which is attacked by free phosphine on its front side or its back side to give the two diastereomers with opposite Fe configurations (Scheme 13) [22]. This epimerization mechanism is another example of how the stereochemical course of reactions at the metal atom can be studied by using optically active labeled compounds.

Scheme 13

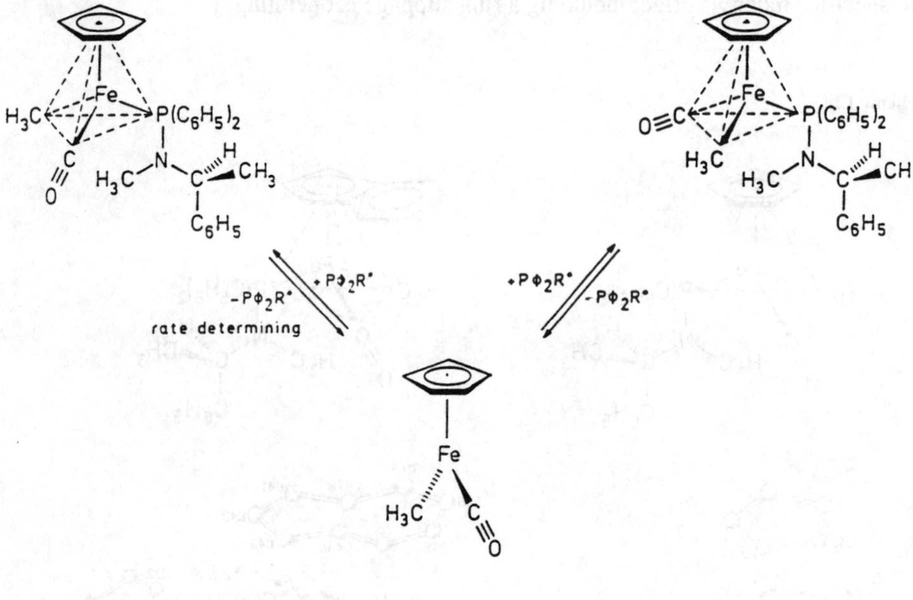

H_3C Fe $P(C_6H_5)_2$ C O N H_3C C H CH_3 C_6H_5

$O{\equiv}C$ Fe $P(C_6H_5)_2$ H_3C N H_3C C H CH_3 C_6H_5

$-P\Phi_2R^*$ $+P\Phi_2R^*$ rate determining

$+P\Phi_2R^*$ $-P\Phi_2R^*$

Fe H_3C C O

The thermal epimerization at the Fe atom in the cyclopentadienyl derivative (Scheme 14, left) occurs at 70 °C with a half life of 70 minutes and in the indenyl derivative (Scheme 14, right) at 50 °C with a half life of 37 minutes. A modern concept in reaction mechanism is electron transfer catalysis [23]. An electron transfer study was carried out with respect to the epimerization of the cyclopentadienyl and indenyl derivatives in Scheme 14. The reductive start of the electron transfer catalysis by sodium amalgam did not lead to an epimerization of the cyclopentadienyl compound at 20 °C, whereas complete epimerization took place in the indenyl compound at -35 °C after one minute, -35 °C being the temperature where Na amalgam becomes a solid and one minute the time necessary for mixing the reactants [22]. Obviously, in the epimerization *via* a 19e species the benzene ring connected to the cyclopentadienyl ligand in the indenyl derivative plays a major role.

On the contrary, the oxidative start of the electron transfer catalysis with $Cp_2Fe^+PF_6^-$ led to complete epimerization of the cyclopentadienyl derivative at 20 °C after five minutes and only partial epimerization was observed for the indenyl analogue at 20 °C

Scheme 14

Epimerisation at the chiral Fe atom via 17e and 19e species

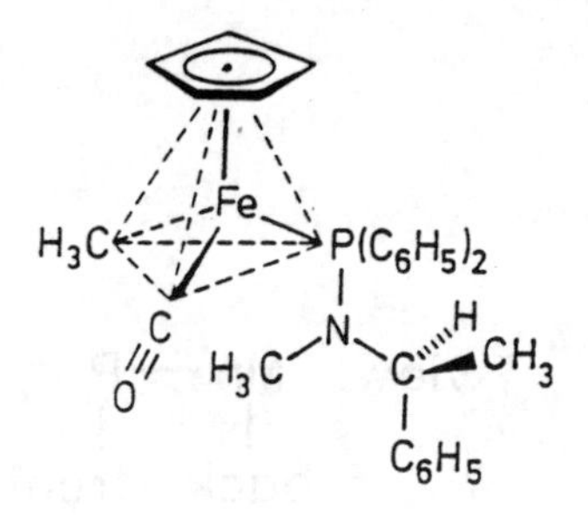

Thermal reaction (C_6D_6):	$\tau_{1/2}$(70°C) = 70 min	$\tau_{1/2}$(50°C) = 37 min
Electron transfer (THF):		
Na-amalgam (reductive)	no epimerisation at 20°C	complete epimerisation at -35°C after 1min
$Cp_2Fe^+ PF_6^-$ (oxidative)	complete epimerisation at 20°C after 5min	partial epimerisation at 20°C after 30min

after 30 minutes. These results show that electron transfer catalysis via 17e and 19e species may drastically accelerate the epimerization at the Fe atom compared to the thermal reaction of the 18e species [22].

Triphenylphosphine is one of the standard ligands in organometallic and coordination chemistry. It is an air stable and inexpensive solid, easy to prepare and to store. In the free ligand and also in its complexes the three phenyl rings of triphenylphosphine form a three-bladed propeller which may have P (plus, right-handed) or M (minus, left-handed) chirality [24]. A well known triphenylphosphine complex is the compound $(C_5H_5)Mn(CO)_2PPh_3$ (Scheme 15). Viewed along the P-Mn bond possible conformations of this molecule are shown in Scheme 15, bottom. In the conformation on the left side, the three phenyl rings adopt a staggered conformation with respect to the C_5H_5 and the CO ligands of the $(C_5H_5)Mn(CO)_2$ fragment. However, due to the right-handed propeller sense, the interactions of the phenyl rings with the C_5H_5 ligand and the CO ligands are different in the conformations of Scheme 15. As far as the cyclopentadienyl ligand

Scheme 15

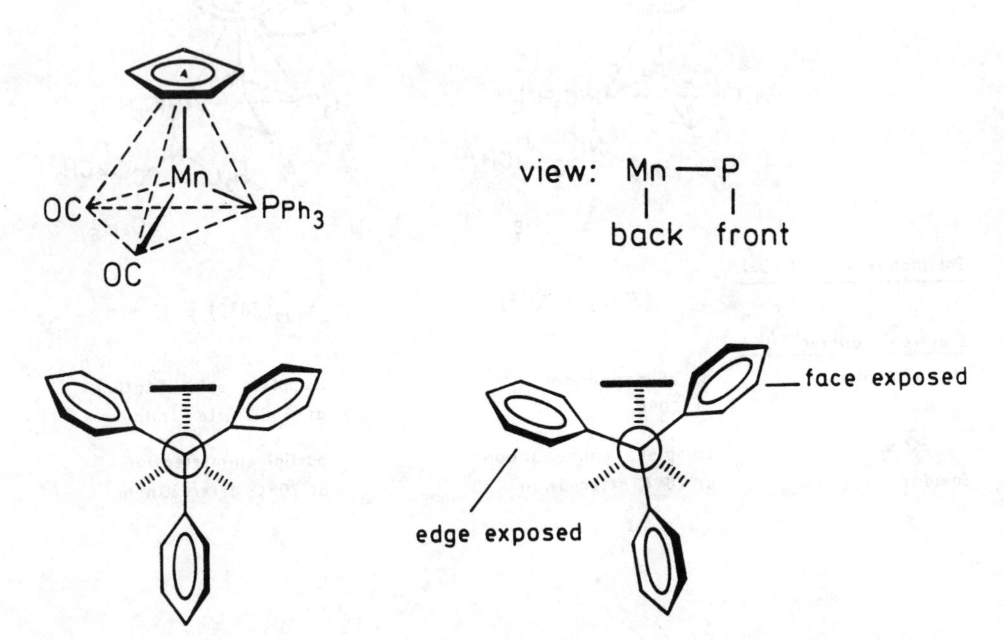

on top of the molecule is concerned the phenyl ring on the left side exposes its ortho-CH bond to the C_5H_5 ligand, whereas the phenyl ring on the right side orients its face to the C_5H_5 ligand [25]. Edge exposure results in an increased steric hindrance compared to face exposure. Therefore, a rotation of the triphenylphosphine propeller around the P-Mn axes with respect to the $(C_5H_5)Mn(CO)_2$ fragment takes place as indicated in the conformation shown on the right side [25]. This rotation of the triphenylphosphine propeller relieves the steric hindrance.

In triphenylphosphine complexes the propeller sense is one element of chirality. If the metal atom in a three-legged piano stool complex is an additional stereogenic center, e.g. $(C_5H_5)Fe(CO)(X)PPh_3$ in Scheme 16, diastereomers arise. In the acetyl complex, X = COMe, the barrier for the propeller inversion in solution was

Scheme 16

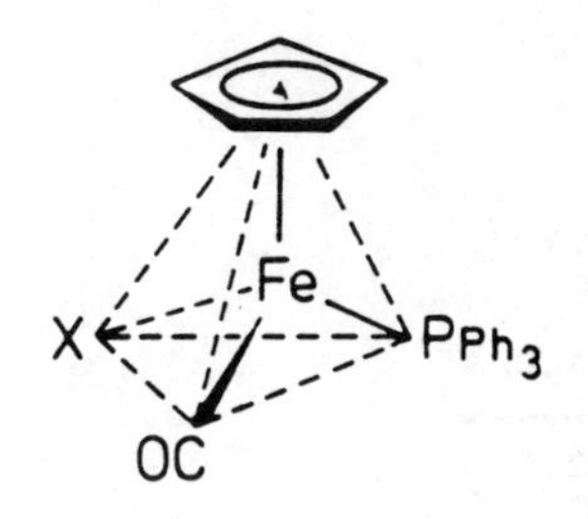

X = $\underline{COCH_3}$, H, CH_3, $CH_2C_6H_5$ ·····

barrier for the propeller inversion ~10 Kcal/mol

only one diastereomer observed in the solid state

X CO

$S_{Fe}P_{PPh_3}$

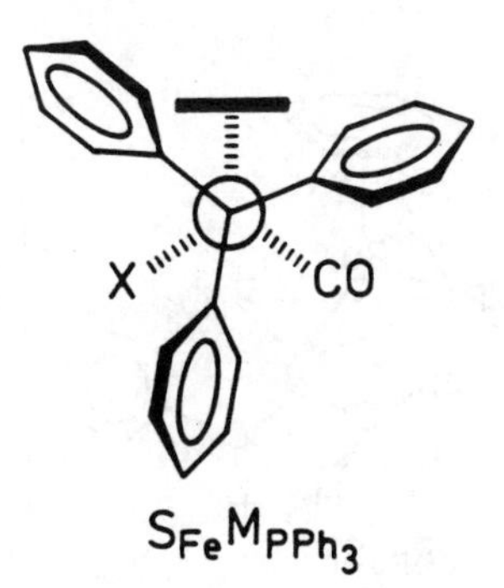

determined to be approximately 10 Kcal/mol [26]. In solution, both diastereomers are present. However, on crystallization, for all cases investigated up to now, only one diastereomer was observed, probably the thermodynamically more stable one.

For the half-sandwich Ru complexes shown in Schemes 17-19 we succeeded for the first time in isolating compounds which only differ in the propeller sense of the triphenylphosphine ligand [27].

In Scheme 17 the starting material is a neutral three-legged piano stool ruthenium complex, which contains a benzene ligand, a chloro ligand and an O-N chelate ligand, derived from salicylaldehyde and the optically active primary amine S-1-phenylethylamine. The Ru atom in this compound is a stereogenic center; only one of the two possible configurations is shown in Scheme 17. Treatment of the chloro compound with $AgPF_6$ removes the chloro ligand from the ruthenium atom. In the

Scheme 17

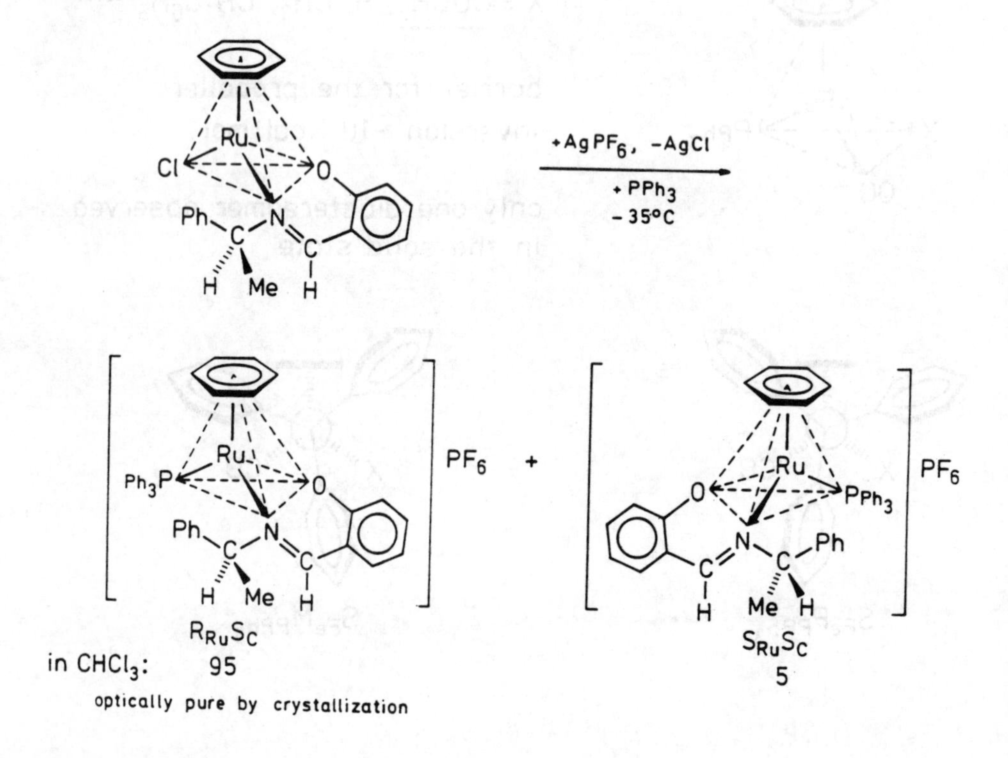

presence of triphenylphosphine at -35 °C the corresponding cationic triphenylphosphine complexes are formed as the PF_6 salts. Their configurations are $R_{Ru}S_C$ and $S_{Ru}S_C$ as indicated below the formulas in Scheme 17. In chloroform solution the diastereomer equilibrium is 95:5 in favor of the $R_{Ru}S_C$ isomer. By crystallization the $R_{Ru}S_C$ isomer can be obtained optically pure [27].

Crystallization of the $R_{Ru}S_C$ isomer, prepared at -35 °C without warming, at 5 °C gives crystals of tetrahedral shape, which by X-ray crystallography turn out to have R configuration at the Ru atom, S configuration at the carbon atom of the chelate ligand and M configuration of the triphenylphosphine propeller. This $R_{Ru}S_C$ M_{PPh3} isomer obviously is formed under kinetic control (Scheme 18, left) [27].

Heating its solution for 50 minutes at 40 °C and crystallization at 5 °C gives another isomer under thermodynamic control. Its crystals form plates. X-ray crystallography shows that is has R configuration at Ru, S configuration at the carbon

Scheme 18

synthesis at -35°C
kinetic control
crystallization at 5°C
$R_{Ru}S_CM_{PPh_3}$

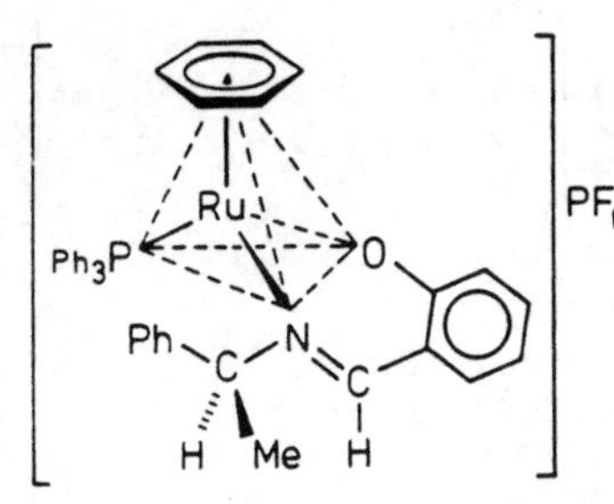

50min at 40°C
thermodynamic control
crystallization at 5°C
$R_{Ru}S_CP_{PPh_3}$
(in $CHCl_3$ solution 5%
$S_{Ru}S_C$)

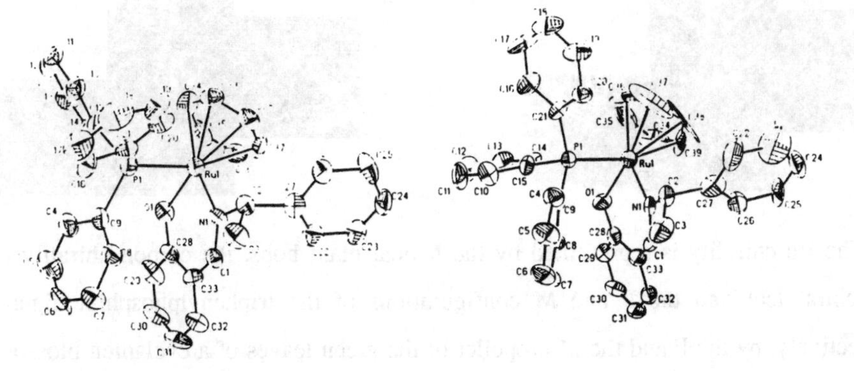

atom of the chelate ligand and P configuration of the triphenylphosphine propeller (Scheme 18, right) [27]. Thus, the two isomers only differ in the propeller sense of the triphenylphosphine ligand.

In Scheme 19 the stereochemistry of the two Ru complexes $R_{Ru}S_{C}M_{PPh3}$ and $R_{Ru}S_{C}P_{PPh3}$ differing only in the triphenylphosphine chirality is "translated" into elements of natural chirality.

Scheme 19

Halfsandwich-Ru-complexes

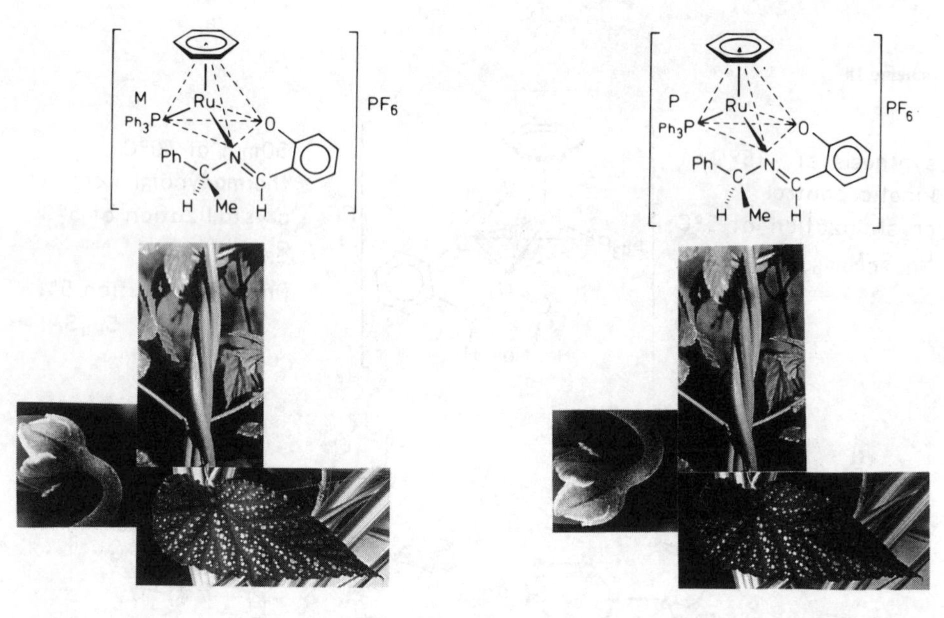

The Ru chirality is represented by the helical plant hops, the carbon chirality by the chiral leaf and the P and M configurations of the triphenylphosphine ligand, respectively, by the P and the M propeller of the green leaves of a cyclamen blossom, the P propeller being the natural one.

References

1. Sommer, L.H. (1965), *Stereochemistry, Mechanism, and Silicon*, McGraw-Hill, New York.
2. Morrison, J.D. (1983-1985), *Asymmetric Synthesis*, Vols. 2-5, Academic Press, New York.
3. Brunner, H. (1991), *From Weak Bosons to the α-Helix*, Janoschek, R. (ed.), Springer-Verlag, Berlin, p. 166.
4. Werner, A. (1911), *Ber. dtsch. chem. Ges.* 44, 1887.
5. Brunner, H. (1971), *Angew. Chem.* **83**, 274; *Angew. Chem. Int. Ed. Engl.* **10**, 249.
6. Brunner, H. (1969), *Angew. Chem.* **81**, 395; *Angew. Chem. Int. Ed. Engl.* **8**, 382.
7. Brunner, H. and Schindler, H.-D. (1970), *J. Organomet. Chem.* **24**, C7.
8. Brunner, H. (1969), *Z. Anorg. Allg. Chem.* **368**, 120.
9. Brunner, H. and Herrmann, W. A. (1973), *J. Organomet. Chem.* **57**, 183.
10. Brunner, H. and Stumpf, A. (1993), *J. Organomet. Chem.* **459**, 139.
11. Brunner, H. and Herrmann, W. A. (1972), *Chem. Ber.* **105**, 3600.
12. Brunner, H. (1980), *Adv. Organomet. Chem.*, Vol. 18, Academic Press, New York, p. 151.
13. Brunner, H. and Schmidt, E. (1969), *J. Organomet. Chem.* **21**, P53.
14. Brunner, H. and Schmidt, E. (1972), *J. Organomet. Chem.* **36**, C18.
15. Brunner, H. and Schmidt, E. (1973), *J. Organomet. Chem.* **50**, 219.
16. Reisner, M.G., Bernal, I., Brunner, H. and Muschiol, M. (1976), Angew. Chem. **88**, 847; Angew. *Chem. Int. Ed. Engl.* **15**, 776.
17. Reisner, M.G., Bernal, A., Brunner, H. and Muschiol, M. (1978), *Inorg. Chem.* **17**, 783.
18. Bernal, I., Brunner, H. and Muschiol, M. (1988), *Inorg. Chim. Acta* **142**, 235.
19. Brunner, H. (1991), *Angew. Chem.* **103**, A-310.
20. Davies, S.G., Dordor-Hedgecock, I.M., Easton, R.J.C., Preston, S.C., Sutton, K.H. and Walker, J.C. (1987), *Bull. Soc. Chim. Fr.* 608.
21. Brunner, H. and Vogt, H. (1981), *J. Organomet. Chem.*, **210**, 223.
22. Brunner, H., Fisch, K., Jones, P.G. and Salbeck, J. (1989), *Angew. Chem.* **101**, 1558; *Angew. Chem. Int. Ed. Engl.* **28**, 1521.
23. Astruc, D. (1988), *Angew. Chem.* **100**, 662; *Angew. Chem. Int. Ed. Engl.* **27**, 643.
24. Cahn, R.S., Ingold, C.K. and Prelog, V. (1966), Angew. Chem. **78**, 413; *Angew. Chem. Int. Ed. Engl.* **5**, 385.
25. Brunner, H., Hammer, B., Krüger, C., Angermund, K. and Bernal, I. (1985), *Organometallics* **4**, 1063.
26. Davies, S.G., Derome, A.E. and McNally, J.P. (1991), *J. Am. Chem. Soc.* **113**, 2854.
27. Brunner, H., Oeschey, R. and Nuber, B. (1994), *Angew. Chem.* **106**, 941.

BIOMIMETIC CHEMISTRY
Bifunctional Binding and Catalysis

R. BRESLOW
Department of Chemistry, Columbia University
New York, NY 10027

Abstract

Mechanistic studies show how the hydrolysis of RNA is catalyzed by imidazole buffer, and this information was used in the design of an effective catalyst for a related process. The geometric preference of this catalyst is evidenced in the greater effectiveness of one isomer, but another isomer is preferred for bifunctional catalysis of enolization and of an internal aldol condensation.

Central to the mechanistic study is the conversion of the normal 3',5'-linked RNA to its 2',5'-linked isomer. We have prepared 2',5'-linked DNA, and compared its properties with those of natural DNA. Binding of substrates into cyclodextrins is much stronger when cyclodextrin dimers are used; they are also the basis of novel effective catalysts.

1. Introduction

Biological models have two different functions: 1) they can assist biochemistry by furnishing the chemical information and analogies needed to explain biological processes; 2) they can expand the scope of chemistry by leading to new chemical inventions inspired by their biological counterparts. We have referred [1] to the latter activity as "Biomimetic Chemistry."

There are many examples of such work, both in our own studies and in those of other laboratories. In this review we will describe novel molecules synthesized to mimic two special features of enzymatic reactions: 1) enzymes normally use at least two functional groups to perform simultaneous bifunctional catalysis; 2) enzymes—and antibodies—normally use at least two interactions with a substrate or antigen to bind it strongly and with good geometric definition. In the course of the work to explore this area, we have discovered novel features of interest about DNA and about the preferred geometries of certain reactions; these features will also be described.

C. Chatgilialoglu and V. Snieckus (eds.), Chemical Synthesis, 113–135.

2. Cleavage of RNA by Buffers

The enzyme ribonuclease A uses two imidazole rings—the sidechains of His-12 and His-119, one acting as a base **B** and the other as an acid **BH+**—as principal catalysts for the cleavage of RNA. We instituted studies some years ago on the cleavage of RNA by imidazole buffer, and found evidence that here too both the **B** and **BH+** forms of the buffer were involved as catalysts. However, in contrast to the situation in the enzyme, our data indicated that there was *sequential* bifunctional catalysis, with the **BH+** acting first to convert the phosphate ester group of RNA to a phosphorane intermediate, and the **B** then catalyzing the cleavage of this intermediate.

This work has been extensively described [2, 3], and therefore will only be summarized here. For the cleavage of RNA itself, one buffer component converts the phosphate monoanion group to a phosphorane monoanion; this protonation state is required to account for the kinetics. However, simple kinetics does not indicate whether the first catalyst is **B** or **BH+**; this was solved by studies of the cleavage of simple dinucleotides, UpU and ApA. With these substrates we saw similar behavior to that of RNA itself with respect to the kinetics of cleavage, but we saw that the buffer catalysts also converted the normal 3',5'-linked UpU or ApA to its isomeric 2',5'-linked uncleaved dimer.

The isomerization of UpU, or of ApA, was catalyzed only by the buffer acid **BH+**. Thus if the isomerization also branched off the same phosphorane intermediate involved in cleavage (Scheme 1), we knew that the first catalyst must be **BH+**, not **B**. The question was whether these two processes had a common intermediate, or whether they had completely independent pathways. Partitioning kinetics showed that they indeed have a common intermediate.

We observed that increasing the concentration of **B**, while holding **BH+** constant, caused a *decrease* in the rate of isomerization, while the cleavage went faster. This is expected if the two processes branch from a common intermediate (Scheme 1). The kinetic equations show this, and physically what happens is that the concentration of the steady state phosphorane intermediate **I** is lower as more **B** sends it along the cleavage path. If the isomerization branches from this intermediate, but not otherwise, such positive catalysis of the cleavage by **B** will lower the rate of isomerization.

Of course changing the **B/BH+** ratio changes the pH, but we tried to correct for this by the normal procedure of extrapolating the buffer concentration to zero, and subtracting the few percent of the rate that was contributed by H^+ or OH^- or water alone [2]. In a later study [3], we avoided the necessity for such corrections. We used a more basic buffer, morpholine, and at high **B/BH+** ratios. We maintained a constant ionic strength and constant pH, and simply increased the concentration of the buffer at a given ratio. We observed that as expected the rate of cleavage of UpU increased as the buffer concentration increased, but with high **B/BH+** ratios the rate of isomerization *decreased* to a plateau, eventually rising again. The decrease at early parts of the curve also indicates a partitioning effect, again supporting the idea that there is a common intermediate for cleavage and isomerization. We saw that such a decrease occurred with high **B/BH+** ratios, but not with a lower one, a further indication that it is the basic component **B** that sends the intermediate along the cleavage pathway, and thus slows the observed isomerization.

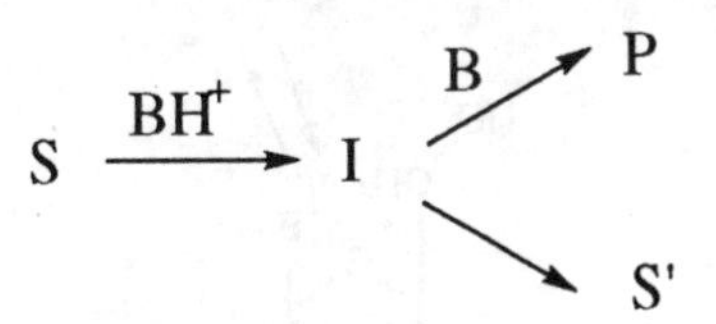

S: 3',5" linked RNA

I: phosphorane intermediate

P: cleavage products

S': 2',5' linked RNA isomer

BH^+: buffer acid

B: buffer base

Scheme 1

In this work [3] there was detailed fitting of the data to theoretical curves for this mechanism, and to some other equations that tested the conclusions. However, the arguments and detailed kinetic treatments are complex, and have not always been fully understood by people who should be experts. Those able to follow complex kinetic treatments have had no trouble with our conclusions.

We described the observation of a slowing of isomerization when the **B** was increased—either in the original work or as part of a **B**-rich buffer in the later work—and called this [2] "negative catalysis." As we have pointed out [3], L. P. Hammett used this same phrase—in his classic book [4] that laid the basis for the field of physical organic chemistry—to describe the slowing of a reaction when a catalyst is added. Despite objections from people not familiar with this history, the phrase is a perfectly good description of the *observation* that a reaction slows when a catalyst is added, although the physical

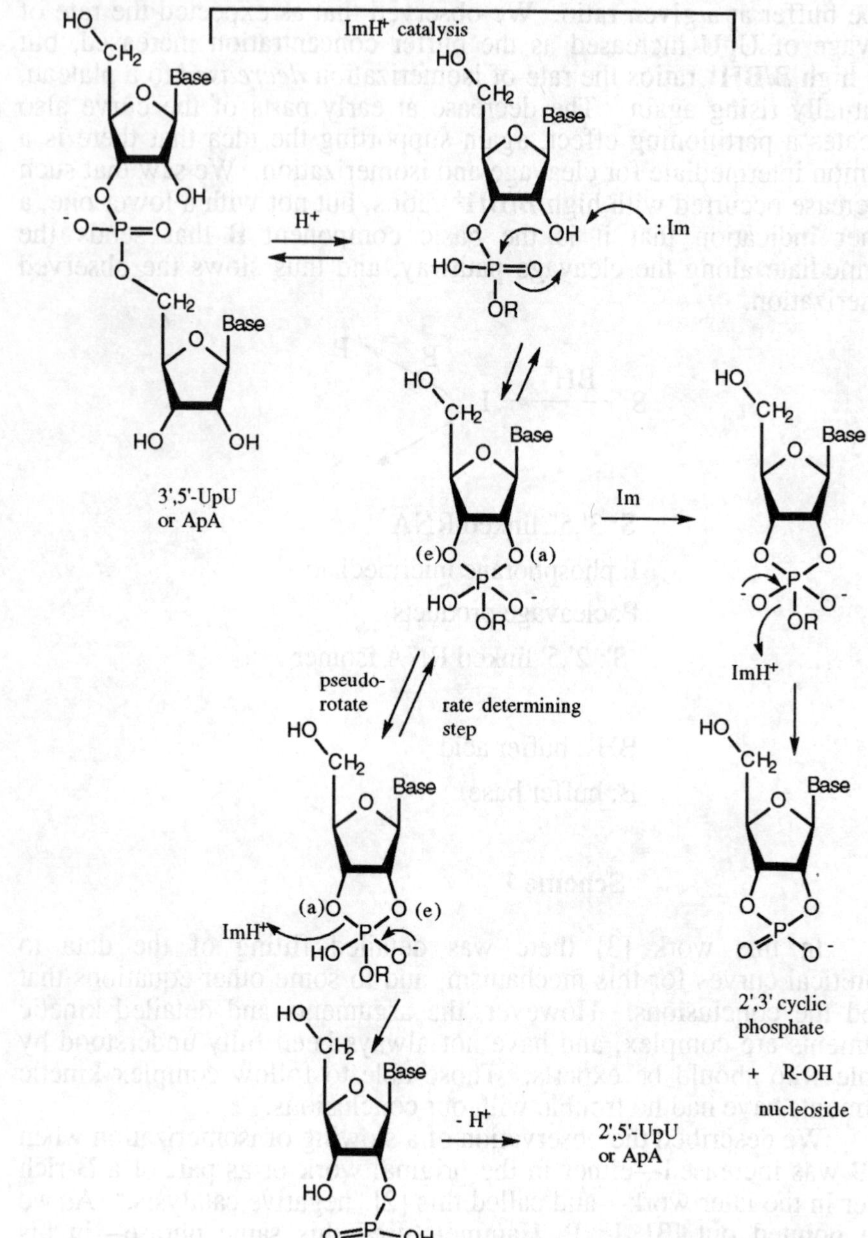

Scheme 2. Buffer-catalyzed RNA cleavage.

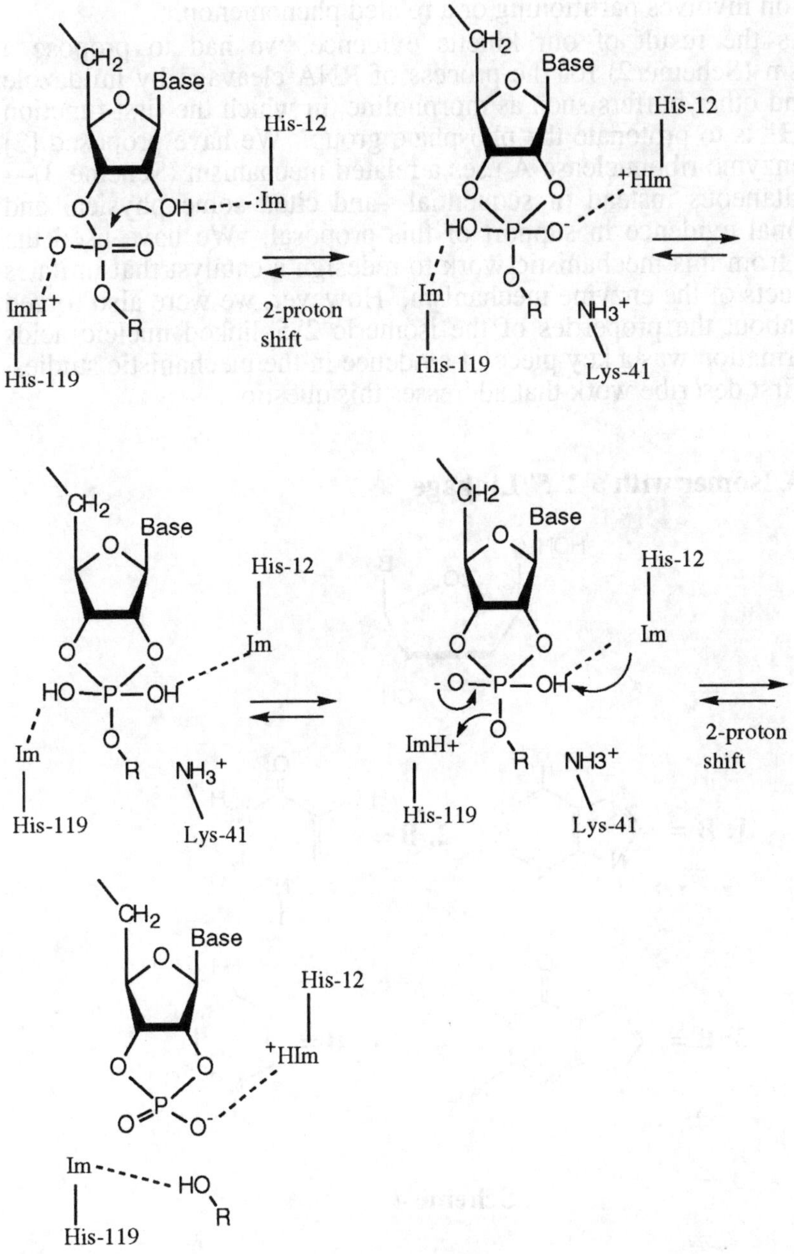

Scheme 3. Our new proposed mechanism for ribonuclease A catalysis.

explanation involves partitioning or a related phenomenon.

As the result of our kinetic evidence, we had to propose a mechanism (Scheme 2) for the process of RNA cleavage by imidazole buffer, and other buffers such as morpholine, in which the first function of the $\mathbf{BH^+}$ is to protonate the phosphate group. We have proposed [2] that the enzyme ribonuclease A uses a related mechanism (Scheme 3)—but simultaneous instead of sequential—and cited some physical and calculational evidence in support of this proposal. We have used the evidence from this mechanistic work to redesign a catalyst that imitates some aspects of the enzyme mechanism. However, we were also forced to think about the properties of the isomeric 2',5'-linked nucleic acids whose formation was a key piece of evidence in the mechanistic studies. We will first describe work that addresses this question.

3. A DNA Isomer with a 2',5' Linkage

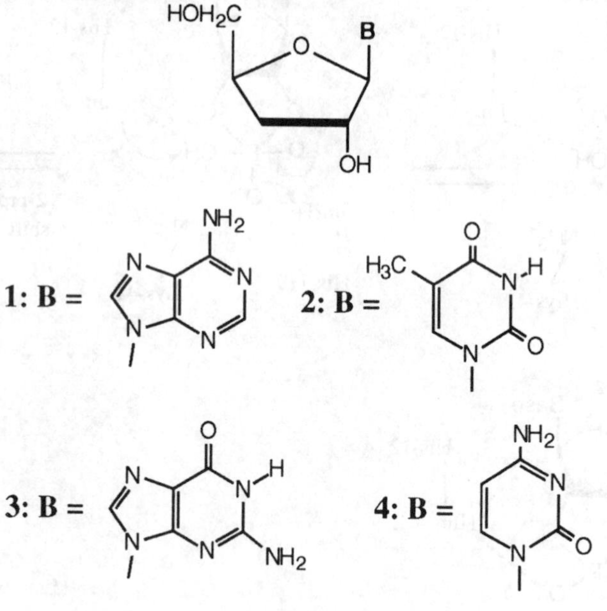

HOH$_2$C B O OH

1: B = NH$_2$ N N N N

2: B = H$_3$C O N H O N

3: B = O N N H N N NH$_2$

4: B = NH$_2$ N O N

Scheme 4

In natural DNA the 2' hydroxyl group of ribose has been removed, and the only possible linkage is from position 3' on one nucleoside to position 5' of another. For other reasons we had become interested in using an isomer in which the 3' hydroxyl of ribose was removed so the linkage had to be 2' to 5', as in our RNA isomers. We have described [5] the synthesis of nucleotides incorporating 3'-deoxyadenosine **1** and the corresponding isomer **2** of thymidine, and in more recent work the related guanine and cytosine analogs **3** and **4** have been prepared and incorporated into polynucleotides (Scheme 4). In early work [5] we saw that the incorporation of one or two A' or T' units (we will use the prime to denote the 3-deoxy isomers) into normal DNA 16-mers,—in which the other nucleotides were normal 2-deoxy A, T, etc.—led to weaker base pairing, but also saw an indication that the A' could base pair with a normal T. However, the most interesting results came from a study of the properties of polynucleotide 16-mers incorporating *only* the A' and T' nucleosides.

In contrast to normal DNA, the isomeric A' and T'—incorporated in mixed sequences into 16-mers—gave a strand that showed *no* evidence of base pairing and helix formation with its complement [5]. There was no hypochromic effect, there was no change in the circular dichroism (CD), and there was no retardation of a radioactive strand by even large excesses of its complement on non-denaturing gel electrophoresis. Computer models showed that it was possible to construct a double helix with our isomeric nucleosides, but the helix differed from that of normal DNA. In particular, the helix was *open*, with a central cavity like that in the center of a spiral staircase, and the bases did not cover each other as well as in a normal DNA double helix. Since the major driving force for double helix formation is base stacking and the resulting coverage of hydrophobic surfaces, the poorer coverage with our isomeric DNA apparently explained the lack of strand association.

However, the situation was quite different when we examined [5] a 16-mer consisting *only* of the A' units, and its interaction with a 16-mer of T' units. Now there was a hypochromic effect in the UV, there was an additional CD when the strands were mixed, and there was retardation in the gel electrophoresis. We pointed out that these contrasting results did not really make sense in terms of a *double* helix, and suggested that a triple or higher-order helix must be involved. Our subsequent collaborative studies confirmed this: the homopolymeric DNA isomeric strands form a *triple* helix only, with one polyA' and two polyT' strands [6].

In the computer model of this triple helix, there is more coverage of the hydrophobic surfaces, since each "step" of the helix is almost 50% longer. This may well be why the triple helix is possible when a double helix does not form under any conditions we examined. The triple helix is of the normal composition, with a central purine and two pyrimidine strands, one with Watson-Crick base pairing and one with Hoogsteen base pairing. Our further work is aimed at a more detailed understanding of this structure in our compounds.

It is clear that no genetic system could have been based on these isomeric nucleotides, in which one does not see the association of *two* strands to transfer and utilize the genetic coding, only three strand association of a special sort. Such studies—related to those by Eschenmoser [7] in which ribose was substituted by other sugars—give us insight into the special chemical properties of natural substances such as DNA that allow them to perform their biological functions.

4. Phosphate Ester Hydrolysis Catalyzed by a Cyclodextrin-based Enzyme Mimic

Cyclodextrins are cyclic oligomers of glucose that have found great use in the construction of enzyme mimics. The most available is β–cyclodextrin, with seven glucose units in a ring, but α–cyclodextrin with six units and γ–cyclodextrin with eight glucoses are also available. In water solution, and to some extent in other polar solvents, these sugar rings bind hydrophobic molecules into their cavities, the binding being driven largely by the hydrophobic effect. Such molecular complexes have many uses; here we will describe their use as models for enzyme-substrate complexes.

Using toluenesulfonyl chloride under appropriate conditions, it is possible to prepare β-cyclodextrin with a tosyl group on the C-6 primary carbon of one glucose residue. This can then be used to attach a single functional group, such as the imidazole ring in **5**, to that position. By the use of bis-sulfonyl chloride reagents with appropriate spacers, it is possible to functionalize two different C-6 groups of a cyclodextrin. Reagents have been described that produce the isomer that is functionalized on neighboring glucose units (the AB isomer), on glucoses separated by one (the AC isomer), or on glucose units as far apart as possible on a seven unit ring (the AD isomer). Using these compounds, we have prepared β-cyclodextrin bis-imidazoles as the AB (**6**), the AC (**7**), and the AD (**8**) isomers [8] (Scheme 5). They have proven to be very interesting bifunctional catalysts.

The phosphate that we first examined [8, 9], and the only one for which we yet have significant evidence, is compound **9**, the cyclic phosphate of 4-tert-butylcatechol (Scheme 6). The t-butylphenyl group binds into β-cyclodextrin with a high affinity in water solution, and the cyclic phosphate group is related to the nucleoside cyclic phosphate that is the product of the first stage of RNA cleavage, and is itself then hydrolyzed by ribonuclease A. Thus the hydrolysis of **9** with catalysis by a cyclodextrin bis-imidazole is a model for the hydrolysis of cytidine-2',3'-cyclic phosphate by ribonuclease, using its own two imidazole groups.

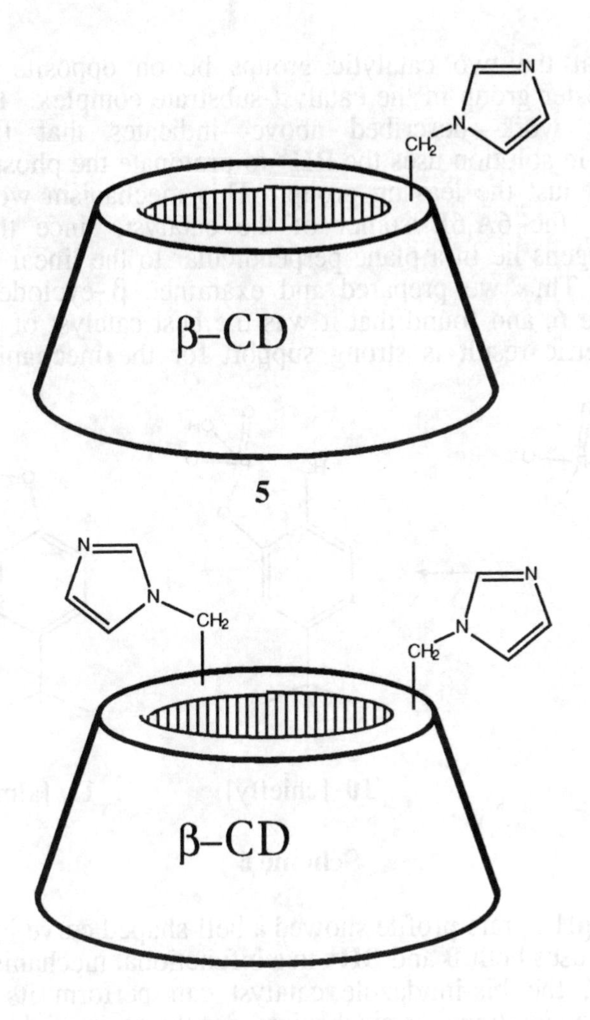

6: on neighboring glucose units
7: on units one apart
8: on units two apart

Scheme 5

In our first studies [9] we used the cyclodextrin 6A,6D-bisimidazole **8** and also the 6A,6C isomer **7**, since by the classic mechanism—written for the enzyme in most textbooks—the base **B** delivers a water to the phosphate group while the acid **BH+** protonates the leaving oxygen atom, in a linear displacement mechanism. This requires that the two catalytic groups be on opposite sides of the phosphate ester group in the catalyst-substrate complex. However, our mechanistic work—described above—indicates that the preferred mechanism in solution uses the **BH+** to protonate the phosphate anionic oxygen, not just the leaving group. This mechanism would be more suitable for the 6A,6B-isomer of the catalyst, since the phosphate anionic oxygens lie in a plane perpendicular to the linear displacement trajectory. Thus we prepared and examined β–cyclodextrin-6A,6B-bisimidazole **6**, and found that it was the best catalyst of the three [8]. This geometric result is strong support for the mechanism we have proposed.

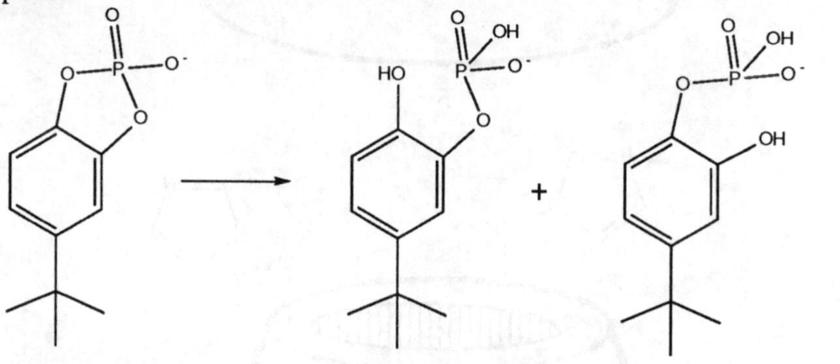

9 **10** [chiefly] **11** [almost none]

Scheme 6

The pH *vs* rate profile showed a bell-shaped curve indicating that this catalyst uses both **B** and **BH+** in a bifunctional mechanism. As with the enzyme, the bis-imidazole catalyst can perform its bifunctional catalysis by a *simultaneous* mechanism, not the sequential mechanism of simple buffer catalysis. We saw that this was indeed the case, as revealed by the tool called "proton inventory." In this technique the reaction is performed in D_2O, in H_2O, and in mixtures of the two. If only one proton that can exchange with D_2O is moving in the transition state, the points all lie on a straight line between the H_2O and slower D_2O points. If two (or more) protons are moving, the line is curved. It had been found for the enzyme ribonuclease A [10] that a curved line was seen corresponding to the movement of two protons, and we also saw a curved plot—with very similar data—for our cyclodextrin-6A,6B-bisimidazole catalyst **6** [11]. Controls established that indeed this was a reliable indication that our system is performing *simultaneous* bifunctional catalysis, just as the enzyme does. In particular, the monoimidazole catalyst **5** showed only *one* proton moving—for monofunctional catalysis—by this test.

The hydrolysis of **9** catalyzed by **6** shows interesting selectivity: it leads to product **10**, with only a few percent of **11** also formed. Thus, in common with the enzyme, this catalyst directs the hydrolysis of a cyclic phosphate essentially in one direction out of the two that are possible; however, the selectivity is not complete, as it is with the enzyme. The pH *vs* rate profile is also similar to that of the enzyme, but the rate is not yet as fast. We are now constructing new related catalysts for which we expect better rates.

5. Enolization and Aldol Condensation Catalyzed by a Bifunctional Enzyme Mimic

The availibility of the three isomers of β–cyclodextrin bisimidazole (**6**, **7**, and **8**) gives us the opportunity to examine other bifunctional catalyses. Geometric preferences among the three isomers should also give us detailed information about the mechanisms, just as it did for the phosphate ester hydrolysis discussed above. The first reactions we have examined (Scheme 7) concern enolization of a ketone and processes that ensue.

We examined exchange of deuterium into the methyl group of ketone **12** catalyzed by the three isomers of cyclodextrin bisimidazole and also by β–cyclodextrin monoimidazole (**5**) in D_2O solution with 14% CD_3OD, buffered at pH 6.2 with phosphate buffer, at 35 °C [12]. We saw essentially no deuterium incorporation into the methyl group over 10 hours with the buffer alone, or the buffer with added simple cyclodextrin. However, all four of the cyclodextrin derivatives (**5**, **6**, **7**, and **8**) catalyzed significant deuterium incorporation into the methyl group over the same period. The AB (**6**) and AC (**7**) isomers were only marginally more effective than was the monoimidazole catalyst, but the AD isomer **8** was clearly better. A pH *vs* rate plot for the reaction catalyzed by the AD isomer showed a bell-shaped curve with a rate maximum near pH 6.2, as expected for bifunctional catalysis, while the reaction catalyzed by cyclodextrin-6-imidazole **5** showed a simple increase to a plateau at pH above 7.5, as expected for monofunctional base catalysis.

Deuterium exchange into the methyl group of a ketone is not a simple process. If the enolization is catalyzed by the **B** and $\mathbf{BH^+}$ of the catalyst, reketonization must also show bifunctional catalysis by a mechanism the reverse of that for enolization. In principle the same proton that was removed by the base **B** should be returned in the ketonization step unless it has had a chance to exchange with the solvent deuterium. Such solvent exchange of a $\mathbf{BH^+}$ may not be instantaneous, but should be fastest under our pH conditions where **B** and $\mathbf{BH^+}$ are in close equilibrium. Thus our observed rate of deuteration is a lower limit of the rate of enolization. As we will discuss later, aldol condensation studies suggest that it is not far off the true rate.

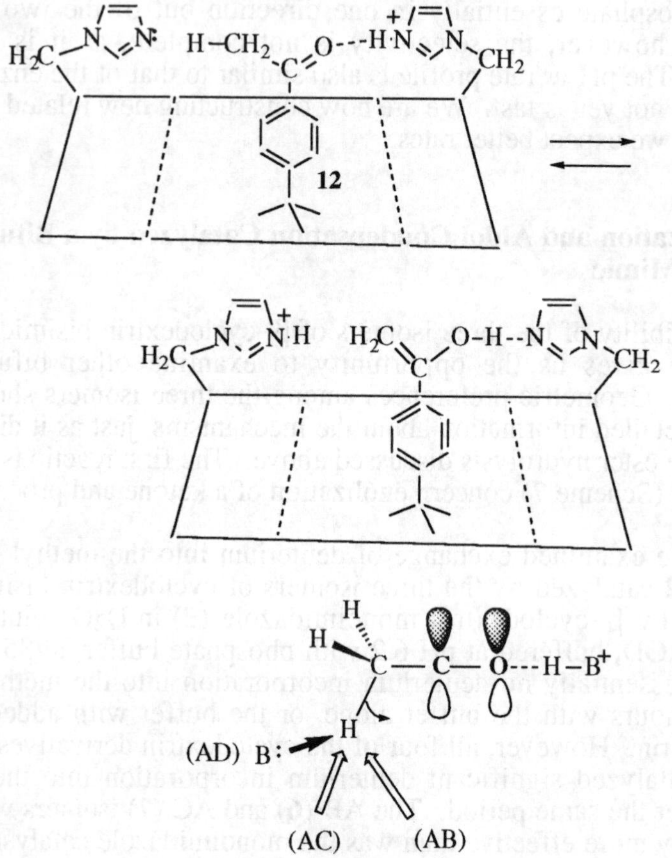

Scheme 7

Molecular models did not make it clear that the AD isomer **8** would be the best. We anticipated that the AC isomer **7** might be preferred, based on a simple idea about the mechanism. If the $\mathbf{BH^+}$ is hydrogen bonded to an unshared electron pair of the ketonic oxygen, the AC isomer apparently—from models—is set up to allow the base **B** to attack the C-H bond in a straight line. From our results and careful thought about what they mean, we now believe that such straight line attack is *not* the preferred direction for removal of the proton in enolization.

As Scheme 7 shows, our results seem to indicate that the preferred attack by base is *off line*, pushing the electrons of the C-H bond not simply towards the carbon but towards the carbon and in a direction required by the developing p orbital; that is, in a direction towards the carbonyl group. This can also be visualized by considering the reverse reaction, ketonization of the enol. By microscopic reversibility, ketonization must follow the enolization pathway but in the reverse direction. As Scheme 8 shows, we propose that the **BH⁺** puts the proton on the enol at the *side* of the carbon p orbital, so the proton will end up in its final position in the re-formed methyl group. If the p orbital electrons displace the **B** from **BH⁺** by an in-line displacement, as they should, then the geometry is as we have deduced. This is a reasonable geometry for proton removal in the course of enolization (or its reverse) but is the first evidence for it.

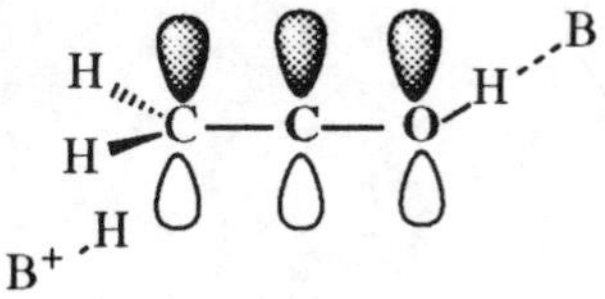

Scheme 8

Enolization is the first step in the aldol condensation, or in the dehydration of β–hydroxyketones. We have observed (Scheme 9) [13] that our cyclodextrin A,D-bisimidazole **8** can catalyze both of these processes. The substrate examined is **13**, a ketoaldehyde that can bind into the cyclodextrin cavity. With the AD bisimidazole catalyst we find that **13** is converted to **14**, in water at room temperature and pH 6.2. Again there is essentially no reaction in the buffer alone, and only at pH 10.0 is the rate of aldol condensation with buffer comparable to that with our catalyst at pH 6.2. Assuming that the aldol condensation without our catalyst is catalyzed by hydroxide ion, this means that our catalyzed rate exceeds 5000 fold over the rate of the spontaneous process at pH 6.2.

We find that the rate of our aldol condensation is comparable to the rate seen above for simple enolization (and deuterium exchange) of ketone **12**. Thus it seems likely that the deuterium exchange was a good measure of the rate of enolization, despite the caveats expressed above.

On further standing, the AD bisimidazole catalyst converts **14** into its dehydrated product **15** [13]. This process also involves enolization, of course, and we are still studying whether our two catalytic groups play other roles as well. For instance, they could promote the addition of the enol to the aldehyde group in the original condensation, or serve to promote the dehydration by using **BH⁺** to protonate the leaving hydroxyl group. The evidence does not yet allow clear choices among these possibilities.

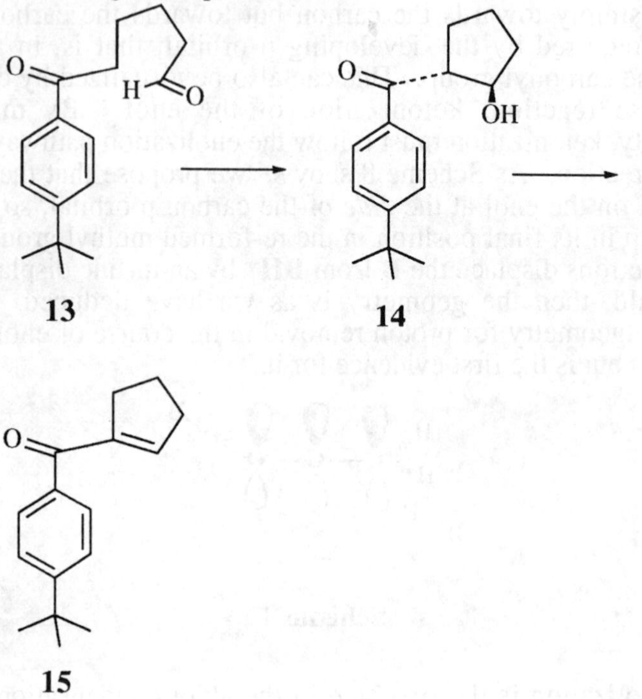

13

14

15

Scheme 9

6. Binding of Appropriate Substrates by Cyclodextrin Dimers

We have written a number of reviews of our early work in this field [14], so we will only summarize it here. In brief, a molecule consisting of two cyclodextrin rings with an intervening linker can bind appropriate substrates much more strongly than monomeric cyclodextrin can. The requirement is that the substrate have *two* hydrophobic regions situated so they can both enter the cyclodextrin cavities.

Our earliest example **16** [15] involved two α–cyclodextrins with a terephthalate ester linkage between the secondary faces (carbons 2 and 3) of the cyclodextrins. However, most of our work has involved β–cyclodextrins, linked on their primary faces at carbon 6. In this section we will discuss their binding ability, while in the next section we will describe catalysts based on such cyclodextrin dimers.

A good substrate for binding to a simple cyclodextrin ring in water, such as substrates **9** and **12** described earlier, has a binding constant of the order of 10^4 M^{-1}. By contrast, we have examples of binding to dimeric cyclodextrins with binding constants exceeding 10^{11} M^{-1}! Several factors come into this difference.

First of all, one might expect that binding two hydrophobic segments into two cyclodextrin cavities could simply double the free energy of binding; if so, a normal 10^4 M^{-1} for monomeric binding would become a 10^8 M^{-1} in the dimeric case. However, there are other considerations as well. For one, the substrate or the dimeric host may have degrees of free rotation around single bonds that must be frozen out for double binding, and these conformational changes can cost energy and lower the binding constant. In the other direction, binding a double ended substrate into a double ended host molecule can involve the entropy advantage often seen in chelate binding—translational entropy does not have to be paid for twice, in contrast to the situation for binding two unattached molecules, so the binding constant can reflect a free energy *greater* than twice that for monomeric binding. As another factor, there may be substrate interactions with the linker used to attach the cyclodextrins to each other, and these can add to or subtract from the binding energy.

Some cyclodextrin dimers that we have prepared are shown in Scheme 10, and some double-ended substrates in Scheme 11. In Table 1 we list a few binding constants for these cases. As the Table shows, in two examples of cyclodextrin dimers with a single flexible link we saw a binding constant of over 10^8 M^{-1} for the rigid cyclopropene substrate **25** . Strong binding was also seen for the ester **23**, but the geometry of **21** is poor and it shows only very weak binding.

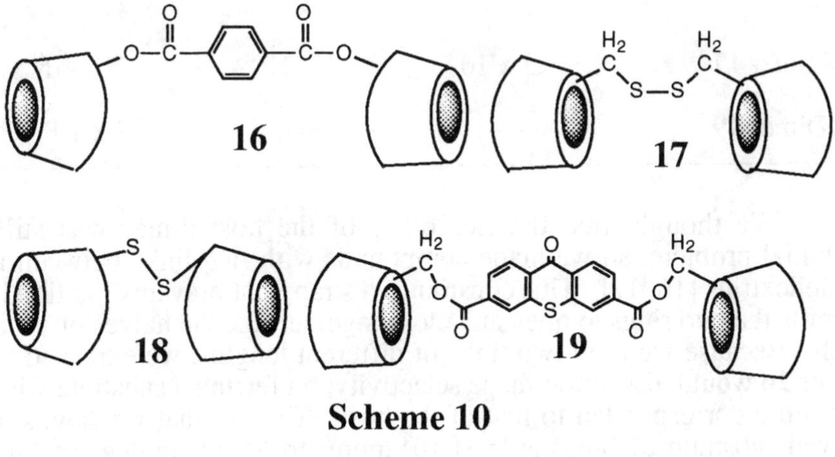

Scheme 10

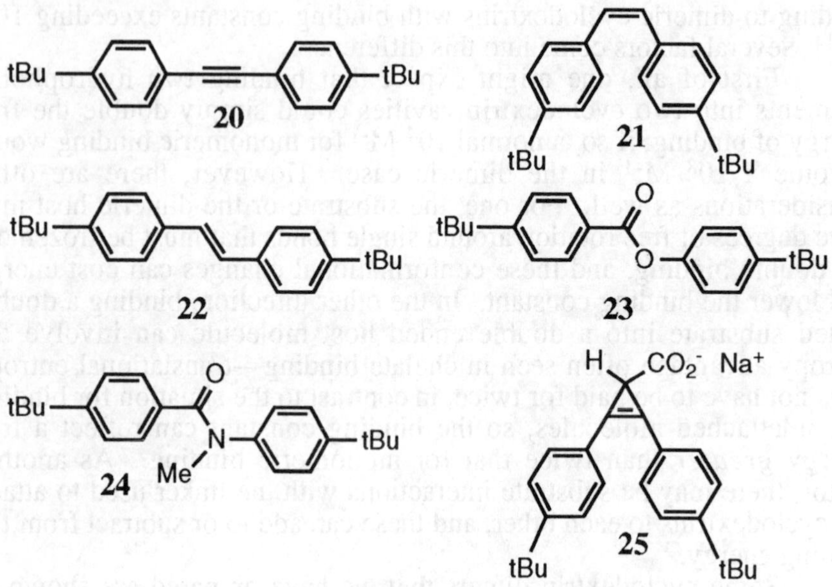

tBu **20** tBu

tBu **21** tBu

tBu **22** tBu

tBu O O **23** tBu

tBu O N Me **24** tBu

H CO_2^- Na^+ tBu **25** tBu

Scheme 11

Table 1. Binding Constants of Compounds **21**, **23**, and **25** to Cyclodextrin Dimers **17** and **19** in Water (M^{-1})

	21	**23**	**25**
To Dimer **17**	$<3 \times 10^3$	1×10^8	3.5×10^8
To Dimer **19**	----	---	7×10^8

We thought that the flexibility of the host dimer was still a potential problem, so we made compounds with *two* links between the cyclodextrins [16, 17]. This constitutes a strap, that prevents rotation but permits the two rings to open and close together like the halves of a clam shell. Because we used two links of different lengths, we expected that dimer **26** would also show shape selectivity, preferring a substrate whose curvature corresponded to that of the host. This is what we saw, since curved substrate **27** binds at least 10^3 more strongly than does its linear analog **28**. (Scheme 12). Such shape selectivity is of considerable potential interest in molecular recognition and catalysis.

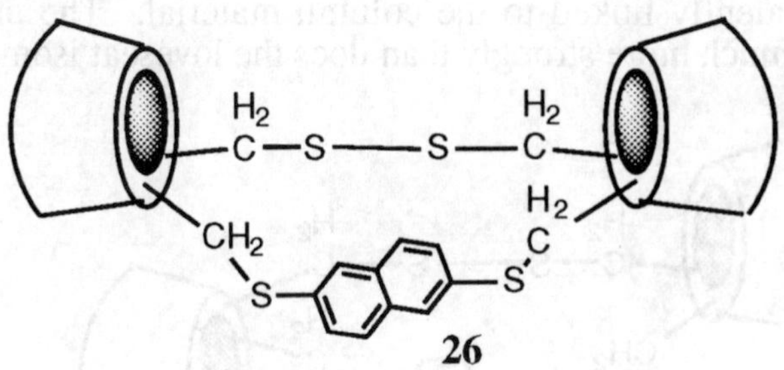

26

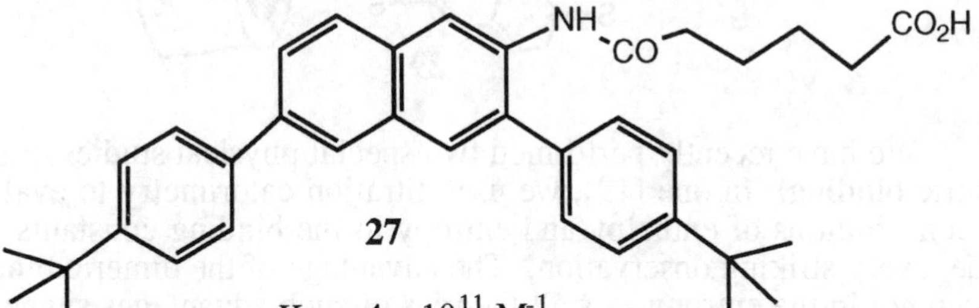

27

$K_a > 4 \times 10^{11}\ M^{-1}$

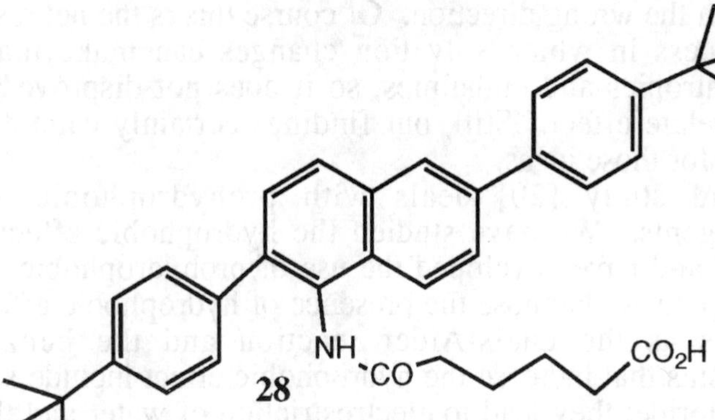

28

$K_a = (1.1 \pm 0.2) \times 10^{8}\ M^{-1}$

Scheme 12

Of course with two links between the cyclodextrins there is another possible isomer, **29**, in which the strap *prevents* the two rings from cooperating in binding a double-ended substrate. We call this the "love seat" isomer. We saw [16] that it has a binding constant only a little above that for simple cyclodextrin itself, the extra affinity perhaps involving some interaction with the linking group but no binding into the second cavity. The two isomers can be separated chromatographically, most strikingly with an affinity column [18] in which **25** is covalently linked to the column material. The clamshell isomer **26** binds much more strongly than does the loveseat isomer **29**.

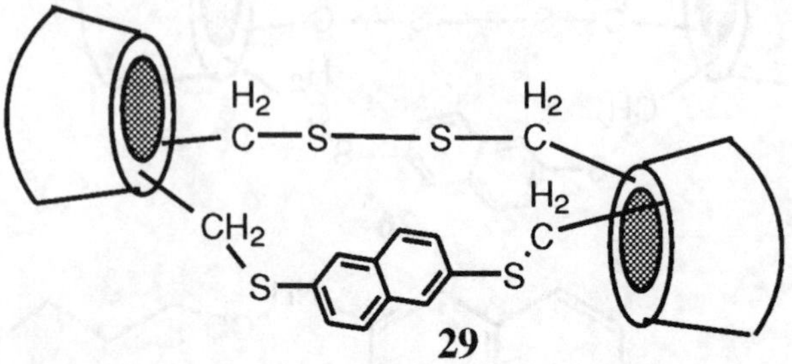

H2 C—S——S—C H2 CH2 S S C H2

29

We have recently performed two special physical studies of such dimeric binding. In one [19], we used titration calorimetry to evaluate the contributions of enthalpy and entropy to the binding constants, and made a very striking observation. The advantage of the dimeric binding is seen not in the entropy—as all theories of such advantages suggest—but in the enthalpy. In fact, in the cases we examined the entropy contribution went in the wrong direction. Of course this is the net result of a complex process in which solvation changes can make major contributions to entropies and enthalpies, so it does not disprove our ideas about the chelate effect. Still, our findings certainly cannot be invoked as support for those ideas.

The second study [20] deals with prohydrophobic and antihydrophobic agents. We have studied the hydrophobic effect in other projects [21] and have developed the use of prohydrophobic and antihydrophobic agents to diagnose the presence of hydrophobic effects in such processes as the Diels-Alder reaction and the benzoin condensation. Agents that increase the hydrophobic effect include salts such as sodium chloride; they lead to electrostriction of water and thus make it more difficult for a non-polar solute to create a cavity in the solvent into which it can fit. Antihydrophobic agents include denaturants such as urea and guanidinium chloride. There had been some ambiguity about how they functioned, but we were able to show that they function to help solvate non-polar solutes in water by bridging between the water and the hydrocarbon solute [21, 22].

In our study with cyclodextrin dimers [20], we examined the magnitudes of the effects of antihydrophobic agents on the binding constants of substrates into cyclodextrin compounds. The major driving force for such binding is hydrophobic, so these agents all led to a decrease in binding in all cases. However, the decrease with dimeric binding was twice or more that for monomeric binding. Thus the magnitudes of these effects can be used as an indicator of the amount of hydrophobic surface that becomes buried in a binding process. Such a tool can be very useful in judging the geometries of complexes or transition states in which hydrophobic effects lead to binding between two components. We are developing this technique further.

7. Catalysis by Cyclodextrin Dimers

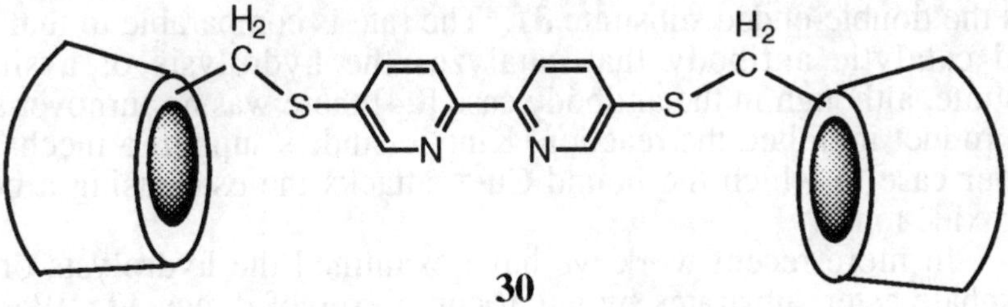

30

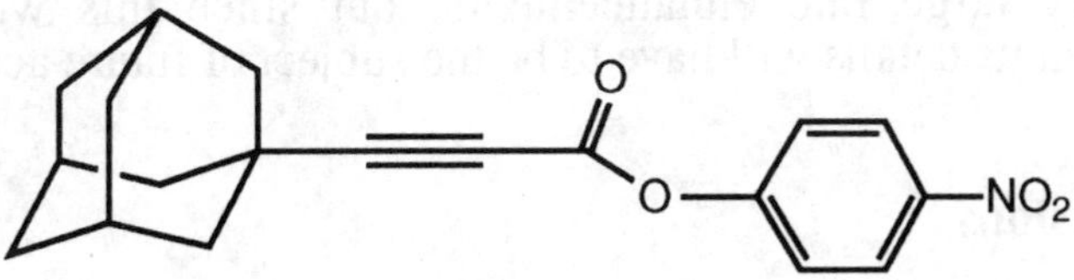

31

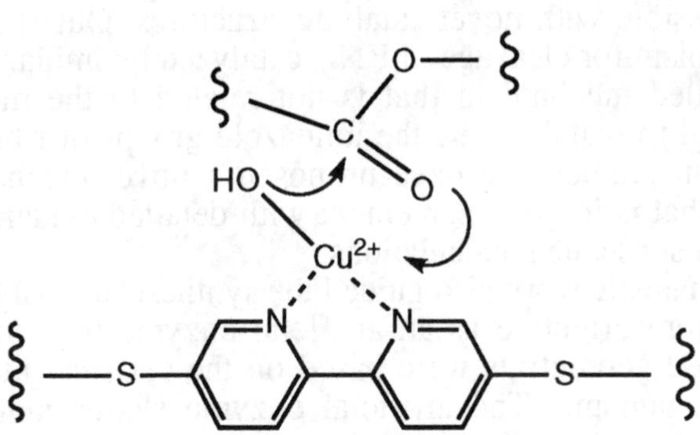

Scheme 13

An interesting feature of the binding of a double-ended substrate into a cyclodextrin dimer is that the substrate is held right on top of the linker group. If that linker carries a catalytic function, one can imagine very effective catalysis of reactions at the middle of the substrate. We have examined a number of such processes, and have achieved some very large catalyses in appropriate cases. Here we will describe catalysis by a cyclodextrin dimer **30** carrying a bipyridyl linker. This linker can bind metal ions, which can then function as catalysts for hydrolysis or oxidation processes.

In our first study [23], we used the Cu^{2+} complex of **30** to catalyze the hydrolysis of ester **31** (Scheme 13). At pH 7.0 and 37 °C we saw a 220,000 fold acceleration of the reaction rate over that for uncatalyzed reaction under the same conditions. There were at least 50 turnovers , since the hydrolysis products bind much more weakly than does the double-ended substrate **31**. The rate is comparable to that for a good catalytic antibody that catalyzes the hydrolysis of a similar substrate, although in the antibody case [24] there was no turnover since the product inhibited the reaction. Kinetic studies support a mechanism for our case in which the bound Cu^{2+} attacks the ester using a bound hydroxide ion.

In more recent work we have examined the hydrolysis of two phosphate ester substrates by metal complexes of dimer **30**. We have found very large rate enhancements, but since this work is still unpublished its details will have to be the subject of future accounts.

8. Conclusions

Mechanistic studies are extremely valuable in guiding and understanding the catalyses possible with novel catalytic structures. Our study of the preferred mechanism for cleavage of RNA catalyzed by imidazole buffer revealed a detailed mechanism that is not related to the mechanism normally invoked for catalysis by the imidazole groups of ribonuclease A. Based on our studies, we have proposed a novel mechanism for ribonuclease A that is in good agreement with detailed evidence on the enzyme and with subsequent calculations.

The mechanistic work also guided the synthesis of a ribonuclease mimic that is more effective as an artificial enzyme than are related compounds whose geometries were based on the classical ideas about the enzyme mechanism. The artificial enzyme shares many of the properties of the natural enzyme.

The availibility of three geometric isomers of our artificial enzyme lets us examine other reactions that can show bifunctional catalysis. Enolization of a ketone—and its addition to an aldehyde group in an aldol condensation—are two cases examined so far in which an isomer of our catalyst is preferred that is not the one that was best in the ribonuclease mimic. The geometric preference indicates something novel about the geometry of enolization reactions.

The mechanistic studies on RNA hydrolysis involved a side reaction in which the 3',5' link of RNA was converted into a 2',5' link. This conversion was critical in proving the mechanism, but it also raises questions about the properties of a 2',5' link in polynucleotides. We have prepared DNA with such linkages—using an unnatural isomer of deoxyadenosine and of thymidine—and explored the properties of this DNA isomer. In contrast to normal DNA, the isomeric DNA does not form a double helix under any available conditions, but it does form a triple helix. The reasons for this unusual behavior are apparent from models.

Binding into cyclodextrin dimers can be very strong with appropriate substrates. Thermochemical studies show that the surprising advantage of such dimers is not entropic, at least in the cases examined. Studies on such dimeric binding with antihydrophobic agents give a quantitative correlation with the amount of hydrophobic surface that is buried in the binding process.

Cyclodextrin dimers have been prepared with a linker that can carry a metal ion next to an ester group of a bound substrate. Very large rates of hydrolysis—and good turnover catalysis—are seen in some examples.

The future of this field is extremely promising [25]. There is every reason to expect that catalysts will be developed that rival enzymes in their effectiveness and selectivities, and that have properties—including stability better than that of proteins—that will make them attractive reagents for chemical transformations.

9. References

1. Breslow, R. (1972) Biomimetic chemistry, Centenary Lecture, *Chem. Soc. Rev.*, **1**, 553-580.
2. (a) Breslow, R. (1991) How do imidazole groups catalyze the cleavage of RNA in enzyme models and in enzymes? Evidence from "negative catalysis", *Accts. Chem. Res.* **24**, 317-324.
   (b) Breslow, R. and Xu, R. (1993) Recognition and catalysis in nucleic acid chemistry, *Proc. Natl. Acad. Sci. USA* **90**, 1201-1207.
3. Breslow, R. and Xu, R. (1993) Quantitative evidence for the mechanism of RNA cleavage by enzyme mimics. Cleavage and isomerization of UpU by morpholine buffers, *J. Am. Chem. Soc.* **115**, 10705-10713.
4. Hammett, L. P. (1940) *Physical Organic Chemistry*; McGraw-Hill, New York; p. 398.
5. Dougherty, J. P., Rizzo, C. J., and Breslow, R. (1992) Oligonucleotides that contain 2',5' linkages: synthesis and hybridization properties, *J. Am. Chem. Soc.* **114**, 6254-6255.
6 Jin, R., Chapman, Jr., W. H., Srinivasan, A. R., Olson, W. K., Breslow, R., and Breslauer, K. J. (1993) Comparative spectroscopic, calorimetric, and computational studies of nucleic

acid complexes with 2',5'- versus 3',5'-phosphodiester linkages, *Proc. Natl. Acad. Sci. USA* **90**, 10568-10572.
7. Eschenmoser, A. & Dobler, M. (1992) Why pentose- and not hexose- nucleic acids? *Helv. Chim. Acta* **75**, 218-259.
8. Anslyn, E. and Breslow, R. (1989) Geometric evidence on the ribonuclease model mechanism, *J. Am. Chem. Soc.* **111**, 5972-5973.
9. Breslow, R., Doherty, J., Guillot, G., and Lipsey, C. (1978) Beta-cyclodextrinyl-bisimidazole, a model for ribonuclease, *J. Am. Chem. Soc.* **100**, 3227-3229.
10. Matta, M. S.; Vo, D. T. (1986) Proton inventory of the second step of ribonuclease catalysis, *J. Am. Chem. Soc.* **108**, 5316-5319.
11. Anslyn, E. and Breslow, R. (1989) Proton inventory of a bifunctional ribonuclease model, *J. Am. Chem. Soc.* **111**, 8931-8932.
12. Breslow, R. and Graff, A. (1993) Geometry of enolization using a bifunctional cyclodextrin-based catalyst, *J. Am. Chem. Soc.* **115**, 10988-10989.
13. Desper, P. and Breslow, R. (1994) Catalysis of an intramolecular aldol condensation by imidazole-bearing cyclodextrins, *J. Am. Chem. Soc.* **116**, 12081-12082.
14. (a) Breslow, R. (1992) Bifunctional binding and catalysis in host-guest chemistry, *Israel Jour. Chem.* **32**, 23-30.
(b) Breslow, R. (1993) Bifunctional binding and catalysis, *Supramolec. Chem..* **1**, 111-118.
15. Chao, Y. (1972) Enzyme models with hydrophobic binding sites, Ph.D. thesis, Columbia University.
16. Breslow, R. and Chung, S. (1990) Strong binding of ditopic substrates by a doubly linked occlusive C_1 "clamshell" as distinguished from an aversive C_2 "loveseat" cyclodextrin dimer", *J. Am. Chem. Soc.* **112**, 9659-9660.
17. Chung, S. (1991) Transformation of binding energy into catalytic energy—implications for amide cleavage, Ph.D. thesis, Columbia University.
18. Halfon, S. (1993) I. Binding and catalysis by beta-cyclodextrin dimers. II. Effects of antihydrophobic agents on binding of beta-cyclodextrin dimers, Ph.D. thesis, Columbia University.
19. Zhang, B. and Breslow, R. (1993) Enthalpic domination of the chelate effect in cyclodextrin dimers, *J. Am. Chem. Soc.* **115**, 9353-9354.
20. Breslow, R. and Halfon, S. (1992) Quantitative effects of antihydrophobic agents on binding constants and solubilities in water, *Proc. Natl. Acad. Sci. USA* **89**, 6916-6918.

21. Breslow, R. (1991) Hydrophobic effects on simple organic reactions in water, *Accts. Chem. Res.* **24**, 159-164.
22. Breslow, R. and Guo, T. (1990) Surface tension measurements show that chaotropic salting-in denaturants are not just water-structure breakers, *Proc. Natl. Acad. Sci. USA* **87**, 167-169.
23. Breslow, R. and Zhang, B. (1992) Very fast ester hydrolysis by a cyclodextrin dimer with a catalytic linking group, *J. Am. Chem. Soc.* **114**, 5882-5883.
24. Wirsching, P., Ashley, J. A., Benkovic, S. J., Janda, K. D., and Lerner, R. A. (1991) An unexpectedly efficient catalytic antibody operating by ping-pong and induced fit mechanisms, *Science* **252**, 680-685.
25. Breslow, R. (1995) Biomimetic chemistry and artificial enzymes: catalysis by design, *Accts. Chem. Res.* **28**, 159-164.

Design of Enzyme Inhibitors

Answering Biological Questions Through Organic Synthesis

Paul A. Bartlett
Department of Chemistry, University of California
Berkeley, California 94720

Abstract. Mechanistic insight and structural information provide the starting points for two contrasting approaches to the design of enzyme inhibitors. Phosphorus-containing peptides are inhibitors of the zinc and aspartic peptidases that mimic key geometric and electronic characteristics of the transition states of these enzymes. The design of compounds to mimic high-energy structures along the reaction path is also an effective strategy for inhibition of two enzymes of the shikimic acid pathway, EPSP synthase and chorismate mutase. For all of these inhibitors, the structural information available from crystallography has been used to interpret their binding behavior, if not for their initial design. Situations in which structural information does play a role in inhibitor design can be divided into three categories, depending on the kind of information available; examples of two of these are presented, along with some computational tools to facilitate the design process. Macrocyclic, constrained inhibitors of thermolysin were designed from the structures of the enzyme complexes of the acyclic transition state analogs, and cyclic hexapeptide mimics of the α-amylase inhibitor tendamistat were designed from the structure of this 74-residue protein. In this connection, the utility of 3-D structural databases and search tools like CAVEAT has stimulated the development of TRIAD and ILIAD, two large databases of computed, minimized structures.

1. The Roles of Mechanistic Insight and Intuition

1.1 INTRODUCTION TO TRANSITION STATE ANALOGS

The idea that you can design a potent enzyme inhibitor by imitating the structure of the transition state was an insightful proposal by Pauling nearly 50 years ago.[1] While this strategy may seem obvious now, in light of our current understanding of enzyme kinetics and mechanism, several decades passed before the idea was implemented.[2-4] At first, a known or naturally occurring inhibitor was designated as a transition state analog because it contained some structural characteristic, often an altered geometry at carbon, associated with the presumed intermediate along the reaction pathway.

During this period, while various strategies for inhibitor design were incubating, we as chemists were content to let the "transition state analog" designation reflect our thought process - how we dreamed up the inhibitor, or what our rationale was for its high

C. Chatgilialoglu and V. Snieckus (eds.), Chemical Synthesis, 137–173.

affinity. With time, designs became more sophisticated, and we developed inhibitors for more complex transformations; however, our basic approach hasn't changed: we look at an enzymatic reaction, we recall what we know about chemical mechanisms – and the structures of the intermediates or transition states involved – and we invent a compound that mimics some aspect of this structure.

In addition to these *qualitative* applications of the "transition state analog" concept, a number of attempts have been made to use it *quantitatively*, either to judge how "good" a mimic of the transition state an inhibitor actually is, or more usefully, to use inhibitor binding affinity as a probe of enzyme mechanism. In the first part of this overview, I describe some of the recent work from my group that illustrates both aspects of the "transition state analog" issue: how we have used this conceptual approach to design inhibitors, and how we have used the binding affinities of such compounds to say something about the mechanisms of the enzymes they inhibit.

1.2 EXAMPLES OF TRANSITION STATE ANALOG INHIBITORS OF THE ZINC AND ASPARTIC PROTEASES

1.2.1 *Potency of Phosphorus-Containing Inhibitors of Carboxypeptidase A*

To put into perspective our ability to "design" potent inhibitors, whether from mechanistic insight or molecular modeling or whatever strategy, I remind myself of the list of inhibition constants in Table 1, for the zinc protease carboxypeptidase A. These phosphorus-containing derivatives of phenylalanine show a progression in inhibitor affinity that is greater than any obtained through structure-based design. By simple extension of the peptide chain, and incorporation of a few side chains that increase hydrophobicity and reduce conformational freedom, a micromolar inhibitor is transformed into a compound whose inhibition constant is 11 **femto**molar, and whose half-life for

TABLE 1. Optimization of Carboxypeptidase A Inhibitors

Compound	K_i (nM)
$CH_3\{PO_2^-$-NH$\}$Phe	1,100
Cbz-Gly$\{PO_2^-$-O$\}$Phe	60
Cbz-Ala-Gly$\{PO_2^-$-O$\}$Phe	0.71
Cbz-Ala-Ala$\{PO_2^-$-O$\}$Phe	0.003
Cbz-Phe-Ala$\{PO_2^-$-O$\}$Phe	0.001
Cbz-Phe-Val$\{PO_2^-$-O$\}$Phe	0.000011

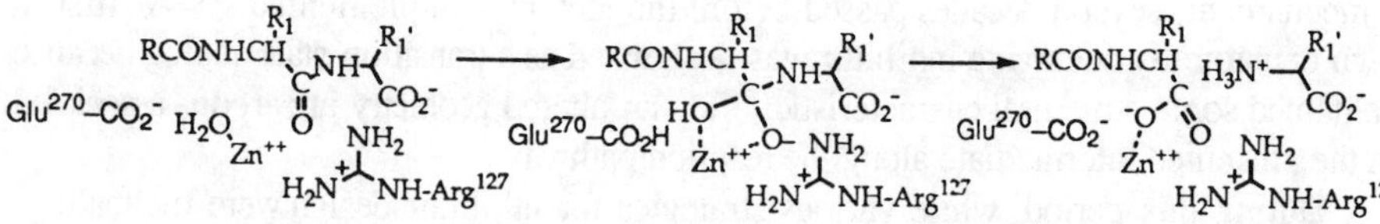

Figure 1. Carboxypeptidase A mechanism

dissociation from the non-covalently bound complex is around 10 years![5] Although we synthesized these compounds with the hope that they would be potent inhibitors, as mimics of the tetrahedral intermediate in the enzyme-catalyzed reaction (Figure 1), we certainly didn't anticipate the vast increases in affinity from simple elaboration of the basic structure.

1.2.2 *Transition State Analogy of Phosphorus-Containing Inhibitors of Thermolysin*

It is important to realize that you can't tell if an inhibitor is a transition state analog simply on the basis of its binding affinity. Many factors that can lead to tight binding have nothing to do with transition state mimicry, and a structural element that reduces binding affinity in one part of the molecule may mask a favorable contribution from this mimicry in another part. However, if the inhibition constants K_i of a series of inhibitors can be compared with the K_m/k_{cat} values of the corresponding substrates, many of these extraneous effects can be removed and a truly rigorous criterion established. As pointed out elsewhere, $\log(K_i)$ is directly proportional to $\log(K_m/k_{cat})$ if the chemical step imitated by the transition state analog is rate limiting for all of the substrates in the series.[4, 6]

We first applied this approach to inhibitors of the zinc endopeptidase thermolysin, eventually studying three series of tripeptide analogs: the phosphonamidates **1-N**, the phosphonates **1-O**, and the phosphinates **1-C**.[6,7] For each series, when we varied the sidechains at the P_2' position, we found exactly this correlation between inhibitor K_i and substrate K_m/k_{cat} values (Figure 2). Note that these correlations hold for the weakly binding phosphonates **1-O** as well as for the tight-binding phosphonamidates **1-N** and phosphinates **1-C**. The three sets of inhibitors have different absolute affinities, as a result of the differing contributions from the Y group, but they show the same dependence on side chain structure at the P_2' position.

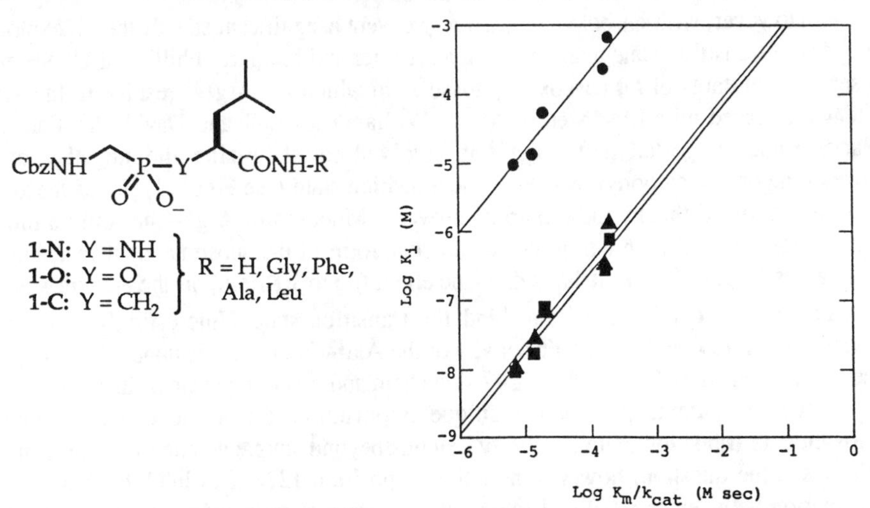

Figure 2. Phosphonamidate (■), phosphinate (▲), and phosphonate (●) inhibitors of thermolysin as transition state analogs

There are a number of insights that we can gain from these correlations. First, the three sets of inhibitors must be binding the same way in the enzyme active site if side chain variation has the same effect in each series. This conclusion was confirmed by structural studies by Brian Matthews and his coworkers at the University of Oregon, who found that the three analogs of Cbz-Gly{PO_2^-Y}Leu-Leu, where Y = NH, O, and CH_2, are bound identically in the thermolysin active site.[8,9] Second, the inhibitors must also be binding like the transition state, at least at the P_2' residue, since variation at this position affects inhibitor and transition state equally, and in a very different manner than they affect ground state binding, as reflected in the substrate K_m values.

It is important to realize that we can't say anything about the "transition state analogy" of parts of the inhibitor structure that are not varied in the correlation. Specifically, we *cannot* conclude from this data that the tetrahedral phosphorus is imitating the tetrahedral center in the transition state structure, although it was with this goal in mind that we made these compounds in the first place. We presume it is the tetrahedral geometry of the phosphorus atom that allows the P_2' residue to fit into the active site the way the transition state does, but there is no way to tell from these experiments whether there is any electronic similarity.

1.2.3 *Varying the Other Component: Binding of Transition State Analogs to Carboxypeptidase A Mutants*

We have also looked at a series of tri- and tetrapeptide phosphonate inhibitors of carboxypeptidase A, with structural variation in the P_2 and P_3 residues.[10] Again, we see a good correlation between the K_i and K_m/k_{cat} values, supporting the view that the mechanisms of this enzyme and thermolysin are similar. This result is also consistent with the notion that a transition state analog motif that is good for one of these zinc peptidases should work the same way for the other.

However, with carboxypeptidase A, we went a significant step further in exploring the idea of transition state analogy. William Rutter and Margaret Phillips at UCSF made a series of mutants of rat carboxypeptidase A in which the Arg^{127} residue in the active site was replaced with Lys, Met, or Ala.[11] William Lipscomb and David Christianson at Harvard had suggested that Arg^{127} is involved in stabilizing the negative charge developing on the carbonyl oxygen in the transition state (see Figure 1),[12] and the altered enzymes allowed this hypothesis to be proved. Mutation of Arg^{127} has only a modest (ca. 10-fold) impact on binding the ground state form of the substrate, as revealed by the K_m values (Figure 3), but it has a dramatic effect (up to 10^5-fold) on the enzyme's ability to catalyze hydrolysis, that is, to bind the transition state. Interestingly, exogenous guanidine can rescue the catalytic activity of the Ala^{127} mutant, presumably by occupying the cavity created by loss of the Arg^{127} side chain and assuming some of its duties.

These mutants provided a unique opportunity for us to examine whether phosphonate transition state similarity extends beyond simple geometric considerations. We posed the question: how will mutation at position 127 affect inhibitor binding? A correlation between $\log(K_i)$ and $\log(K_m/k_{cat})$ across a series of *proteins* would show whether the phosphorus oxyanion, the part of the inhibitor that interacts with the 127-side chain, and the carbonyl oxygen in the transition state respond similarly to changes in the enzyme active site. These mutants were thus the perfect probe for *electronic* similarity between the inhibitors and the transition states.

We could not have asked for more clear-cut results: over 5-orders of magnitude in absolute affinity, mutations that reduce catalytic activity (in terms of K_m/k_{cat}) have the *identical* effect on inhibitor binding (Figure 4).[13] Guanidine even enhances inhibitor binding to the same extent it rescues catalytic activity! The slope of the correlation line is essentially one, meaning that the interactions of the inhibitor and the transition state with

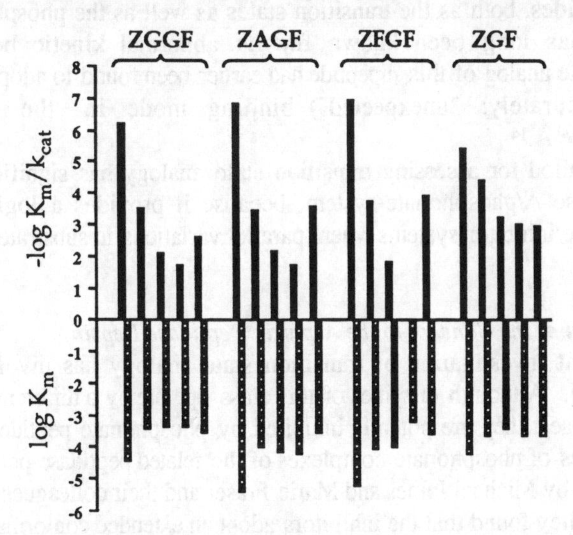

Figure 3. Influence of mutation at Arg[127] on substrate and transition state binding to carboxypeptidase A; each group of bars represents, L to R, the wild type, R127K, R127M, and R127A mutants, and R127A mutant + 0.5 M guanidine•HCl respectively.

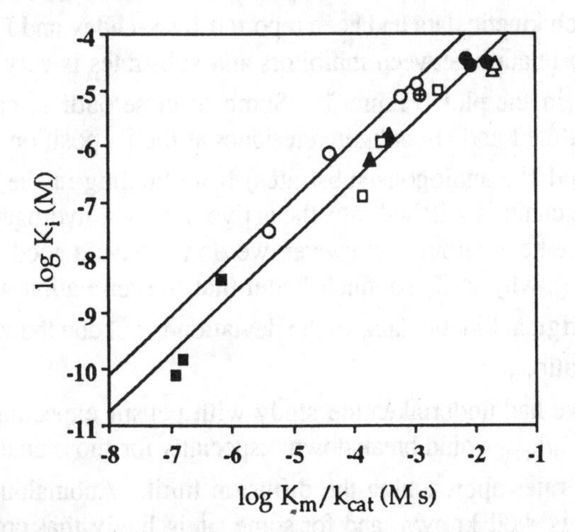

Figure 4. Correlation of inhibitor K_i and substrate K_m/k_{cat} values for Arg[127] mutants of carboxypeptidase A. Tripeptide analogs: ■ = wild-type enzyme, ▲ = R127K, ● = R127M, △ = R127A, □ = R127A in presence of 0.5 M guanidine•HCl. Cbz-Gly-Phe analogs: ○ = with all enzymes, ⊕ = with R127A mutant in presence of 0.5 M guanidine•HCl

the residue at position 127 aren't just similar, they're identical. If either the charge or the position of the phosphorus oxyanion in the active site were different than the carbonyl oxygen in the transition state, we wouldn't have obtained this result.

There was an unanticipated bonus to this study. Notice that the tripeptide analogs generate a different line than the dipeptide Cbz-Gly-Phe. In addition, the effect of Arg^{127} mutation is less for the dipeptide than it is for the tripeptides (275- versus 100,000-fold for Arg^{127} compared to Ala^{127}). These results suggest that Cbz-Gly-Phe binds differently than the tripeptides, both as the transition states as well as the phosphonate inhibitors. Cbz-Gly-Phe has long been known for its abnormal kinetic behavior, and the phosphonamidate analog of this dipeptide had earlier been found to adopt an "anomalous" (or more accurately: "unexpected") binding mode in the active site of carboxypeptidase A.[14]

This method for assessing transition state analogy has significance beyond the carboxypeptidase A/phosphonate system, because it provides a logical approach for studying enzyme/inhibitor systems where parallel variations in substrate and inhibitor are impossible.

1.2.4 *Extension of the Concept to the Aspartic Peptidase Pepsin*

Our most recent investigation of transition state analogy has involved the aspartic peptidase pepsin. Although enzymes of this class operate by a different mechanism than the zinc peptidases, they are potently inhibited by phosphonate peptide analogs as well. Structural studies of phosphonate complexes of the related peptidase penicillopepsin have been carried out by Michael James and Marie Fraser and their colleagues at the University of Alberta.[15] They found that the inhibitors adopt an extended conformation in the active site, with the phosphonate positioned between the two catalytic aspartic acid side chains, as envisioned for the tetrahedral intermediate itself.

We synthesized and evaluated the phosphonate analogs, **2**, of a series of pepsin substrates for which kinetic data had been reported by Sachdev and Fruton in 1970.[16] The K_i vs. K_m/k_{cat} correlation between inhibitors and substrates is very good, although there are a few outliers in the plot (Figure 5). Some of these outliers can be rationalized; for example, N-substituted and D-configured residues at the P_2 position may prevent some of the inhibitors (and the analogous substrates) from binding in the "normal" fashion, in which the peptide chain is stitched into the active site by a hydrogen bonding network to all of the backbone heteroatoms. However, we do not have a good rationalization for the Cbz-Gly-Ile analog: why is K_i so much better that the correlation with K_m/k_{cat} predicts? We verified the original kinetic data, so the deviation is not due the values we used for K_m and k_{cat} or their ratio.

Frankly, we had undertaken the study with pepsin expecting that the correlation between K_i and K_m/k_{cat} would break down, especially for those analogs whose substrates are turned over at rates approaching the diffusion limit. Anomalous kinetic behavior of pepsin substrates is well known, and for some, it is likely that product dissociation, as opposed to bond cleavage, is rate determining.[17] In such circumstances, chemistry is faster, and the chemical transition state bound tighter, than K_m/k_{cat} suggests. This phenomenon may explain the anomalously high affinity for the Cbz-Gly-Ile phosphonate.

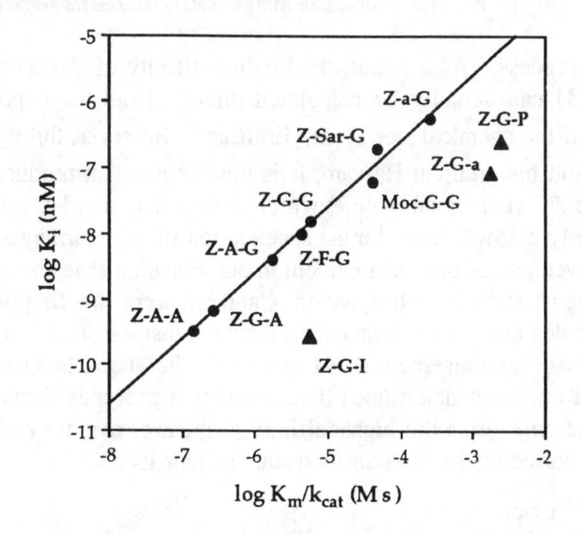

Figure 5. Inhibition of pepsin by phosphonate peptide analogs; the identity of the varied residues in structure **2** is indicated by each point.

1.3 EXAMPLES FROM THE SHIKIMATE-CHORISMATE PATHWAY: MORE INSIGHT, LESS PROOF

The enzymes of the shikimic acid pathway have intrigued bioorganic chemists for some time because of the unusual, in many cases unique, reactions they catalyze.[18] We ourselves have pursued no fewer than nine of these enzymes, designing and synthesizing substrate analogs or inhibitors and hoping to learn something about their mechanisms.[19] These projects represent the subjective aspects of transition state analog design, simply because these enzymes operate on unique substrates and do not lend themselves to the K_i versus K_m/k_{cat} approach.

1.3.1 *Chorismate Mutase*

Our interest in this area began with chorismate mutase, which is responsible for the only sigmatropic rearrangement known in a primary metabolic pathway (Figure 6). It is unique also for the fact that the reaction it catalyzes is unimolecular, and occurs at a measurable rate in aqueous solution by a mechanism that is substantially the same as that

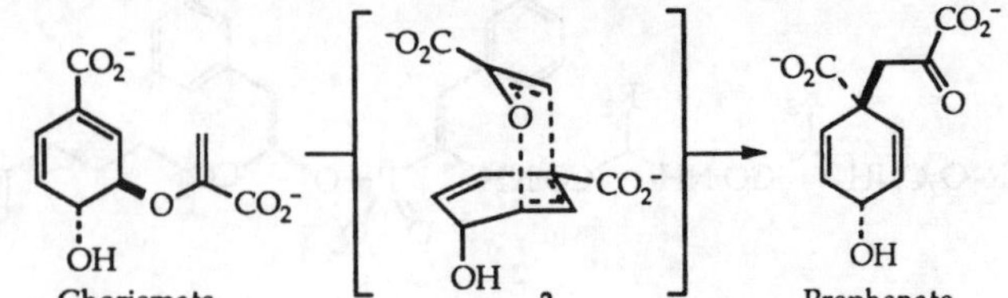

Chorismate 3 Prephenate

Figure 6. The chorismate mutase-catalyzed rearrangement.

of the enzymatic process. As a result, the binding affinity of the chorismate mutase transition state (**3**) can actually be calculated directly from the expression: $K_{TS} = K_S{\cdot}(k_{noncat}/k_{cat})$, if the chemical step is rate limiting.[4] However, through the work of Jeremy Knowles and his group at Harvard, it is now known that product release is the slow step for the *B. subtilis* enzyme,[20] which means that the 3-6 million fold rate enhancement is only a lower limit for the acceleration of the rearrangement step. We hoped to capture even part of this enhancement in our transition state analog designs.

Peter Andrews and his colleagues in Canberra were the first to describe the synthesis of molecules designed to mimic the bicyclic transition state **3** envisaged for the chorismate-prephenate rearrangement.[21] At that time, the stereochemical course of the rearrangement had not been determined (i.e., whether it proceeds through a chair- or boat-like conformation). From the higher affinity of the *exo*- over the *endo*-isomer of **4**, Andrews inferred, correctly, that the transition state is chair-like.

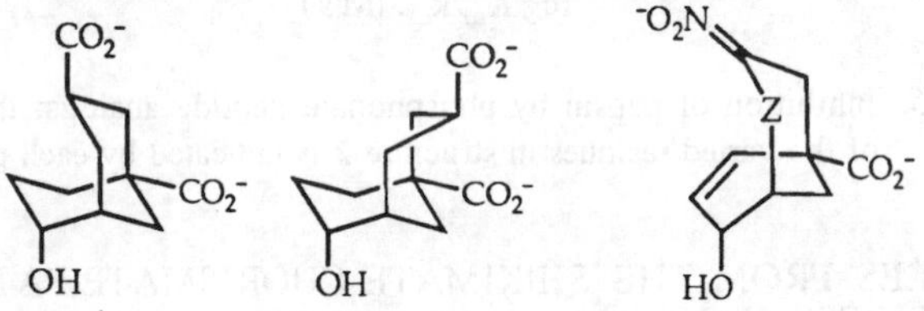

exo-**4** endo-**4** **5-O**: Z = O; **5-C**: Z = CH_2

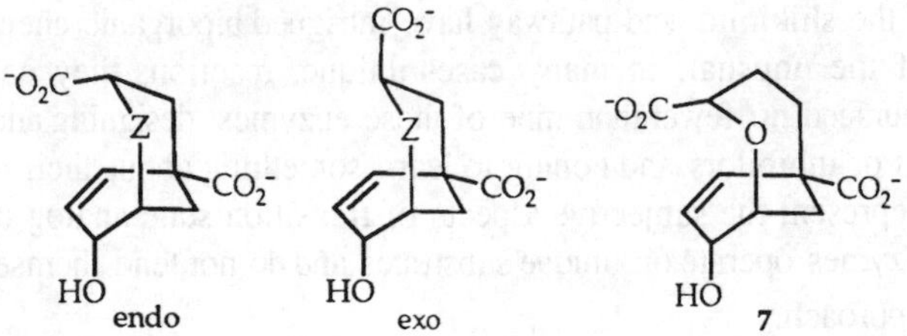

endo exo **7**

6-O: Z = O; **6-C**: Z = CH_2; **6-N**: Z = NH

Our initial designs incorporated more of the structural features of the substrate, including the allylic double bond and the ether linkage, and, in nitronate **5-O**, an sp^2-hybridized carbon in the bridge. The nitronate was supposed to mimic the enolpyruvyl carboxylate in the orientation we imagined it would adopt in the transition state. We never were able to synthesize this nitronate ether, but we did succeed in making the carbabicyclic derivative, **5-C**.[22] In addition, we made various dicarboxylate analogs, both the *endo* and *exo* isomers of the oxa-, aza-, and carbabicyclic frameworks, **6**. We made the carbabicyclic diacids **6-C** to compare with the nitronate we could obtain, and we made the azabicyclic analogs **6-N** on a hunch. Studies of model systems by Gajewski et al.[23] and of the enzymatic process by Knowles and Guilford[24] indicated that the rearrangement involves considerable polarization, with bond cleavage preceding bond formation and an almost ionic transition state. It seemed to us that a good way for the enzyme to assist bond cleavage would be to hydrogen-bond to (or even protonate) the ether oxygen; the transition state would then be more like an enol than an enolate. So, our reasoning went, an amine at this position in the inhibitor would pick up the proton from the enzyme and lead to a favorable ionic or H-bonding interaction.

As the inhibition constants in Table 4 reveal, only the simplest mimic *endo*-**6-O** really worked. The *exo*-isomer presumably puts the carboxylate in the wrong place, and the CH_2 and NH analogs must either lack or have the wrong polarity at the "ether" position. We were particularly disappointed by the weak inhibition of the amine *endo*-**6-N** and the nitronate **5-C**, since we spent a lot of effort in trying to make these compounds! However, we can be pleased that the bicyclic ether *endo*-**6-O** is a good inhibitor of chorismate mutase, binding at least 100-fold more tightly than the substrate (as estimated by the K_m value).[22] Even if it doesn't approach the inhibition constant theoretically attainable for a perfect transition state analog of this enzyme, which would be ca. 3 picomolar, *endo*-**6-O** at least looks like what we imagine the transition state to be.

TABLE 2. Inhibition of chorismate mutase[a] by bicyclic transition state analogs.

Compound	I_{50}/K_m
5-C	120
exo-**6-O**	5.5
endo-**6-C**	2
endo-**6-N**	1.2 (0.2 @ pH 9)
endo-**6-O**	0.008 (K_i = 0.12 μM)

[a] Bifunctional chorismate mutase-prephenate dehydrogenase from *E. coli*; pH 7.5, $K_m \approx 20$ μM.

The inhibitor *endo*-**6-O** has had quite a career since we first described it in 1985. First, in the molecular equivalent of "turn around is fair play", it was used by Peter Schultz and David Jackson and by Don Hilvert and his coworkers to make catalytic antibodies.[25-27] Rearrangement reactions in general, and especially those for which control of the substrate conformation is important, are particularly appropriate candidates for catalytic antibodies since they are unimolecular and seldom require covalent catalysis. The antibody obtained by Schultz and Jackson is reasonably efficient, accelerating the rearrangement 10,000-fold at 0 °C, which is 60% as good as the enzyme in stabilizing the transition state. It is interesting to compare the enzyme and the antibody directly, and their abilities to bind the inhibitor and catalyze the rearrangement, since the results tell us something about design and serendipity.

The enzyme accelerates the reaction 3,000,000-fold over the non-catalyzed, thermal rearrangement, under the conditions of the antibody assay, yet it binds the inhibitor only 250-fold better than substrate (correcting for the racemate).[25] The antibody accelerates the reaction 10,000-fold, and it binds the inhibitor only 60-fold better than substrate. Doesn't it seem like an antibody that is a perfect complement to the inhibitor should only accelerate the rearrangement 250-fold, since that's how "good" a transition state analog it is? Catalysis by this antibody either involves more than molding the right substrate conformation, or its imperfection in complementing the inhibitor has compensated the inhibitor's imperfection in mimicking the transition state - one of the few cases in which two wrongs make a right.

The oxabicyclic inhibitor is making a number of other appearances as well; it has starred in two productions already, and its performance in two more is currently being recorded. The X-ray crystal structures of its complexes with the *B. subtilis* monofunctional chorismate mutase and with the Hilvert catalytic antibody have been reported by William Lipscomb and Yuh-Min Chook at Harvard,[28] and by Ian Wilson and his coworkers at Scripps,[29] respectively, and similar studies with another enzyme and with the Schultz catalytic antibody are underway. When those studies are completed, we will have an unprecedented opportunity to compare the structures of four different proteins that catalyze the same reaction.

The structure of the *B. subtilis* enzyme-inhibitor complex[28] already provides considerable insight into the reasons for the success and shortcomings of our inhibitor designs. This enzyme is a homotrimer, and the three active sites are shared between adjoining monomers. The active site is spectacularly cationic! No fewer than four arginine residues and a lysine are found within 5 Å of the inhibitor. Two arginines bind the carboxylates directly, and a third is sideways to the bridge, H-bonding to the carboxylate and the ether oxygen. This arginine residue must be the proton-donating moiety we postulated would stabilize the pyruvyl enolate and that we had hoped would interact favorably with the amine in *endo*-**6-N**. Too bad it wasn't an aspartic acid side chain...

It is also clear from this crystal structure that *endo*-**6-O** doesn't fill up the active site the way a really good inhibitor ought to. There is quite a gap between the two up by the CH_2 group in the bridge, and a lot of potential van der Waals interaction is lost. It makes sense that the site is larger than the inhibitor, since the transition state has to be bigger as well; after all, the bonds to the enolpyruvyl moiety are mostly cleaved in the transition state, so they are longer than the covalent bonds in the inhibitor. This observation suggests a fairly simple way to improve the inhibitor design: make it bulkier in the vicinity of this CH_2. An inhibitor with 4-carbon bridge (**7**) may not look like a transition state analog if we just consider the Kekulé structure, but it may be a better mimic in three dimensions.

1.3.2 *EPSP Synthase*

5-Enolpyruvylshikimate-3-phosphate (EPSP) synthase is where the money lies. Its identification by Nikolas Amrhein and Hans Steinrücken as the site of action of the blockbuster herbicide glyphosate[30] awakened commercial interest in the shikimate pathway and stimulated a flood of investigations into the mechanisms and structures of its enzymes, and of course into the design of inhibitors. The finer details of the mechanism of EPSP synthase are still emerging, but the overall sequence proposed by Sprinson[31] in the 1960's and depicted in Figure 7 is now firmly established. The actual isolation of the tetrahedral adduct **8** by Karen Anderson and James Sikorski and their coworkers at Monsanto was a milestone in this history,[32] and a tribute to the revolution in mechanistic enzymology that occurred when huge quantities of protein could be obtained through recombinant techniques. A *gram* of enzyme, at steady state equilibrium with substrate, intermediate, and product, was sacrificed in a single denaturation experiment to obtain enough adduct to characterize!

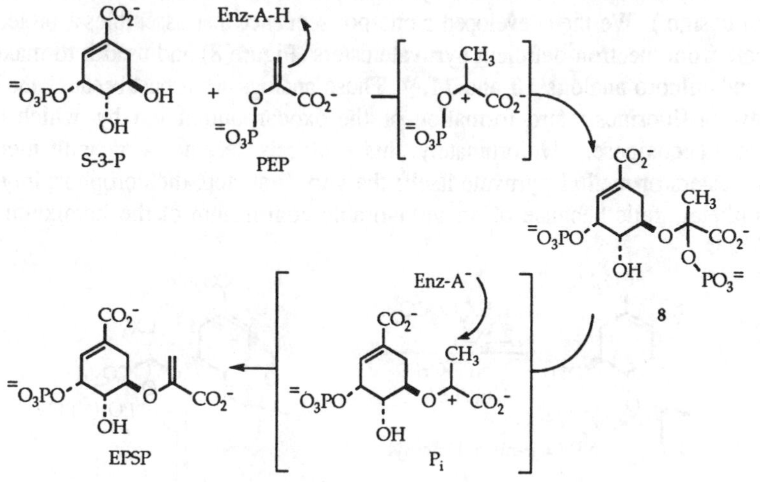

Figure 7. Mechanism of the reaction catalyzed by EPSP synthase.

Ever since the discovery of glyphosate, bioorganic chemists have been fixated on its resemblance to the protonated form of phosphoenolpyruvate (PEP), and rationalized its inhibition of EPSP synthase accordingly. Glyphosate is competitive with PEP and non-competitive with shikimate-3-phosphate (S-3-P), consistent with the ordered mechanism of substrate binding. Not everything about this view makes sense, however. For one, it has not proved possible to improve on glyphosate, in spite of the obvious ways in which its structure could be made to look more like the cation it is supposed to mimic. More

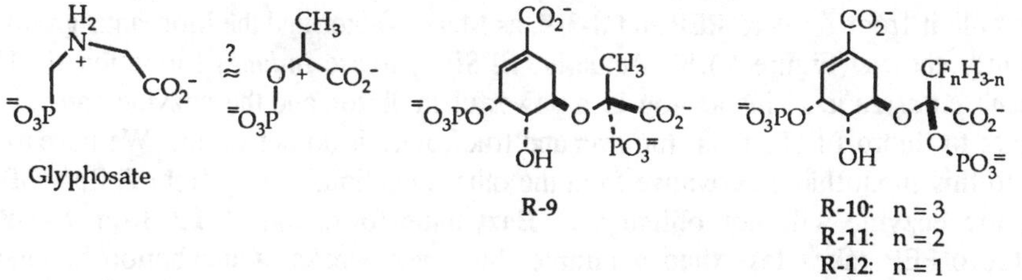

Glyphosate

R-9

R-10: n = 3
R-11: n = 2
R-12: n = 1
S-8: n = 0

definitive evidence has come from recent work of James Sikorski, who finds that glyphosate can bind to the enzyme in a fashion that is inconsistent with its designation as a PEP analog.[33]

Our work has managed to avoid this controversy, focusing on the tetrahedral adduct **8** itself. Although this ketal-phosphate had not been isolated at the time we started our program, we assumed it would be very unstable and very tightly bound to the enzyme; in short, the perfect candidate for analog design.

Mimics of covalent structures are easier to design than transition states, because they have normal bond lengths and angles. Nevertheless, something in the structure of **8** has to be changed to make it stable. We came up with the phosphonates **9** first, because they were easy to make.[34] ("Ease of synthesis" is an essential, if overlooked, element in inhibitor design.) We then developed a one-pot sequence that assembles a protected ketal phosphate from electron-deficient pyruvate esters (Figure 8) and used it to make the trifluoro and difluoro analogs **10** and **11**.[35] These analogs are robust because the electron-withdrawing fluorines retard formation of the oxocarbonium ion by which the ketal phosphates decompose. Unfortunately, this synthesis does not work with monofluoropyruvate esters or methyl pyruvate itself: the very first step, dichlorophosphitylation of the hemiketal, fails because of an unfavorable equilibrium of the hemiketal with its components.

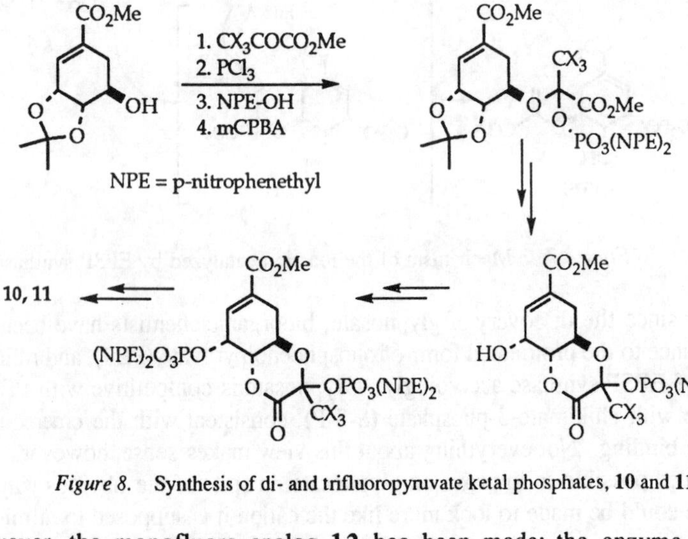

Figure 8. Synthesis of di- and trifluoropyruvate ketal phosphates, **10** and **11**.

However, the monofluoro analog **12** has been made: the enzyme itself will assemble it from Z-fluoro-PEP and S-3-P, as Mark Walker and the Monsanto group have recently shown (Figure 9).[36] Actually, EPSP synthase *induces* formation of **12**, it doesn't *catalyze* it. This analog is a powerful inhibitor and the enzyme can't carry it further to fluoro-EPSP; thus, turnover and true catalysis do not occur. We have tried to get to this monofluoro derivative from the other direction, namely from Z-fluoro-EPSP, but the enzyme will not oblige us. Enzymatic formation of **12** from S-3-P and Z-fluoro-PEP takes less than a minute, but even weeks of incubation of enzyme, phosphate, and Z-fluoro-EPSP give no evidence for formation of **12** or Z-fluoro-PEP.[37]

Does this mean that more cationic character develops in the second step in the enzymatic mechanism than the first?

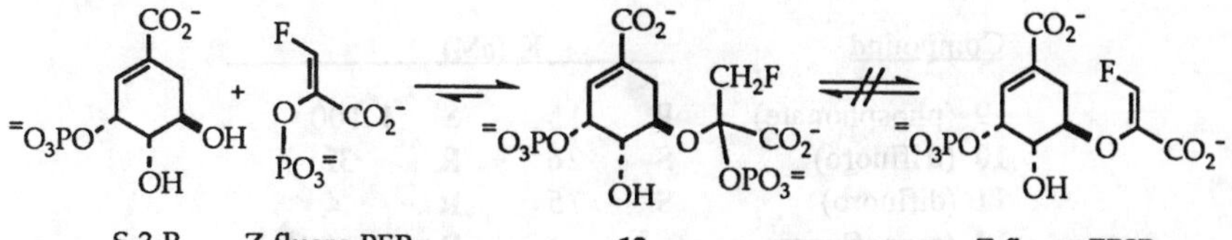

Figure 9. Processing of fluoro*enol*pyruvyl analogs by EPSP synthase.

Although this question is off the subject of inhibitor design, what prospect do we see for a chemical synthesis of **8**, the intermediate itself? We are optimistic that chemistry recently demonstrated by Uday Maitra in a model system can be applied to the appropriate shikimate substrate. He showed that the pyruvate ketal phosphate of 2–phenylethanol (**14**) is formed on tin hydride reduction of dibromide **13**, prepared by the sequence in Figure 8.[38] This reduction is carried out with UV irradiation at 0 °C (in place of the usual radical source and thermal initiation) and is facilitated by a cationic micellar catalyst (Figure 10).

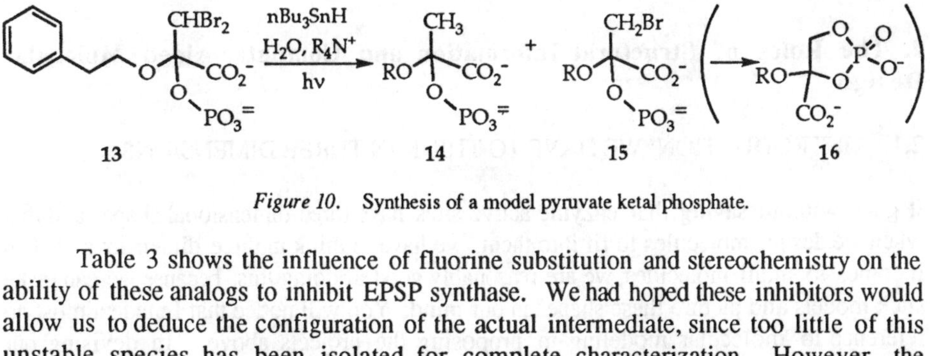

Figure 10. Synthesis of a model pyruvate ketal phosphate.

Table 3 shows the influence of fluorine substitution and stereochemistry on the ability of these analogs to inhibit EPSP synthase. We had hoped these inhibitors would allow us to deduce the configuration of the actual intermediate, since too little of this unstable species has been isolated for complete characterization. However, the phosphonate **9** was a red herring on this trail, even though its diastereomers show the biggest disparity in binding affinity and it was a reasonable assumption that the phosphonate would bind where the phosphate of **8** does. For the trifluoro analog **10**, there is essentially no discrimination between the diastereomers; the trifluoromethyl group may prefer the methyl site, but apparently not well enough to allow the differences between carboxylate and phosphate to be manifested. Removal of one of the fluorines (**11**) increases the binding affinity as well as the difference between diastereomers. Removal of another fluorine (**12**) increases the affinity further, and removal of the last fluorine, to give the actual intermediate, provides the most tightly bound compound of the series, if estimations of this binding constant are correct.[39] It will be very interesting to see what the inhibition constants are for the missing diastereomers of **12** and **8**! (Note that the fluorine-containing analogs and the methyl derivatives with the same 3-dimensional arrangement of groups have the opposite R/S designators.)

TABLE 3. **Inhibition of EPSP synthase by fluorine-containing analogs of the tetrahedral adduct.**

Compound	K_i (nM)			
9 (phosphonate)	R	15	S	1,100
10 (trifluoro)	S	26	R	32
11 (difluoro)	S	75	R	4
12 (monofluoro)			R[a]	0.6[b]
8 (intermediate)			S[a]	[0.05][c]

[a] The stereochemical assignments of **12** and **8** are tentative.
[b] Reference 36. [c] As estimated in reference 32.

Among the most exciting developments in this area are the X-ray crystal structures of a number of enzyme-inhibitor complexes, determined by William Stallings and his colleagues at Monsanto.[40] These structures, which involve several of our inhibitors, will resolve many of the remaining issues of stereochemistry and mechanism when they are completed. For this enzyme as well, the inhibitors will have played a significant role in elucidating structure and mechanism.

2. The Roles of Structural Information and Computer-Aided Molecular Design

2.1 INTRODUCTION: WE HAVE TO THINK IN THREE DIMENSIONS

It goes without saying that enzyme active sites have three-dimensional shape, and that when we design molecules to fit into them, we have to think in three dimensions. When it comes to small molecules, we are reasonably good at doing this, because we can make stick models and picture these shapes in our mind. You will notice that I made almost no reference to molecular modeling in proposing the projects above. In devising our transition state analogs, the information we used was mechanistic, not structural, and the three-dimensional mimicry that we worried about was pretty much at the atomic level: What can we use to mimic a tetrahedral carbon? Should the carboxylate point up or down? etc. While protein crystallography has played an enormous role in understanding how these transition state analogs bind, it did not contribute to their design.

Three-dimensional insight is tremendously seductive, however. Indeed, once you know the shape of a binding site, you can't get it out of your mind when thinking about molecules that might fit into it. We are now in the midst of a virtual explosion in structural information, thanks to advances in X-ray crystallography and NMR methods, and we need to develop strategies for integrating this information into our design of enzyme inhibitors. The problem is challenging because this three-dimensional information is not the kind that we can visualize with a model set or even grasp intuitively. Fortunately, the advances in computation and computer graphics that have contributed to the flood of information also help us to handle it, and we now manipulate protein structures on the computer workstation as readily as we used to flip our Dreiding cyclohexane rings from boat to chair.

There is no single approach to structure-based design: different problems require different kinds of structural information and are likely to be solved in different ways. At the risk of over-simplification, we can describe several common situations and the strategies that have been developed to address them (Table 4). This list is ordered according to my opinion of increasing difficulty.

TABLE 4. Types of structure-based design problems.

Type 1	Information available:	Structure of ligand-receptor complex
	Approach required:	Design analog
	Strategies:	reduce conformational flexibility add favorable binding interactions
Type 2	Information available:	Structure of potent ligand
	Approach required:	Design mimic
	Strategies:	introduce conformational constraint design alternative template
Type 3	Information available:	Structure of receptor
	Approach required:	Design novel ligands *de novo*
	Strategies:	map steric & electronic properties search databases & generate ligands automatically

2.2 STRUCTURE-BASED DESIGN OF CONSTRAINED PEPTIDE ANALOGS: EXAMPLES FROM THERMOLYSIN

The complexes between thermolysin and the phosphonamidate inhibitors **1** provide a good example of a Type 1 system for structure-based design. Their structures provide information not only on the specific interactions between protein and ligand, but also on the bound conformation of a flexible, acyclic peptide. Two straightforward strategies for enhancing binding affinity involve reducing conformational flexibility and incorporating substituents that bring additional binding interaction. Extending a bridging unit from one end of an acyclic inhibitor to the other, thus converting it into a macrocyclic structure, can accomplish the first of these goals, and may also provide an opportunity for the second as well.

We started with the structures of the thermolysin complexes with Cbz-Gly{$PO2^-$-NH}Leu-Leu and Cbz-Phe{$PO2^-$-NH}Leu-Ala.[9,41] In both cases, the inhibitor backbone follows a similar path through the active site from the C-α at P_1 to the C-terminal carboxylate at P_2', with a β-like turn at the P_1'-P_2' linkage (Figure 11). The terminal carboxylate and the side chain at P_1 are close to the surface of the protein, by the entrance to the active site, hence it seemed logical to us to link the α-carbons at P_1 and P_2' and make a macrocycle. There were a few constraints that we needed to deal with, however.

First, all the heteroatoms in these inhibitors are involved in hydrogen-bonding or other polar interactions with active site residues, which we didn't want to disrupt. Second, the entrance to the active site is very narrow, with the side chains of Asn-112 and His-231 coming together from above and below like two jaws. Since both of these side chains are involved in specific interactions with the inhibitors, we figured we had to work our way around them in reaching between the two α-carbons.

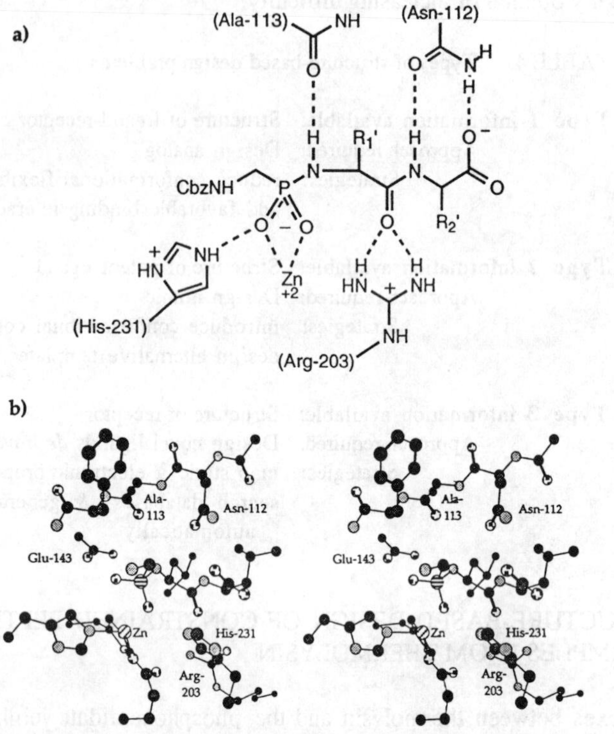

Figure 11. Binding orientation of Cbz-Gly{PO_2^--O}Leu-Leu in the active site of thermolysin. a) Schematic of hydrogen bonding and other polar interactions. b) Stereoview of active site; the Cbz moiety of the inhibitor has been removed in this image.

Our first efforts to design the linking unit took advantage of computer graphics to represent the protein structures, but relied on our own intuition and simple Dreiding model for constructing candidate inhibitors. We became convinced very quickly that a five-atom linkage could span the distance between the C-α carbons (Figure 12). We then fused an aromatic ring at one end to rigidify this chain and because we needed a very slim unit to fit between the teeth of Asn-112 and His-231; the second ring was added to the bridging unit for further rigidification. We put an ether oxygen at the central position of the linker to avoid transannnular interactions arising from hydrogen substituents. Although we incorporated this oxygen into the design purely for steric reasons, it turned out to be a favorable hydrogen-bond acceptor as well, as described below. The structure was then simplified by removal of the Cbz-amino moiety, because this group can adopt a variety of conformations in the P_2-P_3 binding sites[9,41] and because we figured it would not be affected by the macrocyclization strategy anyway. Finally, to enhance the "ease of synthesis" of the target, we added a methyl group to the para position of the aromatic ring; this position projects into solvent, so we reasoned that a substituent here was not going to have any significant affect on binding. The (S,S)-isomer of **17** was thus our target inhibitor.

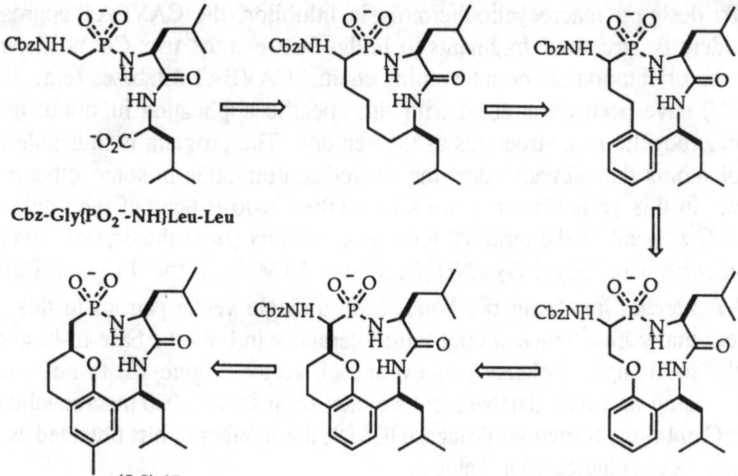

Cbz-Gly{PO_2^--NH}Leu-Leu

(S,S)-17

Figure 12. Evolution of the design of inhibitor **(S,S)-17**.

These design efforts were initiated several years ago and were based almost entirely on intuition, as the discussion above reveals. We used the UCSF computer graphics facility and modeling programs to visualize the thermolysin binding site and gauge whether the designed inhibitors would fit, but we were on our own when it came to thinking up the molecules. However, as we pursued this project, and in particular the design of a tendamistat mimic as described below, we saw a common element in many of the steps that we were taking and we looked for a way to automate them. We thought if we could speed the process up, we would be able to expand the range of our imagination. These ideas are now embodied in our program CAVEAT (Computer-Assisted Vector Evaluation And Target design...).[42,43] Although we didn't have CAVEAT at the time we designed our initial constrained thermolysin inhibitors, it has been interesting to revisit the problem to see what kinds of designs we might have come up with using this program. We conceived of CAVEAT as a tool for identifying structural motifs that can serve as templates in a variety of structure-based design problems. With this program, you can search a database of 3-D structures using a vector relationship between specified bonds in the molecules as the criterion for identifying a hit. As a result, the structures you find are able both to position functional groups in specific locations, as well as orient them in a desired manner relative to each other. CAVEAT searches can accept any number of vectors (bonds), and the relationships between the vectors can be specified flexibly, both with respect to how they are combined and how close a match is required.[43]

In addition to the search algorithm, we have also incorporated a post-search filtering and clustering module into CAVEAT. This utility, called CLASS, screens out hits that contain structural motifs that you don't want or that would encounter steric or conformational difficulties; CLASS also groups the remaining molecules on the basis of their structural similarity. By speeding your evaluation of hit structures, CLASS accelerates the design process significantly, and allows you to open up the search tolerances and net a greater diversity of structural motifs.

To design a macrocyclic thermolysin inhibitor, the CAVEAT approach can be used to identify structural fragments to bridge between the two Cα positions without altering the orientation of the intervening chain. CAVEAT databases (e.g,. the "**intra**" database[43]) have been constructed with this specific application in mind, using bonds within macrocyclic ring structures as the vectors. The program is thus able to identify chains of atoms that already adopt the desired conformation in some other macrocyclic structure. In this particular case, we selected the Cα-to-P bond of the Gly^P residue and the C-to-Cα bond of the terminal leucine as vectors (from the crystal structure of the complex with Cbz-Gly{PO_2^-–NH}Leu-Leu: 5TMN in the Protein Databank[44]). CAVEAT searches involving two bonds (i.e., a single vector-pair, as in this case) typically yield many hits, hence we put tight tolerances in both the base-to-base distance **d** and in the solid angles (tolerance cones) of each vector (Figure 13) to net a manageable number. Using an **intra** database constructed from some 5,000 macrocyclic molecules from the Cambridge Structural Database (CSD), the number of hits obtained as a function of the tolerances is indicated in Table 5.

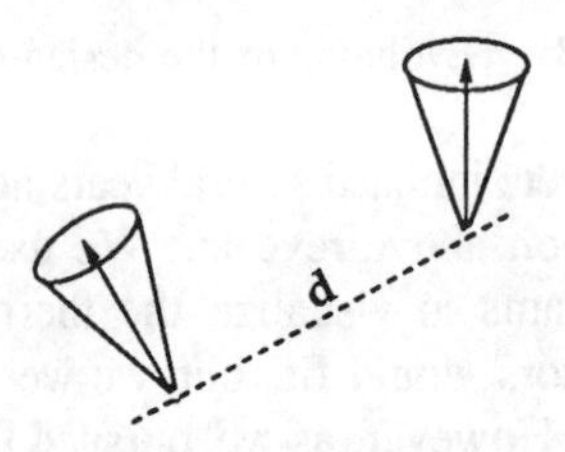

Figure 13. Representation of the tolerance permitted in definition of a vector-pair relationship.

TABLE 5. Relationship between tolerance limits and number of molecules retrieved from the CAVEAT **intra** database.

Tolerance[a]	Number of Hits
(0.03 0.03 0.03)	17
(0.05 0.03 0.03)	21
(0.03 0.05 0.05)	175
(0.05 0.05 0.05)	221
(0.05 0.07 0.07)	748

[a] The three numbers indicate permitted variation in distance between the base atoms of the vector pair (in Å) and in the angular deviations of vector 1 and vector 2, respectively (in radians).

We screened these hits according to a number of criteria. First, we didn't accept any structure if the core of the linking unit was closer than 1.0 Å to the enzyme (the core represents those atoms which require at least two bond-disconnections to be detached from the linking unit[43]). This screen eliminated linking units that bump into the enzyme itself. Second, possible eclipsing interactions across the bonds to be formed between the linking unit and the inhibitor chain were evaluated for each hit, and only those which would result in a staggered conformation were accepted. We also took advantage of the clustering flexibility of CLASS to select a subset of structures that contained cyclic moieties as part of the linking unit, assuming them to be inherently more rigid. A number of the hits that passed these three selection criteria are shown in Figure 14a. While there was no single hit that incorporated a bicyclic structure like that in **17**, the superposition of the various structures clearly suggests the bicyclic linker which we had designed earlier by a considerably more laborious process.

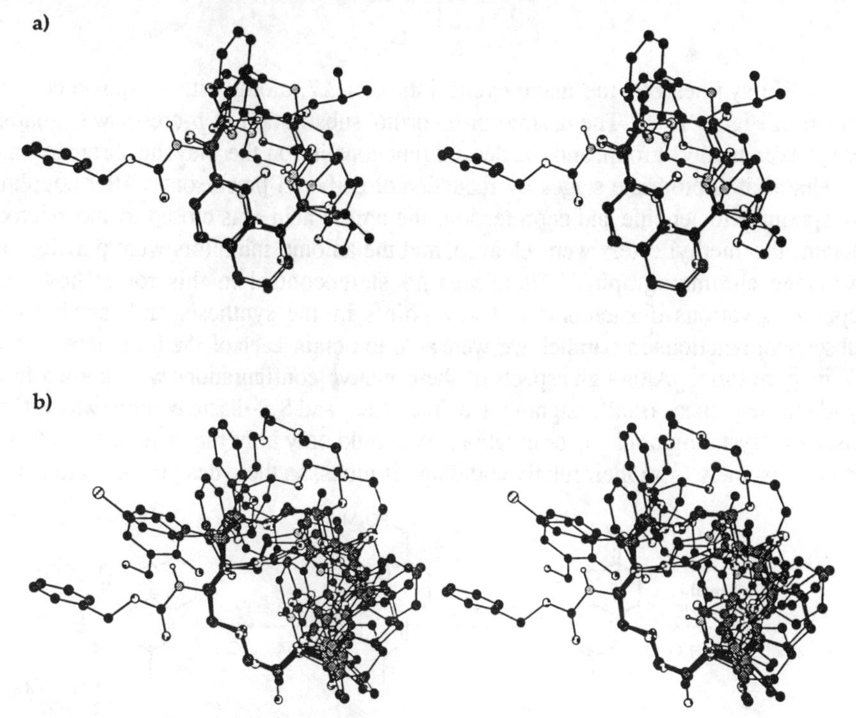

Figure 14. a) Superposition of macrocyclic structures retrieved from the Cambridge Structural Database illustrating potential ring-containing linking units. b) Superposition of a number of crown ethers from the CSD illustrating the linear ether linking group.

Without selecting for ring-containing linkers, and clustering the hits based on back-bone motif, CLASS identifies some 11 crown ethers as one group of hits. The superposition of these molecules shows the usefulness of the clustering approach (Figure 14b). All these molecules represent the same structural motif, and it would be a waste of time to have to look at each one individually in the course of evaluating the hits. Attractive features of this motif are the ether oxygen, which is located at the same position as the earlier bicyclic design **17**, and the fact that it is more accessible synthetically. On the other hand, this linking unit is considerably more flexible than the bicyclic chroman moiety of **17**, and we wondered whether incorporating it into a macrocycle such as **18** would decrease or increase the available conformations.

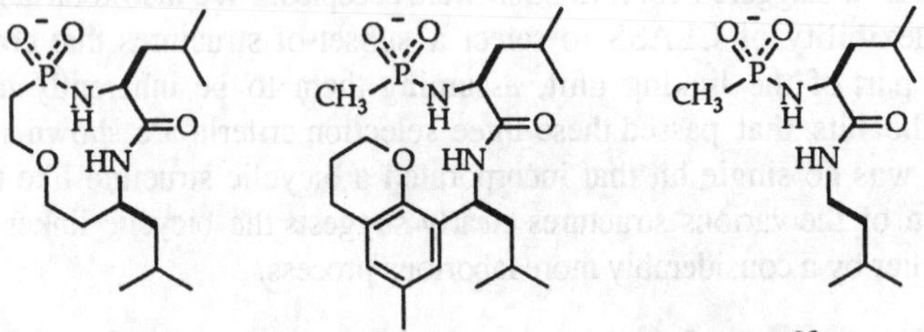

18 **19** **20**

We synthesized the macrocyclic inhibitor **17**, along with its diastereomers, as shown in Figure 15.[45] The desired ortho,ortho' substitution of *p*-cresol was guaranteed by the para methyl group, and the desired functionality on the bicyclic chroman unit was introduced at appropriate stages by reduction of carbonyl precursors. After coupling the phosphonate to leucine and deprotection, the amino acid was closed to the macrocyclic lactam, the methyl esters were cleaved, and the anionic inhibitors were purified by ion exchange chromatography. There was no stereocontrol in this route; however, by separating various diastereomers at key points in the synthesis and carrying out the subsequent reactions in parallel, we were able to obtain each of the four stereoisomers of **17** in pure form. Although aspects of their relative configurations were known from the synthetic sequence, final assignment of the R,R- and S,S-diastereomers was ultimately based on crystallographic determination. We could only infer the assignments of the R,S- and S,R-isomers from their relative binding affinities, so these designations are tentative.

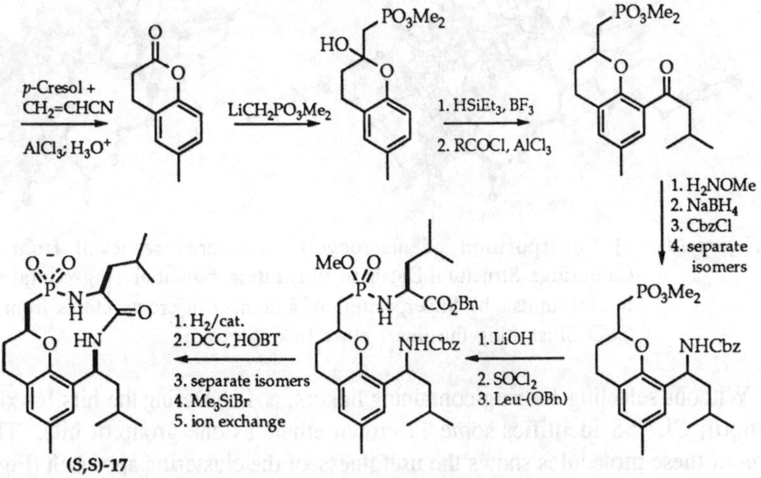

p-Cresol + CH_2=CHCN; $AlCl_3$; H_3O^+ → $LiCH_2PO_3Me_2$ → 1. $HSiEt_3$, BF_3 2. RCOCl, $AlCl_3$ → 1. H_2NOMe 2. $NaBH_4$ 3. CbzCl 4. separate isomers → 1. LiOH 2. $SOCl_2$ 3. Leu-(OBn) → 1. H_2/cat. 2. DCC, HOBT 3. separate isomers 4. Me_3SiBr 5. ion exchange → **(S,S)-17**

Figure 15. Synthesis of thermolysin inhibitor **(S,S)-17**.

The simple macrocyclic ether **18** was synthesized as shown in Figure 16.[46] Chain elongation of L-leucine via the Arndt-Eistert synthesis, followed by reduction, provided the basic 1,3-aminoalcohol building block. The ether linkage was formed by rhodium-catalyzed carbene insertion, and the phosphonate was introduced by an Arbusov reaction on the bromoethyl derivative. Coupling to L-leucine and formation of the macrocyclic lactam proceeded as described above for the more rigid analog.

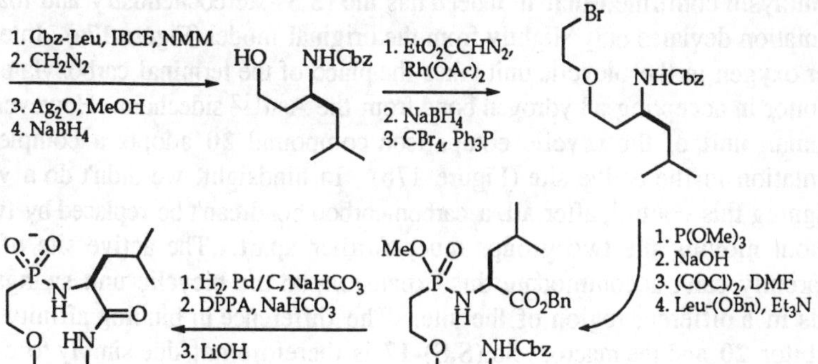

1. Cbz-Leu, IBCF, NMM
2. CH_2N_2
3. Ag_2O, MeOH
4. $NaBH_4$
HO NHCbz
1. EtO_2CCHN_2, $Rh(OAc)_2$
2. $NaBH_4$
3. CBr_4, Ph_3P
Br
O NHCbz
1. $P(OMe)_3$
2. NaOH
3. $(COCl)_2$, DMF
4. Leu-(OBn), Et_3N
MeO O P N H CO_2Bn O NHCbz
1. H_2, Pd/C, $NaHCO_3$
2. DPPA, $NaHCO_3$
3. LiOH
4. ion exchange
O O P N H O O HN
18

Figure 16. Synthesis of the ether-linked inhibitor **18**.

We evaluated the four diastereomers of **17**, as well as the flexible macrocyclic ether **18** and the comparison compounds **19** and **20**, as reversible, competitive inhibitors of thermolysin, and determined the inhibition constants given in Table 6. The stereoisomer of **17** that turned out to have the S,S-configuration as designed proved to be the most potent inhibitor, with a K_i value significantly lower than those of the other isomers, as well as the comparison compounds. The ca. 50-fold enhancement in binding affinity of

TABLE 6. Inhibition of thermolysin by macrocyclic phosphonamidates.

Inhibitor	K_i (nM)[a]
(R,R)-17	>10,000
(R,S)-17[b]	>10,000
(S,R)-17[b]	500
18	500
20	190
19	80
(S,S)-17	4

a Determined in 0.1 M MOPS buffer, 2.5 M NaBr, 10 mM $CaCl_2$, and 2.5% v/v DMF at pH 7.0 at 25 °C.

b Assignments of **(R,S)-1** and **(S,R)-1** are tentative [**(R,S)-1** is epimeric with **(S,S)-1** at the chroman ring position].

this isomer over the bare-bones phosphonamidate **19** indicates that the overall strategy was effective, and the 20-fold enhancement over the intact but acyclic analog **20** suggested that most of this improvement comes from conformational constraint, rather than specific interactions of the chroman unit with the protein.

Brian Matthews and Debra Holland determined the structures of the thermolysin complexes of **(S,S)-17** and **18**, and provided critical insight into both of these inferences. The co-crystal structure of the most potent of the macrocyclic analogs with thermolysin confirmed that it indeed has the (S,S)-stereochemistry and that its binding orientation deviates only slightly from the original model (Figure 17a). Interestingly, the ether oxygen in the bicyclic unit takes the place of the terminal carboxylate in the parent inhibitor in accepting a hydrogen bond from the Asn^{112} sidechain. However, the bicyclic chroman unit of the acyclic comparison compound **20** adopts a completely different orientation in the active site (Figure 17b). In hindsight, we didn't do a very good job designing this control; after all, a carbon-carbon bond can't be replaced by two hydrogens without moving the two groups much further apart. The active site of thermolysin apparently can't accommodate this expansion, so the bicyclic unit swings around and binds in a different region of the site. The difference in binding affinity between this inhibitor **20** and the macrocycle **(S,S)-17** is therefore not due simply to cyclization and conformational constraint.

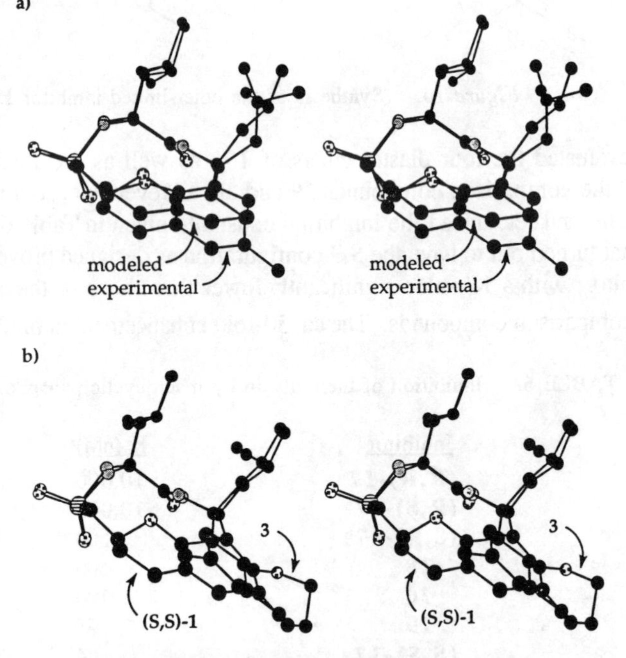

Figure 17. a) Comparison of the modelled and experimentally determined conformations of the chroman-linked inhibitor **(S,S)-17** bound to thermolysin (the modeled inhibitor lacks the methyl substituent on the aromatic ring). b) Comparison between the bound conformations of the cyclic inhibitor **(S,S)-17** and acyclic analog **19**.

The less rigid macrocyclic ether **18** is actually *worse* as an inhibitor than the simple comparison compound without the bridging unit. We don't have a co-crystal structure of this compound in the active site, but we assume that the peptidylphosphonamidate chain fits into the active site in the usual fashion, and that the ether linking unit doesn't do anything unexpected in reaching across the active site. If we knew the low-energy conformation of this macrocycle in solution, and had some assessment of its dynamical behavior, we would be in a better position to determine why **18** is not bound with higher affinity. At this point, we think that the macrocyclic ring is simply too flexible: the entropic advantage of constraining the peptide in a macrocycle is offset by the additional entropy required to fix the conformation of the large ring.

2.3 STRUCTURE-BASED DESIGN OF α-AMYLASE INHIBITORS

The structure of tendamistat, a 74-residue peptidic inhibitor of pancreatic α-amylase,[47] is an example of a Type 2 design problem. A high-resolution crystal structure[48] and the NMR solution structure[49,50] of this inhibitor are available, it is known what region of the inhibitor surface interacts with the enzyme, but the structure of the complex has not yet been refined to the extent that it could be used as the basis for design.[51] Although there is an extensive contact surface between the two proteins, the β-turn region comprising residues Trp^{18}-Arg^{19}-Tyr^{20} is central to this interaction (Figure 18).

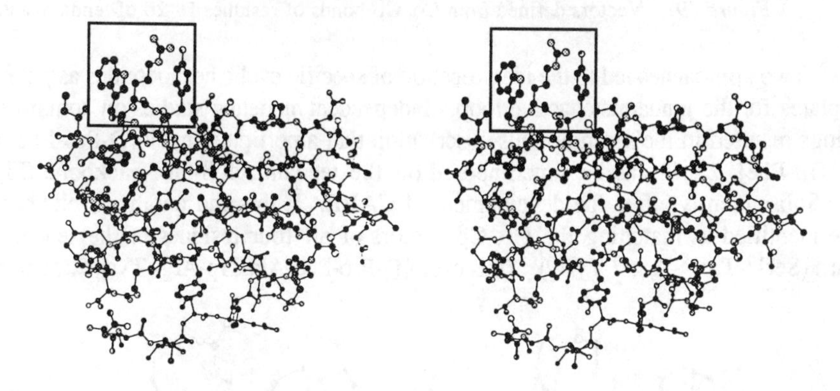

Figure 18. Stereoview of active loop region of tendamistat.

Tendamistat was an attractive target for trying out our ideas on ways to design peptidomimetics. Good structural information was available, and it gave us a chance to see if the binding characteristics of a large protein inhibitor could be embodied in a small mimic. Because tendamistat is a peptide, it presents very different functional groups to the enzyme active site than the normal polysaccharide substrates. A mimic of tendamistat could therefore be an inhibitor of α-amylase designed in a distinctively different manner from the known substrate or transition state analogs.[52]

In approaching this problem, we focused on the peptide backbone (the scaffolding) rather than the functional groups at the ends of the side-chains (the parts that interact directly with the enzyme). The backbone of the active loop of tendamistat is not exposed, so we believe its major role is just to position and orient the side-chains appropriately. If we knew the bound conformation of tendamistat in detail, we could presumably devise a better, more rigid mimic, but without this information, we thought it less dangerous if we focussed on the peptide backbone and looked for a different scaffold or template.

Our next consideration was that this template do more than simply *locate* the Cα atoms of the three residues the correct distance apart; it needed to *orient* the Cα-Cβ bonds of the side-chains in a specific way (Figure 19). We have often expressed this distinction as the difference between a *vector* relationship between bonds, rather than a *Cartesian* relationship between atoms, and it is the underlying principle of CAVEAT as described above.[42, 43]

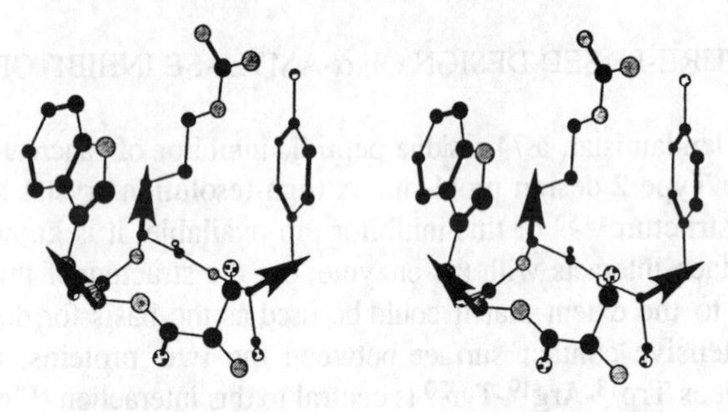

Figure 19. Vectors defined from Cα-Cβ bonds of residues 18-20 of tendamistat.

Two approaches led to the identification of specific cyclic hexapeptides as potential templates for the tendamistat active loop. Independent modeling studies on somatostatin analogs resulted in the serendipitous observation that a portion of cyclo[D-Pro-Phe-Thr-Phe-Trp-Phe], **21**,[53] can be superimposed on the tendamistat β-turn backbone (Figure 20). Subsequently, after the development of CAVEAT, this and another cyclic peptide were identified as matching the Cα-Cβ vectors of all four residues in the tendamistat β-turn (Ser^{17}-Trp^{18}-Arg^{19}-Tyr^{20}). Thus cyclo[D-Pro-Phe-Ala-Trp-Arg- Tyr], **22**, emerged

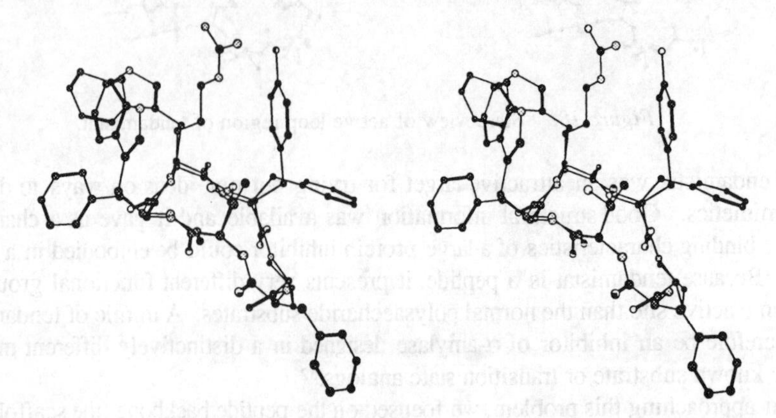

Figure 20. Superposition of cyclo[D-PFTFWF] on residues 17-20 of tendamistat; rms deviation = 0.40Å for the 8-atom match of Cα and Cβ carbons.

as one design for a potential tendamistat mimic. In a second design, cyclo[D-Pro-Phe-Ser-Trp-Arg-Tyr], **23**, alanine is replaced with serine, since this residue at position 17 in tendamistat is also highly conserved in related inhibitors.

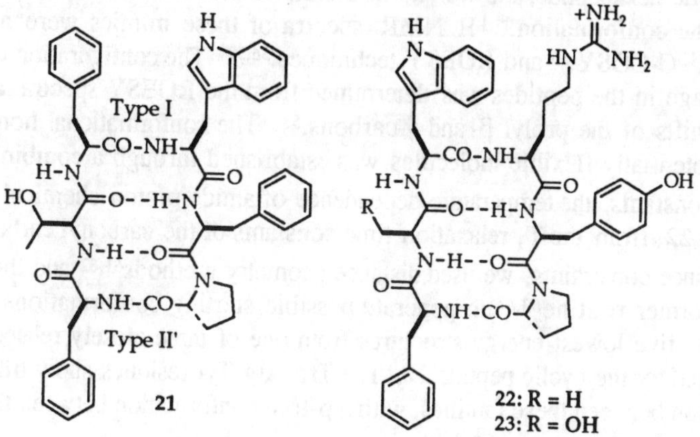

21

22: R = H
23: R = OH

The inhibition constants of these cyclic peptides **22** and **23** and relevant acyclic control compounds are given in Table 7.[54] Comparison between the cyclic (entries 5 & 6) and the linear hexapeptides (entries 3 & 4) indicates the influence of conformational constraint on the binding affinity of the Trp-Arg-Tyr sequence. The linear hexapeptides are even weaker inhibitors than the tripeptides (entries 1 & 2) because the additional three amino acids need to be constrained for this sequence to fit into the active site. The cyclic peptides are clearly better inhibitors than their acyclic comparison compounds, in the case of the alanine-containing derivative **22** by almost a factor of 50. There is less of a gain in binding affinity in the case of the serine analog **23**, which we interpret as the balance between conformational control and solvation: the serine residue may stabilize the correct β-turn conformation by hydrogen bonding interaction with the tyrosine NH group, but it is more hydrophilic and must be desolvated on entering the active site. The relative importance of these two effects is different for the acyclic and cyclic hexapeptides.

TABLE 7. Inhibition constants of tendamistat mimics at pH 7.

Entry	Compound	K_i (μM)
	Tripeptides:	
1	Trp-Arg-Tyr	520 ± 20
2	Ac-Trp-Arg-Tyr-OCH_3	100 ± 3
	Linear peptides:	
3	Ac-Phe-Ala-Trp-Arg-Tyr-D-Pro-NH_2	670 ± 110
4	Ac-Phe-Ser-Trp-Arg-Tyr-D-Pro-NH_2	320 ± 50
	Cyclic peptides:	
5	cyclo[D-Pro-Phe-Ser-Trp-Arg-Tyr], **31**	32 ± 2
6	cyclo[D-Pro-Phe-Ala-Trp-Arg-Tyr], **30**	14 ± 2

To evaluate the success of these mimics from the point of view of design, we needed to know whether the cyclic peptides indeed adopt the intended conformation. For example, the alanine analog **22** was devised from the crystal structure of a different, albeit related, cyclic hexapeptide, and we had no assurance that the different substitution would not alter the conformation. ^{1}H NMR spectra of these mimics were assigned using COSY, $^{1}H^{13}C$-COSY,[55] and ROESY techniques.[56,57] The configuration of the Tyr-Pro amide linkage in the peptides was determined from the ROESY spectra, as well as the chemical shifts of the prolyl β- and γ-carbons.[58] The conformational homogeneity[59,60] of these potentially flexible molecules was established through a combination of $^{3}J_{HH}$ coupling constants, the temperature-dependence of amide-proton chemical shifts, and in the case of **22**, from the T_1 relaxation time constants of the carbonyl carbons. From the NMR distance constraints, we used distance-geometry methods[61,62] and the MacroModel MultiConformer routine[63,64] to generate possible starting conformations.[65] Figure 21 displays the five lowest energy structures from one of three closely related families that we calculated for the cyclic peptide **22**. The Trp-Arg-Tyr residues show little variation in conformation between these families, with a β-turn conformation between the Trp and

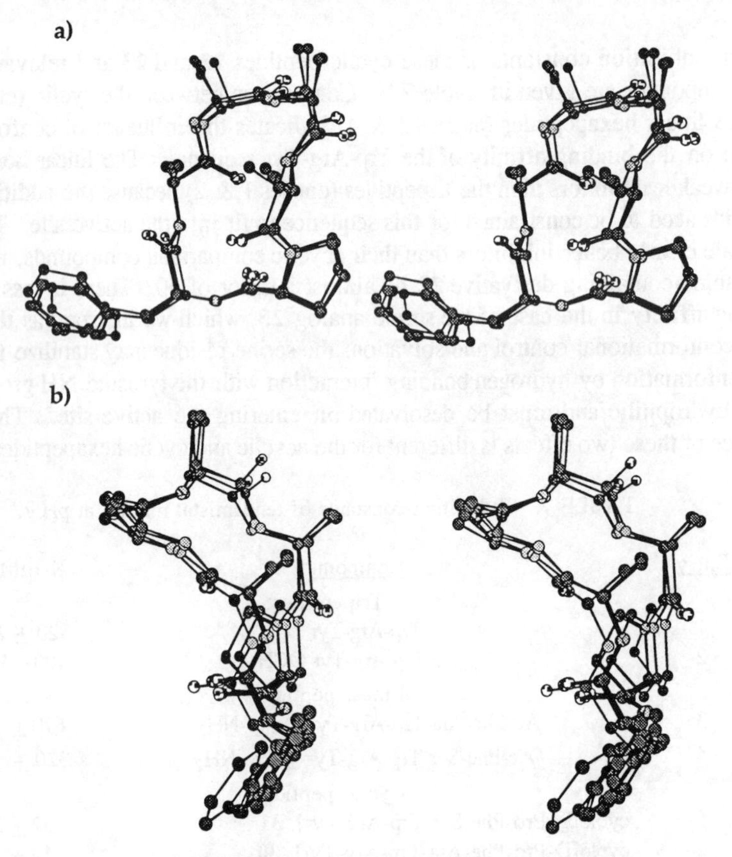

Figure 21. Two views of favored cluster of backbone conformations determined for cyclic peptide **22**.

Arg residues that is intermediate between the canonical Types I and II', in that the carbonyl group points away from the side-chains. It is likely that this amide linkage is conformationally flexible, alternating between the Type I and II' forms. It was particularly interesting to compare this conformation of mimic **22** with the peptide **21** that we used as the template. The rms variance for this comparison was 50° for the angles and 0.53 Å for the distances, less than the angle and distance rmsv among the better structures calculated for **22** itself!

These results support the premise leading to the design of the tendamistat mimics, in that the combination of the key residues and an appropriate scaffolding can imitate the natural inhibitor in a qualitative fashion. Since the Trp-Arg-Tyr sequence is only a small portion of the interaction surface tendamistat, it is not surprising that these analogs do not imitate it quantitatively. However, we are encouraged by the fact that a significant portion of the binding affinity can be captured by a small molecule mimic, and look forward to extending this structure-based approach to inhibitor design.

2.4 COMPUTED DATABASES AS SOURCES OF STRUCTURAL INFORMATION

CAVEAT relies on 3-D databases as its source of structural information; the program does not generate structures or conformations itself, but simply provides a rapid method for searching source databases for the kind of templates or linking units described for the problems above. Currently, the largest source of experimental data on small molecules is the Cambridge Structural Database, which consists of X-ray coordinates for some 100,000 structures that have been deposited with the Cambridge Crystallographic Data Centre. Other databases include the X-ray and NMR structures of biomacromolecules deposited with the Brookhaven Protein Databank, and a number of databases of predicted structures, such as the CAST-3D database distributed by Chemical Abstracts and 3-D databases generated from corporate files.

For any 3-D design problem, you want your database search to give you every reasonable possibility as an answer. Of course, there will be great differences of opinion as to what is "reasonable" or useful, but everyone would agree that the database itself at least ought to be diverse and comprehensive in terms of the structural motifs represented. However, the terms "diverse" and "comprehensive" represent conflicting concepts when they are applied to a limited set of structures. It seems unlikely that it would be possible to generate, let alone search, a database that contains every structure that a chemist could synthesize. Within any bounded subset of possible molecules, a trade-off will have to be made; the database could contain every tricyclic hydrocarbon (within limits) or every macrocyclic polyether (MW below 10,000, say), or some of each, but not both, and everything in between. To my knowledge, no metric for quantitating the diversity or completeness of a database has been used to assess those that are currently available.[66] Most of the current databases, simply because of their size, are not very comprehensive; they have greater or lesser diversity, but they all represent just the tip of an iceberg of possibilities.

It thus seemed to us that a different sort of 3-D database, one that focusses on completeness rather than diversity *per se*, would be complementary to the others that are available, and might therefore useful. We planned to create it computationally, generating and minimizing a collection of structures that would be *comprehensive* within the specification for the database. We set a limit of 10^6 on the number of structures, and then considered what we could get for that number. The first version of this database, which we call TRIAD (for **TRI**cyclics for **A**utomated **D**esign), was thus defined as indicated in Table 8:

TABLE 8. Specification of TRIAD database.

Limitations:

- Tricyclic molecules (if four rings are included, the possibilities explode)
- Hydrocarbons only (if heteroatoms are allowed, there is another combinatorial explosion)
- No ring larger than six-members (this constraint was introduced partly to limit numbers, partly to avoid conformational issues)
- No enantiomers (the search algorithms will take this into account)

Expansions:

- All diastereomers
- All saturation isomers (anthracene, the 5 dihydroanthracenes, the many tetrahydroanthracenes, etc.)
- Zero or one *exo*methylene per ring (introduced to provide a single sp^2-hybridized center within the ring and two additional vectors (the CH bonds) off the ring)

Without the *exo*methylene groups, our back-of-the-envelope calculation suggests that the database would have between 30,000 and 50,000 structures; with this option, the number increases by an order of magnitude.

In gauging the utility of such a database, our most important concern was the relevance of hydrocarbon structures to the sort of problems we plan to address. We do not expect to construct peptidomimetics on tricyclic hydrocarbon frameworks. However, the molecules retrieved in a template search are starting points in the design process; they are going to be substituted and modified, and subjected to refinement and minimization in later steps. We reasoned that hydrocarbon frameworks are similar to various heterocyclic structures (an amide is similar to a double bond, an ether to a methylene, etc. - Figure 22) and, furthermore, that the difference between a hydrocarbon structure and that of a heterocyclic framework is probably going to be less than the difference between the conformation of the original template, hydrocarbon or otherwise, and the final "designed" molecule.

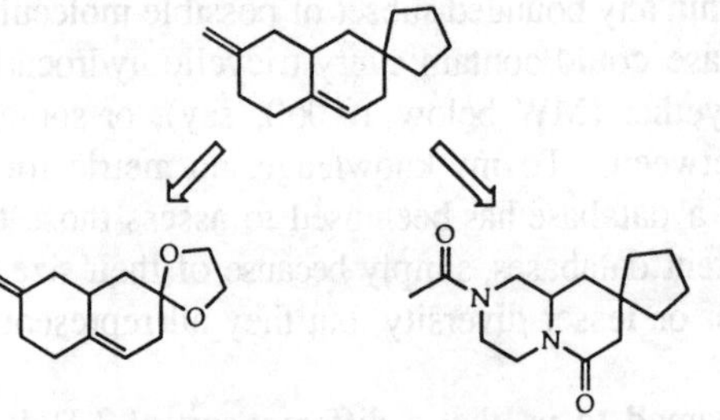

Figure 22. Equivalence of hydrocarbon with heteroatomic structures.

To generate the structures in this database, our first task was to devise an algorithm that would produce all of them. We recognized that there are only two topologies for a tricyclic molecule: one in which a "core" ring can be identified, from which the other two rings emanate as "sidechains", and the other in which there is no core ring, with three sidechains appended to a branched core. Examples of common molecules that represent these topologies are shown in Figure 23.

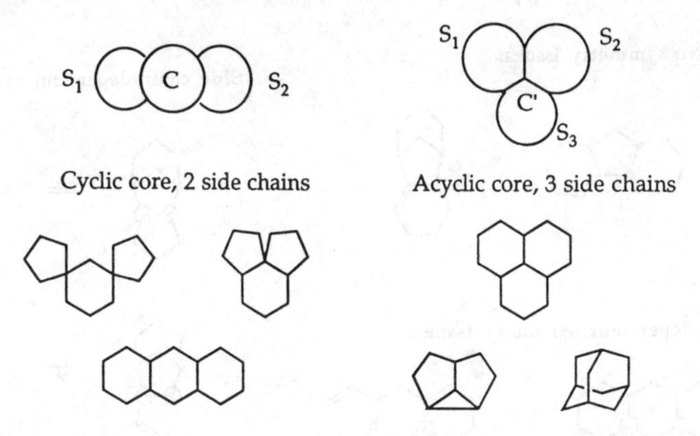

Figure 23. The two topologies of tricyclic molecules.

The task of generating all of the molecules in the database was relatively easy: we simply had to identify all the potential cores, all the potential side chains, and link them together in all the ways possible. We took into account common-sense rules of structure: no *trans*-fused rings of 3- or 4-members, no violations of Bredt's rule or valence, etc. (Figure 24).

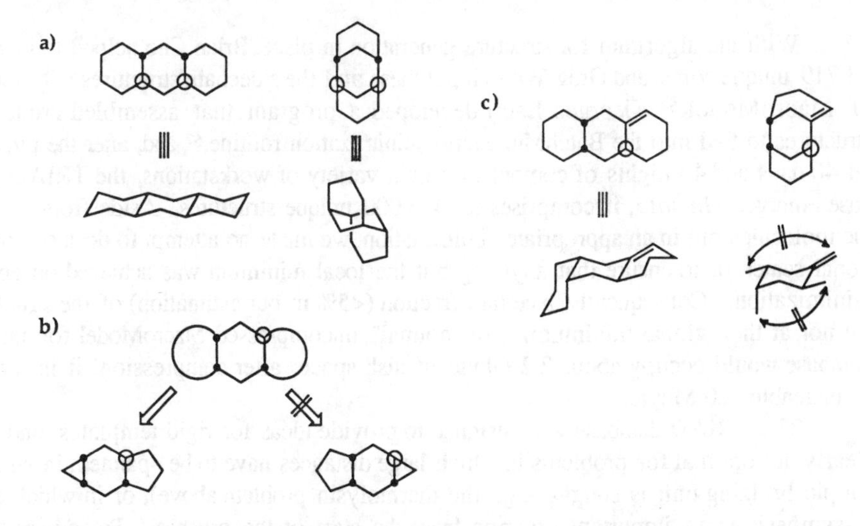

Figure 24. There are two linking patterns for each TRIAD core (a), subject to constraints due to ring-fusion geometry (b) and Bredt's rule, etc. (c).

Our second task was to devise an algorithm that would avoid generating more than a single copy or the enantiomer of any structure. Some of the symmetry issues are straightforward and are inherent in the core; others are dependent on the core-sidechain combinations (Figure 25). Our approach was to assign a symmetry designator to each core and sidechain, and keep track of the overall symmetry of the molecule as the selection of cores, linking pattern, sidechain-1, sidechain-2, (sidechain-3) was made.

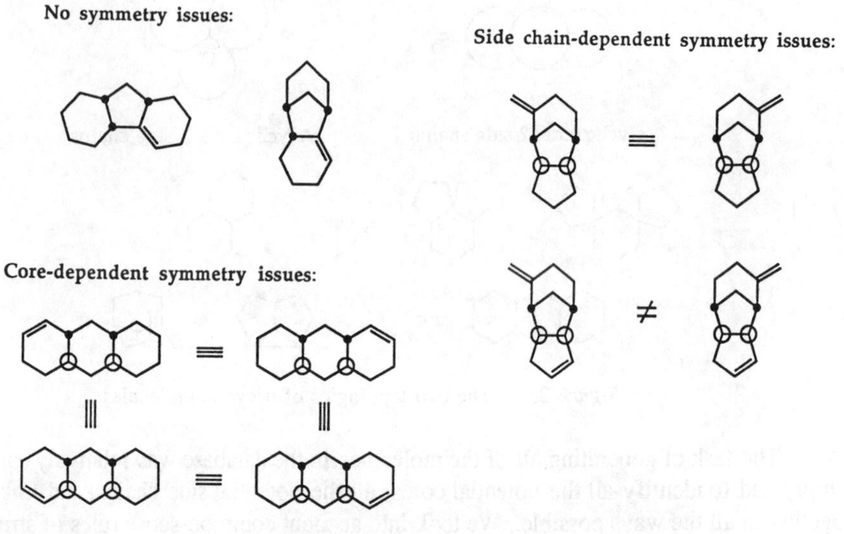

No symmetry issues:

Side chain-dependent symmetry issues:

Core-dependent symmetry issues:

Figure 25. Symmetry issues in selecting sidechain combinations may depend on the core or the sidechains.

With the algorithm for structure generation in place, Brian Cho helped to identify all 719 unique cores, and Greg Weiss input these and the sidechain structures with the aid of MacroModel.[64] Georges Lauri developed a program that assembled crude 3-D structures to feed into the BatchMin energy minimization routine,[67] and, after the proverbial 40 days and 40 nights of computation on a variety of workstations, the TRIAD database emerged. *In toto*, it comprises ca. 403,000 unique structures. Aside from starting the molecules out in an appropriate conformation, we made no attempt to do a conformational search or to ensure that anything but the local minimum was achieved on energy minimization. Consequently, a certain fraction (<5% in our estimation) of the structures are not at their global minimum. In "normal", uncompressed MacroModel format, the database would occupy about 2.2 Gbyte of disk space; after compression, it is a more manageable 220 Mbyte.

The TRIAD database was designed to provide ideas for rigid templates, and it is clearly not optimal for problems in which large distances have to be spanned, in which a simple bridging unit is sought (e.g., the thermolysin problem above), or in which ease-of-synthesis is an important criterion from the start of the process. To address these issues, we created a different database, which we call ILIAD (to rhyme with TRIAD, of course, and rationalized as **I**nter-**LI**nkers for **A**utomated **D**esign...). The molecules in

ILIAD are linear sequences built up from five units, taken from the components listed in Figure 26. The idea that a-hydrocarbon-can-mimic-anything is evident in this design as well, although we incorporate sulfide and disulfide because of their unique bond angles and dihedral angles. The "trimethylenemethane" unit is included as an isostere for urea and carbamate units, and the "cyclopentadienyl" moiety is an isostere for five-membered ring aromatic heterocycles.

H–( )–( )–( )–( )–( )–H, with units selected from:

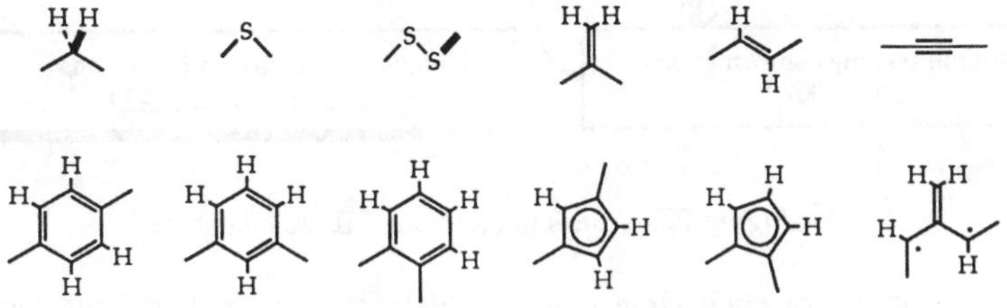

Hydrocarbon - heteroatom equivalence:

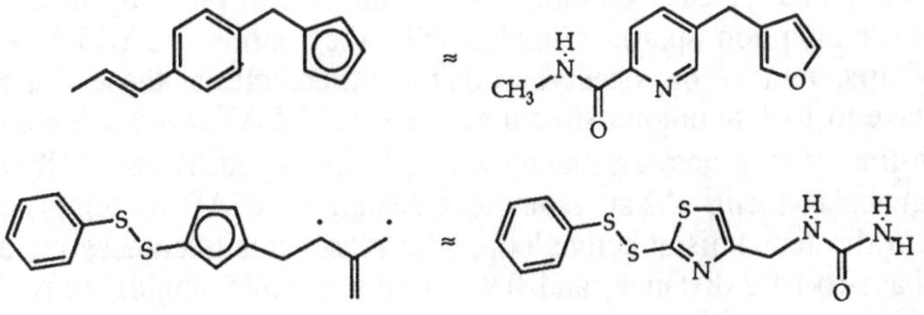

Figure 26. Components of the structures in the ILIAD database and rationale for equivalencies.

In contrast to the TRIAD database, we clearly needed to take conformational heterogeneity into account, so several minimized conformations for each sequence were incorporated. For this database as well, we avoided redundancy by keeping track of both sequences (if A-B-C-D-E, then not E-D-C-B-A), and of conformation as the starting molecules were assembled. A full set of conformations was generated for each molecule in 90° or 60° increments around each of the linking bonds, as appropriate for the hybridizations of the atoms involved. The following process was then followed to produce a minimized set of unique conformations: Self-overlapping starting conformations were eliminated at the outset. Second, every remaining conformation was minimized with MacroModel BatchMin. Third, the minimum-energy conformations were clustered with the help of the CLASS routine in CAVEAT,[43] to identify molecules that had converged to the same conformation. Finally, all unique conformers for each molecule within 10 kcal/mole of the minimum energy conformation were kept in the database. The population of these datasets throughout the process is shown in Figure 27. The final database is comprised of ca. 110,000 unique conformations of 36,000 different molecules.

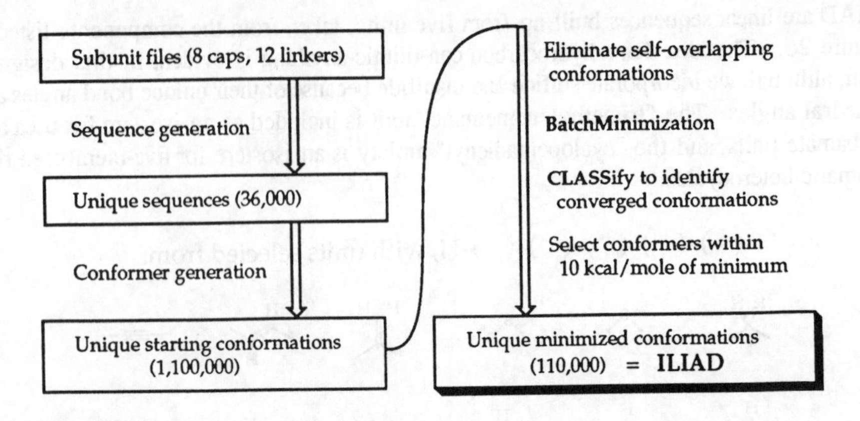

Figure 27. Steps in creation of ILIAD database.

The availability of such large structural databases raises two important issues. How long does it take to search them, and what do you do with the multitude of hits that are found? We have addressed both of these questions in recent developments in CAVEAT. The program is sufficiently fast that even searches in TRIAD can be carried out in real time on a typical SGI Indigo workstation ("real time" for us is a few minutes; I have a short attention span). Coupled with the routine CLASS,[43] which can take hundreds of hits, remove the ones you don't want, and cluster those that remain so that you only have to look at unique structural types, CAVEAT searches are so fast that they are done on-line, giving answers and providing follow-up ideas essentially at the rate that the user can absorb them. As an example, a search in TRIAD for templates for the three sidechains in the tendamistat active loop, with reasonable tolerances around each vector (0.3Å in base-to-base distance, and 0.2 radians in solid angle), provides six unique tricyclic ring systems in an elapsed time of about two minutes (that's clock, not CPU time...; see Figure 28).

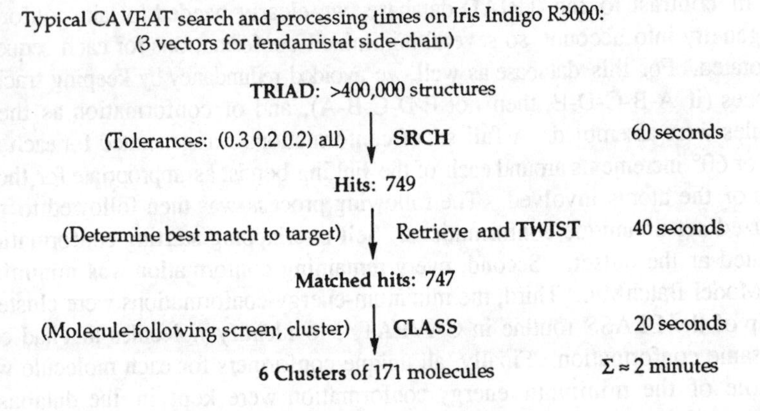

Figure 28. Elapsed time for steps in a typical CAVEAT cycle: search, structure retrieval and overlay, and CLASSification.

The generation of these custom databases is becoming routine, and with disk storage ever less expensive, we can contemplate maintaining a variety of databases for different problems. We may then approach a solution to the diversity/completeness dichotomy, with a diverse set of "complete" databases, adapted to specific problem types. We can even imagine using database construction and computational screening as an element in the "design" of combinatorial libraries,[68,69] especially when favorable structural motifs can be identified;[70] in preliminary work, for example, we have created a comprehensive database of benzodiazepines.[71,72] A real advantage of such databases is the assurance that the molecules identified in a search can be synthesized!

Acknowledgments: The work described above has been supported in its various aspects by grants from the National Institutes of General Medical Sciences (GM-28965-12, GM-30759-19, and GM-46627-16) and by unrestricted gifts. I also want to thank my coworkers who have contributed to these projects. In addition to those whose published work is cited, I want to mention Mark Giangiordano, Mark Hediger, Chris Seto, Uday Maitra, Hyung-Jung Pyun, Brian Cho, Greg Weiss, Marisa Kozlowski, and Georges Lauri. My group has always appreciated and benefitted from our collaborations with a number of structural biology groups; I want to thank in particular David Wemmer, Brian Matthews, William Lipscomb, Michael James, and William Stallings, and their colleagues.

1. Pauling, L. (1946) Molecular Architecture and Biological Reactions, *Chem. Eng. News*, **24**, 1375-1377.
2. Lienhard, G.E. (1972) Transition State Analogs as Enzyme Inhibitors, *Ann. Rep. Med. Chem.*, **7**, 249-258.
3. Wolfenden, R. (1969) Transition State Analogues for Enzyme Catalysis, *Nature*, **223**, 704-705.
4. Wolfenden, R. (1976) Transition State Analog Inhibitors and Enzyme Catalysis, *Annu. Rev. Biophys. Bioeng.*, **5**, 271-306.
5. Kaplan, A.P. and Bartlett, P.A. (1991) An Inhibitor of Carboxypeptidase A with a K_i Value in the Femtomolar Range, *Biochemistry*, **30**, 8165-8170.
6. Bartlett, P.A. and Marlowe, C.K. (1983) Phosphonamidates as Transition State Analog Inhibitors of Thermolysin, *Biochemistry*, **22**, 4618-4624.
7. Morgan, B.P., Scholtz, J.M., Ballinger, M., Zipkin, I. and Bartlett, P.A. (1991) Differential Binding Energy: A Detailed Evaluation of the Influence of Hydrogen-Bonding and Hydrophobic Groups on the Inhibition of Thermolysin by Phosphorus-Containing Inhibitors, *J. Am. Chem. Soc.*, **113**, 297-307.
8. Tronrud, D.E., Holden, H.M. and Matthews, B.W. (1987) Structures of Two Thermolysin Inhibitor Complexes that Differ by a Single Hydrogen Bond, *Science*, **235**, 571-574.
9. Matthews, B.W. (1988) Structural Basis of the Action of Thermolysin and Related Zinc Peptidases, *Acct. Chem. Research*, **21**, 333-340.

10. Hanson, J.E., Kaplan, A.P. and Bartlett, P.A. (1989) Phosphonate Analogs of Carboxypeptidase A are Potent Transition State Analog Inhibitors, *Biochemistry*, **28**, 6294-6305.
11. Phillips, M.A., Fletterick, R. and Rutter, W.J. (1990) Arginine 127 Stabilizes the Transition State in Carboxypeptidase, *J. Biol. Chem.*, **265**, 20692-20698.
12. Christianson, D.W. and Lipscomb, W.N. (1989) Carboxypeptidase A, *Accts. Chem. Research*, **22**, 62-69.
13. Phillips, M.A., Kaplan, A.P., Rutter, W.J. and Bartlett, P.A. (1992) Transition State Characterization: A New Approach Combining Inhibitor Analogs and Variation in Enzyme Structure, *Biochemistry*, **31**, 959-962.
14. Christianson, D.W. and Lipscomb, W.N. (1988) Comparison of Carboxypeptidase A and Thermolysin: Inhibition by Phosphonamidates, *J. Am. Chem. Soc.*, **110**, 5560-5565.
15. Fraser, M.E., Strynadka, N.C.J., Bartlett, P.A., Hanson, J.E. and James, M.N.G. (1992) Crystallographic Analysis of Transition State Mimics Bound to Penicillopepsin: Phosphorus-Containing Peptide Analogues, *Biochemistry*, **31**, 5201-5214.
16. Sachdev, G.P. and Fruton, J.S. (1970) Secondary Enzyme-Substrate Interactions and the Specificity of Pepsin, *Biochemistry*, **9**, 4465-4470.
17. Fruton, J.S. (1976) The Mechanism of the Catalytic Action of Pepsin and Related Acid Proteinases, *Adv. Enzymol. Relat. Areas Mol. Biol.*, **44**, 1-36.
18. Haslam, E. (1993) *Shikimic Acid Metabolism and Metabolites*, John Wiley & Sons, New York.
19. Bartlett, P.A., McLaren, K.L., Alberg, D.G., Fässler, A., Nyfeler, R., Lauhon, C.T. and Grissom, C.B. (1989) Exploration of the Shikimic Acid Pathway: Opportunities for the Study of Enzyme Mechanisms Through the Synthesis of Intermediates and Inhibitors, in L.G. Copping (ed), *Prospects for Amino Acid Biosynthesis Inhibitors in Crop Protection and Pharmaceutical Chemistry*, Society of Chemical Industry, pp. 155-170.
20. Gray, J.V., Eren, D. and Knowles, J.R. (1990) Monofunctional Chorismate Mutase from Bacillus subtilis: Kinetic and ^{13}C NMR Studies on the Interactions of the Enzyme with Its Ligands, *Biochemistry*, **29**, 8872-8878.
21. Andrews, P.R., Cain, E.N., Rizzardo, E. and Smith, G.D. (1977) Rearrangement of Chorismate to Prephenate. Use of Chorismate Mutase Inhibitors to Define the Transition State Structure, *Biochemistry*, **16**, 4848-4852.
22. Bartlett, P.A., Nakagawa, Y., Johnson, C.R., Reich, S. and Luis, A. (1988) Chorismate Mutase Inhibitors: Synthesis and Evaluation of Some Potential Transition State Analogs, *J. Org. Chem.*, **53**, 3195-3210.
23. Gajewski, J.J., Jurayj, J., Kimbrough, D.R., Gande, M.E., Ganem, B. and Carpenter, B.K. (1987) On the Mechanism of Rearrangement of Chorismic Acid and Related Compounds, *J. Am. Chem. Soc.*, **109**, 1170-1186.
24. Guilford, W.J., Copley, S.D. and Knowles, J.R. (1987) On the Mechanism of the Chorismate Mutase Reaction, *J. Am. Chem. Soc.*, **109**, 5013-5019.
25. Jackson, D.Y., Jacobs, J.W., Sugasawara, R., Reich, S.H., Bartlett, P.A. and Schultz, P.G. (1988) An Antibody-Catalyzed Claisen Rearrangement, *J. Am. Chem. Soc.*, **110**, 4841-4842.

26. Hilvert, D., Carpenter, S.H., Nared, K.D. and Auditor, M.-T.M. (1988) Catalysis of Concerted Reaction by Antibodies: The Claisen Rearrangement, *Proc. Natl. Acad. Sci. U.S.A.*, **85**, 4953-4955.
27. Hilvert, D. and Nared, K.D. (1988) Stereospecific Claisen Rearrangement Catalyzed by an Antibody, *J. Am. Chem. Soc.*, **110**, 5593-5594.
28. Chook, Y.M., Ke, H. and Lipscomb, W.N. (1993) Crystal Structures of the Monofunctional Chorismate Mutase from *Bacillus subtilis* and its Complex with a Transition State Analog, *Proc. Nat. Acad. Sci. USA*, **90**, 8600-8603.
29. Haynes, M.R., Sutra, E.A., Hilvert, D. and Wilson, I.A. (1994) Routes to Catalysis: Structure of a Catalytic Antibody and Comparison with Its Natural Counterpart, *Science*, **263**, 646-652.
30. Steinrücken, H.C. and Amrhein, N. (1984) 5-Enolpyruvylshikimate-3-phosphate Synthase of *Klebsiella pneumoniae*. 2. Inhibition by Glyphosate [N-(phosphonomethyl)glycine], *Eur. J. Biochem.*, **143**, 351-357.
31. Bondinell, W.E., Vnek, J., Knowles, P.F., Sprecher, M. and Sprinson, D.B. (1971) On the Mechanism of 5-Enolpyruvylshikimate 3-Phosphate Synthetase, *J. Biol. Chem.*, **246**, 6191-6196.
32. Anderson, K.A., Sikorski, J.A., Benesi, A.J. and Johnson, K.A. (1988) Isolation and Structural Elucidation of the Tetrahedral Intermediate in the EPSP Synthase Enzymatic Pathway, *J. Am. Chem. Soc.*, **110**, 6577-6579.
33. Sikorski, J.A., personal communication.
34. Alberg, D.G. and Bartlett, P.A. (1989) Potent Inhibition of 5-Enolpyruvylshikimate-3-phosphate Synthase by a Reaction Intermediate Analog, *J. Am. Chem. Soc.*, **111**, 2337-2338.
35. Alberg, D.G., Lauhon, C.T., Nyfeler, R., Fässler, A. and Bartlett, P.A. (1992) Inhibition of EPSP Synthase by Analogs of the Tetrahedral Intermediate and of EPSP, *J. Am. Chem. Soc.*, **114**, 3535-3546.
36. Walker, M.C., Jones, C.J., Somerville, R.L. and Sikorski, J.A. (1992) (Z)-3-Fluorophosphoenolpyruvate as a pseudosubstrate of EPSP synthase: enzymatic synthesis of a stable fluoro analog of the catalytic intermediate, *J. Am. Chem. Soc.*, **114**, 7601-7603.
37. Seto, C.T. and Bartlett, P.A., unpublished results.
38. Maitra, U. and Bartlett, P.A., unpublished results.
39. Anderson, K.A. and Johnson, K.A. (1990) "Kinetic Competence" of the 5-Enolpyruvoyl Shikimate-3-phosphate Synthase Tetrahedral Intermediate, *J. Biol. Chem.*, **265**, 5567-5572.
40. Stallings, W., personal communication.
41. Holden, H.M., Tronrud, D.E., Monzingo, A.F., Weaver, L.H. and Matthews, B.W. (1987) Slow- and Fast-Binding Inhibitors of Thermolysin Display Different Modes of Binding: Crystallographic Analysis of Extended Phosphonamidate Transition-State Analogues, *Biochemistry*, **26**, 8542-8553.
42. Bartlett, P.A., Shea, G.T., Telfer, S.J. and Waterman, S. (1989) CAVEAT: A Program to Facilitate the Structure-derived Design of Biologically Active Molecules, in S.M. Roberts (ed), *Molecular Recognition: Chemical and Biological Problems*, Royal Society of Chemistry, pp. 182-196.

43. Lauri, G. and Bartlett, P.A. (1994) CAVEAT: A Program to Facilitate the Design of Organic Molecules, *J. Comp. Aided Mol. Design*, **8**, 51-66.
44. Bernstein, F.C., Koetzle, T.F., Williams, G.J.B., Meyer, E.F., Jr., Brice, M.D., Rogers, J.R., Kennard, O., Shimanouchi, T. and Tasumi, M. (1977) The Protein Data Bank: A Computer-Based Archival File for Macromolecular Structures, *J. Mol. Biol.*, **112**, 535-542.
45. Morgan, B.P., Bartlett, P.A., Holland, D.R. and Matthews, B.W. (1994) Structure-Based Design of an Inhibitor of the Zinc Peptidase Thermolysin, *J. Am. Chem. Soc.*, **116**, 3251-3260.
46. Pyun, H.-J. and Bartlett, P.A., unpublished results.
47. Vértesy, L., Oeding, V., Bender, R., Zepf, K. and Nesemann, G. (1984) Tendamistat (HOE 467), a tight-binding α-amylase inhibitor from *Streptomyces tendae* 4158, *Eur. J. Biochem.*, **141**, 505-512.
48. Pflugrath, J.W., Wiegand, G., Huber, R. and Vértesy, L. (1986) Crystal Structure Determination, Refinement and the Molecular Model of the α-Amylase Inhibitor Hoe-467A, *J. Mol. Biol.*, **189**, 383-386.
49. Kline, A.D., Braun, W. and Wüthrich, K. (1986) Studies by ^{1}H Nuclear Magnetic Resonance and Distance Geometry of the Solution Conformation of the α-Amylase Inhibitor Tendamistat, *J. Mol. Biol.*, **189**, 367-382.
50. Billeter, M., Kline, A.D., Braun, W., Huber, R. and Wüthrich, K. (1986) Comparison of the High-resolution Structures of the α-Amylase Inhibitor Tendamistat Determined by Nuclear Magnetic Resonance in Solution and by X-ray Diffraction in Single Crystals, *J. Mol. Biol.*, **206**, 677-687.
51. Huber, R., personal communication.
52. Laszlo, E., Hollo, J., Hoschke, A. and Sarosi, G. (1978) The active center of amylolytic enzymes. I. A study by means of lactone inhibition of the role of "half-chair" glycosyl conformation at the active center of amylolytic enzymes, *Carbohydr. Res.*, **61**, 387-394.
53. Kessler, H., Bats, J.W., Griesinger, C., Koll, S., Will, M. and Wagner, K. (1988) Peptide Conformations. 46. Conformational Analysis of a Superpotent Cytoprotective Cyclic Somatostatin Analogue, *J. Am. Chem. Soc.*, **110**, 1033-1049.
54. Etzkorn, F.A., Guo, T., Lipton, M.A., Goldberg, S.D. and Bartlett, P.A. (1994) Cyclic Hexapeptides and Chimeric Peptides as Mimics of Tendamistat, *J. Am. Chem. Soc.*, submitted for publication.
55. Bax, A. and Subramanian, S. (1986) Sensitivity-Enhanced Two-Dimensional Heteronuclear Shift Correlation NMR Spectroscopy, *J. Mag. Res.*, **67**, 565-569.
56. Bax, A. (1988) Correction of Cross-Peak Intensities in 2D Spin-Locked NOE Spectroscopy for Offset and Hartmann-Hahn Effects, *J. Mag. Res.*, **77**, 134-147.
57. Bothner-By, A.A., Stephens, R.L., Lee, J., Warren, C.D. and Jeanloz, R.W. (1984) Structure Determination of a Tetrasaccharide: Transient Nuclear Overhauser Effects in the Rotating Frame, *J. Am. Chem. Soc.*, **106**, 811-813.
58. Dorman, D.E. and Bovey, F.A. (1973) Carbon-13 Magnetic Resonance Spectroscopy. The Spectrum of Proline in Oligopeptides, *J. Org. Chem.*, **38**, 2379-2383.
59. Kessler, H. (1982) Conformation and Biological Activity of Cyclic Peptides, *Angew. Chem., Int. Ed. Engl.*, **21**, 512-523.

60. Kessler, H., Griesinger, C., Lautz, J., Muller, A., van Gunsteren, W.F. and Berendsen, H.J. (1988) Conformational Dynamics Detected by Nuclear Magnetic Resonance NOE Values and J Coupling Constants, *J. Am. Chem. Soc.*, **110**, 3393-3396.
61. Blaney, J. and Crippen, G., DGEOM (available from QCPE, University of Indiana, Dept. of Chemistry, Bloomington, IN).
62. Dspace, Dspace v1.0 (available from Hare Research Inc., Woodinville, WA).
63. Lipton, M. and Still, W.C. (1988) The Multiple Minimum Problem in Molecular Modeling. Tree Searching Internal Coordinate Conformational Space, *J. Comp. Chem.*, **9**, 343-355.
64. Mohamadi, F., Richards, N.G.J., Guida, W.C., Liskamp, R., Lipton, M., Caufield, C., Chang, G., Hendrickson, T. and Still, W.C. (1990) MacroModel—An Integrated Software System for Modeling Organic and Bioorganic Molecules Using Molecular Mechanics, *J. Comp. Chem.*, **11**, 440-467.
65. Peishoff, C.E., Dixon, J.S. and Kopple, K.D. (1990) Application of the Distance Geometry Algorithm to Cyclic Oligopeptide Conformation Searches, *Biopolymers*, **30**, 45-56.
66. Blaney, J., COMPARE, unpublished results.
67. Lauri, G., unpublished results.
68. Ohlmeyer, M.H.J., Swanson, R.N., Dillard, L.W., Reader, J.C., Asouline, G., Kobayashi, R., Wigler, M. and Still, W.C. (1993) Complex Synthetic Chemical Libraries Indexed with Molecular Tags, *Proc. Nat. Acad. Sci. USA*, **90**, 10922-10926.
69. Borchardt, A. and Still, W.C. (1994) Synthetic Receptor Binding Elucidated With an Encoded Combinatorial Library, *J. Am. Chem. Soc.*, **116**, 373-374.
70. Simon, R.J., Kania, R.S., Zuckermann, R.N., Huebner, V.D., Jewell, D.A., Banville, S., Ng, S., Wang, L., Rosenberg, S., Marlowe, C.K., Spellmeyer, D.C., Tan, R., Frankel, A.D., Santi, D.V., Cohen, F.E. and Bartlett, P.A. (1992) Peptoids: A Modular Approach to Drug Discovery, *Proc. Nat. Acad. Sci. USA*, **89**, 9367-9371.
71. Bunin, B.A. and Ellman, J.A. (1992) A General and Expedient Method for the Solid-Phase Synthesis of 1,4-Benzodiazepine Derivatives, *J. Amer. Chem. Soc.*, **114**, 10646-10647.
72. DeWitt, S.H., Kiely, J.S., Stankovic, C.J., Schroeder, M.C., Reynolds Cody, D.M. and Pavia, M.R. (1993) "Diversomers": An approach to Nonpeptide, Nonoligomeric Chemical Diversity, *Proc. Nat. Acad. Sci. USA*, **90**, 6909-6913.

ENANTIOSELECTIVE CATALYSIS WITH TRANSITION METAL COMPLEXES

HENRI BRUNNER
Universität Regensburg
Institut für Anorganische Chemie
D-93040 Regensburg (Germany)

Metabolism in man, animal and plant uses optically active compounds, such as L amino acids, D sugars etc., and not racemic mixtures. As a consequence, biology should be approached with optically pure compounds of the correct configuration and not with racemic mixtures. This refers to additives to human food, supplementation of animal food, drugs and agrochemicals, an impressive list of applications which demonstrates the importance of optically active compounds. Therefore, the synthesis of optically active compounds is a challenging problem for the chemical and pharmaceutical industry.

Enzymes are optimized catalysts for the synthesis of the optically active compounds needed in nature. To-day enzymes are also extensively used for the synthesis of optically active compounds needed in industry. However, for substrates which differ from the enzymes' natural substrates usually a drop in reactivity and enantioselectivity is observed. In addition, there are many reactions which are not accessible to enzyme catalysis or in which enzymes are decomposed.

Instead of the natural enzymes also synthetic compounds can be used as enantioselective catalysts. At the beginning of the century this was already

C. Chatgilialoglu and V. Snieckus (eds.), Chemical Synthesis, 175–190.

demonstrated for the cinchona alkaloids. To-day there are many successful examples in organic chemistry. Recently, optically active transition metal catalysts were shown to have a high potential for enantioselective catalysis. The state of the art of this field will be reviewed here in four chapters. The first chapter "Enantioselective Hydrogenation of Dehydroamino Acids" will set the stage. Some historical comments and some examples together with the explanation of the term *in situ* catalyst will serve as a kind of introduction. The next three chapters present the scope of the field. In the chapter "Enantioselective Homo Diels-Alder Reactions" a new reaction type of norbornadiene with monosubstituted acetylenes will be discussed, for which recently high enantioselectivities have been achieved. In the chapter "Stereoselective Hydrogenation of Folic Acid" the target is an important and very sensitive biomolecule, which was not accessible to asymmetric catalysis before. In the last chapter "Optically Active Barbiturates by Allylic Alkylation" it will be shown that this reaction type at present is an unsolved problem as far as the formation of new stereogenic centers at the attacking nucleophile is concerned. An approach to the solution of this problem includes the design of new expanded optically active phosphine ligands.

ENANTIOSELECTIVE HYDROGENATION OF DEHYDROAMINO ACIDS

The hydrogenation of (Z)-α-N-acetamidocinnamic acid to give N-acetylphenylalanine is shown in Scheme 1, top. Rh compounds of the Wilkinson-type catalyze this reaction. With triphenylphosphine as a ligand, the product is formed as a racemic mixture in an achiral catalysis. If triphenylphosphine is replaced by an optically active phosphine, N-acetylphenylalanine is obtained in optically active form in an enantioselective catalysis. In 1968, pioneering studies used optically active phosphines of the Horner-type, such as PMePrPh as ligands [1,2]. Optical inductions, however, remained low with catalysts containing these monodentate phosphines.

Scheme 1

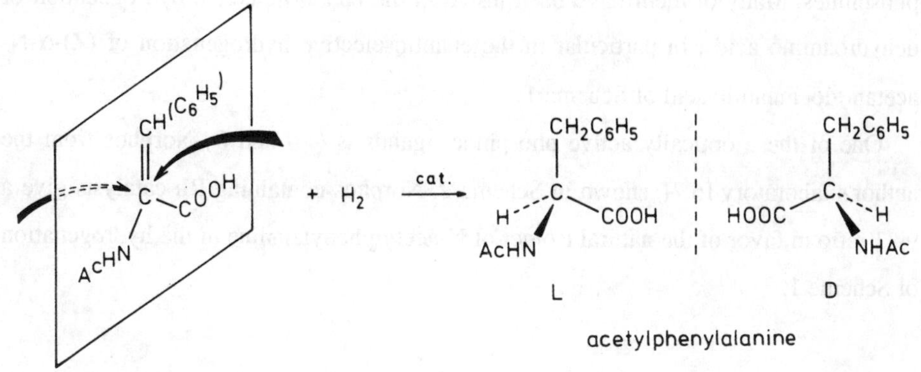

[Rh(cod)Cl]$_2$ / PPhPrMe low ee
[Rh(cod)Cl]$_2$ / Diop 81 % ee

[Rh(cod)Cl]$_2$ / PPhPrMe	low ee
[Rh(cod)Cl]$_2$ / Diop	81 % ee

cat.

L D

acetylphenylalanine

(+) tartaric acid ester

(−) Diop

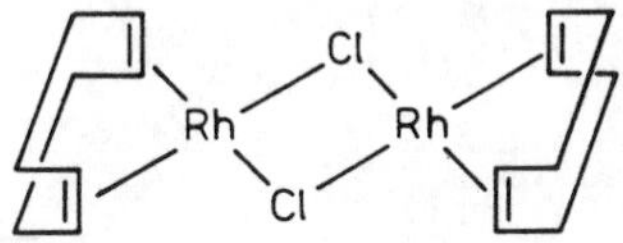

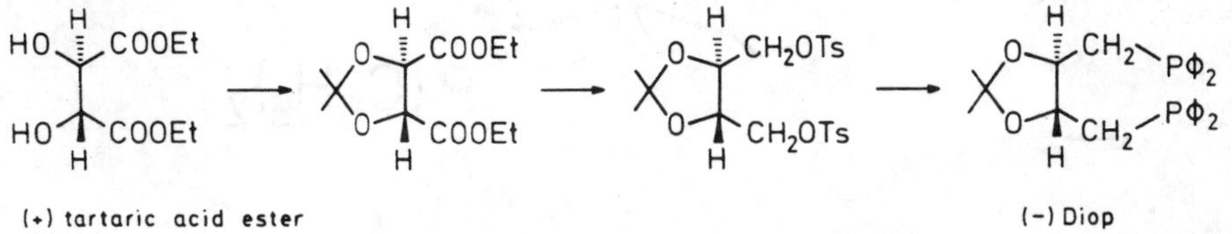

A breakthrough in enantioselective hydrogenation was the design and synthesis of the optically active ligand Diop, accessible from natural tartaric acid by acetalization of the hydroxy groups, reduction of the ester groups, tosylation of the corresponding alcohol and diphenylphosphination, as outlined at the bottom of Scheme 1 [3,4]. Contrary to the monodentate optically active ligand PMePrPh, the bidentate optically active ligand Diop gave high optical inductions of 81 % ee, when used in Rh containing catalysts in the hydrogenation reaction of Scheme 1.

The hydrogenation of dehydroamino acids, such as (Z)-α-N-acetamidocinnamic acid, is a standard reaction in enantioselective catalysis with transition metal complexes since 25 years. Its scope is covered by many reviews, the last of which is the comprehensive data collection in the two-volume "Handbook of Enantioselective Catalysis with Transition Metal Compounds" of the author [5]. This data collection comprises more than 2 000 optically active ligands, most of which optically active phosphines. Many of them have been tested in the enantioselective hydrogenation of dehydroamino acids, in particular in the enantioselective hydrogenation of (Z)-α-N-acetamidocinnamic acid of Scheme 1.

One of these optically active phosphine ligands is (+)- and (-)-Norphos from the author's laboratory [6,7], shown in Scheme 2. Norphos containing Rh catalysts give a 99:1 ratio in favor of the natural isomer of N-acetylphenylalanine in the hydrogenation of Scheme 1.

Scheme 2

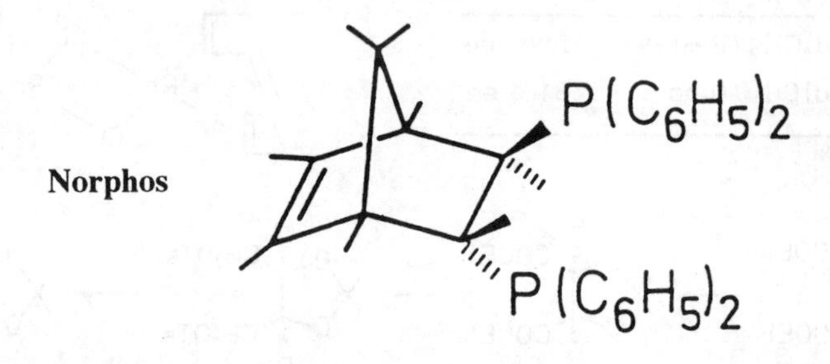

What has been described for Norphos containing Rh catalysts is not unique but typical for the situation at present. A variety of optically active chelating phosphines developed in the last 20 years allows the synthesis of natural and unnatural amino acids by hydrogenation of the corresponding dehydroamino acid derivatives. The hydrogenation of dehydroamino acid derivatives is one of the most successful reaction types in enantioselective catalysis with transition metal compounds [5].

These hydrogenations can be carried out with a catalyst, such as [Rh(cod)P-P]PF_6, in which cod = 1,5-cyclooctadiene and P-P = an optically active chelating phosphine. An alternative is the use of a so-called *in situ* catalyst. Such an *in situ* catalyst consists of two stable precursors, the procatalyst and the cocatalyst (Scheme 3).

Scheme 3

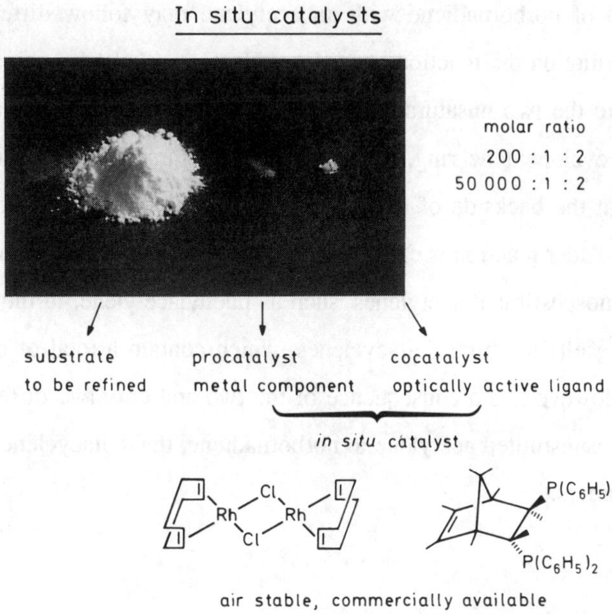

In situ catalysts

molar ratio
200 : 1 : 2
50 000 : 1 : 2

substrate to be refined
procatalyst metal component
cocatalyst optically active ligand

in situ catalyst

Rh Cl Cl Rh
$P(C_6H_5)_2$
$P(C_6H_5)_2$

air stable, commercially available

The procatalyst is the metal component and the cocatalyst is the optically active ligand. Usually, these two components are air stable commercial compounds. At the bottom of Scheme 3 the formulas of the two components of the *in situ* catalyst $[Rh(cod)Cl]_2$/Norphos are shown. In solution these two components combine to give the active enantioselective catalyst. The advantage of an *in situ* catalyst is that no catalyst synthesis is required prior to the reaction which has to be carried out.

The insert in Scheme 3 shows the big pile of the prochiral substrate to be refined in the enantioselective hydrogenation and the small amounts of procatalyst and cocatalyst necessary for this transformation. The relationship demonstrates the elegancy of the approach using an *in situ* catalyst. In laboratory hydrogenations usually molar ratios substrate: procatalyst:cocatalyst (unidentate) of 200:1:2 are applied. On an industrial scale these can be increased up to 50000:1:2 emphasizing the aspect of multiplication of chirality from a small amount of catalyst into a large amount of product.

ENANTIOSELECTIVE HOMO DIELS-ALDER REACTIONS

The reaction of norbornadiene with an acetylene may follow different reaction channels, depending on the reaction conditions. Using a cobalt/phosphine catalyst, an acetylene adds to the two unsaturated front carbon atoms of norbornadiene, giving a five-membered cyclopentene ring. Simultaneously, a three-membered cyclopropane ring is formed at the backside of the molecule. The polycyclic skeleton obtained in this homo Diels-Alder reaction is called the deltacyclene skeleton (Scheme 4).

Use of a monosubstituted acetylenes, such as phenylacetylene, in the homo Diels-Alder reaction results in chiral deltacyclenes, which contain a total of 6 asymmetric carbon atoms. However, as a consequence of the two and only two different addition modes of a monosubstituted acetylene to norbornadiene, the deltacyclene is formed as a pair of enantiomers.

Scheme 4

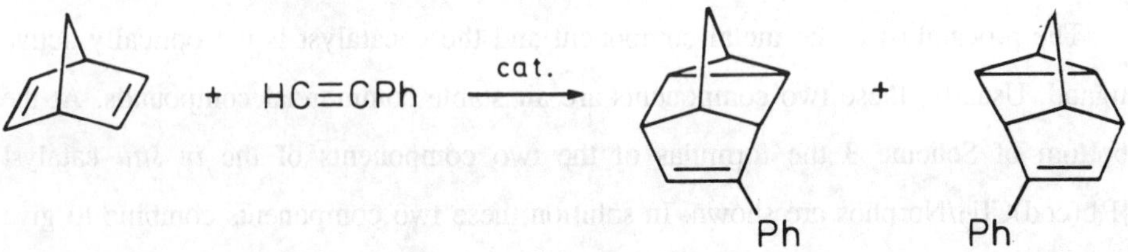

+ HC≡CPh —cat.→ Ph + Ph

cat.: $Co(acac)_3 / PPh_3 / Et_2AlCl$

The reaction in Scheme 4 is catalyzed by an *in situ* catalyst consisting of the procatalyst $Co(acac)_3$ and the cocatalyst triphenylphosphine. Additionally, diethylaluminium chloride is required as a reducing agent to reduce cobalt(III) probably to cobalt(I), which in combination with triphenylphosphine is the active catalyst.

Replacement of the achiral triphenylphosphine by an optically active phosphine renders the reaction enantioselective (Scheme 5). An efficient stereocontrol is achieved with the chelating phosphine Norphos. Phenyldeltacyclene is obtained in an enantiomer ratio of 99.2:0.8. Catalyst quantities of as little as 0.2 mol% are sufficient to yield a quantitative chemical yield after 4 hours reaction time in THF solution at 35 C [8,9].

Scheme 5

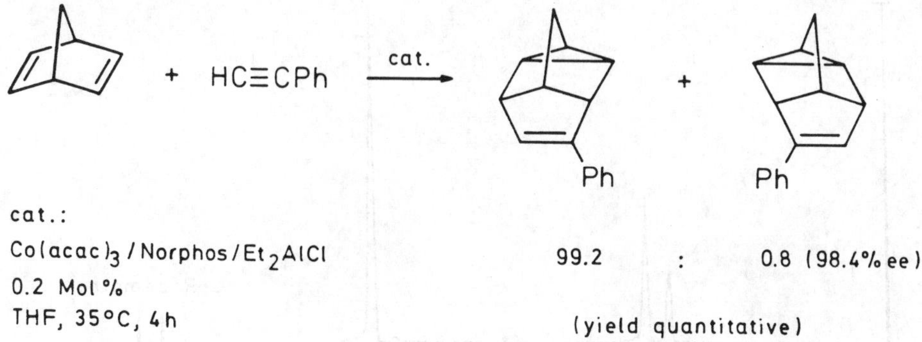

+ HC≡CPh —cat.→ (Ph-substituted deltacyclene) + (Ph-substituted deltacyclene)

cat.:
$Co(acac)_3$ / Norphos / Et_2AlCl
0.2 Mol %
THF, 35°C, 4 h

99.2 : 0.8 (98.4% ee)

(yield quantitative)

The deltacylene formation can be extended from phenylacetylene to other monosubstituted acetylenes. Hexyne-1 yields the *n*-butyl substituted deltacyclene in a 99:1 enantiomer ratio and a quantitative chemical yield with an *in situ* catalyst $Co(acac_3)$/Norphos/Et_2AlCl (Scheme 6). The enantiomer analysis is achieved by gaschromatography using a perpentylated ß-cyclodextrin column (Scheme 6, bottom). The lower trace is obtained with the racemic mixture, the enantiomers being base line separated. The inserts show the 30:70 and 1:99 enantiomer ratios obtained with the optically active ligands Diop and Norphos, respectively [8,9].

In order to explain the high stereoselectivity in the deltacyclene formation in Schemes 5 and 6 a model is used (Scheme 7), which is also successful in predicting the correct product configurations in the hydrogenation of dehydroamino acids [10]. On the left side the chelate skeleton of the ligand Chiraphos is depicted, which also

Scheme 6

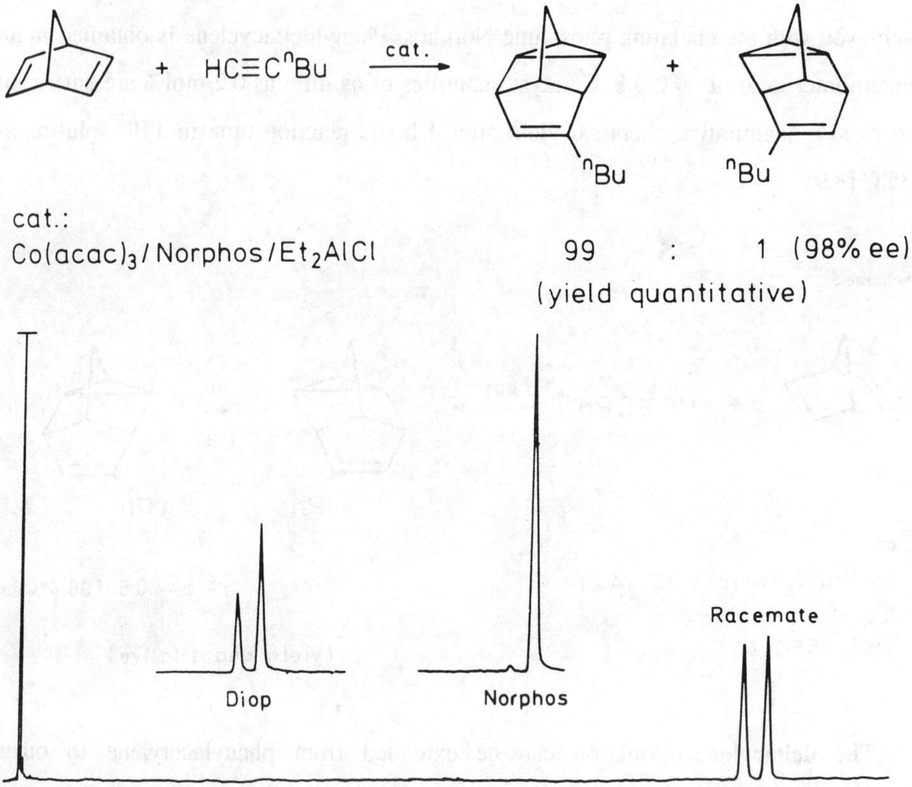

\+ $HC\equiv C^{n}Bu$ —cat.→ ^{n}Bu + ^{n}Bu

cat.:
$Co(acac)_3$/Norphos/Et_2AlCl 99 : 1 (98% ee)
(yield quantitative)

Racemate

Diop

Norphos

induces a high enantiomeric excess in the synthesis of deltacyclenes [9,11]. The two P atoms lie in the plane of the square containing the metal atom, omitted for the sake of clearity, which is perpendicular to the plane of the paper. The stereochemical control derives from the two asymmetric carbon atoms of given chirality above and below this plane, respectively. The puckering of the P-CHMe-CHMe-P skeleton is controlled by the configuration of these two asymmetric carbon atoms, because the large methyl substituents tend to point outwards and the small hydrogen substituents tend to be in axial position. The resulting puckering of the five-membered chelate ring renders the two phenyl substituents at the P atoms different, one being equatorial and one being axial, respectively.

Scheme 7

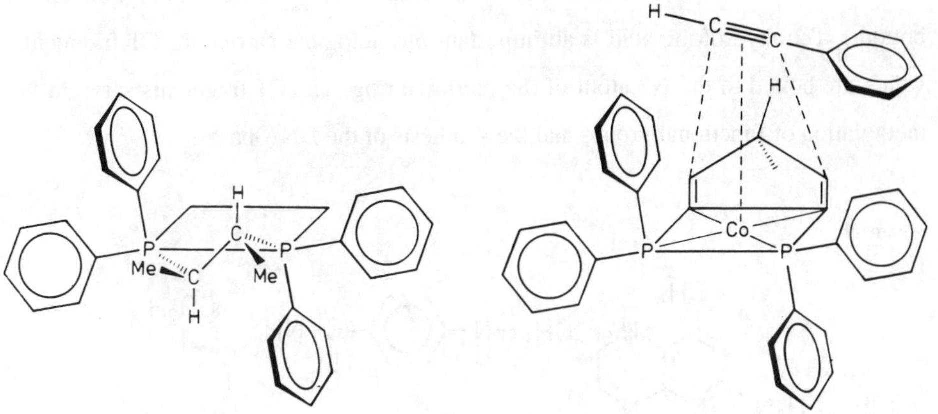

On the right side of Scheme 7 only the equatorial and axial phenyl groups of the chelating phosphine are depicted without the ligand backbone. In addition, on the rear of the cobalt atom half of the norbornadiene substrate is visible. Orientation of the phenyl acetylene in the axial position with respect to the Co atom such that the two C-C bonds to the front carbon atoms of norbornadiene can be formed (dashed lines), makes it obvious, that there is a favored conformation with the large acetylene phenyl substituent on the side of the equatorial phosphine phenyl and the small acetylene hydrogen on the side of the axial phosphine phenyl (Scheme 7). This favored conformation gives rise to the deltacyclene enantiomer formed in 99 % enantiomeric excess. Rotation of the acetylene ligand by 180° results in the unfavorable conformation, from which the other enantiomer is formed in only negligible amounts.

STEREOSELECTIVE HYDROGENATION OF FOLIC ACID

The vitamin folic acid (Scheme 8, top) is an essential food constituent. In the body folic acid is enzymatically reduced to tetrahydrofolic acid. In this reduction the 1,4-diazadiene system of the pteridine ring acquires four reduction equivalents. In the 6-

position of the pteridine system a new asymmetric center is established. In the enzymatic reduction it is exclusively formed in the 6S configuration (Scheme 8, bottom). Tetrahydrofolic acid is an important physiological carrier for C1 fragments, which are bound to the N5 atom of the pteridine ring. The C1 fragments serve in the methylation of functional groups and the synthesis of the DNA-bases.

Scheme 8

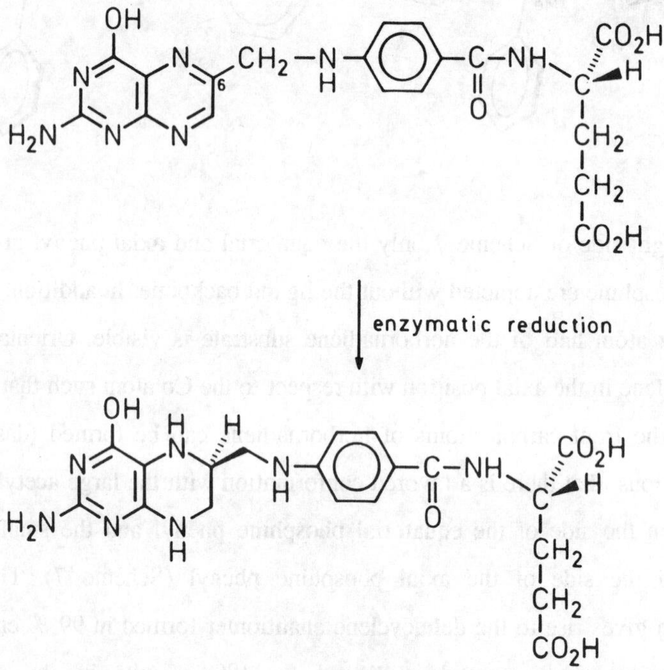

enzymatic reduction

In the therapy of osteosarcomas high doses of methotrexate are used to inhibit the enzyme dihydrofolate reductase resulting in a stop of the DNA synthesis. Fast growing tissues, such as cancers, are particularly hurt by this therapy. To maintain a minimum of metabolism, leucovorine must be administered as a rescue agent. Leucovorine is the Ca salt of 5-formyltetrahydrofolic acid, the formyl group of which is the C1 fragment active in metabolism. Leucovorine is also used together with 5-fluorouracil in the therapy of colon cancers and anemias.

Leucovorine is synthesized in the pharmaceutical industry by reduction of folic acid with $NaBH_4$ or by hydrogenation with Pt and Pd catalysts, respectively. In this reduction the S-configuration of the glutamic acid moiety does not induce an optical induction in the formation of the new stereogenic center in the 6-position of the pteridine system. Therefore, the two diastereomers, the natural 6S,S isomer and the unnatural 6R,S isomer, are obtained in a ratio of approximately 1:1. It is with this 1:1 mixture that medical doctors work even to-day although it is known that the unnatural 6R,S isomer is only slowly metabolized, enriched in the central nervous system and on the long run leads to intoxications. As the separation of the diastereomeric mixture 6S,S/6R,S = 1:1 is extremely difficult, a procedure to synthesize the natural 6S,S-leucovorine by stereoselective hydrogenation of folic acid at present is an important challenge.

There are no reports in the literature describing a successful stereoselective hydrogenation of folic acid. Results from C=N model systems cannot be transferred to folic acid, because folic acid and tetrahydrofolic acid are only soluble in water and not in the usual organic solvents. In addition, the hydrogenation product tetrahydrofolic acid is a biomolecule sensitive to air and acid or base treatment. Therefore, we consider our results concerning the stereoselective hydrogenation of folic acid a success.

We carry out the stereoselective hydrogenation of folic acid [12,13] using an immobilized catalyst prepared by mixing $[Rh(cod)Cl]_2$ and optically active ligands such as (-)-DIOP, (+)-DIOP and (-)-BPPM in methylene chloride. We add silica to the surface of which the catalyst impregnates. In this way, a heterogeneous catalyst is formed which is insoluble in water, the solvent used in the catalytic hydrogenation of folic acid. The best results obtained up to now after hydrogenation and formylation is a leucovorine with a diastereomer ratio of 6S,S/6R,S = 90:10 [13].

Analysis of the 6S,S and 6R,S diastereomers of leucovorine is carried out by HPLC using a BSA column, BSA = bovine serum albumine. Scheme 9 shows two HPLC diagrams, the first of a 6S,S:6R,S = 77:23 mixture and the second of a 6S,S:6R,S = 44:56 mixture.

Scheme 9

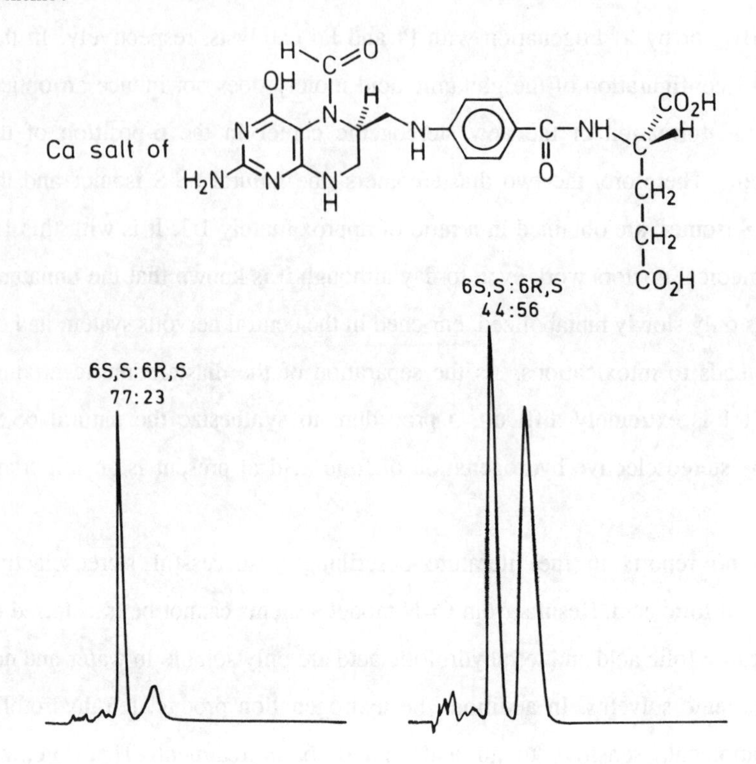

OPTICALLY ACTIVE BARBITURATES BY ALLYLIC ALKYLATION

Optically active barbiturates important in pharmacy, are obtained by classical synthesis and resolution procedures [14]. A new approach to the synthesis of optically active barbiturates is enantioselective catalysis with Pd complexes. An example is shown in Scheme 10 on the right side. The synthesis of the enantiomers of a N-methyl barbiturate containing a quaternary asymmetric carbon atom is achieved by allylation of the corresponding precursor with allyl acetate. Pd/triphenylphosphine complexes are efficient catalysts for this reaction which give a racemic mixture of the product. The reaction proceeds via a π-allyl complex of Pd as an intermediate (Scheme 10, left side). This π-allyl complex is attacked by the nucleophile, the anion of Nmethylbarbiturate, from the side opposite to the metal atom. Formation of the new C-C bond between the nucleophile and the terminal carbon atom of the allyl group establishes the new quaternary stereogenic center. Attempts to render this reaction enantioselective by using optically active phosphines, such as Diop and Norphos in the Pd catalyst were unsuccessful. Obviously, the chiral information in these conventional optically active phosphines is too far away from the position, where the new stereogenic center forms. Therefore, to induce enantioselectivity in syntheses as in Scheme 10 at present is an unsolved problem.

Scheme 10

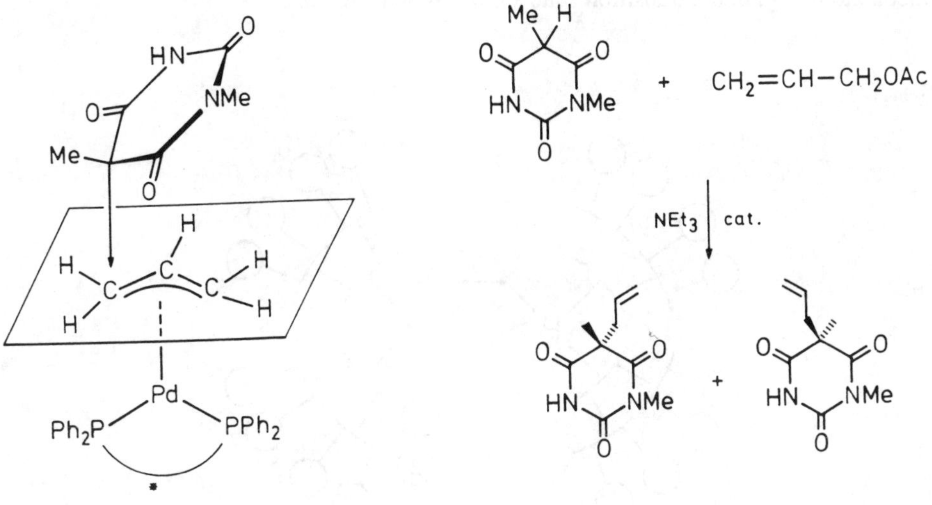

In the following discussion it will be shown how a solution to this problem is attempted by the design of new optically active multi-layer phosphine ligands. Ligands used up to now in enantioselective catalysis are of the conventional type, some of which have been discussed in the preceding chapters. With their phenyl substituent at the phosphorus atoms, these ligands extend only about 5-6 Å into space. Long distance effects are not possible with them. It is not surprising that in a transition state as shown in Scheme 10, the chiral information in conventional optically active ligands does not affect the formation of the new stereogenic center at the incoming nucleophile.

In this situation, the following concept for expanded phosphines was designed (Scheme 11). The center of the ligand consists of a strongly coordinating P-P skeleton. The two additional substituents at each phosphorus atom contain a first, second, third, etc. layer of repeating units, which also may be different from each other. These repeating units can be chiral or achiral. If they contain branching positions, the multi-layer ligand will be space-filling, as indicated in Scheme 11. The last layer is easily made up of optically active terminal groups. With such expanded space-filling ligands, it should be possible to control the formation of asymmetric centers far away from the metal atom, e.g. in the transition state shown in Scheme 10.

Scheme 11

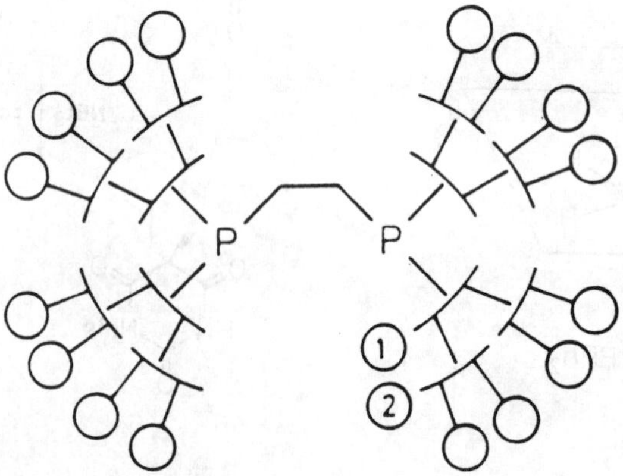

P P 1 2

In Scheme 12 the synthesis of an expanded optically active phosphine is shown. The preparation of the P substituents starts from 1-bromo-3,5-dimethylbenzene which is converted into the bis(bromomethyl)bromoxylene by NBS, depicted in the upper left corner of Scheme 12. Reaction with the Na derivative of menthol forms the ether bonds to the menthyl groups starting from the benzylic positions. In the reaction with BuLi, a metal halogen exchange with the bromine substituent of the aromatic ring is the next step. Addition of $Cl_2P\text{-}CH_2\text{-}CH_2\text{-}PCl_2$ completes the synthesis. The new expanded ligand has a strongly binding PCH_2CH_2P chelating frame and a first layer consisting of building blocks with branching positions. The terminal layer is made up of 8 menthyl substituents. It extends into space by 11-12 Å [15].

Scheme 12

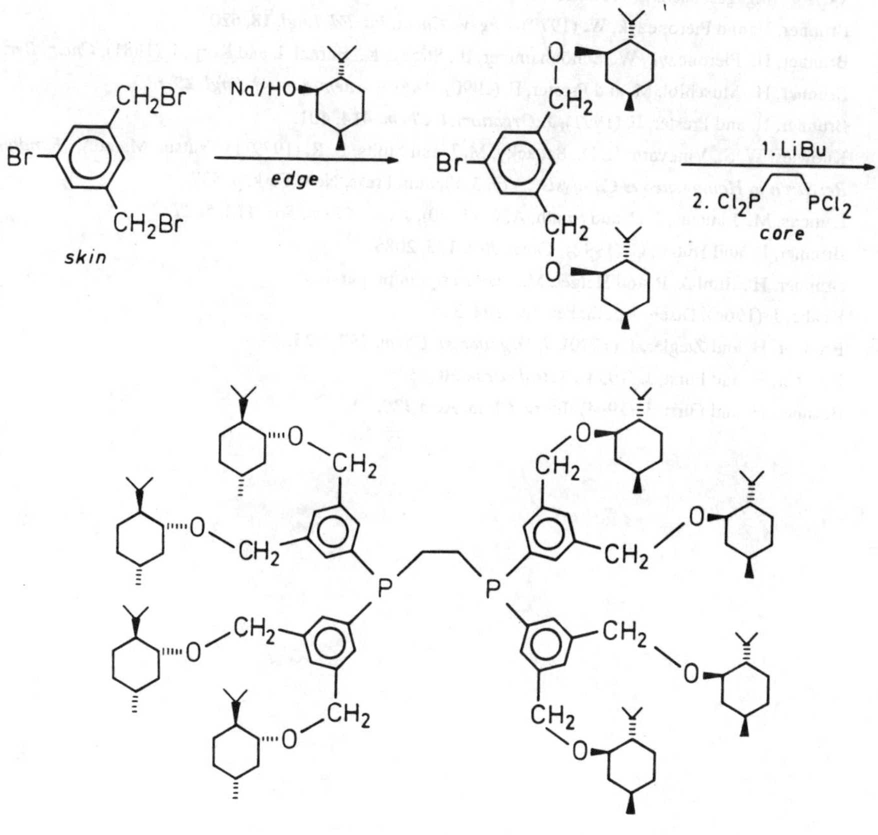

We synthesized a number of other expanded ligands [15-17]. With some of them we already observed enantioselectivities in the bartiturate allylation of Scheme 10 [17], whereas the conventional ligands Diop, Norphos, etc. definitely gave racemic mixtures in this reaction. We do hope that the concept of expanded optically active ligands will allow to solve long range problems in enantioselective catalysis which at the moment are still unsolved.

REFERENCES

1. Horner, L., Siegel, H. and Büthe, H. (1968), *Angew. Chem. Int. Ed. Engl.* **7**, 942.
2. Knowles, W. S., Sabacky, M., J. (1968), *Chem. Commun.* 1445.
3. Dang, T. P. and Kagan, H. B. (1971), *Chem. Commun.* 481.
4. Kagan, H. B. and Dang, T. P. (1972), *J. Am. Chem. Soc.* **94**, 6429.
5. Brunner, H. and Zettlmeier, W. (1993), *Handbook of Enantioselective Catalysis with Transition Metal Compou*
   VCH Verlagsgesellschaft, Weinheim.
6. Brunner, H. and Pieronczyk, W. (1979), *Angew. Chem. Int. Ed. Engl.* **18**, 620.
7. Brunner, H., Pieronczyk, W., Schönhammer, B., Streng, K., Bernal, I. and Korp, J. (1981), *Chem. Ber.* **114**, 113
8. Brunner, H., Muschiol, M. and Prester, F. (1990), *Angew. Chem. Int. Ed. Engl.* **29**, 653.
9. Brunner, H. and Prester, F. (1991), *J. Organomet. Chem.* **414**, 401.
10. Knowles, W. S., Vineyard, B. D., Sabacky, M. J. and Stults, B. R. (1979) in Tsutsui, M. (ed.), *"Fundamental Research in Homogeneous Catalysis"*, Vol. 3, Plenum Press, New York, p. 537.
11. Lautens, M., Lautens, J. C. and Smith, A. C. (1990), *J. Am. Chem. Soc.* **112**, 5627.
12. Brunner, H. and Huber, C. (1992), *Chem. Ber.* **125**, 2085.
13. Brunner, H., Bublak, P. and Helget, M., manuscript in preparation.
14. Knabe, J. (1964), Dtsch. Apotheker-Ztg. **104**, 315.
15. Brunner, H. and Ziegler, J. (1990), *J. Organomet. Chem.* **397**, C25.
16. Brunner, H. and Fürst, J. (1994), *Tetrahedron* **50**, 4303.
17. Brunner, H. and Fürst, J. (1994), Inorg. Chim. Acta **220**, 63.

DIRECTED AROMATIC METALATION: A CONTINUING EDUCATION IN FLATLAND CHEMISTRY

VICTOR SNIECKUS
The Guelph-Waterloo Centre for Graduate Work in Chemistry,
University of Waterloo,
Waterloo, Ontario,
N2L 3G1, Canada

Abstract

Over the past fifteen years, the Directed ortho Metalation (DoM) reaction has emerged as a significant synthetic concept in aromatic chemistry. This article delineates a) the impact of DoM strategies for the regiospecific construction of contiguously substituted aromatics and heteroaromatics using, whenever possible, selected comparisons with conventional methods; b) their increasing and striking utility in large-scale industrial operations; c) the recent evolution of the DoM - transition metal catalyzed cross coupling connection for effective biaryl C-C bond formation and its consequences in the synthesis of condensed aromatics; and d) the early stages of the Directed remote Metalation (DreM) tactic and its complementarity to electrophilic substitution. In this manner, the modern view from the flatlands will hopefully be appreciated.

1. Preamble

Kekuleís recently disputed and controversial dream [1] of the tail-and-hand monkeys provided the structural basis for testing reactivity concepts of the aromatic ring which led, initially by inspiring rationale, [2] and subsequently, by hard-core physical organic studies to the selectivity rules for electrophilic aromatic substitution [3]. This achievement allowed mechanistic rationalization of reactivity patterns of sparsely substituted benzenes and served as a triumph of logic in early instruction of undergraduate organic chemistry [4]. *Tempus non mutantur.* With the exception of the reactions of halonitrobenzenes with nucleophiles, electrophilic substitution occupies by far the largest section in chapters of currently used undergraduate text books [5]. Furthermore, this chemistry is considered so classical that it is given only sparse coverage in the pedagogy of chemical synthesis [6].

The synthesis of aromatics and heteroaromatics is an important endeavor which, as for any class of organic materials, thrives on innovative strategies and economical methods to solve demanding problems in the construction of polysubstituted systems and complex molecules including natural products. From an industrial perspective, the task is amply recognized from the fact that > 80% of today's commercial pharmaceuticals and agrochemicals embody aromatic or heteroaromatic components [7]. Aside from electrophilic substitution, numerous methods, based on aromatics as well as non-

C. Chatgilialoglu and V. Snieckus (eds.), Chemical Synthesis, 191–221.

aromatic SMs, have evolved but are poorly systematized for the student. These processes, presented in highly stylized form in Scheme 1, [8] have been implemented in synthetic aromatic chemistry, with popular emphasis towards electrophilic substitution and accessible, commercial starting points [9].

The emergence of the Directed ortho Metalation (DoM) (Scheme 2) strategy during the late 1970s as a serious contender to the above methodologies, especially electrophilic aromatic substitution, can be traced to concurrent and independent discoveries by Gilman [10] and Wittig, [11] the systematic work of Hauser, [12] and more recently numerous others, in particular Beak, Gschwend, Meyers, Muchowski, Ronald and Slocum and their coworkers [13]. Today, the DoM approach, labeled as die neue Aromatische Chemie, [14] is widely applied in academic and industrial laboratories, on scales from mgs to kgs, for efficient and regioselective synthesis of polysubstituted aromatics. It invariably offers general and usually abbreviated solutions to synthetic problems which are formidable to achieve by more conventional tactics [15].

Scheme 1
Synthetic Approaches to Substituted Aromatics

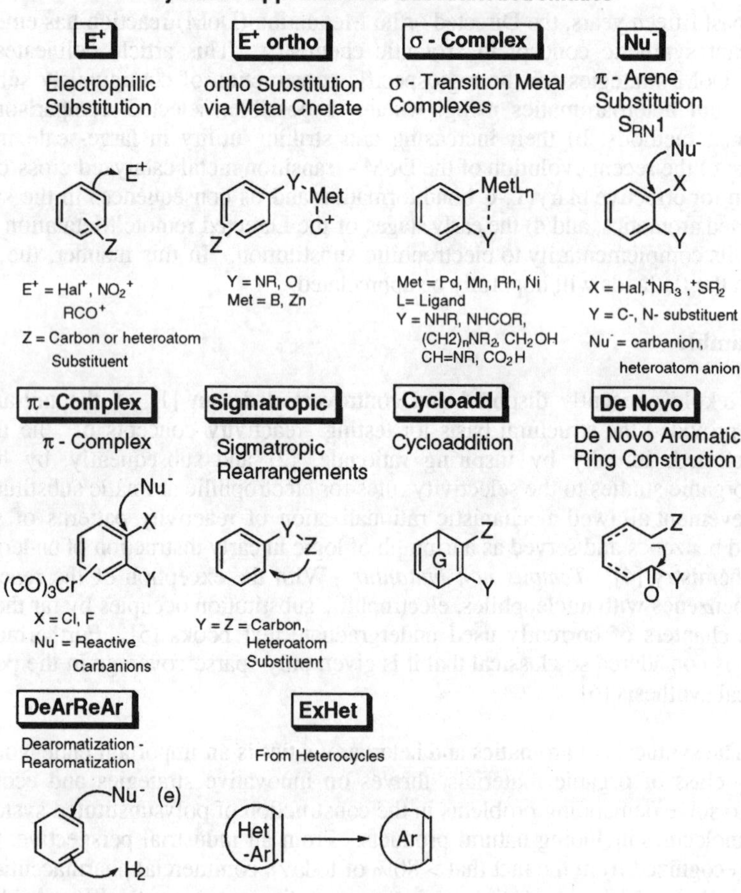

Scheme 2

Directed *ortho* Metalation: The New Aromatic Chemistry

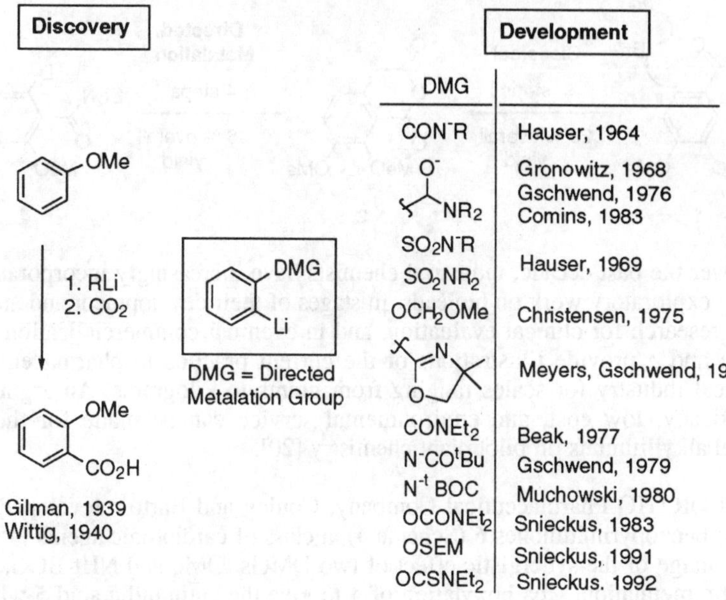

2. Aims of This Article

This account proposes to inform, by comparison wherever possible, how DoM chemistry has made an impact, both conceptually and technically, on the traditional practices of synthetic aromatic chemistry. It will then delineate the recent marriage between DoM and transition metal cross coupling reactions which has added a further dimension to aromatic synthesis. Finally, it will link the ortho metalation and cross coupling to the as yet untrampelled playground of Directed remote Metalation (DreM) and indicate how new opportunities, which again complement electrophilic substitution, are beginning to emerge. In this manner, it is hoped that gnosis to prognosis in the field of flatland chemistry will be adequately addressed.

3. Directed *ortho* Metalation in Methodological Practice

By classical electrophilic substitution, the formation of a contiguously tri- or tetra-substituted and functionalized benzenes is a formidable task which involves several operations including separation of isomers and functional group interconversions. In sharp contrast, reflection based on DoM allows consideration of equally available SMs, rapid focus on the key metalation event (one of several options), and reduction in number and simplification in type of overall manipulations to achieve the same TM with greater efficacy and convenience. This contrasting picture was dramatically revealed to us [16] in 1978 in the preparation of the homophthalic anhydride **2** by classical (from **1**) and metalation (from **3**) routes, results which launched us into DoM chemistry [17].

Scheme 3

Classical vs DoM Route to Homophthalic Anhydrides

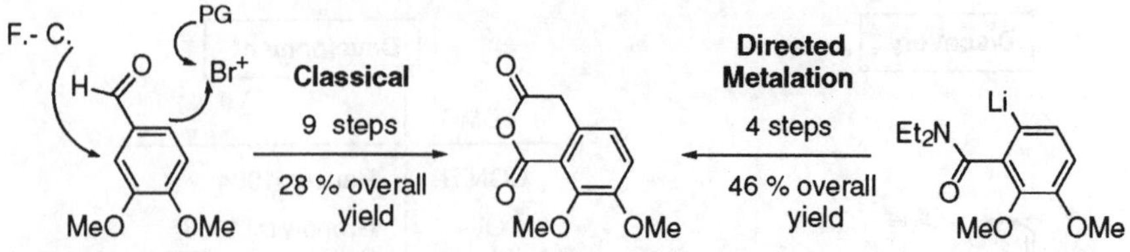

Over the past decade, industrial chemists have increasingly incorporated DoM steps in the exploratory work on bioleads, in stages of their development and analoging, in process research for clinical evaluation, and in eventual commercialization [18,19]. Schemes 4 and 5 provide illustrations of the current practice in pharmaceutical and agrochemical industry for scales ranging from grams to kilograms. An argument for overall efficacy, low cost, and environmental service can be made for the use of commercial alkyllithiums in pilot plant chemistry [20].

At ORTHO Pharmaceutical Company, Conley and Barton developed a DoM route to the benzopyrimidinones **6** (Scheme 4), a class of cardiotonic agents [18]. Thus, taking advantage of the synergistic effect of two DMGs, OMe and NHt-BOC, allowed regiospecific metalation - carboxylation of **4** to give the anthranilic acid **5** which was readily converted in several steps into the target **6**. The indicated scale provides evidence for the industrial viability of DoM chemistry.

Scheme 4

DoM inIndustrial Practice

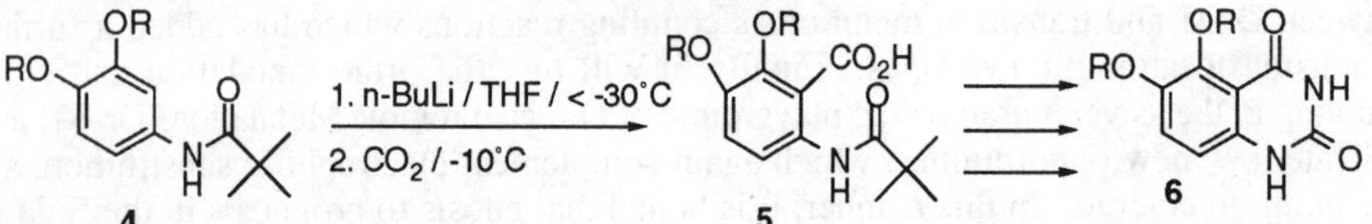

SM	n-BuLi	THF	Scale up	Product	Yield, %
450g	2580mL 1.56M heptane	5340mL	100gal	247g	46

In a more recent development, Hlasta and coworkers at Sterling Winthrop provided a kiloscale DoM procedure for the preparation of the saccharin derivative **9**, an intermediate for an anti-emphysema agent (Scheme 5) [19]. Thus metalation of the readily available benzamide **7** followed by a sequence, established earlier at Dupont Agrochemicals, [21] for the introduction of a sulfonamide functionality, affords **8**. Cyclization using refluxing acetic acid leads to the target **9** in excellent yield thus completing this short synthesis which was carried out batch-wise on the indicated scales. The possibility that the DoM strategy, because of its scope and convenience, can drive the analoging practice is not out of the question.

Scheme 5

DoM in Industrial Practice

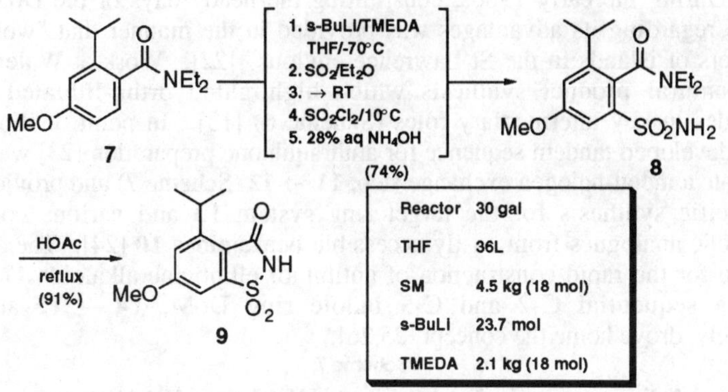

7 → 1. s-BuLi/TMEDA THF/-70°C; 2. SO_2/Et_2O; 3. → RT; 4. SO_2Cl_2/10°C; 5. 28% aq NH_4OH (74%) → 8

8 → HOAc, reflux (91%) → 9

Reactor	30 gal
THF	36L
SM	4.5 kg (18 mol)
s-BuLi	23.7 mol
TMEDA	2.1 kg (18 mol)

Scheme 6

ortho-Lithiated Benzamides: Starting Points for Natural Products

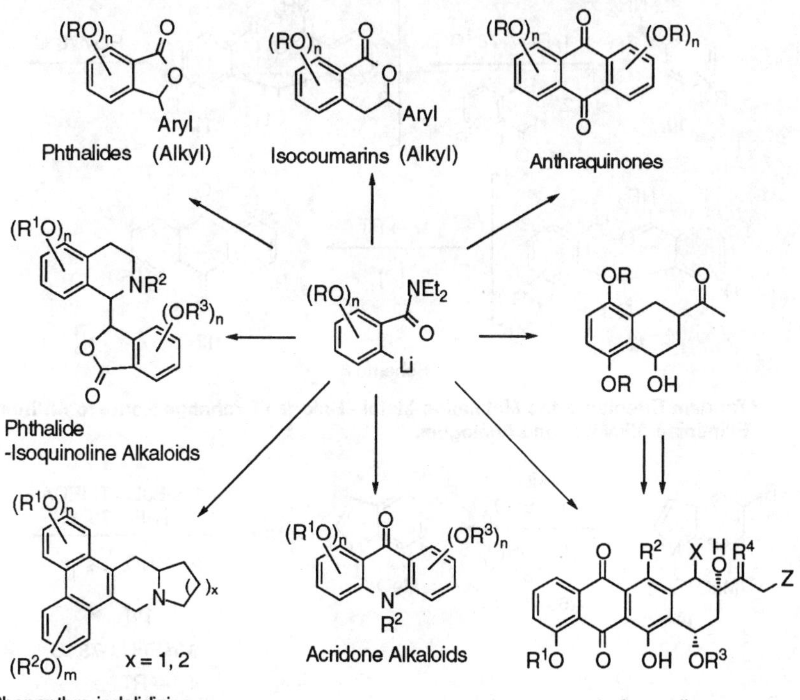

4. DoM as the Key Strategy in Complex Molecules Construction

During the early 1980s, constituting the heady days of the DoM revolution, evidence regarding its advantages was provided in the manner that "would make the discoverers of islands in the St Lawrence envious" [22]. Work at Waterloo provided several natural product synthesis which highlighted ortho-lithiated N,N-diethyl benzamides in key intermediary roles (Scheme 6) [13]. In point of illustration, the initially developed tandem sequence for anthraquinone preparation [23] was modified to incorporate a metal-halogen exchange step, **11** → **12** (Scheme 7) and provided a general, regiospecific synthesis for the target ring system **13** and various condensed and heterocyclic analogues from easily accessible benzamides **10** [24]. The application of this route for the rapid construction of antitumor ellipticine alkaloids **17** (Scheme 8), involving sequential C-2 and C-3 indole ring DoM, **14** → **15** and **15** → **16** respectively, drove home the concept [25,26].

Scheme 7

Anthraquinones by Tandem Directed Metalation - Metal Halogen Exchange

NR_2 · R · **10** · 1. RLi / -78 °C · 2. Br, H, O, R · NR_2 · Br · O · R · OLi · R · **11** · RLi / 78°C

NR_2 · Li · O · R · OLi · R · **12** · 1. → RT · 2. [O] · O · R · R · O · **13**

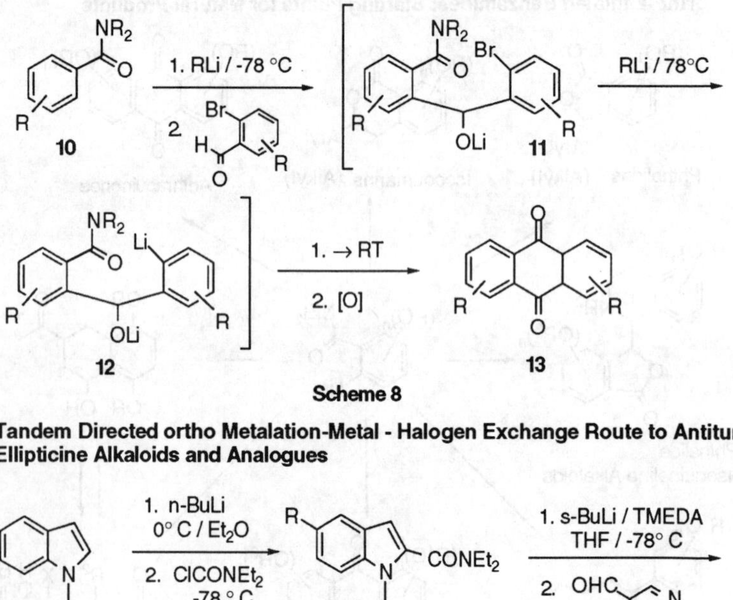

Scheme 8

Tandem Directed ortho Metalation-Metal - Halogen Exchange Route to Antitumor Ellipticine Alkaloids and Analogues

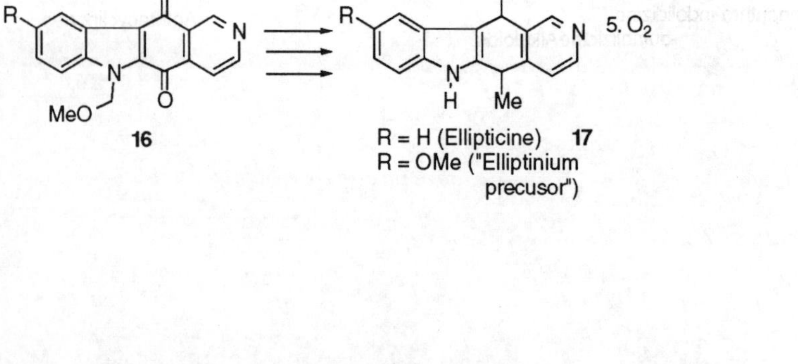

R · N · MeO · **14** · 1. n-BuLi $0° C / Et_2O$ · 2. $ClCONEt_2$ -78 ° C (60-70%) · R · N · $CONEt_2$ · MeO · **15** · 1. s-BuLi / TMEDA THF / -78° C · 2. OHC, N, Br · 3. n-BuLi / -78° C · 4. → RT · 5. O_2

R · O · N · N · O · MeO · **16** · R · Me · N · N · H · Me

R = H (Ellipticine) **17**
R = OMe ("Elliptinium precusor")

Concurrent work in our laboratories provided a slightly lengthier but still efficient route to anthraquinones, including a number of natural products, which brought a fresh approach compared to the classical, historically entrenched Friedel-Crafts based protocol. A comparison of the two in broad brush strokes (Schemes 9 and 10) is instructive [29]. Thus, the normal Friedel-Crafts route has detriments in a) the non-regioselectivity in the initial condensation **18** + **19** → **20** leading to the normal product **21** and b) the treachery of the Hayashi rearrangement of the ortho-benzoyl benzoic acids, of which **20** → **23** → **22** is representative. Consequently, a mixture of anthraquinones, at times with the predominance of the wrong isomer, is a common occurrence.

Scheme 9

Friedel - Crafts Based Retrosynthesis of Anthraquinones

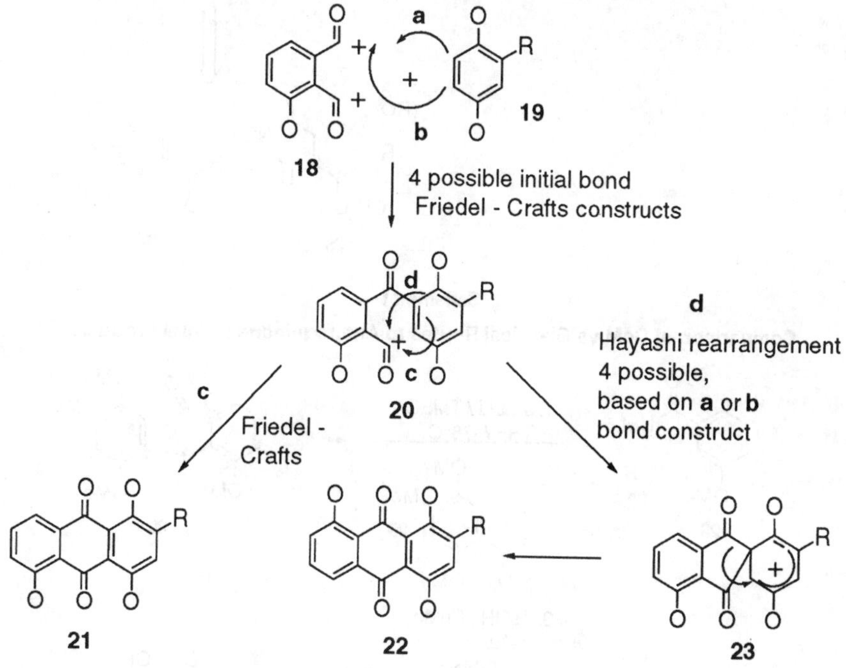

18 + **19** (a, b) → 4 possible initial bond Friedel - Crafts constructs → **20**

20 → (c, Friedel - Crafts) → **21**

20 → (d, Hayashi rearrangement 4 possible, based on **a** or **b** bond construct) → **23** → **22**

On the other hand, DoM protocol (Scheme 10) invites consideration of two dissection modes, **a** and **b**, one of which is shown (**24** ⇒ **25** ⇒ **26** + **27**) which allows double choices of starting materials based, usually, on simple or commercial availability. The intermediate ortho-benzoyl benzoic amide **25** is locked into position for a follow up Friedel-Crafts, directed by the normal electrophilic substitution rules, to give single anthraquinone isomer products. Scheme 11 documents a comparison of DoM and non-DoM approaches to **31**, a one-step (BBr_3) precursor of the natural product, islandicin. In the former, [27] combination of readily available benzamide **28** and benzaldehyde **29** under metalation mediation leads, after TsOH treatment, to the phthalide **30**. Three more conventional steps affords **31** in good overall yield. In an approach based on key photo-Fries (**33** → **34**) and Claisen (**34** → **32**) rearrangement steps [28], islandicin trimethyl ether is obtained in low overall yield and is further hampered by need to prepare the contiguously trisubstituted aromatic **33.** Since these early applications of DoM to rapidly obtain key phthalide intermediates (e.g. **30**) en route to natural products, others have demonstrated their utility in more challenging end results [30].

Scheme 10

Directed ortho Metalation - Based Retrosynthesis of Anthraquinones

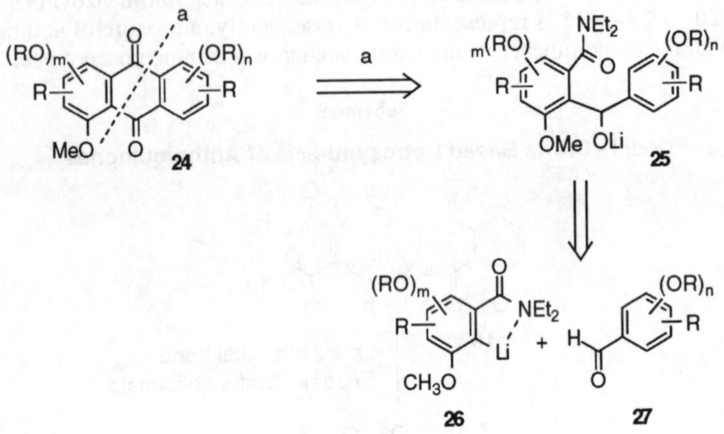

(RO)m (OR)n a R R MeO O O **24** a NEt2 m(RO) (OR)n O R R OMe OLi **25**

(RO)m O NEt2 R Li + H (OR)n R CH3O O **26** **27**

Scheme 11

Comparison of DoM vs Classical Routes to Anthraquinone Natural Products

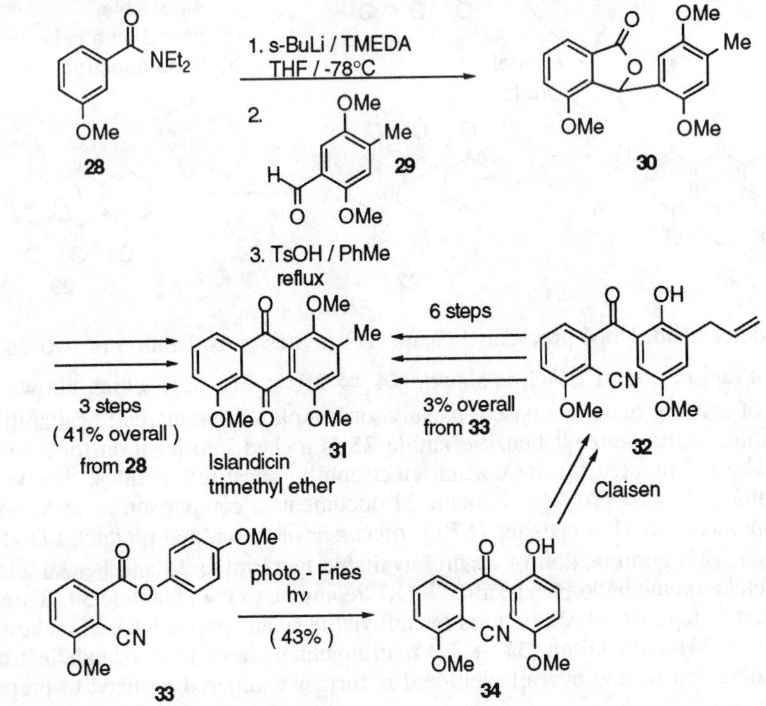

O NEt2 OMe **28** 1. s-BuLi / TMEDA THF / -78°C 2. OMe Me **29** H O OMe 3. TsOH / PhMe reflux O OMe Me O OMe OMe **30**

3 steps (41% overall) from **28** O OMe Me OMe O OMe **31** Islandicin trimethyl ether 6 steps 3% overall from **33** O OH CN OMe OMe **32** Claisen

O O OMe CN OMe **33** photo - Fries hv (43%) O OH CN OMe OMe **34**

The construction of isocoumarin natural products offers another instructive comparison between DoM and more conventional methodologies. Thus, in the synthesis of hydrangenol **37** (Scheme 12), [31a] a five-step, one-pot process directly furnishes the heteroannelated product **36**, a sequence involving a low-temperature DoM of a *dimethyl* amide, an LDA-induced chain extension, and a base-mediated cyclization. BBr_3 treatment leads to the natural product **37** in good overall yield. The chosen classical alternative, [31b] involving Claisen condensation of **39** with **40** to give **38**, is achieved in low overall yield due to the an inefficient preparation of the homophthalic ester **39** (from 3-nitrophthalic anhydride), a memento of an earlier DoM- classical route comparison (Scheme 3).

Scheme 12

Comparison of DoM and Classical Routes to Isocoumarin Natural Products

OMe NMe2 O
35
1. s-BuLi / TMEDA THF / -90°C
2. MeI
3. LDA
4. H O OMe
5. aq. KOH / EtOH
OMe O O
36
OMe
BBr3
(42% overall from **35**)
OMe O O
OH
Hydrangenol **37**
6% overall
OH OH O
MeO2C
OCH2Ph
38
(80%)
DBU / PhH
OH OMe OMe O
39
+
H O
OCH2Ph
40

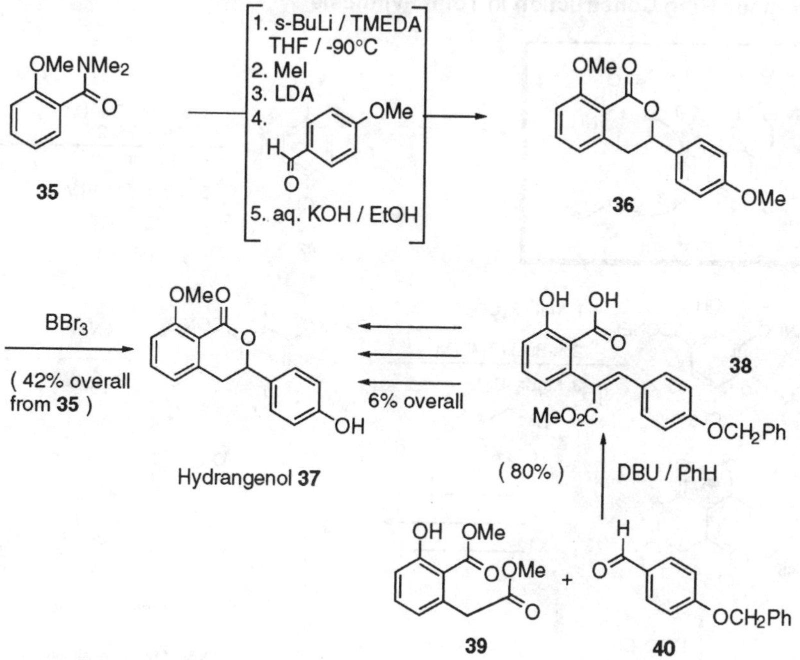

To avoid the *blinkers effect* in organic synthesis, a second illustration of isocoumarin ring construction which is simpler to execute, more economical, and proceeds in better yield by a non-DoM protocol, is examined. The DoM conceptualization **41** (Scheme 13) towards the total synthesis of the toxic fungal metabolite, ochratoxin A (**46**) comprises taking full advantage of the O-aryl carbamate and amide DMGs. Thus, metalation - carbamoyl electrophile quench (step 1), anionic ortho-Fries rearrangement (step 2), and synergistic, two-DMG influenced allylation are invoked to cast the complete substitution pattern of ochratoxin A. In the event, [32a] a one-pot sequence of metalation, carbamoylation, metalation, ortho-Fries rearrangement on **42** gave the isophthalamide **43**. Phenol protection, followed by sequential lithiation, transmetallation to the corresponding Grignard reagent (of necessity, to moderate the basicity in the next reaction), and allylation furnishes product **44**. Treatment of **44** with strong acid effected cyclization, demethylation, and hydrolysis to provide the key coumarin **45** in modest yield [33]. On the non-DoM side of the fence, [32b] the key event is the *de novo* aromatic ring construction (see Scheme 1) of the ortho-toluate **48**. Vinylogous methyl metalation - chain extension produces the isocoumarin **47** which is readily converted into **45** using electrophilic chlorination! Although DoM proponents may find some compensation in arguments that rapid provision of ochratoxin analogues would favor DoM chemistry, clearly they must bow to the *de novo* method (especially since our Honorary Chairman, Professor Prelog contributed to it, - cited in [32b]) for an expedient preparation of **45**. *Audiatur et altera pars.*

Scheme 13

Combined DoM - Anionic *ortho*- Fries Rearrangement *vs de novo* Aromatic Ring Construction in Total Synthesis

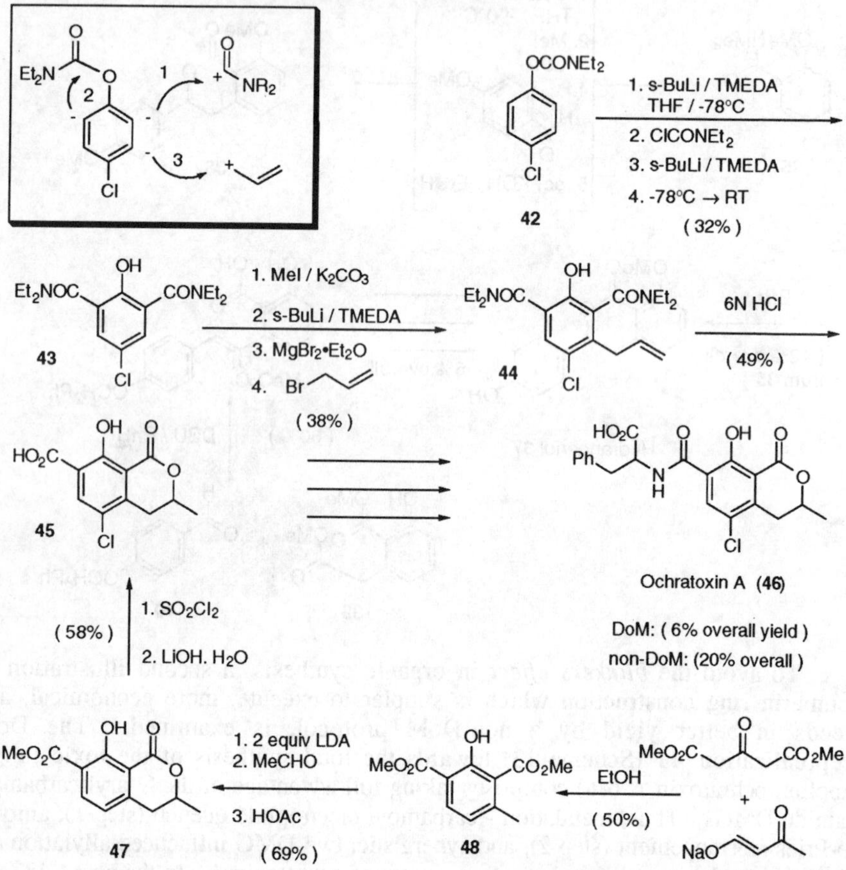

Et_2N O O 1 + O NR_2 2 3 + Cl

$OCONEt_2$ Cl **42**

1. s-BuLi / TMEDA THF / -78°C
2. $ClCONEt_2$
3. s-BuLi / TMEDA
4. -78°C → RT
(32%)

OH Et_2NOC $CONEt_2$ Cl **43**

1. MeI / K_2CO_3
2. s-BuLi / TMEDA
3. $MgBr_2 \cdot Et_2O$
4. Br
(38%)

OH Et_2NOC $CONEt_2$ Cl **44**

6N HCl (49%)

OH O HO_2C O Cl **45**

HO_2C O OH O Ph N H O Cl

Ochratoxin A (**46**)

DoM: (6% overall yield)
non-DoM: (20% overall)

1. SO_2Cl_2
2. LiOH. H_2O
(58%)

OH O MeO_2C O **47**

1. 2 equiv LDA
2. MeCHO
3. HOAc
(69%)

OH MeO_2C CO_2Me **48**

EtOH (50%)

O MeO_2C CO_2Me + O NaO

5. The DoM - Cross Coupling Nexus

With the advantage of 20/20 hindsight, it is arrogant to note that the sterling discovery by Kharash of the cobalt and nickel halide catalyzed homocoupling reactions of aryl Grignard reagents [34] took a long time to be noticed by synthetic chemists. In 1972, independent investigations by Kumada [35] and Corriu [36] demonstrated the first effective Ni(0)-catalyzed aryl metal - aryl halide cross coupling process to produce biaryls (Scheme 14). The findings of aryl zinc - aryl halide coupling by Negishi [37] followed shortly and, in 1981, a truly novel (a word that conjures up patent lawyers) reaction, the cross coupling of aryl boronic acids with aryl halides under Pd(0) catalysis, was reported by Suzuki [38]. The comprehensive work of Beletskaya [39a] and Stille [39b] on the aryl tin - aryl halide and triflate coupling concludes the tetrad of reactions which have gained prominence as new conceptualizations for aryl-aryl bond constructs.

Scheme 14

Prominent Transition Metal Catalysed Cross Coupling Reactions for the Aryl-Aryl Bond

$$Ar^1M + Ar^2X \xrightarrow{NiLn\ or\ PdLn} Ar^1Ar^2$$

M	X		
MgY	Br,I	Corriu Kumada	(1972)[35] (1972)[36]
ZnY	Br,I	Negishi	(1977)[37]
$B(OH)_2$	Br, I	Suzuki	(1981)[38]
SnR_3	OTf	Stille	(1988)[39]

The Suzuki discovery constituted a bold-face arrow in our work since, being cognizant of the oxidative conversion of aryllithiums to phenols via borate intermediates [40], it provided an immediate link to the DoM process [41]. Thus, metalation - boronation of the DMG-bearing substrates **53** followed by aqueous work up afforded the aryl boronic acids **54** which, invariably in crude form [42], were subjected to the originally described [38] conditions to afford biaryls and heterobiaryls **55** with 2-carbon, oxygen, nitrogen, and sulfur substituents in good to excellent yields (Scheme 15) [13,44]. Although aryl bromides and aryl iodides constitute widely used partners, the finding that aryl triflates undergo smooth cross coupling reactions with aryl boronic acids, [45] has not only allowed the use of the more readily available phenol precursors but also provided general complementarity to the Stille process which employs toxic tin cross coupling partners. Gronowitz, who has comprehensive studied the Suzuki reaction for the construction of unusual heterocycles [46], taught us the advantages of using DME, a solvent which, as will be clear in the sequel, is now part of our routine protocol.

Scheme 15

The Directed ortho Metalation - Cross Coupling Connection

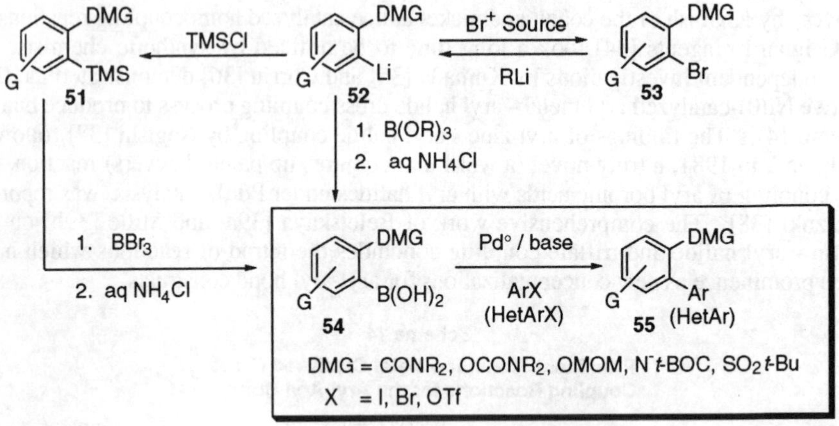

DMG, TMS, G, 51 ← TMSCl ← DMG, Li, G, 52 ⇄ (Br^+ Source / RLi) DMG, Br, G, 53

52 → 1. $B(OR)_3$ 2. aq NH_4Cl → 54

51 → 1. BBr_3 2. aq NH_4Cl → 54

DMG, $B(OH)_2$, G, 54 → (Pd^o / base; ArX (HetArX)) → DMG, Ar (HetAr), G, 55

DMG = $CONR_2$, $OCONR_2$, OMOM, N^-t-BOC, SO_2t-Bu
X = I, Br, OTf

DMG = Directed Metalation Group

Alternative methods for the preparation of aryl boronic acids involve the well established metal-halogen exchange of aryl bromides, **53** → **52** → **54** and the less widely known ipso borodesilylation, **51** → **54** (Scheme 15) [47]. In the former, a boronic acid is formally "stored" in the aryl bromide, to be released at the appropriate time by lithium-halogen exchange, with the obvious understanding of functional group compatibility with or protection against the strongly basic alkyllithium conditions. In the borodesilyation process, the efficient formation of aryl silanes by DoM and their enhanced reactivity to ipso electrophilic substitution may result in better overall yields of **54** compared to the direct method from **52,** although this aspect has not been widely tested [47].

With the initial results of the methodology in hand, opportunities for exploitation towards the construction of polycyclic aromatics, heteroaromatics, and polyphenyls rapidly evolved (Scheme 16). Thus, cross coupling of N-t-Boc aryl boronic acids **59** (DMG = NHt-BOC) with ortho-halobenzaldehydes and benzoates, **58,** R = H and R = OEt respectively, led directly to phenanthridines **56** and phenthridinones **57** [48]. This represents a general, efficient and rapid method for the regiospecific preparation of these heterocycles which has also found application in the synthesis of *Amaryllidaceae* [49] and *Amphimedon* [50] alkaloids. In the latter undertaking (Scheme 17), ortho-iodoaniline (**66**) coupling was best achieved with a 9-BBN-nicotinamide **67** to furnish the azabiaryl **68**. Intramolecular lactam formation under LDA conditions followed by triflation yielded the azaphenanthridine **69**. A Stille cross coupling led to **70** whose conversion to the marine alkaloid, amphimedine (**71**) had been previously accomplished.

Scheme16

The Emergence of the Cross Coupling Reaction

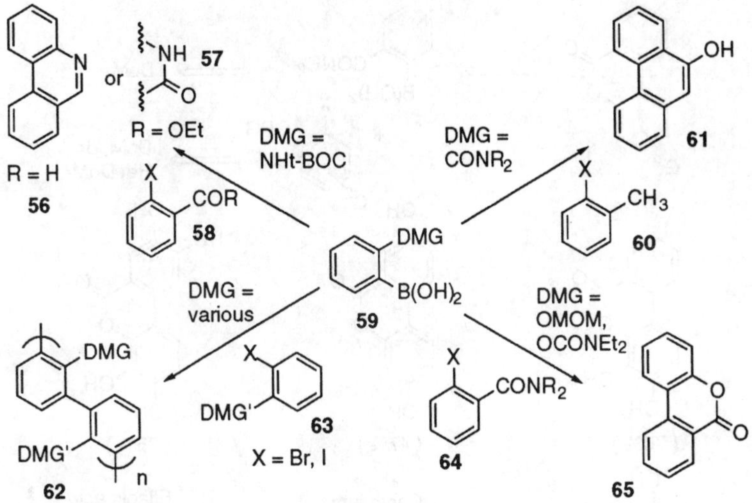

Scheme 17

DoM - Cross Coupling Connections. Formal Synthesis of the Marine Alkaloid Amphimedine, ex *Amphimedon* sp

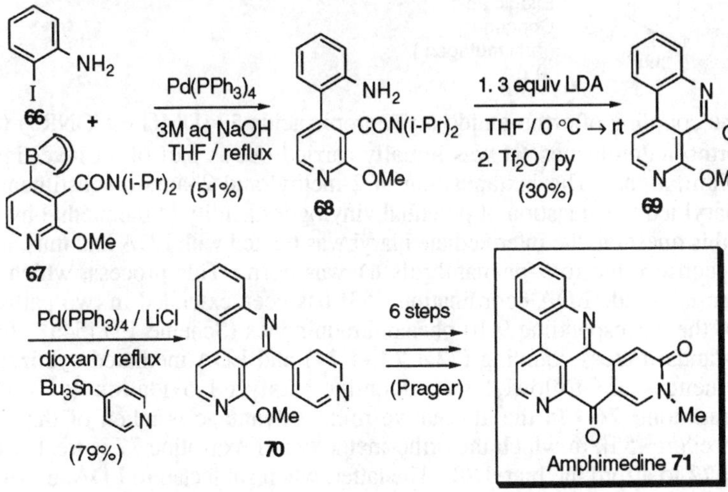

Construction of oxygen heterocycles **65** (Scheme 16) is feasible by sequential coupling of OMOM and $OCONEt_2$ aryl boronic acids **58** (DMG = OMOM, $OCONEt_2$) with ortho-halobenzamides **64** followed by acid-mediated cyclization. This methodology allows access to a variety of dibenzo[b,d]pyranones (Scheme 18) including difficult to prepare systems, natural products, and metabolites of some biological interest [51,52].

Scheme 18

Synthesis of Oxygenated Dibenzo[*b,d*]-pyran-6-ones Illustrated

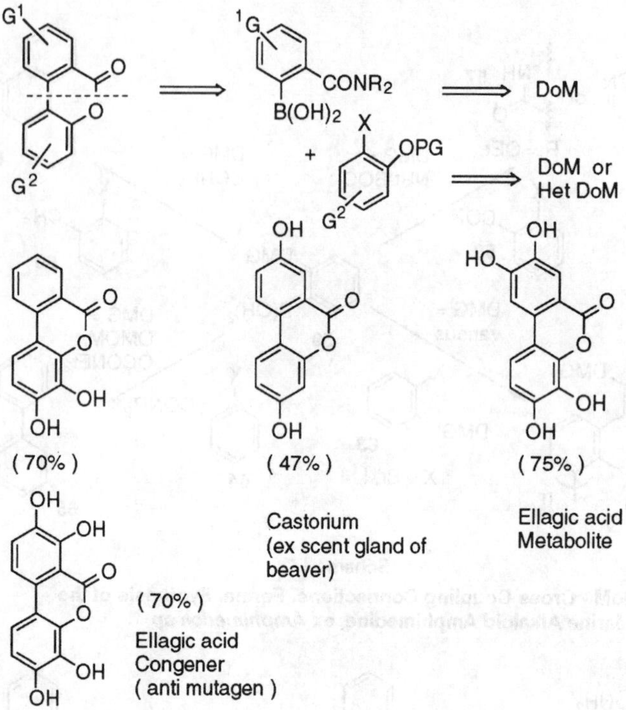

G^1, G^2 ⟹ ^1G, $CONR_2$, $B(OH)_2$ ⟹ DoM

\+ X, OPG, G^2 ⟹ DoM or Het DoM

(70%) (47%) (75%)

Castorium (ex scent gland of beaver)

Ellagic acid Metabolite

(70%) Ellagic acid Congener (anti mutagen)

The coupling of ortho-amido aryl boronic acids **59** (DMG = $CONR_2$) (Scheme 16) with ortho-halotoluenes **60** was initially carried out as part of an investigation of scope and limitations. The juxtaposition of 2-methyl and 2'-amido substituents in the resulting biaryl led to a question of potential vinylogous acidity of the methyl hydrogens. To answer this question, the intermediate biaryl was treated with LDA and thus a general and regiospecific route to 9-phenanthrols **61** was born. This process, which may be driven by initial amide-LDA coordination, [53] has been extended in two paths for the synthesis of the corresponding 9,10-phenanthraquinones (Scheme 19) [54a]. In one of these, the standard cross coupling (**72** + **73** → **74**) and LDA-mediated cyclization (**74** → **75**) sequences are followed by salcomine-catalyzed oxidation to furnish the phenanthraquinone **76**. In the alternative route, advantage is taken of the excellent Meyers procedure [55], in which the ortho-methoxyaryl oxazoline **77** is treated with the Grignard of **72** to afford the biaryl **78**. The latter, when subjected to LDA, undergoes an interesting cyclization - oxazolidine ring opening process, to give the amino ether **79**. Oxidation with Fremy's salt affords the phenanthraquinones **76**. Previous general routes to phenanthridine derivatives involve Pschorr [56] and stilbene [57] photoyclization reactions whose regiochemical outcome depend crucially upon electrophilic aromatic substitution and steric effects respectively.

Scheme 19

Cross Coupling - Remote "*ortho*-Toluyl" Metalation Route to 9,10-Phenathraquinones

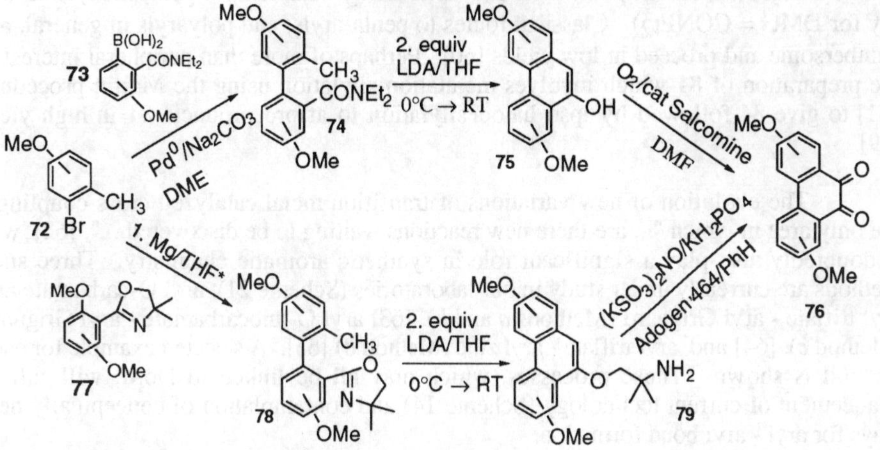

Scheme 20

Regiospecific DoM - Cross Coupling Route to Functionalized Quinquearyls

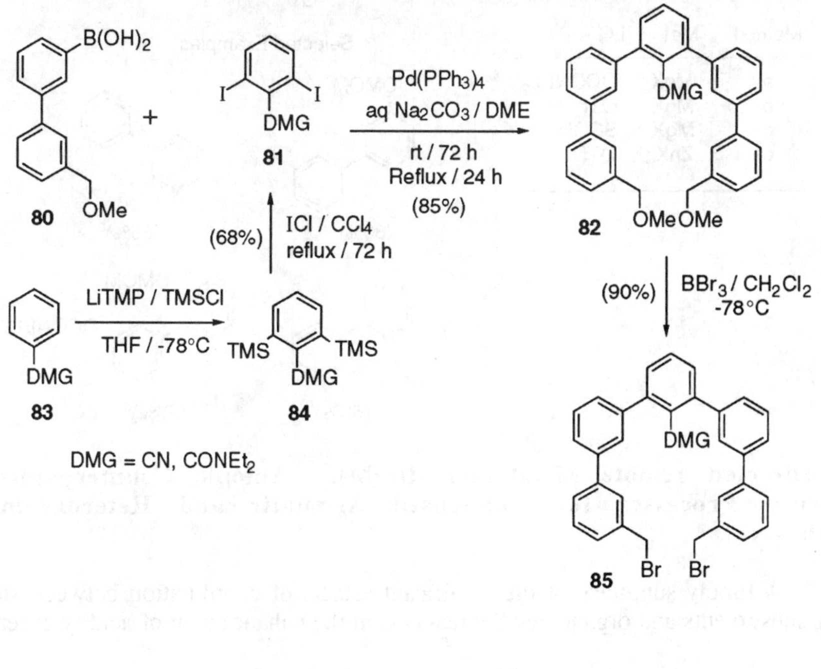

Approaches to polyaryls **59** + **63** → **62**, DMG = various (Scheme 16) via the DoM - cross coupling strategy can take iterative and convergent forms [58]. An illustration of the latter is the construction of a pentaphenyl amide **85** (Scheme 20), of interest as a potential precursor for a macrocyclic host [59].

Thus coupling of the biaryl boronic acid **80**, readily prepared by metal-halogen exchange - boronation of the corresponding bromide, with the 2,6-dihalo benzamide and benzonitrile **81**, DMG = $CONEt_2$ and DMG = CN respectively leads, under the optimum conditions shown, to compounds **82** and, after BBr_3 treatment, to pentaphenyls **85** (X-ray for DMG = $CONEt_2$). Classical routes to penta-aryls, and polyaryls in general, are cumbersome and proceed in low yields [60]. Perhaps of more than peripheral interest is the preparation of **81** which involves metalation-silylation using the Martin procedure [61] to give **84** followed by ipso halodesilylation to afford products **81** in high yield [59].

The evolution of new variations of transition metal catalyzed cross couplings, the only area in which "... are there new reactions waiting to be discovered ...", [62] will undoubtedly also play a significant role in synthetic aromatic chemistry. Three such methods are currently under study in our laboratories (Scheme 21): aryl O-carbamate and aryl triflate - aryl Grignard (Methods *a* and *b*), [63] aryl O-thiocarbamate - aryl Grignard (Method *c*), [64] and aryl triflate - aryl zinc (Method d) [65]. A selected example for each method is shown. These processes, which may all be linked to DoM, will allow ehancement of current technology (Scheme 14) and contemplation of conceptually new ways for aryl - aryl bond formation.

Scheme 21

New Variations of Transition Metal Catalyzed Cross Coupling Reactions

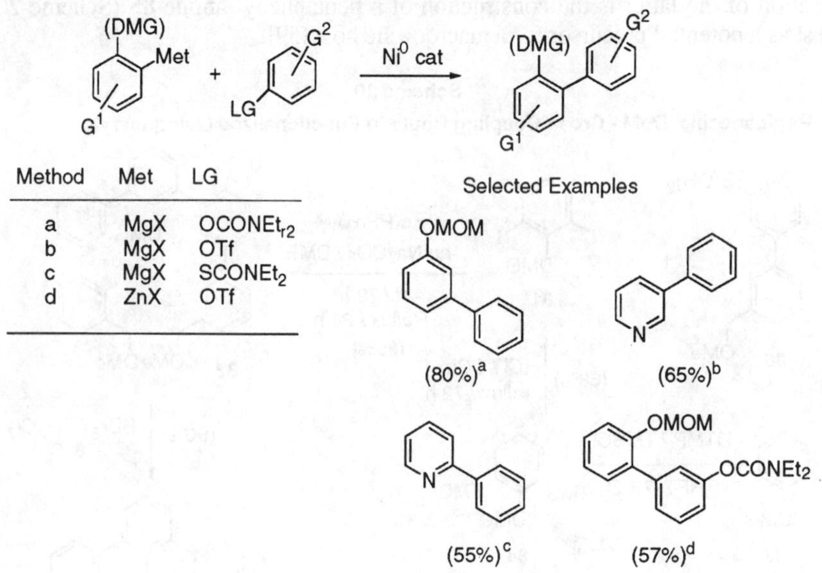

6. Directed remote Metalation (DreM). Anionic Counterpoints to Classical Processes for Condensed Aromatic and Heteroaromatic Synthesis

A timely summary of the significant effects of coordination between strong donor substituents and organometallic reagents in the enhancement of acidity at remote (thermodynamically unacidic) sites (Complex Induced Proximity Effect, CIPE) [53] prompted the deprotonation experiment on the biaryl amide **86** in order to view the consequences of the potential anion *87* electrophile trap and cyclization to fluorenones **88** (Scheme 22). While the former has not been observed, even at very low temperatures [54a], the latter has been developed as a general method and applied to natural product synthesis [66]. Arguably of most significance is the complementarity which this new methodology brings to the Friedel-Crafts reaction. For example, while Lewis acid-mediated trapping of acylium ion derived from **90**, X = OH, OR, Cl is dictated by the overwhelming para directing effect of the electron donating OMe and thereby leads to fluorenone **91**, the corresponding LDA-induced process on **90**, X = NR_2 follows the OMe DMG effect and consequently affords the isomeric **89**. Recently prepared [54b,66], selected fluorenones (Scheme 22, dotted bond is retrosynthetic) are indicative of the further potential of these DReM processes.

Scheme 22

Biaryl Amide DreM. Versatile Anionic Friedel - Crafts Complements for Assemblege of Fluorenones and Azafluorenones

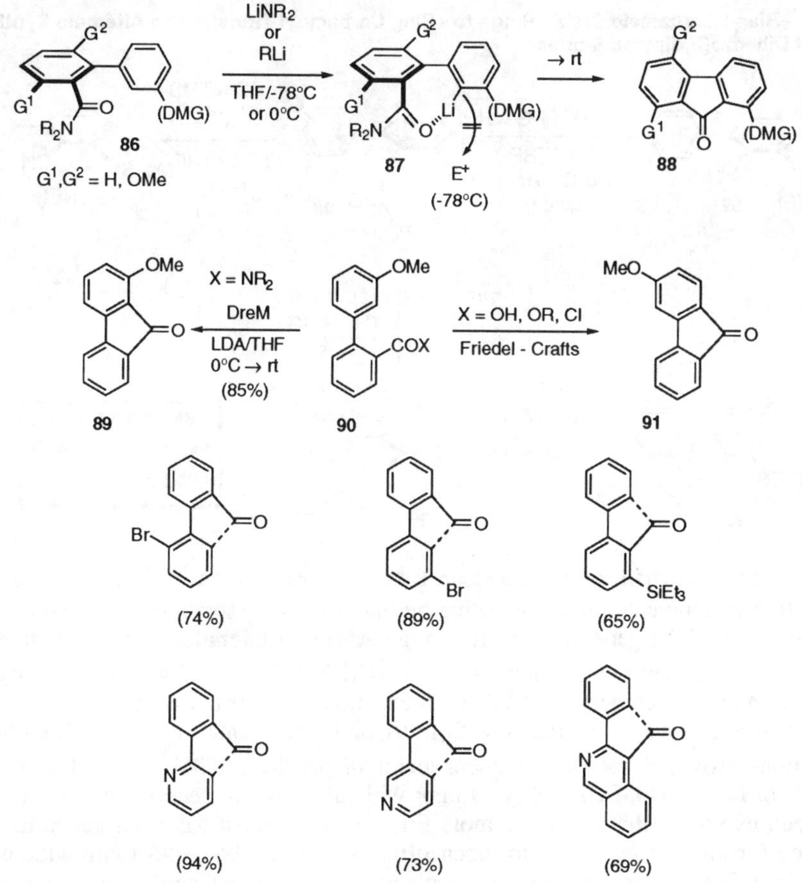

G^2 G^1 O R_2N (DMG) **86** G^1,G^2 = H, OMe

$LiNR_2$ or RLi, THF/-78°C or 0°C

G^2 G^1 R_2N O Li (DMG) **87** E^+ (-78°C)

→ rt

G^2 G^1 O (DMG) **88**

OMe O **89**

X = NR_2 DreM LDA/THF 0°C → rt (85%)

OMe COX **90**

X = OH, OR, Cl Friedel - Crafts

MeO O **91**

Br O (74%)

O Br (89%)

O $SiEt_3$ (65%)

N O (94%)

N O (73%)

N O (69%)

The results of diaryl amide DReM reactivity (Scheme 22) provided the necessary encouragement to test the corresponding diaryl O-carbamates **93** (Scheme 23). Initial trials on **93**, G^1, G^2, PG = H using RLi and LDA bases under a variety of conditions confirmed the suspicion, precedented by our previous work [67], that deprotonation occurs at the site ortho to the powerful carbamate DMG and that the resulting anion is prone to rapid anionic ortho-Fries rearrangement, presumably driven by the steric congestion from the 6-phenyl substituent, to give product **92**. Undaunted by these observations and recalling the TMS protocol to protect the more reactive metalation sites [68], the biaryl O-carbamate **93**, PG = TMS, G^1, G^2 = H was prepared and subjected to LDA treatment. Lack of reactivity at room temperature and impatience prompted *refluxing THF* conditions which caused a temporary breakdown in student-supervisor relationship until he (the supervisor) was shown a spot-to-spot tlc plate. However, the product from this smooth reaction, isolated in good yield, was shown to be **94** [70]. This result has scant precedence but may be rationalized by a CIPE-mediated α-silyl methyl carbanion formation in equilibrium concentrations which is followed by an amide migration, driven again by phenolate leaving group facility.

Scheme 23

O - Biaryl Carbamate DreM. Ring - to - Ring Carbamoyl Transfer and Alternate Synthesis of Dibenzo[b,d]pyran-6-ones

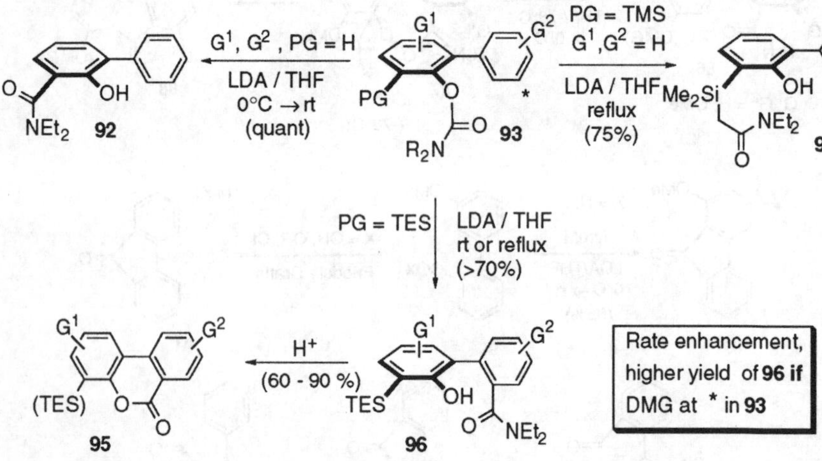

G^1, G^2, PG = H; LDA / THF, 0°C → rt (quant)

PG = TMS, G^1, G^2 = H; LDA / THF reflux (75%)

PG = TES; LDA / THF rt or reflux (>70%)

H^+ (60 - 90 %)

92, 93, 94, 95, 96

Rate enhancement, higher yield of **96** if DMG at * in **93**

The dwarfish intellectual leap from the TMS to the TES (Et_3Si) for ortho-protection circumvented this interesting but undesirable observation and resulted in the development of the general ring - to - ring carbamoyl migration process (Scheme 23) [70]. Thus, the biaryl O-carbamates **93**, PG = TES, G^1, G^2 = H was prepared by cross coupling and low temperature (-90 °C) metalation in order to avoid the anionic ortho-Fries rearrangement. Treatment with LDA under the same vigorous refluxing THF conditions provided good to excellent yields of products **96**, G^1, G^2 = H. Although aspects of further DoM chemistry, with or without phenol protection, on **96** have not as yet been explored, this anionic remote Fries rearrangement has been generalized and adapted for an alternate route to dibenzo[b,d]pyranones, **96** → **95** (with/without TES loss) [51b,70]. Work in progress concerns an additional application towards the construction of the gilvocarcin antibiotics (Scheme 24) [71]. Thus, naphthalene bromocarbamate **97**, when subjected to cross coupling with aryl boronic acid **98** afforded **99** which, upon treatment with LDA led to the carbamoyl migration product **100**. The latter, in refluxing acetic acid, underwent quantitative conversion into **101**.

Scheme 24

Biaryl O-Carbamate DreM. Ring - to - Ring Carbamoyl Transfer Route to a Gilvocarcin Model

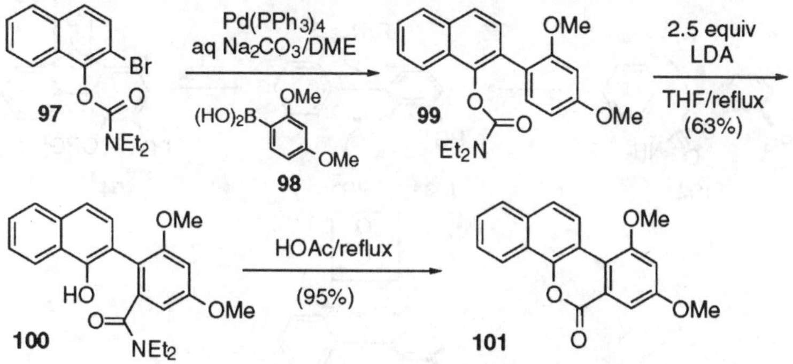

In summary, the DreM anionic Fries rearrangement, **93** → **96** (Scheme 23) provides a) a new strategy, potentially coupled with further DoM, for polysubstituted biaryl construction, b) a route to dibenzo[b,d]pyranones, and, perhaps most significantly, c) opportunity to "piggy-back" a carbon substituent ($CONEt_2$) from one aryl ring to another, thereby allowing the preparation of congested 2,2' - and 2,2',6 -substituted biaryls which are achieved ineffectively by Suzuki coupling. As described in the next section, the latter aspect proved its value in the total synthesis of dengibsin, a fluorenone natural product.

7. Combined Directed remote Metalation (DreM) Processes

What of the feasibility of developing sequential DreM biaryl amide and O-carbamate reactions? One attractive possibility (Scheme 25) involves remote anionic Fries rearrangement, **102** → **103** followed by remote methyl deprotonation and cyclization to a fluorenone, **103** → **104** which conceptually constitutes a reaction of a biaryl 2,2'-dianion with a carbonyl 1,1-dication **105** whereby the latter equivalent is derived from the carbamoyl halide initially incorporated into the starting material **102**. That such a sequence may be practically exploited is illustrated by the last stages of the synthesis of dengibsin (**109**) (Scheme 26), a naturally-occurring fluorenone isolated from the Indian *Dendrobium gibsonii* species [70]. Thus, the biaryl **106**, purposefully differentially protected at oxygen substituents, was subjected to LDA treatment to afford products **107** and **108**. Although the desired **108** is the minor product, the major **107** was recycled to improve the overall conversion. Parenthetically, the formation of **107** may be mechanistically indicative of an LDA-carbamoyl CIPE which, in this case, allows chemoselective deprotection of the 2'-isopropoxy group. Phenol methylation followed by desilyation furnished compound **111** which, when subjected to excess LDA, gave the DReM product **110** in good yield. The synthesis of dengibsin **109** was concluded by selective deisopropylation using BCl_3. The meta-relationship of the oxygen substituents with respect to the strategic aryl - C=O bonds in the target **109** discourages Friedel-Crafts approaches from either precursor biaryl carboxylic acid (dotted lines). In fact, such attempts led mainly, despite various phenol protection regimens, to the reappearance of the old friend dibenzo[b,d]pyranone systems [72].

Scheme 25
Sequential Biaryl Carbamate and Amide DreM Reactions. Regiospecific Route to Fluorenones

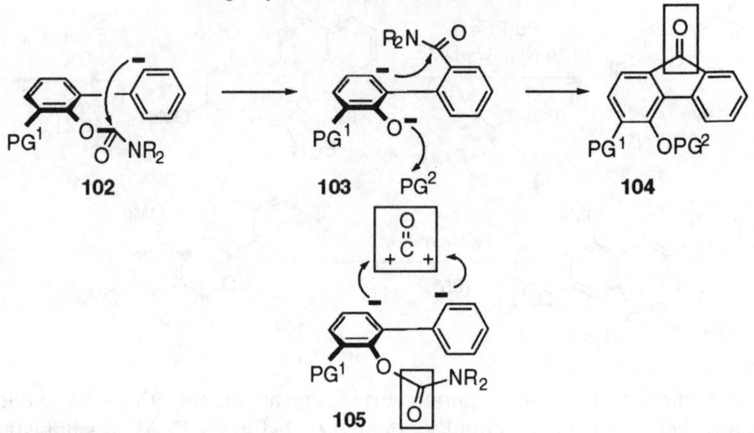

Scheme 26
Combined DoM - Cross Coupling - DreM Strategy to a Naturally Occurring Fluorenone

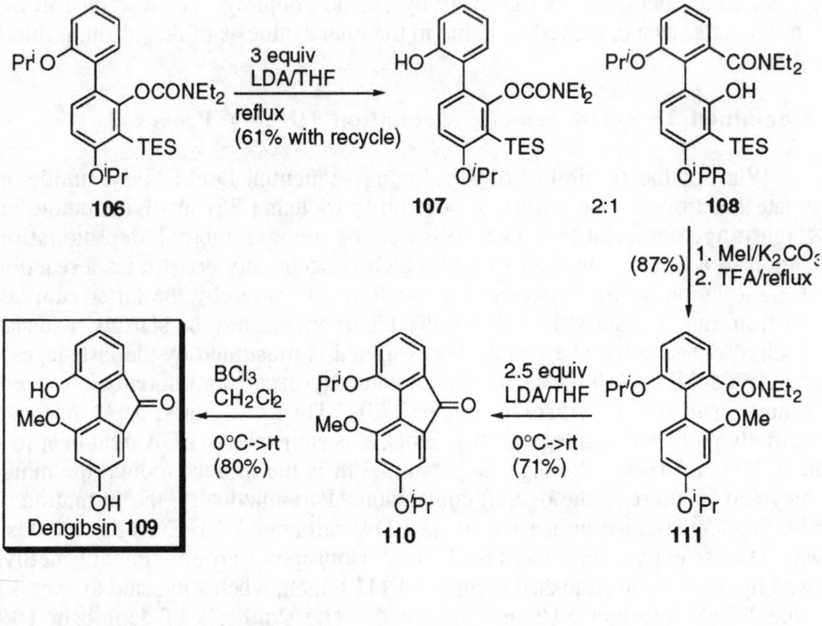

A different sequential DreM pathway (Scheme 27) depicts an amide translocation initiated from a biaryl O-carbamate, **112** → **113** followed by a methyl to amide cyclization, **113** → **114** and formally constitutes bridging a 2,2'-methyl dicarbanion species **115** with a carbonyl dication equivalent which, as before (Scheme 25) is derived from an incorporated carbamoyl group. In an encouraging model study, (scheme 28), the diaryl O-carbamate **116**, upon exposure to excess LDA, gave the oxygenated phenanthrol **117** in good yield [73].

Scheme 27

Sequential Biaryl Carbamate and Vinylogous Amide DreM Reactions. Regioselective Route to Phenanthrols

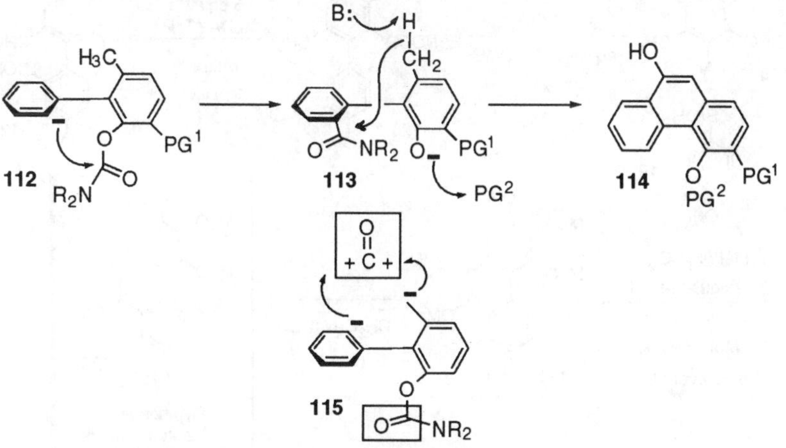

Scheme 28

Sequential DreM Reactions. A Model Study

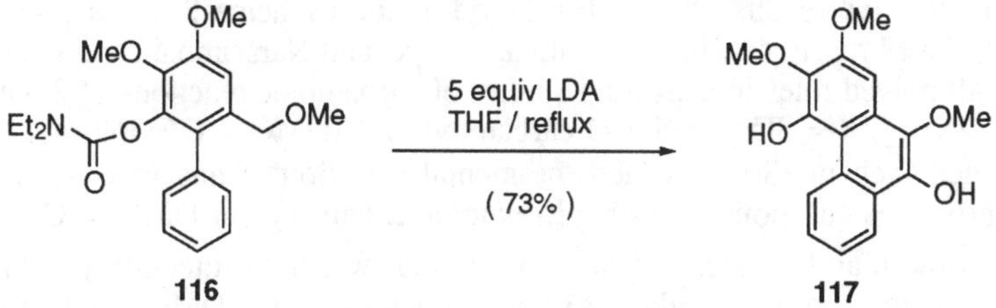

In order to probe this conceptualization in total synthesis, gymnopusin (**123**), (Scheme 29), another natural product isolated from an Indian orchid (*Bulbophyllum gymnopus)* whose structure had ungergone several revisions, was set as a challenging target [73]. Following interesting metal-halogen exchange and DoM chemistry, the key pentasubsituted benzene **118** was obtained and subjected to cross coupling with the easily accessible boronic acid **119** to give the biaryl **120**. LDA-induced ring - to - ring carbamoyl translocation proceeded smoothly to afford **121**. Standard phenol methylation followed by treatment with n-BuLi resulted in the formation of an unstable 9-phenanthrol which was therefore not characterized but subjected to methylation to afford the ether **122** in excellent overall yield. Chemoselective de-isopropylation of **122** using BCl_3

concluded the synthesis of gymnopusin (**123**) [73] leaving no doubt concerning the *final* proposed structure and comparing favorably with that achieved by Sargent via a non-DoM but didactic strategy [74].

Alternative combinations of DOM-DreM processes, yet to be discovered, will undoubtedly provide further synthetic advances in anionic aromatic chemistry.

Scheme 29

Combined DoM - Cross Coupling - DreM Route to a Naturally - Occurring Phenanthrene

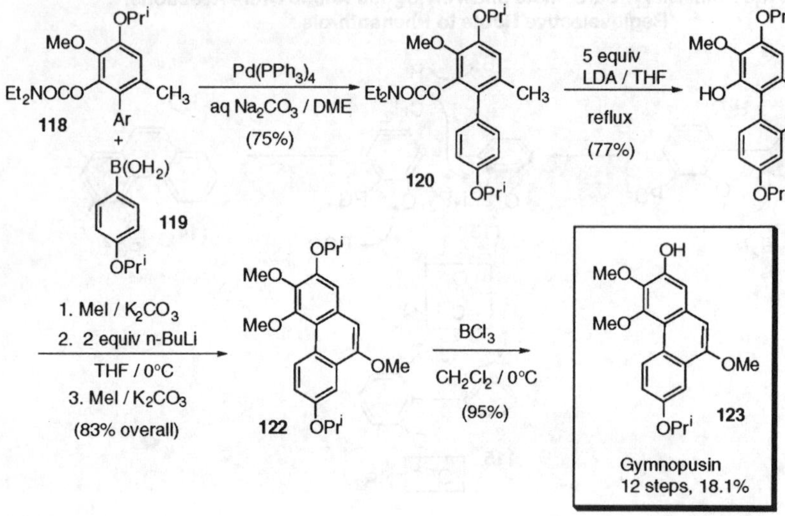

8. Heteroatom-Bridged Biaryl Directed remote Metalation (DreM)?

The positive results observed in biaryl amide (Scheme 22) and O-carbamate (Scheme 23) DreM reactions, chemists Gilman, Truce, and Narasimhan, and the prince of serendipity all played roles in triggering studies of carbanionic reactions of 2-amido **124** (DMG = $CONEt_2$, X = SO_2) and 2-O-carbamoyl **124** (DMG = $OCONEt_2$, X = SO_2) biaryl sulfones (Scheme 30). In fact, the mental insertion of any heteroatom X into biaryl **124** provokes questions regarding its anionic chemistry: for DMG = $CONEt_2$, by analogy with the biaryl amide, one may expect that without protection (PG^1), anion-mediated cyclization to the tricylic **125** may be achieved; for DMG = $OCONEt_2$, by analogy to the corresponding biaryl O-carbamate, one may envisage that, with protection, anionic carbamoyl migration may lead to **126** which, aside from the intrinsic interest of the reaction, has potential for further DoM chemistry. The contributions of Gilman [75], Eaborn [76], and Truce [77], already pointed to the feasibility of diaryl O and S system **127** deprotonation irrespective of a 2-DMG effect. The work of Narasimhan [78] on **128** provided a more appropriate analogy (and hope) for our preliminary studies.

Scheme 30

Heteroatom-Bridged Biaryl DreM

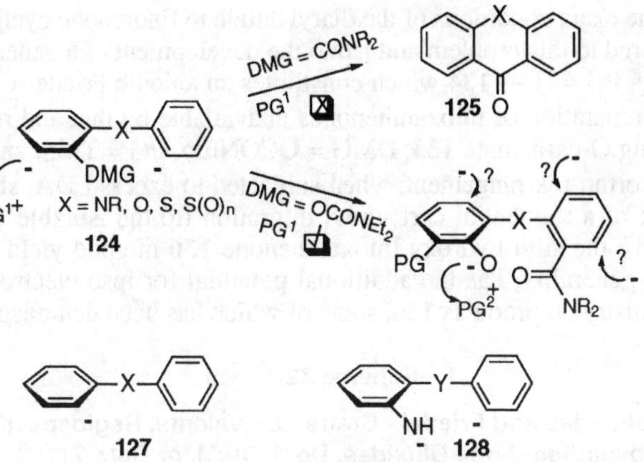

X = O (Gilman, 1939[75]) Y = N, O, S (Narasimhan, 1979[78])

X = S (Gilman, 1939[75]; Eaborn, 1961[76])

X = SO_2 (Truce, 1951[81]; Gilman, 1953[75b])

Serendipity showed its favorable side when we tested the metalation - fluorination of the amido and O-carbamate systems **129** (Scheme 31).

Scheme 31

Serendipitous Synthesis of DMG-Bearing Diaryl Sulfones

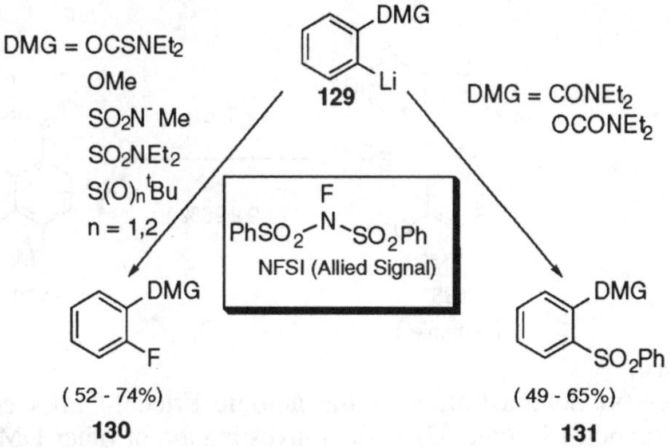

In a study to derive useful methodology for F^+ introduction via DoM tactics, the various ortho-lithiated species **129** were tested in reaction with the Allied Signal N-fluorosulfinimide reagent. With two exceptions, all DMG systems afforded the fluorinated derivatives **130** in modest to good yield [79]. The exceptions, **129**, DMG = $CONEt_2$ and $OCONEt_2$ afforded the corresponding diaryl sulfones **131** for reason(s) currently not appreciated [80] and stimulated the search for alternate routes of diaryl sulfone preparation [81] in order to pursue their metalation chemistry as envisaged

(Scheme 30). Using the excellent literature procedure [81b,c], metalation - phenylsulfonylation of **132** allowed general and less expensive access to the diaryl sulfones **133**, DMG = $CONEt_2$, PG = H and **133**, DMG = $OCONEt_2$, PG = TMS (Scheme 32). The exact conditions of the diaryl amide to fluorenone cyclization (Scheme 22) were transferred to this problem and led to the development of a general reaction **133**, DMG = $CONEt_2$, PG = H → **134** which constitutes an anionic Friedel-Crafts equivalent and allows the preparation of thioxanthenones unavailable by that and related methods. The corresponding O-carbamate **133**, DMG = $OCONEt_2$, PG = TMS, suitably protected to avoid anionic ortho-rearrangement, when subjected to excess LDA, showed the more interesting result of a sequential carbamoyl migration (to the isolable **135**) and amide cyclization to give the silyl hydroxy thioxanthenone **136** in good yield. This reaction, also of apparent generality, has the additional potential for ipso electrophile - induced desilylation chemistry on products **136**, some of which has been demonstrated [82].

Scheme 32

Anionic Remote Fries and Friedel - Crafts Equivalents. Regiospecific Routes to Thioxanthen-9-one Dioxides. DoM, DreM, or Both ?

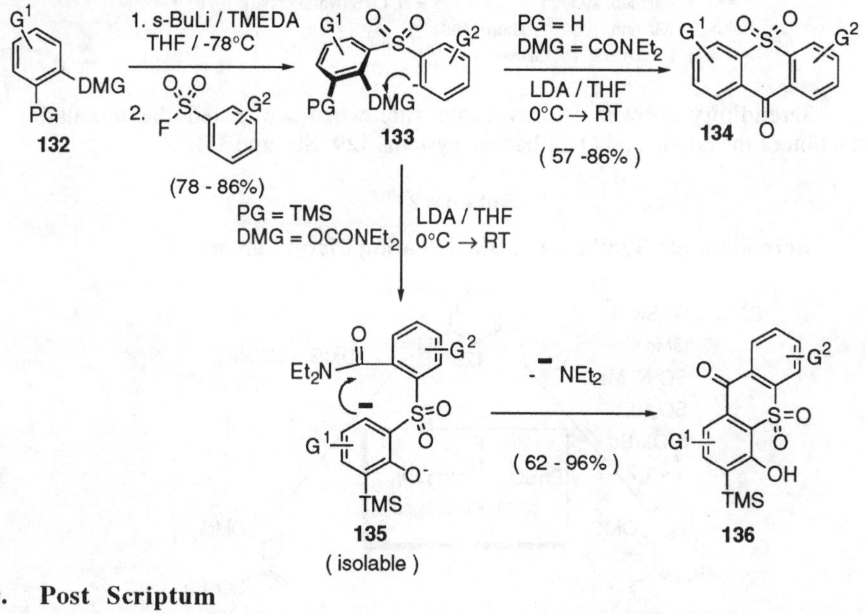

9. Post Scriptum

The successful demonstration of the anionic Friedel-Crafts equivalent for thioxanthone construction (Scheme 32) invited investigation of other DMG-containing heteroatom bridged biaryls (Scheme 30) [83]. In the event, the diphenyl ether amide **138**, prepared from **137** by sequential DoM and Ullmann reactions, underwent smooth LDA-mediated cylization to afford xanthones **140** in good yield (Scheme 33). To take the provocative title of Professor Eschenmoser in our context, "why not phosphorus?" In this event, the prototype phosphonyl derivative **139**, readily prepared by a DoM step from **137**, underwent analogous cyclization to afford the phosphorinone **141** in an equally smooth reaction. Generalization and application of these preliminary results are being pursued [84].

Scheme 33

Anionic Friedel-Crafts Complements. Regiospecific Routes to Xanthen-9-ones and Dibenzo[b,e]phosphorin-10-ones

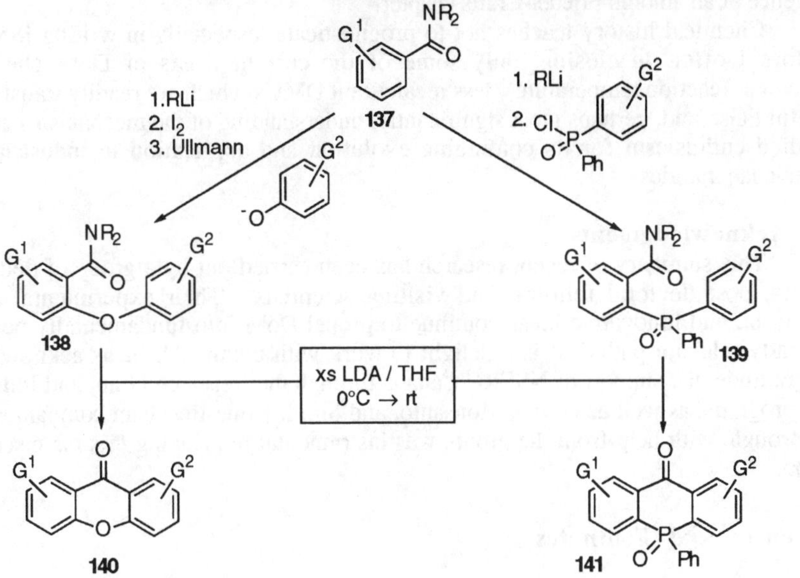

10. Summary and Prognosis

Since the systematic, original work of Hauser, the DoM process has undergone comprehensive development and has found considerable application - yet it is early and treacherous to refer to it as a mature reaction. Hopefully, this review provides convincing evidence for the advantages it lends to the regiospecific preparation of polysubstituted aromatics, the new concepts it offers in total synthesis problems, and the complementarity it shows to classical methodologies. Combinations of modern synthetic methods leads to opportunities which are greater than their individual sum. Such is the case for the DoM - cross coupling link (Scheme 15) which teaches new technologies for the regiospecific construction of biaryls, heterobiaryls, polyaryls (Scheme 20), and various highly condensed architectures, methods which may supersede the traditional, e.g. Ullmann, reactions. Further synthetic consequences of the product biaryls are condensed aromatics (e.g. phenanthrols) and heteroaromatics (e.g. phenenthridines, dibenzopyranones) (Schemes 16 and 18) which also have distinct advantages in natural product synthesis (e.g. Scheme 17).

The discovery of DreM processes pointed to new possibilities. The regiospecific synthesis of fluorenones (Scheme 22) offers a mild anionic Friedel-Crafts equivalent which, by virtue of the DMG effect, may lead to products that are complementary to those dictated by the normal electrophilic substitution rules. The parallel studies on the biaryl O-carbamates (Scheme 23) permits new synthetic approaches to highly substituted biaryls, especially by a "piggy-back" carbamoyl ring-to-ring transfer to circumvent sterically resistant cross coupling reactions, and provides new modes for dibenzopyranone construction (Scheme 24). The broader use of sequential DreM concepts (Schemes 25 and 27), already shown to play key roles in natural product synthesis (Schemes 26 and 29), may be anticipated. The transformations of diaryl sulfone, diaryl ether and diaryl phosphonyl systems into thioxanthone, xanthone, and phosphorinone heterocycles respectively (Schemes 32 and 33) is early promise of an emergence of an anionic Friedel-Crafts chapter.

Chemical history teaches not to prognosticate, especially in writing [85], and therefore I offer, in closing, only some of the current needs of DoM chemistry (convenient reaction temperatures, less recalcitrant DMGs which are readily transformed to useful FGs, and, perhaps most significantly, understanding of the mechanism) and my unbridled enthusiasm for its continuing evolution and application in industrial and academic laboratories.

11. Acknowledgments

This summary of recent research has been carried out by a group of dedicated students, post doctoral fellows, and visiting scientists. Their experimental skills, enthusiasm, and innovative ideas continue to propel DoM into fundamentally new and practically relevant paths. It is a delight to work with them. All of us acknowledge, with gratitude, the support of NSERC Canada through the Research Grant and Industrial Chair programs as well as Glaxo, Monsanto, and Smith Kline Beecham companies. Dr. Paul Brough, with help from the group, was instrumental in bringing this manuscript to fruition.

References and Footnotes

1. McBride, J.M. (1974) *Tetrahedron* **30**, 2009; Wotiz, J.H. and Rudofsky, S. F. (1984) *Chem. Brit.* **20**, 720; Strunz, (1988) *Chem. Unserer Zeit* **23**, 170.
2. Bentley, O.T. (1964) *From Vital Force to Structural Formulas*, Houghton Mifflin Co. Boston, p 101 ff.
3. Stock, L.M. (1968) *Aromatic Substitution Reactions,* Prentice-Hall, Inc. Englewood Cliffs, N.J.
4. See *inter alia*, Roberts, J.D. and Caserio, M.C. (1965) *Basic Principles of Organic Chemistry*, Benjamin, NY, p 799; Cason, J. (1966) *Principles of Modern Organic Chemistry,* Prentice-Hall, Inc. Eglewood, Cliffs, N.J., p 266; Royals, E.E. (1954) *Advanced Organic Chemistry*, Prentice-Hall, Inc. Englewood Cliffs, N.J. p 412.
5. Morrison, R.T. and Boyd, R.N. (1992) *Organic Chemistry*, 6th ed. Prentice Hall, Englewood Cliffs, N.J., p 517.; McMurry, J. (1994) *Organic Chemistry*, 3rd ed. Brooks/Cole, Pacific Grove, CA, p 561; Fox, M.A., Whitesell, J.K. *Organic Chemistry*, Jones and Bartlett, Boston, p 369.
6. Warren, S. (1982) *Organic Synthesis the Disconnection Approach*, Wiley, p 6-15; Norman, R.O.C. *Principles of Organic Synthesis*, 2nd Ed., Chapman and Hall, London, 1993, pp 367-453; Smith, M.B. (1994) *Organic Synthesis*,

McGraw-Hill, Inc. N.Y., pp 183-192. However, for an enlightening, textual "retrosynthetic" treatment before the term was coined, see Hedrickson, J.B.; Cram, D.J.; Hammond, G.S. (1970) *Organic Chemistry*, 3rd Ed., McGraw-Hill, New York, p 683.

7. A hesitant estimate from Laird, T. personal communication; see Katritzky, A.R., Rees, C.W. eds. (1984) *Comprehensive Heterocyclic Chemistry*, Pergamon, Oxford vol. **1**. pp. 143-149; Sneader, W. (1985) *Drug Discovery, the Evolution of Modern Medicines*, Wiley, N.Y.
8. Elaboration and illustration is provided in industrial short courses presented by the author since 1987. For a comprehensive list of references, see ref 13.
9. The 1995 Aldrich catalogue lists 875 monosubstituted benzenes (44 with an alkyl carbon atom and 831 with a carbon other than alkyl or a heteroatom as the direct link); 407 1,2-disubstituted, 243 1,2,3-trisubstituted, and 65 1,2,3,4-tetrasubstituted benzenes. We thank Dr. S.J. Branca, Aldrich Chemical Co. for this information.
10. Gilman, H. and Bebb, R.L. (1939) *J. Am. Chem. Soc.* **61**, 109.
11. Wittig, G. and Fuhrman, G. (1940) *Chem. Ber.* **73**, 1197.
12. Puterbaugh, W.H. and Hauser, C.R. (1964) *J. Org. Chem.* **29**, 853.; Slocum, D.W. and Sugarman, D.I. (1974) *Adv. Chem. Ser.* No. **130**, 227.
13. For leading references to the DoM field, see Snieckus, V. (1990) *Chem. Rev.* **90**, 879.
14. This phrase is likely due to Dieter Seebach although the occasion and date of pronouncement is unknown to the author.
15. The advantageous notwithstanding, DoM reactions have not yet reached undergraduate textbook inclusion. For an exception, see Roberts, J.D. and Caserio, M.C. (1965) *Basic Principles of Organic Chemistry*, Benjamin, N.Y., p 810.
16. A discussion, the first of many, with Peter Beak in 1977 which triggered our work will always be a fond memento.
17. de Silva, O.S., and Reed, J.N. (1978) *Tetrahedron Lett.* **19**, 5099. For a full account with experimental details, see de Silva, S.O., Reed, J.N., Billedeau, R., Wang, X., Norris, D.J., and Snieckus, V. (1992) *Tetrahedron*, **48**, 4863-4878; de Silva, S.O., Ahmad, I., and Snieckus, V. (1979) *Can. J. Chem.* **57**, 1598.
18. Conley, R.A., Barton, D.L. Ortho Pharm. Corp. U.S. Pat 4,617,417 *Chem Abstr.*,1987 **106**, 4980h; U.S. Pat 4,658,054 *Chem. Abstr.* 1988, **107**, 115366u. We thank Dr. Conley for lending early appreciation of the power of

DOM in industry.

19. Hlasta, D.J. *et al* Sterling Winthrop U.S. Pat 5,339; Hlasta, D.J., Court, J.J., and Desai, R.C. (1991) *Tetrahedron Lett*. **32**, 7179. We thank Dr. Hlasta for provision of patent and stimulating discussion.
20. Anderson, R. (1984) *Chem. Ind.(London)*, 205.
21 Cuomo, J., Gee, S.K., and Hartzell, S.C. (1990) in *Synthesis and Chemistry of Agrochemicals*, Baker, D.R., Fenyes, J.G., Moberg, W.K. Eds. *ACS Symposium Series* No. **443**, ACS, Washington, D.C.
22. Manske, R.H.F. (1959) Fifty Years with Alkaloids, *Chemistry in Canada*, June, p 74.
23. Watanabe, M. and Snieckus, V. (1980) *J. Am. Chem. Soc*. **102**, 1457.
24. Wang, X. and Snieckus, V. (1990) *Synlett* 313.
25. Wang, X. (1992) Ph.D. Thesis, University of Waterloo.
26. For an excellent summary of ellipticine alkaloid construction, see Gribble, G.W. (1991) *Synlett*, 289.
27. de Silva, S.O., Watanabe, M., and Snieckus, V. (1979) *J. Org. Chem*. **44**, 4802.
28. Kende, A.S., Belletire, J.L., and Hume, E.L. (1973) *Tetrahedron Lett*. 2935.
29. For an exhaustive and critical discussion of synthetic approaches to anthraquinones, see Krohn, K. (1990) *Tetrahedron* **46**, 291.
30. Bostrycoidin: Watanabe. M., Shinoda. E., Shimizu, Y., Furokawa, S., Iwao, M., and Kuraishi, T. (1987) *Tetrahedron* **43**, 5281; Pyridoanthraquinone precursor to Dynemicin A: Hidenori, C., Porco, J. A. Jr., Stout, T. J., Clardy, J., Schreiber, S. (1991) *J. Org. Chem*. **56**, 1692; Nicolau, K. C., Gross, J.L., Kerr, M.A., Lemus, R.H., Ikeda, K., and Ohe, K. (1994) *Angew. Chem. Int. Ed. Engl*. **33**, 781.
31. a) Watanabe, M., Sahara, M., Kubo, M.;, Furukawa, S., Billedeau, R.J., and Snieckus, V. (1984) *J. Org. Chem*. **49**, 742. b) Naoi, Y., Higuchi, S., Nakano, T., Sakai, K., Nishi, A., and Sano, S. (1975) *Syn. Commun*. **5**, 387.
32. a) Sibi, M.P., Chattopadhyay, S., Dankwardt, J.W., and Snieckus, V. (1985) *J. Am. Chem. Soc*. **107**, 6312. b) Kraus, G.A. (1981) *J. Org. Chem*. , **46**, 201.
33. For another significant use of carbamate DoM chemistry in the assemblage of pancratistatin, see Danishefsky, S. and Lee, J. Y. (1989) *J. Am. Chem. Soc*. **111**, 4829.
34. Kharash, M.S. and Fields, E.K. (1941) *J. Am. Chem. Soc*. **63**, 3216.
35. Corriu, P.J.P. and Masse, J.P. (1972) *J. Chem. Soc. Chem. Commun*. 144.

36. Tamao, K., Sumitani, K., and Kumada, M. (1972) *J. Am. Chem. Soc.* **94**, 4374.
37. Negishi, E.-i., King, A. O., and Okukado, N. (1977) *J. Org. Chem.* **42**, 1821.
38. Miyaura, N., Yanagi, T., and Suzuki, A. (1981) *Synth. Commun.* **11**, 513.
39. a) Beletskaya, I.P. (1983) *J. Organometal. Chem.* **250**, 551. b) Echavarren, A.M. and Stille, J.K. (1987) *J. Am. Chem. Soc.* **109**, 5478; Stille, J.K. (1986) *Angew. Chem. Int. Ed. Engl.* **25**, 508.
40. Kidwell, R.L., Murphy, M., and Darling, S. (1973) *Org. Syn. Coll. Vol.* 5, 918. See also Beak, P. and Brown, R.A. (1982) *J. Org. Chem.* **47**, 34.
41. Sharp, M.J. and Snieckus, V. (1985) *Tetrahedron Lett.* **26**, 5997.
42. Although the intermediate boronic acids, their trimers, and esters have been characterized in many cases [Song, Z.Z., Zhao, Z.Y., Mak, T.C.W., and Wong, H.N.C. (1993) *Angew. Chem. Int. Ed. Eng.* **32**, 432.; Washburn, R.W., Levens, E., Albright, C.F., and Billig, F.A. (1963) *Org Syn, Coll Vol. IV*, p 68; Oh-e, T., Miyaura, N., Suzuki, A. (1990) *Synlett* 221; Sato, M., Miyaura, N., Suzuki, A. (1989) *Chem. Lett.* 1405] they are normally inconvenient to handle and, in some cases, unstable to protodeboronation [Brandao, M.A.F., de Oliveira, A.B. and Snieckus, V. (1993) *Tetrahedron Lett.* **34**, 2437]. Characterization may be achieved via their diethoanolamine adducts [41].
43. See, inter alia, for pharmaceuticals: Reitz. D., Garland, D.J., Norton, M.B., Collins, J.T., Reinhard, E.J., Manning, R.E., Olins, G.M.,Chen, S.T., Palomo, M.A., McMahon, E.G., and Koehler, K.F. (1993) *Bioorg. Med. Chem. Lett.* **3**, 1055.; Santella, III, J.B., Duncia, J.V., Ensinger, C.L., VanAtten, M.K., Carini, D.J., Wexler, R.R., Chiu, A.T., Wong, P.C., and Timmermans, P.B.M.W.M. (1994) *Bioorg. Med. Chem. Lett.* **4**, 2235; Smith, G.B., Dezeny, G.C., Hughes, D.L., King, A.O., and Verhoeven, T.R. (1994) *J. Org. Chem.* **59**, 8151.; for liquid crystals: Reiffenrath, V., Krause, J., Plach, H.J., and Weber, G. EP 0 440 082 A2; DE 40 00 535 A1; *Chem. Abstr.* 1992, **116**, 83917s.
Reitz, D.B., Penick, M.A., Norton, M.B., Reinhard, E.J., Ollins, G.M., Corpus, V.M., Palomo, M.A., McGraw, D.E., and McMahon, E.G. (1994) *Bioorg, Med. Chem. Lett.* **4**, 105.; Reitz, D.B., Penick, M.A., Reinhard, E.J., Cheng, B.K., Olins, G.M., Corpus, V.M., Palomo, M.A., McGraw, D.E., and McMahon, E.G. (1994) *Bioorg, Med. Chem. Lett.* **4**, 99.
44. Quesnelle, C., Iihama, T., Aubert, T., Perrier, H., and Snieckus, V. (1992) *Tetrahedron Lett.* **33**, 2625.

45. First presented in back to back papers at the PACIFICHEM Meeting, Honolulu, Dec. 1989, *Abstr. ORG 499.* See Fu, J.-m. and Snieckus, V. (1990) *Tetredron Lett.* **31**, 1665. and Oh-e, T., Miyaura, N., Suzuki, A. (1990) *Synlett* 221; Oh-e, T., Miyaura, N., and Suzuki, A. (1993) *J. Org. Chem.* **58**, 2201.
46. Review: Martin, A.R. and Yang, Y. (1993) *Acta Chem. Scand.* **47**, 221.
47. Fu, J.-m., Sharp, M.J., and Snieckus, V. (1988) *Tetrahedron Lett.* **29**, 5459.
48. Siddiqui, M.A. and Snieckus, V. (1988) *Tetrahedron Lett.* **29**, 5463.
49. Siddiqui, M.A. and Snieckus, V. (1990) *Tetrahedron Lett.* **31**, 1523.
50. Guillier, F., Nivoliers, F., Godard, A., Marsais, F., Quéguiner, G., Siddiqui, M.A., and Snieckus, V. (1995) *J. Org. Chem.* **60**, 292.
51. a) Alo, B.I., Kandil, A., Patil, P.A., Sharp, M.J., Siddiqui, M.A., Josephy, P.D., Snieckus, V. (1991) *J. Org. Chem.* **56**, 3763.; b) Coelho, A., Costa, P. R.R., and Snieckus, V. unpublished results.
52. Lord, H.L., Snieckus, V., and Josephy, P.D. (1990) *Chem. Res. Toxicol.* **3**, 195.
53. Beak, P. Meyers, A.I. (1986) *Acc. Chem. Res.* **9**, 356; Klumpp, G.W. (1986) *Recl. Trav. Chim. Pays-Bas* . **105**, 1.
54. a) Fu, J.-m. (1990) Ph.D. Thesis, University of Waterloo; b) Zhao, B.-p. (1993) Ph.D. Thesis, University of Waterloo.
55. Reuman, M., Meyers, A.I. (1985) *Tetrahedron* **41**, 837.; Gant, T.G. and Meyers, A.I. (1994) *Tetrahedron* **50**, 2297.
56. Kametani, T. and Fukumoto, K. (1972) *Accts. Chem. Res.* **5**, 212.
57. Mallory, F.B. and Mallory, W.M. (1984) *Orga. React.* **30**, 1.
58. Unrau, C., Campbell, M.G., and Snieckus, V. (1992) *Tetrahedron Lett.* **33**, 2773.
59. Campbell, M.G. Ph.D. Thesis, University of Waterloo, in preparation.
60. For leading literature citations, see ref 58.
61. Krizan, T.D. and Martin, J.C. (1983) *J. Am. Chem. Soc.* **105**, 6155.
62. Seebach, D. (1990) *Angew. Chem. Int. Ed. Engl.* **29**, 1320.
63. Sengupta, S., Liete, M., Raslan, D.S., Quesnelle, C., and Snieckus, V. (1992) *J. Org. Chem.* **57**, 4066.
64. Beaulieau, F. (1994) Ph.D. Thesis, University of Waterloo.; Puumala, K. work in progress.
65. Quesnelle, C., Familoni, O.B. and Snieckus, V. (1994) *Synlett* 349.
66. Fu, J.-m., Zhao, B.-p., Sharp, M.J., and Snieckus, V. (1991) *J. Org. Chem.* **56**, 1683.

67. Sibi, M.P. and Snieckus, V. (1983) *J. Org. Chem.* **48**, 1935.

68. Mills, R.J., Taylor, N.J., and Snieckus, V. (1989) *J. Org. Chem.* **54**, 4372.

69. The temperature at which degradation of THF by LDA occurs has not been measured, but is probably significant above -20 °C (Rathman, T.L. personal communication). For THF degradation by butyl lithium see Bates, R.B., Kroposki, L.M., and Potter D.E. (1972) *J. Org. Chem.* 37, 560; Honeycutt, S.C. (1971) *J. Organometal Chem.* **29**, 1.

70. Wang, W. and Snieckus, V. (1992) *J. Org. Chem.* **57**, 424.

71. James, C. and Snieckus, V. unpublished results.

72. Sharp, M.J. (1986) M.Sc. Thesis, University of Waterloo.

73. Wang, X.; Snieckus, V. (1991) *Tetrahedron Lett.* **32**, 4879.

74. Hughes, A.B. and Sargent, M.V. (1989) *J. Chem. Soc. Perkin Trans.1,* 1787.

75. a) Gilman, H. and Webb, F.J. (1949) *J. Am. Chem. Soc.* **71**, 4062.; b) Gilman, H.; Esmay, D. L. (1953) *J. Am. Chem. Soc.* **75**, 278.

76. Eaborn, C. and Sperry, J. A. (1961) *J. Chem. Soc.* 4921.

77. Truce, W. E. and Amos, M. F. (1951) *J. Am. Chem. Soc.* **73**, 3013.

78. Narasimhan, N. S. and Chandrachood, R. S. (1979) *Synthesis* 589.

79. Snieckus, V., Beaulieu, F., Mohri, K., Han. W., Murphy, C.K., and Davis, F.A. (1994) *Tetrahedron Lett.* **35**, 3465.

80. Differding, E. and Wehrli, M. (1991)*Tetrahedron Lett.* **32**, 3819.

81. a) Truce, R.A. and Ray, W.J., Jr. (1959) *J. Am. Chem. Soc.* **81**, 481; b) Köbrich, G. (1959) *Chem. Ber.* **92**, 2981.; c) Frye, L.L., Sullivan, E.L., Cusack, K.P., and Funaro, J.M. (1992) *J. Org. Chem.* **57**, 697.

82. Beaulieu, F. and Snieckus, V. (1994) *J. Org. Chem.* **59**, 6508.

83. Dicussions with colleagues over real coffee in the garden at Ravello was the final persuasion to experimentally test these ideas upon return to Waterloo.

84. Familioni, O.B., Ionica, I., Gray, M. and Snieckus, V. unpublished results.

85. "Most if not all of the known types of organic derivatives of silicon have been considered ... the prospect of any immediate and important advances in this section of organic chemistry does not seem to be very hopeful." Kipping, F.S. (1937) *Proc. Royal Soc. A.* **159**, 139.

AS WE HEAD INTO THE 21ST CENTURY, IS THERE STILL VALUE IN TOTAL SYNTHESIS OF NATURAL PRODUCTS AS A RESEARCH ENDEAVOR?

CLAYTON H. HEATHCOCK
Department of Chemistry, University of California
Berkeley, CA 94720-1111 USA

1. Introduction

This year (1994) marks the 50th anniversary of the publication of the total synthesis of quinine (**1**) by Woodward and Doering [1]. This accomplishment was the first of a series of increasingly ambitious total syntheses by Woodward and his students and ushered in what has been called the "Woodwardian Era" of organic chemistry. The outstanding synthetic successes that were realized during the 1940s and 1950s, first by Woodward and then by a number of other talented organic chemists, gave birth to a whole new sub-field of chemistry – "Natural Products Total Synthesis".

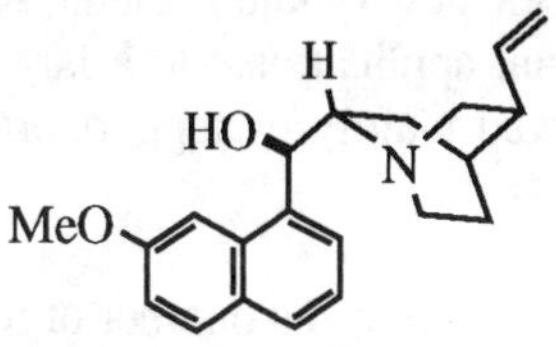

1

The ground rules of the game of Natural Products Total Synthesis were simple. In the beginning of the Woodwardian Era, the very existence of an unsynthesized natural product was sufficient justification for its synthesis and a successful total synthesis was viewed as a final, decisive step in the establishment of the structure of a natural product. However, for this rationale to be valid, it was important that the synthesis be well-planned and that the chemist only employ reliable, well-understood reactions – ones that were not likely to give unexpected results and thereby diminish the value of the synthesis as a proof or confirmation of structure.

The advent of x-ray crystallography, one-dimensional NMR spectrometry, and, perhaps most importantly, two-dimensional NMR spectrometry, has changed this situation. In today's world it is a rare synthesis that contributes to our knowledge of the structure of a natural product. As the value of total synthesis as a tool for structure proof waned, practitioners of the Art sought and found a new *raison d'être*, namely that the practice of

C. Chatgilialoglu and V. Snieckus (eds.), Chemical Synthesis, 223–243.

total synthesis can often yield unexpected results. The case was made effectively in 1963 by Woodward himself in discussing the earlier demise of classical structure elucidation by chemical degradation in a footnote to his classic paper on the synthesis of strychnine [2]:

" Of course, men make much use of excuses for activities which lead to discovery, and the lure of unknown structures has in the past yielded a huge dividend of unsought fact, which has been of major importance in building organic chemistry as a science. Should a surrogate now be needed, we do not hesitate to advocate the case for synthesis."

A corollary of this philosophy was that chemists could be more daring in their choice of synthetic tools. In other words, if the structure of the target is absolutely secure, and the total synthesis will not provide further information on this point, the synthetic chemist can take chances with unknown reactions and unproven strategies. The focus of a total synthesis is shifted to the chemistry used in the synthesis, and the specific target serves to enforce a certain discipline on the chemist. That is, for the synthesis to be successful, the **exact** structure of the target must be synthesized, even if this structure is not easily amenable to state-of-the-art strategies and tactics.

As the field of Natural Products Total Synthesis burgeoned, the related discipline of "Synthetic Methods Development" was born. In this field of research, chemists sought to invent new procedures for accomplishing various synthetic chores, such as introduction and modification of various functional groups, construction of carbon-carbon bonds in sundry ways, control of relative or absolute configuration when new stereocenters are created, etc. Although some of the new synthetic methods that have emanated from these activities have found commercial applications, to a large extent, the field of Synthetic Methods Development has served mainly to support practitioners of Natural Products Total Synthesis.

In the past decade or so, we have witnessed a number of highly impressive total synthesis accomplishments. Notable among these are the epochal conquests of vitamin B_{12} by Woodward and Eschenmoser [3] and palytoxin by Kishi [4]. These signal achievements, and many others, have caused many observers, at least in the United States, to question the value of continued research in the area of synthesis, both total synthesis and methods development. In a previous article, I have commented on the situation as follows [5]:

"If important fields such as medicine, biochemistry, and materials science are to continue to progress, it is essential that we be able to synthesize literally any structure that the imagination can conceive. The goal of research in organic synthesis is to reach a level of sophistication where this ability can be taken for granted. Yet, there is abroad an insidious notion that organic chemists have already become so adept at synthesis that further academic exercises such as that described here are no longer necessary, that there is no longer any need for research in multi-step synthesis. I believe that, in part, this difficulty stems from a confusion of the adjectives effective ["adequate to accomplish a purpose"] [1] and efficient ["performing or functioning in the best possible and least wasteful manner"] [6]. In fact, although chemists **have** over the last half-century become rather adept at constructing small amounts of very complicated molecules, we generally cannot prepare most desired organic compounds in an **efficient**, practical, and truly cost-effective manner. We **can** synthesize almost any given molecule and demonstrate that we have done it. However, this is due more to the development of high-powered separation and analytical tools (hplc, capillary glpc, tlc, FT nuclear magnetic resonance spectrometry) than to some discontinuous change that has come about in our ability to solve synthesis problems. That is, because we can carry out separations and establish structure with minute amounts of material, it is possible to do synthesis on an order-of-magnitude smaller scale. Thus, we can carry out multistep syntheses in a much shorter period of time than was possible ten or twenty years ago, since more time can be devoted to developing new chemistry and less to the time-consuming 'bringing up material.' This maturity should be considered a normal step in the development of a productive science, not a sign that all of the problems have been solved and that the field no longer merits investigation. Problems of synthesis still abound, as is well known by anyone who practices the art. Our textbooks are filled with hundreds of synthetic methods, all of which have limitations that will never be discovered unless the methods are tested in the arena of challenging multifunctional synthesis. Although our approaches to problems **have** matured, we need even more mature strategies of synthesis. There is no reason that organic chemists should not be able to surpass Nature's virtuosity in the synthesis of complex organic structures. In fact, we are still very far from this goal in most cases."

Thus, when confronted with the task of chairing a Panel Discussion on some aspect of organic synthesis for this NATO Workshop, I chose as a topic the rhetorical question that is the title of this chapter. As panelists for the discussion, I selected five distinguished participants in the Workshop: Paul A. Bartlett (Berkeley), Derek H. R. Barton (Texas A&M University), Ronald Breslow (Columbia University), Albert Eschenmoser (ETH), and Stephen Hanessian (Université de Montréal).

At the outset, and to serve as a sub-topical guide for the discussion, several questions were posed to the panelists and audience. A spirited discussion, involving panelists and members of the audience alike, occurred over a period of about three hours on a rainy Thursday afternoon. In the following pages, I attempt to capture the essence of the major points that were touched upon during this lively exchange. Comments are organized along topical lines and not necessarily in the order in which they were made in the discussion. Although the following condensation was made from a tape recording of the actual discussion, statements have been heavily edited and condensed for the purpose of this chapter.

2. Does Total Synthesis Still Have Value as a Method of Structure Determination?

In the Introduction, it was stated that ". . . it is a rare synthesis that contributes to our knowledge of the structure of a natural product." However, there are still many cases in which synthesis is the best way to gain structural information, particularly when stereochemistry is an issue. For example, the synthesis of the "C-40 archaebacterial diol" (**2**) in 1985 did establish both the relative and absolute configurations of this novel natural product, which does not crystallize and is so conformationally mobile that spectroscopic methods cannot readily provide stereochemical information [7].

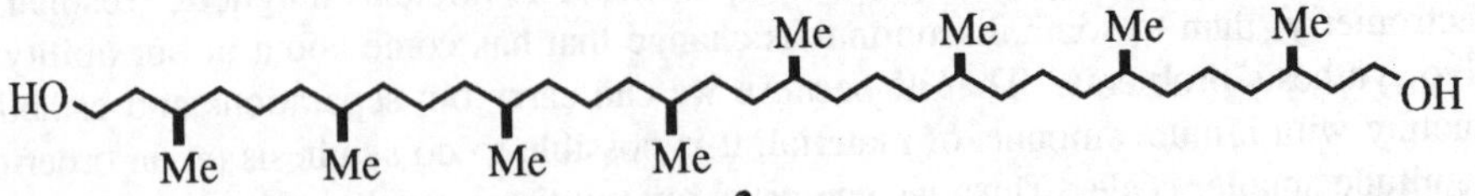

2

Another good example is found in Schreiber's recent synthesis of the immunosuppressant discodermolide, which established the absolute configuration of the natural product to be **3** [8].

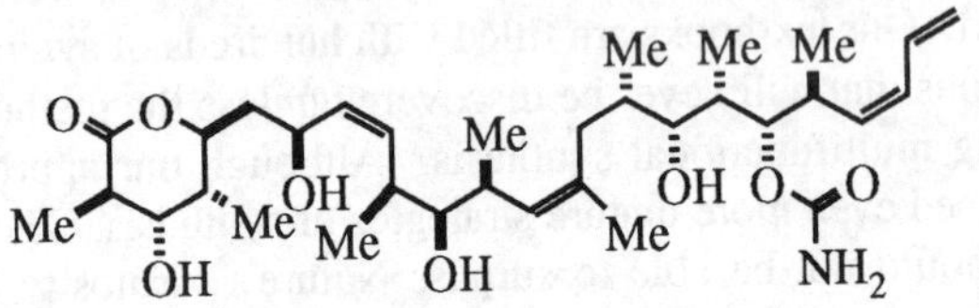

3

This total synthesis also led to a "dividend of unsought fact." By using essentially the same synthetic protocols, Schreiber's group was able to prepare both the natural and the **unnatural** enantiomers of discodermolide. Amazingly, it was found that both enantiomers inhibit cell proliferation, but at different concentrations and by different mechanisms! It is probably fair to say that this remarkable fact would have never been known had not chemists gone to the trouble to carry out a 36-step synthesis of discodermolide and its enantiomer.

3. Is Total Synthesis an Appropriate Vehicle for Educating Students?

This point was not specifically listed as a sub-topic of the discussion because I had assumed that it would be obvious to all that one undisputed value of total synthesis research is that it is an effective means of educating young scientists. In my opening comments and later, I said:

" Now it seems to me that there is no debating the fact that multistep synthesis is a good training vehicle for students, those who are going to go into medicinal chemistry, for example.

" I have trained many graduate students and postdocs who have worked on total synthesis projects, and most of them have gone to work in the pharmaceutical industry. They are highly sought after. The people in industry really value this kind of training, for several reasons. First, multistep synthesis, persisting in order to find an effective path to a specific target molecule, develops experimental skill because you aren't allowed to make many mistakes. That is, you have to get your material through the 10 or 15 or 20 steps without losing it. Second, you are exposed to a diversity (wide variety) of chemistry. Finally, I think that solving a problem of this sort builds character because the student regularly faces barriers and must find ways through or around them."

Albert Eschenmoser commented along similar lines:

" In my experience and in the experience of many young synthetic chemists who have spent Ph.D. time with total synthesis, this kind of research is among the best in giving a young chemist a very broad survey of what is the most important quality organic chemists can bring to natural or molecular science, namely their way of understanding reactivity. The process of preparing a structurally-complex target molecule by chemical synthesis confronts the young practitioner with so many different and difficult reactivity situations that the education that results from the experience is probably the major payoff of the work."

The question was put in a slightly different form for the panel by Dr. W. Rutsch, CIBA-GEIGY:

" Do you think it is fair and right to do natural product synthesis to educate young chemists? Is this a good tool to train them? Do you think that after such training these students are properly equipped to solve the problems that will face them in industry? After all, we in industry and you at the university are not isolated entities. We are affected by public opinion. The chemical industry at the moment has a very **bad** reputation in the public opinion. I don't know about universities. The question should be: What does natural product synthesis contribute to the public opinion about chemistry? Or, in other words, can we justify spending taxpayer's money doing elegant natural product synthesis? Might there be other important problems of society that should be solved before we go into natural product synthesis?"

To this question of "relevance," I replied :

" These students go to Merck or Pfizer or Lilly or Abbott or one of the many other companies and make a big salary and pay lots of taxes. In just a few years they pay back in taxes all of the money that other taxpayers have paid to educate them. Furthermore, for the rest of their lives they continue to pay taxes on their salaries, which are higher than they would be if they had not been so educated. Therefore, I think that total synthesis is both a good tool for education and a good deal for the public. We turn out scientists who are highly qualified and who do good things for society."

Paul Bartlett responded to the relevance issue:

" It's worth elaborating a little bit on this issue of practical application of the research that we do in academe. I am often asked why we don't direct some of our enzyme inhibitor design approaches towards more medicinally relevant enzymes rather than study such common enzymes as thermolysin and carboxypeptidase. After all, we don't really **need** a better inhibitor of our digestive enzymes. For example, why did I not work on the angiotensin-converting enzyme, which is a zinc peptidase just like the ones I have described in my lecture? My response is that I have made a conscious decision to avoid enzymes with which I would be competing with professionals. The goal of industrial research is to come up with practical compounds that can be turned into drugs to earn money for the company. The product of industrial research is profit.

" But profit is not the goal, nor should it be the product, of academic research. Our products are ideas and well-trained students, who can go to industry and solve relevant problems. If we really want to understand how a protein binds a small molecule, and how to design the best inhibitor of an enzyme, thermolysin is just as useful as the angiotensin-converting enzyme.

" Furthermore, commercial drug discovery carries with it a lot of constraints, things such as patent coverage, pharmacokinetics, oral bioavailability, human safety, etc. It doesn't seem to me that it is appropriate for one or two graduate students in my research group to try to compete with 20 chemists at Merck and 15 chemists at Bristol-Myers-Squibb in trying to develop a real drug candidate."

Ronald Breslow introduced another element into the discussion of education:

" Yesterday a student asked me how we can encourage and train people to be creative. Of course, that is an absolutely critical element of education and we simply must do it. I think most American universities require their students to do research proposals. At Columbia we even require research proposals outside the student's research area. But everybody, especially the students at this Workshop, should be listening to the seminars given not just to see if they understand them, but also with the question of what **you** could do in the area of the seminar. After you hear a seminar, you should think about whether you could do something better, or at least use that material in your own work. If you do this and if you attend a lot of seminars that are not just in your own special area, for example, in various parts of biology, you will acquire the breadth that you need. But it has to be a creative breadth, not just an amassing of facts."

The discussion of total synthesis and education was carried in yet another direction by Albert Eschenmoser:

" During my generation total synthesis was heavily executed and large numbers of chemistry students were involved in it, profiting from that quality in education.

However, it is to be admitted that we educators missed one important point – we didn't teach our students how to recognize a research problem. We gave them the target, that was the problem. Therefore, I think educators must pay more attention to teaching their students how to identify significant problems."

Professor Victor Snieckus, University of Waterloo, posed the question :

" How do we go about training students who have the perspective on the one hand and are solidly trained in their specialty on the other hand? On any given day a synthetic organic chemist is thinking so hard about the next step that he usually does not have the time or the possibility to look at the other side of the fence. How do we encourage students to recognize problems that are more general, and still be very good as synthetic organic chemists?"

Ronald Breslow responded to this question:

" That is a very important issue. I think there is no question that the student or postdoc of tomorrow must be a person of many talents. If we just teach chemistry, and only chemistry, if we expose our students only to that discipline, we are doing them a disservice. There has to be at least some exposure, some appreciation, some sensitization to things other than chemistry. I am not saying that chemistry graduate students who are doing a synthesis or doing something that is entirely organic should also be expert pharmacologists or molecular biologists. However, in any University there are other departments and other courses than organic chemistry. Even if each student takes only one or two other courses, such as introductory biochemistry, that could be very useful. Many departments have introduced Special Topics courses. For example, in my department at Columbia, we have introduced a special course entitled 'Topics in Medicinal Chemistry.' We bring in colleagues from industry to give lectures about different subjects. So I think we can't close our eyes and say we are organic chemists and we do what we do and just forget about the rest of the world. I think it is important that we expose our students to as much as possible."

4. What are Some of the Great Unanswered Questions in Organic Chemistry, and can Total Synthesis Play a Role in Finding Answers to These Questions?

Derek Barton began the discussion of this topic with comments about practical challenges that still face us in synthesis:

" The real problem in synthetic chemistry remains as it always was. How do you get 100% yield and 100% stereoselectivity in every reaction? This is the problem that every industrial chemist faces daily. They have a synthesis; they need 100% yield. It solves the pollution problem, because you have nothing to throw away afterwards. It does all sorts of things. What are we doing about getting these 100% yields?"

Ronald Breslow pointed out that chemists have found it difficult to identify field-wide problems, as is commonly the case in other branches of science:

" Chemistry has two unusual problems compared to some other fields. If you are a high energy particle field physicist, you are really trying to find out if the unified field theory is correct or you are trying to find the last unknown boson. That is, the field has just a few central problems and if you are not pursuing one of these few goals you are not considered to be doing serious physics. On the other hand, chemistry is both lucky and unlucky in the sense that we don't have any such central questions. You can imagine formulating some, but they are very general, like what is the relationship between structures and properties. You can't take this seriously as a single target that we all can address, nor could you make a clear statement that you have actually solved the problem. So we have this tremendous diversity and not everyone agrees on what the important problems are. Some people may think that developing new methodology is terribly important; others that it is important to figure out crystal structures. There are all sorts of things people think are important to do in organic chemistry. And that is both our strength and our weakness, because we can never say 'Well, that was clearly a great problem, and now we have solved it!'

" The other problem we have is that we are not a **natural** science, in the sense that more than 95% of all known compounds are compounds chemists have **made**; likewise, chemists have **discovered** most chemical reactions. In this regard, chemists are very different from other scientists. That is, what is the 'synthetic astronomy'? What is the 'synthetic geology'? This aspect of chemistry is very difficult for other scientists to grasp. They ask 'why are you working on this problem? It is not a natural problem. Is it just a problem you have dreamed up? Are you just amusing yourself?' But, of course, we are not. We are really interested in **all** the chemistry that is possible, not just the chemistry that Nature happens to present to us. I think that one of the arguments against working only on the synthesis of natural products is that the richness of chemistry has grown by its extension beyond Nature into other areas, the creation of new polymers, not just biopolymers, the creation of new properties, not just the properties that came along in Nature. It is difficult to explain this to the public because they think we are just inventing problems in order to solve them, like doing crossword puzzles. But we have to explain that this really is not so, we **are** a very special science, and we are different from other sciences, and even other scientists sometimes don't understand what it is we do."

Paul Bartlett expressed some concern about laying out agendas for people to work on:

" I am a little bit worried by the idea that we should try to predict or prescribe what we should do in the future. In hindsight, much of what we do turns out to be useless. The corollary is that ahead of time, we don't always know what is going to turn out to be useless. And whether we are working in organic synthesis, whether we are trying to understand noncovalent interactions, whatever endeavor we undertake, I don't think we are going to be able to tell ahead of time exactly what it is we should do. I think we have to let people's creativity drive them and time will tell what have been the important contributions."

However, Albert Eschenmoser challenged the audience with one grand, central problem:

" In the history of natural science, natural products synthesis has played a very special role, the demystification of nature. It started with the synthesis of urea by Wöhler in 1828 [9]. The effect of that synthesis led to the feeling **we** can do it, namely make compounds that are produced by living matter. Let me take a big jump, to vitamin B_{12} and palytoxin. Again, these Woodwardian Era total syntheses marked a step in demystification, namely that we can synthesize essentially **any** natural product if we just want to. That is, in the Woodwardian Era, the gradient of 'structural complexity' was the direction chemists followed to find their targets. The natural continuation of this gradient would have been chemical synthesis, on a very broad basis, of biopolymers. However, this procedure of identifying targets was vigorously cut off by the discoveries in molecular biology.

" In my opinion, there **is** a problem that is central to organic chemistry alone and in which biologists cannot help us. We all agree in this Workshop that the emphasis in synthetic research is shifting toward the synthesis of **properties**, and not just compounds. Now, the most important property that we can attack by synthesis is the property of living. The problem, rigorously a problem of synthesis, is to study the laws, the rules, and the principles of self-organization of organic matter. We, as synthetic chemists, know that every step in a well-planned synthesis involves, in principle, a self-assembly process. We simply set the stage so that the participants of the synthetic step can assemble themselves. The big problem is to arrange the stage for self-organization leading to a whole cascade of steps without external interference. That is, we should strive to set the stage so that the products resulting from the first event of self-organization are programmed to assemble themselves for a second step, and so on.

" That is, in essence, the problem of self-organization. In this Workshop, we have seen that there are two rather different kinds of self-organization. One is non-covalent self-organization, which biologists strongly depend upon. The other is covalent self-organization, which is a problem of synthesis.

" You may say that understanding how life originated is a dream that is going too far, that we will never know how life originated. However, my belief is that the statement 'we can never know' is nothing more than an opinion. It is our duty, our task, to deal with this uncertainty. I would guess that we can realize in organic chemistry a model for self-organization that led to life within perhaps a few decades. Perhaps in twenty years, organic chemists will be able to present a system in which self-organization happens. We should not continue to leave that problem to the biologists and the physicists. When

I participate in symposia on self-organization, I find physicists and biologists who burn with the question of understanding how organic matter self-organized toward life and both recognize that actually the problem is one of organic chemistry. But there are only a few organic chemists at these symposia. This is unfortunate. We organic chemists have already once abdicated a most beautiful, most important problem of natural product chemistry to non-chemists. I speak of the DNA helix. To recognize how Nature stores information by that base-pairing was probably the most important problem Nature had in store for chemists. But it was biologists who solved this beautiful chemical problem. There is nothing wrong about that; after all, we are just part of science. But I sometimes wonder how students would view organic chemistry today if they could read in their books that organic chemistry solved the problem, if not of the helix itself, at least of the principal of base pairing, which is a rigorously organic process.

" You may say that in order to tackle such a problem we must know what life is. No! It is exactly one of the goals of this kind of research to **find out** what life is. Because Life can be rationalized fully only when we can understand its beginning. This is the opportunity of organic chemistry to contribute to this central question."

5. Given the Past Accomplishments in Total Synthesis, What Kinds of Total Syntheses are Appropriate in the Future? What Kinds of Total Syntheses *should not* be pursued?

As panel moderator, I ventured an opinion on the question of kinds of total synthesis projects that are counter-productive:

" There definitely **are** kinds of total syntheses activities that people should avoid. One of our problems is that many practitioners don't exercise sufficient judgement about whether a synthesis represents an improvement over prior art. It damages our collective image when someone publishes a total synthesis of a compound that has been synthesized several times before, and the new synthesis is twice as long and uses steps that proceed in poorer yield. This kind of total synthesis does not represent an advance, and is really a step backwards.

" These inferior syntheses are not usually created on purpose, but begin with a good idea that just doesn't work out. The chemist then does whatever possible to complete the synthesis, usually involving a much longer and more cumbersome route than was originally envisioned. In the end, we have an inferior synthesis that is published anyway. I think that this is a practice that we should reconsider. It would be better for the field if people would just bury those projects as failures and not carry on just to get a publication that does not represent an advance in our ability to do synthesis.

Professor Carl Johnson, Wayne State University, responded to this point of view :

" I want to make a brief comment on your suggestion that many total syntheses should be buried. I think that is a disservice. When one tries to develop a strategy for the synthesis of a molecule and the strategy doesn't work, sharing this failure can be just as stimulating to other chemists as sharing a success. Publication of a failed idea points out weaknesses in our tools and chemistry can benefit from having these weaknesses identified. Of course, we have to reach a reasonable compromise between these two positions. I do not take the position that **everything** that is tried should be published and I don't think that you really think that every total synthesis that didn't go off exactly like you thought it would, or didn't beat the last one in overall yield or number of steps should be suppressed. In short, I think there is merit in sharing failures."

I responded to Professor Johnson's comments :

" There are two situations in which we synthetic organic chemists give the impression that we are involved in an activity for our own amusement; that is we are not really trying to further science. For example, the following scenario is not uncommon: A chemist whose primary interest is developing new methods has a good idea for a new reaction, tries the reaction, and finds that it does, indeed, work with some generality in simple situations. The chemist then looks around for some natural product target to serve as an advertisement for the new reaction. He finds a compound, say the mating pheromone of the red leaf moth, and proceeds to carry out a multistep total synthesis that features the new synthetic method as a key step. The synthesis requires a total of 15 steps, one of which is the newly-invented reaction. The problem is that someone has published ten years ago a three-step synthesis based on much simpler chemistry. In fact, I think that people who develop methods **should** show that the invention is an advance by solving a problem that cannot be solved as easily using existing knowledge.

' The other common scenario is the synthesis that is based on a novel method or strategy, and which starts out well but bogs down along the way because the protecting groups don't work out or something of that sort. We end up with a long, unwieldy synthesis, one that starts out well but ends like a 'land war in Asia.' This doesn't do the field of synthesis any good. Why not just publish a Note about the novel method or strategy itself. Why slug your way through a 50-step synthesis of strychnine now that Overman has made it in 15 steps. I think it would be best to cut your losses, publish your method and then get onto something that really is new."

" Whatever the reason, when we publish in 1994 a long synthesis that is inferior to a shorter, 1984 synthesis, we appear foolish to those looking from outside into our field. As Professor Breslow put it, we appear to be '. . . just inventing problems in order to solve them'."

Derek Barton addressed this:

" R. B. Woodward used to believe in planned synthesis. He had the kind of mind that could integrate the knowledge of the day better than anyone else and he really could plan a long synthesis and, frequently, execute it. Now days, I don't think that we, in the academic world anyway, should go in for planned synthesis with known reactions, known reagents, and known principles. However, we must also recognize that good things can come by accident, even out of a well-planned synthesis and the orbital-symmetry correlations that came out of the vitamin B_{12} synthesis is a good example of the **unplanned** fruits that can come from a planned synthesis. So let us not be too dogmatic.

" Of course, there are some natural products, particularly those from marine sources, where it is extremely difficult to get hold of enough to do any biology. In this case there would be a justification for total synthesis, even if it is planned along fully predictable lines.

" Furthermore, I think we should make a distinction between what our friends in Industry should be doing and what people in Universities should be doing. I think in Industry a planned synthesis is a very good thing to do. Because if you can execute that synthesis with greater than 90% yield in every step, you will certainly have a satisfactorily planned synthesis. But in the University, we are not supposed to be doing useful things like that because we are supposed to be original. As a University scientist, my general rule would be 'if you know how to do it, you shouldn't do it.' Instead, you should try to find things that you don't know how to do and solve those problems."

Dr. I. Ujváry, Hungarian Academy of Sciences, made the following statement:

" I would like to suggest three criteria for an appropriate total synthesis; 'the three Es of total synthesis.'

1. Essential, in terms of providing essential chemical information, structure information, or biological information.

2. Economical, in terms of yields (100%), use of environmentally-friendly reagents, and requiring few steps.

3. Elegant, in terms of its intellectual content and artistic impact."

Albert Eschenmoser responded as follows:

" Art is always a bonus to synthesis. It is marvelous that we have this bonus, but there is, I think, a kind of obligation of chemists to be serious about the question of what we are doing. An obligation towards the public. We have to seriously analyze whether we are doing right or whether we undergo apparitions that have to be stopped. And therefore

I think that the artistic aspect of a synthesis, beautiful and marvelous as it is, should not be a justification for carrying out a total synthesis. As Einstein put it, 'Let elegance be the affair of tailors and cobblers.' If your problem is truly essential then you don't care about the elegance. The more essential your first E is, the less important your last E becomes. Fortunately, we do often encounter elegance and beauty and art in our profession, but that is not our justification for doing synthesis."

Dr. J. P. Snyder, Institute for Research In Molecular Biology in Rome, addressed the panel as follows:

" I was most excited by the suggestion of Professor Eschenmoser that the really large and unsolved problem of the origin of life might be a target in which to use total synthesis as a tool. However, if synthetic chemists are really to make progress toward solving such a large problem, I think they need to expand their horizons. I work for a small company here in Italy and our goals are to cure diseases, particularly the diseases of osteoporosis, hepatitis, problems of depression, AIDS, and so on. When a young synthetic chemist comes into our company, we no longer describe them as **chemists,** but as **scientists**. Such individuals have some interesting opportunities, because in order to really get one's teeth into a big unsolved problem like the origin of life, one must use ideas and concepts that come from other disciplines and integrate those ideas and concepts with the principles of organic synthesis. And so I would suggest that where total organic synthesis might come to have meaning above and beyond the circle of chemists alone is in situations where there is tight intellectual integration, a creative approach to using these tools to solve truly unsolved problems that affect the lives of large numbers of people. This is applied not only to problems in chemistry but in a vast range of disciplines. But this will require that you no longer identify yourself merely as a chemist, but as a scientist. You must to use all of the tools of science, not just synthesis, but whatever is available to solve the problem. It seems to me that is where synthesis is likely to play the greatest possible role."

Dr. W. Ripka, Corvas International, San Diego, spoke:

" I thought Professor Eschenmoser's statement of purpose was really eloquent. This is the sort of thing we need more of in chemistry, an overall theme of what we are trying to do. A lot of what we talk about tends to be somewhat fragmented. Many people are doing parts of what Professor Eschenmoser said. Certainly, Paul Bartlett is with his work trying to understand why small molecules bind to large ones. Certainly, Professor Breslow's work on artificial enzymes is another component. But I think it is very important to have an overall theme as to why we are doing these things. And I thought Professor Eschenmoser's explanation of that in terms of self-organization is exactly the type of thing that we as chemists should do more of. I should like to ask the panel, all of whom are academics, how well do you think the academic community is doing in terms of setting the kinds of overall goals of the kind Professor Eschenmoser expressed?"

I responded to Dr. Ripka's question as follows:

" I think the academic community is answering the challenge only partly. I do think there is a group of people in academe who are synthesizing for some reason that I can't really categorize other than self-enjoyment and have lost sight of the fact that they should be trying to improve chemistry. There was certainly a time when it was exciting to conquer a molecule, just like it was exciting to climb the Eiger. But it has been demonstrated time and again that we can climb these mountains. We have invented all kinds of very creative crutches like protecting groups and very expensive reagents, like selenium instead of bromine, and things of this sort. We now have to drop some of these crutches and try to get back to using fundamental materials and doing things without protecting groups. That is really the challenge for the academic community. I don't really think it is important that the molecule I work on be biologically active or have importance to materials science. I think it **is** important that I try to find creative new solutions to complex synthetic problems. I think that there is some considerable activity in the academic environment in creating new ideas and new methods, but I also think there is a lot of 'turning-the-crank, mundane, just-do-it-and-publish-it type of research. It is this sort of activity that has brought criticism to the field."

6. To What Extent can the Values of Natural Products Total Synthesis also be Realized if we Direct our Efforts Instead Toward the Synthesis of 'Unnatural' Products?

This subject came up many times and in many forms during the discussion. A clear statement of the situation was made by Ronald Breslow in his opening remarks:

" I would like to put the question in the most negative way I can and then I will come back to a more positive view of it. Let's imagine that you are explaining to someone who is not a chemist what you do when you are a synthetic chemist. They say 'Oh, that's very exciting You make all sorts of new molecules, and it must be very interesting to find out their properties.' You say, 'No, we only consider it to be synthesis if we make known compounds. If we make new compounds, that doesn't count'. And there is a very serious rule that exists in many quarters that only the synthesis of natural products is a true synthesis, because it is the only one that has the challenge that you don't invent your target.' So then the person says, 'Well, that's interesting. But if the material is fairly rare, even if already known, perhaps your synthesis contributes to the supply.' You say 'No, we often use relays, and we consume more of the final product than we actually make, so the result of the synthesis is a **decrease** in the world's supply of the material.'

" Now, this makes total synthesis sound fairly crazy, so what **is** the argument in favor of it? I think there is a value, and that is in having a **specific**, predetermined target. We are not permitted to put in an extra methyl group or to leave a methyl group out just to make the synthesis easier. We must make **exactly** what's there, and if we don't know how to do it, we had better figure out how to do it.

" Since this is a principle value of total synthesis research, it follows that there is no great argument for saying that once a compound has been made the problem is solved. Morphine and strychnine have each been synthesized several times, and each new synthesis that contributes new insights and new chemistry is as valuable as the first synthesis. That is, the value of a total synthesis is not just that we have conquered another mountain.

" Thus, natural products synthesis **does** have value because of the discipline that is enforced by the necessity of preparing the target in exact detail. My research group does a great deal of synthesis. However, whereas our synthetic targets generally have structural details that are highly desired, other parts can be varied without much effect on the desired property. Thus, we can modify our target somewhat to take advantage of available methodology or to avoid serious synthetic problems. When I am feeling particularly self-effacing, I contrast what we do in synthesis with what natural product synthetic people do. They have a specific target and they fire an arrow at it and try to hit the bulls-eye; we shoot an arrow into the wall and then we paint a bulls-eye around it.

" However, I do strongly object to the idea that the **only** valid synthesis is the synthesis of a naturally-occurring target. That definition is too restrictive. There are many other perfectly good synthesis targets, and synthetic work directed at these targets will not only advance biology or materials science, but will also advance chemistry. We are not yet a service science; at least that is not our only function. We have serious things to do **within** chemistry itself. Important molecules that will answer serious theoretical questions in organic chemistry remain to be made."

I addressed the audience on this point:

" I would like to reiterate Professor Breslow's observation that there is value in the enforced discipline that comes naturally in the synthesis of a natural product. I call your attention to one slide in Professor Bartlett's lecture about rational design of enzyme inhibitors. You will recall that, using his CAVEAT program, Professor Bartlett designed a molecule that he thought might fit nicely into a certain enzyme active site. The problem was then to synthesize this complicated potential inhibitor and test the theory. However, you will recall that the actual compound that Professor Bartlett synthesized had an extra methyl group on the benzene ring. Now, this methyl group was predicted to be totally out of the picture in the enzyme-inhibitor complex, so it really shouldn't matter if it is there or not. Professor Bartlett jokingly told us that this methyl group was introduced because of the 'EOS factor,' meaning 'ease of synthesis.' That is, it was a lot easier to synthesize the molecule with the superfluous methyl group than the simpler structure without it.

" This situation exactly illustrates why natural product synthesis is more valuable **for the purpose of furthering synthesis as a tool** than synthesis of equally complex structures that we just make up."

Dr. J. C. Orr, Health Science Center at the Memorial University of Newfoundland, spoke on this issue:

" I would like to suggest that the editors here might devote a little section of their journals as a sort of 'chemistry bazaar,' wherein those who are biochemists or biologists and need specific compounds could get together with synthetic organic chemists who are looking for projects. For example, I work in a medical school and quite often people come to me and say, 'I would love to have this compound to test for certain biological properties. Do you know how it can be synthesized?' Often I am just too busy to sit down and do this."

I responded to Dr. Orr:

" That is an interesting suggestion. However, recall Professor Barton's statement that if an academic chemist knows **exactly** how to synthesize a given compound, he shouldn't be doing it. I agree that we want to have organic synthesis contribute new tools every time we do carry out a multistep synthesis. Therefore, if there were such a 'dating-service' connecting biologists who want specific compounds with chemists who are looking for 'relevant' synthesis projects, I would hope chemists would only tackle problems in which there is the opportunity to develop some synthetic method or strategy, or in some other way to add to our knowledge about how to do synthesis. Furthermore, I think it is important that chemists who really want to contribute to the improvement of synthesis as a tool avoid using the EOS factor, as Professor Bartlett put it. We must discipline ourselves to synthesize **exactly** the structure desired, and not make changes just to make the synthesis easier."

Ronald Breslow made the following observation on this question:

" Sometimes there **is** a tendency, particularly from those in the pharmaceutical industry, to say that anyone who has not been synthesizing natural products is not suited for work in that industry. I think that attitude is a mistake and one I worry about. An awful lot of what I do is synthetic organic chemistry. I wouldn't let a student graduate who had not made a new molecule, because synthesis is the one unique tool that only chemists have. Physicists and biologists can **talk** about molecules, but we can **make** them. We must never forget that we have this powerful ability, or leave it out of the training of our students.

" On the other hand, people should recognize that there is a significant amount of synthesis that goes on in many areas of organic chemistry, usually not identified specifically as synthetic chemistry. For example, in the molecular recognition business, students probably spend at least 80% of their time making the molecules they are going to study. I think that is an activity that ought to be recognized."

7. To What Extent has Computer-Designed Synthesis Reduced the Need for Total Synthesis Research?

Although not specifically advertised as a sub-topic for the discussion, this question was brought up by Professor D. Arigoni, ETH:

" I agree to a very large extent if not totally with Professor Barton's statement that if know you can do it, you should not do it. But I wonder from this point of view, Derek, how you (or other members of the panel) feel about computer-assisted total synthesis?"

Derek Barton responded:

" Well, I think that if it's in the computer, it's a known fact. So, if the computer predicts how you can do a synthesis, it will be a summation of known facts. That's fine. People in industry should certainly use that tool. I don't think it has value for people in the academic world because it just limits them to the use of known facts, known reagents, known reactions, and known principles."

Professor U. Jordis, Technical University of Vienna, disagreed with this view:

" I would like to disagree with Professor Barton's view of the computer synthesis planning. In my view, the computer allows you to ask new questions that you could not otherwise ask"

Ronald Breslow then made some observations about the value of computer-designed synthesis planning:

" Often when we devise a synthetic plan, we are so interested in the methodology or using a certain key reaction that we lose sight of the starting material. We tell ourselves we are going to find it in Aldrich or Fluka. Then, at the moment of truth, we realize that we had better see what sort of starting material we have after all. This is where computers might help. As human beings, we are limited with our sense of vision. We can't see everything even for a fraction of a second. Perhaps we can see it but then we lose it. And if there is a tool out there, be it a computer or a machine or something that can actually make me see things and freeze those frames on a computer screen to tell me here are some possibilities for you to choose from, without invading my brain, without telling me the first reagent is going to be this, the second is that and the third is going to be that. Just to give me a choice of starting materials, then I think that would be very interesting."

8. What About 'Biomimetic' Synthesis?

Strictly speaking, 'biomimetic synthesis' implies that one prepares a compound by the same method used by Nature for the biosynthesis of compound. However, chemists commonly use the term in a more general way--meaning that a synthesis is inspired by a probable or even possible biosynthetic construction of bonds. That is, it is often possible to deduce from the structure of a natural product what the starting material must be, and by application of sound mechanistic ideas about electron movement, it is often possible to propose a sequence of steps that **may be** used by Nature. When a chemist is inspired by such considerations to design a synthesis, and then proceeds to reduce the plan to practice by experiment, this is often called a 'biomimetic' synthesis, even if there has been absolutely no experimental work directed at determining the actual biosynthetic pathway.

Although the subject of biomimetic synthesis was really allocated to another panel discussion, the subject did come up several times. It was first introduced by Albert Eschenmoser as follows:

"I think it would be very helpful if Professor Heathcock would expose his two types of synthesis of the squalenoid alkaloids [10]. What these syntheses show us is that we can greatly reduce the number of steps in huge synthetic projects if we pay attention to biomimetic ideas. That is, you can really go much too long a distance when you try to use vigorously unnatural chemistry to make a natural compound. Nature did not have so much freedom in producing her natural products. Indeed, Nature has apparently been restricted to a surprisingly large extent, and we should not start to synthesize really complex natural products without checking whether by adapting to the natural channels followed by Nature, we might be able to plan or achieve an optimal synthesis."

I replied to these comments:

"Let me just make one observation about this because I think it is an example of the value of doing total synthesis of natural products. In the project that Professor Eschenmoser is referring to I spent probably five years trying to force designed chemistry using carbanions and other well-behaved organic reagents onto a structure. And it was only after four or five years, that some light dawned and we saw how Naturemight do this using some very simple chemistry and it turned out that this simple chemistry works in the laboratory very well.

"In a way, planned syntheses using conventional reagents and biomimetic syntheses are related in the same way as the two methods for finding enzyme inhibitors, as described for us by Professor Bartlett in his lecture on Monday. You recall that he talked about 'rational design', but he also described the 'combinatorial' approach. When we approach a synthesis biomimetically, we are taking advantage of the biodiversity of Nature. Nature has had lots of time to be a process chemist and find the most efficient and most economical way to solve some of these big problems. Much of this chemistry is not known to us. At least it is not obvious to us, and it is only by trying to solve these problems in the laboratory that we will perhaps discover some of this very nice, simple, environmentally-friendly chemistry."

On the subject of biomimetic synthesis, I would also like to quote a passage that was authored by Albert Eschenmoser in the abstract of his classic article on the origin of the molecular structure of vitamin B_{12} [3c]:

"The goal is to arrive experimentally at a perception of the biomolecule's intrinsic potential for structural self-assembly. This potential, together with the specific type of reactivity related to the biological function, is considered to be responsible for the biomolecule having been chosen by natural selection. The chemical rationalization of the structure of biomolecules is an objective of organic natural product chemistry. The field of natural product synthesis provides appropriate conceptual and methodological tools to approach this objective experimentally."

9. Summary

It is not possible to summarize a discussion such as the one held in Ravello on Thursday afternoon, May 12, 1994, but we can at least review some of the cogent points that emerged.

There was general agreement that total synthesis still does have intrinsic value as a method of structure proof in many cases, particularly where stereochemistry is involved. It was also recognized that it is sometimes the only way to obtain sufficient amounts of rare natural products with which to carry out biological experiments.

Furthermore, there seemed to be agreement that this form of research is an excellent way to train students, particularly those headed for careers in pharmaceutical chemistry or the biological sciences. However, it was pointed out by several that we educators need to do more to teach our students just how to identify significant problems, and that we need to do more to encourage the kind of breadth that is increasingly necessary in the modern world of science.

The case was strongly made that, in spite of the recent significant achievements in the arena of total synthesis, we still have far to go before we can accomplish practical syntheses of any desired structure, no matter how complicated. We have come a long way from Wöhler's synthesis of urea to the Woodward-Eschenmoser synthesis of vitamin B_{12} and Kishi's synthesis of palytoxin. However, even these monumental synthetic feats are only big steps along the long road toward synthetic perfection. It will probably take another 150 years before chemists will be able to prepare non-biological compounds of comparable complexity in a truly practical manner. So there is continuing value in our trying to solve larger and larger problems by simpler and simpler means.

However, the point was also made that we must be aware of opinion in the community. Practitioners of multistep synthesis must continually question what they are doing; they must carefully evaluate their synthetic approaches to assure that each synthesis really does have the potential to teach us something new, be it a new method or a new strategy of synthesis. Professor Eschenmoser eloquently made the point that total synthesis has played an important role in the history of organic chemistry – the demystification of Nature. It would seem that the time has past when we need to do fully planned synthesis solely for the purpose of convincing ourselves or the world that we can do it, that we can make anything if we are just willing to work hard enough.

The point was made that synthesis is a unique tool that chemists have, that this ability to make things to study, rather than just studying what Nature provides for us, sets chemists apart from other scientists. It was also pointed out that chemistry, unlike some of the other sciences, does not have an agenda of 'big problems.' Along these lines, Professor Eschenmoser challenged us to use synthesis to address a really big problem – understanding how life began! This bold suggestion, particularly his ambitious estimate that we might achieve a model for the kind of self-organization that may have led to the beginning of life in just a few decades, clearly caught the imagination of many participants in the discussion and no doubt provided an inspiration that many of us took with us as we returned home to our own laboratories.

References

1. Woodward, R. B. and Doering, W. E. (1944) The Total Synthesis of Quinine, *J. Am. Chem. Soc.* **66**, 849. For the full account, see: Woodward, R. B. and Doering, W. E. (1945) The Total Synthesis of Quinine, *J. Am. Chem. Soc.* **67**, 860.

2 Woodward, R.B., Cava, M.P., Ollis, W.D., Hunger, A., Daeniker, H.U. and Schenker, K. (1963) The Total Synthesis of Strychnine, *Tetrahedron* **19**, 247.

3. (a) Woodward, R. B. (1973) The Total Synthesis of Vitamin B_{12} *Pure Appl. Chem.* **33**, 145; (b) Eschenmoser, A. and Wintner, C. (1977) Natural Product Synthesis and Vitamin B_{12} *Science* **196**, 1410, and the many previous references cited in these two publications. See also (c) Eschenmoser, A. (1988) Vitamin B_{12}: Experiments Concerning the Origin of Its Molecular Structure *Angew. Chem., Internat. Edn. Engl.* **27**, 5.

4. Armstrong, R. W., Beau, J.-M., Cheon, S. H., Christ, W. J., Fujioka, H., Ham, W.-H., Hawkins, L. D., Jin, H., Kang, S. H., Kishi, Y., Martinelli, M. J., McWhorter, Jr., W. W., Mizuno, M., Nakata, M., Stutz, A. E., Talamas, F. X., Taniguchi, M., Tino, J. A., Ueda, K., Uenishi, J.-i., White, J. B. and Yonaga, M. (1989) *J. Am. Chem. Soc.* **111**, 7530.

5. Heathcock, C. H. (1992) The Enchanting Alkaloids of Yuzuriha, *Angew. Chem.* **104**, *675; Angew. Chem. Int. Ed. Engl.* **31**, 665.

6. Stein, J. (1979) *The Random House Dictionary of the English Language*, Random House, New York.

7. Heathcock, C. H., Finkelstein, B. L., Aoki, T. and Poulter, C. D. (1985) Stereostructure of the Archaebacterial C_{40} Diol, *Science, Washington, D.C.* **229**, 862.

8. Nerenberg, J. B., Hung, D. T., Somers, P. K. and Schreiber, S. L. (1993) Total Synthesis of the Immunosuppressive Agent (–)-Discodermolide *J. Am. Chem. Soc.* **115**, 12621.

9. Wöhler, F. (1828) Ueber künstliche Bildung des Harnstoffs *Ann. Physik* **12**, 253.

10. Professor Eschenmoser refers here to the *Daphniphyllum* alkaloids. See ref. 5.

CONVERGENT STRATEGIES FOR SYNTHETIC RECEPTORS

P. TIMMERMAN, W. VERBOOM and D. N. REINHOUDT
Laboratory of Organic Chemistry
University of Twente
P. O. Box 217
7500 AE Enschede
The Netherlands

ABSTRACT. Combination of medium-sized building blocks such as calix[4]- or resorcinarenes comprises a new strategy for the synthesis of receptor molecules with unique complexation properties. Calix[4]arenes bridged at the lower rim with a polyglycol chain show very high association constants for K^+ (8.9×10^9 M^{-1} in $CDCl_3$). Kinetically stable Rb^+-complexes (half-life time in $CDCl_3$ of 180 days) can be prepared with calix[4]arenes bridged at the lower rim with a *m*-terphenyl moiety. The combination of calix[4]arenes with a uranyl containing salophen unit is used for the synthesis of lipophilic carriers for selective transport of urea through supported liquid membranes. Connecting the salophen unit via the upper rim of a calix[4]arene carrying four ester groups at the lower rim a ditopic receptor able to selectively complex NaH_2PO_4 is synthesized. Calix[4]arenes can also be incorporated in carcerands via combination with resorcinol-based cavitands. Inclusion of dissymmetrical solvent molecules introduces a new type of diastereoisomerism with potential application in the preparation of molecular switches. The combination of calix[4]arenes with resorcinol-based cavitands is also used in the synthesis of the first holand, a molecule with a rigid organized cavity of nanosize dimensions.

1. Introduction

Supramolecular chemistry is a relatively new field in which specific interactions between molecules constitute the central theme. Its main source of inspiration is found in biological life which is almost entirely governed by very specific interactions.

Nature constructs biological receptors by the combination of a limited number of components. Amino acids are combined to proteins, nucleosides to DNA or RNA and

C. Chatgilialoglu and V. Snieckus (eds.), Chemical Synthesis, 245–261.

monosaccharides to carbohydrates. The linear structures fold in a three-dimensional way in order to form a recognition site which size and dimensions largely depend upon the sequence of the monomeric components. Systematic variation of the monomer sequence gives access to an almost infinite number of different receptors. However, this is achieved at the expense of a high molecular weight.

One of the aims of supramolecular chemistry is the mimicry of biological processes with synthetic molecules. Since the discovery by Pedersen in 1968 that certain crown ethers show a high affinity for alkali metal cations [1,2], the field of molecular recognition has developed rapidly.

The first studies mainly concerned the complexation of cations which is based on the relatively strong ion-dipole interactions. Lehn and Cram introduced their cryptands [3] and spherands [4] which show substantially higher affinities for alkali and alkaline-earth metal cations compared to simple crown ethers. With these new macrocyclic host molecules they exemplified the importance of *preorganization* in the host and *complementarity* between host and guest [5].

Complexation of organic molecules requires a totally new set of electronical and structural demands for the synthesis of appropriate receptor molecules, because in most cases *neutral* guest molecules are involved that are generally larger and less symmetrical than cations. Much weaker interactions, like hydrogen bonding, π-π stacking and (induced) dipole - (induced) dipole interactions, account for the complexation but the larger size of the guest molecules permits multiple binding-site interactions which usually act in a cooperative way, resulting in a high *selectivity* in the binding of structurally related guest species.

Most synthetic receptors are prepared via *de novo* synthesis using modern synthetic methodologies, which allow almost unlimited variation. The strategy focusses on the complementarity of functional groups between receptor and guest species and aims for minimal reaction steps and molecular weights. The drawback is that for each individual guest molecule a new synthetic pathway has to be developed; the learning process is not efficiently accumulated.

A new strategy for the synthesis of artificial receptor molecules which is a compromise between the two extremes described above, starts from medium-sized molecules that can be used as platforms to which functional groups for intermolecular interactions can be attached [6]. Several of these components can be combined to build up larger structures. The components must have a well-defined shape, be readily available from cheap starting materials and easily functionalizable. Examples of such molecular building blocks are calix[4]arenes (**1**) [7] and resorcinarenes (**2**) [8].

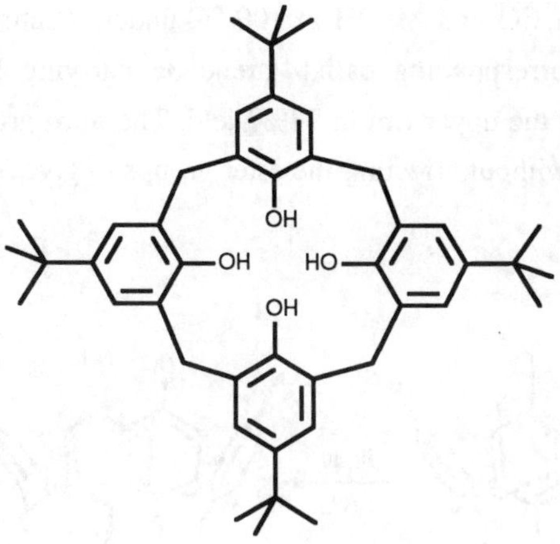

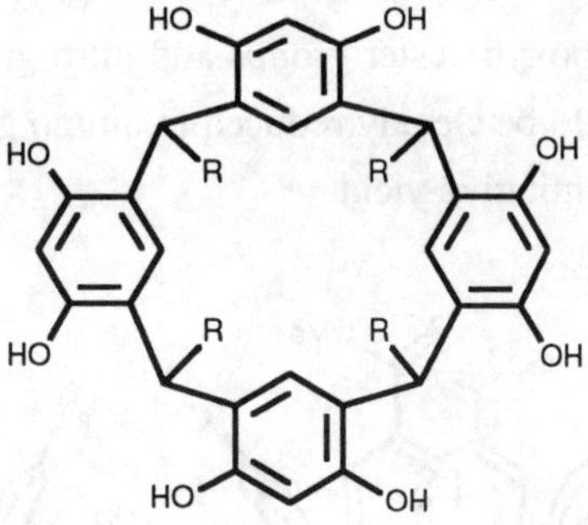

HO OH HO R R OH HO R R OH HO OH 2

OH OH HO OH 1

In this article we will illustrate our new strategy with a few examples of receptor molecules composed of calix[4]arenes in combination with several other building blocks that exhibit unique complexation properties. Moreover we have found new routes for the selective introduction of functional groups both in calix[4]arenes and resorcinarenes. One of these methods comprises the key step in the synthesis of calix[4]arene-based carcerands in which guest molecules are permanently encapsulated in an asymmetric environment. Finally, the first example of a holand, a molecule with a large rigid cavity of nanosize dimensions, will be described.

2. Selective Functionalization

2.1. SELECTIVE FUNCTIONALIZATION OF CALIX[4]ARENES

Calix[4]arenes (**1**) represent a class of synthetic molecules that has been a subject of research in our group for the past decade. Several methods have been developed for the selective functionalization of calix[4]arenes both at the lower and the upper rim; most of them have been summarized in a review article [9]. In this section only a recent example will be described.

Nitrocalix[4]arenes iodinated at the upper rim via treatment with CF_3COOAg/I_2 are excellent starting materials for the synthesis of calix[4]arenes carrying a combination of carboxylic ester, amino and nitro groups at the upper rim [10]. Treatment of 1,3-diiodo-2,4-dinitrocalix[4]arene **2a** with CO and MeOH at 100 °C under 10 atm for 24 h using NEt_3 as a base gave the corresponding calix[4]arene **3a** carrying both carboxylic ester groups and nitro groups at the upper rim in 74% yield. The nitro groups could be cleanly reduced to amino groups without affecting the ester groups to give **3b** in quantitative yield.

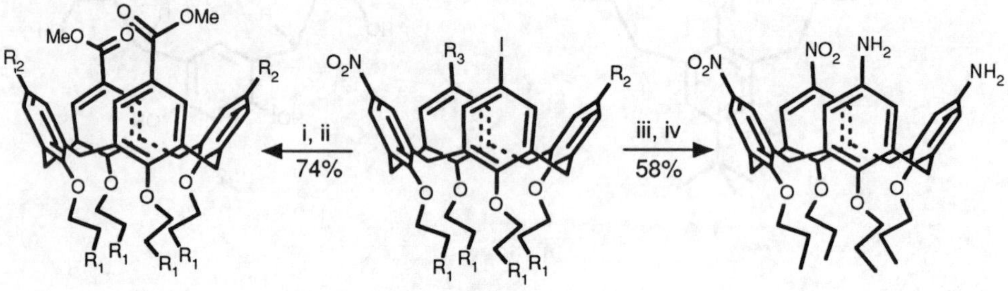

a R_1=OCH_2CH_3, R_2=NO_2 **a** R_1=OCH_2CH_3, R_2=NO_2, R_3=I

b R_1=OCH_2CH_3, R_2=NH_2 **b** R_1=CH_3, R_2=I, R_3=NO_2

i) CO (10 atm), MeOH, NEt_3, $Pd(PPh_3)_2Cl_2$, anisole, 100 °C, 24 h; ii) $NH_2NH_2.H_2O$, Raney Ni, MeOH, reflux, 2 h; iii) phthalimide, $Cu(I)_2O$, collidine, 190 °C, 24 h; iv) $NH_2NH_2.H_2O$, EtOH, reflux, 2 h, conc. HCl, reflux, 0.5 h.

Scheme 1

Reaction of iodonitrocalix[4]arenes with phthalimide in the presence of $Cu(I)_2O$ followed by treatment with hydrazine and conc. HCl provides an easy route into calix[4]arenes carrying both nitro and amino groups at the upper rim. In this way 1,2-diiodo-3,4-dinitrocalix[4]arene **2b** was converted to the corresponding aminonitro-calix[4]arene **4** in 58% yield.

2.2. SELECTIVE FUNCTIONALIZATION OF RESORCINARENES

Tri-bridged resorcinarene **5**, prepared in 54% yield by reaction of the corresponding resorcinarene with CH_2BrCl in DMSO, could be selectively debrominated at two of the four aromatic rings by treatment with 5 equiv of *n*-BuLi at -70 °C for 15 seconds followed by quenching with H^+ to give **6a** in 76% yield [11]. When the reaction was quenched with $B(OMe)_3$ followed by oxidative workup, tetrol **6b** could be obtained in 47% yield. After incorporation of the fourth methylene bridge in **6a**, the two remaining bromo atoms could be substituted by carboxylic ester, cyano or hydroxyl groups to give compounds **7** in 60-95% yield.

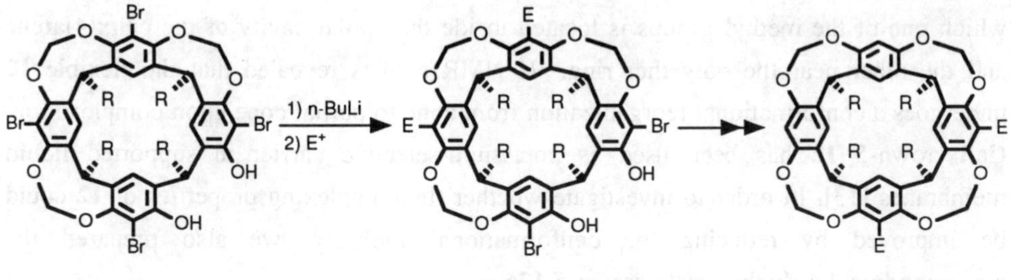

5 → 1) n-BuLi 2) E^+ → **6** → **7**

6: **a** E=H (76%); **b** E=OH (47%)

7: **a** E=C(O)OMe (>95%); **b** E=CN (60%); **c** E=OH (62%)

Scheme 2

3. Combination of Calix[4]arenes with Other Building Blocks

For the design of new receptor molecules with unique complexation properties calix[4]arenes have been combined with several other building blocks making use of the methods for their selective functionalization that were developed. In this paragraph the combination of calix[4]arenes with glycol chains (**8**), a terphenyl (**9**), and a uranyl containing salophen moiety (**10**) will be discussed.

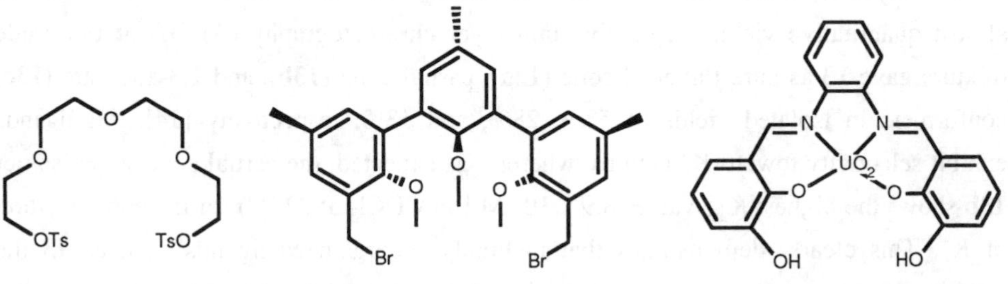

8 **9** **10**

3.1. CALIXCROWN ETHERS

Reaction of calix[4]arene **1** with tetraethylene glycol ditosylate (**8**) gave calixcrown ether **11** in 53% yield. Subsequent methylation of **11** afforded the 2,4-dimethoxycalixcrown-5 **12** which shows a surprisingly high K^+/Na^+ selectivity in extraction experiments [12]. The preferred conformation for binding in **12** appeared to be a flattened partial cone, in which one of the methyl groups is located inside the apolar cavity of the calix[4]arene and the other near the polyether ring. ^{1}H NMR studies revealed that the flexible **12** undergoes a conformational reorganization from cone to partial cone upon complexation. Calixcrown-5 **12** has been used as potassium-selective carrier in supported liquid membranes [13]. In order to investigate whether the complexing properties of **12** could be improved by reducing the conformational mobility, we also prepared the corresponding 2,4-diethoxycalixcrown-5 **13a**.

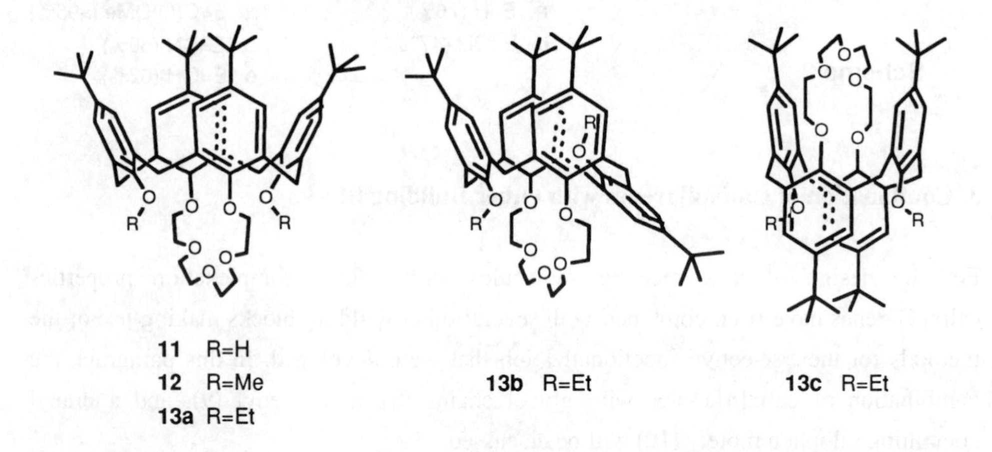

2,4-Dihydroxycalixcrown-5 **11** was dialkylated with ethyl iodide in THF/DMF in almost quantitative yield. Preparative thin layer chromatography (Al_2O_3) of the crude mixture gave **13** as pure flattened cone (**13a**), partial cone (**13b**), and 1,3-alternate (**13c**) conformers in isolated yields of 53%, 28%, and 13%, respectively [14]. All ligands exhibit selectivity toward K^+ cations, whereas, as expected, the partial cone stereoisomer **13b** shows the highest K_{ass} value (8.9 x 10^9 M^{-1} in $CDCl_3$ at 22 °C) for the complexation of K^+. This clearly demonstrates that in highly preorganized ligands such as **13** the stability of the complexes can be strongly affected by subtle changes in the geometry around the binding region. The superiority of the partial cone stereoisomer **13b** in K^+/Na^+ selectivity has also been found in chemically modified field effect transistor (CHEMFET) and membrane ion selective electrode (ISE) measurements [15].

3.2. CALIXSPHERANDS

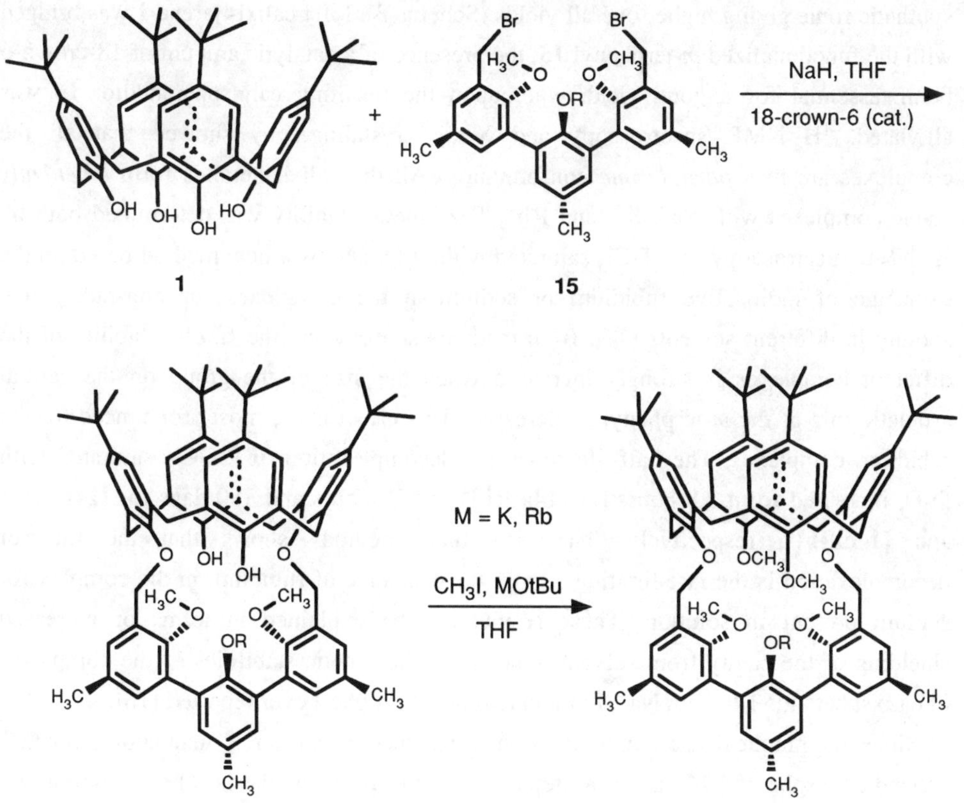

Br Br H_3C CH_3 O O OR NaH, THF 18-crown-6 (cat.) + H_3C CH_3 OH OH HO OH CH_3

1 **15**

M = K, Rb

CH_3I, MOtBu

THF

O OH OH O H_3C CH_3 O O OR H_3C CH_3 CH_3

O OCH_3 OCH_3 O H_3C CH_3 O O OR H_3C CH_3 CH_3

16 (50-80%)

14 (>80%)
a R=Me
b R=Et
c R=i-Pr

Scheme 3

Some years ago we reported the synthesis of calixspherand **14a** by reaction of 26,28-dimethoxy-*p-tert*-butylcalix[4]arene with *m*-terphenyl **15** in a yield of less than 30% [12]. This calixspherand forms complexes with Na^+ and K^+ which are kinetically stable, with decomplexation half-life times at room temperature of 3.7 and 2.2 years, respectively. With Rb^+ a complex is formed with a much lower kinetic stability; the half-life time of decomplexation at room temperature is only 2.8 h. Since we are particularly interested in the formation of kinetically stable Rb^+-complexes with the ultimate goal to immobilize rubidium for organ imaging, we have prepared calixspherands having a more shielded cavity [16]. Therefore we developed a new synthetic route giving higher overall yields (Scheme 3). First calix[4]arene **1** was bridged with the functionalized *m*-terphenyl **15**, the presence of a catalytic amount of 18-crown-6 being essential for a good yield, whereupon the resulting calixspheranddiol **16** was alkylated. ^1H NMR spectroscopy and X-ray crystallography showed that all the complexes are in a *partial cone* conformation. All the calixspherands form *kinetically* stable complexes with Na^+, K^+, and Rb^+. The kinetic stability was determined both by ^1H NMR spectroscopy, in $CDCl_3$ saturated with D_2O, and by a new method based on the exchange of radioactive rubidium or sodium in the complexes for non-radioactive sodium in different solvents [17]. Both methods showed that the *kinetic* stability of the different complexes is strongly increased when the size of the group on the central aromatic ring of the *m*-terphenyl is increased. This effect is the most pronounced for the rubidium complexes. The half-life times for decomplexation, in $CDCl_3$ saturated with D_2O, increased from 2.8 hours for **[14a.Rb]$^+$** to 139 hours and 180 days for **[14b.Rb]$^+$** and **[14c.Rb]$^+$**, respectively. The "exchange method" shows that the rate of decomplexation is the rate-limiting step in the exchange of rubidium in the complex for sodium present in solution. These results can be explained in terms of increased shielding of the cavity from solvent molecules. The kinetic stabilities of the complexes of calixspherands **14c** with Na^+, K^+, and Rb^+ are the highest ever reported [16].

Since for practical use coupling to an organ-specific carrier is mandatory, functionalized calixspherand **17** has been prepared in a multi-step synthesis. The formation of a conjugate between **17** and the low-molecular-weight protein (LMWP) lysozyme was confirmed by ion-spray mass spectrometry [18].

Recently, we found that calixspherands **14b** and **14c** form kinetically stable complexes with Ag^+, in $CDCl_3$ saturated with D_2O, with half-life times of decomposition of 51 and 131 h, respectively [19].

17

3.3. CALIX SALOPHEN CROWN ETHERS

Macrocycles that contain a uranyl containing salen or salophen unit are very good complexing agents for urea and other neutral molecules containing a nucleophilic center [20]. The synthesis of calix salophen crown ethers **19** is outlined in Scheme 4 [21]. The key step involves the macrocyclization of dialdehyde **18** by addition of 1 equiv of *cis*-1,2-cyclohexanediamine to a solution of **18** and 2 equiv of $Ba(OTf)_2$, which serves as a template, in THF. Subsequent addition of uranyl acetate gave the crude product which after purification afforded pure calix salophen crown ethers **19** in 78-88% yield. Complexation of neutral molecules has been demonstrated by the X-ray structures of the H_2O, MeOH, and DMSO complexes. Due to the presence of the calix[4]arene unit receptors **19** are highly lipophilic, making them useful as carriers for urea in supported liquid membranes [22].

3.4. CALIX[4]ARENE BASED DITOPIC RECEPTOR

Previously we reported that neutral metalloclefts and metallomacrocycles containing both an immobilized Lewis acidic UO_2-center and amido C(O)NH units as additional binding sites are excellent receptors for anions with a high selectivity for dihydrogen phosphate $H_2PO_4^-$ [23,24]. By combination of anionic and cationic binding sites in one molecule, a bifunctional receptor can be obtained which is able to complex

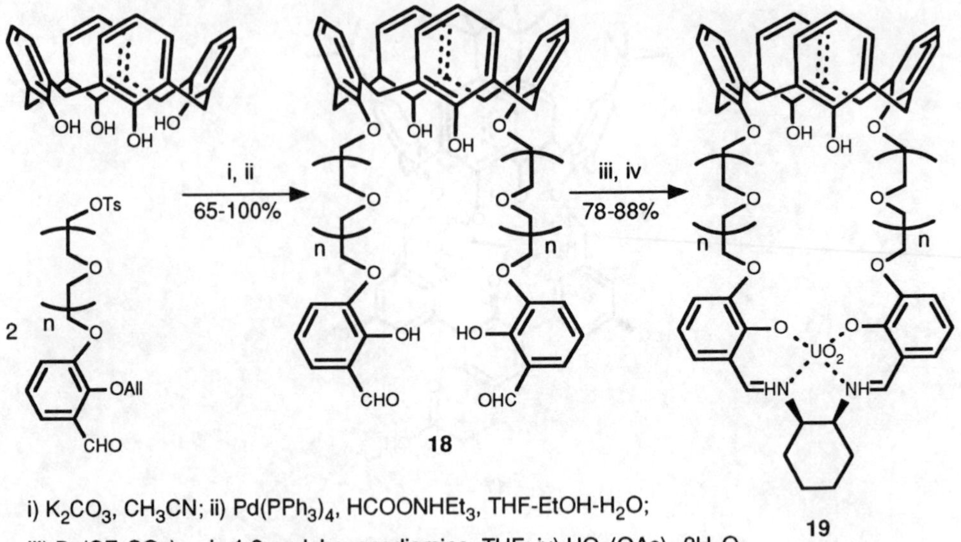

i) K_2CO_3, CH_3CN; ii) $Pd(PPh_3)_4$, $HCOONHEt_3$, THF-EtOH-H_2O;
iii) $Ba(CF_3SO_3)_2$, cis-1,2-cyclohexanediamine, THF; iv) $UO_2(OAc)_2.2H_2O$

Scheme 4

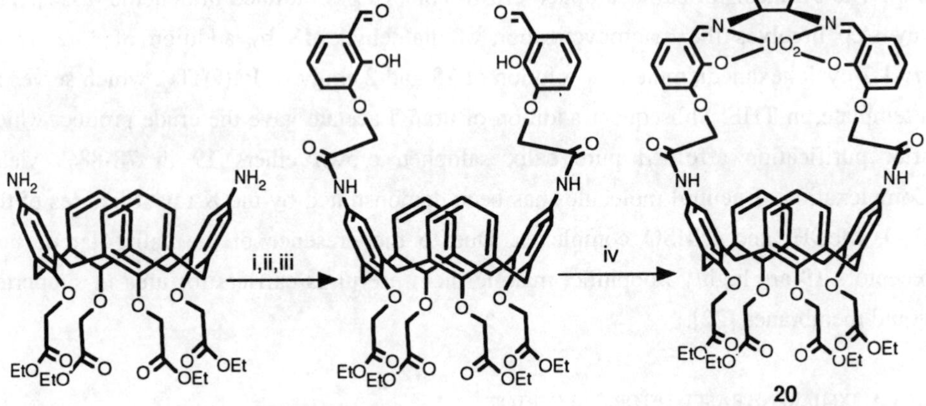

i) $ClCH_2C(O)Cl$, NEt_3, CH_2Cl_2 ii) 2-allyl-3-hydroxybenzaldehyde, K_2CO_3, KI, MeCN(67%)
iii) $Pd(OAc)_2$, PPh_3, HCOOH, NEt_3, EtOH (aq) (90%) iv) cis-1,2-diaminocyclohexane
$UO_2(OAc)_2.2H_2O$, EtOH (20%)

Scheme 5

simultaneously anionic and cationic species [25]. It is well-known that calix[4]arenes containing four preorganized ester fragments at the lower rim complex alkali metal cations with a high selectivity for Na^+ [26]. Using calix[4]arene as a molecular platform we attached both four ester groups and a uranyl containing salen moiety to give the calix[4]arene based bifunctional receptor **20**, the synthesis of which is summarized in Scheme 5 [27]. The complexation of NaH_2PO_4 has been proven by FAB mass spectrometry. ^{1}H NMR dilution experiments showed a selectivity of $H_2PO_4^-$ over Cl^-, HSO_4^-, and ClO_4^- anions.

4. Combination of Calix[4]arenes with Resorcinol-based cavitands

The work on the combination of calix[4]arenes and resorcinol-based cavitands was initiated primarily for the synthesis of dissymmetrical carcerands with ellipsoidal shape that are able to encapsulate guest molecules with a dipole moment in two different orientations giving rise to a new type of diastereomerism. Therefore the coupling of upper rim functionalized calix[4]arenes to resorcinol-based cavitands was investigated. We developed a stepwise route for the synthesis of calix[4]arene-based carcerand **21** because the direct coupling analogously to the synthesis of previously reported symmetrical carcerands [28] was not successful.

4.1. RECEPTOR MOLECULES WITH EXTENDED SURFACES FOR THE COMPLEXATION OF PREDNISOLON ACETATE

First, we studied the reaction between upper rim 1,2-functionalized calix[4]arene **22a**, prepared in 72% overall yield by reduction of the corresponding 1,2-dinitro compound [29] and subsequent reaction with two equivalents of α-chloroacetyl chloride, and cavitand **23**. When this reaction was performed in a 1:1 ratio of **22a** and **23** the diastereomeric products **24a** and **b** in which the calix[4]arene moiety is coupled to the cavitand in a 1,2- (proximal) fashion with the *endo* and *exo* stereochemistry were isolated in 20 and 32% yield, respectively. In addition to these 1:1 reaction products small amounts of the three possible isomeric 2:1 products **25a-c** were formed. Products in which the calix moiety is coupled in a 1,3- (distal) fashion to the cavitand could not be detected. When the reaction between **22a** and **23** was carried out in 2:1 ratio only the 2:1 products **25** were isolated in an almost statistical ratio of endo-endo (**25a**), endo-exo

Cs_2CO_3, KI

CH_3CN

22

23

a R_1=OCH_2CH_3, R_2=H

b R_1=CH_3, R_2=NO_2

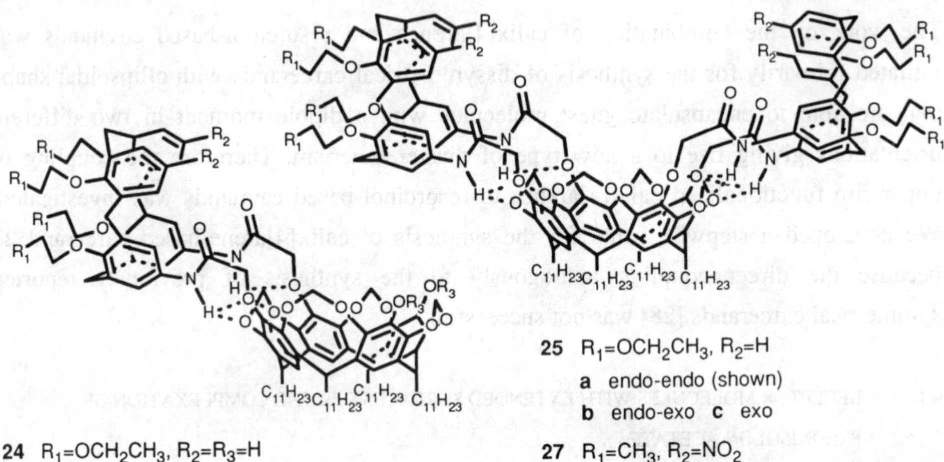

25 R_1=OCH_2CH_3, R_2=H

a endo-endo (shown)

b endo-exo **c** exo

24 R_1=OCH_2CH_3, R_2=R_3=H

a endo (shown) **b** exo

26 R_1=CH_3, R_2=NO_2, R_3=H endo

28 R_1=CH_3, R_2=NO_2, R_3=$Si(CH_3)_2C(CH_3)_3$ endo

29 R_1=CH_3, R_2=$NHC(O)CH_2Cl$

R_3=$Si(CH_3)_2C(CH_3)_3$ endo

27 R_1=CH_3, R_2=NO_2

a endo-endo (shown) **b** endo-exo

30 R_1=CH_3, R_2=$NHC(O)CH_2Cl$ endo-endo

Scheme 6

(**25b**) and exo-exo (**25c**) in a total yield of 64% [30]. These 2:1 products show very selective complexation of the corticosteroid prednisolon acetate (K_{ass}=6.0-8.5 x 10^2 M^{-1} in $CDCl_3$ at 25 °C), a compound that is often used because of its anti-inflammatory effect [31].

Apparently in the reaction between **22a** and **23** there is a slight preference for the formation of the 1:1 exo product **24b**. A possible way to stimulate the formation of the 1:1 endo product, the desired precursor for calix[4]arene-based carcerand **21**, is the introduction of functional groups at the calix[4]arene fragment that favorably interact with the cavitand moiety in the transition state. Reaction of a 1:1 mixture of calix[4]arene **22b**, prepared by acylation of **4** (Scheme 1) with 2 equiv of α-chloroacetyl chloride, and cavitand **22** exclusively gave the 1:1 endo isomer **26** together with small amounts of the 2:1 products **27a** (endo-endo) and **27b** (endo-exo). Because of its instability, **26** was isolated after silylation of the free hydroxyl groups as **28** in 41% yield. The nitro groups in **28** could easily be reduced to amino groups using Raney Ni/hydrazine [30] and after reaction with α-chloroacetyl chloride the bis(2-chloro-acetamido) derivative **29** was isolated in quantitative yield.

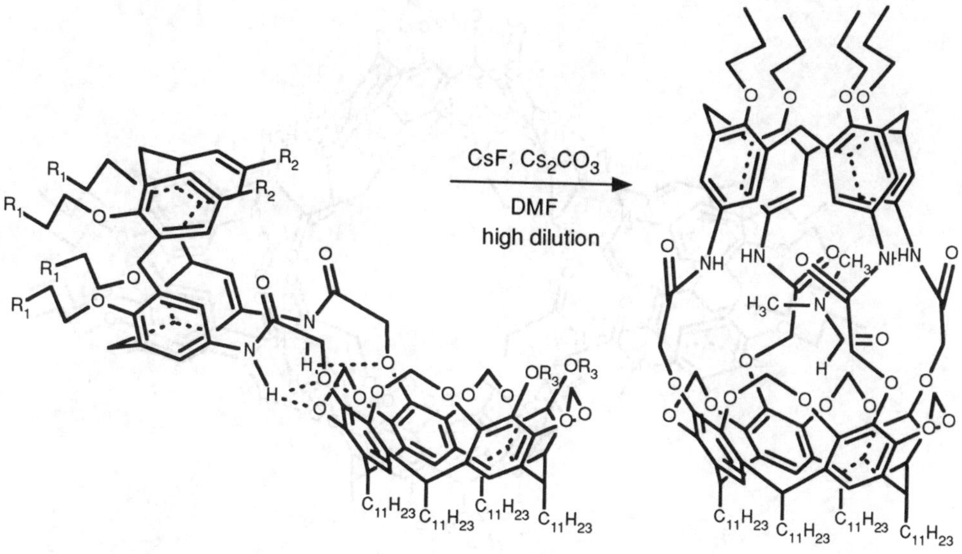

CsF, Cs_2CO_3
DMF
high dilution

29 $R_1=CH_3$, $R_2=NHC(O)CH_2Cl$
$R_3=Si(CH_3)_2C(CH_3)_3$ endo

21

Scheme 7

When compound **29** was reacted under high dilution conditions with CsF and Cs_2CO_3

in DMF at 70 °C for 24 h calix[4]arene-based carcerand **21**, with one molecule of DMF inside its interior, was isolated in essentially quantitative yield [32]. The presence of the incarcerated solvent molecule is evident from both MS FAB [M^{+}_{obs} = 2126, ($M+DMF+Na^{+}$, 100%)] and ^{1}H NMR spectroscopy showing two different singlets for the two methyl groups of DMF at 0.66 and -0.86 ppm, respectively. Even after heating for 1 h at 100 °C in DMF-d_7 no exchange of the incarcerated DMF was observed. Compound **21** is the first example of a dissymmetrical carcerand. According to variable ^{1}H NMR and NOE experiments the incarcerated solvent molecule has a preferred orientation in the cavity, but exchange with other orientations is too fast to be measured even at low temperatures [32]. Studies are underway to complex solvent molecules that have such a high rotation barrier inside the cavity that different diastereomers can be isolated.

4.3. SYNTHESIS OF A HOLAND; a molecule with a shielded hole of nanosize dimensions.

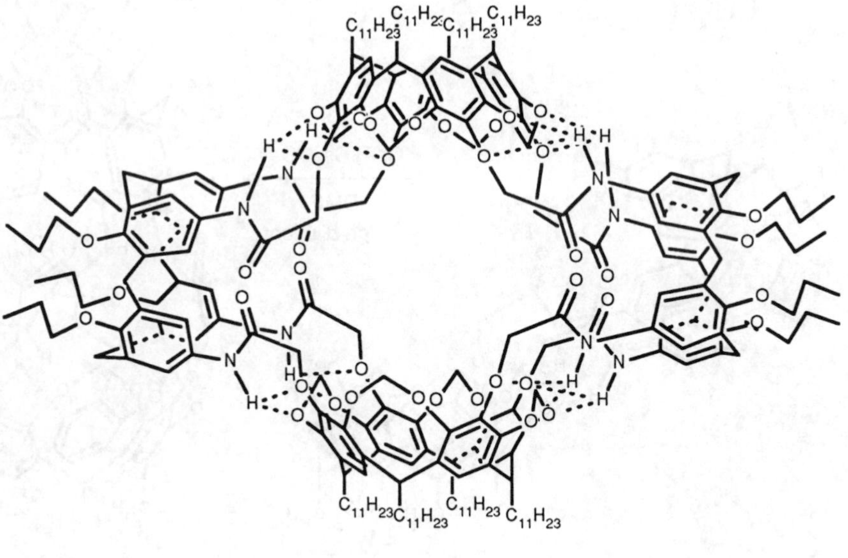

31

When **29** was desilylated with CsF in DMF for 2 h at 70 °C and subsequently reacted with Cs_2CO_3/KI in a concentration of 5 mM, beside calix[4]arene-based carcerand **21** (27%) a second compound, characterized as holand **31**, was formed in 26% yield. The latter compound could also be synthesized by dropwise addition of an equimolar solution of **30**, obtained from **27a** in quantitative yield in a similar way as **29** was obtained starting from **28**, and **23** in DMF to a suspension of Cs_2CO_3 and KI in DMF to give **31** in 35% yield. Holand **31** has a large rigid organized cavity. The rigidity is partly a result of a bifurcated hydrogen bond between the amide hydrogen atom and both the oxygen atoms in the spacer itself and an oxygen atom in the methylenedioxy bridge as indicated in **31** with dashed lines. Holand **31** contains a cavity of nanosize dimensions having, according to CPK-models, axes of about 1.5 and 2.0 nm long. The calculated internal volume is approximately 1.0 nm^3 (1000 $Å^3$). Holand **31** is expected to have unique complexation properties. The size of the cavity permits complexation of host molecules which themselves are good complexing agents.

5. Conclusions

An efficient strategy for the synthesis of new receptor molecules which exhibit unique complexation properties was developed by the covalent linkage of only a limited number of medium-sized molecules in many different ways. Several examples have been described showing the validity of the concept.

6. Acknowledgements

The authors are indebted to their colleagues, whose names appear in the references, for their contribution included in this paper. Part of this work has been performed in collaboration with the University of Parma (Professors Pochini and Ungaro). Financial support from the Technology Foundation (STW), Technical Science Branch of the Netherlands Organization for Scientific Research (NWO), and the EEC Science Program is gratefully acknowledged.

7. References

1. Pedersen, C. J. *J. Am. Chem. Soc.* **1967**, *15*, 153.
2. Pedersen, C. J. (Nobel Lecture) *Angew. Chem.* **1988**, *100*, 1053.
3. Lehn, J.-M. (Nobel Lecture) *Angew. Chem.* **1988**, *100*, 91.

4. Cram, D. J. (Nobel Lecture) *Angew. Chem.* **1988**, *100*, 1041.
5. Cram, D. J.; Cram, J. M. *Selectivity, a goal for synthetic efficiency*, Bartmann, W.; Trost, B. M., Eds.; Verlag Chemie: Weinheim, 1984, p 43.
6. Groenen, L. C.; Reinhoudt, D. N. *Calix[4]arenes, molecular platforms for supramolecular structures,* in *Supramolecular Chemistry*, Balzani, V.; De Cola, L., Eds.; Kluwer Academic Publishers: Dordrecht, 1991.
7. a) Gutsche, C. D. *Calixarenes, monographs in supramolecular chemistry*; Stoddart, J. F., Ed.; Royal Society of Chemistry: Crambridge, 1989; Vol. 1; b) Vicens, J.; Böhmer, V., Eds. *Calixarenes: a versatile class of macrocyclic compounds*; Kluwer Academic Press: Dordrecht, 1991.
8. a) Erdtman, H.; Högberg, S.; Abrahamsson, S.; Nilsson, B. *Tetrahedron Lett.* **1968**, 1679; b) Högberg, A. G. S. *J. Am. Chem. Soc.* **1980**, *102*, 6046; c) Högberg, A. G. S. *J. Org. Chem.* **1980**, *45*, 4498.
9. Van Loon, J.-D.; Verboom, W.; Reinhoudt, D. N. *Org. Prep. Proc. Int.* **1992**, *24*, 437.
10. Timmerman, P.; Verboom, W.; Reinhoudt, D. N.; Arduini, A.; Grandi, S.; Sicuri, A. R.; Pochini, A.; Ungaro, R. *Synthesis* **1994**, 185.
11. a) Timmerman, P.; van Mook, M. G. A.; Verboom, W.; van Hummel, G. J.; Harkema, S.; Reinhoudt, D. N. *Tetrahedron Lett.* **1992**, *33*, 3377; b) Timmerman, P.; Boerrigter, H.; Verboom, W.; van Hummel, G. J.; Harkema, S.; Reinhoudt, D. N. submitted for publication in *J. Incl. Phenom.*
12. Dijkstra, P. J.; Brunink, J. A. J.; Bugge, K.-E.; Reinhoudt, D. N.; Harkema, S.; Ungaro, R.; Ugozzoli, F.; Ghidini, E. *J. Am. Chem. Soc.* **1989**, *111*, 7567.
13. Nijenhuis, W. F.; Buitenhuis, E. G.; De Jong, F.; Sudhölter, E. J. R.; Reinhoudt, D. N. *J. Am. Chem. Soc.* **1991**, *113*, 7963.
14. Ghidini, E.; Ugozzoli, F.; Ungaro, R.; Harkema, S.; El-Fadl, A. A.; Reinhoudt, D. N. *J. Am. Chem. Soc.* **1990**, *112*, 6979.
15. Brzozka, Z.; Lammerink, B.; Reinhoudt, D. N.; Ghidini, E.; Ungaro, R. *J. Chem. Soc., Perkin Trans. 2* **1993**, 1037.
16. Iwema Bakker, W. I.; Haas, M.; Khoo-Beattie, C.; Ostaszewski, R.; Franken, S. M.; Den Hertog, Jr., H. J.; Verboom, W.; de Zeeuw, D.; Harkema, S.; Reinhoudt, D. N. *J. Am. Chem. Soc.* **1994**, *116*, 123.
17. Iwema Bakker, W. I.; Haas, M.; den Hertog, Jr., H. J.; Verboom, W.; de Zeeuw, D.; Reinhoudt, D. N. *J. Chem. Soc., Perkin Trans. 2* **1994**, 11.
18. Iwema Bakker, W. I.; Haas, M.; den Hertog, Jr., H. J.; Verboom, W.; de Zeeuw,

D.; Bruins, A. P.; Reinhoudt, D. N. *J. Org. Chem.* **1994**, *59*, 972.

19. Iwema Bakker, W. I.; Verboom, W.; Reinhoudt, D. N. *J. Chem. Soc., Chem. Commun.* **1994**, 71.
20. For a few recent examples, see: a) van Doorn, A. R.; Schaafstra, R.; Bos, M.; Harkema, S.; van Eerden, J.; Verboom, W.; Reinhoudt, D. N. *J. Org. Chem.* **1991**, *56*, 6083; b) van Doorn, A. R.; Verboom, W.; Reinhoudt, D. N. *Recl. Trav. Chim. Pays-Bas* **1992**, *111*, 421; c) Reichwein, A. M.; Verboom, W.; Reinhoudt, D. N. *Recl. Trav. Chim. Pays-Bas* **1993**, *112*, 358; d) Reichwein, A. M.; Verboom, W.; Reinhoudt, D. N. *Recl. Trav. Chim. Pays-Bas* **1993**, *112*, 595.
21. Reichwein, A. M.; Verboom, W.; Harkema, S.; Spek, A. L.; Reinhoudt, D. N. *J. Chem. Soc., Perkin Trans. 2* in press.
22. van Straaten-Nijenhuis, W. F.; van Doorn, A. R.; Reichwein, A. M.; de Jong, F.; Reinhoudt, D. N. *J. Org. Chem.* **1993**, *58*, 2265.
23. Rudkevich, D. M.; Stauthamer, W. P. R. V.; Verboom, W.; Engbersen, J. F. J.; Harkema, S.; Reinhoudt, D. N. *J. Am. Chem. Soc.* **1992**, *114*, 9671.
24. Rudkevich, D. M.; Verboom, W.; Brzozka, Z.; Palys, M. J.; Stauthamer, W. P. R. V.; van Hummel, G. J.; Franken, S. M.; Harkema, S.; Engbersen, J. F. J.; Reinhoudt, D. N. *J. Am. Chem. Soc.* in press.
25. Rudkevich, D. M.; Brzozka, Z.; Palys, M.; Visser, H.; Verboom, W.; Reinhoudt, D. N. *Angew. Chem., Int. Ed. Engl.* **1994**, *33*, 467.
26. Arnaud-Neu, F.; Collins, E. M.; Deasy, M.; Ferguson, G.; Harris, S. J.; Kaitner, B.; Lough, A. J.; McKervey, M. A.; Marques, E.; Ruhl, B. L.; Schwing-Weill, M. J.; Seward, E. *J. Am. Chem. Soc.* **1989**, *111*, 8681.
27. Rudkevich, D. M.; Verboom, W.; Reinhoudt, D. N. accepted for publication in *J. Org. Chem.*
28. Sherman, J. C.; Knobler, C. B.; Cram, D. J. *J. Am. Chem. Soc.* **1991**, *113*, 2194.
29. van Loon, J.-D.; Heida, J. F.; Verboom, W.; Reinhoudt, D. N. *Recl. Trav. Chim. Pays-Bas* **1992**, *111*, 353.
30. Timmerman, P.; Verboom, W.; van Veggel, F. C. J. M.; van Hoorn, W. P.; Reinhoudt, D. N. accepted for publication in *Angew. Chem.*
31. Timmerman, P.; Brinks, E. A.; Verboom, W.; Reinhoudt, D. N. unpublished results.
32. Timmerman, P.; Verboom, W.; Reinhoudt, D. N. unpublished results.

THE ADVANCEMENT OF RADICAL-BASED SYNTHETIC STRATEGIES: FROM REAGENT DESIGN TO POLYMER MODIFICATION

C. CHATGILIALOGLU
I.Co.C.E.A. - Consiglio Nazionale delle Ricerche
Via P. Gobetti 101, 40129 Bologna, Italy

1. Introduction

The uses of radicals in organic synthesis have increased enormously within the last decade and now they span all the areas of organic chemistry. For example, at the beginning of the 1980's the application in natural product total synthesis was limited to a few functional group transformations and now hundreds of sophisticated applications exist [1]. The majority of radical reactions of interest to synthetic chemists are chain processes and generally they take place under reducing conditions [2]. In this article, we will deal with some recent applications of free radical chain reactions to organic synthesis and provide information for the design of new radical-based synthetic reactions.

2. A Comparison between Bu_3SnH and $(TMS)_3SiH$

In the last three decades organotin compounds have increased in importance for performing free radical reactions. In fact, four of the most important classes of radical reactions, i.e., reduction [3], intermolecular C-C bond formation [4], intramolecular C-C bond formation [5] and allylation [4], use either organotin hydride or allyltin derivatives. Conceptually, all these reactions are very similar. The Bu_3Sn• radical, for example, abstracts an atom (i.e., Cl, Br, I) or a group (i.e., OC(S)OPh, CN, SePh, NO_2) from an organic molecule to produce a carbon-centered radical. This radical can either abstract a hydrogen from Bu_3SnH to give the desired product and regenerate Bu_3Sn• (reduction), or add to a C-C double bond followed either by H-transfer (C-C bond formation) or the ejection of the Bu_3Sn• radical by β-elimination (allylation).

However, it is well known that there are several problems associated with triorganotin compounds. The main drawback consists of difficulties in the complete removal of the highly toxic tin by-products from the final products. For this reason, although in fine chemical synthesis the majority of free radical applications deal with tri-*n*-butyltin hydride, it is inappropriate to extend these reactions to industrial and medicinal chemistry. It is, therefore, of great importance to find new reagents which are suitable from the toxicological point of view and are analogous to organotin hydrides in their chemistry, so that the knowledge we gained in the last twenty years in the field of free radicals can be further applied.

Pioneers in organotin chemistry like W.P. Neumann, recently developed a polymer-supported organotin reagent based on cross-linked polystyrene of large pore size, where the tin moiety is separated from the aromatic ring of polystyrene by two methylene units [6-9]. The application of this polymeric material (**1**) has some practical advantages, like the regeneration of starting material, as well as some disadvantages, like the absence of

C. Chatgilialoglu and V. Snieckus (eds.), Chemical Synthesis, 263–276.

control in the hydrogen donation. However, even in this case, the desired reaction products are contaminated by tin-containing by-products and again this is the main drawback of the reagent.

$CH_2CH_2Sn(H)Bu_2$

n

1

A few years ago it seemed that an alternative approach based on organosilanes would be more suitable. We know that trialkylsilanes are poor reducing agents under free-radical conditions, but from our own work and the work of others it emerged that the incorporation of appropriated thermodynamic forces could change significantly the reactivity of the SiH moiety. The rationalization of the physico-chemical data of the silane/silyl radical systems indicated tris(trimethylsilyl)silane [10,11], a compound reported by Gilman *et al.* [12] in 1965 and then almost totally ignored for the next 20 years. $(TMS)_3SiH$ has indeed quickly proven to be a valid alternative to tin hydride for the majority of its radical chain reactions, although in a few cases the two reagents can complement each other [13]. For example, $(TMS)_3SiH$ rivals Bu_3SnH in efficiency for the reduction of halides, chalcogen groups, thioesters (deoxygenation of secondary alcohols), isocyanides (deamination of primary amines). $(TMS)_3SiH$ adds across the double bonds of a variety of olefins and ketones under free-radical conditions to give adducts. It has been shown that $(TMS)_3SiH$ and its silicon-containing by-products are less toxic than the corresponding tin compounds [14].

The fact that $(TMS)_3SiH$ reacts with alkyl radicals ca. 10 times slower than Bu_3SnH supported the expectation that it would be more efficient in cyclization reactions. Scheme 1 shows three examples which represent the key steps in the total synthesis of (–)-slaframine[15], (–)-zearalenone[16] and (±)-tacamonine[17], respectively.

Like the stannyl radicals [18], $(TMS)_3Si\bullet$ radicals are known to add to a double bond reversibly and therefore, to isomerize alkenes [19,20]. That is, these radicals add to (*Z*)- or (*E*)-alkene to form radical **2** or **3**, respectively. Interconversion between the two radical adducts by rotation around the carbon-carbon bond, followed by β-scission can then lead to the formation of either (*Z*)- or (*E*)-alkene, depending on the radical-alkene combination.

$Si(TMS)_3$ Z H Y H — 180° ⇌ — $Si(TMS)_3$ H Z Y H

2 **3**

Scheme 1

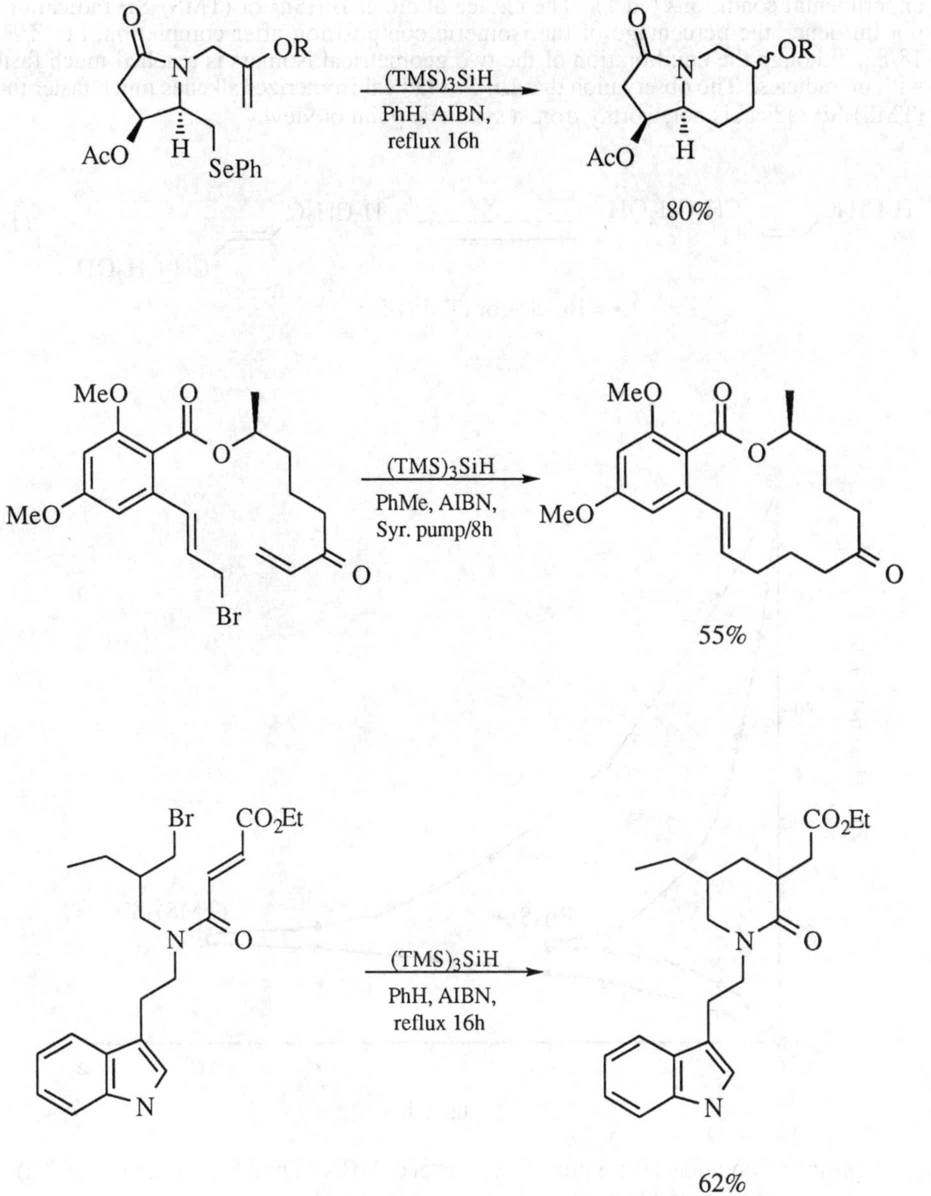

O, OR, N, AcO, H, SePh — $(TMS)_3SiH$ / PhH, AIBN, reflux 16h → O, N, OR, AcO, H

80%

MeO, O, O, MeO, Br, O — $(TMS)_3SiH$ / PhMe, AIBN, Syr. pump/8h → MeO, O, O, MeO, O

55%

Br, CO_2Et, N, O, N — $(TMS)_3SiH$ / PhH, AIBN, reflux 16h → CO_2Et, N, O, N

62%

For comparison we report in Figure 1 the reaction profile for the interconversion of (*E*)- to (*Z*)-3-hexen-1-ol and *vice versa* by $(TMS)_3Si\bullet$ and $Bu_3Sn\bullet$ radicals, under identical experimental conditions (eq 1). The choice of either $Bu_3Sn\bullet$ or $(TMS)_3Si\bullet$ radical does not influence the percentage of the isomeric composition after completion, i.e. Z/E = 18/82, although the equilibration of the two geometrical isomers is reached much faster with tin radicals. The observation that $Bu_3Sn\bullet$ radical isomerizes alkenes much faster than $(TMS)_3Si\bullet$ radical is noteworthy from a synthetic point of view.

$$H_3CH_2C\text{–CH=CH–}CH_2CH_2OH \;\underset{}{\overset{X\bullet}{\rightleftharpoons}}\; H_3CH_2C\text{–CH=CH–}CH_2CH_2OH \quad (1)$$

$$X\bullet = Bu_3Sn\bullet \text{ or } (TMS)_3Si\bullet$$

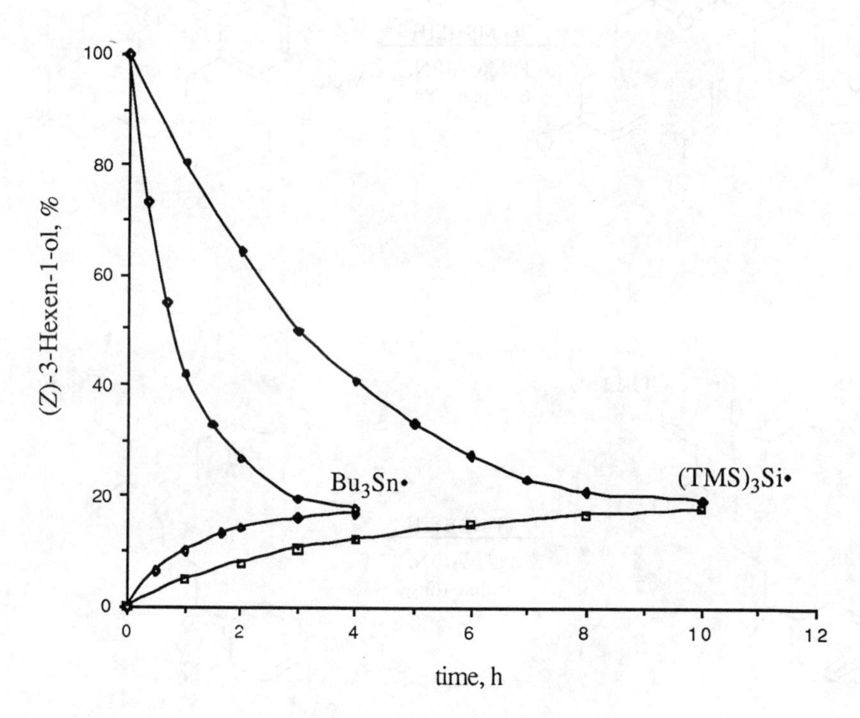

Figure 1. Conditions: 0.5 equiv of hydride and AIBN (5 mol % at intervals of 2 h) in refluxing benzene.

The isomerization can occurr *in situ*, while accomplishing other reactions. Johnson and Poulos [21] in synthesising (*Z*)-methyl triacont-21-enate (**4a**), *via* the reduction of the thiocarbonate **4b** or iodo **4c** derivatives with $(TMS)_3SiH$, have obtained (*E*)-methyl triacont-21-enate. That is, during the reduction process a (*Z*)-(*E*) isomerization occurred, by the addition-elimination of $(TMS)_3Si\bullet$ radical.

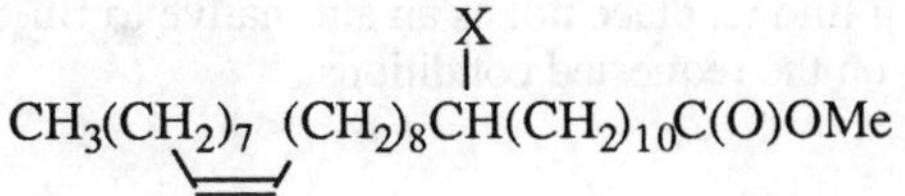

$$CH_3(CH_2)_7CH{=}CH(CH_2)_8CH(X)(CH_2)_{10}C(O)OMe$$

4a X = H
4b X = OC(S)OPh
4c X = I

Radical deoxygenation of *vic*-diols to olefins by reaction of Bu_3SnH or $(TMS)_3SiH$ with the corresponding bisdithiocarbonate derivatives has been performed on complex biologically active molecules like the potent immunosuppressive agents FK 506 and FR 900520, in which the region C(9)-C(11) is known to influence the binding to proteins [22]. The *E*/*Z* ratio strongly depends on the reagent and condition used and with the choice of suitable reaction temperature, reaction time and reducing agent high stereoselectivity can be obtained. Although the complexity of the substrates does not allow any satisfactory mechanistic interpretation (two possible reaction paths were discussed), the experimental data were reported in some detail, so we can again suggest that the interconversion of the two geometrical isomers may occur *via* an addition-elimination mechanism involving either $(TMS)_3Si\bullet$ or $Bu_3Sn\bullet$ radicals.

Lowinger and Weiler [23] have recently shown that the intramolecular cyclization of secondary alkyl radicals with α,β-alkynyl esters **5** proceeds with high stereoselectivity to give predominantly either (*E*)- or (*Z*)-exocyclic alkenes depending upon the reaction conditions and reducing agents (Scheme 2). It has been suggested that the main factor controlling the formation of these products is the ability of $(TMS)_3Si\bullet$ and $Bu_3Sn\bullet$ radicals to isomerize the product alkene. That is, the (*E*)-(*Z*) isomerisation, under Bu_3SnH reduction conditions is confirmed whereas no such transformation is observed with $(TMS)_3SiH$.

Scheme 2

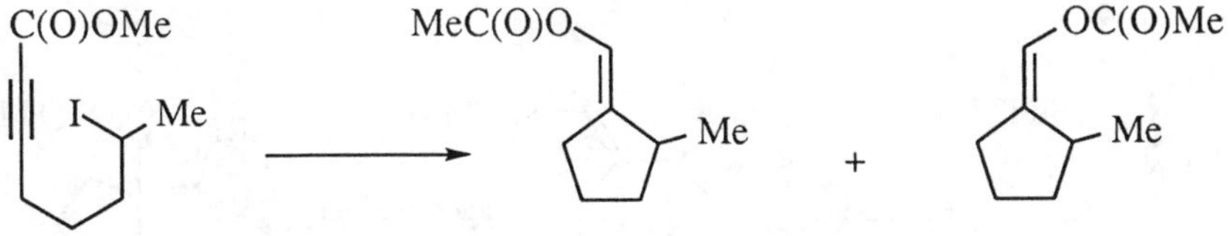

C(O)OMe, I, Me — **5** → MeC(O)O, Me (*E*) + OC(O)Me, Me (*Z*)

Bu_3SnH, AIBN, 80°C — *E*/*Z* = 98/2 (82%)

$(TMS)_3SiH$, Et_3B/O_2, -78°C — *E*/*Z* = 11/89 (85%)

Recently, Pattenden and co-workers have reported that treatment of the acetylene derivative **6** with $(TMS)_3SiH$ leads, in one pot, to the bicyclic compound **7** [24,25]. The proposed mechanism is based on the addition-elimination concept. That is, the $(TMS)_3Si\bullet$ radical adds to the triple bond to form an adduct radical followed by a cascade of radical cyclization-fragmentation-transannulation- ring expansion and termination *via* ejection of the $(TMS)_3Si\bullet$ radical to afford the bicyclic product (Scheme 3). As the addition-elimination methodology becomes more and more important in programming synthetic strategies, it is expected that $(TMS)_3SiH$ will find its place not as an alternative to Bu_3SnH but as a complementary reagent, depending on the requested conditions.

Scheme 3

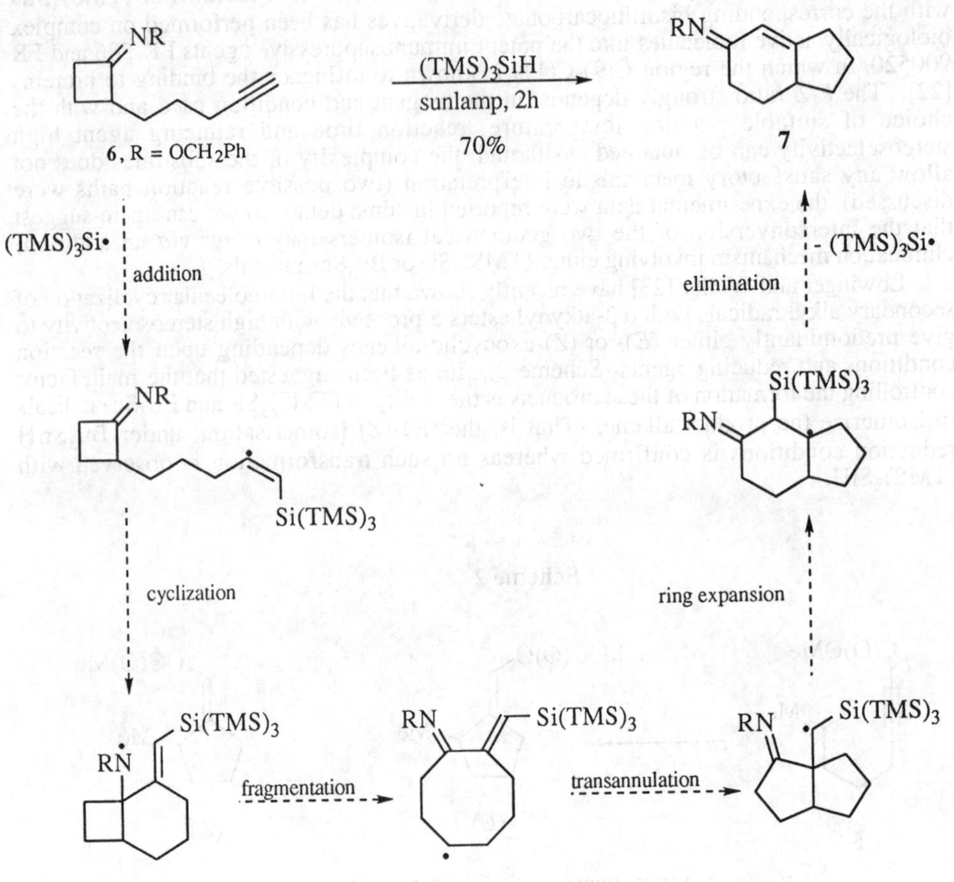

NR

6, R = OCH_2Ph

$(TMS)_3SiH$

sunlamp, 2h

70%

RN

7

$(TMS)_3Si\bullet$

addition

- $(TMS)_3Si\bullet$

elimination

NR

Si(TMS)$_3$

Si(TMS)$_3$

RN

cyclization

ring expansion

Si(TMS)$_3$

RN

RN

Si(TMS)$_3$

RN

Si(TMS)$_3$

fragmentation

transannulation

3. Silanes with Hydrogen Donating Abilities Lower than $(TMS)_3SiH$

Following the success of $(TMS)_3SiH$, it was thought that other organosilanes might be capable of sustaining analogous radical chain reactions and that, therefore, it might be possible to modulate the hydrogen donating abilities of silanes by simply changing the nature of the substituents. In particular, we investigated in some detail the use of silanes **8**, **9** and **10** as radical-based reducing agents for a variety of organic substrates [26-28]. The reactions, which proceed via radical chain mechanisms, have been found to occur in high yields with all these silanes. The experimental conditions for the reaction with silanes **8** and **10** are similar to those with $(TMS)_3SiH$; that is, in toluene or benzene at ca. 80°C and in the presence of AIBN as radical initiator. Reactions with silane **9** require excess of the appropriate silane and preferably *tert*-butyl perbenzoate as initiator at 125°C. Triethylsilane and α-phenyl substituted silanes have also been used for the deoxygenation of secondary alcohols via thiono esters with success [29]. However, some of these last approaches are considered "mechanistically adventurous" and, therefore, of poor synthetic utility [30].

$(TMS)_2Si(H)Me$	$(TMS)_{3-n}SiMe_nSi(H)Me_2$	$(RS)_3SiH$
8	**9** (n = 0,2,3)	**10**

Figure 2 reports rate constants for hydrogen abstraction from a variety of silanes by primary alkyl radicals at room temperature [31]. The rate constants cover a range of three orders of magnitude and, as we expected, the hydrogen donating abilities in organosilanes can be modulated by substituents. As the hydrogen abstraction step becomes too slow, other reaction paths will compete [32]. Therefore, if a synthetic strategy requires a slower hydrogen donor, $Me_3SiSi(H)Me_2$ is at the limit of practical value.

Kinetic data are not yet available for the reaction of tris(alkylthio)silanes with alkyl radicals. However, the bond dissociation energy of the Si-H bond in **10** is ca. 4 kcal/mol higher than that in $(TMS)_3SiH$ suggesting that the former may be poorer hydrogen donors [28].

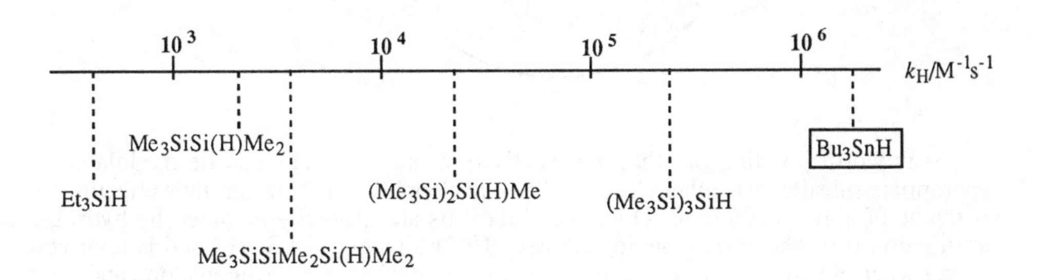

Figure 2. Rate constants for hydrogen abstraction from a variety of reducing agents by primary alkyl radicals at 25°C.

4. Silanes with Hydrogen Donating Abilities Higher than $(TMS)_3SiH$

It is well-known that thiols are rather good hydrogen donors toward carbon-centered radicals and that the corresponding thiyl radicals are poor atom (or group) abstracting agents and therefore do not support chain reactions analogous to $(TMS)_3SiH$. The facts that thiols are good H-atom donors toward alkyl radicals and that silyl radicals are among the most reactive known species for abstraction and addition reactions suggest that any class of compounds, which allows the transformation of a thiyl to a silyl radical *via* a fast intramolecular rearrangement, will potentially be a good radical-based reducing agent (Scheme 4). This turned out to be the case for tris(trimethylsilyl)silylthiol [33].

Scheme 4

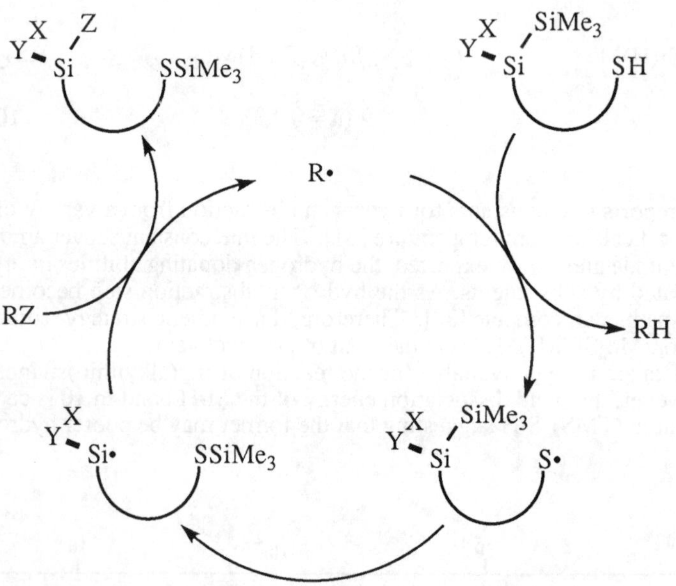

Y X Z Si SSiMe3 — Y X SiMe3 Si SH — R• — RZ — RH — Y X Si• SSiMe3 — Y X SiMe3 Si S•

It is worth pointing out that the reactivity of silyl radicals can be modulated by appropriate substituents, either electrophilic or nucleophilic in character, thus affecting the outcome of a given reaction. Organosilyl radicals are generally obtained by hydrogen abstraction from the corresponding silanes. Unfortunately, the Si-H bond is relatively strong when electron withdrawing substituents are attached to the silicon atom, and up to now, this approach is largely limited to electron donating substituents (see above). Therefore, appropriately substituted organosilanes (see Scheme 4) could overcome this problem and offer, for the first time, the ability to have either electrophilic or nucleophilic silyl radicals.

5. Polymers Containing Repeating Cycloketonic Units

The copolymers of olefins with carbon monoxide are of considerable importance from an industrial point of view. It is due, for example, to the facts that carbon monoxide is particularly plentiful and inexpensive, the carbonyl chromophore makes these copolymers photo degradable, and polyketones are good starting materials for other types of functionalized polymers with specialized properties. These copolymers have been synthesized either through radical-initiated or metal-catalyzed polymerization. The former methodology produces *random* olefin-carbon monoxide copolymers (olefin:CO>1) [34]. The problems associated with this procedure include the requirement for elevated temperatures and pressures. In addition, the CO content rarely exceeds 25 mol%, except for the C_2H_4:CO copolymer. The resulting copolymers with transition metal catalysts have a regular structure with the *alternating* olefin and carbon monoxide units (eq 2) (olefin:CO=1) [35].

$$H_2C{=}CH_2 \quad + \quad CO \longrightarrow \left(\;\; \right)_n \qquad (2)$$

In recent years, chemically modified polymers have gained an increasing importance in the manufacture of rubbers and plastic materials. Unsaturated polymers are particularly suitable for such transformations. It seemed to us in 1990 that a complementary approach to radical-initiated copolymerization of ethylene-carbon monoxide would be the reaction of polybutadiene with carbon monoxide under free radical conditions (eq 3). Due to the entropy factors, which are favorable in unimolecular reactions, it was expected that mild experimental conditions would be suitable, i.e. relatively low reaction temperature and pressure. Furthermore, it was hoped to find some special properties in this material containing polycyclopentanonic units. From the chemical point of view, the expectation turned out to be partially correct.

$$\left(\;\; \right)_n \xrightarrow{CO\,/\,I_R} \left(\;\; \right)_n \qquad (3)$$

A new class of polymers with repeating cycloketonic units along their chain can be prepared by reaction of polydiene containing adjacent structural units derived from 1,4-*cis* polymerization of conjugated dienes with carbon monoxide. This is best achieved in the presence of a free radical initiator and, preferably, a compound capable of acting as a hydrogen donor [36]. Infrared studies clearly indicate that the carbonyl moiety is incorporated in the forms of cyclopentanones and cyclohexanones. The content of carbonyl units as well as the ratio of cyclopentanones/cyclohexanones depends strongly on the experimental conditions. The simple mechanistic Scheme 5 for the chain modification of the polymer is not adequate to explain such a behavior.

Scheme 5

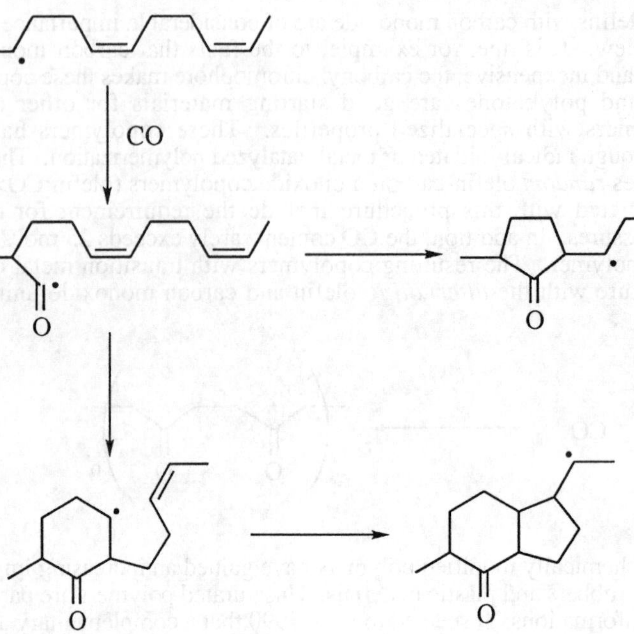

A search in the literature gave us a mechanistic picture for the cyclization of acyl radicals that is reported in Scheme 6 [37]. This disconcerting picture is derived from the contribution of several groups, some of whom are leaders in the field of free radical chemistry, in the last thirty years. However, the total absence of quantitative measurements together with a careful evaluation of the data that do exist, persuaded us to study in some detail the elementary steps of our polymer modification.

First, methods of generation of acyl radicals and their kinetics were studied at 80°C (eqs 4 and 5) [38,39]. Then, rate constants for the reactions of acyl radicals with Bu_3SnH and $(TMS)_3SiH$ were measured using the decarbonylation of primary acyl radicals as timing devices [40]. The values of 1.1×10^6 and 7.2×10^4 M^{-1} s^{-1} derived at 80°C for Bu_3SnH and $(TMS)_3SiH$, respectively, were used to obtain the rate constant of 3.7×10^5 s^{-1} for the decarbonylation of secondary acyl radicals. With these data in our hands we were able to design and carry out a series of experiments which allowed us to determine rate constants (at 80°C, temperature of the polymer carbonylation) for each elementary step of interest, as it is reported in Scheme 7 [41,42].

$$RC(O)Cl + (TMS)_3Si\bullet \xrightarrow{7 \times 10^5\ M^{-1}s^{-1}} R\dot{C}O + (TMS)_3SiCl \quad (4)$$

$$RC(O)SePh + (TMS)_3Si\bullet \xrightarrow{2 \times 10^8\ M^{-1}s^{-1}} R\dot{C}O + (TMS)_3SiSePh \quad (5)$$

Scheme 6

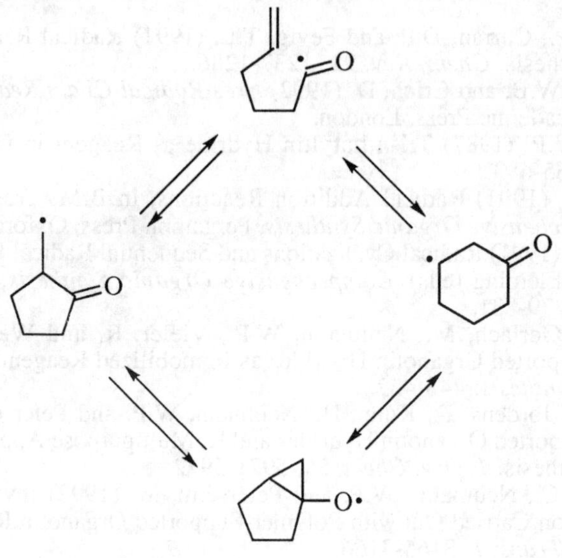

Scheme 7

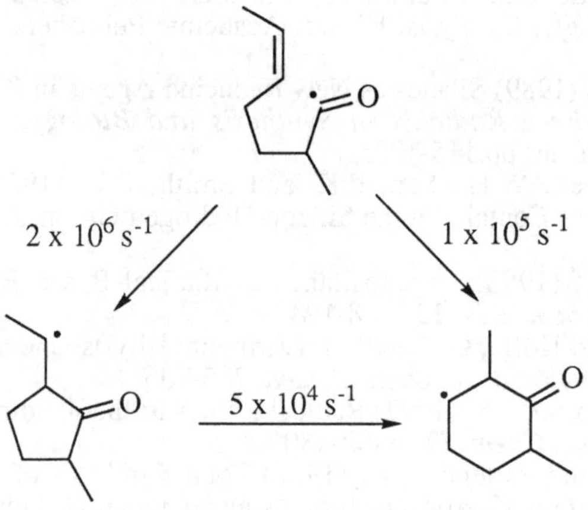

The above described polymers show a good combination of chemical, mechanical and process properties, and are particularly useful, either alone or as mixtures, as high-tenacity thermoplastic polymers and as elastomers or elastomeric modifiers. One can control both the incorporated amount of carbon monoxide and the ratio of cyclopentanones/cyclohexanones and, consequently, the properties of the desired polymer. This makes such a class of new materials of great potential interest.

6. References

1. Jasperse, C.P., Curran, D.P. and Fevig, T.L. (1991) Radical Reactions in Natural Product Synthesis, *Chem. Rev.* **91**, 1237-1286.
2. Motherwell, W.B. and Crich, D. (1992) *Free Radical Chain Reactions in Organic Synthesis*, Academic Press, London.
3. Neumann, W.P. (1987) Tri-*n*-butyltin Hydride as Reagent in Organic Synthesis, *Synthesis*, 665-683.
4. Curran, D.P. (1991) Radical Addition Reactions, in B.M. Trost and I. Fleming (eds), *Comprehensive Organic Synthesis*, Pergamon Press, Oxford, pp. 715-777.
5. Curran, D.P. (1991) Radical Cyclizations and Sequential Radical Reactions, in B.M. Trost and I. Fleming (eds), *Comprehensive Organic Synthesis*, Pergamon Press, Oxford, pp. 779-831.
6. Gerigk, U., Gerlach, M., Neumann, W.P., Vieler, R. and Weintritt, V. (1990) Polymer-Supported Organotin Hydrides as Immobilized Reagents for Free Radical Synthesis, *Synthesis*, 448-452.
7. Gerlach, M., Jördens, F., Kuhn, H., Neumann, W.P. and Peterseim, M. (1991) A Polymer-Supported Organotin Hydrides and Its Multipurpose Application in Radical Organic Synthesis, *J. Org. Chem.* **56**, 5971-5972.
8. Bokelmann, C., Neumann, W.P. and Peterseim, M. (1992) Investigations on the Giese Reaction Carried Out with Polymer-Supported Organotin Reagents, *J. Chem. Soc., Perkin Trans. 1*, 3165-3166.
9. Ruel, G., The, N.K., Dumartin, G., Delmond, B. and Pereyre, M. (1993) Un Nouvel Hydride Organostannique Greffé sur Support Insoluble, *J. Organomet. Chem.* **444**, C18-C20.
10. Chatgilialoglu, C. (1989) Recent Advances in the Chemistry of Silyl Radicals, in M. Chanon, M. Julliard, and J. Poite (eds), *Paramagnetic Organometallic Species in Activation/Selectivity, Catalysis*, Kluwer Academic Publishers, Dordrecht, pp.119-129.
11. Chatgilialoglu, C. (1989) Silanes as New Reducind Agents in Organic Synthesis, in F. Minisci (ed), *Free Radicals in Synthesis and Biology*, Kluwer Academic Publishers, Dordrecht, pp.115-123.
12. Gilman, H., Atwell, W.H., Sen, P.K. and Smith, C.L. (1965) Branched-Chain Organic Polysilanes Containing the Silicon-Hydrogen Group, *J. Organomet. Chem.* **4**, 163-167.
13. Chatgilialoglu, C. (1992) Organosilanes as Radical-Based Reducind Agents in Synthesis, *Acc. Chem. Res.* **25**, 188-194.
14. Schummer, D. and Höfle, G. (1990) Tris(trimethylsilyl)silane as a Reagent for the Radical Deoxygenation of Alcohols, *Synlett*, 705-706.
15. Knapp, S. and Gibson, F.S. (1992) Radical Routes to Indolizidines. Synthesis of (–)-Slaframine, *J. Org. Chem.* **57**, 4802-4809.
16. Hitchcock, S.A. and Pattenden, G. (1992) Total Synthesis of the Mycotoxin (–)-Zearalenone Based on Macrocyclization using a Cinnamyl Radical Intermediate, *J. Chem. Soc., Perkin Trans. 1*, 1323-1328.
17. Ihara, M., Setsu, F., Shohda, M., Taniguchi, N. and Fumumoto, K. (1994) A Total Synthesis of (±)-Tacamonine (Pseudovincamone I) Through Radical Cyclization, *Heterocycles* **37**, 289-292.
18. Pereyre, M., Quintard, J.-P. and Rahm, A. (1988) *Tin in Organic Synthesis*, Butterworths, London.
19. Ferreri, C., Ballestri, M. and Chatgilialoglu, C. (1993) An Unexpected Stereochemical Control of Alkene Formation by the Choice of Radical Initiator. The

Reversible Addition of $(TMS)_3Si$• Radicals to Alkenes, *Tetrahedron Lett.* **34**, 5147-5150.

20. Chargilialoglu, C., Ballestri, M., Ferreri, C. and Vecchi, D. (1995) (*Z*)-(*E*) Interconversion of Olefins by the Addition-Elimination Sequence of the $(TMS)_3Si$• radical, *J. Org. Chem.* **60**, 3826-3831.
21. Johnson, D.W. and Poulos, A. (1992) Tris(trimethylsilyl)silane in the Deoxygenation of Long Chain Unsaturated Alcohols Converts cis to trans Isomers, *Tetrahedron Lett.* **33**, 2045-2048.
22. Emmer, G. and Weber-Roth, S. (1992) Synthesis of Derivatives of FK 506 and FR 900520: Modification at the Binding Domain, *Tetrahedron* **48**, 5861-5874. (1993) corrigenda, *Tetrahedron* **49**, 291.
23. Lowinger, T.B. and Weiler, L. (1992) Stereoselective Formation of *E*- or *Z*-Exocyclic Alkenes via Radical Cyclization Reactions of Acetylenic Esters, *J. Org. Chem.* **57**, 6099-6101.
24. Pattenden, G. and Schulz, D.J. (1993) Cascade Radical Reactions in Synthesis. A New Radical Mediated Double Ring Expansion-Cyclization Process with Oxime Ethers, *Tetrahedron Lett.* **34**, 6787-6790.
25. Hollingworth, G. J., Pattenden, G. and Schulz, D. J. (1995) Cascade Radical Cyclization-Fragmentation-Transannular-Ring Expansion Reactions Involving Oximes. A New Approach to Synthesis of Angular Triquinanes, *Aust. J. Chem.* **48**, 381-399.
26. Chatgilialoglu, C., Guerrini, A. and Lucarini, M. (1992) The Me_3Si Substituent Effect on the Reactivity of Silanes. Structural Correlations between Silyl Radicals and Their Parent Silanes, *J. Org. Chem.* **57**, 3405-3409.
27. Ballestri, B., Chatgilialoglu, C., Guerra, M., Guerrini, A., Lucarini, M. and Seconi, G. (1993) Organosilanes as Radical-Based Reducing Agents with Low Hydrogen Donating Abilities, *J. Chem. Soc., Perkin Trans. 2*, 421-425.
28. Chatgilialoglu, C., Guerra, M., Guerrini, A., Seconi, G., Clark, K.B., Griller, D., Kanabus-Kaminska, J. and Martinho-Simões, J.A. (1992) A Study on the Reducing Abilities of Tris(alkylthio)silanes, *J. Org. Chem.* **57**, 2427-2433.
29. Barton, D.H.R., Jang, D.O. and Jaszberenyi, J.Cs. (1993) The Invention of Radical reactions. Part XXIX. Radical Mono- and Dideoxygenations with Silanes, *Tetrahedron* **49**, 2793-2804.
30. Chatgilialoglu, C. and Ferreri, C. (1993) Progress of the Barton-McCombie Methodology: From Tin Hydrides to Silanes, *Res. Chem. Intermed.* **19**, 755-775.
31. Chatgilialoglu, C. (1995) Structural and Chemical Properties of Silyl Radicals, *Chem. Rev.* **95**, 1229-1251.
32. Chatgilialoglu, C., Ferreri, C. and Lucarini, M. (1993) A Comment on the Use of Triethylsilane as a Radical-Based Reducing Agent, *J. Org. Chem.* **58**, 249-251.
33. Ballestri, B., Chatgilialoglu and Seconi, G. (1991) $(TMS)_3SiSH$: A new Radical-Based Reducing Agent, *J. Organomet. Chem.* **408**, C1-C4.
34. Sen, A. (1986) The Copolymerization of Carbon Monoxide with Olefins, *Adv. Polym. Sci.* **73/74**, 125-145.
35. Sen, A. (1993) Mechanistic Aspects of Metal-Catalyzed Alternating Copolymerization of Olefins with Carbon Monoxide, *Acc. Chem. Res.* **26**, 303-310.
36. Sommazzi, A., Cardi, N., Garbassi, F. and Chatgilialoglu, C. (1994) Polymers Containing Repeating Cycloketonic Units and Method for Obtaining Them, *US Patent* 5,369,187.
37. Dowd, P. and Zhang, W. (1993) Free-Radical-Mediated Ring Expansion and Related Annulations, *Chem. Rev.* **93**, 2091-2115.

38. Ballestri, M., Chatgilialoglu, C., Cardi, N. and Sommazzi, A. (1992) The Reaction of Tris(trimethylsilyl)silane with Acid Chlorides, *Tetrahedron Lett.* **33**, 1787-1790.
39. Chatgilialoglu, C. and Lucarini, M. (1995) Rate Constants for the Reaction of Acyl Radicals with Bu_3SnH and $(TMS)_3SiH$, *Tetrahedron Lett.* **36**, 1299-1302.
40. Chatgilialoglu, C., Ferreri, C., Lucarini, M., Pedrielli, P. and Pedulli, G. F. (1995) Rate Constants and Arrhenius Parameters for the Reaction of Acyl Radicals with Bu_3SnH and $(Me_3Si)_3SiH$, *Organometallics* **14**, 2672-2676.
41. Chatgilialoglu, C., Ferreri, C., Lucarini, M. and Zavitsas A. (submitted) Mechanistic Insights into the Free Radical Mediated One-Carbon Ring Expansion in Cyclopentanones, *J. Am. Chem. Soc.*
42. Chatgilialoglu, C., Ferreri, C., Venturini, A. and Ballestri, M. (submitted) Mechanistic Studies on the Cyclization of Hex-5-enoyl Radicals, *J. Am. Chem. Soc.*

TEMPLATED SYNTHESIS OF ENZYME MIMICS: HOW FAR CAN WE GO?

SALLY ANDERSON AND JEREMY K. M. SANDERS
Cambridge Centre for Molecular Recognition
University Chemical Laboratory, Lensfield Road
Cambridge CB2 1EW, United Kingdom

1. Introduction

Watson and Crick discovered the DNA double helix in 1953 and realised immediately that its replication involved a templated synthesis.[1] Their work inspired Todd[2] to issue this challenge in 1956:

> *"The use of one molecule as a template to guide and facilitate the synthesis of another.........has not hitherto been attempted in laboratory synthesis, although it seems probable that it is common in living systems. It represents a challenge which must, and surely can, be met by organic chemistry."*

In retrospect, it is clear that metal ion templated syntheses of phthalocyanines had been observed as early as 1932,[3] but it is only with Busch's pioneering and systematic work in the 1960s that different modes of templating were seriously studied and classified.[4,5] The recent emergence of supramolecular chemistry in solution and in the solid state has brought together the historically separate organic and inorganic strands of template chemistry so that metal-ligand binding, hydrogen bonding and π–π interactions can now be exploited to allow the synthesis of ever-larger structures with a remarkable degree of control. Templating has been reviewed very recently[6–8] so we mainly concentrate in this article on our own recent work with porphyrin oligomers; these are of interest as enzyme mimics, as described in the accompanying article.

A template intervenes in the macroscopic geometry of the reaction rather than in the chemistry; it provides instructions for the formation of a single product from a substrate or substrates which otherwise have the potential to assemble and react in a variety of ways. Changing the template should result in a different substrate assembly and consequently a different product. In general, after the template has directed the formation

C. Chatgilialoglu and V. Snieckus (eds.), Chemical Synthesis, 277–291.

of the product, it is removed to yield the template-free product, although this feature is not found where the template is an integral part of the structure it helps to form.

Our interest in building enzyme mimics based on porphyrins led us to synthesise cyclic porphyrin oligomers using Glaser coupling.[9] This reaction oxidatively combines two terminal acetylenes to give a butadiyne link:

$$\mathrm{Ar{-}C{\equiv}C{-}H\ +\ H{-}C{\equiv}C{-}Ar\ \rightarrow\ Ar{-}C{\equiv}C{-}C{\equiv}C{-}Ar}$$

The linkage has the attractions of being linear and relatively rigid, and the reaction can be made essentially quantitative: under optimized conditions, there is little side reaction and no unreacted starting material remains. The nature and yields of the products are determined solely by geometrical considerations and by the kinetics of the encounter between reactive ends. Thus, reaction of the monomeric porphyrin **1** (shown in cartoon form; the actual substitution pattern is shown in Fig. 9) gives a complex mixture of cyclic oligomers. The dimer **2** and the trimer are relatively easy to isolate, but the tetramer is very difficult to separate from the small quantities of even higher oligomers formed in the reaction. The separated oligomers proved to have very different ligand-binding characteristics, tempting us to try using these differences to influence the synthesis by templating. After much optimization of conditions, we arrived at a very successful cyclic dimer synthesis using 4,4'-bipyridyl (**Bipy**) as the template, CuCl–TMEDA–dry air as the oxidant, and dichloromethane solvent.[9]

The first step in this templated reaction must be the combination of two porphyrin units to yield linear dimer **5** (Fig 1). Then either an intramolecular cyclisation occurs or reaction with a further monomer porphyrin yields linear porphyrin trimer. The latter process is rendered less important by the **Bipy** template which induces the reactive ends of the linear dimer intermediate to come into close proximity and so increases the rate of intramolecular cyclisation. This reaction therefore proceeds in the same way as classical metal-cation templated macrocylisations. The yield of cyclic dimer was increased by a factor of three in the presence of **Bipy**.

The templated synthesis of cyclic trimer, using the complementary ligand tripyridyltriazine, **Py_3T**, was less successful, improving the yield from 34% to around 50%, while at the same time drastically reducing the yield of dimer. Again, the first step must be the coupling of monomer units to yield linear dimer, but the conformation adopted when this is bound to **Py_3T** has the reactive ends held apart (Fig. 1); intramolecular cyclisation is now disfavoured, leaving the way open for intermolecular reaction to take place. Bifunctional ligands which bind with the same geometry, such as dipyridylpyridine **Py_2Py**, give a similar product distribution, confirming that the template in the synthesis of cyclic trimer plays a mainly preventative role.

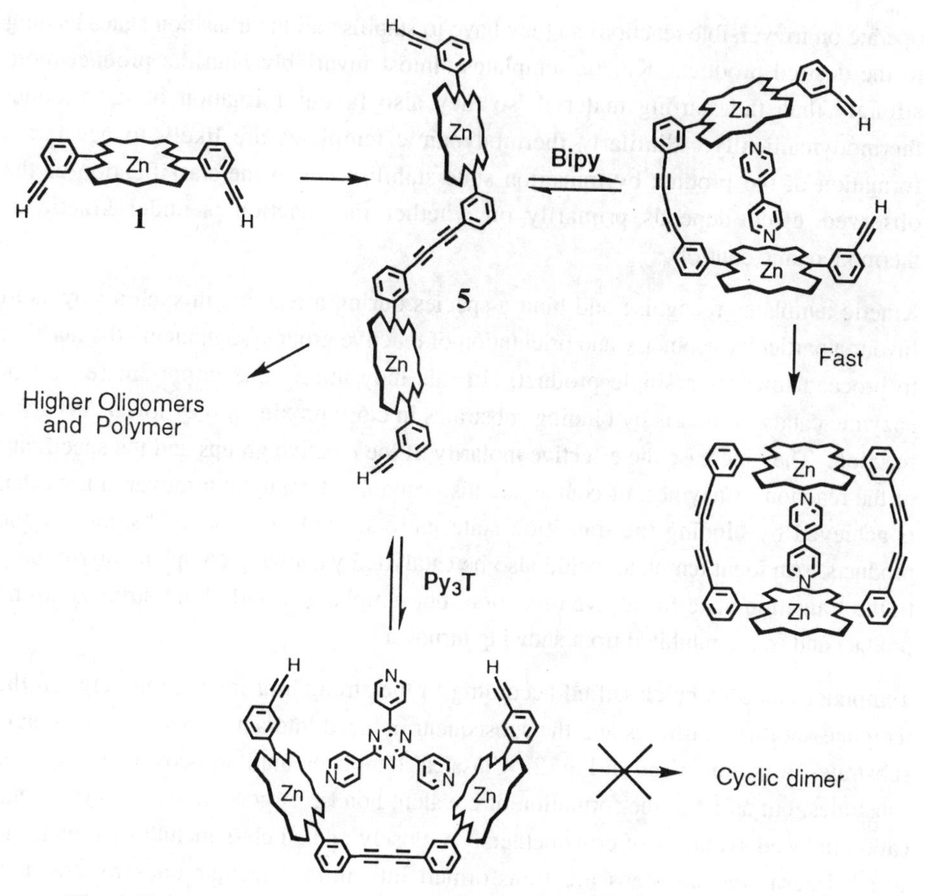

Fig. 1 Templated Glaser coupling of bis-alkynyl porphyrin **1** in the presence of **BiPy** and **Py3T**.

Our detailed analysis of these differences in the templating abilities of **BiPy** and **Py3T** led to efficient syntheses of porphyrin tetramers and an octamer, and the understanding of template principles described below and in our review.[6]

2. Classification of Templates

Templates can be classified as either thermodynamic or kinetic.[4] Thermodynamic templating occurs in reversible reactions under thermodynamic control: the template, when added to a reaction mixture at equilibrium, binds to one of the products and shifts the equilibrium towards this species. Thermodynamic templating can result in very high yields because the template only has to stabilise the product. Kinetic templates operate on irreversible reactions so they have to stabilise all the transition states leading to the desired product. Kinetic templates almost invariably bind the product more strongly than the starting material, so they also favour formation of the product thermodynamically. Similarly thermodynamic templates are likely to accelerate formation of the product by transition state stabilisation so the classification of the observed effect depends primarily on whether the reaction is under kinetic or thermodynamic control.

Kinetic templates recognise and bind a species during a reaction in such a way as to favour a particular geometry and orientation of reactive groups, so inducing the reaction to proceed towards a single product. In this they mimic one important feature of enzymic catalysis, that is by binding substrates in close proximity they lower $\Delta S^{\ddagger}$ for a reaction. They increase the effective molarity of the reactive groups and the specificity of the reaction. Enzymes, of course, are also capable of catalytic turnover, a trick that is achieved by binding the transition state more strongly than the substrates or the products. An ideal template would also be catalytic by showing complementarity only to the transition state for a given reaction, but templates usually bind strongly to the product and so are inhibited from showing turnover.

Templates can also be classified according to the strength of interaction between the template and the substrates and the subsequent induced interaction between the bound substrates, as summarised in Fig. 2. Classical templates bind non-covalently to their substrates, but assist in the formation of covalent bonds. A good example is the metal cation induced synthesis of crown ethers. A closely related class includes templates in which intermolecular steps are transformed into intramolecular ones by covalent bonding between template and substrates.[10] 'Self-assembly' processes could also be classified as aggregation-templated, a non-covalent interaction between template and substrates inducing a second non-covalent interaction.[6, 7]

Finally, templates can be classified according to their topology, as shown in Fig. 3. We now explore the features of cyclisation, linear and interweaving templates.

2.1 Cyclisation templates. The simplest type of template in this category, and perhaps the most widely exploited, is the metal cation which induces macrocyclisation. As mentioned above this same kind of effect is exhibited in our templated synthesis of cyclic porphyrin dimer from monomer.[9] In this latter example the linear intermediate was produced in situ and not isolated. In order to study the cyclisation step in isolation we synthesised and isolated the linear porphyrin dimer **5**.[11, 12] Using 0.2 mM **5** and no template, the yields of cyclic dimer and higher oligomers were 21% and 79% respectively.[12] The presence of **Bipy** increased the yield of cyclic dimer by four fold

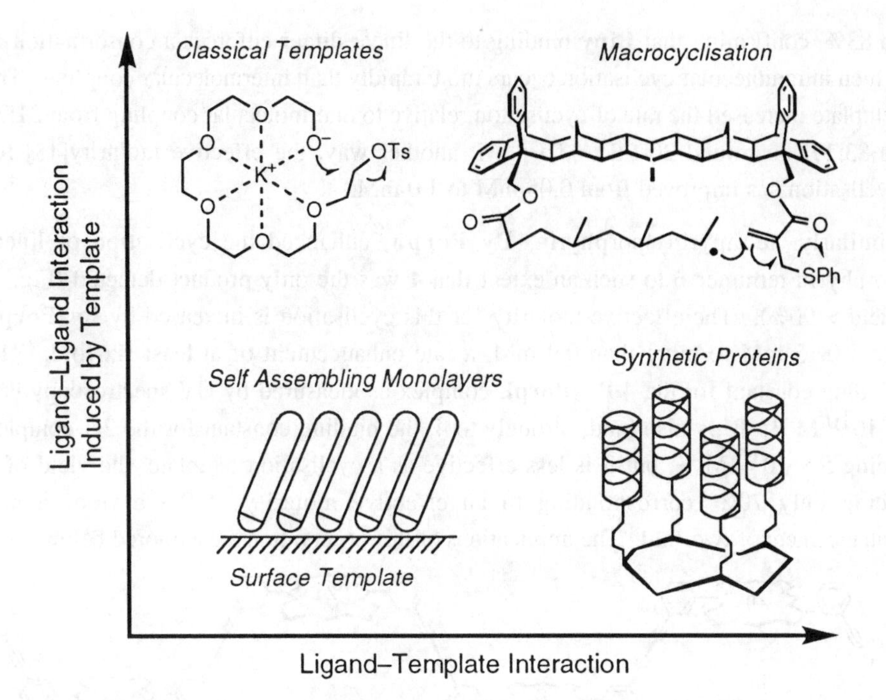

Fig. 2 Templating and self-assembly: templating systems range from those where non-covalent substrate–template interactions induce non-covalent interactions between substrates (bottom left), to those where covalent interactions induce further covalent interactions (top right).

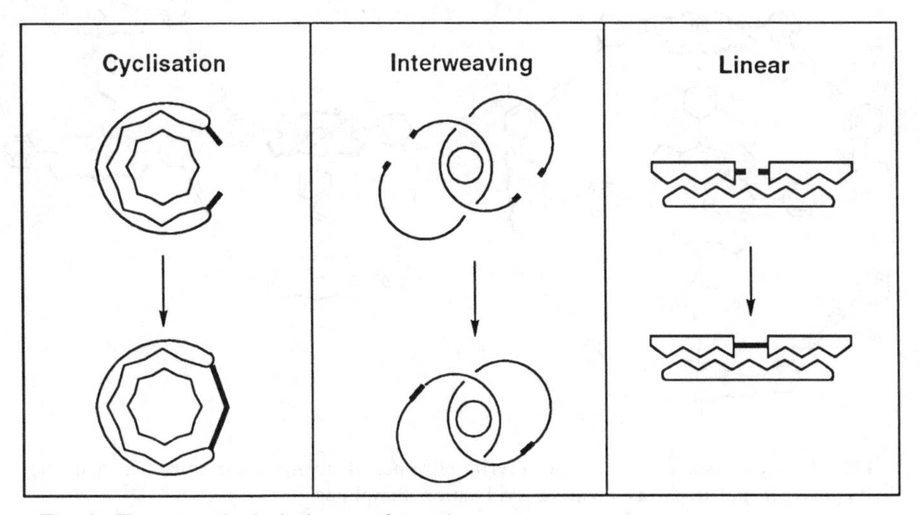

Fig. 3 Three topological classes of template.

to 83%, confirming that **Bipy** binding to the linear dimer enforces a conformation in which intramolecular cyclisation occurs more rapidly than intermolecular coupling. The template increased the rate of cyclisation relative to intermolecular coupling from 21:79 to 83:17, or around 20-fold. To put it another way, the effective molarity[13] for cyclisation has improved from 0.05 mM to 1.0 mM.

Similarly, tetrapyridylporphyrin, **Py_4Porph**, enhanced the cyclisation of linear porphyrin tetramer **6** to such an extent that **4** was the only product detected (Fig. 4; yield > 90%). The effective molarity for this cyclisation is increased by **Py_4Porph** from 0.15 mM to more than 0.9 mM, a rate enhancement of at least six-fold. The binding constant for the **4•Py_4Porph** complex as measured by UV spectroscopy is 2×10^{10} M^{-1}. **Bipy** also binds strongly to **4**, the binding constant for the 2:1 complex being 5×10^{16} M^{-2}, but it is less effective as a cyclisation template, the yield of **4** being only 70%, corresponding to an effective molarity of 0.3 mM or a rate enhancement of two-fold. The implications of this observation are explored below.

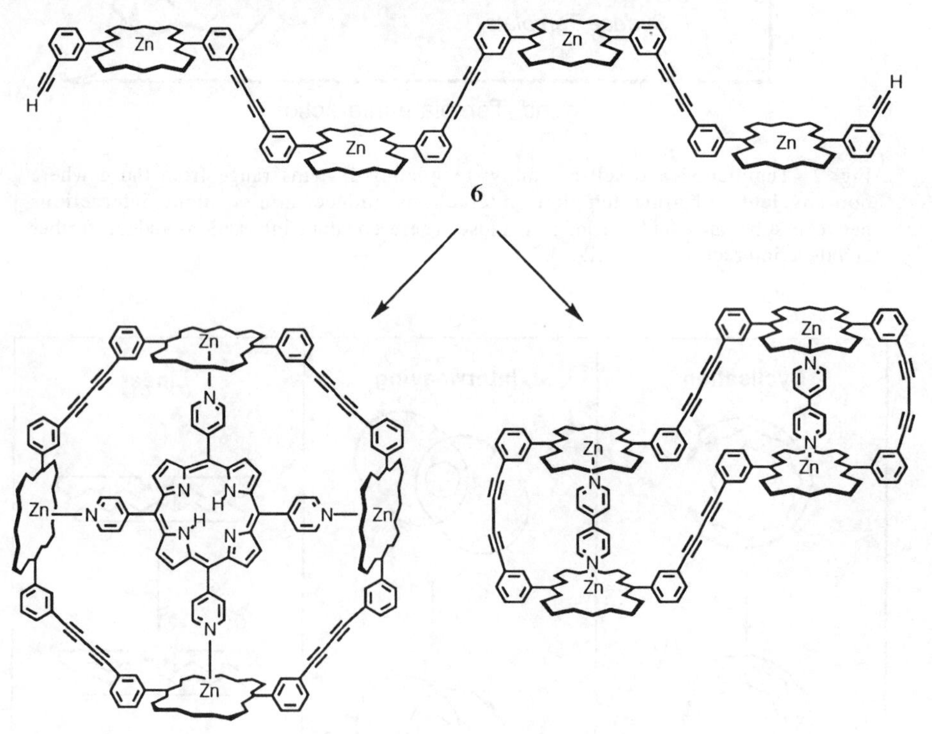

Fig. 4 Synthesis of cyclic porphyrin tetramer **4** from linear tetramer **6** using **Py_4Porph** and **Bipy** as positive cyclisation templates.

2.1.1. Scavenging. Cyclisation templates can be used to scavenge cyclisable material in a reaction mixture and thus facilitate the formation of linear oligomers.[11] Using the strategy for synthesis of linear oligomers in Fig. 5(a), dimerisation of the partially-deprotected material can only be carried out efficiently in the absence of fully deprotected molecules because the latter can couple with mono-protected material to generate a new reactive oligomer and ultimately a complex mixture. This problem can be avoided by separation of the doubly-reactive material before coupling, but this becomes more difficult with increasing chain length. Scavenger templates overcome the problem by enforcing intramolecular reaction of doubly-reactive molecules (Fig. 5 (b)); mono-protected molecules have no choice but to couple to each other. The separation is left until after coupling, when it has become much easier because the desired linear compound is twice as massive as either the starting material or the cyclic by-product.

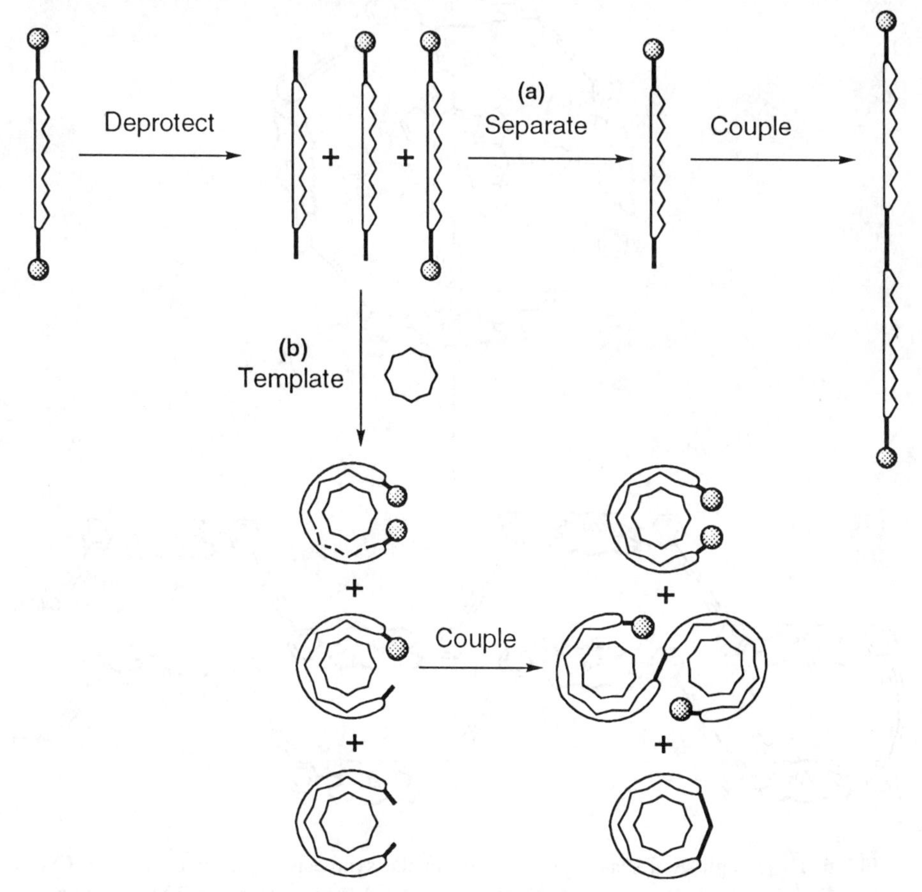

Fig. 5 Strategies for the synthesis of linear oligomers from a symmetrically protected building block. Hatched circles represent protecting groups.

We used this scavenging strategy to synthesise linear tetramer **6.TMS$_2$** from dimer **5.TMS** (not shown) and a linear octamer from tetramer **6.TMS** (Fig. 6), using **Bipy** and **Py$_4$Porph** respectively as templates. In both cases the yields of desired product are close to the theoretical maximum of 50%; the doubly-protected starting material can be recycled and the cyclic by-product is useful in its own right. In the absence of template, both reactions do give significant amounts of the desired linear coupling product, but there are so many other coupling products that separation of pure material on a preparative scale is difficult and inefficient. In the presence of a scavenger template, only three products are formed, greatly easing the isolation process; this advantage is as important as the improvement in absolute yield.

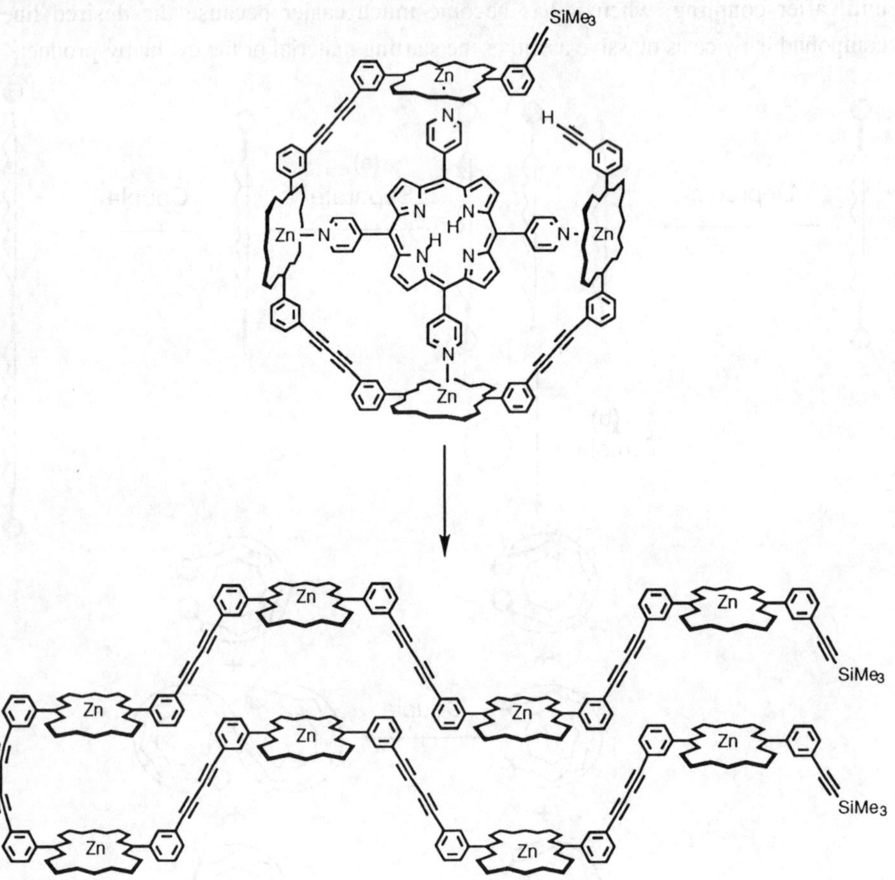

Fig. 6 **Py$_4$Porph** as a scavenger template in the synthesis of linear octamer **7.TMS$_2$**. For clarity, only the coupling of the mono-protected linear tetramer **6.TMS** is shown.

2.2 Linear Templates. These template the reaction between two bound substrates rather than between two ends of the same substrate. The classic example of this type of templating is replication of DNA and RNA. In both cases a single strand of DNA is used as a template, the base sequence on the template strand being complementary to the sequence produced on the new strand. Linear templates are not restricted to the formation of a single bond: a Diels–Alder reaction[14] has been effectively templated by connecting diene and dienophile covalently through a phenylboronic acid, so transforming the reaction from intermolecular to intramolecular. This is a templated reaction in the sense that the phenylboronate linker is readily removed at the end of the reaction. The Diels–Alder reaction accelerated by binding of diene and dienophile in the cavity of our cyclic porphyrin trimer and described in the accompanying article is also an example of linear templating, as are Mock's[15] cucurbituril-catalysed 4 + 2 cycloaddition, Feldmann's[10] system in which several bonds are formed in series, and the increasing number of self-replicating systems.[16]

2.3 Interweaving templates. Sauvage's synthesis of catenanes uses copper(I) as a template to induce two phenanthroline ligands to adopt a tetrahedral conformation around the metal (Fig. 7). The template forces the ligands to remain interwoven while coupling with the linking group takes place. Removal of the metal yields the catenane. In this synthesis the template does not increase the efficiency of cyclisation of any individual ring so it is not a cyclisation template, but it does increase the probability of the product having an interesting topology. This strategy has been extended to the synthesis of complex knots.[17] Stoddart, Hunter and Vögtle have prepared catenanes and rotaxanes[6, 7] using π–π interactions to enhance both cyclisation and interweaving, leaving the template as an integral part of the final product, while the massive objects recently constructed with DNA are also strictly catenanes.[18]

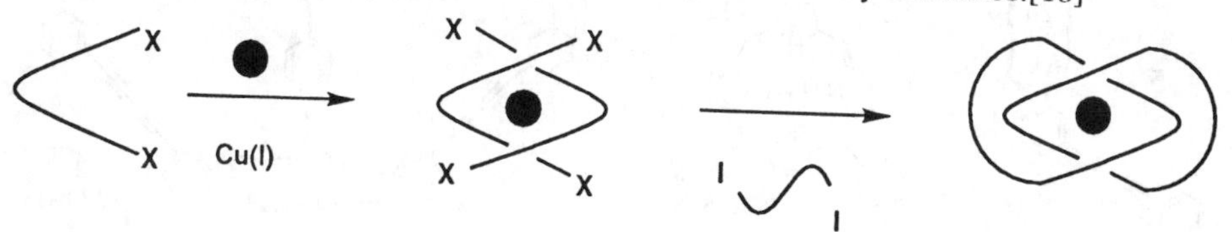

Fig. 7 Copper(I) as an interweaving template in the synthesis of a catenane.

2.4 Negative templates The templates discussed so far favour reaction between bound substrates; they are positive templates. It is also possible for a template to disfavour reaction between bound substrates; they are then negative templates. A negative template disfavours the formation of a given product not by accelerating a competitive reaction but by specifically disfavouring the formation of that product. As we show below, the negative template may have a second function as a positive template during a subsequent step. In principle any system which can be induced to adopt a particular molecular architecture by a positive template can be prevented from doing so by a negative template. For example the Glaser cyclisation of linear dimer **5** can be inhibited by the negative cyclisation template mentioned above, **Py2Py**, the yield of cyclic dimer being halved, i.e. the effective molarity for cyclisation is reduced from 0.05 mM to 0.03 mM. Clearly the ligand binds so that linear dimer adopts a conformation in which the reactive groups are held apart so as to discourage intramolecular cyclisation to cyclic dimer. This same effect was observed in the **Py3T** templated reaction of monomer **1** (Figure 1). To summarize the results described earlier, the template binds to the linear dimer intermediate, preventing it from cyclising, and then acts as a modest cyclisation template, improving the yield of cyclic trimer. The major contribution to this increase is derived from the suppression of cyclic dimer.

A more satisfying example of negative templating is the efficient one-step synthesis of cyclic tetramer **4** from linear dimer **5** templated by **Py4Porph** (Fig. 8).[6] This is currently our best route for production of cyclic tetramer. Here the template inhibits cyclic dimer formation by acting negatively, and positively accelerates cyclisation of the intermediate linear tetramer **6**.

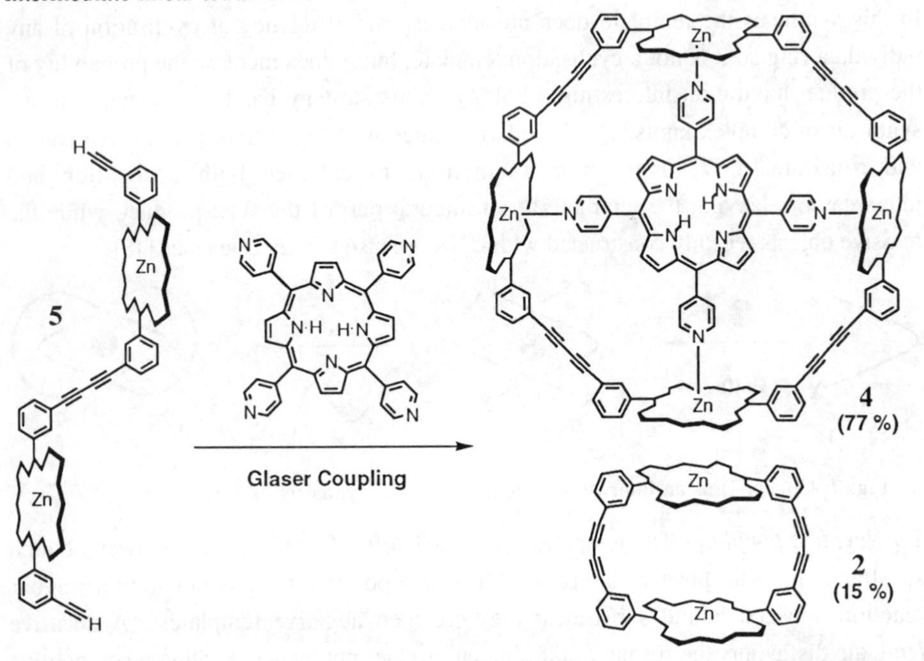

Fig. 8 Synthesis of cyclic porphyrin tetramer **4** from linear dimer **5** using **Py4Porph** as template.

2.5 Multiply-acting templates We have seen that **Py4Porph** acts as both a positive cyclisation and a negative template. Does it also accelerate the coupling of two dimer units to yield the linear tetramer? Under our standard conditions the Glaser coupling is too fast to allow easy observation of the linear tetramer intermediate so we looked at the Glaser coupling of a mixture of free base and metallated linear dimers which were blocked at one end (Fig. 9): in this experiment, coupling is necessarily terminated at the linear tetramer stage, and the question becomes whether the distribution of metallated and free base sites in the coupling products is influenced by the presence of **Py4Porph**. In the absence of template, one would expect a statistical mixture of free base, bis-zinc and tetra-zinc species. In the presence of an effective template, one would expect a larger yield of tetrazinc species. Experimentally it proved easiest to use a double-labelling strategy, the free base porphyrins carrying methyl ester side chains and the zinc porphyrins carrying *iso*-decyl esters; this allowed us to remove the labile metal labelling and use the esters as a permanent, and chromatographically visible label.

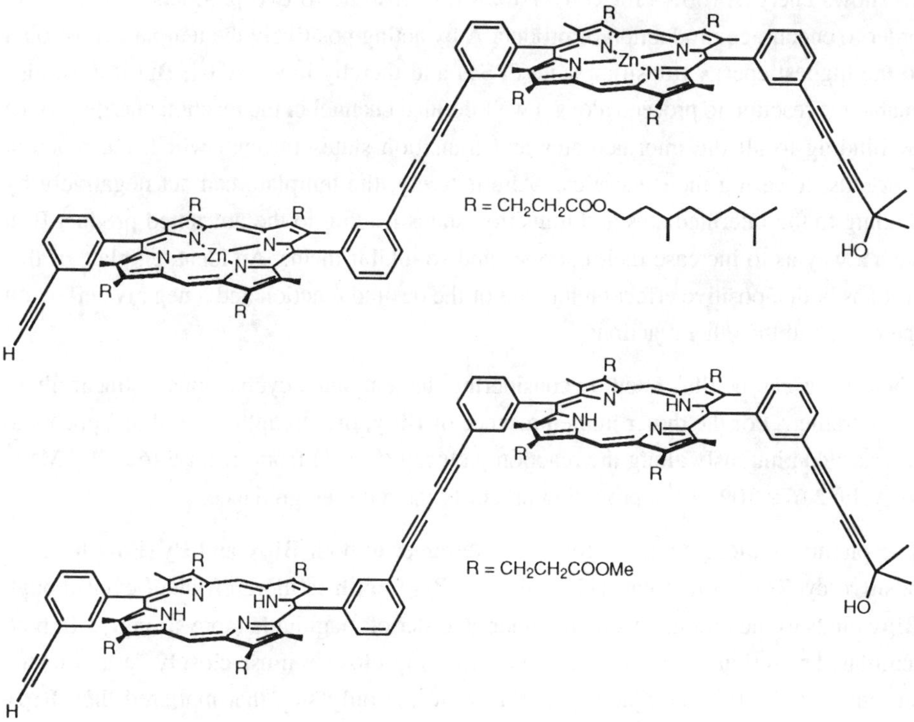

R = CH_2CH_2COO-(iso-decyl)

R = CH_2CH_2COOMe

Fig. 9 "Doubly labelled", blocked linear dimers used for detection of templated coupling on **Py4Porph**.

We actually observe a non-statistical distribution of tetramers even in the absence of template as a result of preferential activity of one of the dimers, and the standard templating conditions gave no significant change in product distribution. When we changed the copper chelating agent from TMEDA to 2,2′-bipyridyl, the coupling process slowed substantially, and it was now possible to detect an acceleration of linear dimer coupling by the template, corresponding to an effective molarity of 0.5 mM. Under these new, slower coupling conditions we were able to follow the formation and consumption of linear tetramer in the actual synthesis of cyclic tetramer and show directly that **Py_4Porph** indeed carries out all three roles: suppressing dimer cyclisation, accelerating dimer dimerisation, and accelerating linear tetramer cyclisation.

3. Free energy profiles for templates

Many of the ideas discussed above can be usefully summarised in energy profiles. Fig. 10 shows energy profiles for conversion of a substrate to two products A and B. In order to encourage production of product A by acting positively the template must bind to the highest energy transition state (**TS_1**) and thereby lower $\Delta G^{\ddagger}$. But it must also enable the reaction to proceed along a well-defined channel in the reaction energy surface by binding to all the intermediates and transition states through which the reaction proceeds, lowering their energies. Alternatively, the template can act negatively by binding to the intermediates and transition states leading to the unwanted product B in such a way as to increase their energies and so inhibit them. An ideal template of this sort has both a positive effect on the rate of the desired reaction and a negative effect on the rate of all the other reactions.

These ideas can be illustrated by considering the templated cyclisations of linear dimer and tetramer. For the dimer in the presence of **Bipy**, the strength of binding appears to increase continuously along the reaction pathway (Fig. 1) from linear **5** (6×10^6 M^{-1}) to cyclic **2** (6×10^8 M^{-1}), providing precisely the required guidance.

The picture is more complex for linear tetramer **6**; both **Bipy** and **Py_4Porph** act as positive cyclisation templates (Fig. 4), but **Py_4Porph** is more efficient even though **Bipy** binds to the product cyclic tetramer an order of magnitude more strongly. In both templated reactions the transition state for ring-closure must closely resemble the structure of the final complex so if this were the only step that mattered then **Bipy** would probably be a better template than **Py_4Porph**. However, the template must lead the initial substrate through a maze of alternative reaction pathways to reach the final transition state; its efficiency depends both on its ability to prevent linear tetramer from undergoing intermolecular reaction and on its ability to bring the two reactive ends together. It appears that **Bipy** is less efficient than **Py_4Porph** at preventing linear tetramer from undergoing intermolecular reaction early in the reaction pathway.

Returning to the templated formation of cyclic trimer in Fig. 1 we can now see that the template **Py_3T** must undertake a wide variety of roles if it is to be successful: complementarity to the trimer cyclisation transition state may be important, but unless the reaction is effectively guided to this transition state most of the substrate will have undergone irreversible side reactions before it is reached. It is pointless making a deep well in the reaction energy surface if a large proportion of the substrate never reaches it. What is required is a well defined channel along which the reaction can proceed.

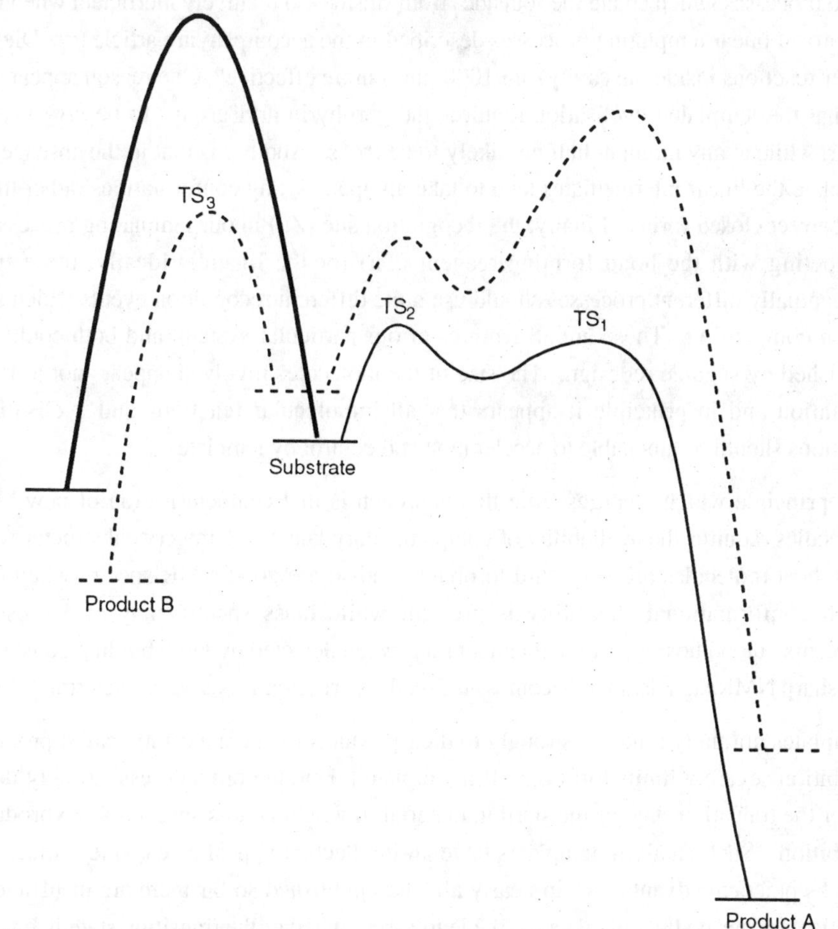

Fig. 10 Energy profiles for reactions which are untemplated (dashed line), positively templated (solid line) and negatively templated (bold line).

4. Discussion and prospects

We have prepared an array of linear and cyclic porphyrin oligomers using the Glaser coupling as the only carbon–carbon bond-forming reaction, and have achieved control through templating. Relatively small templates have been used to construct remarkably extended molecular architectures around them: the octamer is a monodisperse species with molecular weight of over 7400, but was prepared in just three conceptual steps from monomer. The question arises as to how far one could go using this approach. It should be born in mind that the templated processes described here are actually rather inefficient: effective molarities of 0.1–1 mM mean that no template effects would have been seen at all if our syntheses had been carried out at higher concentrations. Why are these processes which create the 'outside' from 'inside' so relatively inefficient when the reciprocal linear templating processes described in the accompanying article (i.e. Diels–Alder reactions inside the cavity) are 1000 times more effective? One reason appears to be that the templated cyclisation requires the porphyrin aryl groups to be *cis* to each other, while at any moment half are likely to be *trans*. Another is that in the absence of template the linear intermediates tend to take up open, wrong conformations rather than the correct closed forms. Finally, the recognition site (Zn) in our templating process is competing with the bond forming reagent (Cu) for the ligand. Ideally, these two conceptually different processes should use quite different recognition events which are not in competition. These are all features of our particular system, and both could be abolished by suitable redesign. The size of the molecules involved appears not to be a limitation and in principle it appears that all bimolecular reactions and cyclisation reactions should be amenable to acceleration and control by templates.

One principle which emerges from this approach is that characterisation of new host molecules requires the availability of complementary guests. X-ray crystal structures of large host molecules are very hard to obtain, and so are good NMR spectra when too much conformational flexibility is present, while mass spectra may also present problems. Good host–guest complementarity, when detected by large binding constants and sharp NMR signals, then becomes an excellent criterion for structure determination.

Templates normally bind too strongly to their products to operate catalytically: product inhibition severely limits turnover. If a template bound the product less strongly than either the transition state or the starting material, it would be less susceptible to product inhibition. Such catalytic templates have an intellectual appeal as enzyme mimics as well as practical advantages. In nearly all cases published so far there are insufficient kinetic and thermodynamic data available to assess whether the transition state is bound more or less strongly than the product. It will be fascinating to see whether it is possible to make templates which bind transition states sufficiently selectively to operate efficiently as catalysts, capable of rivalling natural enzymes. The application of templates appears limited only by our imagination and ingenuity.

5. Acknowledgements

We thank Dr Harry Anderson for his pioneering contributions to the early parts of this work and for his continuing advice. This work was supported by the Science and Engineering Research Council (UK), Rhône-Poulenc-Rorer Ltd, Magdalene College Cambridge and Trinity College Cambridge.

6. References

1. Watson, J. D. and Crick, F. H. C. (1953) *Nature*, **171**, 737–738.
2. Todd, A. R. (1956) *Perspectives in Organic Chemistry*; A. R. Todd: Interscience Publishers Ltd. London, p 263.
3. Byrne, G. T.,Linstead, R. P., and Lowe, A. R. (1932) *J. Chem. Soc.*, 1017–1022.
4. Thompson, M. C. and Busch, D. H. (1964) *J. Am. Chem. Soc.*, **86**, 213–217.
5. Curtis, N. F. (1960) *J. Chem. Soc.* 4409–4413.
6. Anderson, S., Anderson, H. L., and Sanders, J. K. M. (1993) *Acc. Chem. Res.*, **26**, 469-475.
7. Hoss, R. and Vögtle, F. (1994) *Angew. Chem. Int. Ed. Engl.*, **33**, 375–384.
8. Thomas, J. M. (1994) *Nature*, **368**, 289–290.
9. Anderson, H. L. and Sanders, J. K. M. (1990) *Angew. Chem. Int. Ed. Engl.*, **29**, 1400–1440.
10. Feldmann, K. S., Bobo, J. S., and Tewalt, G. L. (1992) *J. Org. Chem.* **114**, 4573–4574.
11. Anderson, S., Anderson, H. L., and Sanders, J. K. M. (1992) *Angew. Chem. Int. Ed. Engl.*, **31**, 907–910.
12. Anderson, S. and Sanders, J. K. M. Unpublished results.
13. Kirby, A. J. (1980) *Adv. Phys. Org. Chem.*, **17**, 183–278.
14. Gillard, J. W., Fortin, R., Grimm, E. L., Maillard, M., Tjepkema, M., Bernstein, M. A., and Glaser, R. (1991) *Tetrahedron Lett.*, **32**, 1145–1148.
15. Mock, W. L., Irra, T. A., Wepsiec, J. P., and Adhya, M. (1989) *J. Org. Chem.* **54**, 5302–5308.
16. Rebek, J. Jr (1994) *Chemistry in Britain*, **30**, 286–290.
17. Dietrich-Buchecker, C.; Sauvage, J-P. (1992) *New J. Chem.*, **16**, 277–285.
18. Zhang, Y and Seeman, N. C. (1994) *J. Am. Chem. Soc.*, **116**, 1661–1669.

TOWARDS A CHEMICAL ETIOLOGY OF THE NATURAL NUCLEIC ACIDS' STRUCTURE

ALBERT ESCHENMOSER
Swiss Federal Institute of Technology (ETH)
Laboratory of Organic Chemistry
Universitätstrasse 16
CH-8092 Zürich (Switzerland)

This report provides a short introduction *) into the general topic of the two lectures given, followed by copies of the complete set of slides shown in these lectures (Lecture 1, Figs. 1-43; Lecture 2, Figs. 44-78), whereby each slide figure is complemented by a short commentary.

An etiology of life will have to be primarily a *chemical* etiology and, quite probably, one concerned mainly with the origin of the type of molecular structure which we encounter today in Nature's nucleic acids. A general concept by which the problem of the origin of this - or any other - type of biomolecular structure can be approached *experimentally* is the systematic study of the chemistry of *structural alternatives*, molecular structures which - according to chemical reasoning - could have, but have not, been chosen by Nature to become (or to survive as) biomolecules. The structure type of an alternative is derived from a *chemical hypothesis* on the constitutional self-assembly that could have given rise to the biomolecule's origin and is chosen according to two criteria: first, whether the structure of the alternative has a potential for constitutional self-assembly comparable to that of the biomolecule itself and, second, whether its chemical properties might be such that the alternative could in principle fulfil the same type of biological function as the actual biomolecule. By chemically synthesizing such an alternative structure and comparing its re-

---

*) Reprinted with permission from *Chemistry and Biology* Volume 1, pages iv-v, 15th April, 1994.

C. Chatgilialoglu and V. Snieckus (eds.), Chemical Synthesis, 293–340.

levant chemical properties with those of the actual biomolecule we can expect to learn about the reasons why the latter, and not the alternative, has been chosen by Nature to become a biomolecule. Such information will deepen our understanding of the structural basis of the actual biomolecule's functioning and, if we are lucky, we may come across molecular struct ures which reveal themselves as candidates for having been intermediates (if there were any) along the evolutionary path towards the biomolecule we know today.

"Why pentose- and not hexose-nucleic acids?": this was the question that stood at the outset of a comprehensive experimental study in our laboratory on alternative nucleic acid structures [1] [2]. The question had emerged from our investigation of the aldomerization chemistry of glycolaldehyde phosphate, where we observed rac-ribose-2,4-diphosphate to be the kinetically favored product member of the pentose family and - in the absence of formaldehyde - rac-allose-2,4,6-triphosphate of the hexose family, the latter forming with comparable ease and selectivity as the former [3]. In a sequence of investigations hexopyranosyl-(6' $\rightarrow$ 4')-oligonucleotides derived from 2',3'-dideoxy-D-glucose (building block of "homo-DNA"), D-allose, 2'-deoxy- and 3'-deoxy-D-allose, D-altrose and D-glucose were synthesized [4] and their pairing properties compared with those of corresponding DNA-oligonucleotides. These studies presented us with a cascade of surprises and, consequently, of insights: whereas hexopyranosyl-(6' $\rightarrow$ 4')-oligonucleotides in the model system homo-DNA show Watson-Crick purine-pyrimidine pairing that is uniformly *stronger* than the pairing in the DNA series and, in addition, display unprecedented purine-purine pairing to duplexes in the reverse-Hoogsteen mode [5] [6], corresponding oligonucleotides derived from the *natural* hexopyranoses D-allose, D-altrose and D-glucose show pairing that is in some respects similar, but in others drastically different from and, above all, uniformly much weaker than the pairing in homo-DNA. It can be convincingly argued that the reason for this divergence is intrastrand steric hindrance in the pairing conformation. These $(CH_2O)_6$ hexopyranose sugars (and, foreseeably, also the four remaining diastereomers) are too bulky to serve as building blocks of efficient pairing systems. The shortcut answer to the question "why pentose- and not hexose-nucleic acids" that emerges from these studies seems simply to be: "*too many atoms*!" [7].

A comprehensive experimental involvement in the problems of a chemical etiology of the nucleic acids' structure would require a systematic extension of the study into hexo- and pentopyranosyl (as well as hexo- and pentofuranosyl) oligonucleotide systems which have their phosphodiester link between positions *other* than the (6' $\rightarrow$ 4')- or the (5' $\rightarrow$ 3')-

link of the structures investigated so far. Screening the pyranosyl section of this structure space by qualitative conformational analysis predicts the existence of a variety of so far experimentally untouched pairing systems and, above all, foresees a *pyranosyl isomer* of RNA ("p-RNA") that contains the phosphodiester linkage between positions C-2' and C-4' of neighboring ribopyranosyl units and is expected to show purine-pyrimidine and purine-purine (Watson-Crick) pairing comparable in strength to that observed in homo-DNA. *Experimentally* [8], not only does Watson-Crick pairing in p-RNA turn out to be *stronger* than in both RNA and DNA, but p-RNA also appears to be the most *selective* oligonucleotide pairing system known so far. This statement is based on the observation that homo-oligomers p-Ribo (A_n) and p-Ribo (G_n) (n up to 10) show neither reverse-Hoogsteen nor Hoogsteen self-pairing, in sharp contrast to homo-DNA and, with respect to guanine, also to DNA and RNA. These (together with other, so far conjectured) properties render this *constitutional isomer* of RNA of special interest in the context of the problem of RNA's origin [7].

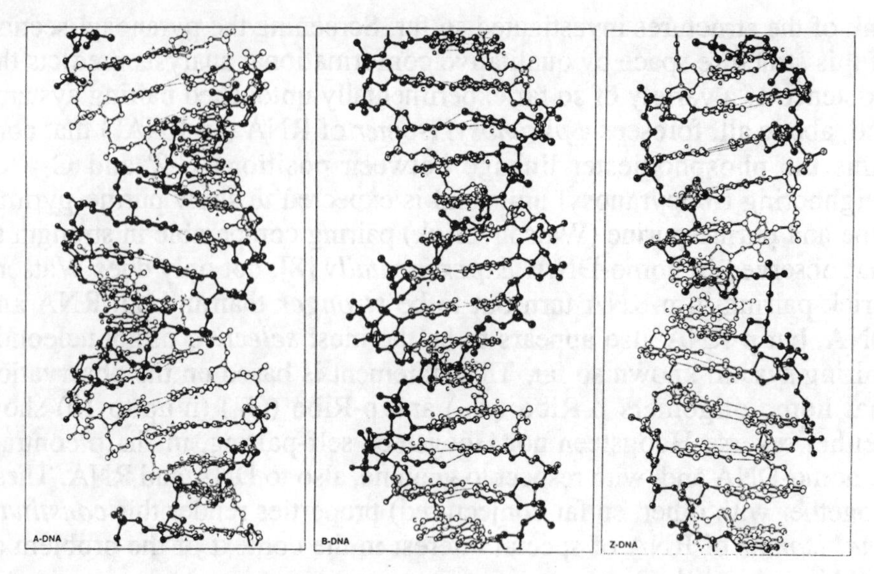

Figure 1. DNA: The three types of duplex conformation, A, B, and Z (from l. to r.) [9].

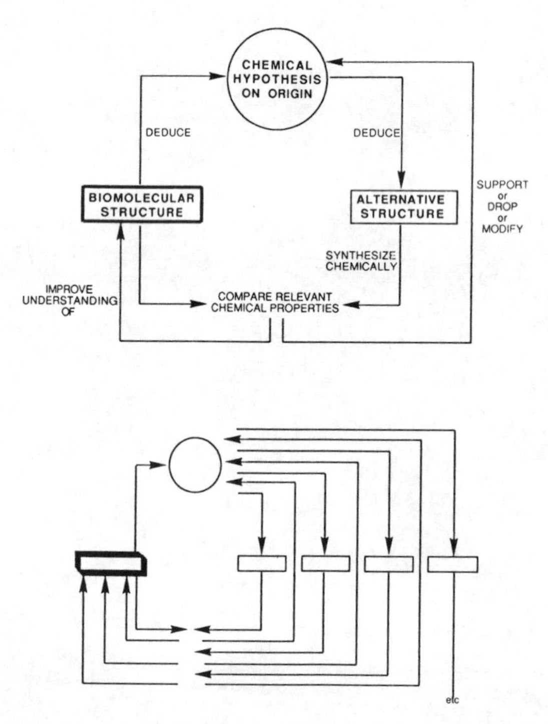

Figure 2. The study of structural alternatives, a general concept for experiments directed at a chemical etiology of biomolecular structures.

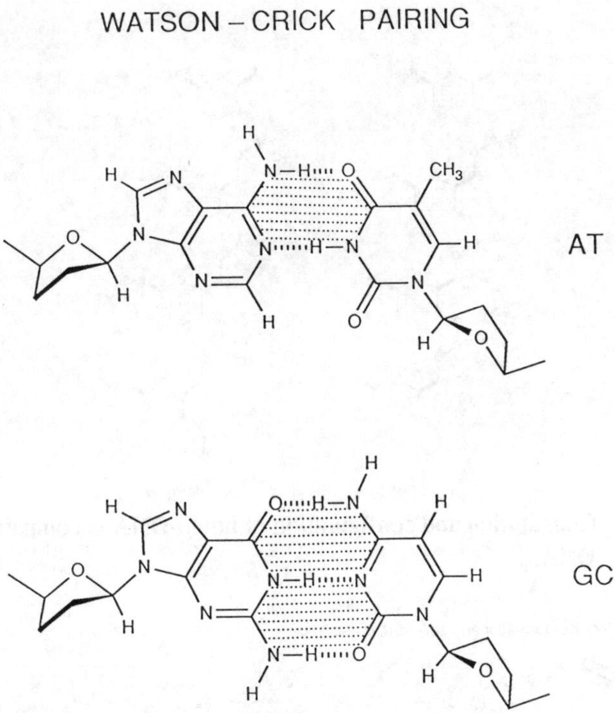

WATSON – CRICK PAIRING

AT

GC

Figure 3. The classical Watson-Crick base pairing in DNA and RNA.

Why purines and pyrimidines	and not .....
Why a sugar	and not .....
Why a pentose	and not a hexose
Why ribose	and not another pentose
Why ribofuranose	and not ribopyranose
Why phosphate	and not .....

Figure 4. Cascade of "why-questions" directed at a chemical etiology of the natural nucleic acids' structure. Such "why-questions" must define alternatives in order to define experiments.

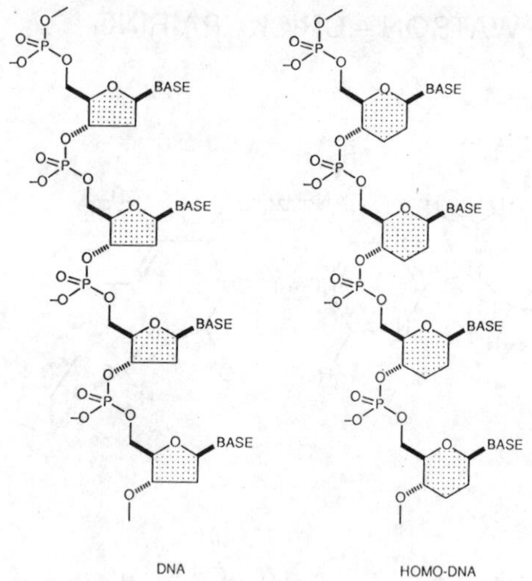

DNA

HOMO-DNA

Figure 5. Constitution and configuration of homo-DNA in comparison to DNA.

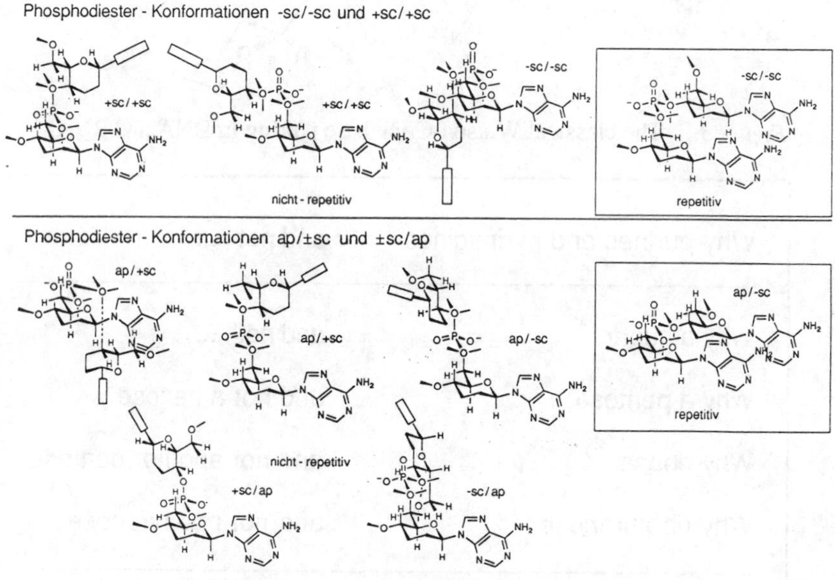

Phosphodiester - Konformationen -sc/-sc und +sc/+sc

nicht - repetitiv

repetitiv

Phosphodiester - Konformationen ap/±sc und ±sc/ap

nicht - repetitiv

repetitiv

Figure 7. With synclinal/synclinal phosphodiester bond arrangements there are four least strained (idealized) conformations in a homo-DNA nucleotide unit; only one of them is conformationally repetitive and, therefore, a pairing conformation. If one allows for synclinal/ antiperiplanar phosphodiester bond arrangements, then there are six more, including one further pairing conformation [2] [5].

HOMO – DNA : QUALITATIVE CONFORMATIONAL ANALYSIS

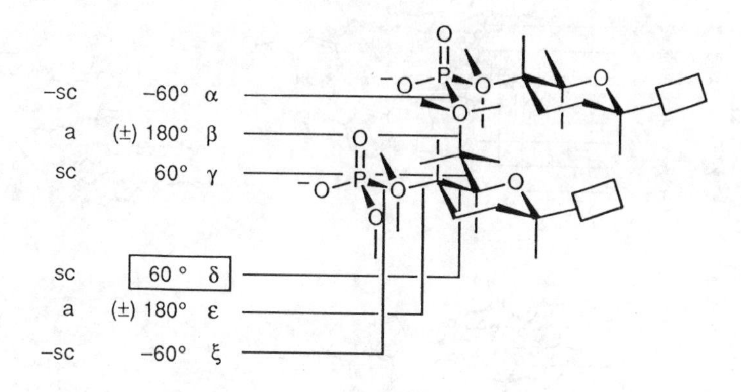

AMONG A TOTAL OF 486 (= 2·3^5) POSSIBLE, IDEALIZED CONFORMATIONS OF A HOMO – DNA MONOMER BACKBONE UNIT THERE ARE **FOUR** STERICALLY LEAST UNFAVORABLE CONFORMATIONS *).

ONLY **ONE** OF THEM IS REPETITIVE AND THEREFORE PRONE TO BASE PAIRING IN AN OLIGOMER DUPLEX.

THIS **UNIQUE** CONFORMATION GIVES RISE TO A HOMO – DNA BACKBONE THAT IS **LINEAR** (IN ITS IDEALISED FORM).

AT THE SAME TIME, IT BELONGS TO THE **TYPE** OF CONFORMATION THAT OCCURS IN DUPLEXES OF NATURAL A – DNA.

THE HELICAL SHAPE OF THE LATTER APPEARS AS A DIRECT CONSEQUENCE OF THE **FIVE-MEMBEREDNESS** OF THE SUGAR FURANOSE RING ($\delta > 60°$).

*) ACCORDING TO (ONLY) THREE QUALITATIVE CONFORMATIONAL CRITERIA:
— SINGLE BONDS STAGGERED THROUGHOUT
— 1,5–REPULSIONS MINIMIZED
— PHOSPHODIESTER CONFORMATIONS ACCORDING TO ANOMERIC EFFECT

Figure 6. Qualitative conformational analysis of the homo-DNA single strand backbone based on idealized conformations. Three simple criteria predict one single pairing conformation of a nucleotide unit, provided that the phosphodiester groups have a synclinal/synclinal bond arrangement [2].

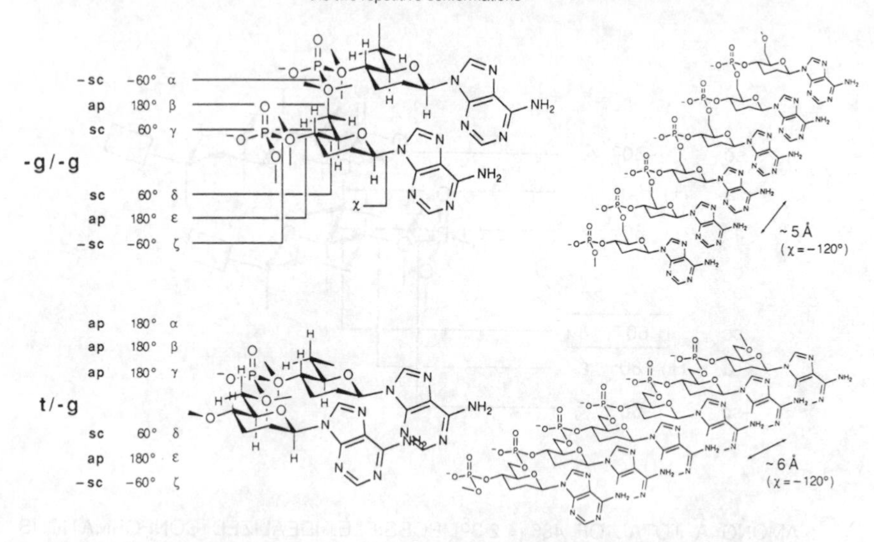

Figure 8. The two pairing conformations of the homo-DNA nucleotide units correspond to oligonucleotide single strands which are not helical, but linear (in their idealized conformation).

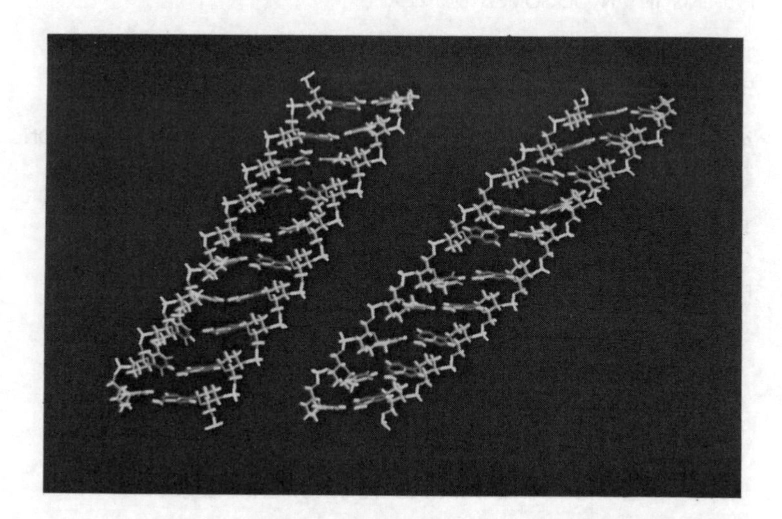

Figure 9. Structure models of the self-complementary homo-DNA duplex dd(A-A-A-A-A-T-T-T-T-T)$_2$ according to NMR-spectroscopy. The NMR-data suggest a fast equilibrium between the two quasi-linear models whose conformational types correspond to those of the idealized models shown in fig. 8 [6].

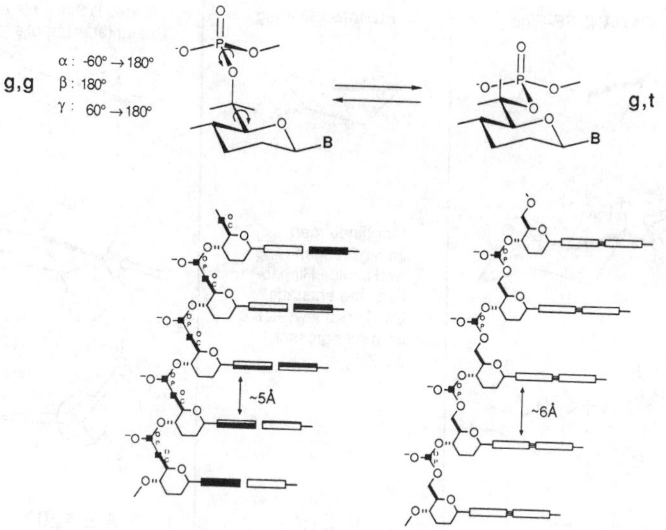

g,g α : -60° → 180° β : 180° γ : 60° → 180°

g,t

~5Å

~6Å

Figure 10. The two models of fig. 9 can easily interconvert into each other by two countercurrent bond rotations without interruption of base pairing.

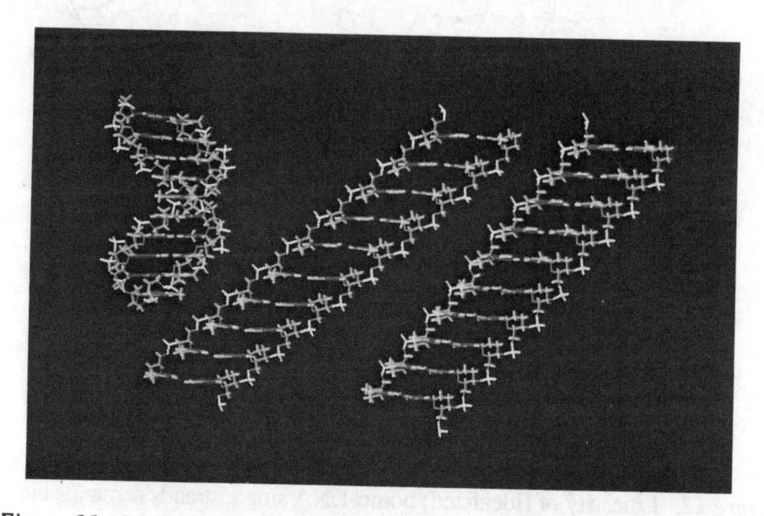

Figure 11. Models of the homo-DNA duplex dd(A-A-A-A-A-T-T-T-T-T)$_2$ juxtaposed with the model of the corresponding DNA duplex. Whereas the distances between base pairs in the homo-DNA models are too large for optimal base pair stacking, these distances can be optimal in the DNA model as a consequence of the model's helicality.

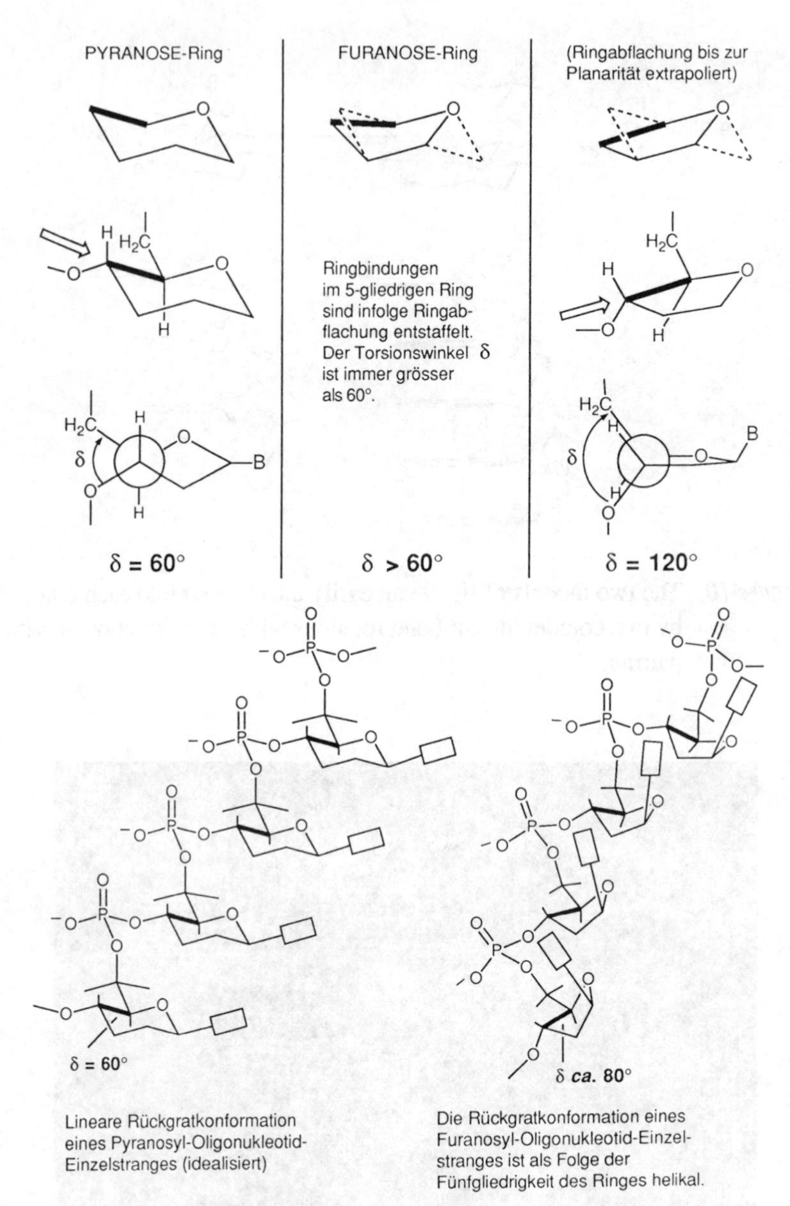

PYRANOSE-Ring
FURANOSE-Ring
(Ringabflachung bis zur Planarität extrapoliert)
Ringbindungen im 5-gliedrigen Ring sind infolge Ringabflachung entstaffelt. Der Torsionswinkel δ ist immer grösser als 60°.
δ = 60°
δ > 60°
δ = 120°
δ = 60°
δ ca. 80°
Lineare Rückgratkonformation eines Pyranosyl-Oligonukleotid-Einzelstranges (idealisiert)
Die Rückgratkonformation eines Furanosyl-Oligonukleotid-Einzelstranges ist als Folge der Fünfgliedrigkeit des Ringes helikal.

Figure 12. Linearity of (idealized) homo-DNA single strands demands the endocyclic torsion angle δ in the pyranose rings to have a value of 60°. In the furanose rings this torsion angle must always be greater than 60° for geometric reasons; this leads - ceteris paribus - to a right-handed (D-series) helix for the DNA single strand. The helical shape of DNA duplexes is a direct consequence of the five-memberedness of the furanose ring [2].

MODELLING EXPERIMENT (PURELY GEOMETRICAL) :
IDEALIZED NA - BACKBONE GEOMETRY AS A FUNCTION
OF TORSION ANGLE δ
(ALL OTHER TORSION ANGLES CONSTANT AND OF
A - DNA TYPE)

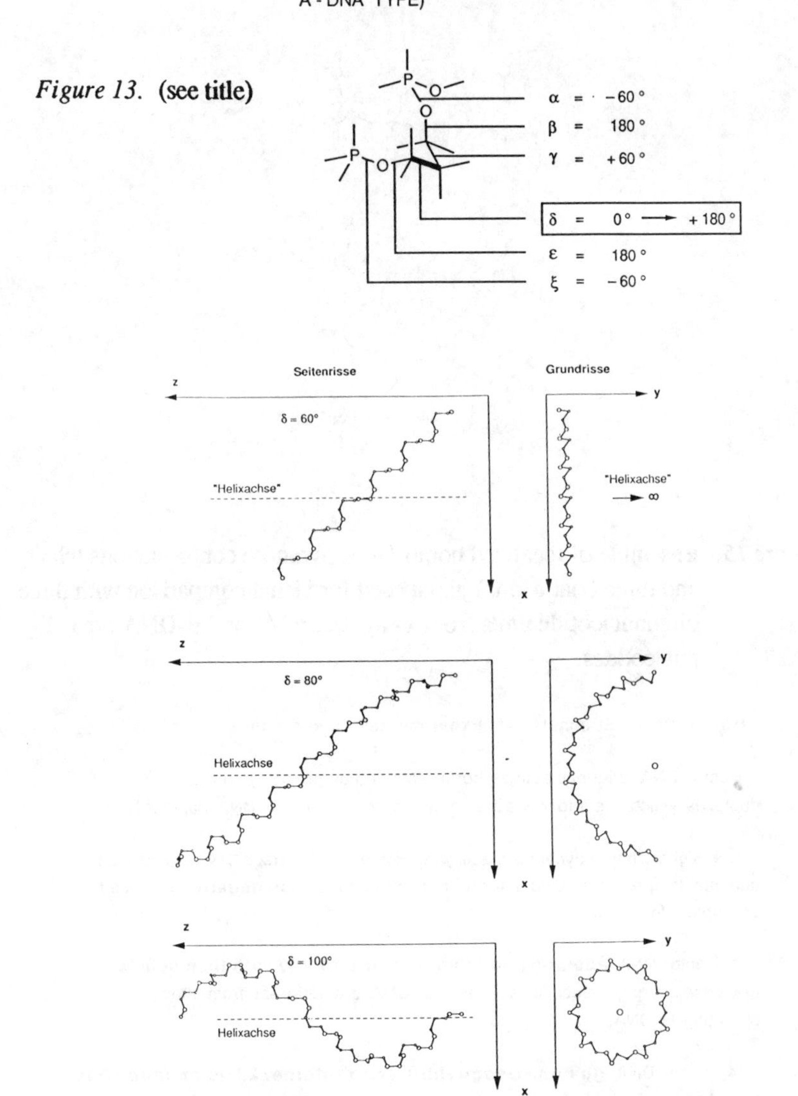

Figure 13. **(see title)**

Figure 14. **Top view (right) and side view (left) of a homo-DNA backbone in idealized conformation where torsion angle δ is changed from 60° to 80° and 100° (torsion axis δ perpendicular to projection plane of top view). δ = 60°: linear strand; δ = 100°: helical strand, helix diameter ca. 13 Å [2].**

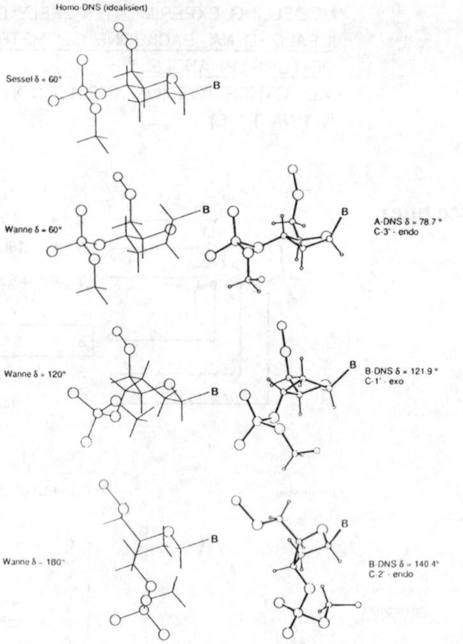

Figure 15. **Ensemble of idealized homo-DNA pyranose conformations (chair and three boat forms) juxtaposed for visual comparison with three oligonucleotide units from x-ray data of A- and B-DNA type oligonucleotides.**

Homo-DNA: Summary of Experimental Observations

- Homo-DNA oligonucleotides form antiparallel purine-pyrimidine duplexes which are **more stable** than the corresponding DNA duplexes.

- The higher thermodynamic stability of Homo-DNA versus DNA duplexes is due not to greater binding energy, but rather to a **less negative entropy** of duplex formation.

- In Homo-DNA **adenine and guanine pair strongly with themselves**: the base pairing selectivities in Homo-DNA are different from those operating in DNA.

- In Homo-DNA **guanine/isoguanine** and **xanthine/2,6-diaminopurine** form base pairs which are comparable in strength to guanine/cytosine.

- Complementary base sequences of Homo-DNA and DNA do not pair: **Homo-DNA is an autonomous artificial pairing system.**

Figure 16. **Homo-DNA: Summary of Experimental Observations. For the synthesis of homo-DNA oligonucleotides see [4].**

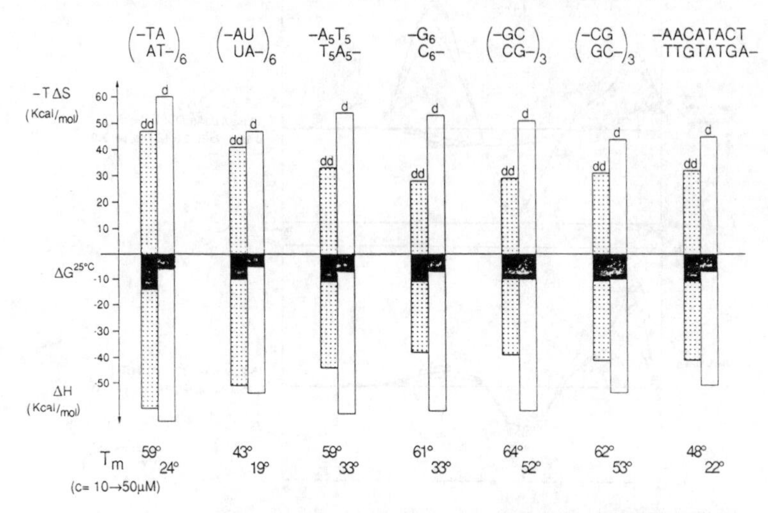

Figure 17. **Comparison of thermodynamic data of homo-DNA and corresponding DNA duplexes. Homo-DNA duplexes are more stable (relative to their single strands) than DNA duplexes. This difference is entropic in origin [5].**

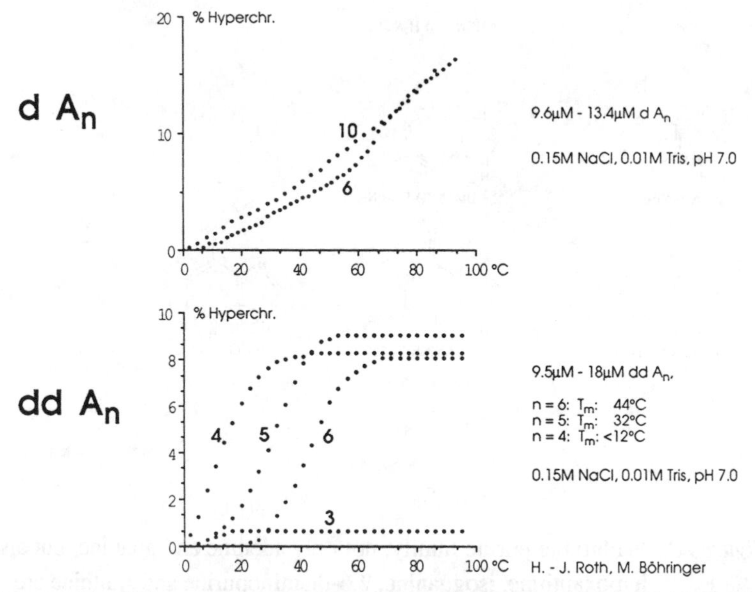

Figure 18. **UV-spectroscopic observation of adenine-adenine pairing in the homo-DNA series dd(A_n). There is no such adenine self-pairing in duplexes of the DNA series.**

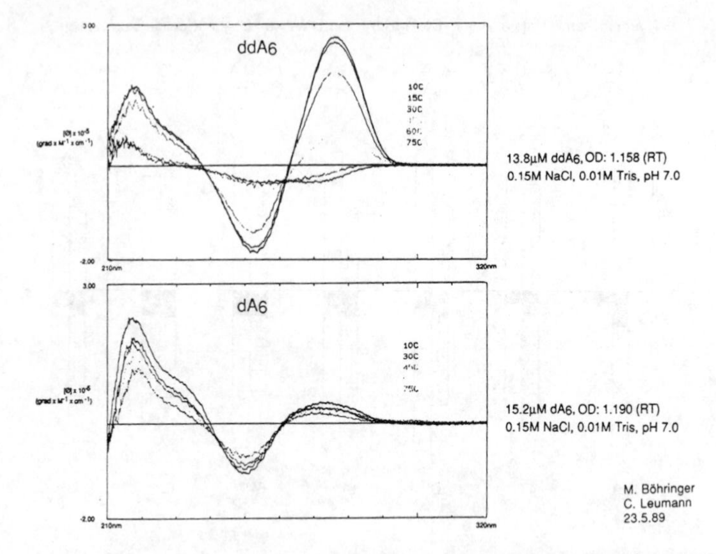

Figure 19. CD-spectroscopic observation of adenine-adenine pairing in the homo-DNA hexamer dd(A_6) (in contrast to the DNA hexamer d(A_6)).

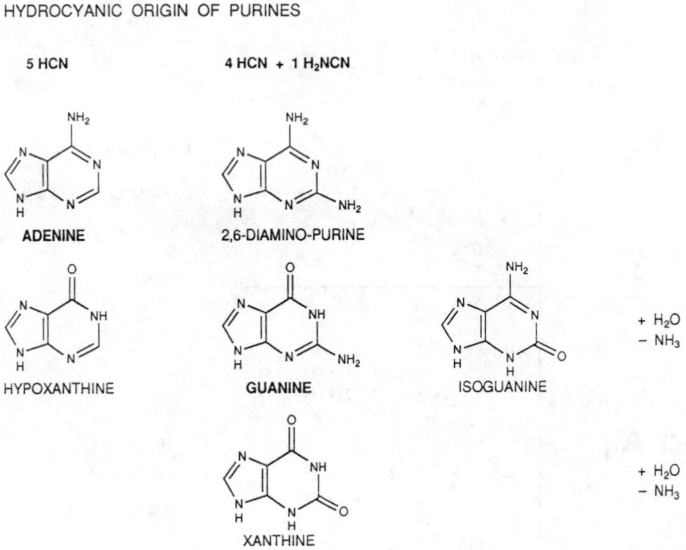

HYDROCYANIC ORIGIN OF PURINES

5 HCN

4 HCN + 1 H_2NCN

ADENINE

2,6-DIAMINO-PURINE

HYPOXANTHINE

GUANINE

ISOGUANINE

+ H_2O
− NH_3

XANTHINE

+ H_2O
− NH_3

Figure 20. Within the purine family, not only adenine and guanine, but also hypoxanthine, isoguanine, 2,6-diaminopurine and xanthine are potentially prebiological structures. Therefore, the pairing properties of these bases must also be investigated in the context of the present study.

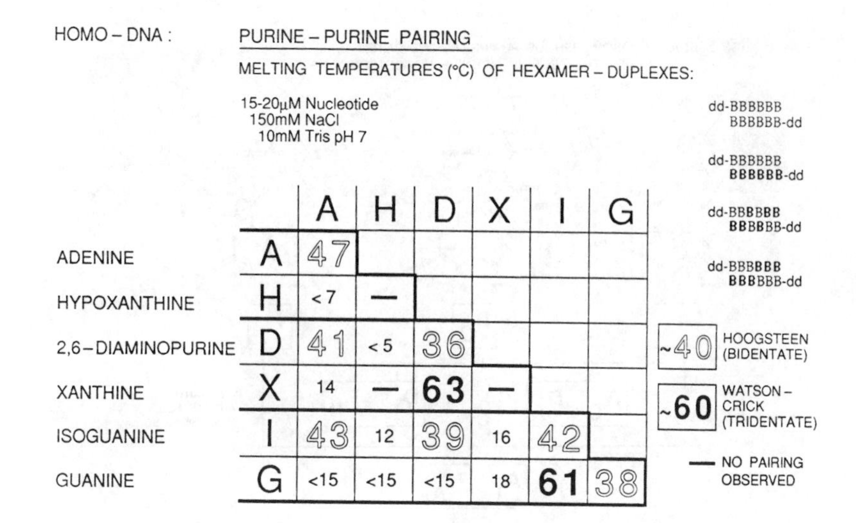

HOMO – DNA :
PURINE – PURINE PAIRING
MELTING TEMPERATURES (°C) OF HEXAMER – DUPLEXES:
15-20μM Nucleotide
150mM NaCl
10mM Tris pH 7
dd-BBBBBB
BBBBBB-dd
A H D X I G
ADENINE A 47
HYPOXANTHINE H <7 —
2,6–DIAMINOPURINE D 41 <5 36
XANTHINE X 14 — 63 —
ISOGUANINE I 43 12 39 16 42
GUANINE G <15 <15 <15 18 61 38
~40 HOOGSTEEN (BIDENTATE)
~60 WATSON – CRICK (TRIDENTATE)
— NO PAIRING OBSERVED
KATRIN GROEBKE, MARKUS BOEHRINGER, HANS-JÖRG ROTH, JÜRG HUNZIKER, ULF DIEDERICHSEN,
DR. WILLIAM FRASER, DR. CHRISTIAN LEUMANN

Figure 21. **Melting temperatures of oligonucleotide hexamers demonstrating strong purine-purine pairing in the homo-DNA series. A = adenine, H = hypoxanthine, D = 2,6-diaminopurine, X = xanthine, I = isoguanine, G = guanine [10].**

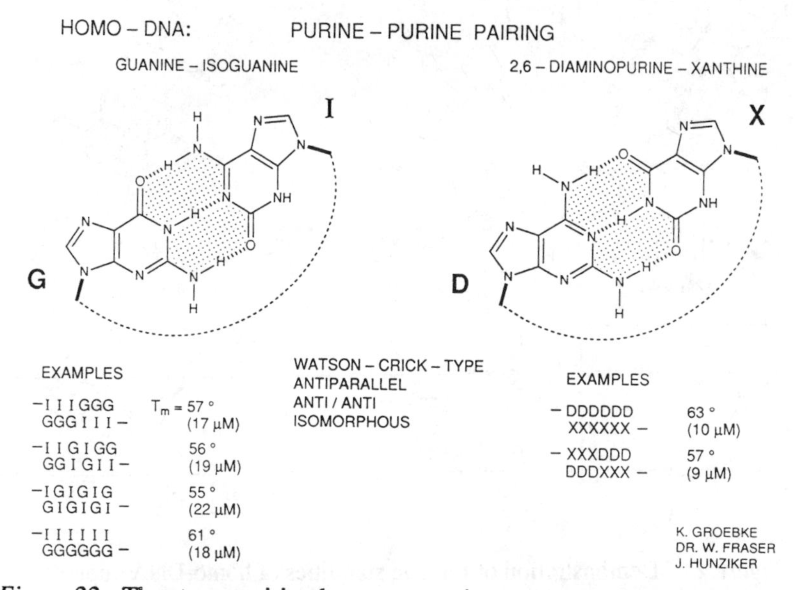

Figure 22. **The strong pairing between guanine and isoguanine as well as between 2,6-diaminopurine and xanthine in the homo-DNA series is assigned the Watson-Crick mode [10].**

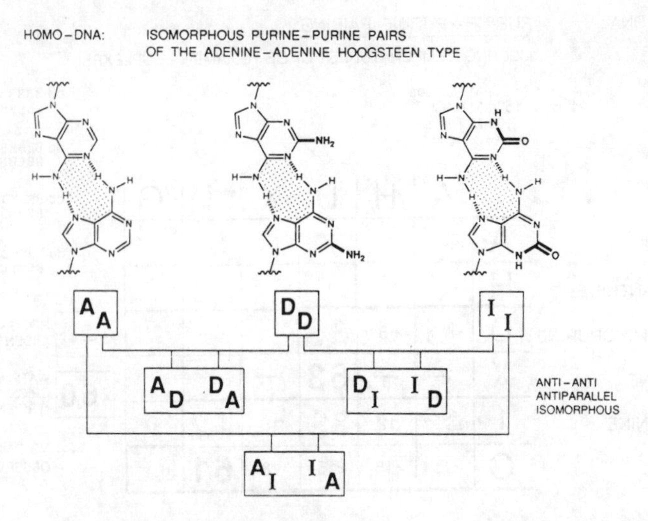

Figure 23. **Within the "adenine family" (adenine, 2,6-diaminopurine and isoguanine) there are six possible combinations of purine-purine pairing in the reverse-Hoogsteen mode. This prediction is fully corroborated by experiment (see fig. 21).**

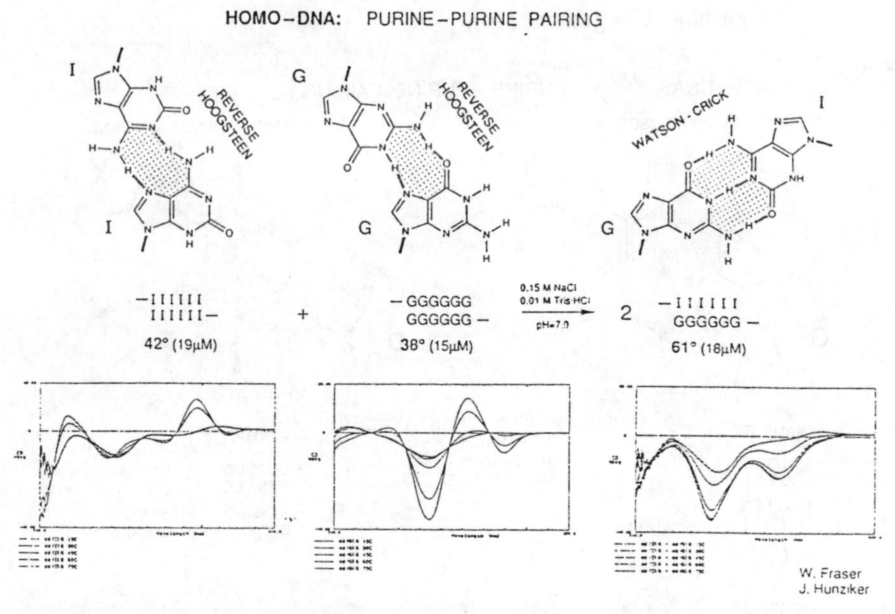

Figure 24. **Demonstration of relative stabilities of homo-DNA duplexes with reverse-Hoogsteen versus Watson-Crick pairing [11].**

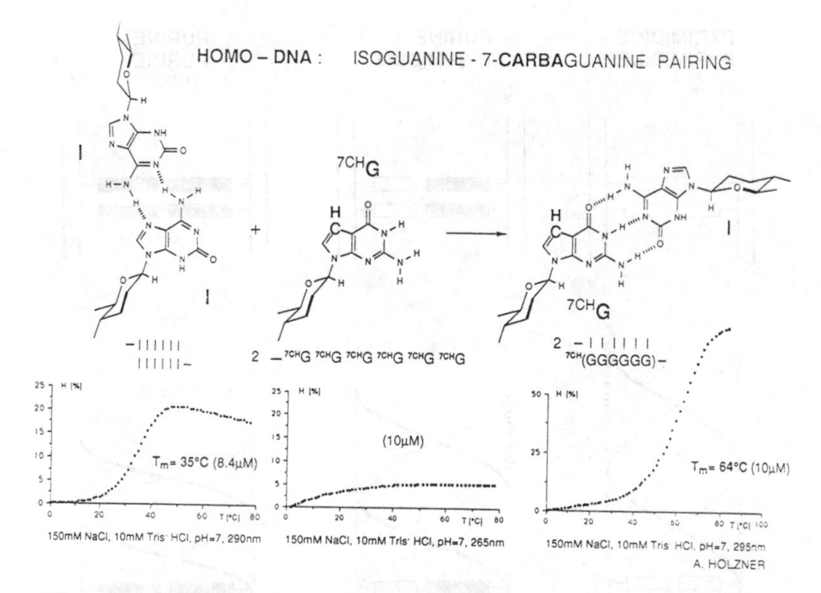

Figure 25. In the hexamer dd(G_6) in which all guanines are replaced by 7-carba-guanine, there is no longer self-pairing, but still cross-pairing with isoguanine, corroborating constitutional assignments of pairing modes [12].

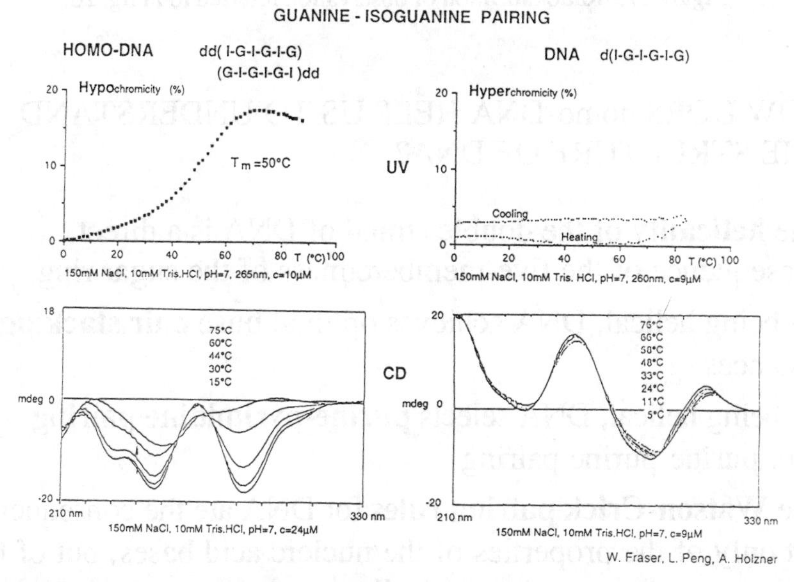

Figure 26. In sharp contrast to the homo-DNA series, there is no (cumulative) guanine-isoguanine Watson-Crick pairing in the DNA series [12].

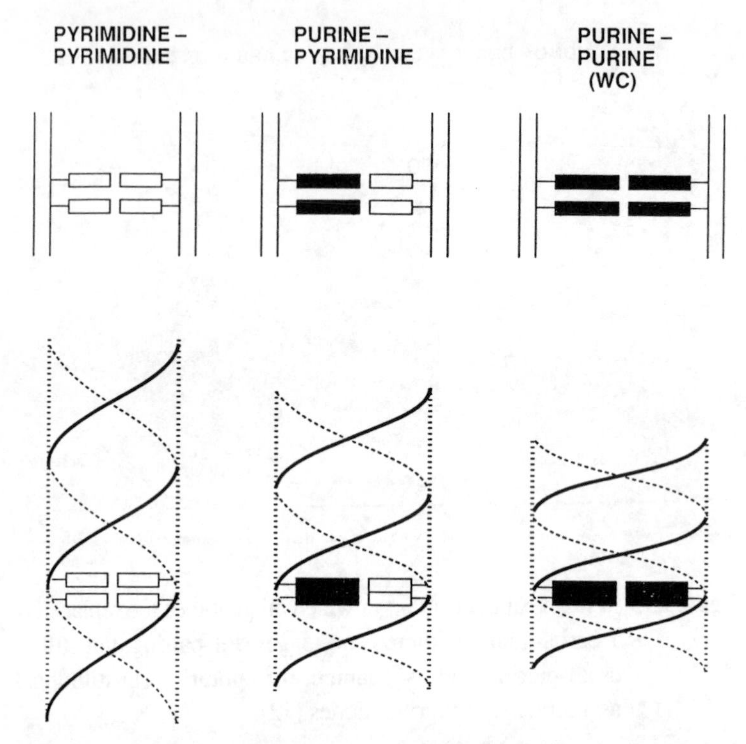

Figure 27. Rationalization of observation referred to in fig. 26.

HOW DOES homo-DNA HELP US TO UNDERSTAND THE STRUCTURE OF DNA?

- The **helicality** of the double strand of DNA is a direct consequence of the five-memberedness of the sugar ring
- By being helical, DNA achieves optimal **base pair stacking** distances
- By being helical, DNA selects **purine-pyrimidine pairing** over purine-purine pairing
- The **Watson-Crick pairing rules** for DNA are the consequence not only of the properties of the nucleic acid bases, but of the sugar backgone structure as well

Figure 28. (see title)

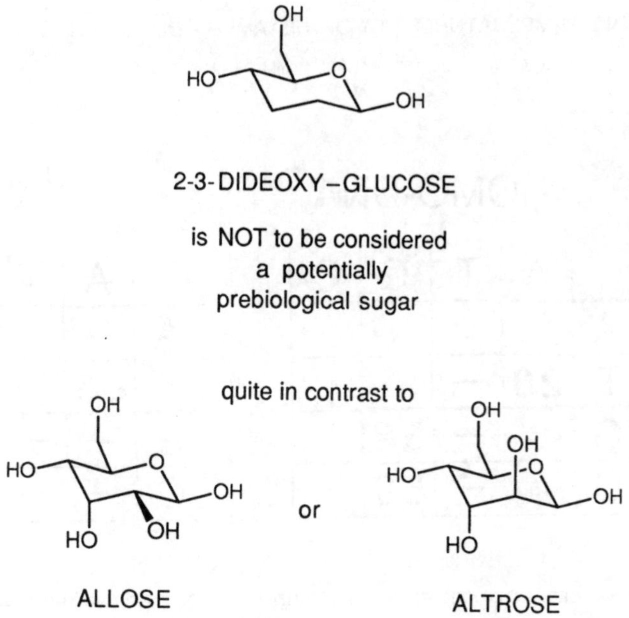

2-3-DIDEOXY–GLUCOSE

is NOT to be considered
a potentially
prebiological sugar

quite in contrast to

ALLOSE or ALTROSE

Figure 29. **(see text of the figure)**

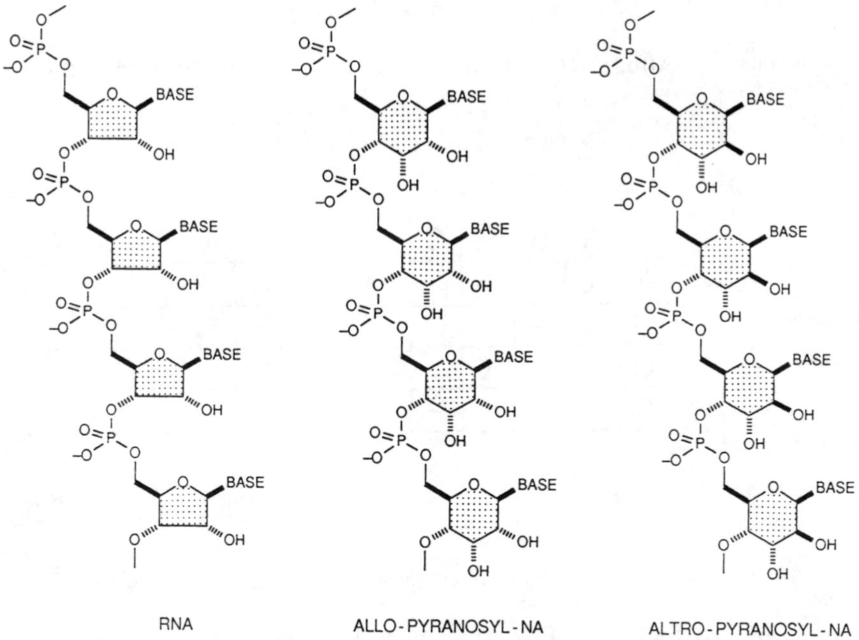

RNA ALLO-PYRANOSYL-NA ALTRO-PYRANOSYL-NA

Figure 30. Constitution and configuration of allo- and altropyranosyl-(6' → 4')- oligonucleotides in comparison to RNA.

MELTING TEMPERATURES (°C) OF HEXAMER – DUPLEXES

15 - 20 μM Nucleotide, 150 mM NaCl, 10 mM Tris pH 7

dd-BBBBBB
BBBBBB-dd
dd-BBBBBB
BBBBBB-dd

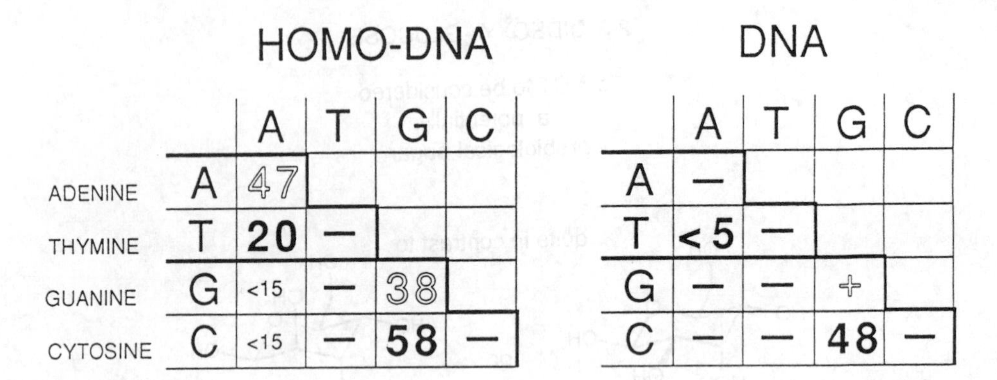

HOMO-DNA

		A	T	G	C
ADENINE	A	47			
THYMINE	T	20	—		
GUANINE	G	<15	—	38	
CYTOSINE	C	<15	—	58	—

DNA

	A	T	G	C
A	—			
T	< 5	—		
G	—	—	+	
C	—	—	48	—

MARKUS BOEHRINGER, HANS-JÖRG ROTH, JÜRG HUNZIKER, FREDI GIGER, DR. CHRISTIAN LEUMANN

Figure 31. **Relative strength of pairing in the homo-DNA versus the DNA series (for comparison with fig. 32).**

Melting Temperatures (°C) of Duplexes

Allose-NA (Octamers)

5-10 μM
150 mM NaCl
10 mM TRIS; pH 7

		A	U	G	C
ADENINE	A	16			
URIDINE/THYMINE	U	<0	—		
GUANINE	G			13	
CYTOSINE	C			<10	—

HOMO-DNA (Hexamers)

15-20 μM
150 mM NaCl
10 mM TRIS; pH 7

	A	T	G	C
A	47			
T	20	—		
G		—	38	
C	<15	—	58	—

40 HOOGSTEEN (BIDENTATE) 60 WATSON-CRICK (TRIDENTATE)

Markus Böhringer, Hans-Jörg Roth, Jürg Hunziker, Andreas Helg, Reto Fischer, Alfred Giger, Dr. William Fraser, Dr.Christian Leumann

Figure 32. **Relative strength of pairing in the homo-DNA versus the allo-pyranosyl-(6' → 4')-oligonucleotide series. Pairing in this latter series is drastically weaker [13] [14].**

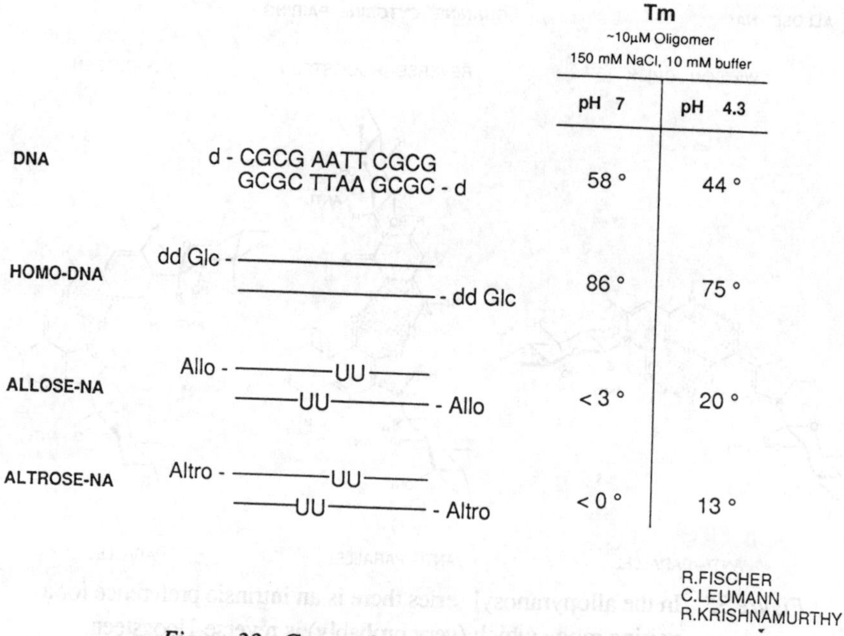

		Tm ~10μM Oligomer 150 mM NaCl, 10 mM buffer	
		pH 7	pH 4.3
DNA	d - CGCG AATT CGCG GCGC TTAA GCGC - d	58 °	44 °
HOMO-DNA	dd Glc - ——— ——— - dd Glc	86 °	75 °
ALLOSE-NA	Allo - ———UU——— ———UU——— - Allo	< 3 °	20 °
ALTROSE-NA	Altro - ———UU——— ———UU——— - Altro	< 0 °	13 °

R.FISCHER
C.LEUMANN
R.KRISHNAMURTHY

Figure 33. **Comparison of pairing strength [13].**

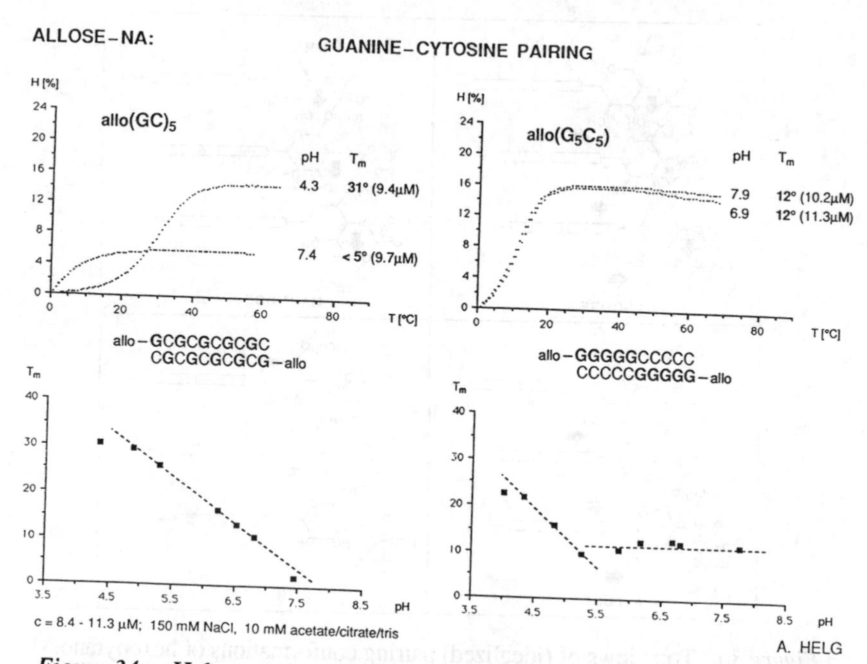

Figure 34. **pH dependence of guanine-cytosine pairing in the allopyranosyl series, and its dependence on the base-pair sequence [14].**

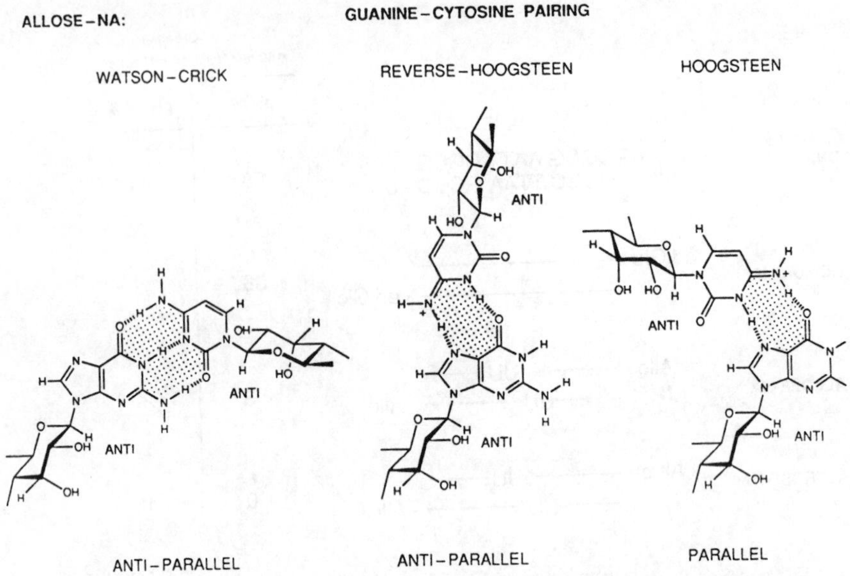

ALLOSE – NA: GUANINE – CYTOSINE PAIRING

WATSON – CRICK | REVERSE – HOOGSTEEN | HOOGSTEEN

ANTI – PARALLEL | ANTI – PARALLEL | PARALLEL

Figure 35. In the allopyranosyl series there is an intrinsic preference for a pairing mode which (very probably) is reverse-Hoogsteen.

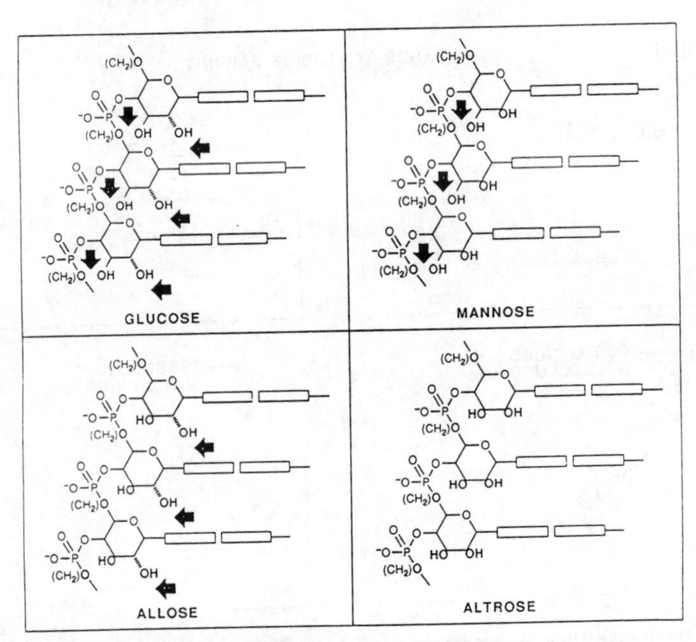

Figure 36. Top views of (idealized) pairing conformations of hexopyranosyl systems. Arrows indicate sources of severe steric hindrance in the pairing conformation.

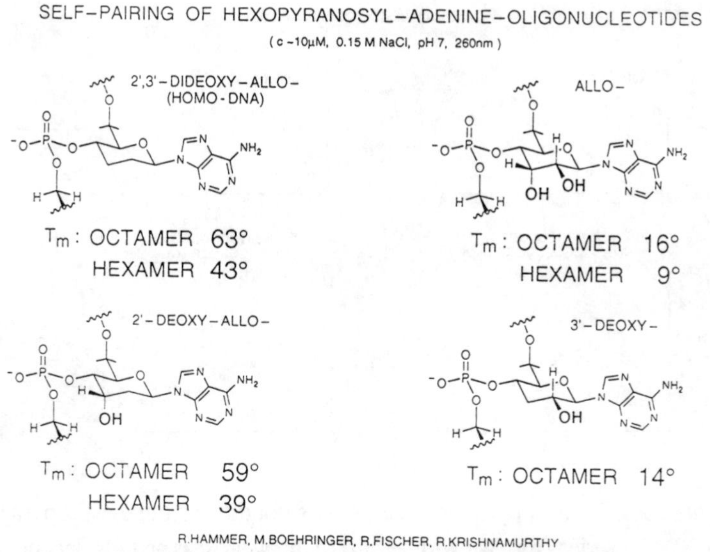

SELF-PAIRING OF HEXOPYRANOSYL-ADENINE-OLIGONUCLEOTIDES

(c ~10μM, 0.15 M NaCl, pH 7, 260nm)

2',3'-DIDEOXY-ALLO- (HOMO-DNA)

T_m: OCTAMER 63°
HEXAMER 43°

ALLO-

T_m: OCTAMER 16°
HEXAMER 9°

2'-DEOXY-ALLO-

T_m: OCTAMER 59°
HEXAMER 39°

3'-DEOXY-

T_m: OCTAMER 14°

R.HAMMER, M.BOEHRINGER, R.FISCHER, R.KRISHNAMURTHY

Figure 37. **Comparison of melting temperatures show that, in the allopyranosyl series, it is the equatorial hydroxyl group at position C-2' (and not the axial hydroxyl group at C-3') which is responsible for the much weaker (as compared to homo-DNA) self-pairing of adenine [15].**

Self-Pairing of Hexopyranosyl-Oligonucleotides: Thermodynamic Data

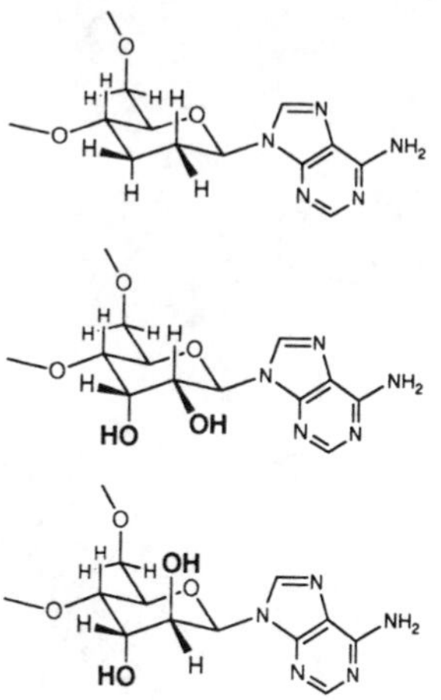

	T_m (°C) (10μM)	ΔH (Kcal/mol)	TΔS (Kcal/mol)	$\Delta G^{25°C}$ (Kcal/mol)
dd(A_6) 2',3'-Dideoxyallo- –AAAAAA AAAAAA–	43°	-39.4	-30.1	-9.3
allo(A_{12}) –AAAAAAAAAAAA AAAAAAAAAAAA–	29°	-67.8	-60.1	-7.7
altro(A_{12}) –AAAAAAAAAAAA AAAAAAAAAAAA–	39°	-47.9	-39.0	-8.9

R.Fischer
M.Böhringer
K.Groebke

Figure 38. Adenine self-pairing in the altropyranosyl series is also very much weaker than in the homo-DNA series [10].

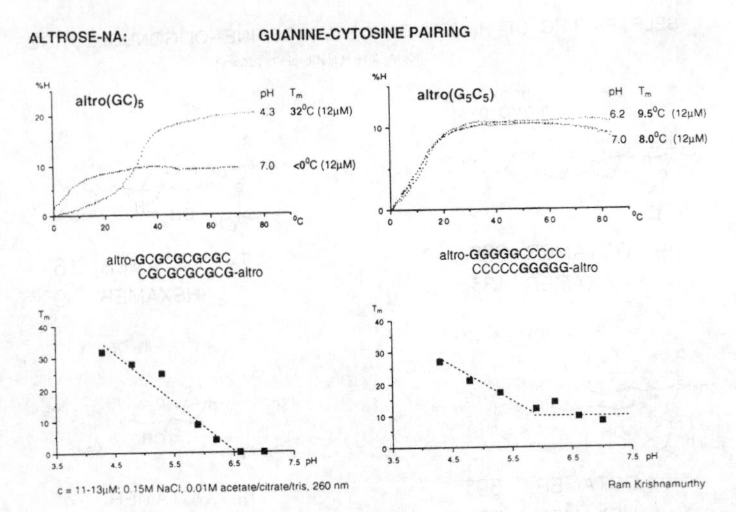

Figure 39. The altropyranosyl series does not differ from the allopyranosyl series (fig. 34) with regard to the weakness and pH dependence of the guanine-cytosine pairing [16].

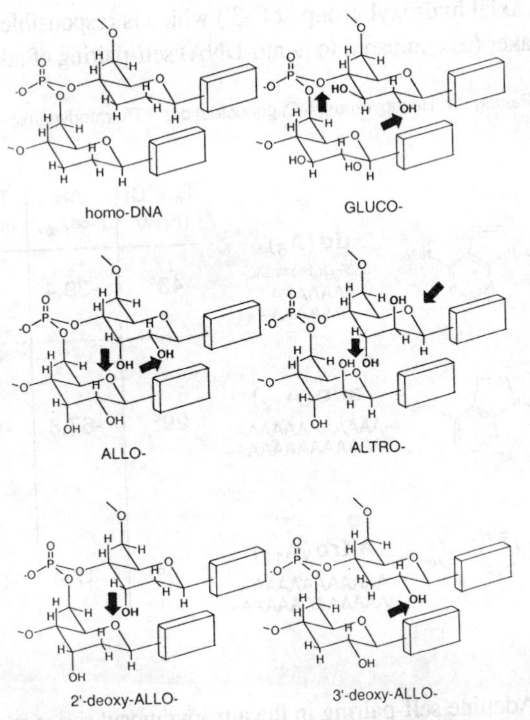

Figure 40. Arrows indicate sources of severe steric hindrance which hamper base pairing in hexopyranosyl oligonucleotide systems.

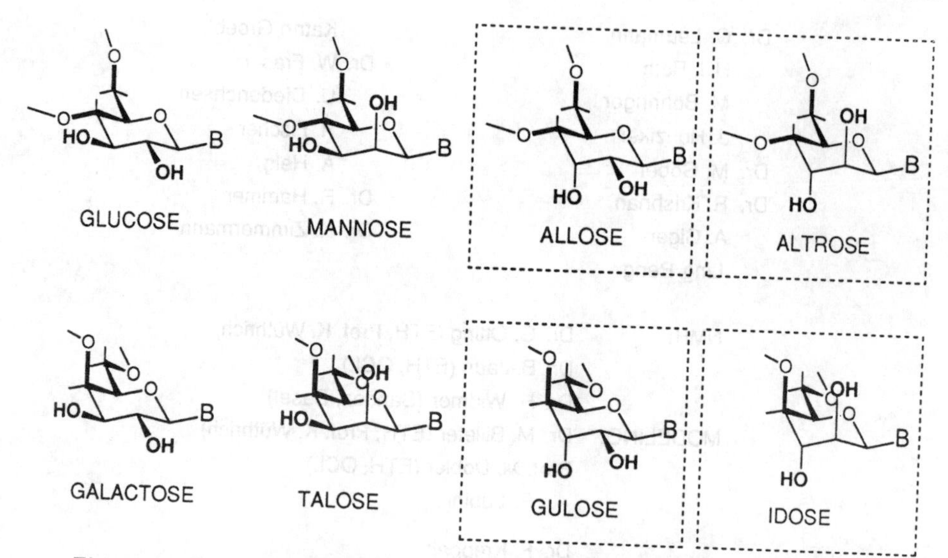

Figure 41. The observations made with allo-, altro-, and glucopyranosyl-(6' → 4')-oligonucleotides can be extrapolated to hexopyranosyl-(6' → 4')-oligonucleotide systems based on other hexopyranoses.

Why PENTOSE- and not HEXOSE-nucleic acids ?

CONCLUSION (preliminary) based on observations of the pairing behaviour of allo-, altro- and glycopyranosyl-, as well as 2'-deoxy-, 3'-deoxy- and 2',3'-dideoxy-(6' → 4')-oligonucleotides (up to dodecamers):

Whereas base pairing in homo-DNA is stronger than in DNA, hexopyranosyl-(6' → 4')-analogues of RNA derived from fully hydroxylated $(CH_2O)_6$-hexoses are pairing systems drastically inferior to RNA. The reason is intrastrand steric hindrance in the pairing conformation.

"TOO MANY ATOMS"

Figure 42. An unexpectedly simple and general answer to the title question emerges from the systematic experimental study of the pairing properties of hexopyranosyl-(6' → 4')-oligonucleotides.

Dr. C. Leumann
H.J. Roth
M. Böhringer
J. Hunziker
Dr. M. Göbel
Dr. R. Krishnan
A. Giger
Ling Peng

Katrin Groebke
Dr. W. Fraser
U. Diederichsen
R. Fischer
A. Helg
Dr. R. Hammer
Dr. K. Zimmermann

NMR: Dr. G. Otting (ETH, Prof. K. Wüthrich)
Dr. B. Jaun (ETH, OCL)
Dr. H. Widmer (Sandoz, Basel)

MODELING: Dr. M. Billeter (ETH, Prof. K. Wüthrich)
Prof. M. Dobler (ETH, OCL)
P. Lubini

Dr. F. Kreppelt

Figure 43. **Names of doctoral and postdoctoral collaborators who participated in the hexopyranosyl oligonucleotide study.**

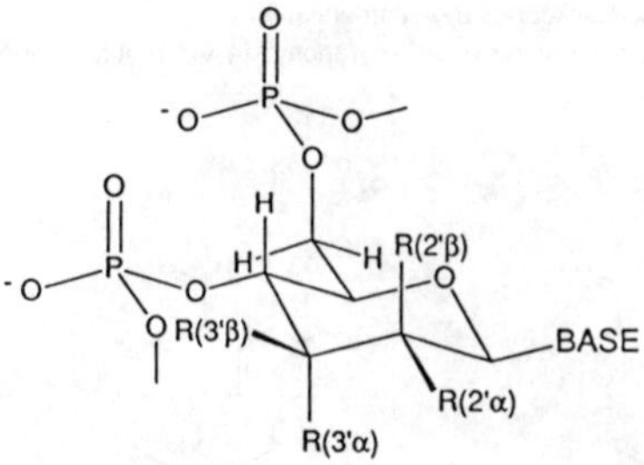

O
-O—P—O—
O
H
H
H R(2'β)
O
-O—P—O
O R(3'β)
BASE
R(2'α)
R(3'α)

Oligonucleotides	R(2'α)	R(2'β)	R(3'α)	R(3'β)
2',3'-dideoxy-β-D-glucopyranosyl ("homo - DNA")	H	H	H	H
β-D-allopyranosyl	OH	H	OH	H
β-D-altropyranosyl	H	OH	OH	H
2'-deoxy-β-D-allopyranosyl	H	H	OH	H
3'-deoxy-β-D-allopyranosyl	OH	H	H	H
β-D-glucopyranosyl	OH	H	H	OH

Figure 44. **All hexopyranosyl oligonucleotide systems investigated so far (cf. lecture 1) have their phosphodiester groups between the position C-6' and C-4' of the hexopyranose units.**

Constitutionally Isomeric Oligonucleotide Backbones
(Phosphodiester junctions between sugar postions)

	PYRANOSES		FURANOSES	
HEXO–	2'→6' ■ = 2'→4' ■ = 2'→3' 3'→6' 3'→4' [4'→6' ■ =]	6' 4' O 3' 2'	2'→6' 2'→5' 2'→3' 3'→6' 3'→5' 5'→6'	6' 5' O 3' 2'
PENTO–	2'→4' ■ = 2'→3' 3'→4'	4' O 3' 2'	[2'→5'] 2'→3' [3'→5' RNA]	5' O 3' 2'

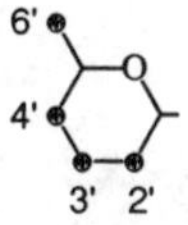

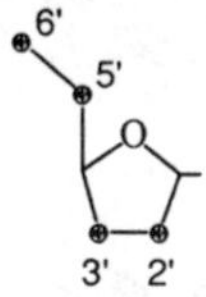

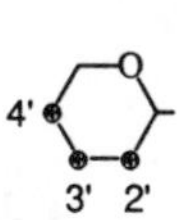

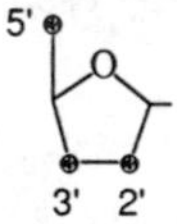

■ retrosynthetically derivable via aldomerization pathway

= cooperative base pairing predicted by qualitative conformation analysis

☐ experimentally studied (so far)

Figure 45. **Survey of the constitutions of (formally) conceivable oligonucleotide systems derived from aldohexoses and aldopentoses, and predictions about their pairing propensities.**

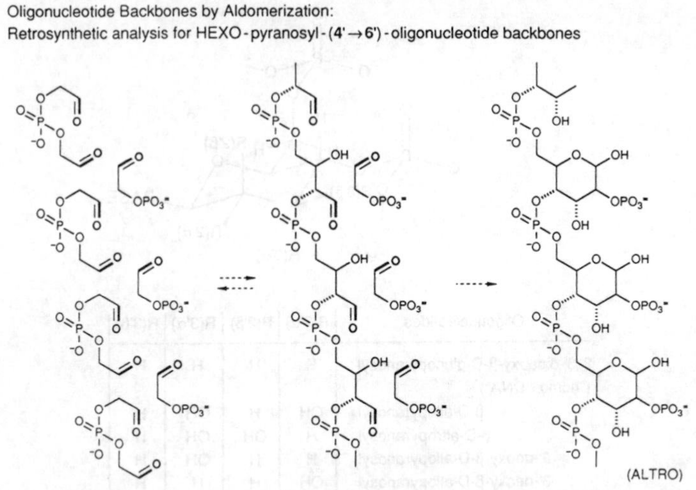

Oligonucleotide Backbones by Aldomerization:
Retrosynthetic analysis for HEXO-pyranosyl-(4'→6')-oligonucleotide backbones

(ALTRO)

Figure 46. *Formal* derivation of the constitution of hexopyranosyl-(6' → 4')-oligonucleotides via the aldomerization pathway.

Potential Aldomerization Pathways for Constitutional Self-Assembly of HEXO- and PENTO-Pyranosyl-Oligonucleotide Backbones

4'→6' Altropyranosyl- Allo- Gulo-

2'→6' Allopyranosyl- Altro-

2'→4' Talopyranosyl- Manno-

2'→4' Ribopyranosyl- Lyxo-

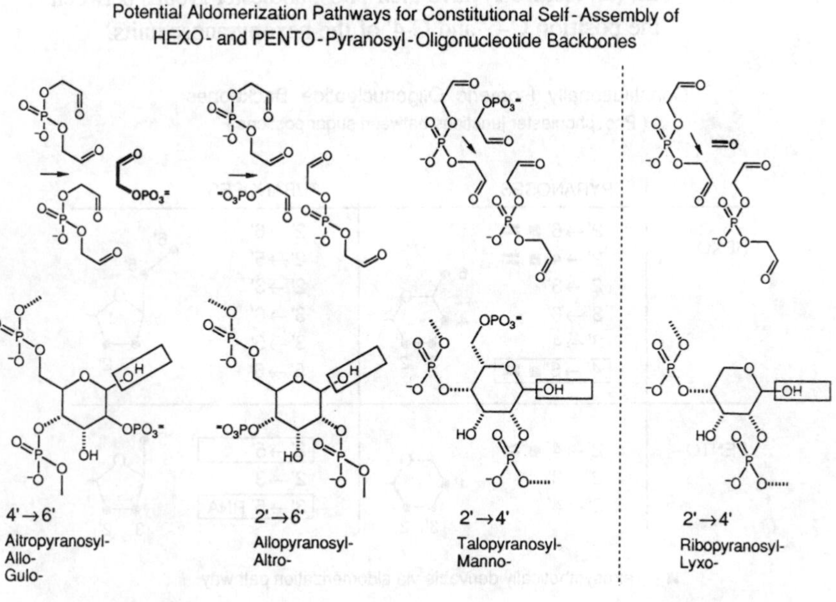

Figure 47. Not only (6' → 4'), but also (6' → 2'), and (4' → 2') *hexo*-pyranosyl oligonucleotide systems can be formally derived via the aldomerization pathway. The backbones of the diastereomeric (4' → 2') systems in the hexopyranosyl series are all sterically hindered in their pairing conformation. This is not necessarily so for the (4' → 2') *pento*pyranosyl systems.

Oligonucleotide Backbones by Aldomerization:
Retrosynthetic analysis for PENTO-pyranosyl-(2'→4')-oligonucleotide backbones

(RIBO)

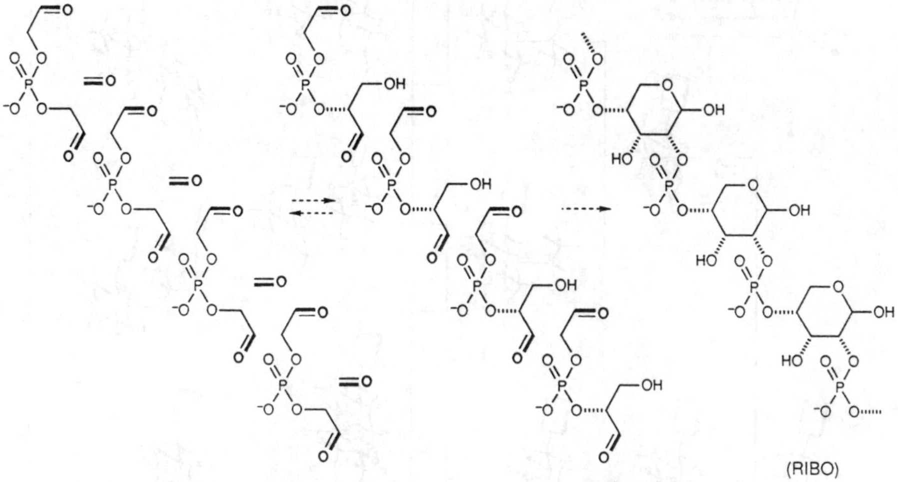

Figure 48. *Formal* derivation of the constitution of *pento*pyranosyl-(4' → 2')-oligonucleotides by the aldomerization pathway.

β-RIBO-PYRANOSYL-(4'→2')-OLIGONUCLEOTIDES
"PYRANOSYL - RNA", an ISOMER of RNA
a target of chemical synthesis and of studies on the constitutional self-assembly of potentially self-replicating systems :

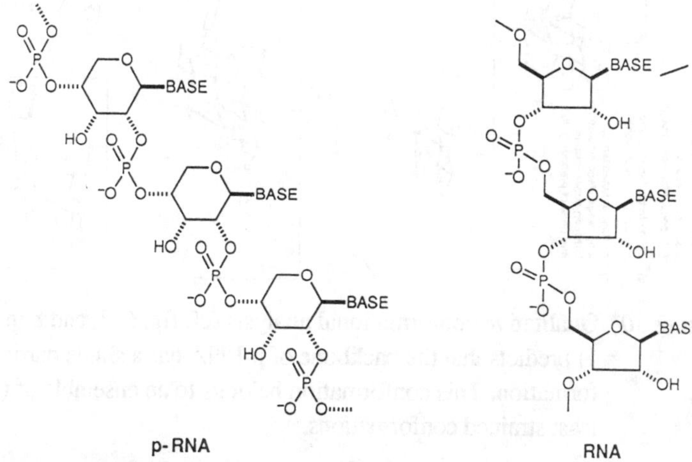

p-RNA

RNA

Figure 49. Constitution and configuration of the ribopyranosyl-(4' → 2')-oligonucleotide system ("Ribopyranosyl-RNA", "p-RNA") [17].

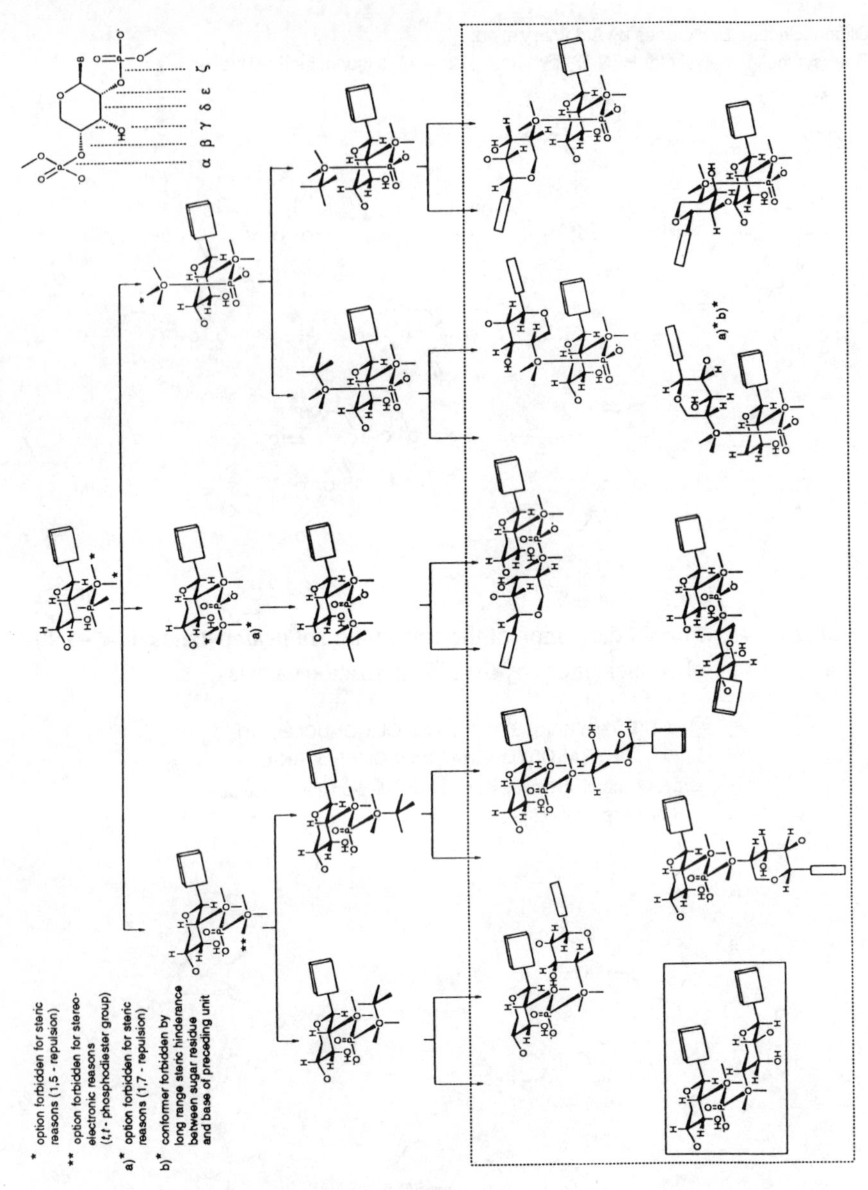

Figure 50. Qualitative conformational analysis (cf. fig. 6, 7, and 8 in lecture 1) predicts that the backbone of p-RNA has a single pairing conformation. This conformation belongs to an ensemble of (nine) least strained conformations.

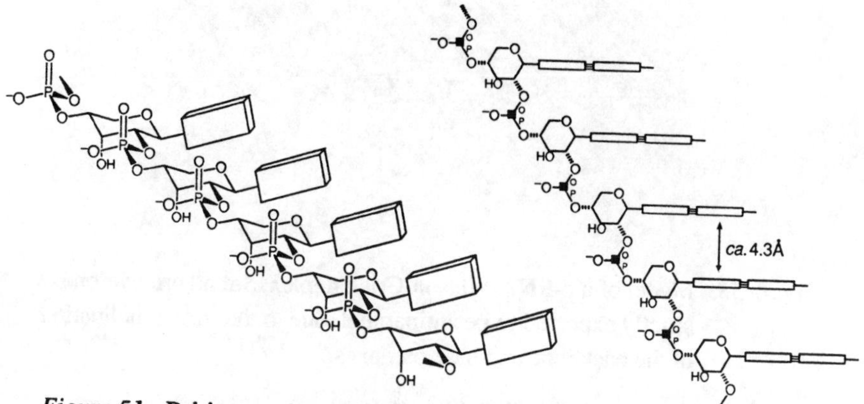

Figure 51. **Pairing conformation of p-RNA single strand. It was predicted, that p-RNA should form duplexes comparable in stability to homo-DNA duplexes.**

HEXO- versus **PENTO**-pyranosyl-(4'→2')-oligonucleotides

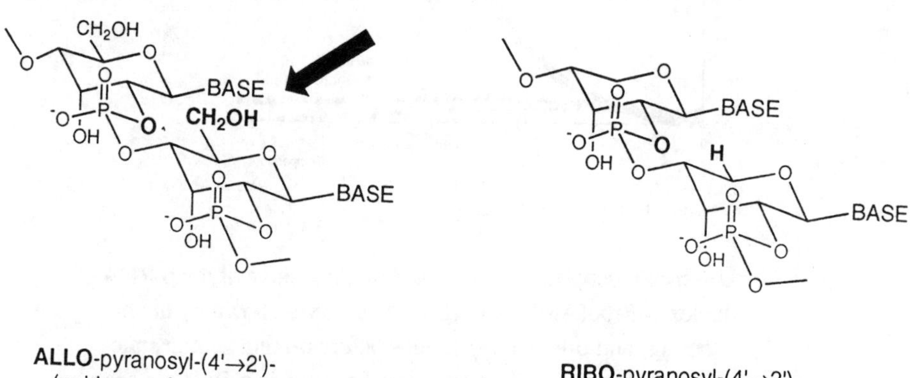

Figure 52. **Juxtaposition of the backbones of p-RNA and its allopyranosyl analogue in their idealized pairing conformations. The latter, in contrast to the former, would suffer prohibitive steric hindrance in this conformation.**

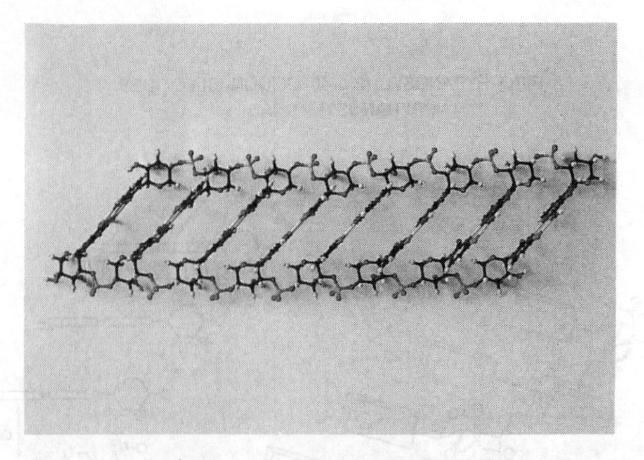

Figure 53. Model of a p-RNA Watson-Crick duplex. Strand orientation is strictly expected to be antiparallel, due to the strong inclination of the backbone and base pair axes.

p-Ribo (U - U - U - U - U - U - U - U)
(A - A - A - A - A - A - A - A) p-Ribo

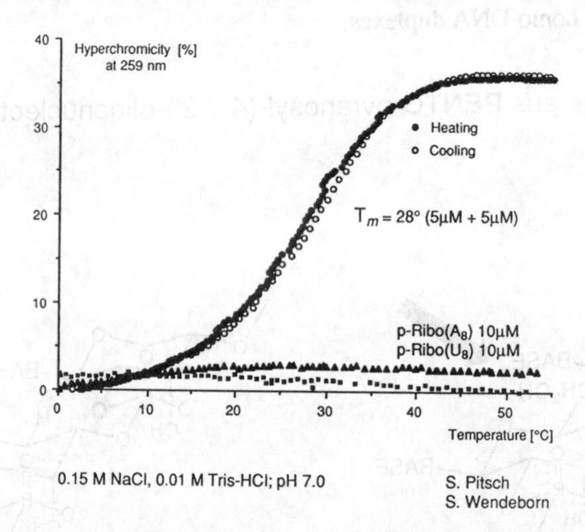

Figure 54. UV-spectroscopically determined melting curve of the p-RNA duplex p-Ribo(A_8)/p-Ribo(U_8). The curves shown by the p-Ribo(A_8) and p-Ribo(U_8) strands before mixing demonstrate that there is no adenine-adenine self-pairing in p-RNA, in sharp contrast to homo-DNA (cf. fig. 18 in lecture 1). For the synthesis of p-RNA oligonucleotides see [8].

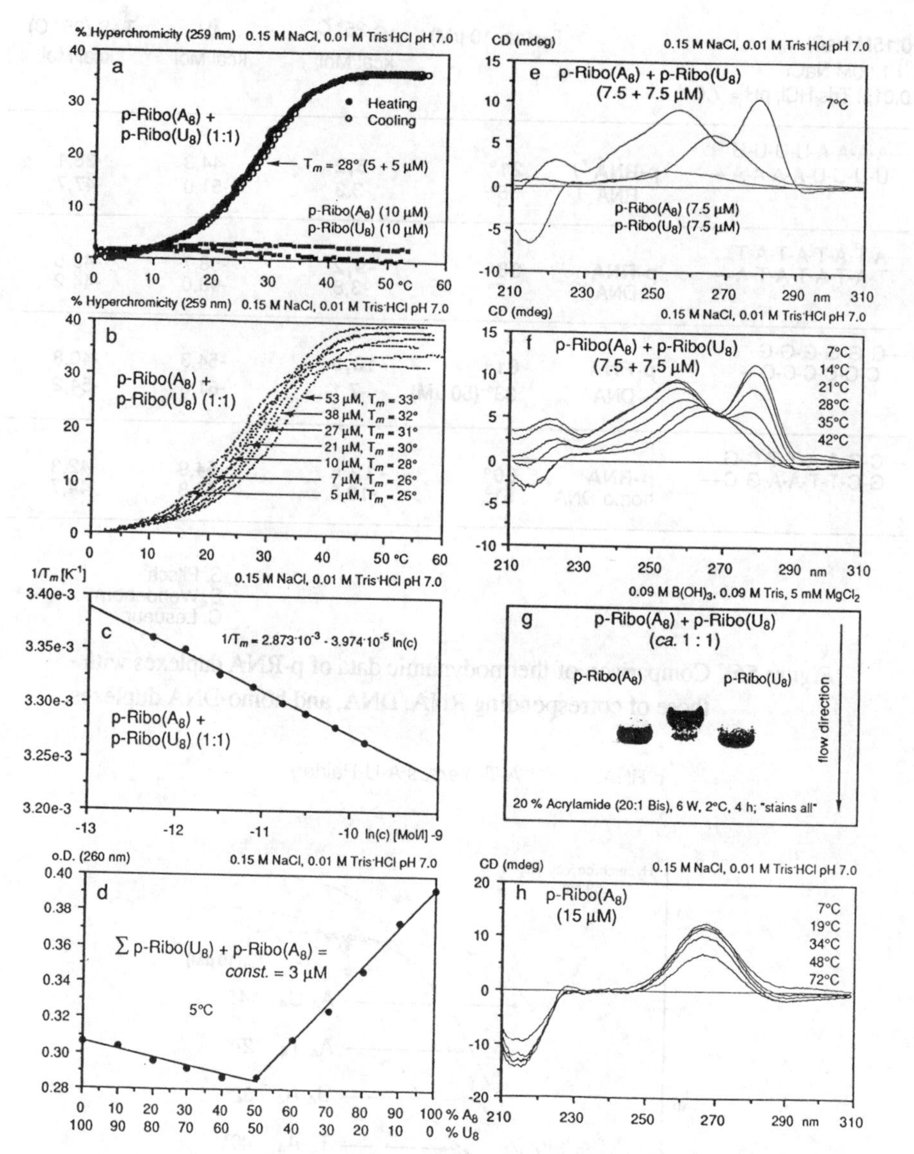

Figure 55. **Characterization of the p-Ribo(A_8)/p-Ribo(U_8) duplex by temperature dependent UV-spectroscopy, temperature dependent CD-spectroscopy and (non-denaturing) gel electrophoresis [8].**

0,15M NaCl *) 1,00M NaCl 0,01M Tris·HCl, pH = 7,0		T_m (ca. 10 μM)	ΔG 25° C kcal/Mol	ΔH kcal/Mol	TΔS (25° C) kcal/Mol
—A-A-A-A-U-U-U-U U-U-U-U-A-A-A-A—	**p-RNA** *)	**21°**	**-6,2**	-44,3	-38,1
	RNA *)	5°	-3,3	-51,0	-47,7
—A-T-A-T-A-T-A-T T-A-T-A-T-A-T-A—	**p-RNA**	**38°**	**-9,2**	-58,7	-49,5
	DNA	7°	-3,8	-48,0	-44,2
—G-G-G-G-G-G C-C-C-C-C-C—	**p-RNA**	**61°**	**-13,5**	-54,3	-40,8
	DNA	33° (50 μM)	-7,1	-61,3	-54,2
—C-G-A-A-T-T-C-G G-C-T-T-A-A-G-C—	**p-RNA**	**60°**	**-12,6**	-54,9	-42,3
	homo-DNA	51°	-10,6	-55,3	-44,7

S. Pitsch
S. Wendeborn
C. Lesueur

Figure 56. **Comparison of thermodynamic data of p-RNA duplexes with those of corresponding RNA, DNA, and homo-DNA duplexes.**

p-RNA: A-T- *versus* A-U-Pairing

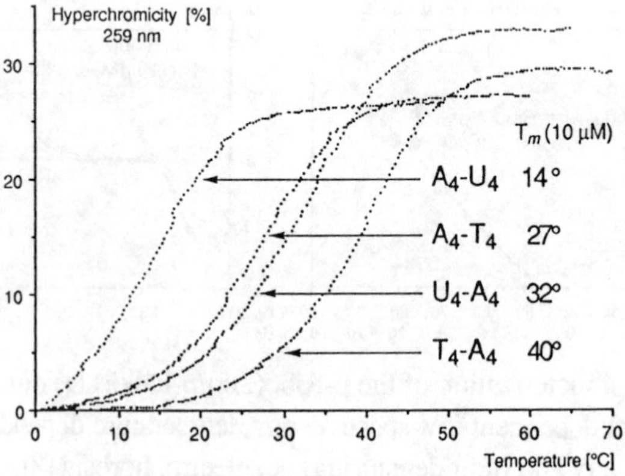

Hyperchromicity [%]
259 nm

T_m (10 μM)

A_4-U_4 14°

A_4-T_4 27°

U_4-A_4 32°

T_4-A_4 40°

0 10 20 30 40 50 60 70
Temperature [°C]

0.15 M NaCl; 0.01 M Tris-HCl, pH 7.0

S. Pitsch
S. Wendeborn

Figure 57. **Comparison of adenine-uracil with adenine-thymine pairing.**

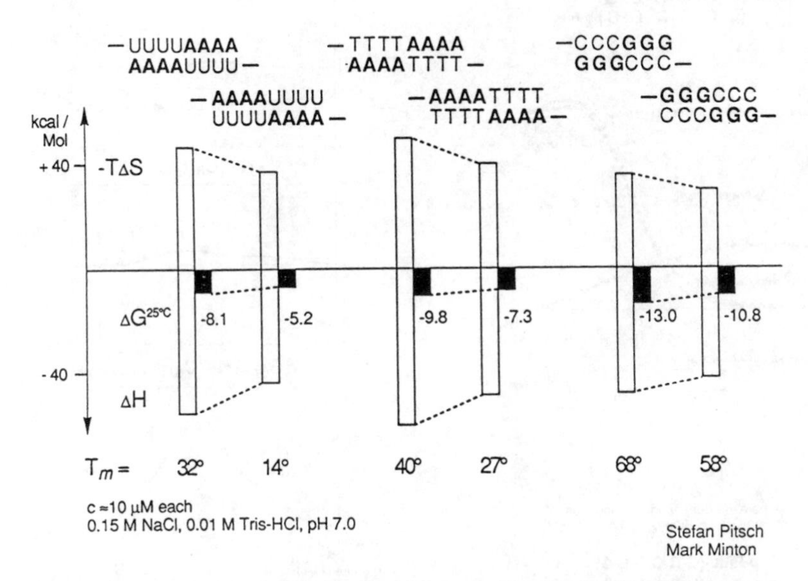

Figure 58. Base-pair-sequence dependence of purine-pyrimidine pairing in p-RNA duplexes.

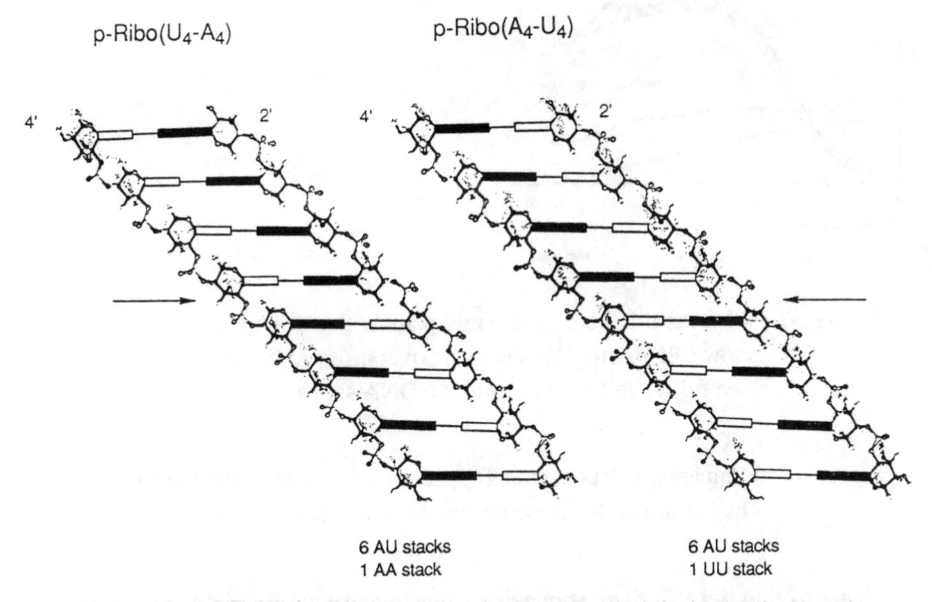

Figure 59. Models of p-Ribo-octamer duplexes (pyrimidine)$_4$-(purine)$_4$ and (purine)$_4$-(pyrimidine)$_4$. There is more purine-purine stacking in the former.

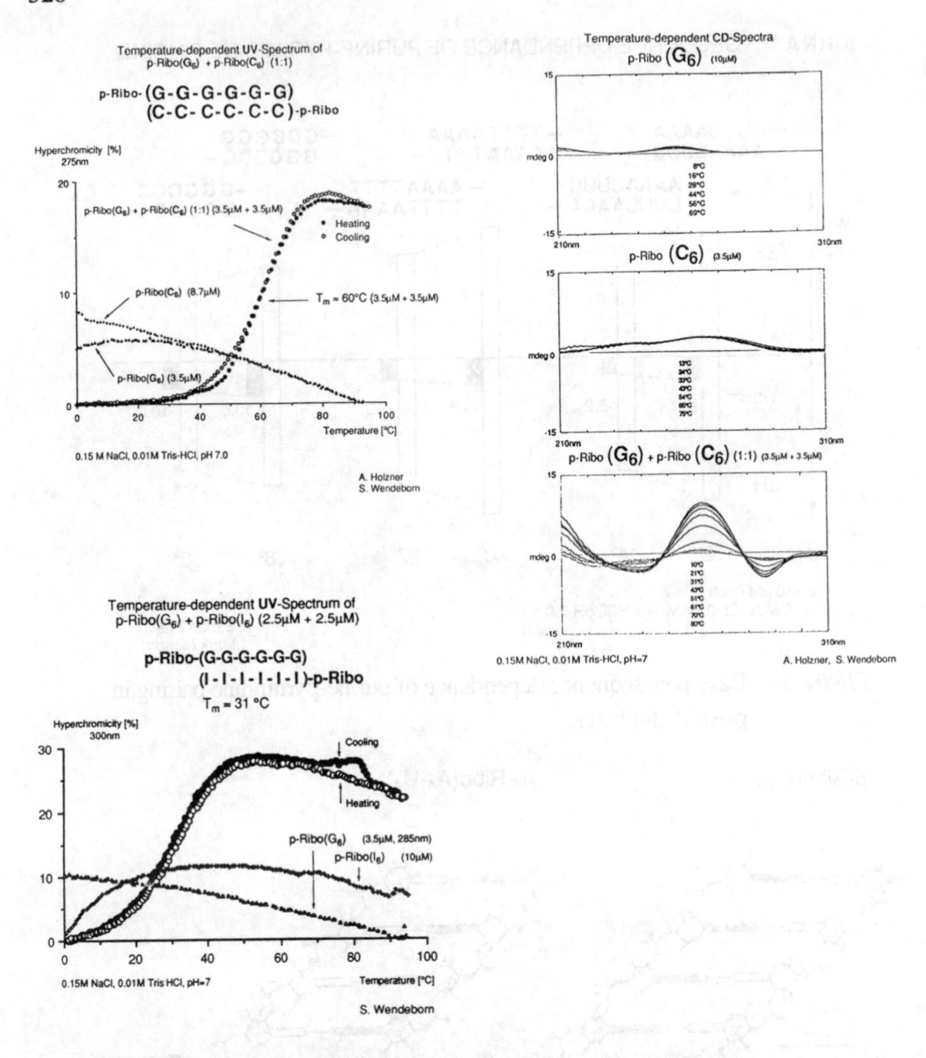

Figure 60. Whereas p-Ribo(G_6) strongly pairs with p-Ribo(C_6), it does not show self-pairing; this is in sharp contrast to both the homo-DNA (see fig. 21 in lecture 1) and the DNA series.

Figure 61. Temperature dependent CD-spectroscopy corroborates the conclusion drawn from the observation of fig. 60.

Figure 62. In p-RNA, there is guanine-isoguanine (Watson-Crick) pairing as in homo-DNA (see fig. 22 in lecture 1) but no reverse-Hoogsteen self-pairing of isoguanine.

p-RNA: SUMMARY OF EXPERIMENTAL OBSERVATIONS

- **Purine-pyrimidine pairing** (Watson-Crick, antiparallel strand orientation) is **stronger than in RNA,** comparable to homo-DNA
- Strong **purine-purine Watson-Crick pairing** between guanine and isoguanine (as in homo-DNA, but in contrast to DNA)
- **No** reverse-Hoogsteen purine-purine **selfpairing** (in contrast to homo-DNA)
- **No** Hoogsteen **selfpairing** of guanine (in contrast to DNA and RNA)

Figure 63. (see title)

p-RNA is not only the **strongest**, but also the **most selective** oligonucleotide pairing system known so far.

Figure 64. The most interesting property of p-RNA so far discovered is its remarkable pairing selectivity. The non-pairing of guanine with itself is especially significant in context of the question, whether p-RNA base sequences can replicate themselves by template chemistry.

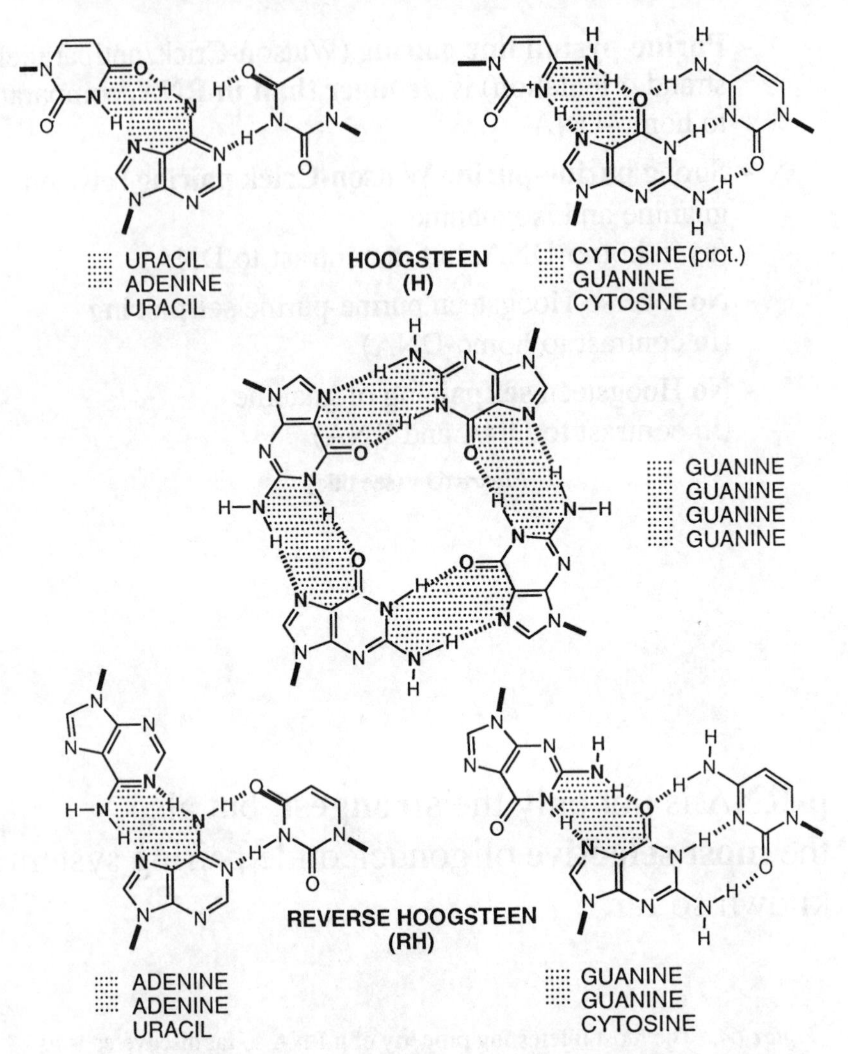

Figure 65. Hoogsteen and reverse-Hoogsteen pairing modes encountered in DNA triplexes. None of them has been observed (so far) in the p-RNA series (in sharp contrast to the homo-DNA series).

BASE PAIRING		DNA (RNA)	Homo-DNA	p-RNA
PURINE-PYRIMIDINE				
WC	A - T	+	+	+
WC	G - C	+	+	+
H	A - T	+ *	(+) *	(–)
H	G - C^{H+}	+ *	(+) *	(–)
PURINE-PURINE				
WC	G - I	–	+	+
WC	D - X	(–)	+	(+)
RH	A - A	+ *	+	–
RH	G - G	+ *	+	–
H	G - G	+ *		–
RH	I - I		+	–
RH	D - D		+	(–)

+ Pairing in Duplexes
+ * Pairing in Tri(or Tetra)plexes
( ) inferred as against observed

Figure 66. List of observed pairing modes in the DNA, homo-DNA, and p-RNA series.

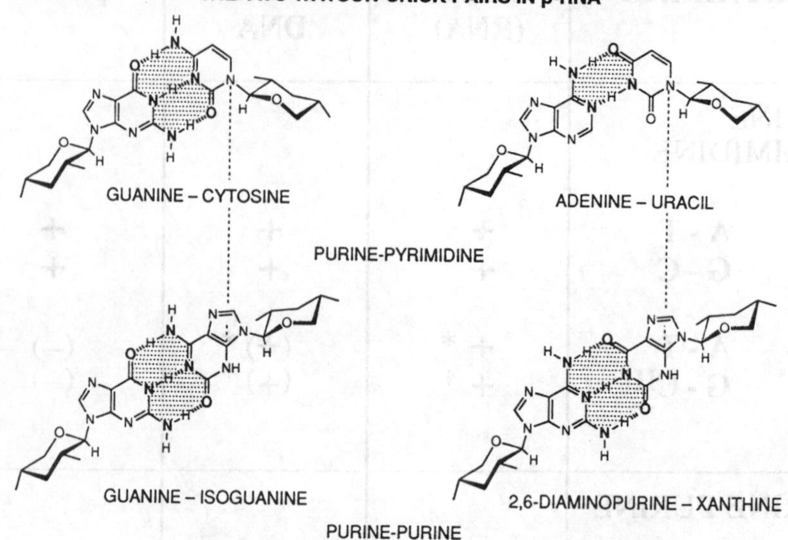

Figure 67. The pairing observed so far in the p-RNA series is (believed to be) exclusively Watson-Crick. The figure shows the two purine-pyrimidine and the two purine-purine pairs (the diaminopurine-xanthine pair is expected, but not yet observed, to occur).

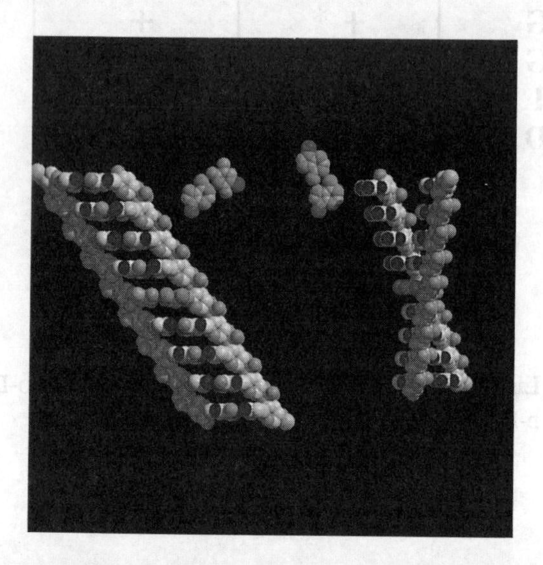

Figure 69. Deviation of p-RNA backbones (in their idealized conformation) from colinearity: left: purine-pyrimidine, Watson-Crick; right: adenine-adenine, reverse-Hoogsteen.

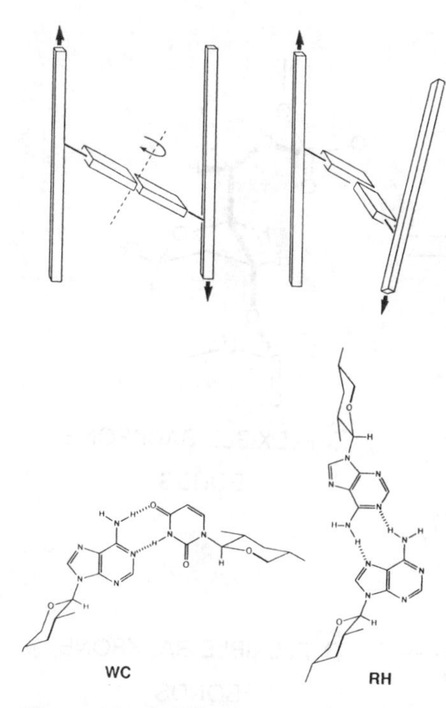

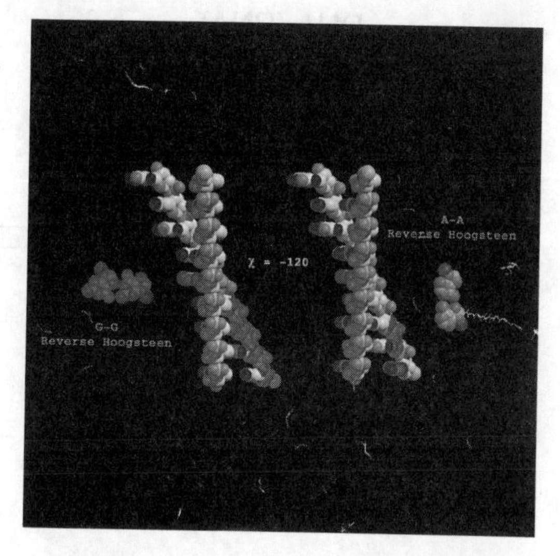

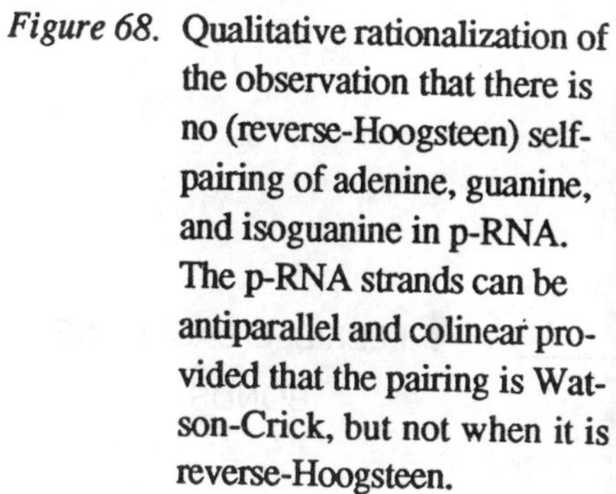

Figure 68. Qualitative rationalization of the observation that there is no (reverse-Hoogsteen) self-pairing of adenine, guanine, and isoguanine in p-RNA. The p-RNA strands can be antiparallel and colinear provided that the pairing is Watson-Crick, but not when it is reverse-Hoogsteen.

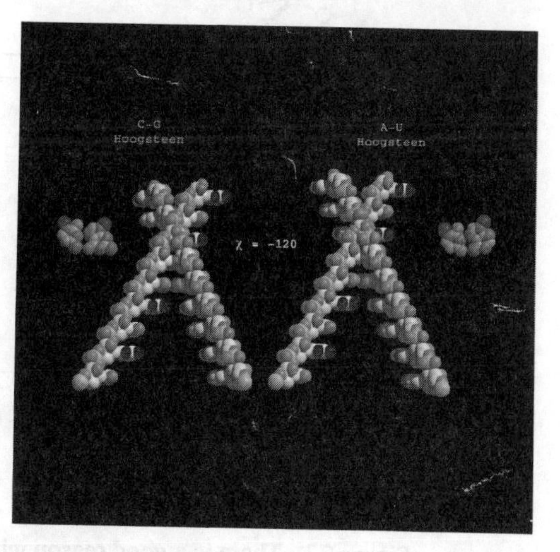

Figure 70. Right: adenine-adenine, reverse-Hoogsteen; left: guanine-guanine, reverse-Hoogsteen.

Figure 71. Right: adenine-uracil, Hoogsteen; left: guanine-(protonated) cytosine, Hoogsteen. In order to pair in a Hoogsteen (or reverse-Hoogsteen) mode, the p-RNA backbone would have to adjust by conformational deformation.

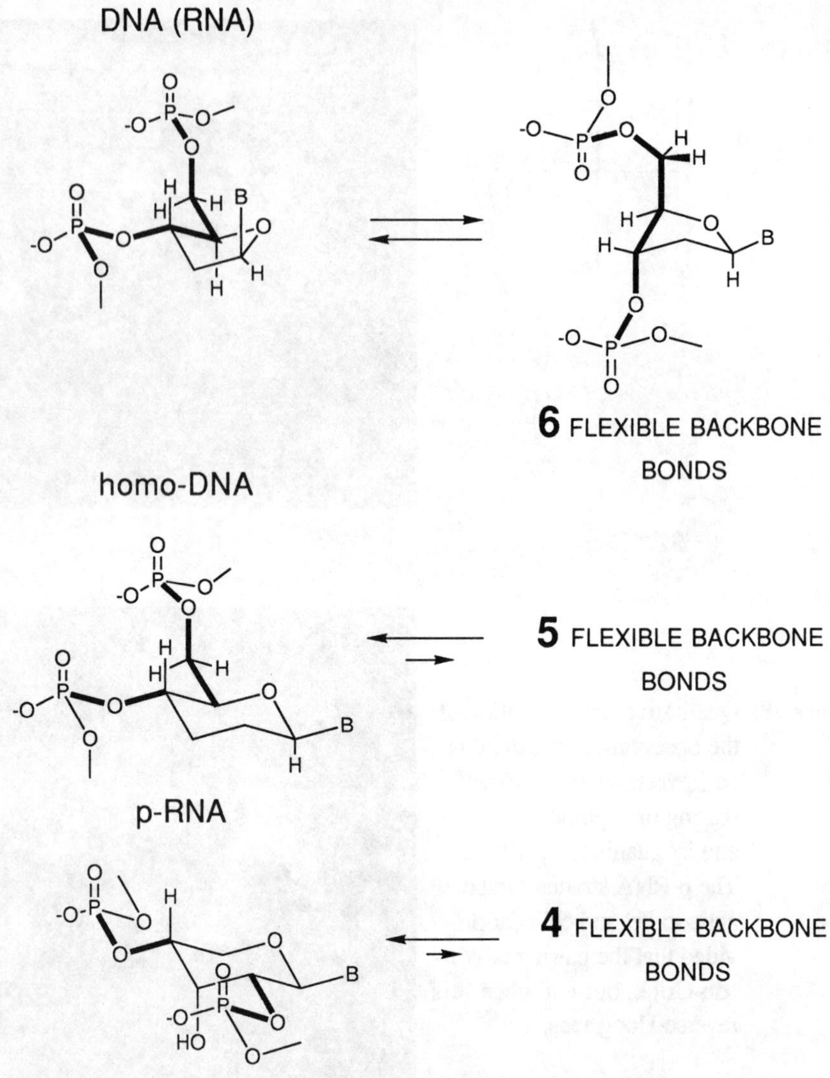

DNA (RNA)

6 FLEXIBLE BACKBONE BONDS

homo-DNA

5 FLEXIBLE BACKBONE BONDS

p-RNA

4 FLEXIBLE BACKBONE BONDS

Figure 72. **There is a good reason why the p-RNA backbone should have less conformational freedom than the DNA or the homo-DNA backbone (see text of the figure).**

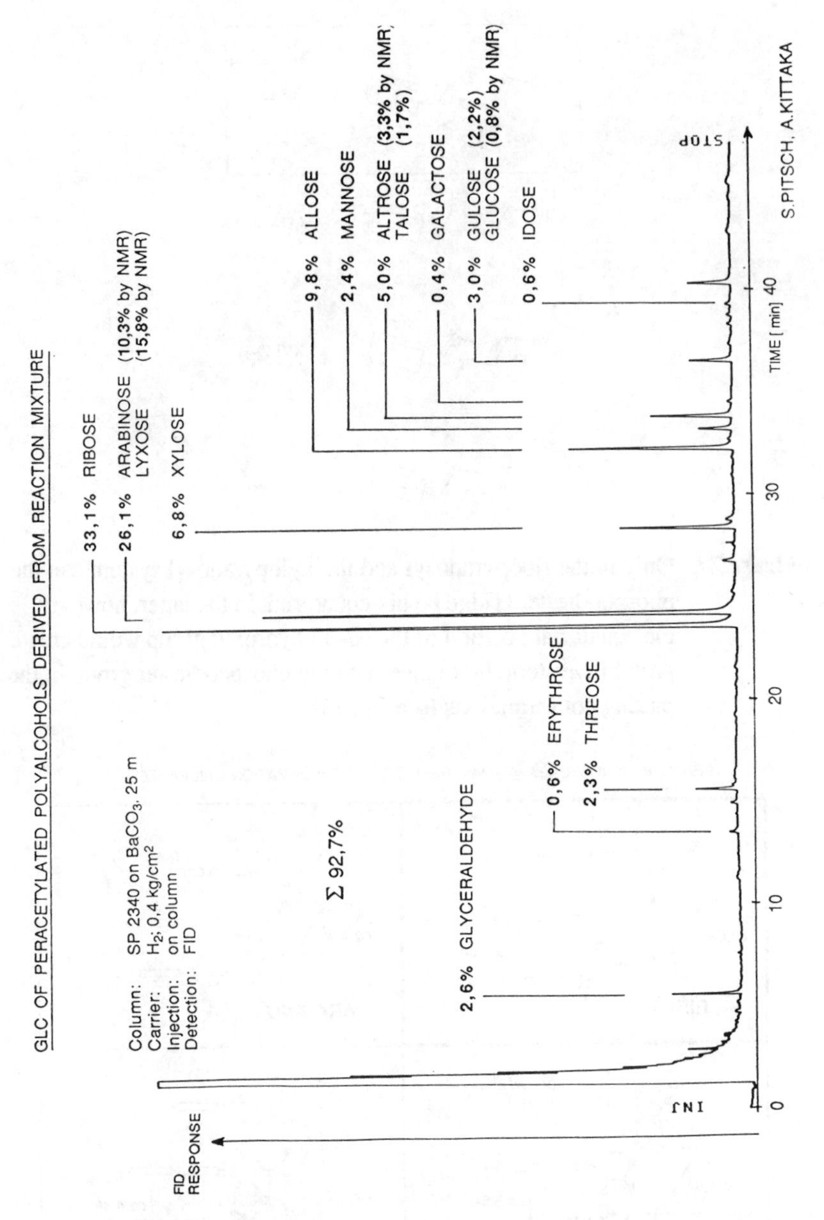

Figure 73. If a pentose and not a hexose, why then ribose and not another pentose? The gas chromatogram (of polyalcohol acetates) characterizing the reaction mixture produced by aldolization of glycolaldehyde phosphate in the presence of half an equivalent of formaldehyde shows ribose to be formed as the major product among all (racemic) tetrose-, pentose-, and hexose-phosphates [3].

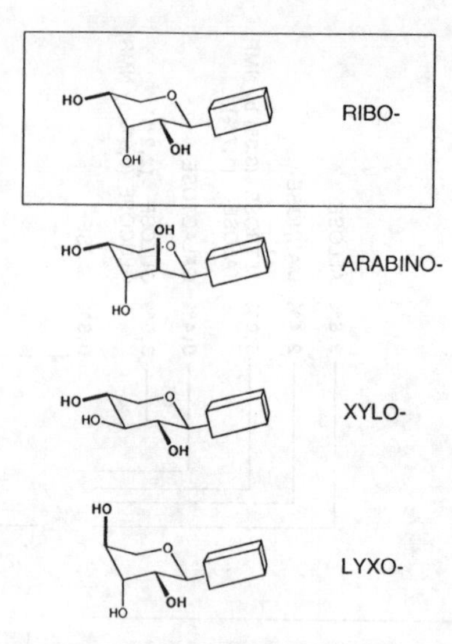

Figure 74. Only in the ribopyranosyl and the xylopyranosyl system can the phosphodiester bridge be bis-equatorial; in the latter, however, the equatorial position of the (C-3') hydroxyl group would create prohibitive steric hindrance with the phosphodiester group in the pairing conformations (see fig. 75).

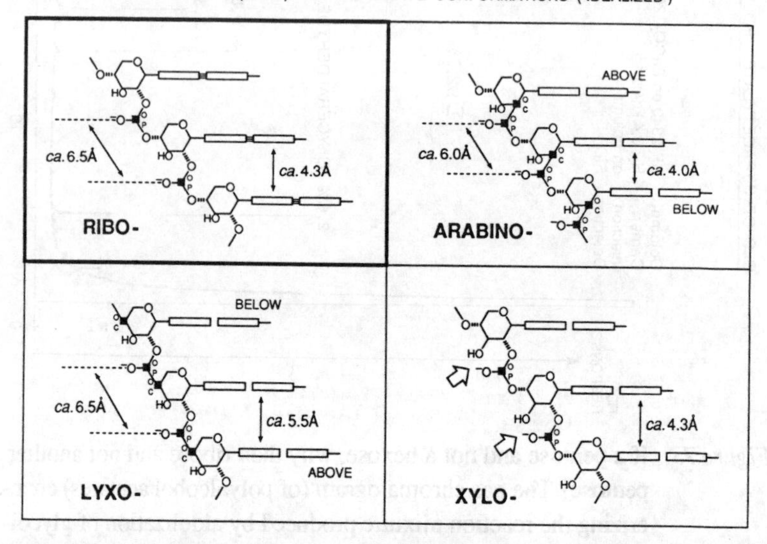

Figure 75. An analysis of the idealized pairing conformations suggests that only the ribopyranosyl system can adopt unhampered colinearity of antiparallel strands in duplexes.

p-RNA: CONJECTURES, to be dealt with experimentally:

- p-RNA has a greater potential for **constitutional self-assembly** in a natural environment than RNA
- p-RNA has a greater potential for **non-enzymic self-replication** than RNA
- **Entropy** favors RNA over p-RNA
- An intraduplex **p-RNA → RNA transition** with retention of base-sequence information is mechanistically feasible

Figure 76. (see title)

Figure 77. Sculpture by Edward Class on the campus of the University of British Columbia in Vancouver, Canada (photography by the author).

ETH		**Frankfurt**
Stefan Pitsch		Christian Miculka
Sebastian Wendeborn	(1992-93)	Norbert Windhab
Armin Holzner	(1992-93)	
Georg Issakides	(1992-93)	
Guillermo Delgado	(1992-93)	
Rama Krishnamurthy		
Mark Minton		
Catherine Lesueur		
Christoph Spinner		

Figure 78. Names of postdoctoral colleagues who have done (or are still doing) the work on the chemistry of p-RNA.

Acknowledgment

The author expresses his deep appreciation and gratitude to his group of outstanding doctoral and postdoctoral collaborators who have carried out the work presented here; their names are listed in figures 43 and 78. The research was supported by the Ciba AG, Basle, Firmenich & Cie., Geneva, the Schulleitung of ETH Zürich, and the Swiss National Science Foundation. A special thank goes to my colleague Duilio Arigoni, whithout whose administrative help and generosity the work on p-RNA could not have been carried out.

References

1. Eschenmoser, A. (1991) Warum Pentose- und nicht Hexose-Nucleinsäuren?, *Nachr. Chem. Tech. Lab.* **39**, 795-806.
2. Eschenmoser, A. and Dobler, M. (1992) Why Pentose and Not Hexose Nucleic Acids? Part I. Introduction to the Problem, Conformational Analysis of Oligonucleotide Single Strands Containing 2',3'-Dideoxyglucopyranosyl Building Blocks ('Homo-DNA'), and Reflections on the Conformation of A- and B-DNA, *Helv. Chim. Acta* **75**, 218-259.
3. Müller, D., Pitsch, S., Kittaka, A., Wagner, E., Wintner, C.E., and Eschenmoser, A. (1990) Chemie von α-Aminonitrilen. Aldomerisierung von Glykolaldehyd-phosphat zu racemischen Hexose-2,4,6-triphosphaten und (in Gegenwart von Formaldehyd) racemischen Pentose-2,4-diphosphaten: rac-Allose-2,4,6-triphosphat und rac-Ribose-2,4-diphosphat sind die Reaktionshauptprodukte, *Helv. Chim. Acta* **73**, 1410-1468.
4. Böhringer, M., Roth, H.-J., Hunziker, J., Göbel, M., Krishnan, R., Giger, A., Schweizer, B., Schreiber, J., Leumann, C., and Eschenmoser, A. (1992) Why Pentose and Not Hexose Nucleic Acids? Part II. Preparation of Oligonucleotides Containing 2',3'-Dideoxy-β-D-glucopyranosyl Building Blocks, *Helv. Chim. Acta* **75**, 1416-1477.
5. Hunziker, J., Roth, H.-J., Böhringer, M., Giger, A., Diederichsen, U., Göbel, M., Krishnan, R., Jaun, B., Leumann, C., and Eschenmoser, A. (1993) Why Pentose- and Not Hexose-Nucleic Acids? Part III. Oligo(2',3'-dideoxy-β-D-glucopyranosyl)nucleotides. ('Homo-DNA'): Base-Pairing Properties, *Helv. Chim. Acta* **76**, 259-352.
6. Otting, G., Billeter, M., Wüthrich, K., Roth, H-.J., Leumann, C., and Eschenmoser, A. (1993) Why Pentose- and Not Hexose-Nucleic Acids? Part IV. 'Homo-DNA': ^{1}H-, ^{13}C-, ^{31}P,- and ^{15}N-NMR-Spectroscopic Investigation of ddGlc (A-A-A-A-A-T-T-T-T-T) in Aqueous Solution, *Helv. Chim. Acta* **76**, 2701-2756.
7. Eschenmoser, A. (1993) Toward a Chemical Etiology of the Natural Nucleic Acids' Structure, Proc. R.A. Welch Foundation Conferences on 'Chemical Research XXXVII: 40 Years of DNA Double Helix', Houston, Texas, pp. 201-235.
8. Pitsch, S., Wendeborn, S., Jaun, B., and Eschenmoser, A. (1993) Why Pentose- and Not Hexose-Nucleic Acids? Part VII. Pyranosyl-RNA ('p-RNA'), *Helv. Chim. Acta* **76**, 2161-2183.
9. Dickerson, R.E. (1985) Helix Polymorphism and Information Flow in DNA, Proc. R.A. Welch Foundation Conferences on 'Chemical Research XXIX. Genetic Chemistry: The Molecular Basis of Heredity', pp. 39-75.
10. Groebke, K. (1993) Über Purin-Purin-Paarungen bei Hexopyranose-Nukleinsäuren, Thesis No. 10149, ETH Zürich.

11. Fraser, W. (1992) unpublished work, ETH Zürich.
12. Holzner, A. (1993) unpublished work, ETH Zürich.
13. Fischer, R.W. (1992) Allopyranosyl-Nukleinsäure: Synthese, Paarungseigenschaften und Struktur von Adenin-/Uracil-haltigen Oligonukleotiden, Thesis No. 9971, ETH Zürich.
14. Helg, A. (1994) Allopyranosyl-Nukleinsäure: Synthese, Paarungseigenschaften und Struktur von Guanin-/Cytosin-enthaltenden Oligonukleotiden, Thesis No. 10464, ETH Zürich.
15. Hammer, R. (1992) unpublished work, ETH Zürich.
16. Krishnamurthy, R. (1994) unpublished work, ETH Zürich.
17. Eschenmoser, A. (1993) Hexose nucleic acids, *Pure & Appl. Chem.* **65**, 1179-1188.

PRINCIPLES OF ANTIBODY CATALYSIS

DONALD HILVERT
Departments of Chemistry and Molecular Biology
The Scripps Research Institute
10666 North Torrey Pines Road
La Jolla, California 92037 USA

Abstract

A wide range of chemical transformations can be catalyzed by antibody molecules elicited with rationally designed transition state analogs. The development of catalytic antibodies consequently represents one of the most versatile and general strategies for creating new enzymes to emerge in the last several years. Recent advances in the production and characterization of these agents are reviewed.

1. Introduction

The mammalian immune system is an unparalleled source of tailored receptor molecules. In response to antigen challenge, antibody molecules are produced over the course of several weeks or months that exhibit extraordinarily high affinity and specificity for their ligands [1-3]. It is estimated that more than 10^8 binding specificities are accessible in the immune system's primary repertoire, and this population can be further expanded and refined through a microevolutionary process involving somatic mutation and selection [4,5]. Harnessing this remarkable protein-generating system for the production of tailored antibody catalysts has emerged in recent years as a versatile and powerful approach for creating novel enzymes [6].

All antibodies share a common scaffold (Figure 1) [7]. Two heavy and two light chains combine to form a homodimeric, Y-shaped molecule of Mr 160,000. The antigen-binding sites are located in the N-terminal variable

C. Chatgilialoglu and V. Snieckus (eds.), Chemical Synthesis, 341–359.

domains of the protein and are formed by six peptide loops contributed by the heavy (H1 to H3) and the light (L1 to L3) chains. Antibody diversity derives from the hypervariability in length and sequence of these so-called complementarity determining regions (CDRs). The combinatorial arrangement of six CDR loops allows an extensive range of shapes to be constructed — from deep pockets, clefts, and grooves to flatter, more modulating surfaces — in response to a wide array of structurally diverse antigens [7,8].

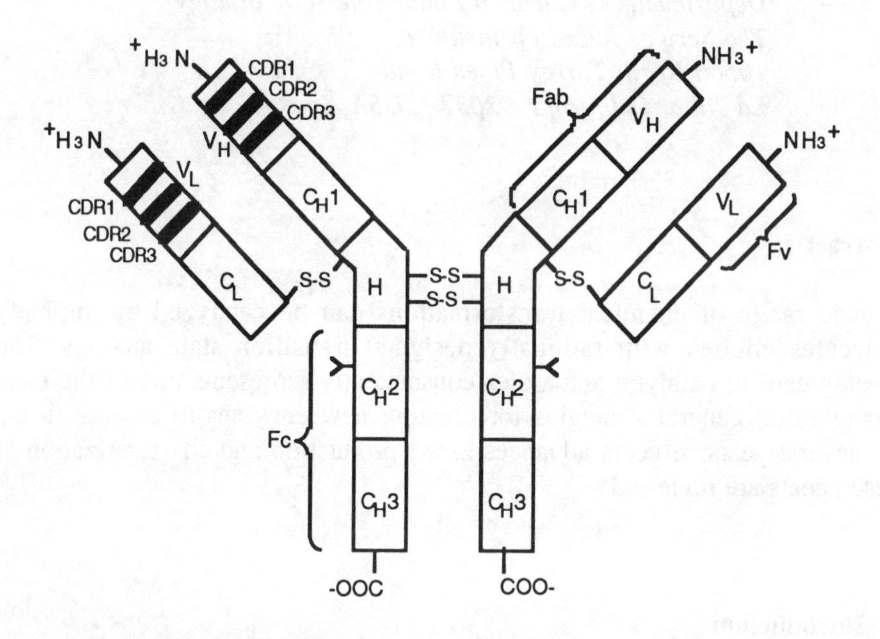

Figure 1. Schematic drawing of an antibody molecule. The active site is formed by the complementarity determining regions (CDRs) of the variable domains of the heavy and light chains.

The notion that antibodies might serve as highly selective catalysts originated with Jencks in 1969 [9]. Enzymes and antibodies exhibit many common features [10]. Not only are the basic sizes and shapes of their binding pockets similar, but ligand recognition is mediated in both cases by a combination of hydrogen bonding, electrostatics, dispersion, and hydrophobic interactions. However, whereas enzymes have evolved for maximum stabilization of the ephemeral transition state of the reaction they catalyze, antibodies are raised in response to a ground state molecule. Jencks' key insight [9] was that antigens mimicking the essential stereoelectronic features of the transition state for a reaction might elicit antibody binding sites with properties suitable for catalysis.

This strategy works remarkably well. Well over 50 chemical transformations have been promoted by antibodies since 1986 [6]. These include acyl transfers, eliminations, pericyclic processes, isomerizations, and redox reactions. Antibody catalysts not only accelerate these reactions 10^2 to 10^6-fold over background, they also exhibit a high degree of substrate specificity, regio- and stereoselectivity. Because these properties correlate well with the structural features of the transition state analog, the development of catalytic antibodies provides the means to create tailored agents for a wide range of applications in chemistry and biology. Furthermore, detailed mechanistic and structural studies of individual catalytic antibodies have the potential to reveal a great deal about the fundamental principles of protein catalysis.

Much effort is currently directed toward defining both the scope and limitations of catalytic antibody technology. Expansion of the repertoire of reactions amenable to antibody catalysis is one important objective. Chemical transformations that cannot be achieved efficiently or selectively with available chemical methods are particularly attractive targets. Elucidation of the structural basis of antibody catalysis is another major goal. Improved design of transition state analogs and optimization of the catalytic efficiency of first-generation catalytic antibodies via protein engineering will require a detailed understanding of the structure-function relationships that govern catalysis. Structural studies are also important for determining what if any limitation antibody structure places on the nature and efficiency of the reactions that can be carried out within the combining site. As outlined below, considerable progress has recently been made on each of these fronts.

2. Structural basis of antibody catalysis

The rearrangement of chorismate **1** into prephenate **3** is a useful transformation for illustrating the basic strategies for creating catalytic antibodies and the lessons that can be learned through detailed study of their structure and properties. This simple unimolecular reaction is formally a Claisen rearrangement and does not require general acid-base or nucleophilic catalysis. It is, moreover, a rare example of a biologically relevant pericyclic reaction: in plants and lower organisms, it is the branchpoint in the shikimate pathway leading to the essential aromatic amino acids tyrosine and phenylalanine. *In vivo* this step is catalyzed by the enzyme chorismate mutase with a rate acceleration $> 10^6$ over background [11].

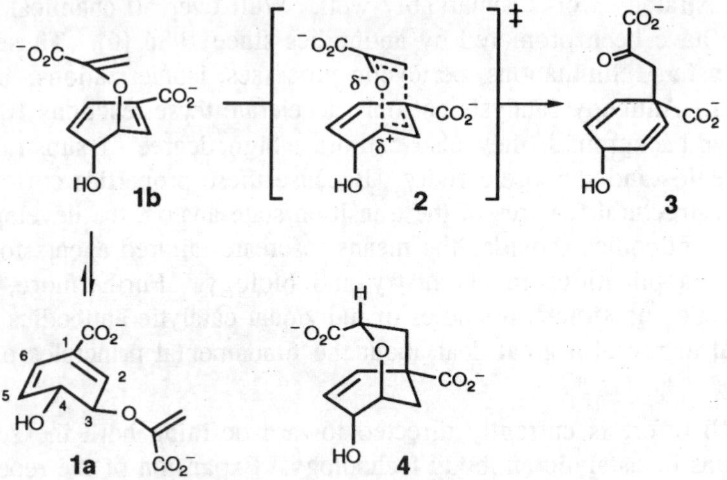

How chorismate mutase speeds the rearrangement has been a matter of debate for some time. Labeling studies and inhibitor experiments indicate that both the catalyzed and spontaneous reactions proceed via a conformationally restricted chair-like transition state **2** [12,13]. Solvent and kinetic isotope effects show that the reaction is asynchronous with carbon-oxygen bond cleavage preceding carbon-carbon bond formation [14,15]. Some of the observed rate acceleration may therefore result from stabilization of charge separation in the transition state through specific hydrogen bonding or electrostatic interactions at the active site [16]. However, entropic contributions must also be important. The flexible chorismate molecule adopts an extended pseudo-diequatorial conformation in solution (**1a**) and must undergo an energetically unfavorable conformational change to the diaxial conformer **1b,** in which the enol pyruvate side chain is oriented properly for the ensuing reaction [15]. Given this, the enzyme might act as an 'entropy trap' [17] to stabilize **1b** and freeze out the rotational degrees of freedom of the side chain in the transition state [11,18]. The favorable activation entropy of the enzyme-catalyzed process compared with the spontaneous thermal rearrangement ($\Delta\Delta S^{\ddagger}$ = 13 cal K^{-1} mol^{-1}) is consistent with this notion [18].

Strong inhibition of chorismate mutase by the conformationally restricted oxabicyclic transition state analog **4** [19] also suggests that the enzyme has an active site congruent to the compact transition state structure **2**. Compound **4** binds ≈200-times more tightly to the enzyme than chorismate [19], capturing a significant fraction of the energy available for stabilizing the metastable species. This compound has also proved effective as a template for producing antibodies with substantial chorismate mutase activity [20,21]. Because small molecules are generally not immunogenic, it was necessary to couple **4** to a carrier protein to generate an effective immune response. The coupling strategy is dictated by the need for recognition of both the enolpyruvate side chain and the cyclohexadiene ring of chorismate. Attachment of **4** to keyhole limpet hemocyanin via its secondary alcohol ensures this by making the two carboxylates the primary determinants for immune recognition.

This approach has yielded antibodies that catalyze the chorismate to prephenate reaction in a highly enantioselective fashion with rate accelerations in the range 10^2-10^4 fold over background [20-23]. The best catalyst (11F1-2E9) is only two orders of magnitude less efficient than natural chorismate mutase, and like the natural enzyme, has a very favorable entropy of activation compared to the thermal rearrangement [23]. Surprisingly, the less active antibody catalyst (1F7) achieves its 200-fold rate acceleration entirely by lowering the enthalpy of activation [20]. The entropy of activation in this case is actually 10 cal K^{-1} mol^{-1} less favorable than the uncatalyzed reaction. While it is difficult to interpret activation parameters even under the best of circumstances, the differences exhibited by 1F7 and 11F1-2E9 illustrate the capacity of the immune system to solve a specific chemical problem in several ways.

The unfavorable entropy of activation observed for 1F7 raises interesting questions regarding the extent to which the antibody populates and stabilizes the diaxial substrate conformer **1b**. This issue has been investigated by NMR spectroscopy [24]. Diagnostic transferred nuclear Overhauser effects (TRNOEs) between protons on the enol pyruvate side chain and protons on the cyclohexadiene ring demonstrate that a significant fraction of the chorismate bound at the active site of 1F7 does adopt the reactive geometry 'programmed' by the conformationally restricted transition state analog. Although additional, smaller TRNOE signals suggest that the antibody may not completely reduce the rotational degrees of freedom of the bound substrate, the NMR results rule out simple preorganization of the substrate as the major contributor to the high entropy barrier for the antibody-catalyzed reaction.

Crystallographic studies of 1F7 complexed with the transition state analog **4** [25] complement and extend the spectroscopic experiments. As shown in Figure 2, hapten recognition in 1F7 is achieved through a combination of hydrophobic interactions, hydrogen bonds and electrostatics. Indeed, the combining site of the antibody exhibits excellent overall shape and charge complementarity to **4**. It is this complementarity that presumably dictates preferential binding of the correct substrate enantiomer in a conformation appropriate for reaction. Moreover, the orientation of the transition state analog within the active site reflects the coupling strategy

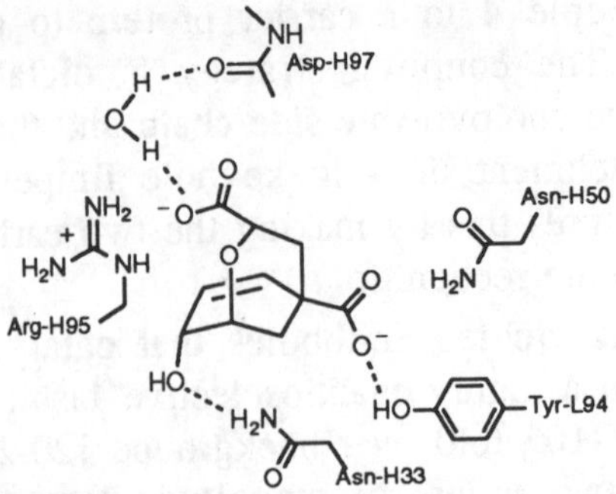

Figure 2. Schematic drawing illustrating the hydrogen bonding and electrostatic interactions between the chorismate mutase antibody 1F7 and the transition state analog **4**.

adopted for the purposes of immunization: the two hapten carboxylates are the primary recognition elements, while the C-4 alcohol group sticks out of the cavity. The antibody's structural properties thus accurately mirror those of the transition state analog.

Why is 1F7 10^4-times less active than chorismate mutase? The antibody molecule's distinctive architecture does not appear to impose intrinsic structural limitations on catalysis. Indeed, comparison of the active sites of the antibody and the monofunctional enzyme from *Bacillus subtilis* [26] suggests that the differences between them are more a matter of degree than of kind. Upon complex formation, the hapten is buried to a similar extent in both proteins, and similar types of interactions are available for orienting the flexible substrate correctly for reaction. Furthermore, the enzyme and the antibody seem to promote the rearrangement of chorismate via the same concerted transition state as the uncatalyzed reaction. Other formal mechanistic possibilities, such as a two-step heterolytic process assisted by an enzymic nucleophile, can be ruled out by the lack of appropriate functional groups in the respective active sites [25,26].

Unlike 1F7, chorismate mutase appears to have more extensive interactions with the hapten's alcohol moiety than with the tertiary carboxylate [26], but this is unlikely to account for the observed differences in rate. In conjunction with the overall shape complementarity of the protein pocket, recognition of either the C-1 carboxylate or the C-4 alcohol of chorismate should be sufficient for correctly positioning the cyclohexadiene ring for reaction. Consistent with this idea, antibody 11F1-2E9, which was elicited in the same way as 1F7, has remarkably high catalytic activity [21]. Detailed structural data for 11F1-2E9 are eagerly awaited.

A more plausible reason for the relatively low efficiency of 1F7 is that the antibody is a poorer 'entropy trap' than chorismate mutase. The antibody exploits far fewer specific hydrogen bonding and electrostatic interactions for ligand recognition than the enzyme [25]. This point is illustrated by the interactions available for fixing the enol pyruvate side chain in place. In the enzyme, the carboxylate moiety of the side chain is hydrogen bonded to a tyrosine and two arginine residues [26]. In the antibody, this group binds loosely in a relatively hydrophobic pocket. ArgH95 is probably important for charge neutralization, but it is too far away to donate a hydrogen bond to either of the carboxylate oxygens. A water molecule takes on this role, serving as a bridge between the carboxylate and the protein backbone. However, the water molecule is likely to provide only poor control over the placement of the side chain in the flexible substrate during the reaction. In addition to being an ineffective means of freezing out rotational degrees of freedom in the bound substrate, fixation of a water molecule from bulk solvent will be entropically quite costly [27].

In contrast to the enzyme, the antibody also appears relatively poorly equipped to accommodate charge separation in the transition state [25]. The enzyme possesses a dipolar binding site near the substrate carbon-oxygen bond that breaks during rearrangement [26]: Arg95 is poised to stabilize the partial negative charge that accumulates on the ether oxygen, and the side chains of Glu78, Cys75, and possibly even Phe57 may help to neutralize any positive charge that develops on the cyclohexadiene ring. The side chain of ArgH95 occupies a similar position in the antibody as Arg95 in the enzyme, but given its role in neutralizing a full negative charge on the substrate, it will be much less effective in accommodating additional negative charge on the ether oxygen at the transition state. The antibody has no groups analogous to Glu78 or Cys75. Of course, the lack of a dipolar binding site in the antibody is not surprising. Although hapten **4** mimics the conformationally restricted geometry of **2** very well, it models the polarized character of the transition state poorly.

If 1F7's relatively low catalytic efficiency is due to its short evolutionary history and imperfect transition state analog design, rather than to an intrinsic architectural bias against catalysis, rational modifications of the strategy that yielded this antibody may increase the probability of finding better catalysts. Expanded screening of anti-**4** antibodies is one possibility. A relatively small number of antibodies that bind **4** ($\approx$ 50) have been tested for chorismate mutase activity [20,21], representing but a tiny fraction of the immune response. By assaying more antibodies, even better catalysts than 11F1-2E9 might be found. Because the hapten was racemic, it may also be possible to identify catalysts that preferentially rearrange the unnatural isomer of chorismate. Recently developed methods for displaying antibody fragments on phage [28], in conjunction with powerful biological screening and selection techniques, may facilitate such a search by providing improved access to the diversity within the immunological repertoire.

While more extensive screening can increase the chances of finding rare but highly active catalysts, improvements in hapten design could substantially increase the presence of such catalysts in the population of candidate clones. Knowledge of the critical interactions between chorismate mutase and **4** [26] will be invaluable in this regard. For example, derivatives of **4** in which the C-4 alcohol is replaced with an amine or in which the secondary carboxylate is replaced with a phosphonate might be much better enzyme inhibitors because of improved electrostatic complementarity with the active site. For the same reason, they might also be more effective at producing antibodies that resemble natural chorismate mutase.

The availability of structural information on the antibody-hapten complex also makes possible the redesign of the first generation catalyst by protein engineering. Thus, engineering a dipolar binding site into 1F7 via site-directed mutagenesis is conceivable. Alternatively, the water molecule that makes a crucial interaction with the hapten's secondary carboxylate might be replaced with an amino acid side chain through redesign of the CDR3 loop of the heavy chain. These tasks will be greatly aided by the availability of a biological selection assay that has been developed for 1F7. This catalyst has been shown to function *in vivo,* conferring a substantial growth advantage to a chorismate mutase-deficient yeast strain under auxotrophic conditions [29]. The ability to select for this phenotype provides the means to identify 1F7 derivatives with enhanced potency. Large numbers of variants, generated by random mutagenesis, can be assayed directly for their ability to complement the chorismate mutase deficiency of the host organism. Analogous experiments in engineered strains of *Escherichia coli* lacking chorismate mutase appear feasible as well (P. Kast, unpublished results).

How general are the lessons learned through study of 1F7? Although investigation of the mechanistic and structural basis of antibody catalysis is only beginning, striking parallels between a number of antibodies and analogous enzymes are apparent [30,31]. Moreover, to the extent they have been investigated, the mechanisms of action of most antibody catalysts reflect the structural properties of the transition state analog quite well. In some cases, the mechanisms are actually more complex than envisaged. For example, a remarkably active hydrolytic antibody (43C9) appears to operate via a two-step mechanism involving transient acylation of the catalyst reminiscent of the serine proteases [32], even though it was generated with a simple phosphonamidate transition state analog [33]. As more crystal structures of catalytic antibodies become available, rules that make one hapten better than another may become apparent. Elucidation of such rules will be essential if highly efficient first-generation antibody catalysts are to be created as a routine exercise.

3. Controlling chemical reactivity and selectivity with antibodies

The power of the catalytic antibody approach for creating new enzymes is its generality. Virtually any reaction should be susceptible to antibody catalysis, as long as it can be carried out in a protein microenvironment in aqueous buffer and its energy barrier for converting substrates to products does not exceed the ca. 10 to 20 kcal mol^{-1} of binding energy available from an antibody. As in enzymes, substrate destabilization, proximity effects, nucleophilic and general acid-base catalysis, and catalytic cofactors can all be exploited to help reduce the activation barrier for many different reaction types.

The importance of substrate destabilization is illustrated by the antibody-catalyzed decarboxylation of 3-carboxybenzisoxazoles (**5**) to give salicylonitriles (**7**) [34]. This reaction is extraordinarily sensitive to its solvent microenvironment, with rate enhancements up to 10^8-fold observed upon transfer of the reactant from aqueous buffer to aprotic dipolar solvents [35]. Desolvation of the negatively charged carboxylate group greatly destabilizes the substrate, while the charge delocalized transition state (**6**) may be stabilized by dispersion interactions with solvent. Similar factors are believed

5 **6** **7** **8**

to contribute to the efficiency of a number of decarboxylase enzymes. The 1,5-naphthalene disulfonate hapten **8** was used to generate antibodies capable of catalyzing the decarboxylation of 5-nitro-3-carboxybenzisoxazole [34]. Rate accelerations up to 2 x 10^4 over background were observed, and because the model reaction is completely insensitive to general acid-base chemistry, the catalytic advantage provided by these proteins can be ascribed entirely to medium effects. In essence, the antibody provides a 'rigid' and chemically well-defined solvent microenvironment for investigating how dipolarity, electrostatics and hydrogen bonding influence chemical reactivity. Kinetic and spectroscopic studies [34, 36] suggest that the hapten's large aromatic ring and negatively-charged substituents yielded an appropriately-sized apolar pocket containing positively-charged residues for sequestering the anionic species that occur along the reaction coordinate.

B: H N O 9 — [B H N O 10]‡ → CN O⁻ 7 + BH⁺ H₃C H₃C H N⁺ NH₂ N COOH 11

Many important chemical transformations, including aldol condensations, S_N2 substitutions, E2 eliminations, and even some hydrolytic reactions, are solvent sensitive and should be highly susceptible to antibody catalysis. The base-promoted deprotonation of benzisoxazoles **9** to give salicylonitriles **7** [37] has recently been used as a model system to probe how microenvironmental factors influence general base catalysis in an antibody active site [38]. The benzimidazolium hapten **11** (pK_a = 7.8) was designed to capitalize on charge complementarity, which has been exploited in a number of instances to induce general acids or bases in an antibody binding pocket. Because it is planar, **11** also correctly mimics all reacting bond geometries of the transition state **10**. Hapten **11** was an unusually effective hapten, eliciting highly active antibody catalysts. The activity of antibody 34E4 depends on an active site base, which chemical modification studies have identified as a carboxylate with a perturbed pK_a. The rate acceleration observed for this catalyst is in excess of 10^8 $[(k_{cat}/K_m)/k_{AcO^-}]$, with an effective molarity for the active site carboxylate of ca. 40,000 M. By way of contrast, effective molarities rarely exceed 10 M for intramolecular general base catalysis in model systems or for general catalysis in other antibodies. Moreover, 34E4 achieves more than 10^3 turnovers per active site as a consequence of the dissimilarity in shape and charge of product **7** and hapten **11**. The extraordinary activity of 34E4 shows that very large effects can be achieved by strategic use of haptenic charge.

Bringing two molecules together in the proper orientation for reaction — catalysis by approximation — is one of the easiest things an antibody can do. As the benzisoxazole elimination shows, large rates can be achieved by juxtaposing a substrate and a catalytic group. Bimolecular Diels-Alder reactions are also susceptible to proximity effects. Antibody catalysis of cycloadditions [39-41] represents a particularly significant accomplishment, given that analogous enzymes are as yet unknown. The boatlike transition state geometry for these processes has been successfully mimicked with bicyclic compounds. For example, the hexachloronorbornene derivative **17** resembles the transition state for the cycloaddition between **12** and **13** as well as the transition state for the subsequent chelotropic elimination of SO_2. An anti-**17** antibody (1E9) catalyzes this two-step transformation with high efficiency (effective molarities > 10^2 M) and multiple turnovers [39]. The large geometrical change accompanying the reaction minimizes problems associated with product inhibition. This feature was 'programmed' into the catalyst through the design of the transition state analog, which binds to 1E9 five to six orders of magnitude more tightly than substrate (**13**) or products (**15** and **16**), respectively (K. Hill, unpublished results). While structure-function relationships must await detailed crystallographic studies, sequence data show that the antibody is closely related to a family of steroid-binding antibodies [42]. Models based on the structure of an anti-progesterone antibody (whose structure is available [43]) show a hydrophobic active site with multiple polar interactions that could be used to orient the diene and dienophile and facilitate breakdown of the unstable intermediate **14**. The development [43] of a powerful immunoassay (catELISA) for this reaction in which catalytic activity can be detected directly in tissue culture supernatant has led to the discovery of even more potent catalysts than 1E9 and provides the means to systematically screen large numbers of alternative scaffolds.

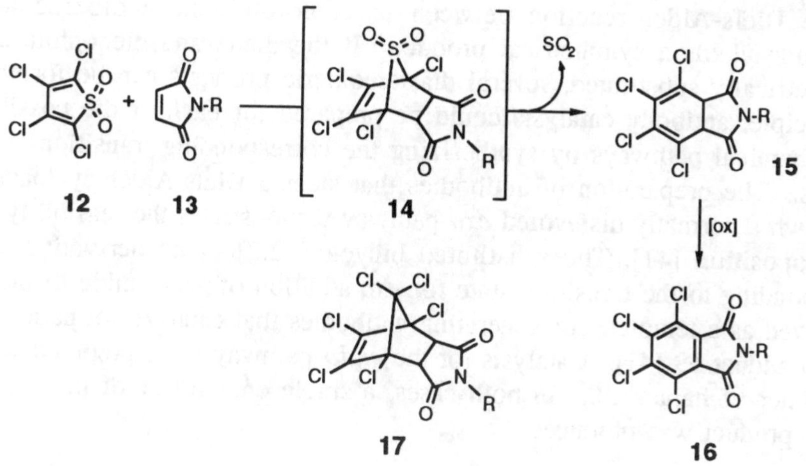

The Diels-Alder reaction between tetrachlorothiophene dioxide and maleimide yields a symmetrical product. If the diene and dienophile are asymmetrically substituted, several diastereomeric products can be formed. In principle, antibody catalysts could be prepared for each of the possible stereochemical pathways by synthesizing the corresponding transition state analogs. The preparation of antibodies that steer a Diels-Alder cycloaddition down a normally disfavored *exo* pathway demonstrates the feasibility of this proposition [41]. The substituted bicyclo[2.2.2]octane derivative **20**, corresponding to the transition state for *exo* addition of acrylamide to diene **18**, served as a template for generating antibodies that catalyze formation of the *exo* adduct **19** [41]. Catalysts for the *endo* pathway were prepared with the epimer of hapten **20**. In both cases, a single enantiomer of the cyclohexene product was produced.

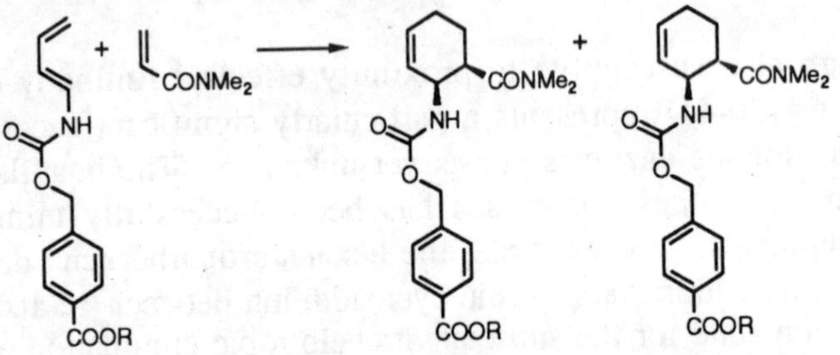

18 **19** *endo* **19** *exo*

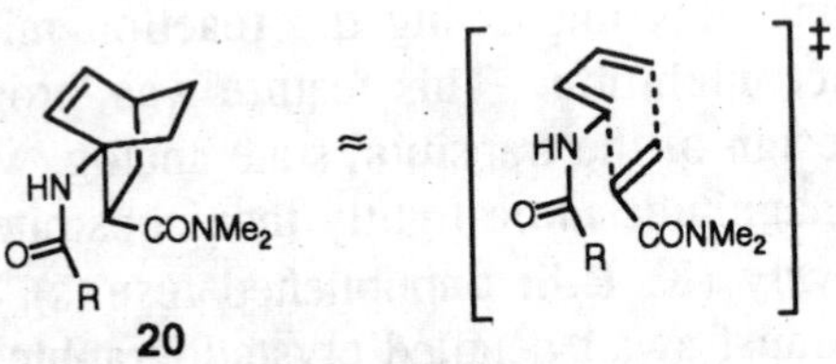

20

The control of chemical selectivity is a classical problem in organic chemistry and an antibody's ability to route reactions via disfavored pathways, as in the asymmetric Diels-Alder reaction, has important practical ramifications. Not surprisingly, extension of this approach to the catalysis of other unselective or energetically demanding reactions has become an important focus of research. Syn eliminations [45], cationic olefin cyclizations [46], and enol ether hydrolyses [47] are among the processes that have been investigated. For the purposes of the current discussion, however, the regio- and stereoselective reduction of the diketone **21** [48] illustrates what can be accomplished. In the presence of an antibody raised against *N*-oxide **23**, **21** is reduced by $NaCNBH_3$ in high yield and greater than 175:1 regioselectivity at the *p*-nitrobenzyl-substituted ketone. Comparable selectivity would be difficult or impossible to achieve with more standard chemical methods. The cyclization of oxirane **24** provides another impressive example of how an antibody can alter the balance between competing reaction pathways [49]. In the absence of catalyst, tetrahydrofuran **26** is the exclusive cyclization product, but an antibody raised against the *N*-oxide hapten **27** overwhelms the normal preference for formation of five-membered rings to give enantiomerically pure **25**. Calculations [50] suggest that the catalyst must stabilize the disfavored 6-*endo* transition state by 3.6 kcal/mol more than the 5-*exo* transition state.

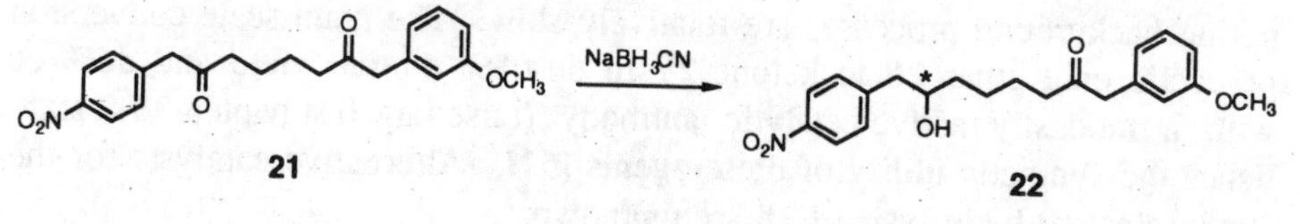

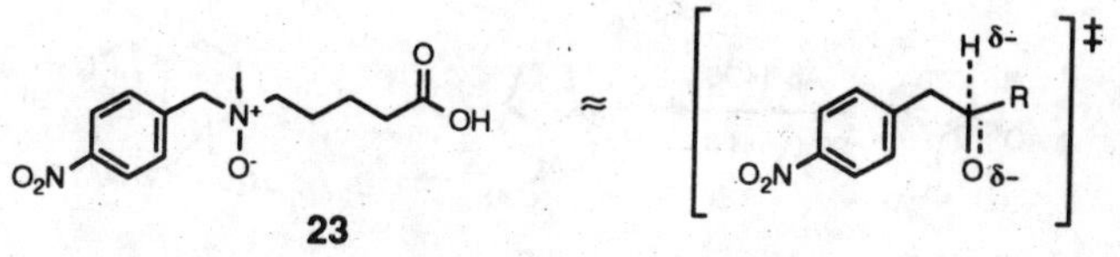

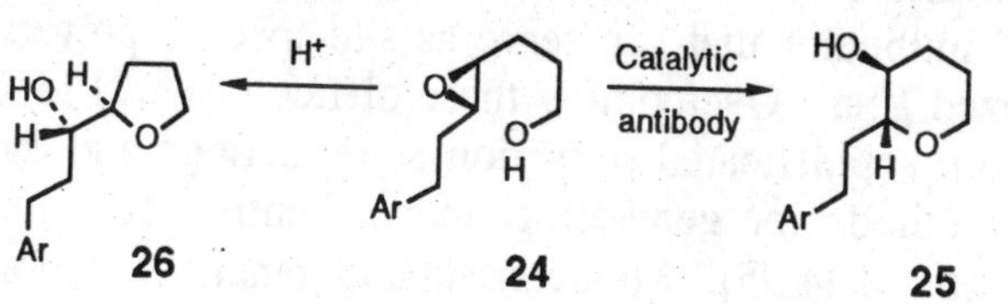

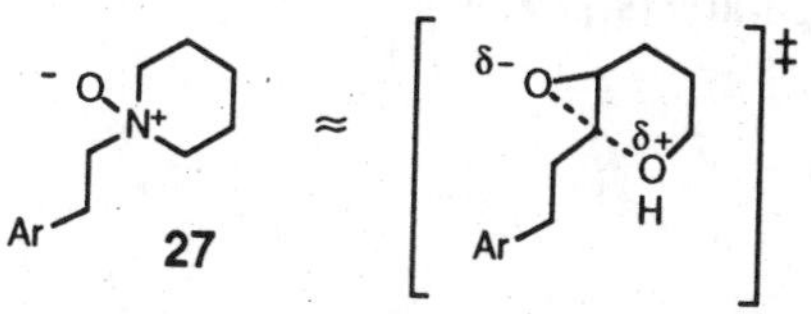

While antibodies are clearly capable of controlling chemical reactivity and selectivity, the low specific activities (compared to enzymes) that are typically observed is of concern and could limit practical applications. However, an antibody-catalyzed process need not be particularly fast to be useful to the synthetic chemist, as long as it is highly selective and the competing background processes are relatively slow. The gram-scale conversion of cyclic enol ether **28** to ketone **29** in 60-65% overall yield and 86% ee with a modestly active catalytic antibody (raised against hapten **30**) establishes the synthetic utility of these agents [51]. Alternative catalysts for the stereoselective hydrolysis of **28** are unknown.

MeO Ar **28** — Ab 14D6, 20°C, pH 6.0 → O Ar **29**

NHR, N+, H_3C, CH_2Ar **30**

The scope of antibody catalysis has clearly expanded dramatically since the discovery of immunoglobulins with modest esterolytic activity in 1986 [52,53]. Nevertheless, many challenges remain. Aldol condensations, S_N2 displacements, and phosphoryl and glycosyl group transfers are among the important reaction types that have not yet been catalyzed by antibodies. Moreover, the strategies employed in the past to generate catalysts for some reactions, like amide hydrolysis, may not be easily generalized. For example, despite the availability of excellent transition state analogs, only two monoclonal antibodies with amidase activity have been prepared to date [33,54]. The difficulties encountered may reflect the intrinsic stability of the amide bond and the low probability of eliciting an effective constellation of catalytic groups within the antibody combining site. Whatever the reason, the production of antibodies that can serve as site-specific proteases remains an as yet unrealized goal. Overcoming these difficulties will require the development of more sophisticated transition state analogs and the implementation of new methods for generating and screening large immunoglobulin libraries [see e.g. 28,44,55]. Mutagenesis and genetic selection, on the other hand, are potentially powerful tools for optimizing the activity of first-generation antibody catalysts [29,56].

4. Outlook

1994 was the centennial anniversary of Emil Fischer's famous 'lock and key' description of enzyme action [57-60]. This beautiful metaphor can also be applied to the field of catalytic antibodies in which a rationally designed transition state analog is used as a key to unlock the catalytic potential of the mammalian immune system. This simple yet elegant approach to tailored catalysts complements other strategies for creating new enzymes, including the *de novo* synthesis of proteins, the reengineering of existing enzymes through site-directed mutagenesis and chemical modification, ribozyme selection, and the chemical synthesis of low molecular weight macrocyclic or cleft-like molecules. The generation of catalytic antibodies is an inherently multidisciplinary undertaking, involving chemical synthesis, immunology, molecular biology and genetics. By successfully integrating these various disciplines, increasingly active catalysts for increasingly difficult chemical transformations may become available on demand.

5. Acknowledgments

Research in the author's laboratory has been generously supported by grants from the National Institutes of Health, the American Cancer Society, the National Science Foundation, the Army Research Office, and the Office of Naval Research.

6. References

1. Pressman, D.; Grossberg, A. (1968) *The Structrual Basis of Antibody Specificity*, Benjamin, New York.
2. Nisonoff, A., Hopper, J., and Spring, S. (1975) *The Antibody Molecule*, Academic Press, New York.
3. Kabat, E.A. (1976) *Structural Concepts in Immunology and Immunochemistry*, Holt, Reinhart and Winston, New York.
4. Alt, F.W., Blackwell, T.K., and Yancopoulos, G.D. (1987) Development of the primary antibody repertoire, *Science* **238**, 1079.
5. Rajewsky, K., Förster, I., and Cumang, A. (1987) Evolutionary and somatic selection of the antibody repertoire in the mouse, *Science* **238**, 1088.
6. Lerner, R.A., Benkovic, S.J., and Schultz, P.G. (1991) At the cross-roads of chemistry and immunology: Catalytic antibodies, *Science* **252**, 659.
7. Davies, D.R., Padlan, E.A., and Sheriff, S. (1990) Antibody-antigen complexes, *Ann. Rev. Biochem.* **59**, 439.
8. Wilson, I.A. and Stanfield, R.L. (1993) Antibody-antigen interactions, *Current Opinion in Struct. Biol.* **3**, 113.

9. Jencks, W.P. (1969) *Catalysis in Chemistry and Enzymology*; McGraw-Hill: New York, p. 288.
10. Pauling, L. (1948) Chemical achievement and hope for the future, *Am. Sci.* **36**, 51.
11. Andrews, P.R., Smith, G.D., and Young, I.G. (1973) Transition-state stabilization and enzymic catalysis. Kinetic and molecular orbital studies of the rearrangement of chorismate to prephenate, *Biochemistry* **12**, 3492.
12. Sogo, S.G., Widlanski, T.S., Hoare, J.H., Grimshaw, C.E., Berchtold, G.A., and Knowles, J.R. (1984) Stereochemistry of the rearrangement of chorismate to prephenate: Chorismate mutase involves a chair transition state, *J. Am. Chem. Soc.* **106**, 2701.
13. Andrews, P.R., Cain, E.N., Rizardo, E., and Smith, G.D. (1977) Rearrangement of chorismate to prephenate. Use of chorismate mutase inhibitors to define the transition state structure, *Biochemistry* **16**, 4848.
14. Addadi, L., Jaffe, E.K., and Knowles, J.R. (1984) Secondary tritium isotope effects as probes of the enzymic and nonenzymic conversion of chorismate to prephenate, *Biochemistry* **22**, 4494.
15. Copley, S.D. and Knowles, J.R. (1987) The conformational equilibrium of chorismate in solution: Implications for the mechanism of the non-enzymic and the enzyme-catalyzed rearrangement of chorismate to prephenate, *J. Am. Chem. Soc.* **109**, 5008.
16. Severence, D.L. and Jorgensen, W.L. (1992) Effects of hydration on the Claisen rearrangement of allyl vinyl ether from computer simulations, *J. Am. Chem. Soc.* **114**, 10966.
17. Westheimer, F.H. (1962) Mechanisms related to enzyme catalysis, *Adv. Enzymol.* **24**, 441.
18. Görisch, J. (1978) On the mechanism of the chorismate mutase reaction, *Biochemistry* **17**, 3700.
19. Bartlett, P.A., Nakagawa, Y., Johnson, C.R., Reich, S.H., and Luis, A. (1988) Chorismate mutase inhibitors: Synthesis and evaluation of some potential transition-state analogues, *J. Org. Chem.* **53**, 3195.
20. Hilvert, D., Carpenter, S.H., Nared, K.D., and Auditor, M.-T.M. (1988) Catalysis of concerted reactions by antibodies: The Claisen rearrangement, *Proc. Natl. Acad. Sci. USA* **85**, 4953.
21. Jackson, D.Y., Jacobs, J.W., Sugasawara, R., Reich, S.H., Bartlett, P.A., and Schultz, P.G. (1988) An antibody-catalyzed Claisen rearrangement, *J. Am. Chem. Soc.* **110**, 4841.

22. Hilvert, D. and Nared, K.D. (1988) Stereospecific Claisen rearrangement catalyzed by an antibody, *J. Am. Chem. Soc.* **110**, 5593.
23. Jackson, D.Y., Liang, M.N., Bartlett, P.A. and Schultz, P.G. (1992) Activation parameters and stereochemistry of an antibody-catalyzed Claisen rearrangement, *Angew. Chem. Int. Ed. Engl.* **31**, 182.
24. Campbell, A.P., Tarasow, T.M., Massefski, W., Wright, P.E., and Hilvert, D. (1993) Binding of a high-energy substrate conformer in antibody catalysis, *Proc. Natl. Acad. Sci. USA* **90**, 8663.
25. Haynes, M.R., Stura, E.A., Hilvert, D., and Wilson, I.A. (1994) Routes to catalysis: Structure of a catalytic antibody and comparison with its natural counterpart, *Science* **263**, 646.
26. Chook, Y.M., Ke, H., and Lipscomb, W.N. (1993) Crystal structures of the monofunctional chorismate mutase from *Bacillus subtilis* and its complex with a transition state analog, *Proc. Natl. Acad. Sci. USA* **90**, 8600.
27. Dunitz, J. (1994) The entropic cost of bound water in crystals and biomolecules, *Science* **264**, 670.
28. Garrard, L.J. and Zhukovsky, E.A. (1992) Antibody expression in bacteriophage systems: The future of monoclonal antibodies? *Current Opinion in Biotechnology* **3**, 474.
29. Tang, Y., Hicks, J.B., and Hilvert, D. (1991) *In vivo* catalysis of a metabolically essential reaction by an antibody, *Proc. Natl. Acad. Sci. USA* **88**, 8784.
30. Stewart, J.D., Liotta, L.J., and Benkovic, S.J. (1993) Reaction mechanisms displayed by catalytic antibodies, *Acc. Chem. Res.* **26**, 396.
31. Golinelli-Pimpaneau, B., Gigant, B., Bizebard, T., Navazá, J., Saludjian, P., Zemel, R., Tawfik, D.S., Eshhar, Z., Green, B.S., and Knossow, M. (1994) Crystal structure of a catalytic antibody Fab with esterase-like activity, *Structure* **2**, 175.
32. Stewart, J.D., Roberts, V.A., Thomas, N.R., Getzoff, E.D., and Benkovic, S.J. (1994) Site-directed mutagenesis of a catalytic antibody: An arginine and a histidine residue play key roles, *Biochemistry* **33**, 1994, and references therein.
33. Janda, K.D., Schloeder, D., Benkovic, S.J., and Lerner, R.A. (1988) Induction of an antibody that catalyzes the hydrolysis of an amide bond, *Science* **241**, 1188.
34. Lewis, C., Krämer, T., Robinson, S., and Hilvert, D. (1991) Medium effects in antibody-catalyzed reactions, *Science* **253**, 1019.

35. Kemp, D.S. and Paul, K.G. (1975) The physical organic chemistry of benzisoxazoles. III. The mechanism and the effects of solents on rates of decarboxylation of benzisoxazole-3-carboxylic acids, *J. Am. Chem. Soc.* **97**, 7305.
36. Tarasow, T.M., Lewis, C., and Hilvert, D. (1994). Investigation of medium effects in a family of decarboxylase antibodies. *J. Am. Chem. Soc.* **116**, 7959.
37. Casey, M.L., Kemp, D.S., Paul, K.G., and Cox, D.D. (1973) The physical organic chemistry of benzisoxazoles. I. The mechanism of the base-catalyzed decomposition of benzisoxazoles, *J. Org. Chem.* **38**, 2294.
38. Thorn, S. N., Daniels, R. G., Auditor, M.-T. M. and Hilvert, D. (1995). Large rate accelerations in antibody catalysis by strategic use of haptenic charge. *Nature* **373**, 228.
39. Hilvert, D., Hill, K.W., Nared, K.D., and Auditor, M.-T.M. (1989) Antibody catalysis of a Diels-Alder reaction, *J. Am. Chem. Soc.* **111**, 9261.
40. Braisted, A.C. and Schultz, P.G. (1990) An antibody-catalyzed bimolecular Diels-Alder reaction, *J. Am. Chem. Soc.* **112**, 7430.
41. Gouverneur, V.E., Houk, K.N., Pascual-Teresa, B., Beno, B., Janda, K.D., and Lerner, R.A. (1993) Control of the exo and endo pathways of the Diels-Alder reaction by antibody catalysis. *Science* **262**, 204.
42. Hilvert, D. (1993). Antibody catalysis of carbon-carbon bond formation and cleavage. *Acc. Chem. Res.* **26**, 552.
43. Arevalo, J.H., Stura, E.A., Taussig, M.J., and Wilson, I.A. (1993) Three-dimensional structure of an anti-steroid Fab' and progesterone-Fab' complex, *J. Mol. Biol.* **231**, 103.
44. MacBeath, G. and Hilvert, D. (1994) Monitoring catalytic activity by immunoassay: Implications for screening, *J. Am. Chem. Soc.* **116**, 6101.
45. Cravatt, B.F., Ashley, J.A., Janda, K.D., Boger, D.L., and Lerner, R.A. (1994) Crossing extreme mechanistic barriers by antibody catalysis: Syn elimination to a cis olefin, *J. Am. Chem. Soc.* **116**, 6013.
46. Li, T., Janda, K.D., Ashley, J.A., and Lerner, R.A. (1994) Antibody catalyzed cationic cyclization, *Science* **264**, 1289.
47. Sinha, S.C., Keinan, E., and Reymond, J.-L. (1993) Antibody-catalyzed reversal of chemoselectivity, *Proc. Natl. Acad. Sci. USA* **90**, 11910.
48. Hsieh, L.C., Yonkovich, S., Kochersperger, L., and Schultz P.G. (1993) Controlling chemical reactivity with antibodies, *Science* **260**, 337.

49. Janda, K.D., Shevlin, C.G., and Lerner, R.A. (1993) Antibody catalysis of a disfavored chemical transformation, *Science* **259**, 490.
50. Na, J., Houk, K.N., Shevlin, C.G., Janda, K.D., and Lerner, R.A. (1993) The energetic advantage of 5-exo versus 6-endo epoxide openings: A preference overwhelmed by antibody catalysis, *J. Am. Chem. Soc.* **115**, 8453.
51. Reymond, J.-L., Reber, J.-L., and Lerner, R.A. (1994) Enantioselective, multigram-scale synthesis with a catalytic antibody, *Angew. Chem. Int. Ed. Engl.* **33**, 475.
52. Tramontano, A., Janda, K.D., and Lerner, R.A. (1986) Catalytic antibodies, *Science* **234**, 1566.
53. Pollack, S.J., Jacobs, J.W., and Schultz, P.G. (1986) Selective chemical catalysis by an antibody, *Science* **234**, 1570.
54. Iverson, B.L. and Lerner, R.A. (1989) Sequence-specific peptide cleavage catalyzed by an antibody, *Science* **243**, 1184.
55. Tawfik, D.S., Green, B.S., Chap, R., Sela, M., and Eshhar, Z. (1993) catELISA: A facile general route to catalytic antibodies, *Proc. Natl. Acad. Sci. USA* **90**, 373.
56. Lesley, S.A., Patten, P.A., and Schultz, P.G. (1993) A genetic approach to the generation of antibodies with enhanced catalytic activities. *Proc. Natl. Acad. Sci. USA* **90**, 1160.
57. Fischer, E. (1894) Einfluss der Configuration auf die Wirkung der Enzyme, *Ber. Dt. Chem. Ges.* **27**, 2985.
58. Eschenmoser, A. (1994) One hundred years lock-and-key principle. *Angew. Chem. Int. Ed. Engl.* **33**, 2363.
59. Lichtenthaler, F.W. (1994) 100 Years "Schlüssel-Schloss-Prinzip": What made Emil Fischer use this analogy? *Angew. Chem. Int. Ed. Engl.* **33**, 2364.
60. Koshland, D.E., Jr. (1994) The Key-Lock Theory and the Induced Fit Theory. *Angew. Chem. Int. Ed. Engl.* **33**, 2375.

New Homocalixarenes and Catenanes

- From Molecular Recognition to Mechanical Bonds

S. MEIER, S. OTTENS-HILDEBRANDT, G. BRODESSER and F. VÖGTLE*
Institut für Organische Chemie und Biochemie der Universität Bonn, Gerhard-Domagk-Straße 1, D-53121 Bonn, Germany

1. Introduction

Molecular recognition is one of the most essential phenomena in biological processes and it is a particular interest of the life sciences to understand such processes. Supramolecular chemistry is a tool for gaining the knowledge about molecular recognition by tailoring synthetic host molecules, which could act as model compounds (perhaps as enzyme mimics). A supramolecular building block concept makes various host architectures feasible and allows the endowment of host molecules with selected functions for rendering novel host properties. Investigations of their guest binding abilities give insight into the correlations of structure and function.
Two types of such host architectures are introduced in this contribution: homocalixarenes and catenanes.

2. Homocalixarenes and -Calixpyridines

2.1. CONCEPT

A host type, which is expected to fulfill demands for enzyme mimic is the well known calixarene [1, 2, 3]. Guided by success and shortcomings of calixarenes, we aimed to design a host which promises the advantages of high selectivity, through cooperative binding, for a spectrum of relevant guests: **Homocalixarenes** [4].
Whereas calixarenes (**I**) are based on a $[1_n]$metacyclophane skeleton, *all*-homocalixarenes (**II**) contain an expanded (c.f.: Sessler:"Expanded Porphyrins" [5]) cavity, which is achieved by synthesis of $[2_n]$metacyclophanes, and yields one additional CH_2-group compared to calixarenes in all their bridges (Fig. 1). In addition, homocalixarenes allow use of pyridines as aromatic units in their ring skeleton: **Homocalixpyridines (III)**.

C. Chatgilialoglu and V. Snieckus (eds.), Chemical Synthesis, 361–379.

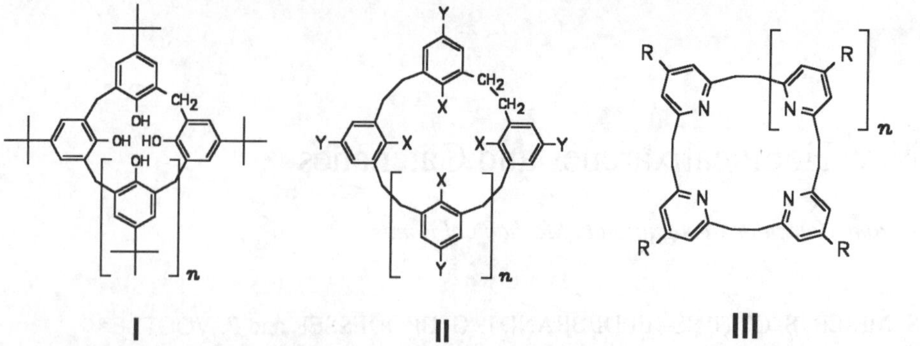

Fig. 1. Structural features of calixarenes (I) versus homocalixarenes (II) and homocalixpyridines (III).

For the development of homocalixarenes, we attempted to incorporate the following requirements:

- high guest selectivity by tailoring:
  appropriate cavity size ⇒ for steric host/guest fit
  different binding sites ⇒ for electronic fit of charged guests and neutral/ acidic/basic guests
  ⇒ appropriate parameters for charge, dipole moment, hydrogen bonds, lipophilicity
- conformative flexibility correlated with preorganization
- good solubility in organic solvents
- stability in reaction environment
- anchor groups for immobilization and units which give characteristic signals

Using these criteria, a chemically stable, large ring of appropriate cavity size is constructed whose functional groups may be modified even by the use of vigorous conditions. For tailoring, the host molecule is endowed with various binding sites inside or outside the cavity (Fig. 2). The near to planar ring skeleton is intended to provide greater solubility compared to the calixarenes, which is possible by an alternating arrangement of rigid aromatic and flexible aliphatic moieties.

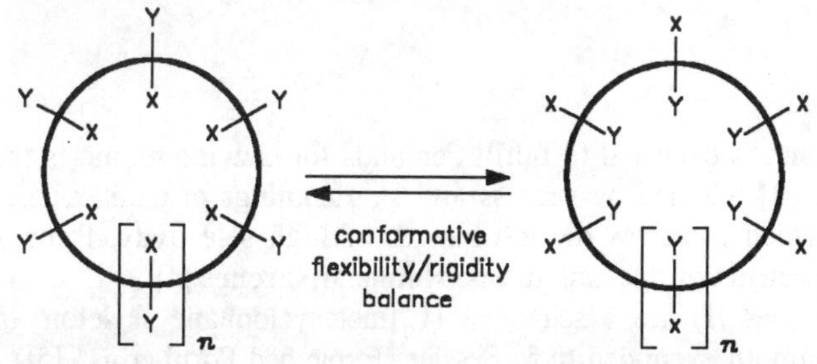

Fig. 2. Carbocyclic *all*-homocalixarenes with variable ring size and functionalities X, Y inside and outside (schematic).

Unfavourable rigid conformers (which perhaps could not enclose guests) are thereby avoided as mixtures of stable conformers which would require characterization.
Although cavity size is generally defined by the number of aromatic units, ligand arms fixed inside the cavity may be tailored to assist cooperative guest binding. This may be

carried out by adjusting steric and electronic host/guest fit in a manner precedented by natural ionophores (Fig. 3).

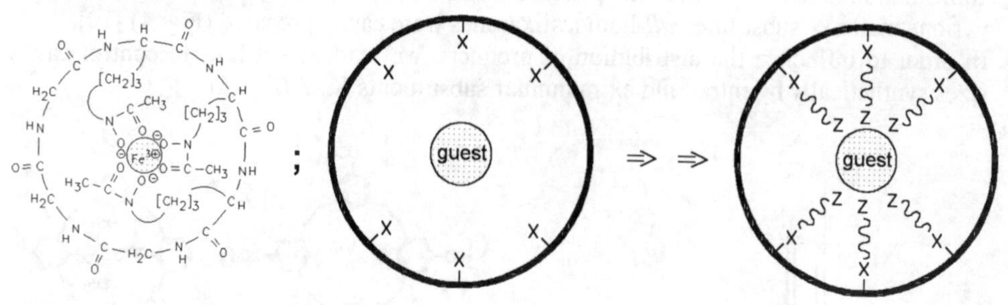

Fig. 3. Ferrichrome Fe $^{3\oplus}$ complex as example from nature for lariat type complexon (left side), adjustment of steric and electronic host/guest complementarity in *all*-homocalixarenes (right side).

2.2. SYNTHESIS

Of the multitude of available cyclization methods, that of Müller-Röscheisen [6] constitutes the method of choice because of its versatility. As shown, with this method, both $[2_n]$metacyclophanes [7] and $[2_n]$pyridinophanes [8, 9] with various ring sizes are accessible (Fig. 4).

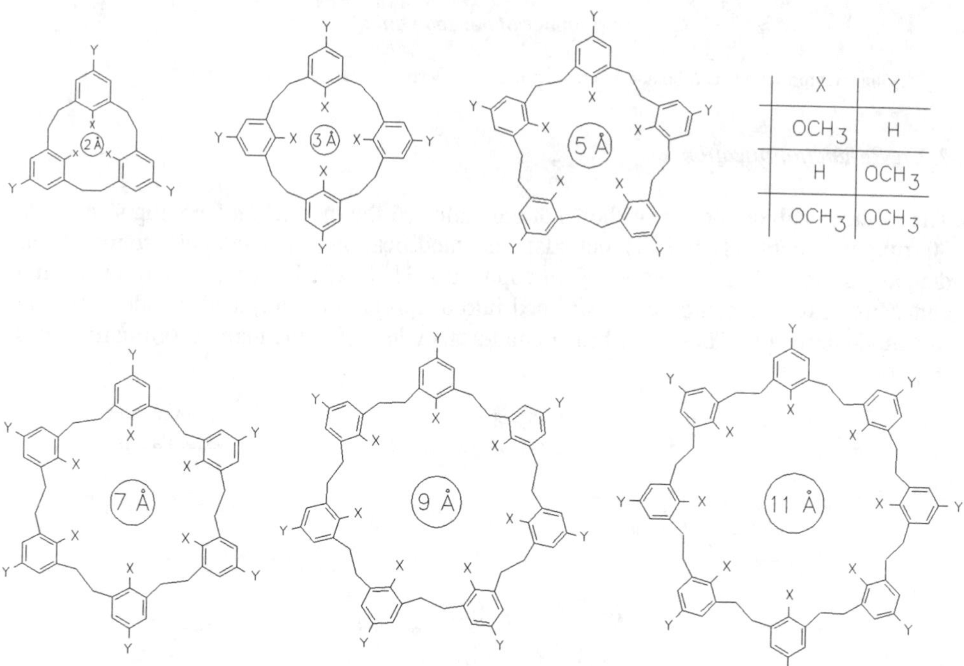

X	Y
OCH_3	H
H	OCH_3
OCH_3	OCH_3

Fig. 4. A spectrum of eligible cavity sizes feasible by Müller-Röscheisen method (longest O-O distances estimated for X = OCH_3, Y = H).

As starting compounds for introducing additional heteroatoms for individual coordination sites for guest complexation from different bis(bromomethyl)benzenes, various methoxy substituted *all*-homocalixarenes were easily prepared (Fig. 4) [10].
In order to influence the distribution of products, we worked out how to control cavity sizes synthetically by intra- and extraannular substituents X, Y (Fig. 5).

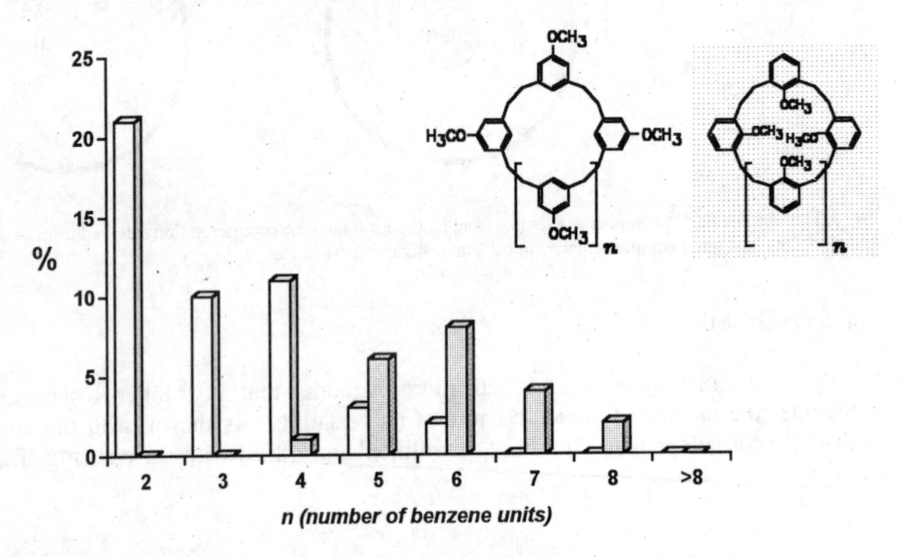

Fig. 5. Oligomer formation in *all*-homocalixarenes synthesis is steered by intraannular substituents.

2.2.1. Refunctionalization

The *all*-homocalixarene type of host not only allowed the modification of ring size (10 to 80 ring members, cf. Fig. 4) but also the modification of functional groups. Their demethylation led to a series of oligophenols [11] which,. aside from exhibiting endoacidic cavities, could be transformed into oxapropionic ester, acid, amide [12], and thioamide derivates. These now bear ligand-arms which provide many coordination sites (Fig. 6).

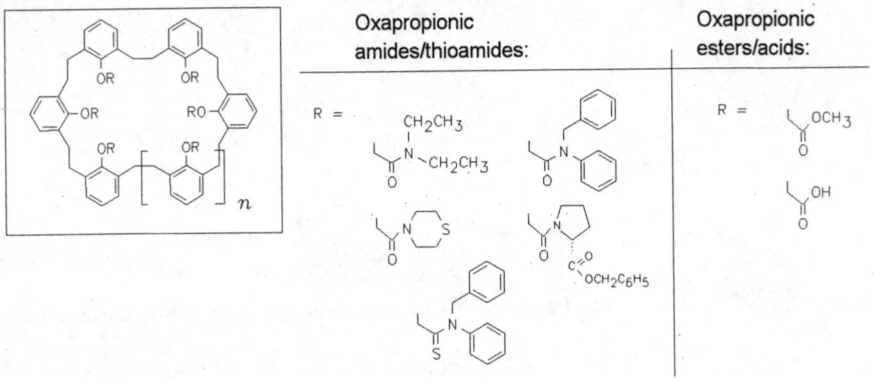

Fig. 6. Introduction of various binding sites into *all*-homocalixarenes.

2.3. *ALL*-HOMOCALIXPYRIDINES

In contrast to calixarenes, *all*-homocalixarenes allow a choice between endoacidic and endobasic cavities (Fig. 7), because cyclization of pyridine units is possible, too. The binding strength of the hosts containing 4-OR substituted pyridines are increased over the unsubstituted systems.

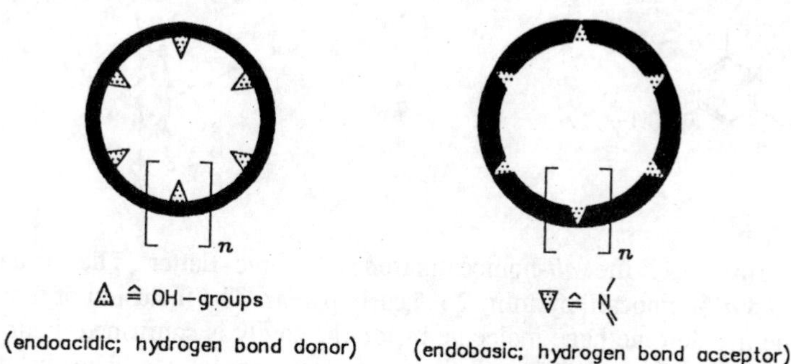

Fig. 7. Schematic diagram of endoreceptors fitted with acidic or basic donor sites.

Distribution of products in the case of cyclization of 2,6-bis(bromomethyl)-4-methoxypyridine depends mainly on dilution and reaction temperature. At temperatures about -100 to -90°C formation of open-chained by-products is decreased. Using high dilution leads to odd-numbered ring major products like trimer **1** (12%) and pentamer **2** (2%) [13]. Increase of concentration produces a broader spectrum of cyclooligomers.

1 **2**

2.4. CRYSTALLOGRAPHIC STUDIES

X-ray structure analyses have not been obtained for unsubstituted $[2_n]$metacyclophanes and $[2_n]$pyridinophanes. Evidence for conformational characteristics was only revealed by NMR-data. Single crystals suitable for crystallographic studies were obtained from *all*-homocalixarenes as well as from *all*-homocalixpyridines of various ring sizes.

In the trimeric, extraannular trimethoxy *all*-homocalixarene and -calixpyridine, the aromatic rings do not form a plane but show the 'partial cone' conformation (Fig. 8). The distances of the nitrogen atoms projecting into the 15-membered ring of **1** range from 408.4 pm to 356.1 pm.

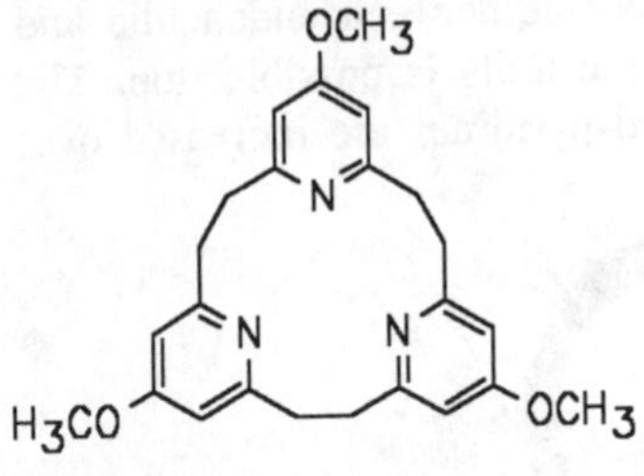

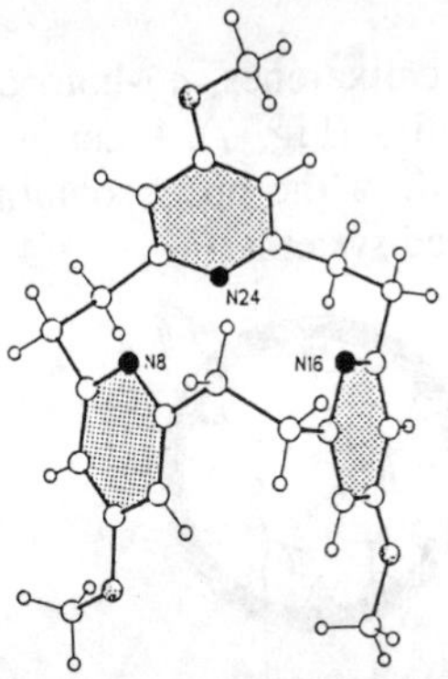

Fig. 8. Crystal structure of **1**.

With increasing ring size, the *all*-homocalixarenes become flatter. The pentameric methoxy substituted *all*-homocalixpyridine **2** is nearly planar. The inclusion of five water molecules and one trichloromethane molecule inside the cavity is confirmed. Four of the water molecules form hydrogen bridges between two nitrogen atoms. The distance of N1A···N1D with a value of 856 pm is twice that of the largest N-N distance in trimer **1** (Fig. 9).

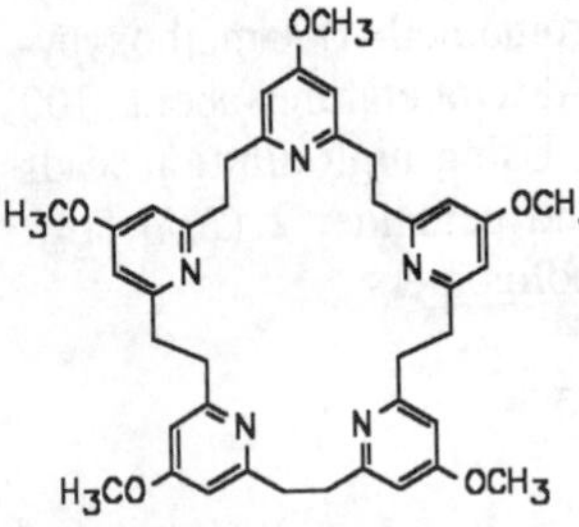

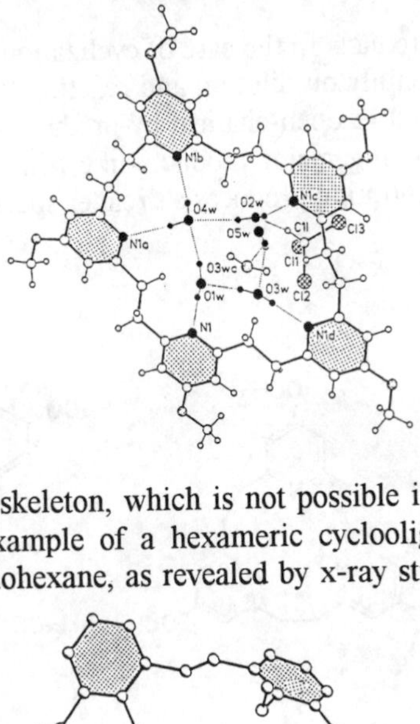

Fig. 9. Crystal structure **2**.

The planarity of this macrocyclic ring skeleton, which is not possible in the calixarene series, is also demonstrated by the example of a hexameric cyclooligomer **3**, which unfolds allowing guest binding of cyclohexane, as revealed by x-ray structural analysis (Fig. 10).

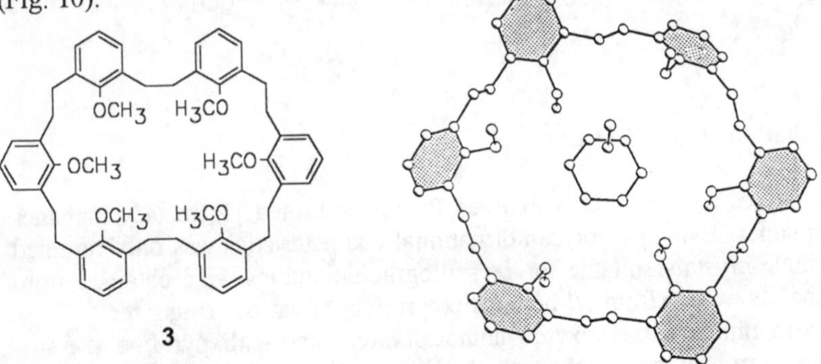

Fig. 10. Crystal structures of a near to planar *all*-homocalix[6]arene 3.

Hitherto no x-ray structure analysis of any heptameric *all*-homocalixarene has been obtained. The highest cyclooligomer studied by crystallography is the octamer **4**. All hydroxy groups are oriented inwards (Fig. 11). Each *all*-homocalixarene molecule encapsulates one ethanol and one cyclohexane molecule. Whereas four of the hydroxy groups of one host molecule are free, the other four form hydrogen bridges with a disoriented ethanol molecule.

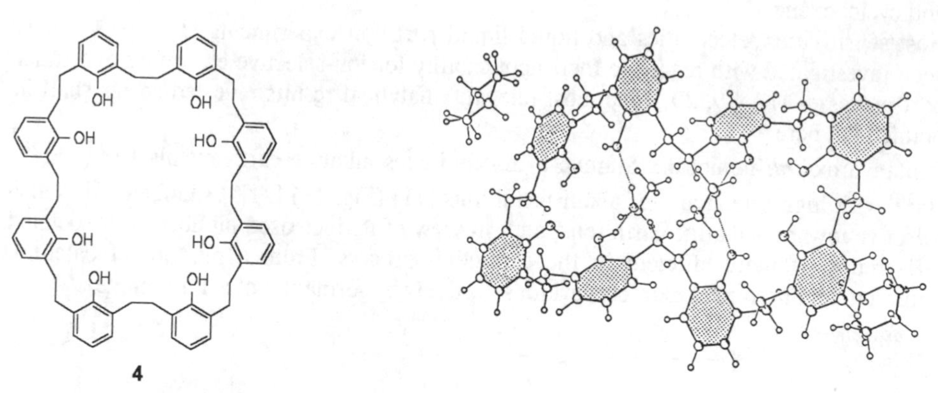

Fig. 11. Crystal structure of **4**.

2.5. RIGID *ALL*-HOMOCALIXARENES

Conformational flexibility is important in particular for transport processes where fast kinetics of complexation/decomplexation is desired. However, the rigidity of the *all*-homocalixarenes may be tailored by capping the cavity in order to fine tune them for specific guests.

Triphenol **5** was capped with tribromo compound **6** by intermolecular reaction in DMF under high dilution conditions and yielded the capped *all*-homocalixarene **7** [14].

Preliminary results of the x-ray structure analysis are shown in Fig. 12. A trichloromethane molecule is oriented over one of the benzene rings in a distance of 338 pm in the crystal [14].

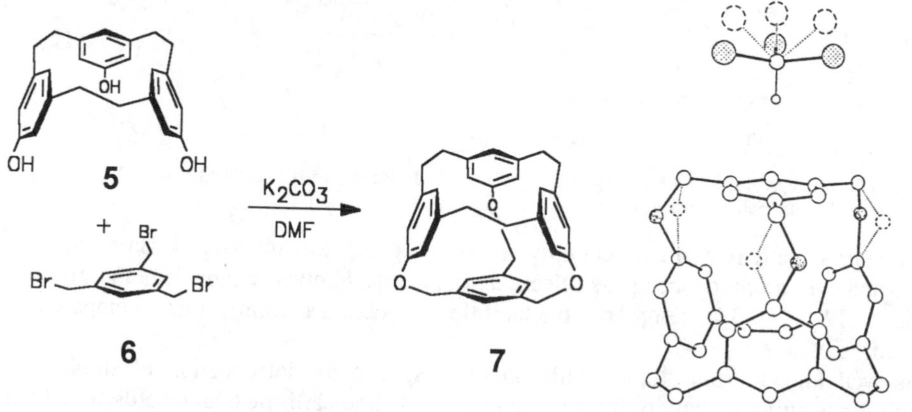

Fig. 12. Preliminary crystal structure of **7**.

2.6. HOST PROPERTIES

Some decisive indications concerning *all*-homocalixarenes and -calixpyridines to bind guests are already given by the results of x-ray structure analyses. The cluster type inclusion of water by the pentameric pyridine macrocycle **2** via multiple hydrogen bridges seems an exceptional case. Oligophenol **4** allows guest binding towards ethanol and cyclohexane.

Host sensitivities/selectivities and liquid-liquid partition experiments [15, 16, 17] have been investigated with regard to their applicability for ion-selective electrodes and transport processes [18, 19, 20]. The most relevant analytical results revealed so far shall be pointed out here.

Pentahydroxy-*all*-homocalix[5]arene **8** discriminates alkaline earth metals $Ca^{2\oplus}$, $Sr^{2\oplus}$, $Ba^{2\oplus}$ with high selectivity for alkali metal ions [11] (Fig. 13) [21]. Octaphenol **4** is also selective towards alkaline earth ions, and, in view of its increased number of dissociated OH-groups, is more efficient in the extraction process. From a mixture of calcium, strontium and barium, **4** extracts barium selectively by formation of a 1:1 complex.

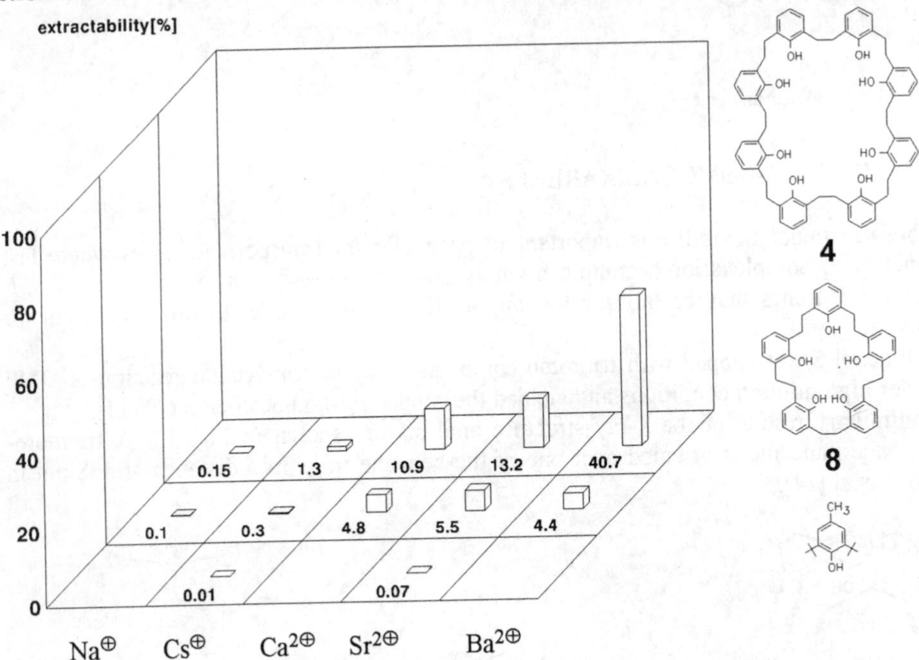

Fig. 13. Extraction properties of **4** and **8**. $[M(NO_3)_n] = 1x10^{-4}$ M; $[NaNO_3] = 0.1$ M; $[NaOH] = 0.1$ M; $[ligand] = 5x10^{-3}$ in trichloromethane.

To favour selectivity and extractability, an amido group was introduced. $Ba^{2\oplus}$ ions are extracted with a pronounced peak selectivity in contrast to other monovalent and bivalent cations [12] (Fig. 14). (Higher extractabilities are obtained from toluene compared to trichloromethane).

It is well known from studies with calixarenes, that the introduction of sulphur as binding site shifts selectivity from alkaline and alkaline earth metals towards transition metals [22]. Recent results provide evidence for selectivity towards $Pd^{2\oplus}$ and $Hg^{2\oplus}$ with a thiomorpholine substituted hexaamido *all*-homocalixarene.

all-Homocalixpyridines have also been investigated regarding their host properties. With pyridine nitrogen atoms as a preorganized binding site [23], *all*-homocalixpyridines show a commendable selectivity towards transition metals, in particular $Ag^{\oplus}$, $Hg^{2\oplus}$, $Cu^{2\oplus}$, $Pd^{2\oplus}$, which may be tuned by cavity size and by 4-substituents in the pyridine ring. Liquid-liquid extraction results compared to those of analogous open chained ligands gave evidence for a "macrocyclic effect".

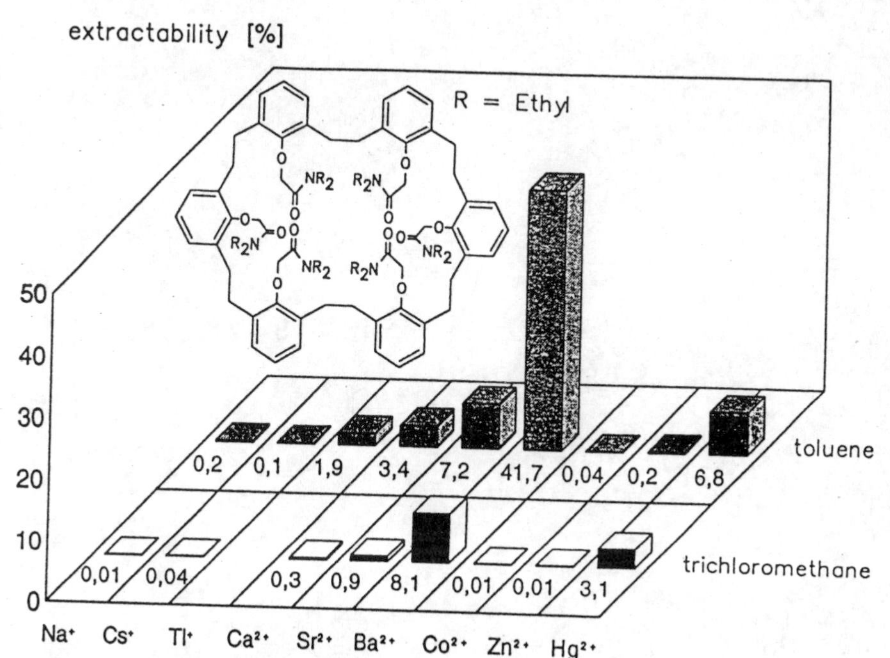

Fig. 14. Extractability of metal ions with *all*-homocalixarene **9**. $[M(NO_3)_n] = 1x10^{-4}$ M; [picric acid] $= 5x10^{-3}$ M; [ligand] $= 1x10^{-3}$ in trichloromethane or toluene.

3. Catenanes

During the last thirty years, many attemps have been made to synthesize aesthetically attractive molecules and topologically novel compounds. Frisch and Wassermann introduced the concept of topological isomerism and coined the term catenanes for molecules composed of interlocking rings [24].

The catenanes and related interlinked compounds have attracted considerable interest, not only because of the stereochemistry, but as important building units for the synthesis of large molecular assemblies.

The early attemps to synthesise catenanes relied on statistical or covalent templating routes, which resulted in relatively poor yields and made catenanes freaks of the laboratory. Later they became preparatively accessible through the use of template effects [25].

3.1 COVALENT TEMPLATE

The first efficient catenane synthesis was developed by Schill and Lüttringhaus who used covalent bonds to direct formation of the interlocked ring system [26].
In the precursor **9** the cyclisation can only occur in a predetermined way to give the catenane **10**.

MeO CHO MeO CHO HO HO $[CH_2]_{25}$ $Cl-[CH_2]_{12}$ $Cl-[CH_2]_{12}$ O O R $[CH_2]_{25}$

: R = H

: R = NO_2

9 : R = NH_2

$[CH_2]_{12}$ O O N $[CH_2]_{12}$ $[CH_2]_{25}$ RO RO N $[CH_2]_{12}$ $[CH_2]_{25}$ C=O $[CH_2]_{12}$

O HO O $[CH_2]_{25}$ HN $[CH_2]_{12}$ $[CH_2]_{12}$ C=O AcO AcO OAc $[CH_2]_{25}$ AcN $[CH_2]_{12}$ $[CH_2]_{12}$ C=O

10

An analogous approach allowed Schill et al. to obtain the first [3]-catenane in moderate yield [27].

3.2 METAL ION TEMPLATE

Sauvage et al. introduced the first template directed catenane synthesis using transition metals. Transition metal ions have the specific capability to coordinate molecules with suitable functional groups. This orthogonalisation is responsible for the three-dimensional template effect [28].
This effect favors the catenane formation with remarkable yields and allowed Sauvage et al. to synthesize impressive types of catenanes and even molecular knots.

Examples in which the synthesis is based on the formation of a complex between copper (I) and two phenanthroline ligands are shown in structures **IV** and **V**.

+ $Cu(CH_3CN)_4^{\oplus}$ +

Cs_2CO_3/DMF

IV

V

3.3 π–DONOR-π-ACCEPTOR INTERACTION TEMPLATE

A molecular template effect was developed by Stoddart et al.. This approach is based on weak intermolecular forces which are used for many extremely efficient syntheses in nature like biological self-assembly and self-replication [29].

In this case, the forces are induced by electron donor-acceptor interactions between a synthetic receptor molecule like bis(phenylene) crown ether (ring) and its substrate in the form of appropriate substituted paraquats (thread).
In the synthesis of Stoddart and coworkers depicted below, the crown ether ring **11** and the paraquat thread **12** "assemble" themselves. Subsequently, the supramolecular complex undergoes cyclisation via a bridging unit to the catenane **13**.

11

12

2 $PF_6^{\ominus}$

Br

Br

4 $PF_6^{\ominus}$

13

3.4 HYDROGEN BONDING AND π-STACKING TEMPLATE

A new template-directed catenane synthesis was published by Hunter in 1992 [30]. Starting from the intermediate **14** and isophthalic acid dichloride **15**, he succeeded in interlocking two identical macrocyclic tetraamides to the catenane type **16** (R=H) in remarkable yield (29 %).

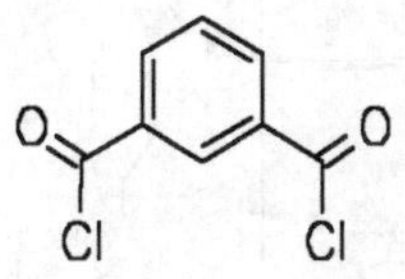

14

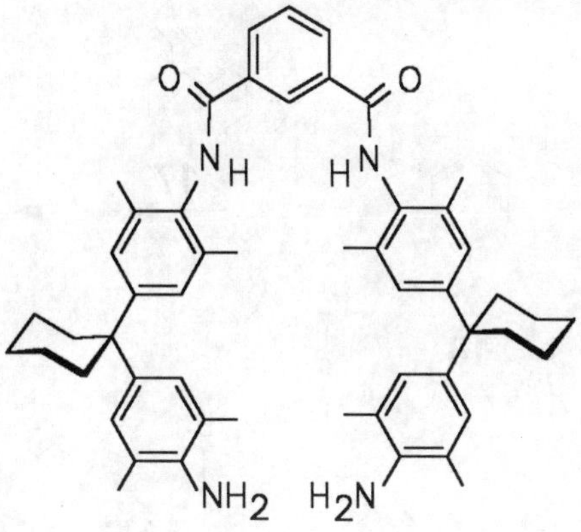

15

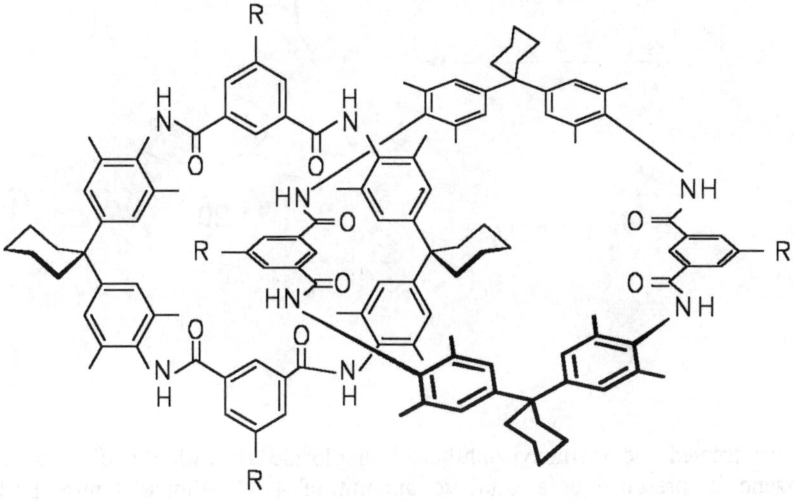

16 (R = H)

Whereas Hunter did not obtain the catenane **16** directly from **14** and **17**, but only after preparation of the intermediate **15**, we independently succeeded in the same year in a one-step synthesis of a substituted representative of the same catenane type **16** (R = OCH_3) starting from the simple components **17** and **18** [31]. In connection with earlier syntheses of basket-shaped host molecules [32], we had applied the cyclohexylidene-substituted diamine **17** [33] instead of unsubstituted diphenylmethane, p-phenylene and naphthalene units in order to increase the solubility of macropolycyclic lactams of type **19**. The OCH_3-substituents introduced via **18** are intended to be cleaved to OH-groups after the cyclization process. This should be useful for further bridging and, in particular, for attaching the bottom of the basket (spacer in **20**).

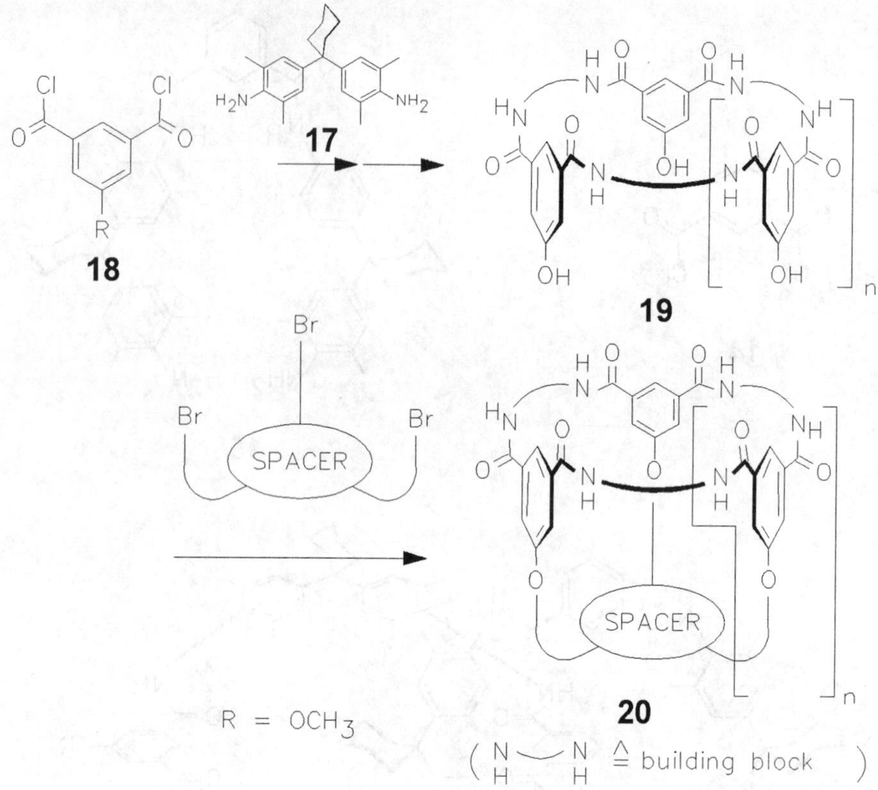

In 1991, we treated the methoxyisophthaloyl dichloride **18** with the diamine **17** in chlorobenzene in presence of a catalytic amount of 4-(*N*,*N*-dimethylamino)pyridine (DMAP) at 50-60°C using high dilution conditions. From this reaction, we isolated the sparingly soluble dimethoxy-substituted macromonocycle **21** (12.4 % yield) along with the 72-membered octalactam **22** (10 % yield). Its isomer **16** (R = OCH_3), a catenane, according to FAB mass and ^1H and ^{13}C NMR spectra, was obtained in 8.4 % yield. The spectroscopic properties of **16** (R = OCH_3) are in full accord with those of the unsubstituted catenane reported by Hunter [30]. Thus the mass spectrum of **16** shows the characteristic fragmentation due to the loss of one of the catenane rings. The NMR spectra of **16** are more complicated than those of its isomer **22**.

16

17

a: R = H
b: R = OCH_3
c: R = NO_2

21 : n = 1
22 : n = 3

The four methoxy groups give rise to two singlets. One explanation is that neither the cyclohexylidene units nor the methoxybenzene groups pass easily through the mechanically bound second macrocycle, if at all. Thus, **16** has a relatively fixed conformation which has two inner and two outer methoxy groups. As a consequence, **16** should have two enantiomeric forms **VI** and **VII**, which are expected to be more stable than the enantiomers of the unsubstituted catenane. H,H-COSY and NOESY spectra also support the catenane structure.

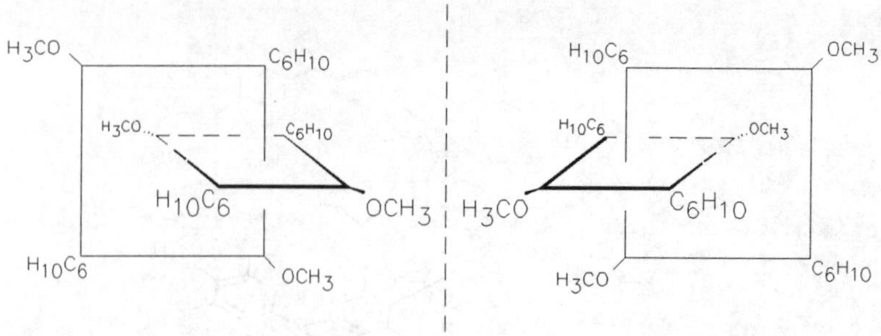

VI VII

The template assistance in the reaction, which appears to be necessary for the smooth formation of the catenane, may be due to a) hydrogen-bonding interactions between the carbonyl group of the isophthaloyl unit and two NH-groups of the cyclic amide (cf. **23**), and b) the "steric fit" and π-π interactions between the benzene rings of the spacer units and the isophthalic acid. The readily soluble, bulky diamine **17**, which shows a preference for forming large rings, seems to favor the catenane formation.

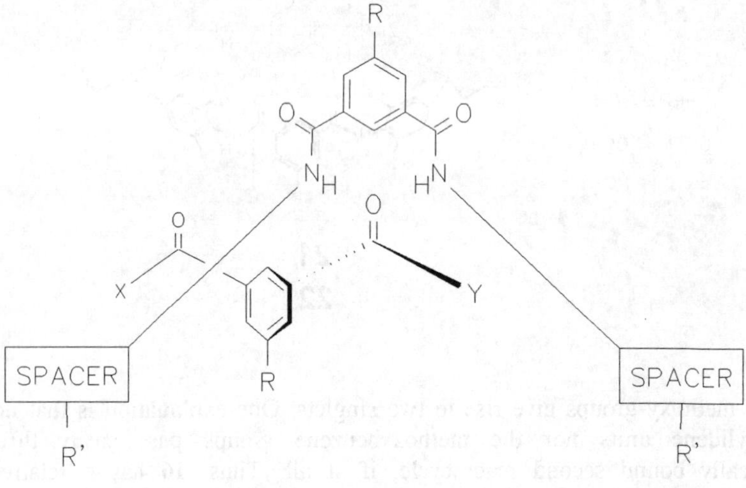

23 : X,Y = Cl

The analogous cyclization of the nitro substituted **18** (R = NO_2) with **17** yielded the dimer **21** (11 % yield) as the main product and a tetramer (7 % yield). The FAB-MS spectrum of the latter hints at the macromonocycle **22** and the isomeric catenane **16** (R = NO_2) has not been detected so far. In the template interaction here, the NO_2-groups perhaps compete with the carbonyl groups in forming hydrogen bonds with the NH groups.

The compounds with extraanular methoxy substitutents have the advantage that they can be converted into the hydroxy compounds according to known procedures. Hydroxy-substituted catenanes should be suitable for further cyclisations and for linkage to oligomeric catenanes. By affixing large extra- and intra-anular groups, the conformational mobility of the catenane rings be manipulated.
When other carbonyl compounds, for example, the substituted terephthalic acid dichloride **24**, were employed, catenanes were not formed, but rather the 34- and 68-membered macromonocycles **25** (<1%) and **26** (34 % yield), respectively were obtained. In this case, the arrangement of the amino groups is apparently not complementary to the carbonyl groups, so that templated catenane formation is not possible.

24 R = OCH_3

25 : n = 1

26 : n = 3

As shown by the last two examples, a directed catenane synthesis requires exact planning of the host-guest intermediates (cf. **23**). Many macromonocycles can also be integrated in such synthetic strategies. Other catenanes with isomeric structures [34] and containing sulfone amide and thioamide groups replacing the amide bonds as well as the first rotaxanes of this amide type have also been synthesized. The first x-ray analysis was achieved in the case of a difurano catenane [35].

4. ACKNOWLEDGEMENT

Thanks are due to DECHEMA e.V., Frankfurt (Dr. K. Wagemann, Dr. D. Tiebes) and BMFT for financial assistance. We also thank Prof. Dr. K. Gloe, Dr. H. Stephan, O. Heitzsch, Technische Universität Dresden, and Prof. Dr. K. Cammann, Dipl.-Chem. W. Hasse, Dipl.-Chem. B. Ahlers, Universität Münster for fruitful cooperation.

5. REFERENCES

1. C. D. Gutsche: *Calixarenes*, Monographs in Supramolecular Chemistry, Vol.1, Ed. J. F. Stoddart, The Royal Society of Chemistry, Cambridge (1989).
2. J. Vicens and V. Böhmer (Eds.): *Calixarenes: A Versatile Class of Macrocylic Compounds*, Kluwer Academic Publishers, Dordrecht (1991).
3. C. D. Gutsche: *Acc. Chem. Res.* **16**, 161 (1983).
4. Cf.: G. Brodesser, R. Güther, R. Hoss, S. Meier, S. Ottens-Hildebrandt, J. Schmitz, and F. Vögtle: *Pure & Appl. Chem.* **65**, 2325 (1993); G. Brodesser, J. Schmitz, F. Vögtle, K. Gloe, O. Heitzsch, and H. Stephan: *Book of Abstr.* 'Second Workshop on Calixarenes and Related Compounds' in Kurume, Fukuoka, Japan, **IL-8** (1993); G. Brodesser, R. Güther, S. Meier, S. Ottens-Hildebrandt, J. Schmitz, and F. Vögtle: *Book of Abstr.*: 'XVIII. Int. Symposium on Macrocyclic Chemistry' in Enschede, the Netherlands, **P-12** (1993); F. Vögtle: *Ann. Quim.* **89**, 29 (1993).
5. J. L. Sessler and A. K. Burrell: *Top. Curr. Chem.* **161**, 177 (1991).
6. E. Müller and G. Röscheisen: *Chem. Ber.* **90**, 543 (1957).
7. K. Burri and W. Jenny: *Helv. Chim. Acta* **50**, 1978 (1967).
8. W. Jenny and H. Holzrichter: *Chimia* **22**, 306 (1968).
9. W. Jenny and H. Holzrichter: *Chimia* **23**, 158 (1968).
10. F. Vögtle, J. Schmitz, and M. Nieger: *Chem. Ber.* **125**, 2523 (1992).
11. J. Schmitz, F. Vögtle, M. Nieger, K. Gloe, H. Stephan, O. Heitzsch, H.-J. Buschmann, W. Hasse, and K. Cammann: *Chem. Ber.* **126**, 2483 (1993).
12. J. Schmitz, F. Vögtle, M. Nieger, K. Gloe, H. Stephan, O. Heitzsch, H.-J. Buschmann: *Supramol. Chem.* **4**, 115 (1994).
13. F. Vögtle, G. Brodesser, M. Nieger, K. Rissanen: *Recl. Trav. Chim. Pays-Bas* **112**, 325 (1993).
14. H.-B. Mekelburger, J. Groß, J. Schmitz, M. Nieger, and F. Vögtle: *Chem. Ber.* **126**, 1713 (1993).
15. Cooperation with Prof. Dr. K. Gloe, Dresden.
16. K. Gloe, H. Stephan, O. Heitzsch, J. Schmitz, and F. Vögtle: *Book of Abstr.*: 'XVIII. Int. Symposium on Macrocyclic Chemistry' in Enschede, the Netherlands, **A-70** (1993).
17. H. Stephan, K. Gloe, O. Heitzsch, G. Brodesser, and F. Vögtle: *Book of Abstr.*: '24. GDCH-Hauptversammlung', in Hamburg, Germany, **SUP 9** (1993).
18. Cooperation with Prof. Dr. K. Cammann, Münster.
19. J. Reinbold, B. Ahlers, W. Hasse, K. Cammann, G. Brodesser, and F. Vögtle: *Book of Abstr.*: 'XVIII. Int. Symposium on Macrocyclic Chemistry' in Enschede, the Netherlands, **B-55,56** (1993).
20. W. Hasse, K. Cammann, F. Vögtle, and G. Brodesser: *Book of Abstr.*: 'Eurosensor VII', in Budapest, Hungary, **AP-24, 4430-1** (1993).
21. Cf.: R. M. Izatt, J. D. Lamb, R. T. Hawkins, P. R. Brown, S. R. Izatt, and J. J. Christensen: *J. Am. Chem. Soc.* **105**, 1782 (1983); S. R. Izatt, R. T. Hawkins, J. J. Christensen, and R. M. Izatt: *J. Am. Chem. Soc.* **107**, 63 (1985).
22. R. Perrin and S. Harris: in *Calixarenes: A Versatile Class of Macrocylic Compounds*, J. Vicens and V. Böhmer (Eds.), Kluwer Academic Publishers, Dordrecht, p.241 (1991).

23. Cf.: S. Shinkai, T. Otsuka, K. Araki, and T. Matsuda: *Bull. Chem. Soc. Jpn.* **62**, 4055 (1989).
24. E. Wasserman *J. Am. Chem. Soc.* **82**, 4433-4434 (1960); H. L. Frisch, and E. Wasserman, *J. Am. Chem. Soc.* **83**, 3789-3795 (1961); for recent reviews on the synthesis of catenanes see i. e. G. Schill, *Catenanes, Rotaxanes, and Knots*, Academic Press (1971), New York; C. O. Dietrich-Buchecker, and J. P Sauvage, *Bioorganic Chemistry Frontiers*, 195-248 (1991/2); D. Philp, and. J. F Stoddart, *Synlett* 445-458 (1991).
25. F. Vögtle, and R. Hoss, *Angew. Chem.* **106**, 389 (1994); *Angew. Chem. Int. Ed. Engl.* **33**, 375-384; S. Anderson, H. L. Anderson, and J. K. M. Sanders, *Angew. Chem.* **104**, 921-924 (1992); *Angew. Chem. Int. Ed. Engl.*, **31**, 907.
26. G. Schill, and A. Lüttringhaus, *Angew. Chem.* **76**, 567-568 (1964).
27. G. Schill, and C. Zürcher, *Angew. Chem.* **81**, 996-997 (1969).
28. C. O. Dietrich-Buchecker, B. Frommberger, I. Lüer, J. P. Sauvage, and F. Vögtle, *Angew. Chem.* **105**, 1526-1529 (1993); *Angew. Chem. Int. Ed. Engl.*, **32**, 1434 and references cited therein.
29. D. Armspach, P. R. Ashton, C. P Moore, N.Spencer, J. F. Stoddart, T. J. Wear, and D. J. Williams, *Angew. Chem.* **105**, 944-948 (1993); *Angew. Chem. Int. Ed. Engl.* **32**, 854; D. B. Amabilino, P. R. Ashton, M. S. Tolley, J. F. Stoddart, and D. J. Williams, *Angew. Chem.* **105**, 1358-1362 (1993); *Angew. Chem. Int. Ed. Engl.* **32**, 1297 and references cited therein.
30. C. A. Hunter, *J. Am. Chem. Soc.* **114**, 5303-5311 (1992).
31. F. Vögtle, S. Meier, R. and Hoss, *Angew. Chem.* **104**, 1628-1631 (1992); *Angew. Chem. Int. Ed. Engl.* **31**, 1619-1622.
32. L. Wambach, and F Vögtle, *Tetrahedron Lett.*, 1483-1486 (1985); B. Dung, and F. Vögtle, *J. Incl. Phenom.* **6**, 429-442 (1988); J. Breitenbach, K. Rissanen, U. Wolf, F. and Vögtle, *Chem. Ber.* **124**, 2323-2327 (1991); C. Seel, and F. Vögtle, in A. F. Williams, C. Floriani, A. E. Merbach (eds) *Perspectives in Coordination Chemistry*, VCH, Weinheim, S. 31 (1992). DFG-project Nr. Vo 145/38-1 (01.03.1990); we thank the Deutsche Forschungsgemeinschaft for the support of these studies (till 31.03.92).
33. C.A. Hunter, *J. Chem. Soc., Chem. Commun.*, **11**, 749 (1991).
34. S. Ottens-Hildebrandt, S Meier, W. Schmidt and F. Vögtle *Angew. Chem.* **1994**, *106*, 1818; *Angew. Chem. Int. Ed. Engl.* **1994**, *33*, 1767; F. Vögtle, M. Händel, S. Meier, S. Ottens-Hildebrandt, F. Ott and T. Schmidt, Chem. Ber., submitted.
35. F. Vögtle, S. Ottens-Hildebrandt, M. Nieger, K. Rissanen, J. Rouvinen, S. Meier and G. Harder, *J. Chem. Soc., Chem. Comm.*, submitted; cf. Lectures at the European Research Conference on molecular Chemistry at Mainz (11. - 16.08.94) and at the RSC UK Makrocycle Group Annual Meeting at Newcastle (04. - 05.01.95).

SELF-ASSEMBLY IN CHEMICAL SYNTHESIS

STEVEN J. LANGFORD, J. FRASER STODDART
School of Chemistry
University of Birmingham
Edgbaston, Birmingham B15 2TT, UK

1. Introduction

More and more chemists are beginning to realise that the conventional synthetic methodology of making compounds group-by-group or molecule-by-molecule, employing reagents or catalysts to make or break covalent bonds and so manipulate functional groups and transform molecular structures, is insufficient by itself to construct the materials they would like to make in the future. Along its many synthetic trails, nature does not rely upon the inefficient use of protecting groups and the complexity of reagents according to the usual manner and practice of the synthetic organic chemist in a traditional sense. Indeed, one of the keys to the efficient operation of biological systems is their ability to self-assemble,[1] self-organise[2] and self-replicate.[3]

The use of self-assembly in nature has many different guises and generates a wide diversity of thermodynamically stable structures at both the cellular and sub-cellular levels.[4-8] The overall superstructures, which characterise biological systems, *e.g.*, the cell, are derived from the mutual recognition displayed by the molecular components such that they interact physically and chemically in defined ways with a level of specificity and architectural control that is unparalleled in modern chemical synthesis. The term self-assembly[1,9] relates to the construction of discrete molecular assemblies and supramolecular arrays *via* the use of either weak (yet specific) noncovalent bonding interactions, reversible covalent bonding (*e.g.*, disulfide linkages, imines), or a mixture of both. Using these means, molecular assemblies and supramolecular arrays have programmed into them a way by which their three-dimensional architectures and functions can be controlled with remarkable ease and efficiency.

Only recently have the scientific community started to take advantage of the synthetic paradigms of nature, borne out with clarity in the structural elucidation of DNA by Watson and Crick,[10] to construct unnatural molecular and supramolecular systems that are designed to exploit the use of noncovalent bonding interactions.[3,11-18] One of the main driving forces behind the development of supramolecular chemistry[19] has been the urge to understand how natural systems operate to form supramolecular arrays and then to extend this understanding to wholly synthetic molecular and supramolecular structures with perhaps novel functions, such as chemical sensing or information processing.[20]

C. Chatgilialoglu and V. Snieckus (eds.), Chemical Synthesis, 381–401.

The production of synthetic systems displaying molecular recognition characteristics through self-assembly processes is now beginning to be realised.[1,19] This is not to say, that, in the everyday life of the chemist, self-assembly has not been used often. Indeed, crystallisation is a prime example of self-assembly, wherein molecules are discriminated between because of their shape, size, bonding and electronic properties, giving rise to long range three-dimensional structures. Molecules that are unable to form the maximum number of bonding interactions or that disrupt a crystalline lattice are readily excluded during the crystallisation process. Other well-known examples of self-assembly include liquid crystalline mesophases and the formation of membranes, micelles, and vesicles.

2. Preamble

In most of the self-assembly pathways[1] found in nature, structurally complex architectures are usually constructed from small, relatively simple sub-units. These simple sub-units contain the information to promote their own assembly into arrays. They allow nature's systems to excercise a remarkable economy with respect to the amount of information they require in order to produce the superstructures. Very precise recognition features in each sub-unit renders the assembly processes self-checking. The use of relatively weak noncovalent bonds, or reversible covalent bonds, allows a degree of control over the construction processes, such that an alteration or correction to the superstructure can be performed without encountering the potential difficulties associated with irreversible bond formation. The construction processes are synthetically efficient under mild conditions. This regime is a direct result of the self-checking nature of the processes and the mechanistically simple reaction types that are happening.

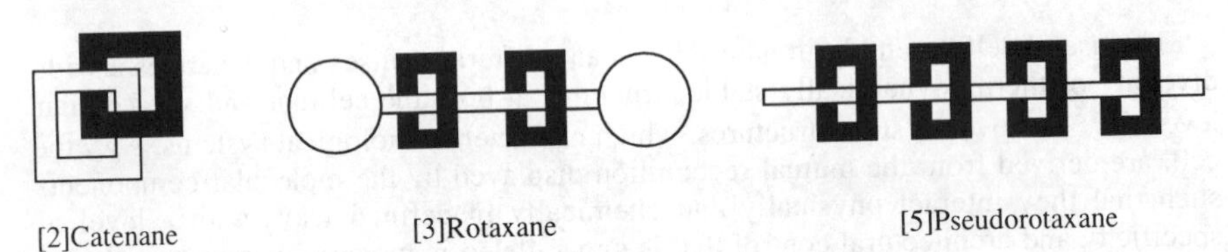

Figure 1. An assortment of schematically drawn mechanically interlocked structures

The synthetic strategy we have adopted relies upon the use of mechanically interlocked and intertwined molecules and complexes (***Figure 1***) in the shape of the so-called catenanes, rotaxanes, and pseudorotaxanes.[12,21-26] The choice of crown ethers,[27] which are ideally suited to address the challenges posed by these mechanically entangled structures, was an obvious one. We have argued that the knowledge and experience gained by carrying out the self-assembly of these relatively small molecular and supramolecular systems in the solution state, will establish the fundamental rules for the elaboration ultimately of larger polymolecular structures and superstructures. As such, the template-directed synthesis of catenanes and rotaxanes illustrates the growing potential of molecular self-assembly processes in the realm of organic synthesis. The ultimate objective of research in this field is to be able to control the construction, the form, and the function of synthetic nanometer-scale structures[28] from the appropriate substrates with the same degree of precision that is displayed by nature, without the need for reagent control or catalysis.

3. Evolving a Synthetic Strategy

The strategy for the molecular self-assembly of catenanes and rotaxanes emerged from parallel observations carried out on two complementary host-guest systems. During the early 1980s, we became interested in developing efficient, synthetic receptors for the important bipyridinium herbicides,[29] Diquat $[DQT]^{2+}$ and Paraquat $[PQT]^{2+}$. The most successful receptors that emerged were dibenzo-30-crown-10[27,30] for $[DQT]^{2+}$ and the constitutional isomer, bisparaphenylene-34-crown-10 (BPP34C10),[31] for $[PQT]^{2+}$. The synthesis of BPP34C10 can be achieved[12] from the diol BHEEB[32] - obtained from the base-promoted reaction of hydroquinol with 2-(2-chloroethoxy)-ethanol - and its derivatised bistosylate, BTEEB (***Scheme 1***). BPP34Cl0 forms a deep orange-coloured 1:1 inclusion complex with $[PQT][PF_6]_2$ in acetone at room temperature.

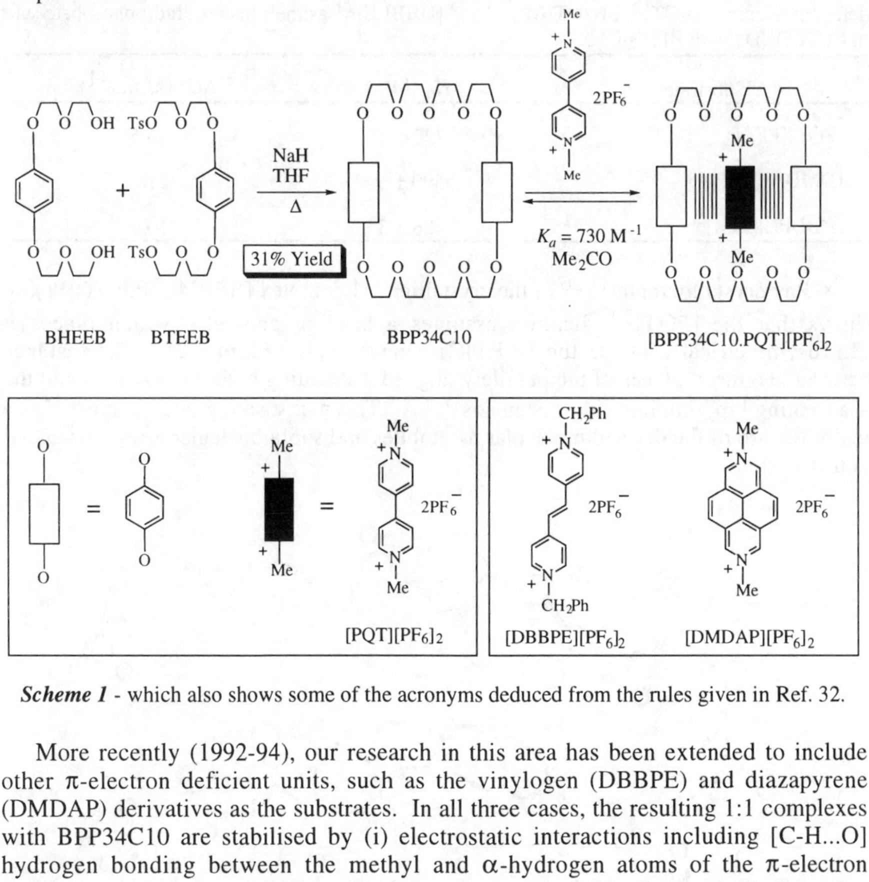

Scheme 1 - which also shows some of the acronyms deduced from the rules given in Ref. 32.

More recently (1992-94), our research in this area has been extended to include other π-electron deficient units, such as the vinylogen (DBBPE) and diazapyrene (DMDAP) derivatives as the substrates. In all three cases, the resulting 1:1 complexes with BPP34C10 are stabilised by (i) electrostatic interactions including [C-H...O] hydrogen bonding between the methyl and α-hydrogen atoms of the π-electron deficient guests and some of the polyether oxygen atoms of the host and (ii) dispersive forces, including π-π stacking[33] and charge transfer interactions between the π-electron deficient units in the guests and the π-electron rich hydroquinol rings in the host. It is these charge-transfer interactions which give rise to the deep orange colour exhibited by the complexes. The difference in the stabilities (***Table 1***) of the three 1:1 complexes can be explained in terms of the extent of π–π interactions occurring between the π-electron deficient and π-electron rich components and the "snugness" of the fit of the dications within the cavity of the receptor - a factor which will influence hydrogen bonding interactions. The more extensive the π-systems, the greater will be the stabilities of the 1:1 complexes. As a corollary, the greater the separation of the pyridinium cations within the π-electron deficient unit (as exemplified by [DBBPE][PF_6]$_2$), the smaller will be the occupying area by the active recognition sites (*e.g.*, the α-hydrogen atoms and Me groups for hydrogen bonding and the pyridinium rings for π–π interactions) of the substrate molecules in the cavity. The consequence of this modified geometry is to reduce the chance for optimal hydrogen bonding to the polyether oxygen atoms, as well as impair the extensive π–π interactions between the two aromatic units, which, in turn, render the complex less stable than it might otherwise have been.

Table 1. A comparison between the free energies of complexation ($\Delta G°$) for the π-electron deficient dications $[PQT]^{2+}$, $[DMDAP]^{2+}$ and $[DBBPE]^{2+}$ as their bishexafluorophosphate salts in CD_3COCD_3 with BPP34C10.

Substrate	K_a (M^{-1})	$\Delta G°$ (kJ mol^{-1})
[PQT][PF_6]$_2$	730 ± 23	16.5
[DMDAP][PF_6]$_2$	5500 ± 500	21.6
[DBBPE][PF_6]$_2$	86 ± 3	10.9

X-Ray crystallography[11,26] of the crystalline 1:1 complex [BPP34C10.PQT][PF_6]$_2$, shows that the $[PQT]^{2+}$ dication assumes a binding geometry which places it centrosymmetrically inside the BPP34C10 macrocycle (***Figure 2***). The distance between the mean planes of the parallely-aligned π-donating hydroquinol rings and the π-accepting bipyridinium ring systems is 3.7 Å. This π-π stacking interaction[24,33] is a major feature of the derived molecular assemblies and supramolecular arrays discussed in this article.

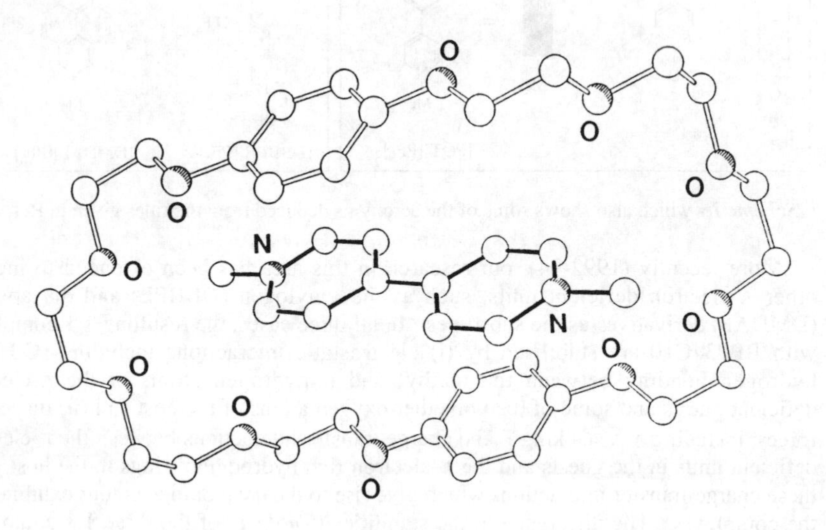

Figure 2. The X-ray crystal structure of [BPP34C10.PQT][PF_6]$_2$

The binding geometries and strengths are sensitive to a range of reaction conditions, such as solvent, pH, and absolute, as well as relative, concentrations. This situation can be exemplified by noting the following observations. When the binding study between $[PQT][PF_6]_2$ and BPP34C10 was performed in acetone, an association constant of 730 M^{-1} was obtained.[26] When the same experiment was carried out in acetonitrile, a much lower (250 M^{-1}) K_a value was found.[34] Furthermore, the addition of water (10 equivalents) to a solution of BPP34C10 and $[DMDAP][PF_6]_2$ in acetonitrile caused the two components to remain dissociated in solution.[35]

With the demonstration of the existence of a successful host-guest system between BPP34C10 and $[PQT][PF_6]_2$, the next question to be addressed was the reversal of the roles of the host and the guest, *i.e.*, the formation of a receptor incorporating bipyridium units, which would complex with a neutral hyrdroquinol derivative as the substrate. To this end, the tetracationic cyclophane $[BBIPYBIXYCY][PF_6]_4$ was synthesised (***Scheme 2***) from the known[36] $[BBIPYXY][PF_6]_2$ - obtained from 4,4'-bipyridine and 1,4-bis(bromomethyl)benzene (BBB) after counterion exchange (NH_4PF_6/H_2O) - and 1,4-bis(bromomethyl)benzene. The final ring closure step affords only a 12% yield of the desired tetracationic cyclophane.[37]

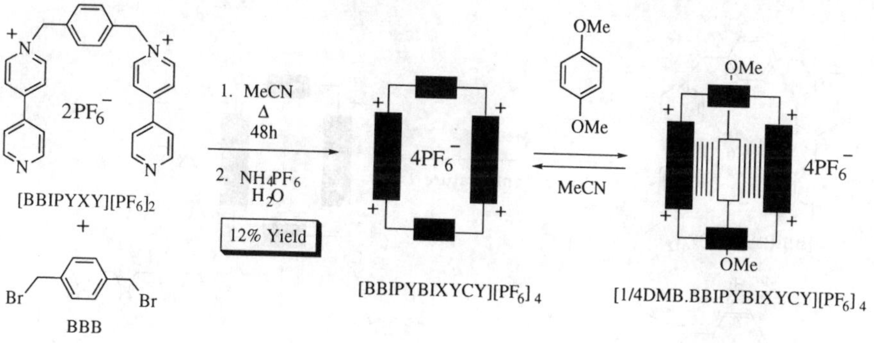

Scheme 2

The ability of the tetracationic cyclophane[37,38] to form 1:1 complexes[38,39] with π-donor aromatic substrates - 1,2- and 1,4-dimethoxybenzene (DMB),[40] as well as tetrathiafulvalene,[41] for example, is a result of stabilising dispersive forces that include (i) π-π-stacking and charge transfer interactions between the π-electron-rich guests and the π-electron-deficient bipyridinium units in the host and (ii) electrostatic "T-type" edge-to-face interactions involving the aromatic guests and the paraphenylene units in the tetracationic cyclophane has led us to examine the complexing properties of a wide range of substrates. The T-type interactions, although quite weak, play a significant organising role in the molecular assembly processes involving the $[BBIPYBIXYCY]^{4+}$ tetracation.

The rotaxane-like orientations observed in the solid state for the centro-symmetrically-located guests in the middle of the $[BBIPYBIXYCY]^{4+}$ cavities conjure up a highly suggestive situation. The marriage of the molecular recognition features from these two host-guest systems, combined with further covalent modification, leads to proposals of quite intricate molecular assemblies of a rotaxane and catenane nature in which two or more molecular components are linked mechanically, but not covalently.

4. Making a [2]Catenane to Order- Learning Lessons from Nature

When [BBIPYXY][PF_6]$_2$ is mixed with an excess of BPP34C10 in acetonitrile solution, there is no visible evidence for any appreciable interaction between the dication and the crown ether. The addition, however, of BBB to this solution, followed by stirring at room temperature overnight, resulted[12,42] in the deposition of a highly coloured red precipitate. After counterion exchange and recrystallisation, the catenane {[2]-[BPP34C10]-[BBIPYBIXYCY]catenane}[PF_6]$_4$ was isolated (***Scheme 3***) in a quite remarkable yield of 70%! Confirmation of the self-assembly of the [2]catenane followed from FABMS, ^{1}H and ^{13}C NMR spectroscopies, and X-ray crystallography.[12] Variable temperature NMR spectroscopy led us to the conclusion that the structural order observed in the solid state was also evident in solution. In fact, the dynamics of the system in solution was illustrated by the fact that the neutral ring is circumrotating through the charged ring 300 times a second at +25 °C while pirouetting around it 2000 times a second at the same temperature.[42]

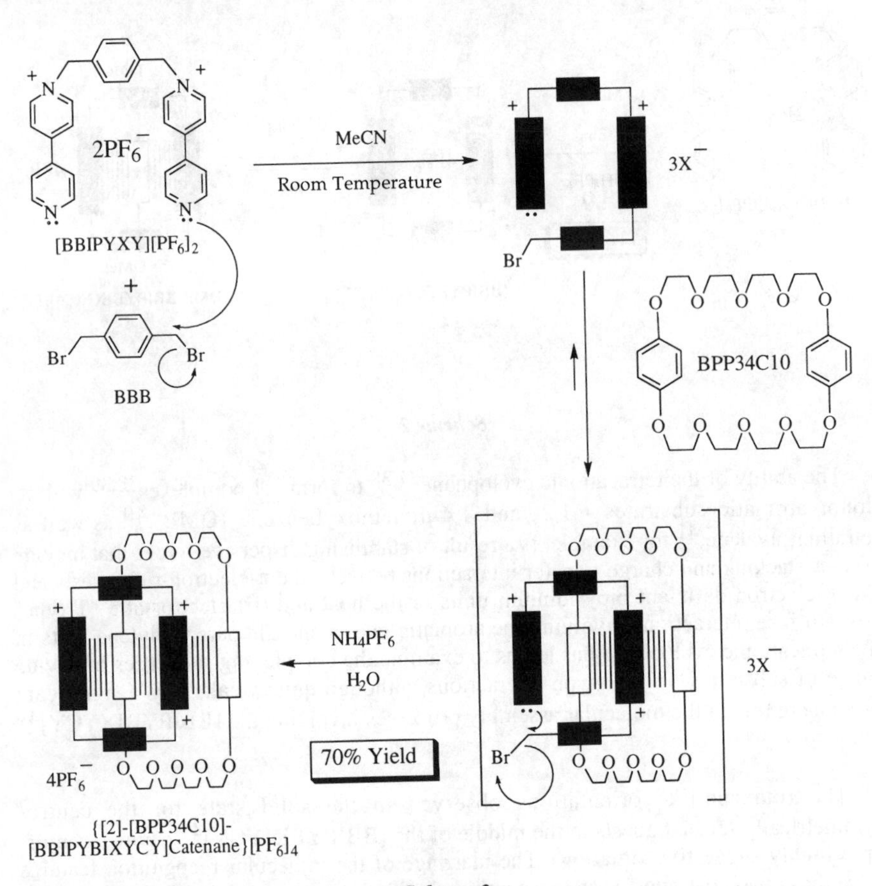

Scheme 3

The spectacularly high yield of the [2]catenane is a testament to the efficiency of the template-directed route to its synthesis. In acetonitrile, the $[BBIPYXY]^{2+}$ dication and BPP34C10 exist in a noninteractive state. However, the addition of BBB results in the formation of a highly coloured solution. The reason is as follows. Once the first *inter*molecular S_N2 reaction has occurred to form the trication (***Scheme 3***), the mutual recognition (for example, by the emerging $\pi-\pi$ stacking interactions) between the components ensures that the second *intra*molecular S_N2 reaction to afford the [2]catenane, proceeds rapidly *via* a favoured transition state to the cooperatively assembled molecule. The relatively fast rate of the final cyclisation step is illustrated by the fact that the 1:1 complex between the tricationic species and BPP34C10 has not been detected in the reaction mixture.

From the practical point of view, the formation of the precise three-dimensional structure occurs because only those molecules having the correct positioning of functional groups can fit together to form the maximum number of noncovalent bonding interactions. The small free-energy increment in the association step is sufficient to induce the sharp transition from producing the interactive components to the assembling of a highly ordered structure involving mechanically-linked components. Indeed, this is an example of a *template-directed* synthesis[1] wherein the template ends up being another component of the molecular structure. The template-directed synthesis of the catenane is a simple example of noncovalent self-assembly, followed by post-assembly modification involving covalent bond formation to give an aesthetically pleasing and physically intruiging molecule. In essence, the whole process can be described as an *irreversible* self-assembly.[1] Finally, it could be added that, because the need for reagent control or catalysis has been eliminated, the formation of the [2]catenane also constitutes an example of a *structure-directed* synthesis.[43]

So far, we have witnessed in the context of self-assembly that molecular and supramolecular structures can be constructed from the appropriate components, provided that we build into them the required geometries in order to favour the self-assembly process. But what happens if we relax the initial degree of preorgansation to its bare minimum? When the reaction between *five* "molecules" - *two* of BP, *two* of BBB, and *one* BPP34C10 - was performed (***Scheme 4***) in dimethylformamide at room temperature, the [2]catenane was isolated in 18% yield. When a similar mixture was subjected to ultra-high pressure (12 kbar), the isolated yield[37] rose to a staggering 42%.

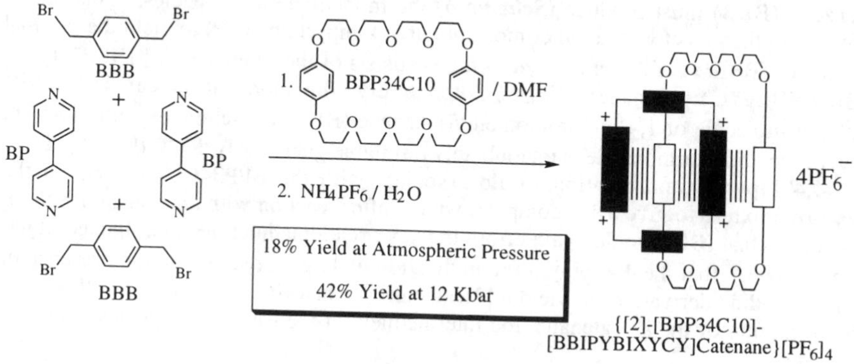

Scheme 4

In a similar vein, the tetracationic cyclophane $[BBIPYBIXYCY]^{4+}$ is self-assembled (***Scheme 5***) from *two* molecules of each of BP and BBB in dimethylformamide at room temperature, in the presence of bishydroxyethoxyethoxybenzene (BHEEB) as the template in a modest 35% yield after counterion exchange.[37]

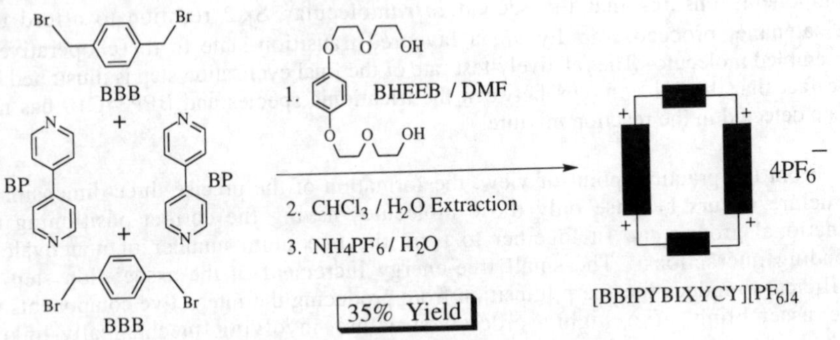

BBB + BP + BP + BBB

1. BHEEB / DMF

2. $CHCl_3$ / H_2O Extraction

3. NH_4PF_6 / H_2O

35% Yield

$4PF_6^-$

$[BBIPYBIXYCY][PF_6]_4$

Scheme 5

During the template-directed synthesis, the BHEEB template organises the collection of π-electron deficient ligands, and in so doing favours *intra*molecular cyclisations to *inter*molecular reactions, *i.e.*, polymerisations. In the absence of π-electron rich templates and certainly at higher concentrations, reactions of this type would tend to yield polymers. The two-fold increase in yield of the tetracationic cyclophane in the presence of the alkoxybenzene is a direct result of the dramatic organising effect of this unit when it performs its role as a template. This *molecular imprinting* is an example of a *directed* self-assembly whereby a temporary template participates as a structural element in the assembly process but does not appear in the final assembled product. We rely upon liquid-liquid extraction, which exploits the rapid equilibrium between the inclusion complex, to separate the template away from the newly formed tetracationic cyclophane.

Let us return to the question of how the [2]catenane self-assembles from *five* molecules. The initial reaction between 4,4'-bipyridine (BP) and 1,4-bis(bromomethyl)-benzene (BBB) must produce (***Scheme 6***) the monoquaternary species. This monocation can then react with another molecule of 4,4'-bipyridine (BP) to yield the original precursor $[BBIPYXY]^{4+}$ employed in the synthesis of the catenane {[2]-[BPP34C10]-[BBIPYBIXYCY]catenane}$[PF_6]_4$. Altenatively, the monocation can react with another molecule of 1,4-bis(bromomethyl)benzene (BBB) to yield a 4,4'-bipyridinium derivative. Such a molecule is undoubtably a suitable guest for BPP34C10.[31,44] Thus, the 4,4'-bipyridinium dication would associate with the BPP34C10 present in the reaction mixture to give a 1:1 complex, which, after reaction with another molecule of 4,4'-bipyridine (BP), would again afford the key tricationic intermediate. Indeed, it has been shown[45] that the 4,4'-bipyridinium dication undergoes reaction in the presence of 4,4'-bipyridine derivatives to yield a [2]catenane. Whatever mechanism is followed to the formation of the [2]catenane, the intermediacy of the tricationic species is clearly indicated.

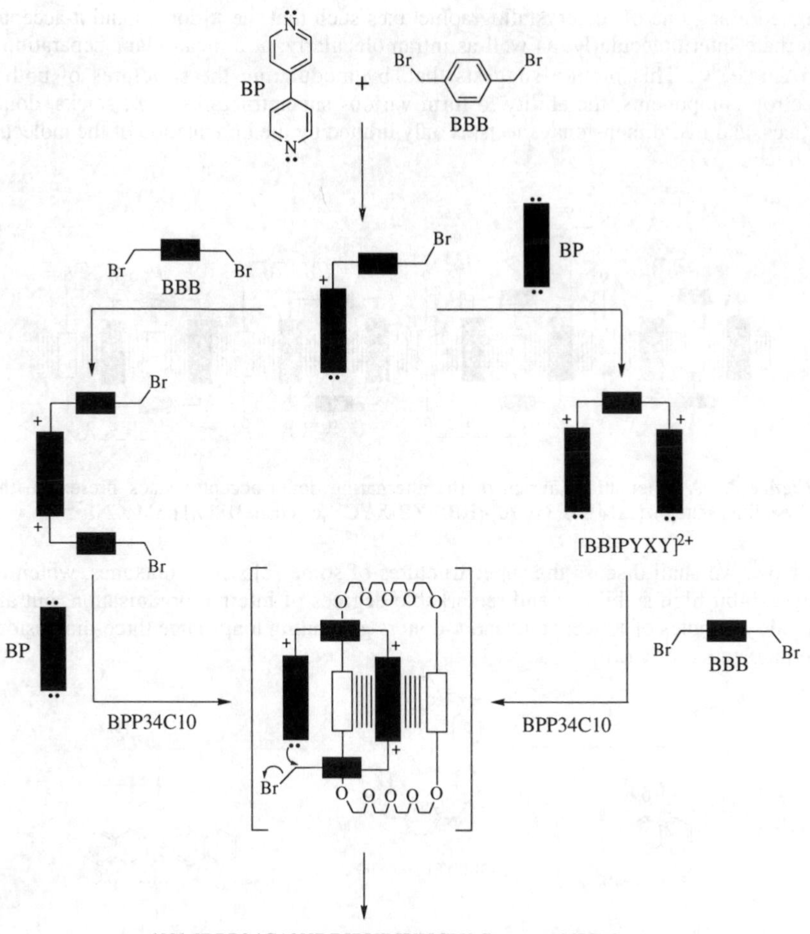

{[2]-[BPP34C10][BBIPYBIXYCY}Catenane}$[PF_6]_4$

Scheme 6

5. Supramolecular Arrays

The recent surge of interest in the reversible control of supramolecular arrays in the solid state has led researchers to recognise the importance of weak noncovalent intermolecular forces.[46] The reliance on noncovalent bonding interactions during crystallisation to form long-range three-dimensional structures has a close parallel with the processes found in nature. The reversibility allows for error-checking, such that the unwanted units are rejected and excluded from the crystalline lattice. This facet of supramolecular chemistry has been centred principally on the use of hydrogen bonding[47] as the mediator to the molecular recognition involved in the organisation of one-, two-, and three-dimensional superstructures.

The X-ray crystal structure of {[2]-[BPP34C10]-[BBIPYBIXYCY]catenane} $[PF_6]_4$.5MeCN reveals the pattern of continuously stacked [2]catenanes (***Figure 3***) aligned along one of the crystallographic axes such that the π-donors and π-acceptors alternate intermolecularly, as well as intramolecularly, at a mean-plane separation of 3.5Å.[12,42,48] This picture suggests that, by modulating the structures of both π-electron components, the ability to form various superstructures, *e.g.*, stacks, double helices, and two-dimensional sheets, is only limited by the imagination of the molecular architect.

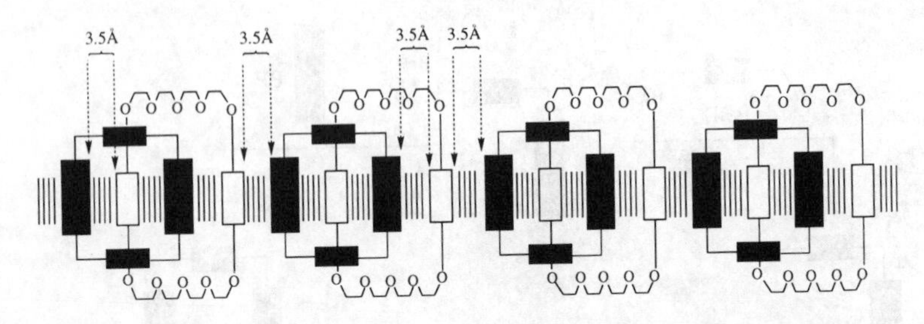

Figure 3. A schematic diagram of the alternating donor-acceptor stack present in the crystalline state of {[2]-[BPP34C10]-[BBIPYBIXYCY]catenane}$[PF_6]_4$.5MeCN[12,48,50]

Now, we shall discuss the superstructures of some [2]pseudorotaxanes, which not only exhibit high stabilities and remarkable degrees of internal organisation, but also provide examples of π-acceptors and π-donors generating long-range three-dimensional supramolecular order.

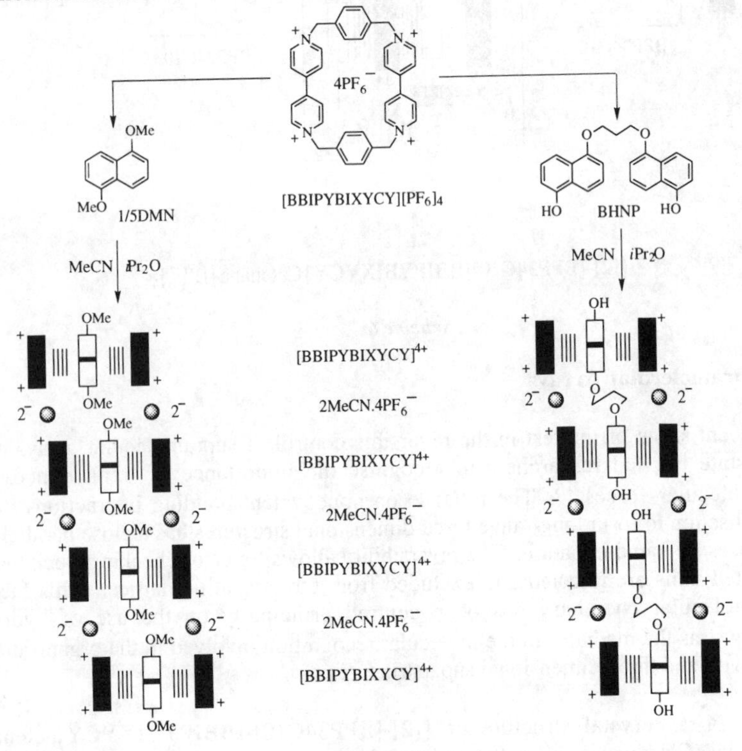

Scheme 7

The picture of a molecular abacus has emerged from the solid state structure of the 1:1 complex formed between [BBIPYBIXYCY][PF_6]$_4$ and 1,5-dimethoxynaphthalene (1/5DMN) (***Scheme 7***). The packing within the crystal lattice of both components is such as to produce an alternating stack in one crystallographic direction with the 1/5DMN molecules aligning themselves along the free channel that runs through the centre of the cyclophane stack. This observation led to the design of a prototype of a molecular abacus in the shape of a 2:1 complex (or a [3]pseudorotaxane) between [BBIPYBIXYCY][PF_6]$_4$ and 1,3-bis(5-hydroxy-1-naphthoxy)propane (BHNP) whose crystal structure is shown schematically in ***Scheme 7***.[49] The dominant feature of this structure is once again the channel-like arrangement of both the cyclophane and the four hexafluorophosphate counterions down one crystallographic axis. Inside the channel is included the BHNP guest in such a way that each naphthalene unit of the BHNP lies within the centre of the tetracationic cyclophane.

When BBPEEEEN is added to a solution of [BBIPYBIXYCY][PF_6]$_4$ in acetonitrile, it spontaneously threads itself (***Scheme 8***) through the centre of the tetracationic cyclophane,[50] producing a highly ordered 1:1 complex whose X-ray crystal structure indicates that, whilst the central naphthalene ring of the guest is located inside the tetracationic cyclophane, the terminal naphthalene rings are positioned against the sides of the bipyridinium units of the cyclophane. In the crystalline lattice, the [2]pseudorotaxanes form continuous two-dimensional sheets, sustained by a combination of the π–π stacking and aromatic [CH...π] interactions that extend to form a grid-like pattern. The hexafluorophosphate counterions are located between the sheets.

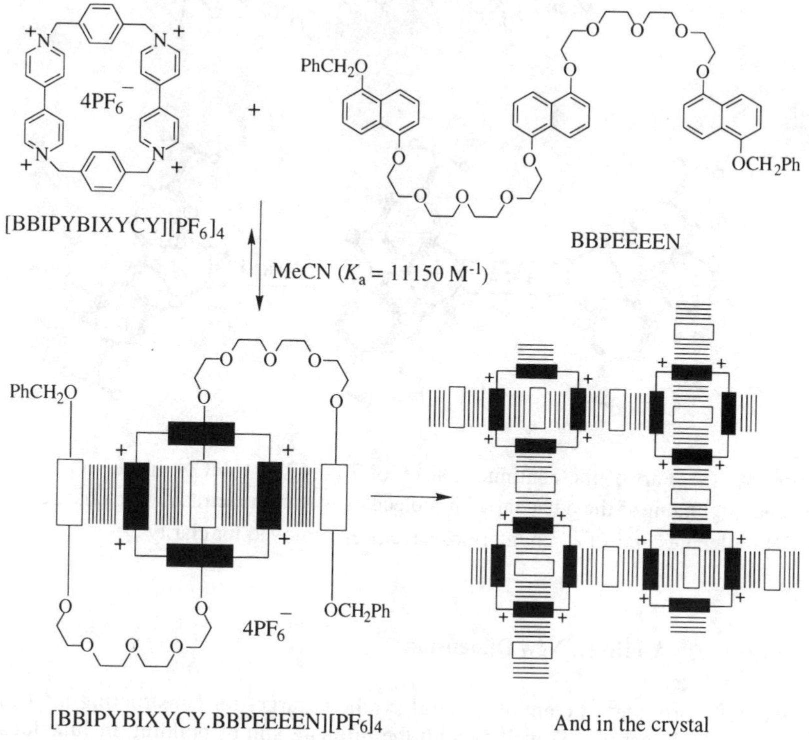

Scheme 8

The 1:1 complex between the tetracationic cyclophane and tetrathiafulvene (TTF) forms continuous stacks with no hinderance of the central channel by either solvent molecules or the counterions (***Figure 4***).[41] The TTF molecules have their long axis almost perfectly aligned along this crystallographic direction with a separation of 4.9 Å between the tertiary alkenic carbon atoms in adjacently located TTF molecules.[41] This supramolecular structural arrangement also demonstrates the potential of the tetracationic cyclophane to form polyrotaxane-like assemblies with suitably linked π-donor compounds. Furthermore, its inclusion inside $[BBIPYBIXYCY]^{4+}$ illustrates its potential for the production of molecular "shuttles" and "switches".

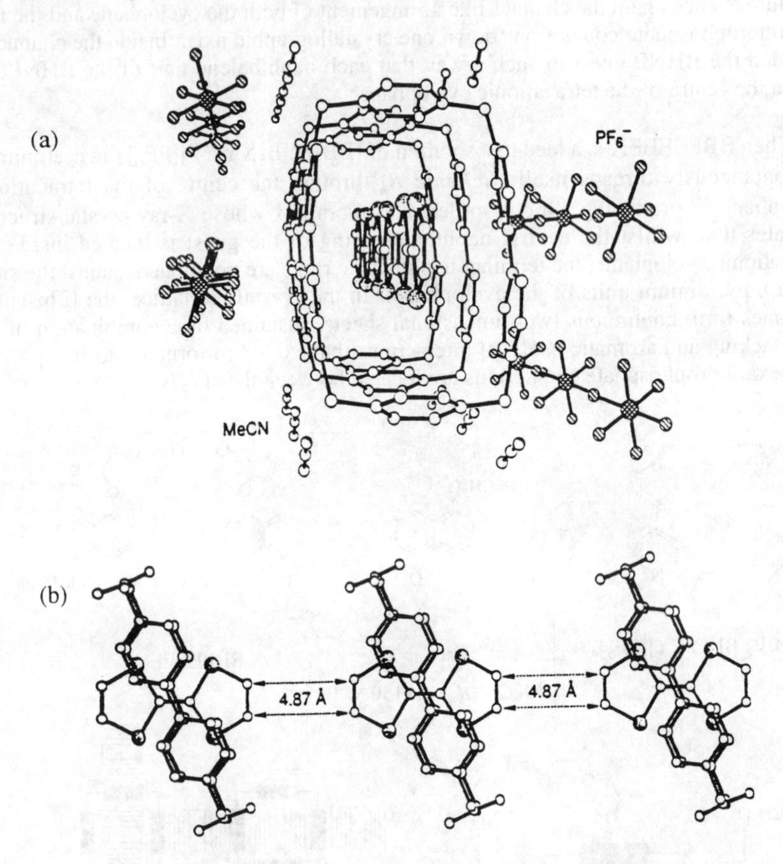

Figure 4. (a) Part of the continuous stack of $TTF.4PF_6.4MeCN$ in the solid state.[41] (b) The alignment of the adjacent TTF molecules within the tetracationic channel. The MeCN molecules and PF_6^- counterions have been removed for clarity.

6. Rotaxanes - Adding a New Dimension

Rotaxanes[51] provide a potentially useful synthetic target for constructing molecules with novel architectures as well as with the ultimate aim of building in functions in order to create nanoscale devices. In chemical terms, this type of molecule contains a linear component encircled by a macrocyclic component. To prevent the wheel from leaving the axle *readily*, the linear component must be terminated at both ends by large stoppers, *i.e.*, blocking groups. Traditionally, there have been two different approaches to the formation of rotaxanes. The *threading* approach (***Figure 5***) relies on statistical or noncovalent bonding directed association of a thread-like component with a bead-like macrocyclic component.[12] This assembly is then trapped mechanically by the covalent attachment of blocking groups which are of sufficient size to stop the dissociation of the components.[12] In the *clipping* approach (***Figure 5***), the bead-like macrocyclic component is assembled around a preformed dumbbell-shaped component to form the [2]rotaxane. To date, rotaxanes made by this approach have been limited to the synthesis of the tetracationic cyclophane as the bead component.

Recently, a third process has been identified which constitutes an example of *strict* self-assembly.[22,23] The so-called *slippage* approach to rotaxane synthesis (***Figure 5***) is based upon size complementarity between the macrocyclic polyether, BPP34C10, containing two π-electron rich recognition sites and the stoppers comprised of suitable tetra-arylmethane groups at the ends of dumbbell-shaped molecules incorporating at least one 4,4'-bipyridinium unit.[22,23]

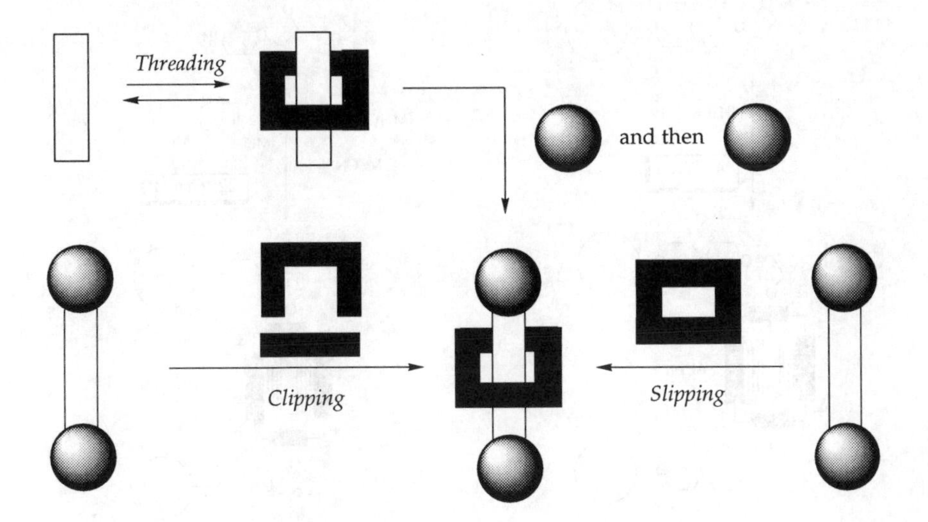

Figure 5. The methods of rotaxane synthesis

The threading and clipping methods used to self-assemble [2]rotaxanes are also limited in the uses for stoppers and threads (***Scheme 9***). The base sensitive nature of the tetracationic cyclophane makes it difficult to find a suitable synthetic route for the addition of stoppers to the included thread-like molecule in the *threading* approach. The *clipping* approach, while commendable in yield as a self-assembling process, has the disadvantage of being relatively low yielding for use as a preparative method to [2]rotaxanes incorporating one station. We must remember that, if we add in a second recognition site comparable to the first, then the clipping approach is the method of choice.[12,52]

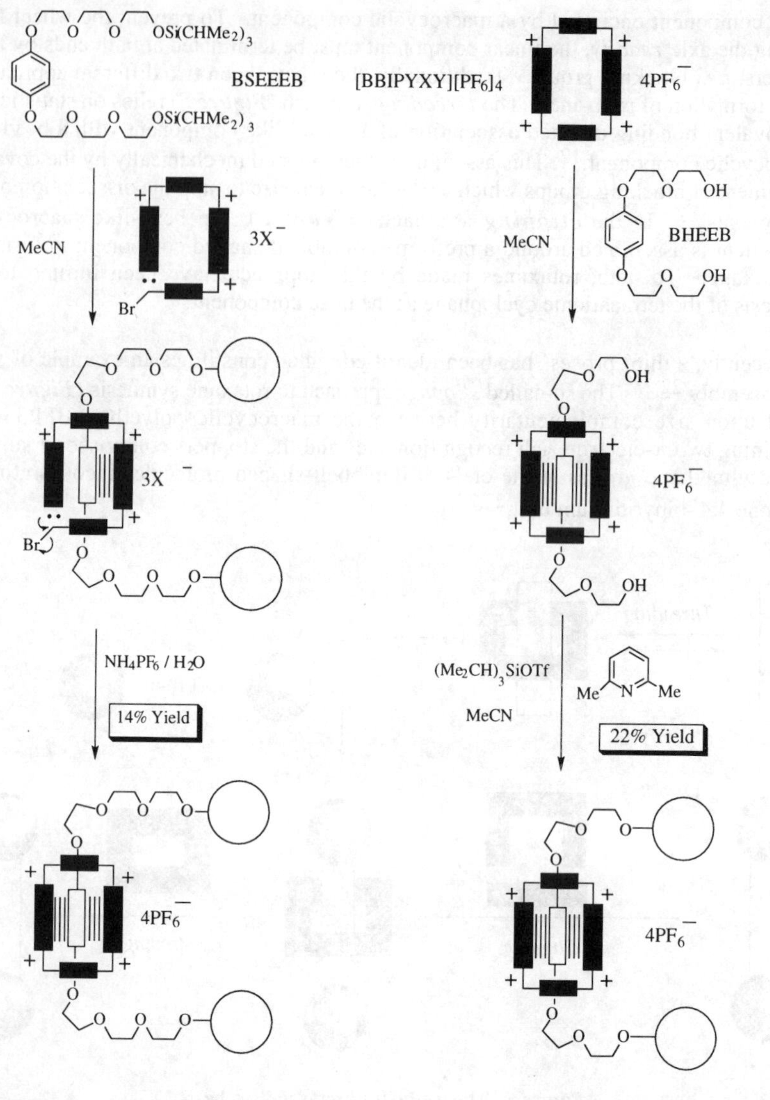

Scheme 9

Before the advent of slippage, the formation of [2]rotaxanes included a very limited array of choices for the macrocyclic bead and the rod. In essence, the use of macrocyclic polyethers, like BPP34C10 as the bead component, is only beneficial if there is a mutual recognition between the π-electron rich hydroquinol rings contained within the crown and a π-electron poor site on the dumbbell component. This route was limiting in that it requires *two* 4,4'-bipyridinium dications to be present in order that *one* BPP34C10 ring can be incorporated into the assembly (***Scheme 10***).

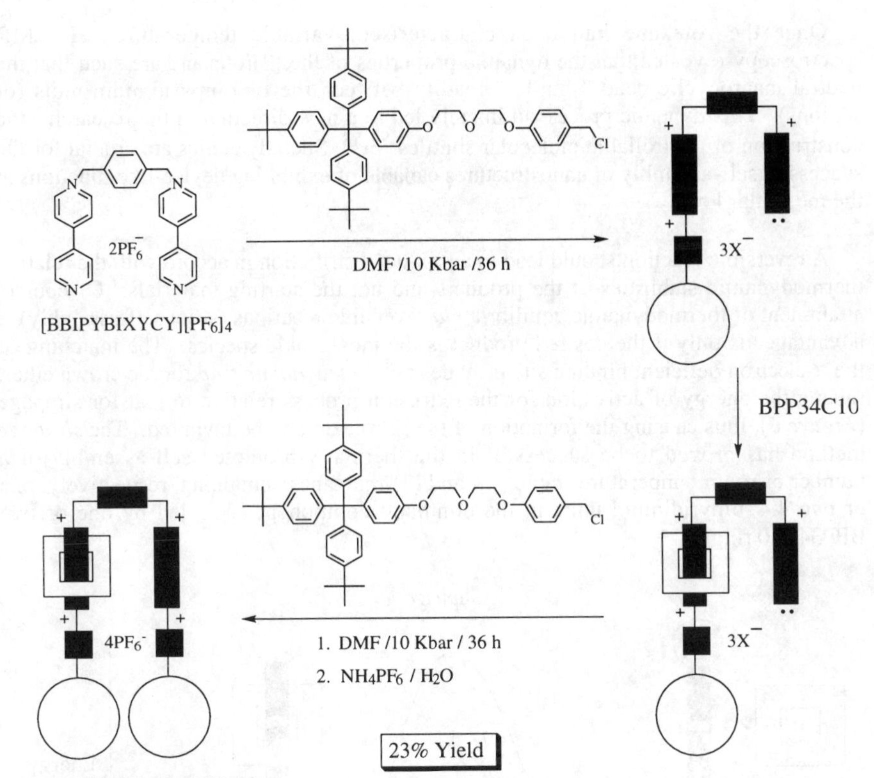

$2PF_6^-$

[BBIPYBIXYCY][PF_6]$_4$

DMF /10 Kbar /36 h

$3X^-$

BPP34C10

$4PF_6^-$

1. DMF /10 Kbar / 36 h

2. NH_4PF_6 / H_2O

$3X^-$

23% Yield

{[2][BPP34C10][TtBuPhMPEEBz]$_2$
[BBIPYXY]Rotaxane}[PF_6]$_4$

Scheme 10

The different components of the [2]rotaxane may be regarded as "intelligent" insofar as they hold the information necessary for their mechanical interlocking (***Scheme 11***). The addition of a stopper to a rod component forms a recognition site onto which a bead component spontaneously threads. The addition of a second stopper to the rod forms a second recognition site and, provided the stoppers are large enough, prevents the macrocyclic bead from leaving the rod.

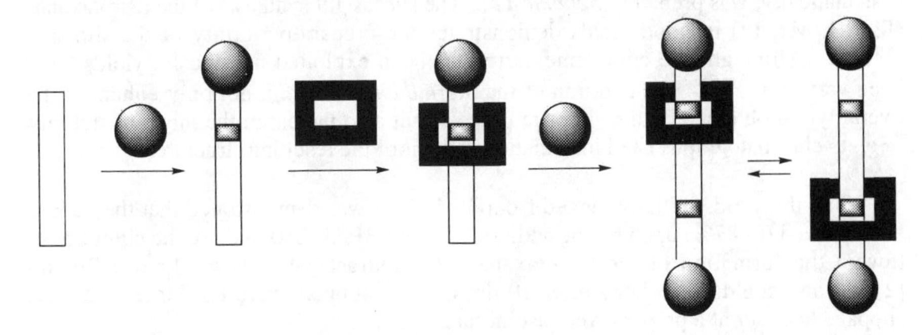

Scheme 11

Once the rotaxane had been characterised, variable temperature ^{1}H NMR spectroscopy revealed that the dynamic properties of the [2]rotaxane are such that the neutral macrocyclic bead "shuttles" readily between the two bipyridinium units (or stations). This dynamic process ultimately led to a new direction in the research - the construction of controllable molecular shuttles.[52,53] Such dynamics are crucial for the successful self-assembly of nanostructures capable of exhibiting device-like functions at the molecular level.

A reversible reaction should lead to a product distribution in accord with the relative thermodynamic stabilities of the products and not the starting materials. Of course, attainment of thermodynamic equilibria *via* reversible reactions (strict self-assembly) is advantageous only if the desired product is the most stable species. The matching of the π-electron deficient binding site provides a *thermodynamic trap* for the crown ether, raising the energy of activation for the extrusion process relative to that for slippage (***Figure 6***), thus causing the formation of the [2]rotaxane to be favoured. The *slippage* method has proved to be successful in the thermally-promoted self-assembly of a number of room temperature stable [2]- and [3]-rotaxanes containing, respectively, one or two 4,4'-bipyridinium units in the dumbbell component encircled by one or two BPP34C10 rings.

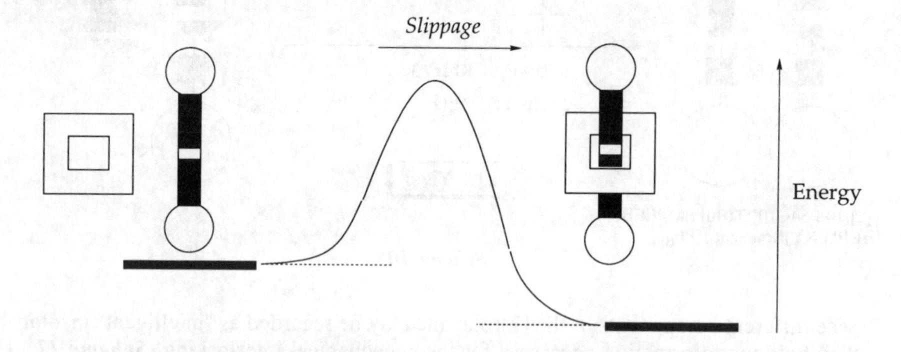

Figure 6. Energy diagram showing slippage

In order to investigate slippage as a synthetic route, a range of dumbbell-shaped compounds containing bipyridium units, where the size of the stoppers was varied systematically, was prepared (***Scheme 12***). The successful synthesis of the [2]rotaxanes (R = H, Me, Et) in good yields demonstrates the preparative utility of the slippage method. Although size complementarity has been exploited previously, yields have been very low.[54,55] The addition of the *thermodynamic trap*, not only enhances the overall yield obtained by the slippage process, but also increased the inherent stability (*e.g.*, to chromatography) and information content of the resulting structures.

While the yields initially ranged from 45-52%, it was demonstrated that they could be improved (to 87%) upon by the addition of more BPP34C10 to drive the equilibrium toward the formation of the [2]rotaxane. By contrast, when R = *i*-Pr or *t*-Bu, no [2]rotaxane could be isolated after 10 days. The stoppers were too large to permit slippage to occur at a preparatively useful rate.

Extrusion of the macrocycle from the [2]rotaxane was observed by ^{1}H NMR spectroscopy in CD_3SOCD_3 at +100°C.[22,23] The rates of formation of the [2]-

rotaxanes (R = H, Me, Et) are almost identical. However, when R = *i*-Pr, the rate is ten times less. These features, together with the sheer synthetic simplicity of the experimental approach, recommend *slippage* as a viable and alternative synthetic procedure for the construction of the larger and more highly "intelligent" rotaxanes.

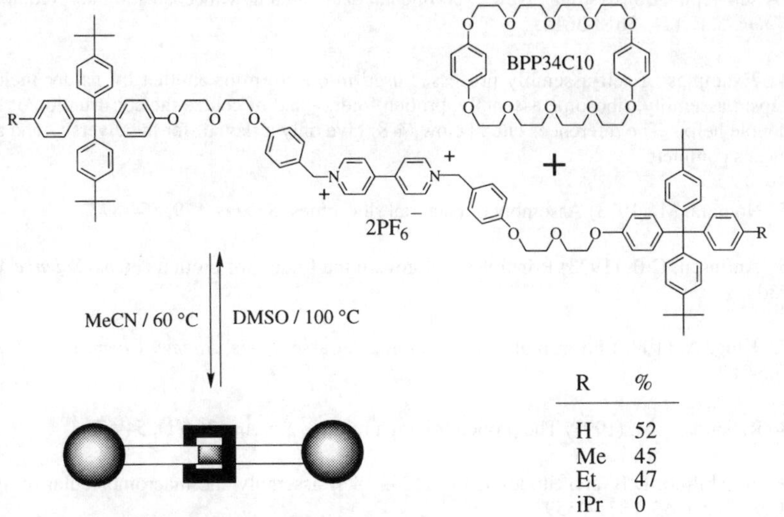

BPP34C10

$2PF_6^-$

MeCN / 60 °C

DMSO / 100 °C

R	%
H	52
Me	45
Et	47
iPr	0

{[2]-[B(MPhtBuPhM)PEEBz][BPP34C10]Rotaxane}$[PF_6]_2$

Scheme 12

7. Some Reflections

We continue to be amazed by the extent and efficiency of self-assembly processes in chemical synthesis. There is an enormous potential for the construction of a diverse range of molecular and supramolecular structures. Now that the foundation has been laid, our explorations into just how far structurally (the form) we can take these systems begins. Furthermore, the addition of photoactive or electrochemically active components could very well open the way for control of the function of these systems. One of the messages that has been noted throughout this account is that it is often easier to make the molecular assemblies than to prepare the molecular components by themselves. There could hardly be a stronger recommendation for the utility of self-assembly in chemical synthesis.

References

1. Lindsey, J.S. (1991) Self-assembly in synthetic routes to molecular devices. Biological principles and chemical perspectives : a review, *New J. Chem.* **15**, 153-180.

2. Ringsdorf, H., Scharb, B., Venzmer, J. (1988) Molecular architecture and function of polymeric orientated systems: Models for the study of organisation, surface recognition, and dynamics, *Angew. Chem. Int. Ed. Engl.* **27,** 113-158. Muller, W., Ringsdorf, H., Rump, E., Wildburg, G., Zhang, X., Angermaier, L., Knoll, W., Liley, M., and Spinke, J. (1993) Attempts to mimic docking processes of the immune system, *Science* **262**, 1706-1708.

3. Tjivikva, T., Ballester, P., and Rebek Jr.J. (1990) Self-replicating systems, *J. Am. Chem. Soc.* **112**, 1249-1250. Rebek Jr.J. (1994) A template for life, *Chem. Br.* **30**, 286-290. Li, T. and Nicolaou, K.C. (1994) Chemical self-replication of palindromic duplex DNA, *Nature* **396**, 218-221. Sievers, D. and von Kiedrowski, G. (1994) Self-replication of complementary nucleotide-based oligomers, *Nature* **369**, 221-224. Menger, F.M., Eliseev, A.V., and Khanjin, N.A. (1994) "A self-replicating system": New experimental data and a new mechanistic interpretation, *J. Am. Chem. Soc.* **116**, 3613-3614.

4. Examples of self-assembly processes used in one form or another by nature include viral capsid assembly, ribosome assembly, protien folding and of course the construction of the DNA double helix. The references cited below (4-8) give only a flavour for the diversity and are by no means complete.

5. Nomura, M. (1973) Assembly of bacterial ribosomes, *Science* **179**, 864-873.

6. Anfinsen, C.B. (1973) Principles that govern the folding of protien chains, *Science* **181**, 223-230.

7. Klug, A. (1983) From molecules to biological assemblies, *Angew. Chem. Int. Ed. Engl.* **22,** 565-582.

8. Richards, F.M. (1991) The protien folding problem, *Sci. Am.* **264**(1), 54-57.

9. Amabilino, D.B. and Stoddart, J.F. (1993) Self assembly and macromolecular design, *Pure Appl. Chem.* **65**, 2351-2359.

10. Watson, J.D. and Crick, F.H.C. (1953) Molecular structure of nucleic acids, *Nature* **171**, 737-738.

11. Anelli, P.L., Ashton, P.R., Spencer, N., Slawin, A.M.Z., Stoddart, J.F., and Williams, D.J. (1991) Self-assembling [2]pseudorotaxanes, *Angew. Chem. Int. Ed. Engl.* **30**, 1036-1039. Stoddart, J.F. Chirality in drug design and synthesis, C. Brown (Ed.), Academic Press, London, 1990, p. 53-81.

12. Anelli, P.L., Ashton, P.R., Ballardini, R., Balzani, V., Delgado, M., Gandolfi, M.T., Goodnow, T.T., Kaifer, A.E., Philp, D., Pietraszkiewicz, M., Prodi, L., Reddington, M.V., Slawin, A.M.Z., Spencer, N., Stoddart, J.F., Vicent, C., and Williams, D.J. (1992) Molecular meccano. 1: [2]Rotaxanes and a [2]catenane made to order, *J. Am. Chem. Soc.* **114**, 193-218.

13. Lehn, J-M. (1990) Perspectives in supramolecular chemistry, *Angew. Chem. Int. Ed. Engl.* **29**, 1304-1319. Koert, U., Harding, M.M., and Lehn, J.-M. (1990) DNH deoxyribonucleo-helicates: Self-assembly of oligonucleosidic double-helical metal complexes, *Nature* **346**, 339-342.

14. Rao, T.V.S. and Lawrence, D.S. (1990) Self-assembly of a threaded molecular loop, *J. Am. Chem. Soc.* **112**, 3614-3615. Manka, J.S. and Lawrence, D.S. (1990) Template-driven self-assembly of a porphyrin-containing supramolecular complex, *J. Am. Chem. Soc.* **112**, 2440-2242.

15. Dietrich-Buchecker, C.O. and Sauvage, J.P. (1989) A synthetic molecular trefoil knot, *Angew. Chem. Int. Ed. Engl.* **28**, 189-192.

16. Seto, C.T. and Whitesides, G.M. (1990) Self-assembly based on the cyanuric acid-melamine lattice, *J. Am. Chem. Soc.* **112**, 6409-6411. Seto, C.T. and Whitesides, G.M. (1991) Self-assembly of a hydrogen bonded 2+3 supramolecular complex, *J. Am. Chem. Soc.* **113**, 712-713.

Whitesides, G.M., Mathias, J.P., and Seto, C.T. (1991) Molecular self-assembly and nanochemistry: A chemical strategy for the synthesis of nanostructures, *Science*, **254**, 1312-1319.

17. Saalfrank, R.W., Stark, A., Bremer, M., and Hummel, H.V. (1990) Formation of a tetranuclear chelate (4-) ions of divalent metals (Mn, Co, Ni) with idealised T symmetry by spontaneous self-assembly, *Angew. Chem. Int. Ed. Engl.* **29**, 311-314.

18. Anderson, H.L. and Sanders, J.K.M. (1990) Amine-template-directed synthesis of cyclic porphyrin oligomers, *Angew. Chem. Int. Ed. Engl.* **29**, 1400-1403

19. Lehn, J.-M. (1988) Supramolecular chemistry - scope and perspective, *Angew. Chem. Int. Ed. Engl.* **27**, 89-112.

20. Amabilino, D.B. and Stoddart, J.F. (1994) Molecules that build themselves, *New Scientist* Vol 141, No 1913, 19 Feb, p. 25-29.

21. Ashton, P.R., Brown, C.L., Chrystal, E.J.T., Goodnow, T., Kaifer, A.E., Parry, K.P., Philp, D., Slawin, A.M.Z., Spencer, N., Stoddart, J.F., and Williams, D.J. (1991) The self-assembly of a highly ordered [2]catenane, *J. Chem. Soc., Chem. Commun.* 634-639.

22. Ashton, P.R., Belohradsky, M., Philp, D., and Stoddart, J.F. (1993) Slippage - an alternative method for assembling [2]rotaxanes, *J. Chem. Soc., Chem. Commun.*, 1269-1274.

23. Ashton, P.R., Belohradsky, M., Philp, D., and Stoddart, J.F. (1993) The self-assembly of [2]- and [3]-rotaxanes by slippage, *J. Chem. Soc., Chem. Commun.* 1274-1277.

24. Ashton, P.R., Philp, D., Spencer, N., and J. F. Stoddart (1991) The self-assembly of [n]pseudorotaxanes *J. Chem. Soc., Chem. Commun.* 1677-1679.

25. Ashton, P.R., Philp, D., Reddington, M., Slawin, A.M.Z., Spencer N., Stoddart, J.F., and Williams, D.J. (1991) The self-assembly of complexes with [n]pseudorotaxane structures, *J. Chem. Soc., Chem. Commun.* 1680-1683.

26. Anelli, P.L., Ashton, P.R., Goodnow, T.T., Slawin, A.M.Z., Spencer N., Stoddart, J.F., and Williams, D.J. (1991) Self-assembling [2]pseudorotaxanes, *Angew. Chem., Int. Ed. Engl.* **30**, 1036-1039.

27. Pedersen, C.J. (1967) Cyclic polyethers and their complexes with metal salts, *J. Am. Chem. Soc.* **89**, 2495-2496. Pedersen, C.J. (1967) Cyclic polyethers and their complexes with metal salts, *J. Am. Chem. Soc.* 89, 7017-7036.

28. Feynman, R.P. (1960) The wonders that await the microscope, *Sat. Rev.* **43**, 45-47.

29. Summers, L.A. (1980) *The Bipyridinium Herbicides*, Academic Press: London.

30. Helgeson, R.C., Tarnowski, T.L., Timko, J.M., and Cram, D.J. (1977) Host-guest complexation. 6. The [2.2]paracyclophanyl structural unit in host compounds, *J. Am. Chem. Soc.* **99**, 6411-6418.

31. Allwood, B.L., Spencer, N., Shahriari-Zavareh, H., Stoddart, J.F., and Williams, D.J. (1987) Complexation of paraquat by a bisparaphenylene-34-crown-10 derivative, *J. Chem. Soc., Chem. Commun.* 1064-1066.

32. It will be convienient at this point to describe the acronyms, which are composed of letters, that identify the neutral and charged compounds displayed throughout this essay. Compounds such as 1,4-dihydroxybenzene and 1,4-dimethoxybenzene are abbreviated to 1/4DHB and 1/4DMB, respectively, and bis-*p*-phenylene-34-crown-10 to BPP34C10. The other acronyms can be deduced from the following rules: B stands for bis when at the beginning, for benzyloxy when in the middle and benzene when at the end of the name. E, H, P, S, and T stand for ethoxy, hydroxy, phenoxy, triisopropylsilyl, and tosyloxy units, respectively. CY, TU, XY, and BIXY represents cyclophane, trioxaundecane, xylylene, and bisxylylene units, respectively. In addition, BIPY stands for a bipyridinium ring system with fromal charges being indicated in the usual way. BBB stands for bis(bromomethyl)benzene. The neutral molecules are unshaded, whereas they are shaded in the case of the positively charged organic species with the formal charges positioned appropriately.

33. Hunter, C.A., and Sanders, J.K.M. (1990) The nature of π–π interactions, *J. Am. Chem. Soc.* **112**, 5523.

34. Langford, S.J. and Stoddart, J.F. (1994) University of Birmingham, Unpublished results.

35. Langford, S.J., Raymo, F.M., and Stoddart, J.F. (1994) University of Birmingham, Unpublished results.

36. Geuder, W., Hunig, S., and Suchy, A. (1983) Phanes with two 4,4'-bipyridinium moieties - A new class of compound, *Angew. Chem. Int. Ed. Engl.* **22**, 489-490. Geuder, W., Hunig, S., and Suchy, A. (1986) Single and double bridged viologenes and intramolecular pimerization of their cation radials, *Tetrahedron*, **42**, 1665-1672.

37. Brown, C.L., Philp, D., and Stoddart, J.F. (1991) The template directed synthesis of a rigid tetracationic cyclophane receptor, *Synlett*, 462-464.

38. Odell, B., Reddington, M.V., Slawin, A.W.Z., Spencer, N, Stoddart, J.F., and Williams, D.J. (1988) Cyclo(paraquat-*p*-phenylene). A tetracationic multipurpose receptor, *Angew. Chem. Int. Ed. Engl.* **27**, 1547-1550.

39. Goodnow, T.T., Reddington, M.V., Stoddart, J.F., and Kaifer, A.E. (1991) Cyclo(paraquat-*p*-phenylene). A novel synthetic receptor for amino acids with electron-rich aromatic moieties, *J. Am. Chem. Soc.* **113**, 4335-4337.

40. Ashton, P, Odell B., Reddington, M.V., Slawin, A.W.Z., Stoddart, J.F., and Williams, D.J. (1988) Isostructural, alternately-charged receptor stacks, *Angew. Chem. Int. Ed. Engl.* **27**, 1550-1553.

41. Philp, D., Slawin, A.M.Z., Spencer, N., Stoddart, J.F., and Williams, D.J. (1991) The complexation of tetrathiafulvalene by cyclobis(paraquat-*p*-phenylene), *J. Chem. Soc., Chem. Commun.* 1584-1586.

42. Ashton, P.R., Goodnow, T.T., Kaifer, A.E., Reddington, M., Slawin, A.M.Z., Spencer N., Stoddart, J.F., Vicent, C., and Williams, D.J. (1989) A [2]catenane made to order, *Angew. Chem., Int. Ed. Engl.* **28**, 1396-1399.

43. Philp, D. and Stoddart, J.F. (1991) Self-assembly in organic synthesis, *Synlett* 445-458.

44. Reddington, M.V. (1989) Non-covalent bonding interactions: cyclophanes, catenanes, and rotaxanes, PhD. Thesis, University of Sheffield.

45. Ashton, P.R., Ballardini, R., Balzani, V., Gandolfi, M.T., Marquis, D.J.-F., Pérez-Garcia, L., Prodi, L., Stoddart, J.F., and Venturi, M., (1994) The self-assembly of controllable [2]catenanes, *J. Chem. Soc., Chem. Commun.* 177-180.

46. G.R. Desiraju (1989) *Crystal Engineering. The design of organic solids,* Elsevier, Amsterdam.

47. Pimentel, G.C. and McClellan, A.L. (1960) *The Hydrogen Bond*, Freeman, San Fransisco. Jeffrey, G.A. and Saenger, W. (1991) *Hydrogen Bonding in Biological Structures*, Springer-Verlag, Berlin. Etter, M.C., Urbanczyk-Lipkowska, Z., Jahn, D.A., and Frye, J.S. (1986) Solid-state structural characterisation of 1,3-cyclohexanedione and of a 6:1 cyclohexane dione: benzene cyclamer, a novel host-guest species, *J. Am. Chem. Soc.* **108**, 5871-5876. Etter, M.C., Urbanczyk-Lipkowska, Z., Zia-Ebrahimi, M., and Panunto, T.W. (1990) Hydrogen bond-directed cocrystallisation and molecular recognition properties of diarylureas, *J. Am. Chem. Soc.* **112**, 8415-8426. Etter, M.C. and Reutzel, S.M. (1991) Hydrogen bond-directed cocrystallisation and molecular recognition. properties of acyclic imides, *J. Am. Chem. Soc.* **113**, 2586-2598. Gorbitz, C.H. and Etter, M.C. (1992) Hydrogen bonds to carboxylate groups. Syn/anti distributions and steric effects, J. Am. Chem. Soc. **114**, 627-631. Etter, M.C., Reutzel, S.M., and Choo, C.G. (1993) Self-organisation od adenine and thymine in the solid state, *J. Am. Chem. Soc.* **115**, 4411-4412.

48. Ashton, P.R., Brown, C.L., Chrystal, E.J.T., Goodnow, T.T., Kaifer, A.E., Parry, K.P., Philp, D., Slawin, A.M.Z., Spencer, N., Stoddart, J.F., and Williams, D.J. (1989) The self-assembly of a highly ordered [2]catenane, *J. Chem. Soc., Chem. Commun.* 634-639.

49. Reddington, M.V., Slawin, A.M.Z., Spencer, N., Stoddart, J.F., Vicent, C., and Williams, D.J. (1991) Towards a molecular abacus, *J. Chem. Soc., Chem. Commun.* 630-634.

50. Ashton, P.R., Philp, D., Spencer, N., Stoddart, J.F., and Williams, D.J. (1994) A self-organised layered superstructure of arrayed [2]pseudorotaxanes, *J. Chem. Soc., Chem. Commun.* 181-184.

51. The name rotaxane is derived [Schill, G. (1971) *Catenanes, rotaxanes and knots*, Academic Press, New York] from the Latin words *rota* meaning wheel and *axis* meaning axle.

52. Anelli, P.L., Spencer, N., and Stoddart, J.F. (1991) A molecular shuttle, *J. Am. Chem. Soc.* **113**, 5131-5133.

53. Ashton, P.R., Philp, D., Spencer, N., and Stoddart, J.F. (1992) A new design strategy for the self-assembly of molecular shuttles, *J. Chem. Soc., Chem. Commun.* 1125-1128.

54. Cram, D.J., Blando, M.T., Park, K.. and Knobler, C.B. (1992) Constrictive and intrinsic binding of a hemicarcerand containing four portals, *J. Am. Chem. Soc.* **114**, 7765-7772.

55. Harrison, I.T. (1972) The effect of ring size on threading reactions of macrocycles, *J. Chem. Soc., Chem. Commun.* 231-232.

OLIGONUCLEOTIDE-DIRECTED RECOGNITION OF DOUBLE-HELICAL DNA

Claude HÉLÈNE and Thérèse GARESTIER
Laboratoire de Biophysique
INSERM U.201 - CNRS URA 481
Muséum National d'Histoire Naturelle
43 rue Cuvier
75231 Paris Cedex 05 (France)

1. Introduction

The rational design of sequence-specific ligands of nucleic acids is an active field of research with several goals i) to provide molecular biologists with new tools to investigate the function of specific genes and the role of specific sequences in the control of gene expression, ii) to develop sequence-specific artificial nucleases that could cleave long DNA fragments at selected sites, e.g., for mapping genes on chromosomes, iii) to provide a rational basis for the development of new therapeutic agents based on selective modulation of gene expression.

During the past few years, oligonucleotides and analogues have received considerable attention as a very versatile class of reagents for sequence-specific recognition and modification of nucleic acids. Several strategies have been developed (1)

1. In the antisense strategy, the oligonucleotide is targeted to a messenger or viral RNA where it forms a double helix through Watson-Crick hydrogen bonding interactions. It can block translation of mRNA and replication or reverse transcription of viral RNA.
2. Ribozymes constitute a special class of oligoribonucleotides which not only bind but also induce cleavage of a target RNA (Fig 1).
3. In the antigene strategy, the oligonucleotide binds to double-helical DNA and forms a triple helix. It can block transcription of a targeted gene. This strategy will be further discussed below (3,4).
4. In the sense strategy, the oligonucleotide (single-stranded or double-stranded) binds and traps a protein involved in the control of gene

C. Chatgilialoglu and V. Snieckus (eds.), Chemical Synthesis, 403–417.

expression. A high selectivity of action is expected when the target protein is of viral or parasitic origin. If a cellular transcription factor is trapped several genes are expected to be affected because transcription factors are usually involved in controlling the expression of several genes (1).

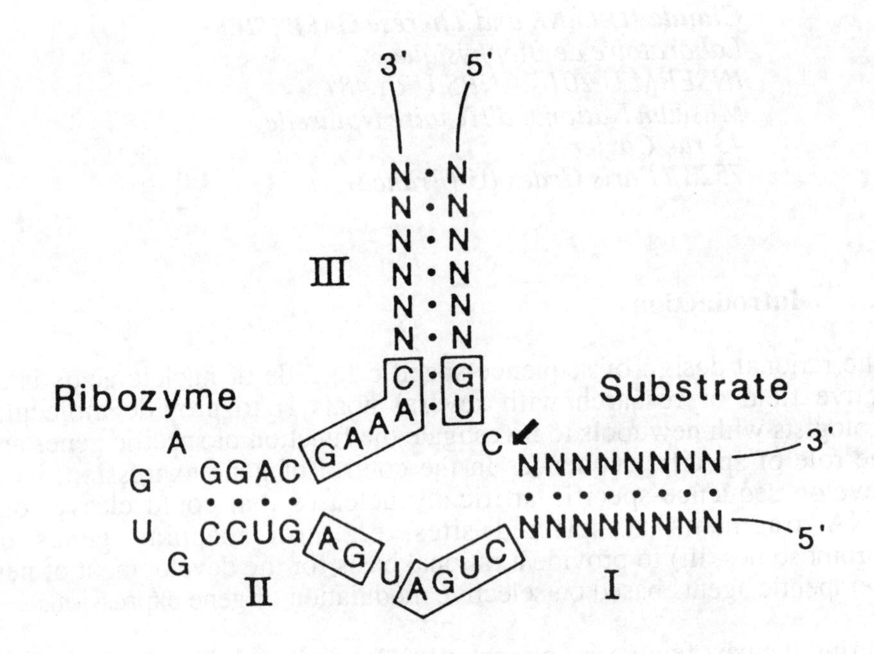

Figure 1 : Ribozyme : a ribozyme is an RNA fragment which binds to a target RNA (messenger or viral RNA) via Watson-Crick base pair formation (indicated by N.N) and induces cleavage of the target RNA at the site indicated by the arrow. The only requirement on the target site is the GUN trinucleotide sequence with N = U,C,A (not G). The boxed nucleotides in the ribozyme sequence cannot be changed to any other base to maintain the cleavage activity. A three-dimensional structure of a ribozyme has been recently published (26). A ribozyme acts as a catalyst if it can dissociate after cleavage of the targets and bind to a new substrate. This requires that the base-paired region (N.N) are sufficiently short to dissociate after cleavage. Ribozymes can be produced within living cells by transcription of appropriate DNA constructs.

5. Oligonucleotides can be selected so as to bind a protein which is not normally involved in any nucleic acid interaction under physiological conditions. These "aptamers" represent a new class of potential inhibitors of enzymes, hormones, receptors, channels (5).

The present review will focus on recognition of double-helical DNA by triple helix formation and its applications in both the antigene and antisense strategies.

2. Triple Helix-Formation

Hydrogen bonding interactions are responsible for holding the two strands of the DNA double helix together, through formation of Watson-Crick base pairs (two hydrogen bonds in an A.T base pair, three hydrogen bonds in a G.C base pair). The base pairs themselves still possess several hydrogen bond donor and acceptor groups. They can be recognized by nucleic acid bases to form base triplets (Fig 2). The following abbreviation will be used : a base triplet designated X.YxZ means that X.Y is the Watson-Crick base pair and Z is the base in the third strand which recognizes the purine Y in the second strand of DNA. Short oligonucleotides can therefore bind to long DNA molecules by forming hydrogen bonds with Watson-Crick base pairs (3,4,6). However, base triplet formation involves recognition of only the purine bases of Watson-Crick base pairs. Therefore formation of a triple helix requires that all purines of the DNA target be on the same strand. Under such conditions, the oligonucleotide wraps around the DNA within the major groove of the double helix. In order to form two hydrogen bonds with G in a C.G base pair, cytosine has to be protonated on N_3. The pK_a of cytosine is 4.1 but due to the polyelectrolyte effect the apparent pK_a in oligonucleotides and triple-helical complexes is raised. Triple helices involving cytosines in the third strand exhibit a pH-dependent stability which varies with the number and distribution of cytosines. All other natural bases in the third strand do not require protonation to form two hydrogen bonds with purines of Watson-Crick base pairs. Base analogues such as pseudoisocytosine and 8-oxo-adenine have been introduced into oligonucleotides to recognize C.G base pairs without any requirement for protonation (Fig 3). Methylation of cytosine at position C_5 enhances the stability of triple-helical complexes, as a result of hydrophobic interactions of these methyl groups with those of thymines (also at position C_5 on the pyrimidine ring).

The orientation of the third strand depends on its base sequence : oligonucleotides containing C and T form Hoogsteen hydrogen bonds and bind parallel to the oligopurine sequence of DNA ; Oligonucleotides containing G and A form reverse Hoogsteen hydrogen bonds and bind antiparallel to the oligopurine sequence. Oligonucleotides containing G and T can bind in both orientations depending on the number of 5'-GT-3' and 5'-TG-3' steps. This result arises from two opposite constraints : on the one hand, the intrinsic stability of the C.GxG base triplets is higher in the Hoogsteen (parallel) rather than in the reverse Hoogsteen (antiparallel) configuration while there is no large difference for T.AxT

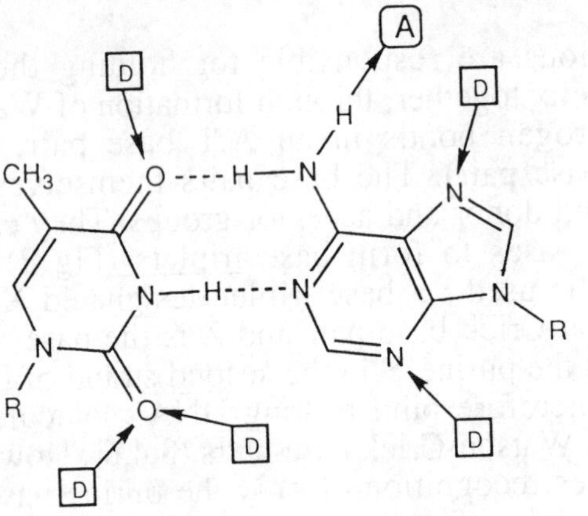

THYMINE - ADENINE

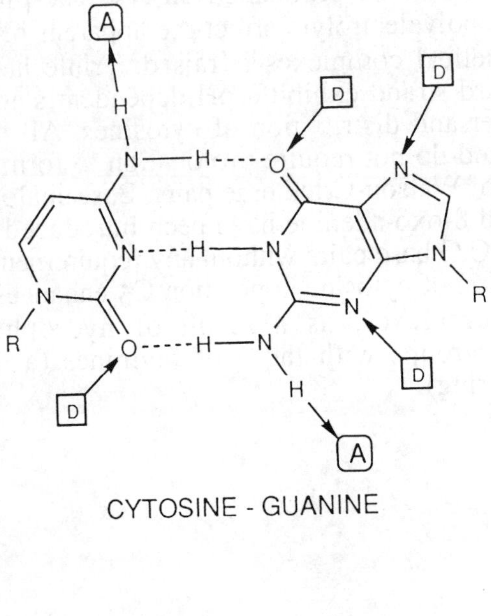

CYTOSINE - GUANINE

Figure 2 :

Watson-Crick base pair formation. Hydrogen bond donor (D) and acceptor (A) sites within both the major and the minor groove are indicated.

base triplets ; on the other hand, the distortion of the third strand backbone when moving from G to T or T to G (in the 5'-3' direction) along the third strand is higher in the Hoogsteen than in the reverse Hoogsteen configuration (in other words the deviation from isomorphism of the C.GxG and T.AxT base triplets is higher in the Hoogsteen than in the reverse Hoogsteen configuration). The net result of these two conflicting constraints is that orientation will depend on the number of -GT- and -TG- steps. A small number of such steps will favor a parallel orientation, a large number, an antiparallel orientation. The total length of the oligonucleotide as well as the distribution of -TG- and -GT- steps might also play a role in determining the oligonucleotide mode of binding.

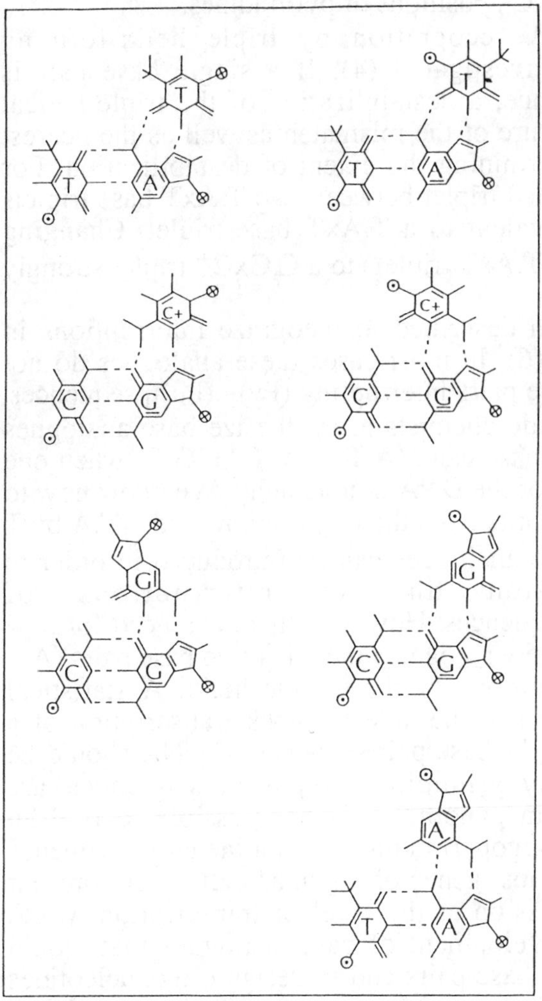

Figure 3 :

Hydrogen-bonding interactions in base triplets formed by natural nucleotides. The left column corresponds to Hoogsteen-like and the right column to reverse Hoogsteen configurations. These configurations correspond to different positions of the glycosidic bond of the third strand with respect to that of the polypurine strand. The orientation of the third strand with respect to the polypurine strand is indicated by a dot (parallel orientation) or a cross (antiparallel) at the position of the C_1' atom of the third strand assuming that all nucleotides adopt the *anti* glycosidic conformation. Non-natural bases can also be used to recognize base pairs in DNA (4).

The two orientations of (G,T)- containing oligonucleotides and the opposite orientations of (C,T)- and -(G,A)-containing oligonucleotides can be exploited to extend the range of recognition sequences for triple helix formation. An oligonucleotide can be designed to recognize two oligopurine sequences, one on each strand of DNA. Alternatively, two oligonucleotides of the same family (C,T or G,A or G,T) which recognize oligopurine sequences that alternate on the two strands can be linked by their 3'-ends or their 5'- ends. For example, binding of two (C,T)-containing oligonucleotides on alternate strands will bring 3'-ends facing each other if the target sequence is 5'(Pu)n(Py)m3' and 5'- ends if the target sequence is 5'(Py)m(Pu)n3'. Different chemical linkages between terminal 3'-OH(or 5'-OH) groups have been designed. Alternatively the linkage can take place via the terminal bases using, for example, a linker between the C_5 positions of pyrimidines.

The specificity of DNA recognition by triple helix-forming oligonucleotides has been investigated (4). If a single base pair is changed in the target sequence, a destabilization of the triple-helical complex is observed. The nature of the mismatch as well as the nearest neighbors play a role in determining the extent of destabilization. For example, an A.TxG mismatched triplet between two T.AxT base triplets gives a stability nearly equivalent to a T.AxT base triplet. Changing either one of the neighboring T.AxT triplets to a $C.GxC^+$ triplet strongly destabilizes the complexes.

Base analogues have been described to recognize interruptions in polypurine target sequences (6). In most cases these analogues do not distinguish A.T from G.C base pairs interrupting $(Py)_n.(Pu)_n$ sequences. It is still a challenge to organic chemists to synthesize base analogues that would recognize all four base pairs (A.T, T.A, C.G, G.C) when one moves along one the grooves of the DNA double helix. We know how to recognize two of these four motifs from the major groove side (T.A by T or A, C.G by C^+ or G). Base analogues can be introduced in order to limit the destabilization due to interruptions of polypyrimidine.polypurine sequences. However, there is a need for new derivatives that would selectively recognize the other two base pairs (A.T and G.C) within the major groove of the double helix. A statistical calculation shows that in order to be able to block transcription of a single gene in human cells, a 17 base pair sequence of DNA should be targeted. Even though polypyrimidine.polypurine sequences are overrepresented in the human genome, it is not possible to find an appropriate (~ 17 base pairs) oligopyrimidine.oligopurine target sequence for triple helix formation in most genes of a human cell. Therefore our ability to control gene expression at the level of transcription would strongly benefit from the development of base analogues that would allow us to recognize all four base pairs and to design oligonucleotides that would bind selectively any sequence of base pairs along the DNA double helix.

3. Oligonucleotide-Intercalator Conjugates

Oligonucleotide-intercalator conjugates have been synthesized by attachment of intercalating agents to the ends of oligonucleotides or to the nucleic acid bases within the sequence (4,7). The purpose of these syntheses was three-fold :

i) to stabilize triple-helical complexes by intercalation at the triplex-duplex junction (e.g., acridine derivatives) (7) or within the triple-helical region (e.g., pyridoindole derivatives) (8) ;

ii) to induce sequence-specific cleavage of long DNA fragments at selected triple helix-forming sites by using intercalating agents that can be activated either chemically (e.g., phenanthroline-Cu chelate) or photochemically (e.g., ellipticine derivatives) (9) ;

iii) to induce sequence-specific cross-linking of the oligonucleotide to its target sequence (e.g., psoralen derivatives) (10).

3.1 STABILIZATION OF TRIPLE-HELICAL COMPLEXES

Acridine derivatives intercalate in DNA in a non-sequence specific way. It was surmised that attachment to the 5' or 3'-end of a triple helix-forming oligonucleotide would allow the acridine ring to intercalate at the triplex-duplex junction (Fig 4). Since intercalation is favored at 5' Py-Pu 3' dinucleotide steps, attachment to the 5'-end of (C,T)-containing oligonucleotides should allow intercalation at the 5' Py-Pu 3' step preceding the target oligopurine stretch. This prediction was borne out by experiments which showed that the melting temperature of a triple-helical complex formed by a (C,T)-containing oligonucleotide with an oligopurine.oligopyrimidine sequence of double-helical DNA was strongly enhanced by attachment of an acridine derivative to the 5'-end of the third strand (7). Stabilization was significantly less upon attachment to the 3'-end.

We have recently described triple-helix specific intercalating agents (8) (Fig 5). Upon binding, they strongly stabilize triple-helical complexes. The equilibrium : duplex + single strand $\rightleftarrows$ triplex, is strongly shifted to the right. Consequently, the melting temperature of the triplex is markedly increased. Two families of molecules were shown to exhibit these properties : benzopyridoindole and benzoquinoxaline derivatives. They prefer to intercalate between T.AxT base triplets. When attached covalently to triple-helix-forming oligonucleotides, they strongly stabilize triple-helical complexes. That intercalation takes place within the triple-helical structure and not at the triplex-duplex junction was demonstrated by photochemical reactions. Upon UV irradiation, benzopyridoindoles form adducts with nucleic acid bases. Modified bases can be identified and shown to be located within the triple-helical region, especially on the oligopyrimidine strand of the DNA target.

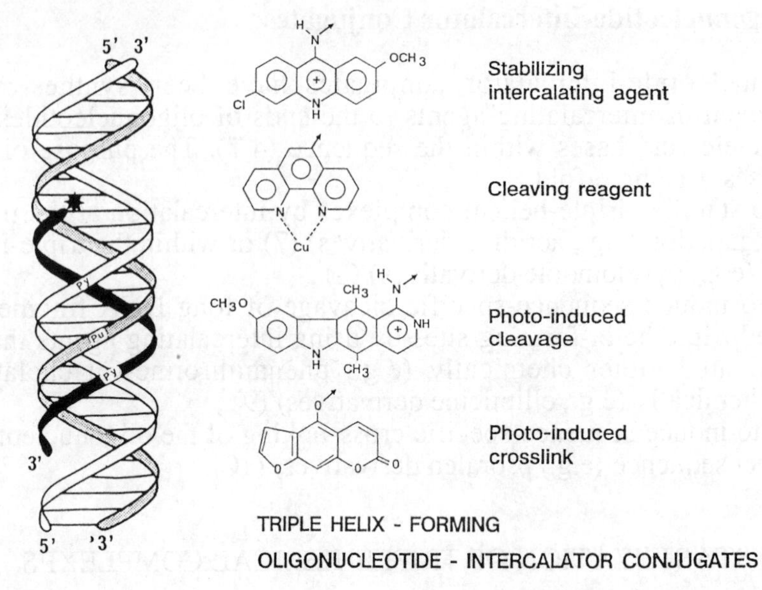

Figure 4 : A triple helix formed by a pyrimidine oligonucleotide (black ribbon) covalently linked to an intercalating agent (star). The intercalating agent binds at the triplex-duplex junction and can i) stabilize the triple-helical structure (acridine derivative, 7), ii) induce double strand cleavage (phenanthroline, 12), iii) photo-induce cleavage (ellipticine derivative, 13), iv) photo-induce cross-linking of the two DNA strands (psoralen, 10).

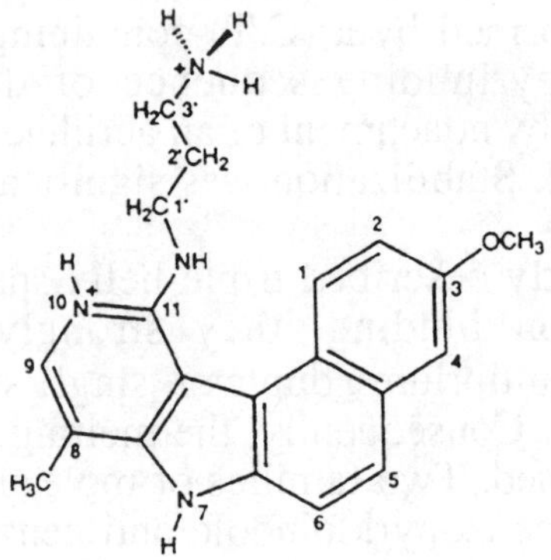

BePI

Figure 5 : Structure of benzo[e]pyridoindole which gives a very efficient stabilization of the triple helix.

3.2 SEQUENCE-SPECIFIC CLEAVAGE

Triple helix-forming oligonucleotides may be armed with an intercalating agent which may be activated to induce cleavage reactions in the DNA target (9). Phenanthroline covalently attached to the 5'-end of the oligonucleotide can intercalate at the triplex-duplex junction. In the presence of copper ions, molecular oxygen and a reducing agent, a single double-strand cut can be observed on a long DNA fragment containing the target sequence for triple helix formation by the oligonucleotide. The yield of the reaction reached ~ 70 % for a phenanthroline-11-mer oligonucleotide conjugate on SV40 (a simian virus) DNA (12). There is a clear need for better cleaving reagents that could be used in living cells to cleave DNA at specific sites, for example within the sequence of a viral DNA. The AIDS virus, HIV, is an RNA virus, whose RNA is first reverse transcribed into DNA which then integrates as a provirus within the genomic DNA of the infected cell. This proviral DNA contains two identical oligopurine.oligopyrimidine sequences of 16 base pairs which are good targets for a phenanthroline-16-mer oligonucleotide conjugate. Cleavage occurs very selectively at the expected sites, excising about 4000 base pairs of viral DNA from the cellular DNA. Clearly the development of a better DNA cleaving reagent that could be tethered to the triple helix-forming 16-mer oligomer and which could be used *in vivo* would provide an interesting way of eliminating nearly half of the viral genome from infected cells and inactivate the viral functions.

An ellipticine derivative attached to the 5'-end of a triple helix-forming oligonucleotide was shown to induce cleavage of the targeted DNA upon photochemical activation in the UV range at wavelengths ($\lambda >$ 300 nm) where nucleic acid bases do not absorb light (13). Unfortunately, the yield of this cleavage reaction was very low. In addition, ellipticine was photochemically cross-linked to the DNA target. The development of an efficient photochemically-activatable cleaving reagent would be of interest as an analytical tool, e.g., for mapping genes on long DNA fragments.

3.3 SEQUENCE-SPECIFIC CROSS-LINKING OF THE TWO DNA STRANDS

Psoralen contains two double bonds that undergo [2+2] photocycloaddition reactions with the 5,6-double bond of thymines (Fig 2). When intercalated at a 5'TA3' step in DNA, the two strands of DNA become covalently cross-linked, upon UV irradiation, with psoralen forming cyclobutane rings with two thymines, one on each strand. The structure of this bis-adduct has been determined by NMR studies and shows that the C_8 atom of psoralen is located on the major groove side of DNA for the intercalation complex that leads to the cross-linking reaction. Therefore the psoralen ring was covalently attached to the 5'-end of triple helix-forming oligonucleotides via the C_8 position. When the target sequence contains a 5'TA3' step on the 5' side of the oligopurine sequence, the psoralen-oligonucleotide conjugate can be photochemically cross-linked to both strands of DNA (10). This photo-induced cross-linking reaction may be used to selectively inhibit gene expression in cell cultures (14).

Alkylating agents may be used to cross-link an oligonucleotide to its target DNA. However, most alkylating agents react with only one strand of the double helix.

4. Alpha-anomeric Oligonucleotides

Natural nucleic acids make use of the β-anomers of nucleotide units as building blocks. It is possible to synthesize the α-anomeric monomers and use them to produce α-oligonucleotides. There is no anomeric effect on the yield of the coupling reactions. α-Oligonucleotides bind to complementary single-stranded sequences through Watson-Crick hydrogen bonding interactions. In contrast to a natural β-oligonucleotide, an α-oligonucleotide binds parallel to its complementary natural sequence. An exception is provided by α-$(dT)_n$ which binds in an antiparallel orientation with respect to a complementary ribonucleotide sequence (poly(rA)) even though it binds parallel to a complementary deoxyribonucleotide sequence (poly(dA)).

α-Oligonucleotides can bind to double-helical DNA to form triple helices (15). The stability of these triple-helical complexes is somewhat less than that of β-oligomers. The α-oligonucleotide binds with an opposite orientation to that of a β-oligonucleotide of the same sequence. One exception is an α-$(dT)_n$ which binds in a parallel orientation with respect to the oligo$(dA)_n$ sequence of the target DNA, as does β-$(dT)_n$.

α-Oligonucleotides are much more resistant to nucleases than β-oligonucleotides. Their uptake by living cells is similar to that of β-oligomers. They can be substituted by intercalating agents to improve binding to their target DNA sequence or to induce irreversible reactions.

5. Oligonucleotide Clamps

An oligopurine sequence on a single-stranded nucleic acid can be recognized by a complementary pyrimidine oligonucleotide. The double-helical complex may in turn be recognized by a second oligonucleotide which can bind and form a triple helix. The two oligonucleotides can be linked together by either an oligoethylene glycol (16) or an oligonucleotide linker (17) (Fig 6). The chimeric molecule can bind to the target oligopurine sequence by both Watson-Crick and Hoogsteen (or reverse Hoogsteen) hydrogen bonding interactions. The guanine is recognized by five hydrogen bonds (3 on the Watson Crick side and 2 on the Hoogsteen side). Adenine is recognized by four hydrogen bonds. Therefore recognition is highly sequence specific : a single base change in the target sequence will be recognized as a mismatch from both the Watson-Crick and Hoogsteen sides. The binding free energy is high as compared to separate oligonucleotides both for enthalpic and entropic reasons.

An intercalating agent can be attached to the 5'-end of the (C,T)-containing oligonucleotide portion which forms Hoogsteen hydrogen bonds (17). If a triple helix-specific intercalator is used (see above), an appropriate linker allows it to fold back and intercalates within the triple-helical region. Alternatively, the Watson-Crick part of the chimeric oligomer can be increased in length by one or two bases to provide a triplex-duplex junction where intercalation can take place (Fig 6). When psoralen is used as an intercalating agent, the two strands of the Watson-Crick part of the complex can be photochemically cross-linked. Therefore the complex is irreversibly locked in place and can block nucleic acid-processing enzymes such as polymerases. Even in the absence of a cross-linking reaction, the chimeric oligonucleotide-intercalator conjugate acts as a clamp that can block biological processes.

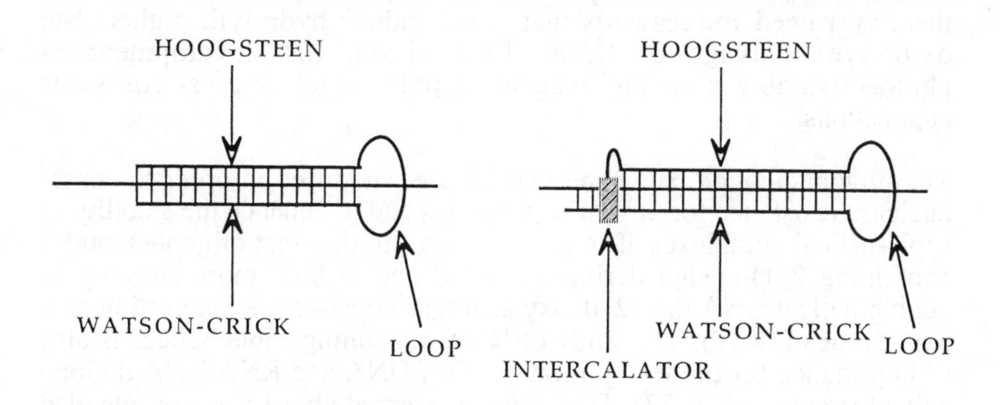

Figure 6 : Oligonucleotide clamps. An oligonucleotide can bind to an oligopurine single-stranded target via both Watson-Crick and Hoogsteen hydrogen bonds (left). An intercalating agent attached to the 5'-end can intercalate at the triplex-duplex junction provided the Watson-Crick part is lengthened by one or two nucleotides (right).

There are several possibilities of linking two oligonucleotides in a triple-helical complexes, depending on the base composition (and therefore orientation) of the third strand. Only very few of these possibilities have been explored to date. Further studies will provide more information on oligonucleotide clamps as a new class of reagents to control gene expression.

6. Conclusion

Oligonucleotides and their analogues represent a very versatile class of sequence-specific DNA ligands. Only purines of purine.pyrimidine Watson-Crick base pairs are recognized when natural bases are used. The formation of two hydrogen bonds per base pair provides a high selectivity of binding. It has been calculated (1) that, on a statistical basis, an oligonucleotide of about 17 bases should be able to recognize a single site within the human genome. This corresponds to the formation of 34 hydrogen bonds. However other hydrogen bonding sites are still available within the major groove and could be used if base analogues were introduced in the third strand. Alternatively, one might combine an oligonucleotide with another ligand binding either the minor groove or the new groove that lies between the third strand and the pyrimidine-rich Watson-Crick strand. Another area of interest involves derivatization of the oligonucleotide third strand to confer new properties to the triple-helical complex such as chemical modification or cleavage of the target double-helical DNA. Some possibilities have been described above. But there is a need for reagents that could induce hydrolytic rather than oxidative cleavage of DNA. In addition, the development of photoactivatable cleaving reagents would be of interest for some applications.

The oligonucleotide backbone may also be modified in order to confer nuclease resistance for *in vivo* applications and to enhance the stability of triple-helical complexes. The recent demonstration that oligonucleotides containing 2'-O-methyl derivatives of C and U bind more strongly to double-helical DNA than 2'-deoxy analogues opens some interesting new possibilities (18,19). The study of RNA-containing triple helices is also of importance for the recognition of RNA.DNA and RNA.RNA double-helical structures (20-22). The survey presented above was not intended to cover all aspects of oligonucleotide-directed triple helix formation. The kinetics of triple helix formation involving oligopyrimidines as third strands has been shown to be much slower than that of double helix formation (23). Changes in association/dissociation kinetics must be investigated when new backbones and/or new bases are introduced in the third strand. The above discussion was limited to target sequences where all purines belong to the same strand of the double helix. There have been several attempts to extend the range of recognition sequences for triple helix formation. A nucleotide analogue bas been recently described which allows for recognition of oligopurine sequences which are interrupted by a pyrimidine (24). It is possible to design oligonucleotides which recognize oligopurine sequences that alternate on the two strands of DNA (25). However, it still remains a challenge to chemists to design base analogues that will recognize all four base pairs (A.T, T.A, C.G, G.C) when one moves along the major groove of the DNA double helix. This research area is of primary importance not only to provide new and selective tools for molecular and cellular biology but also to provide a rational basis for the development of a new therapeutic approach aimed at selectively controlling the expression of specific genes (at the transcriptional level) rather than the activity of the products of their expression.

7. References

1. Hélène, C., Toulmé, J.J. (1990) Specific regulation of gene expression by antisense, sense and antigene nucleic acids. *Biochimica et Biophysica Acta, **1049**, 99-125.*

2. Ojmang J.O., Hampel, A., Looney, D.J., Wong-Staal, F., & Rapaport, J. (1992) Inhibition of HIV-1 expression by a hairpin ribozyme, *Proc. Nat. Acad. Sci,* USA, **89**, 10802-10806.

3. Hélène, C. (1991) The anti-gene strategy : control of gene expression by triplex-forming-oligonucleotides, *Anti-cancer Drug Design*, **6**, 569-584.

4. Thuong, N.T., & Hélène, C. (1993) Sequence-specific recognition and modification of double-helical DNA by oligonucleotides, *Angew. Chem. Internat. Ed. Engl.,* **32**, 666-690.

5. Bock, L.C., Griffin, L.C., Latham, J.A., Vermaas, E.H., & Toole, J.J. (1992) Selection of single-stranded DNA molecules that bind and inhibit human thrombin. *Nature*, **355**, 564-566.

6. Sun, J.S., and Hélène, C. (1993) Oligonucleotide-directed triple-helix formation, *Current Opinion in Structural Biology*, **3**, 345-356.

7. Sun, J.S., Giovannangeli, C., François, J.C., Kurfurst, R., Montenay-Garestier, Th., Saison-Behmoaras, T., Thuong N.T., & Hélène, C. (1991) Triple-helix formation by α-oligodeoxynucleotides and α-oligodeoxynucleotide-intercalator conjugates, *Proc. Natl. Acad. Sci.*, USA, **88**, 6023-6027.

8. Mergny, J.L., Duval-Valentin, G., Nguyen, C.H., Perrouault, L., Faucon, B., Rougée, M., Montenay-Garestier, Th., Bisagni, E., & Hélène, C. (1992) Triple helix specific ligands, *Science*, **256**, 1681-1684.

9. Hélène, C. (1993) Sequence-selective recognition and cleavage of double-helical DNA, *Current Opinion in Biotechnology*, **4**, 29-36.

10. Takasugi, M., Guendouz, A., Chassignol, M., Decout, J.L., Lhomme, J., Thuong, N.T., & Hélène, C. (1991) Sequence-specific photo-induced cross-linking of the two strands of double-helical DNA by a psoralen covalently linked to a triple helix-forming oligonucleotide, *Proc. Natl. Acad. Sci.*, USA, **88**, 5602-5606.

11. Strobel, S.A., and Dervan, P.B. (1992) Triple helix-mediated single-site enzymatic cleavage of megabase genomic DNA, *Methods in Enzymology,* **216**, 309-321.

12. François, J.C., Saison-Behmoaras, T., Barbier, C., Chassignol, M., Thuong, N.T., & Hélène, C. (1989) Sequence-specific recognition and cleavage of duplex DNA via triple-helix formation by oligonucleotides covalently linked to a phenanthroline-copper chelate, *Proc. Natl. Acad. Sci.*, USA, **86**, 9702-9706.

13. Perrouault, L., Asseline, U., Rivalle, C., Thuong, N.T., Bisagni, E., Giovannangeli, C., Le doan, T., & Hélène, C. (1990) Sequence-specific artificial photo-induced endonucleases based on triple helix-forming oligonucleotides, *Nature*, **344**, 358-360.

14. Grigoriev, M., Praseuth, D., Guieysse, A.L., Robin, P., Thuong, N.T., Hélène, C., & Harel-Bellan, A. (1993) Inhibition of gene expression by triple helix-directed DNA cross-linking at specific sites, *Proc. Natl. Acad. Sci.*, USA, **90**, 3501-3505.

15. Sun, J.S., François, J.C., Montenay-Garestier, Th., Saison-Behmoaras, T., Roig, V., Thuong, N.T., & Hélène, C. (1989) Sequence-specific intercalating agents : intercalation at specific sequences on duplex DNA via major groove recognition by oligonucleotide-intercalator conjugates, *Proc. Natl. Acad. Sci.*, USA, **86**, 9198-9202.

16. Giovannangeli, C., Montenay-Garestier, Th., Rougée, M., Chassignol, M., Thuong, N.T., & Héléne, C. (1991) Single-stranded DNA as a target for triple-helix formation, *J. Am. Chem. Soc.*, **113**, 7775-7777.

17. Giovannangeli, C., Thuong, N.T., & Hélène, C. (1993) Oligonucleotide clamps arrest DNA synthesis on a single-stranded target, *Proc. Natl. Acad. Sci.*, USA, **90**, 10013-10017.

18. Shimizu, M., Konishi, A., Shimada, Y., Inoue, H. & Otsuka, E. (1992) : Oligo (2'-0-methyl) ribonucleotides. Effective probes for duplex DNA, *FEBS Lett,* **302**, 155-158.

19. Escudé, C., Sun, J.S., Rougée, M., Garestier, Th., & Hélène, C. (1992) : Stable triple helices are formed upon binding of RNA oligonucleotides and their 2'-0-methyl derivatives to double-helical DNA, *C.R. Acad. Sci., Paris, Série III*, **315**, 521-525.

20. Roberts, R.W. and Crothers, D.M. (1992) Stability and Properties of Double and Triple Helices : Dramatic Effects of RNA or DNA Backbone Composition, *Science*, **258**, 1463-1466.

21. Han, H. and Dervan, P.B. (1993) Sequence-specific recognition of double-helical RNA and RNA-DNA by triple helix formation, *Proc. Natl. Acad. Sci.*, USA, **90**, 3806-3810.

22. Escudé, C., François, J.C., Sun, J.S., Ott, G., Sprinzl, M., Garestier, Th. and Hélène, C.(1993) Stability of triple helices containing RNA and DNA strands : experimental and molecular modeling studies, *Nucleic Acids Research*, **21**, 5547-5553.

23. Rougée, M., Faucon, B., Mergny, J.L., Barcelo, F., Giovannangeli, C., Montenay-Garestier, T. & Hélène, C. (1992) : Kinetics and thermodynamics of triple helix formation : effects of ionic strength and mismatches, *Biochemistry*, **31**, 92699278.

24. Stilz, H.U. & Dervan, P.B. (1993) : Specific recognition of C.G base pairs by 2'-deoxynebularine within the purine•purine•pyrimidine triple helix motif, *Biochemistry*, **32**, 2177-2185.

25. Beal, P.A. & Dervan, P.B. (1992) : Recognition of double helical DNA by alternate strand triple helix formation, *J. Am. Chem. Soc.*, **114**, 4976-4982.

26. Pley, H.W., Flaherty, K.M. & Mc Kay, D. (1994) : Three-dimensional structure of a hammerhead ribozyme, *Nature*, **372**, 68-74.

CAN ENZYME MIMICS COMPETE WITH CATALYTIC ANTIBODIES?

CHRISTOPHER J. WALTER, LINDSEY G. MACKAY
AND JEREMY K. M. SANDERS
Cambridge Centre for Molecular Recognition
University Chemical Laboratory, Lensfield Road
Cambridge CB2 1EW, United Kingdom

1. Introduction

Many scientists are trying to create synthetic host molecules which mimic some of the recognition and catalytic properties of real enzymes. Enzymes catalyse chemical reactions with an efficiency that is awe-inspiring: they can bring together two unreactive substrate molecules, induce them to react, and then release the product at an astonishing speed. They are also subtle, usually forming just one of the many products that would result from a simple reaction carried out by a chemist. We know in general terms that enzymes achieve their catalysis by binding two substrate molecules in close proximity, by using binding energy and conformational changes to facilitate the approach to the transition state, and by using their own functional groups to intervene chemically, but we do not know all the rules. In particular we do not understand the balance between structural rigidity leading to pre-organised binding sites, and flexibility which allows the site to recognise and respond to the shape and size of the bound substrate.

So why should anyone want to build mimics which are bound to be inferior to the real thing? Precisely because we want to discover the rules that we don't know! As Kirby points out,[1] "our current level of understanding fails....the severe practical test...of designing and making artificial enzymes systems with catalytic efficiencies which rival those of natural enzymes." Two quite different strategies for developing enzyme mimics are emerging:

- *Design* In this approach, one decides on the target molecule or system to be made, makes it, and then studies its binding and catalytic properties. The molecule may be made from natural protein components, i.e. amino acids;[2] from other natural components such as carbohydrates or steroids;[3–5] or from totally synthetic

C. Chatgilialoglu and V. Snieckus (eds.), Chemical Synthesis, 419–428.

components.[6] In any case, it will be necessary to use large functional building blocks, ideally in a highly convergent manner, in order to assemble big molecules rapidly. This approach, which we are using as described below, has the attraction that one knows what one is trying to make, and that it is possible to build in systematic rational changes to the system. It has the obvious disadvantage that we are quite likely to fail with any given design because we don't know the rules.

- *Selection* In this approach, one generates a vast number of related structures chemically or biologically by some kind of random or combinatorial process, and then selects the most active. Catalytic antibodies are the classic examples of this approach, but ribozymes provide another promising avenue.[7, 8] Selection from a large 'library' has the advantage of improving the chances of success, but several drawbacks: the selection and identification of a catalytically-active species can be very difficult; imposing rational or systematic change may be difficult; and understanding the details of binding may also be difficult or impossible. Recently the creation of tailored cavities within solid polymers has emerged as a variant of the selection approach with its own attractions and problems.[9]

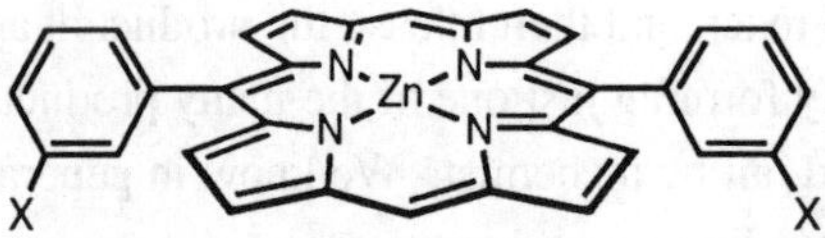

1 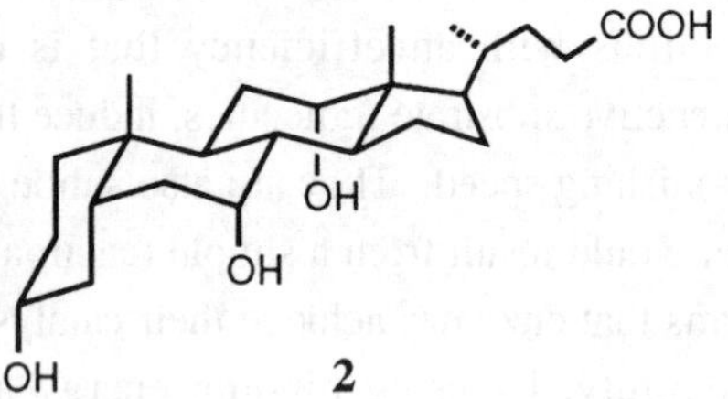 **2**

This article describes the results of our design-based approach. We use large building blocks such as diaryl porphyrins, **1**, and cholic acid, **2**, which can be assembled into macrocyclic host structures that enclose a cavity capable of binding two or more guest molecules in close proximity. Such hosts should catalyse reactions simply by virtue of their binding properties. Cholic acid has the attractions that it is available cheaply and as a single enantiomer, it has a rigid concave surface with inwardly-facing hydroxyl groups for binding and catalysis, and it has terminal functional groups for oligomerisation. Cholate-based receptors have many interesting properties[4, 5] but they are not discussed further here. Diarylporphyrins have the attractions that they are easy to synthesise on a gram scale, they provide a versatile framework for the construction of molecular architectures, they have a delocalised π-system which is spectroscopically eloquent both in UV/visible and NMR spectroscopy, and they possess a central metal ion which provides the binding site. To date we have concentrated on using zinc because it is easy to put in and take out, is diamagnetic and therefore spectroscopically benign, and exclusively four- or five-coordinate in porphyrins. Other metals may, of course, have advantages which we have yet to explore.

Our starting point is the butadiyne-linked porphyrin trimer schematically illustrated in Fig. 1. The ligand-binding properties and synthesis of this molecule and its relatives are described in the accompanying article on templating; the key feature is that the binding constant for pyridine–porphyrin binding at ambient temperature in chlorinated solvents is around 1000 M^{-1}. When two or three pyridine ligands are bound within the cavity their effective concentration is dramatically increased while at the same time their range of relative orientations is limited by the geometry of Zn–N coordination. The trimer should, therefore, act as an 'entropic trap' and accelerate any reaction whose transition-state geometry matches the orientation of bound ligands.

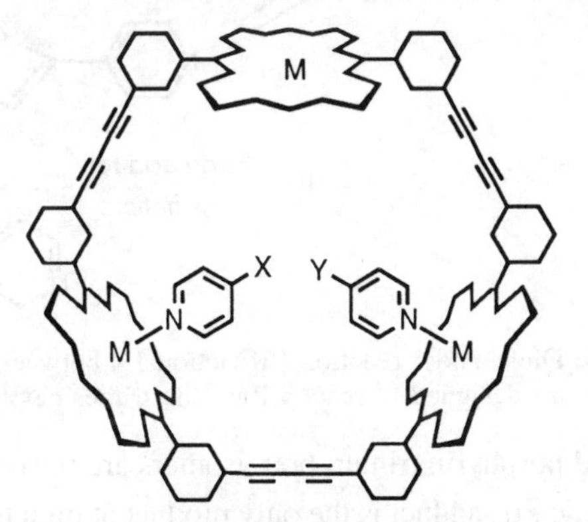

Fig. 1 Schematic view of a porphyrin trimer with two ligands bound within the cavity.

2. Accelerating the Diels–Alder Reaction

The Diels–Alder reaction attracted us because it has an interesting stereo- and regiochemistry, a stringent geometrical requirement, no need for external reagents, and it has been the subject of related studies using catalytic antibodies and cyclodextrins. It also offers the possibility that the stereo- and regiochemistry might be altered through geometrical control within a cavity. We chose to use furan as a diene as its Diels–Alder reaction is reversible: this feature offers the opportunity of studying the kinetics (and therefore the approach to the transition state) in both the forward and reverse direction. The precise reaction studied is shown in Fig. 2; this is a practical synthetic reaction at high concentrations, but at millimolar concentrations the equilibrium lies on the diene–dienophile side. Model building suggested that the *endo*-adduct would fit less well into the trimer cavity than the *exo*-adduct. The measured binding constants confirm this, the *exo*-adduct binding around 15 times more strongly in $C_2H_2Cl_4$ at 30° C (9×10^6 M^{-1} *vs* 6×10^5 M^{-1}) and six times more strongly at 60°.[10]

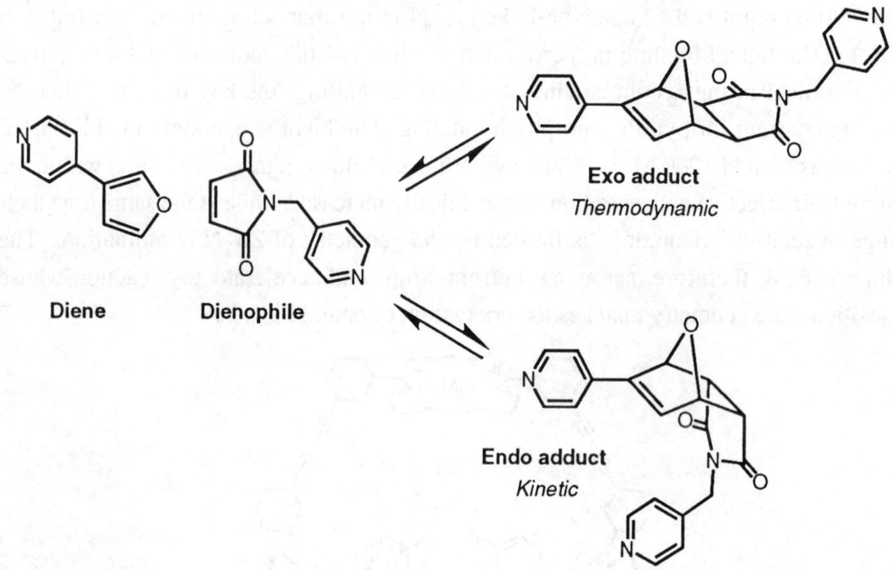

Diene

Dienophile

Exo adduct
Thermodynamic

Endo adduct
Kinetic

Fig. 2 Reversible Diels–Alder reaction ('Reaction 1') between a diene and dienophile which are designed to react within the trimer cavity.

In the absence of added porphyrin trimer, both isomers are formed at room temperature while the thermodynamic *exo*-adduct is the only product at high temperatures. Addition of one equivalent of trimer to a dilute solution of the two reactants (0.9 mM each in tetrachloroethane, 30° C) accelerates the initial observed rate of the forward Diels–Alder reaction around 1000-fold and yields the *exo*-adduct as the only detectable product.[4, 11] Stoichiometric amounts of trimer are required because the products bind strongly and so inhibit the trimer from further activity. The initial reaction rate in the presence of trimer is almost temperature independent under the experimental conditions because as the temperature is raised the binding of diene and dienophile to the trimer decreases; this almost exactly offsets the intrinsic increase in rate of the reaction within the porphyrin cavity.[4] Many control experiments involving other oligomers or competitive inhibitors indicate that the reaction is indeed occurring within the cavity and that the *exo*–transition state is well recognised as shown in Fig. 3.

Measurements of the forward and reverse reaction rates in the absence and presence of trimer allow construction of the complete free energy profile for the reaction at 60° as shown in Fig 4. This profile identifies three separate sources of the *exo*-selectivity in the trimer-accelerated reaction: (i) the *endo*-adduct is intrinsically less stable thermodynamically — this is well known qualitatively, but measurements of the reverse reaction rates of the separated adducts, and an estimate of the forward *endo* rate allow us to quantify the difference as *ca* 13 kJ mol^{-1}; (ii) the *endo*-adduct is less well recognised (bound), as described above; and (iii) experimentally, the trimer slows the reverse reaction of the *endo*-adduct, demonstrating that the *endo*-transition state is less well recognised than the *endo*-adduct; by contrast the reverse reaction of the *exo*-adduct is not slowed by trimer, so the *exo*-transition state is as well recognised as the *exo*-adduct. These three factors combine to place the *endo*-transition state within the trimer cavity some 14 kJ mol^{-1} above the *exo*-transition state, corresponding to around 150-fold ratio for the formation rates.

One consequence of these energetics is that the trimer actually accelerates conversion of the *endo*-adduct to its *exo*-isomer: the reverse reaction is marginally slowed, but once the diene and dienophile have been created they bind within the cavity and are converted to *exo*-adduct. Thus, a 0.9 mM solution of *endo*-adduct in the absence of trimer isconverted over a period of hours to a largely dissociated diene–dienophile pair, while in the presence of trimer it is largely converted instead into *exo*-adduct.

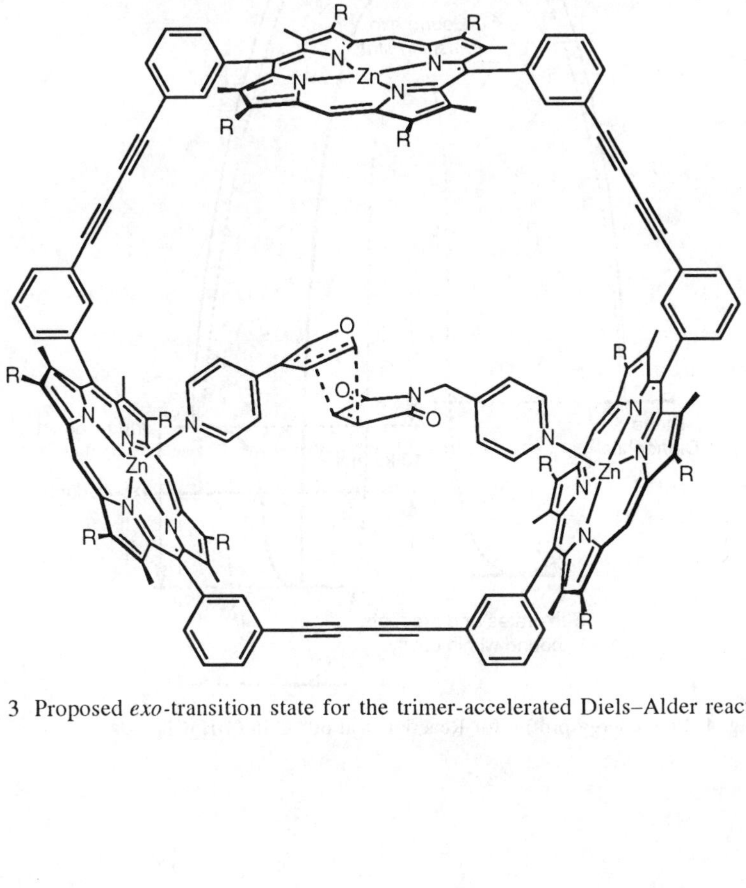

Fig. 3 Proposed *exo*-transition state for the trimer-accelerated Diels–Alder reaction.

One measure of how good a host is at accelerating a reaction is the Effective Molarity (EM).[12] The larger this number, the better acceleration, and natural enzymes can achieve values in excess of 10^{10} M. Computer simulation of our trimer-accelerated Diels–Alder reaction suggests effective molarities of around 20 M at 30°, decreasing at higher temperatures, presumably as a result of greater host flexibility. An X-ray crystal structure of the host[13] confirms that there is a good deal of flexibility, and synthetic efforts are presently being made to stiffen the host in systematic ways. It will be interesting to see whether such hosts are better catalysts because they explore less conformational space or are worse catalysts because they are less well able to distort towards the transition state geometry.

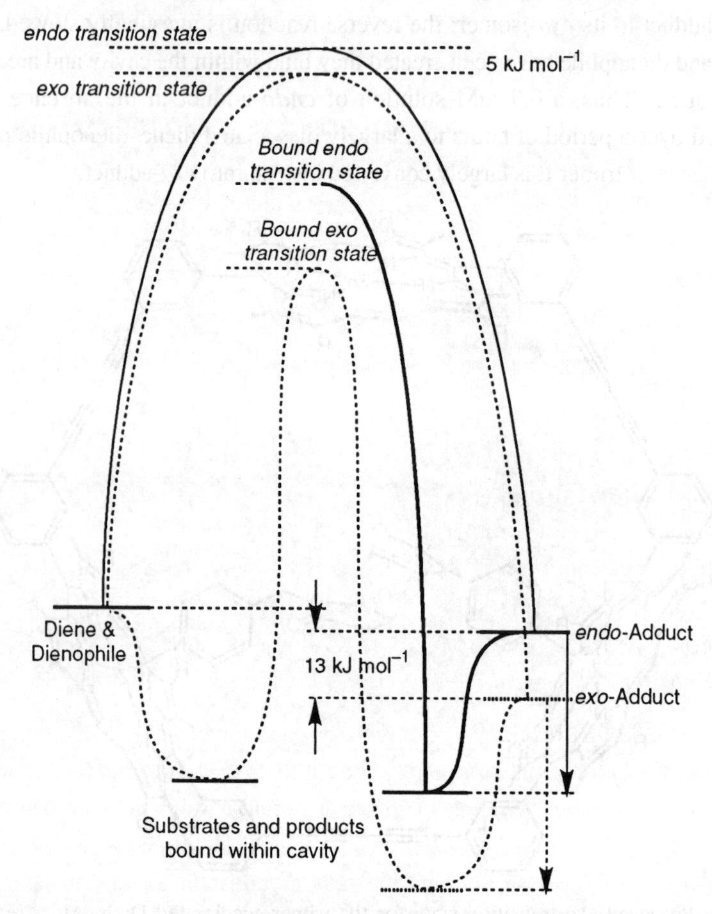

Fig. 4 Free energy profile for Reaction 1 at 60° C in $C_2H_2Cl_4$.

The design of host and guests in this reaction are closely matched, and the question arises as to how important that design actually is. We therefore studied the isomeric Diels–Alder reaction shown in Fig. 5. In this 'Reaction 2', the maleimide portion is the same as in Reaction 1 (Fig. 2), and the porphyrin-binding pyridine is also the same, but their geometrical relationship is different: the maleimide is now attached to the pyridine is at the 3-position. Models suggested a worse fit inside the trimer cavity for the new Diels–Alder adducts, and indeed the binding of the *exo*-adduct is 20–30 times weaker and the acceleration is now only around one tenth that of the first reaction.

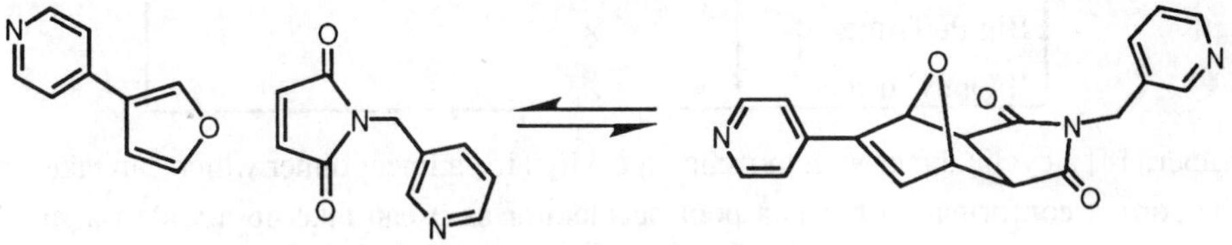

Fig. 5 'Reaction 2', differing only in the geometry of the dienophile.

However models suggested that the 'stretched' dimer **3** would be a better host for this Reaction 2 than for Reaction 1, and that prediction too is borne out by experiment: it does not detectably accelerate the first reaction but it accelerates the second reaction around 50-fold at 30°.[10] Quantitatively, this last result is disappointing, as molecular models suggest a rather good a fit, but qualitatively it is pleasing that we can predict the relative order of reactivities for different geometries.

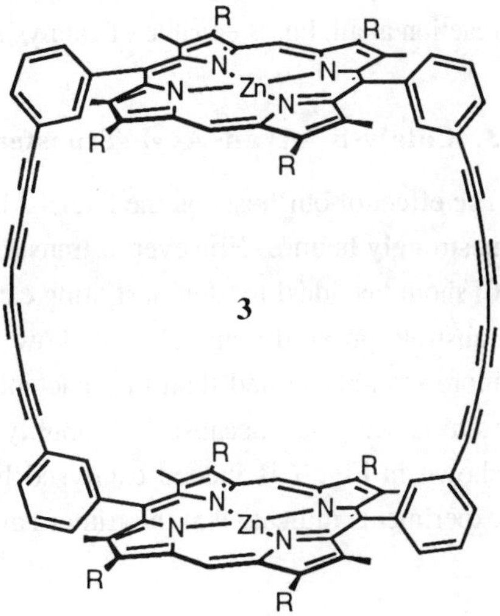

Table 1 summarises the results for both reactions and a variety of different hosts using an NMR assay to screen reaction mixtures; with 0.9 mM substrates the background reaction is so slow that accelerations of less than around 10-fold are not reliably detected and are marked by the symbol ×. Where substantial acceleration is seen (✓), there is a good correlation between the rate and the binding constant for the *exo*-adduct. The hosts studied include a platinum-linked trimer which has a larger cavity than the normal

TABLE 1. Which host can accelerate which reaction?

Host	Reaction 1	Reaction 2
Monomer	×	×
Cyclic Dimer	×	×
Stretched dimer	×	✓
Linear dimer	✓	×
Cyclic Trimer	✓	✓
Big Pt-Trimer	×	
'Floppy' trimer	×	

trimer;[14] a cyclic dimer with too small a cavity;[15] a linear dimer which can take up the correct conformation but is a poor accelerator as it can take up a wide range of strain-free conformations, most of which are unproductive;[15] and, perhaps most intriguing, a trimer which has the butadiyne linkages hydrogenated to give a 'floppy' tetramethylene chain.[4] This floppy trimer is unable to accelerate the Diels–Alder reaction at all, but is capable of catalysing the acyl transfer described below.

3. Catalysis of an Acyl Transfer Reaction

The effect of our hosts on the Diels–Alder reaction is not catalytic because the product is strongly bound. However, a transfer reaction of the type A + BC → AB + C (Fig. 6) should be ideal for demonstrating catalysis and turnover: it should be accelerated by substrate proximity and should show efficient turnover because the products are no more strongly bound than the reactants. Furthermore, the intermediate or transition state is stabilised because it is doubly-bound to the host. The acyl transfer reaction shown in Fig. 7 is indeed catalysed by the same porphyrin trimer,[16] and control experiments indicate that the trimer cavity is a crucial component. In accordance with

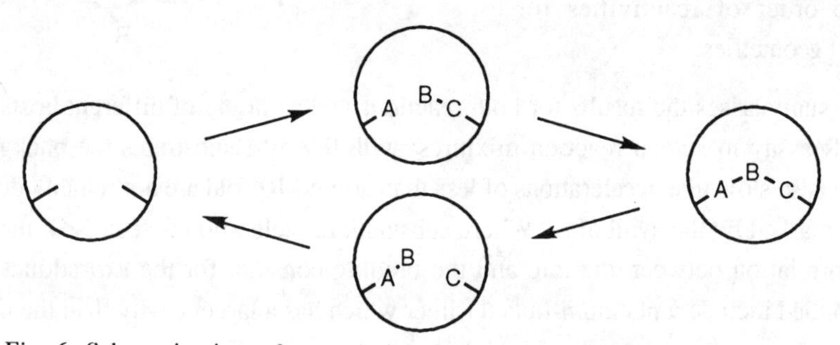

Fig. 6 Schematic view of a proximity-catalysed transfer reaction.

expectation, the trimer is effective catalytically rather than stoichiometrically. We would like to believe that the tetrahedral intermediate shown in Fig. 7 is strongly bound within the trimer cavity, but do not yet have any direct evidence for it.

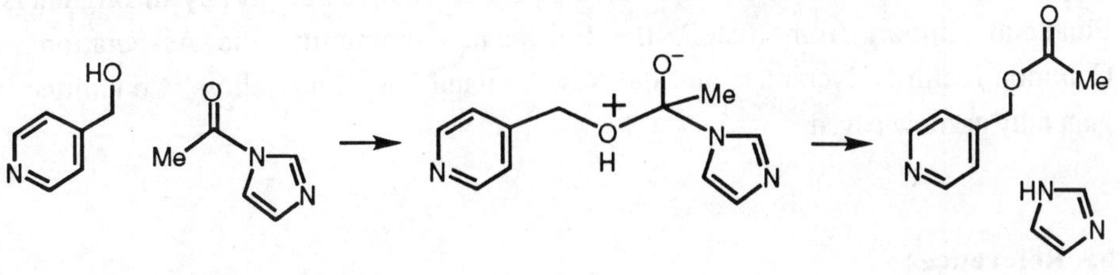

Fig. 7 Acyl transfer reaction catalysed by porphyrin trimer. The precise nature of the tetrahedral intermediate, and the timing of the proton transfer to imidazole, are not known.

We have tested the same range of porphyrin hosts as for the Diels–Alder reaction, and find almost exactly the same spectrum of activity as for reaction 1; this is hardly surprising as both were designed for the same trimer. However, the flexible methylene-linked trimer is still a catalyst for this reaction, even though it is ineffective in the Diels–Alder reaction; we do not yet understand this difference in behaviour but are looking in more detail into the mechanism and scope of the catalysis.

4. Conclusion

As explained in the Introduction, our present design approach is only one of many that are currently being explored. Each approach will have its own advocates, merits and drawbacks, but the field is at an early stage of development and it is too soon to predict how it will turn out. The question posed in the title is deliberately provocative, and the answers will depend on context and what one is trying to achieve, but it is nevertheless worth asking. It is clear, for example, that we can design hosts which will accelerate Diels–Alder reactions with efficiencies which are comparable with those reported for catalytic antibodies. The synthetic design approach allows us to build up a repertoire of closely-related hosts, each of which can be screened for activity in any new reaction we choose to study. Furthermore, we can make rational and systematic design changes which should allow us to factor out the various contributions to catalysis and ultimately to learn the elusive rules. We have a long way to go before we have synthetic enzymes worthy of the name, but we do at least know that it can be done.

5. Acknowledgements

We thank Harry Anderson, Sally Anderson, Richard Bonar-Law, Valèrie Marvaud and Stephen Wylie for valuable help, discussion and samples of porphyrin oligomers. Financial support from SERC, the European Community, the Association of Commonwealth Universities, and the New Zealand Vice-Chancellors' Committee is gratefully acknowledged.

6. References

1. Kirby, A. J. (1994) *Angew. Chemie Intl. Edn.*, **33**, 551–553.
2. Johnsson, K., Alleman, R. K., Widmer, H., and Benner, S. A. (1993) *Nature*, **365**, 530–533.
3. Breslow, R (1991) *Acc. Chem. Res.*, **24**, 318–324.
4. Bonar–Law, R. P., Mackay, L. G., Walter, C. J., Marvaud, V., and Sanders, J. K. M. (1994) *Pure Appl. Chem.*, **66**, 803–810.
5. Bonar-Law, R.P. and Davis, A. P. (1993) *Tetrahedron*, **49**, 9829–9854; Bonar-Law, R.P. and Sanders, J. K. M. (1991) *J. Chem. Soc. Chem. Commun.*, 574–577.
6. (a) Diederich, F. and Lutter, H.-D. (1989) *J. Amer. Chem. Soc.*, **111**, 8438–8446; (b) Mock, W. L., Irra, T. A., Wepsiec, J. P., and Adhya, M. (1989) *J. Org. Chem.*, **54**, 5302–5308; (c) Kelly, T. R., Bridger, G. J., and Zhao, C. (1990) *J. Amer. Chem. Soc.*, **112**, 8024–8034; (d) Nowick, J. S., Feng, Q., Tjivikua, T., Ballester, P., and Rebek, J. Jr (1991) *J. Amer. Chem. Soc.*, **113**, 8831–8839; (e) Schneider, H.-J., Kramer, R. and Rammo, J. (1993) *J. Amer. Chem. Soc.*, **115**, 8980–8984.
7. For reviews see the special issue of *Accounts of Chemical Research*, August 1993.
8. (a) Benner, S. (1993) *Science*, **261**, 1402–1404 (b) Bartel, D. P. and Szostak, J. W. (1993) *Science*, **261**, 1411–1418.
9. Flam, F. (1994) *Science*, **263**, 1221–1222; Heilmann, J. and Maier, W. F. (1994) *Angew. Chemie Intl. Edn.*, **33**, 471–473.
10. Walter, C. J. and Sanders, J. K. M. (1995) *Angew. Chemie Intl. Edn.*, **34**, 217–219.
11. Walter, C. J., Anderson, H. L., and Sanders, J. K. M. (1993) *J. C. S., Chem. Commun.*, 458–460.
12. Kirby, A. J. (1980) *Adv. Phys. Org. Chem.*, **17**, 183–278.
13. Anderson, H. L., Bashall, A., Henrick, K., McPartlin, M., and Sanders, J. K. M. (1994) *Angew. Chemie Intl. Edn. Engl.*, **33**, 429–431.
14. Mackay, L. G., Bonar-Law, R. P., and Sanders, J. K. M. (1992) *J. C. S., Chem. Commun.*, 43–44.
15. Anderson, S., Anderson, H. L., and Sanders, J. K. M. (1993) *Accounts of Chemical Research*, **26**, 469–475.
16. Mackay, L. G., Wylie, R. S., and Sanders, J. K. M. (1994) *J. Amer. Chem. Soc.*, **116**, 3141–3142.

SYNTHESIS OF GIANT MODULAR STRUCTURES

Panel Discussion II: Supramolecular Assemblies and Well-Defined Macromolecules

JOSEF MICHL
Department of Chemistry and Biochemistry,
University of Colorado
Boulder, CO 80309-0215
U.S.A.

1. Introduction

The first of our panel discussions dealt with the total synthesis of natural products, and the second with the relation of organic synthesis to the life sciences. The panelists for the third and last discussion, Professors Lehn [Louis Pasteur University, Strasbourg] , Reinhoudt [University of Twente], and Whitesides [Harvard University], with Prof. Eschenmoser [ETH Zürich] lurking in the background, will consider an even more complex subject, the synthesis of large objects composed of molecular modules.

1.1 FUNDAMENTAL CONCEPTS

As in most relatively young areas of science, in modular chemistry terminology is not yet settled in a definitive fashion. Before launching into the subject it may therefore be wise to agree on the use of several common terms that might otherwise mean somewhat different things to different people. The proposal outlined in the following reflects a personal view and may not be optimal, but should serve the present purpose. As is usual in chemistry, borderlines between different categories defined in modular chemistry are not very sharp, and a fair amount of good will and understanding on part of the audience was expected and received at the three-hour session.

C. Chatgilialoglu and V. Snieckus (eds.), Chemical Synthesis, 429–452.

1.1.1 *Modules*

In a sense, all chemistry is modular, with atoms being the building modules. Presently, however, we restrict the term module to molecular units of at least moderate complexity, certainly not just atoms, and preferably not even very simple molecules. The modules normally possess a fair degree of structural and stereochemical rigidity, and impart some rigidity and shape to the large objects whose synthesis will be the subject of the present panel discussion. Their use makes the synthesis of very large objects practical, since without such subunits the number of individual synthetic steps required to achieved the desired degree of complexity would be overwhelming.

Typically, a synthetic target will contain some passive modules, which serve the function of mere spacers, and some active modules, such as recognition sites, light absorbers or emitters, electron conductors, donors, or acceptors, etc.

1.1.2 *Synthetic Targets*

The modules can be assembled and held together by non-covalent forces, in which case we refer to the resulting large multi-molecule structures as *supramolecular assemblies*. A naturally occurring example is a lipid bilayer, and an artificial one is a Langmuir-Blodgett monolayer. Alternatively, the large structure may be a *single well-defined giant molecule*, constructed from the modules by the use of covalent bonds. A naturally occurring example is a protein, and an artificial one is a well-defined dendritic polymer.

In order to avoid confusion, I propose to reserve the verb "to assemble" for non-covalent synthesis and the verb "to construct" for covalent synthesis. The verb "to synthesize" will then be more general and applicable to either case.

1.1.3 *Synthetic Procedures*

Commonly used concepts that I find particularly difficult to define are self-synthesis, i.e., *self-assembly* or *self-construction* on the one hand, and *actively enforced assembly* or *construction* on the other. The former occurs spontaneously once the right ingredients are present in a reaction vessel, and requires only such external intervention as went into the choice of the reactants and reaction conditions. An obvious example is the assembly of a monolayer of an alkanethiol on a gold suface [1]. There are few good examples of the latter, such as the assembly of a Langmuir-Blodgett film [2] or the as yet mostly hypothetical synthesis with the tip of a scanning tunneling microscope. Multi-step syntheses of large objects also fall into this category, although each individal step may be of the self-construction kind.

In a sense, all ordinary one-pot chemical reactions that produce large molecules from smaller modules without human intervention other than the choice of the initial conditions are self-synthesis, but we shall use the term only for if the starting modules are sufficiently complicated. It will not be used for trivial processes such as the preparation of a hexacyanoferrate from a Fe(III) salt and cyanide.

Although self-assembly and self-construction are spontaneous by definition, the amount of human input that goes into them is usually enormous, and takes the form of instructions that are built into the structures of the starting modules and into the reaction conditions. These then determine the course of the reactions and the structure of the final product.

1.1.4 *Function and Structure*

The synthesis of large structures is ultimately driven by the desire to attain some useful function, such as sensing, information storage, energy conversion, etc., even though some large and complex objects appear to be worthy synthetic targets already because of their purely esthetic appeal. It is hoped that the "designer materials" presumably attainable through modular chemistry will revolutionize material science, microelectronics, and similar disciplines, but these goals still remain to be achieved.

Although the desired function imposes restrictions on the types of structures that can be used for any particular purpose, an overwhelming number of target structures can be considered: biomimetic or entirely artificial, organic or inorganic, linear (one-dimensional), two-dimensional, or three-dimensional, etc. The size of the target structures can range from nanoscopic (scale of nm) to microscopic (scale of μm) and macroscopic (scale of mm and larger).

1.2 DISCUSSION TOPICS AND AIMS

The discussion of the synthesis of very large chemical structures at Ravello developed two extensive themes, one encompassing the general and the other the more specific aspects of the field.

The general part of the panel discussion dealt with the identification of the goals and opportunities, with public support and funding, and with relations to other disciplines. Problems associated with nomenclature were also addressed.

In subsequent discussion attention was turned to the specific issues involved in the synthesis of very large chemical structures, and to a lesser degree, in their function.

2. General Aspects

2.1 ULTIMATE GOALS

2.1.1 *Where Are the Opportunities?*

Professor Whitesides opened the discussion by emphasizing the importance of identifying and analyzing the ultimate goals. Chemists have a unique ability to make molecules and to make materials from them, and constantly exercise it. After all, this is what uniquely distinguishes them from other scientists, such as physicists. The latter are frequently utterly amazed at the chemists' ability to make bonds between atoms in a controlled fashion. Given this ability, how can it be put to use?

There are two ways in which this can be done. One is to look at problems that are presently pressing and interesting, and start from the top down. The other is to proceed in the opposite direction, identifying interesting new ideas and concepts in chemistry and figuring out what problems they might be able to help us solve.

Professor Whitesides did not hide his conviction that there ought to be an element of utility, or potential utility, in what chemists do. He finds it amusing that apparently as a result of this, his colleagues at Harvard will occasionally call him a chemical engineer, a label he does not mind.

2.1.2 *The Top Down Approach*

Professor Whitesides selected the top down approach and asked, what are the current areas of importance? One that used to be important, and is no longer, are national security issues. We are less concerned with dropping explosives on cities from ballistic trajectories, and there is less need for the wonderful "stealth" technology. The preeminent position of this concern has now been assumed by another concern, namely manufacturing, since the production of consumer goods is the plane at which nations now compete. Other problems that remain very much with us are health, information and communications, and environment.

Each of these offers some stunning opportunities to synthetic chemists. Yet, the only one to which organic chemistry has made a significant contribution to date is medicine. In the eyes of society, the existence of synthetic organic chemists is justified by the fact that they prepare beneficial and improved drugs. But how about the wonderful opportunities in the areas of manufacturing, information, and environment, based on new structures leading to new functions?

Traditional organic synthesis has been primarily involved with natural products and drugs. However, nowadays, there may well be more interesting types of target structures to consider, and it is time to move beyond the traditional objectives. With a new set of targets, the most appropriate techniques may also be different, typified perhaps by an electron beam irradiation rather than an aldol condensation?

Given the large number of possibilities listed in the introduction, such as covalent and non-covalent interactions, self-assembly and directed assembly, and so on, one should be able to make good progress. This requires first, an identification of a goal, which should not be just a publication of a paper in a prestigious journal, and second, the knowledge of all or many of these techniques. Since they are so numerous, this is an interesting problem in its own right.

Once one takes a truly entrepreneurial view of the issues that are of current interest in the general area of science and technology, the list of problems to whose resolution the chemists' unique ability to make bonds between atoms can contribute is seen to be virtually endless.

2.1.3 *The Bottom Up Approach*

It seems to me that the top down approach to the choice of a research problem, just discussed at some length by Professor Whitesides, and the bottom up approach, in which an unexpected opportunity comes up and subsequently is exploited for some originally unanticipated application, go in hand in hand. It so happens that my own entry into the field of synthesis of giant molecules is of the bottom up kind. An accidental formation of inert rod-like molecules (functionalized oligomers of [1.1.1]propellane) in a reaction intended to produce a precursor for a completely unrelated spectroscopic study prompted me to consider what such molecular rods might be good for. Ever since, my group has pursued the notion of a molecular-scale analogue of a children's construction kit (i.e., of Tinkertoys), based on rigid rods and connectors. By now we have identified potential practical applications as our immediate goals, and it could be argued that our present research is applied science.

2.1.4 *Applied versus Basic Science: Is It a Meaningful Distinction?*

Compared to the opinions described by Professor Whitesides, Professor Lehn placed less emphasis on the utilitarian aspects of what chemists do and spoke in favor of basic science. Why not do research and explore things just because they are beautiful, or because it is enjoyable to do so, as long as we can convince others that they are worth doing and exploring? After all, they may ultimately be "useful", some perhaps in a practical sense, and others perhaps only because they contribute to mankind's understanding of the universe.

Prof. Eschenmoser stressed that as a member of his generation he is struck by the intensity with which the ultimate applications are emphasized by many chemists who are today at the frontiers of their discipline. One might almost become pessimistic about the future of chemistry as a scientific discipline unless one remembered that chemistry has been application-driven since its very beginnings, and certainly during the entire period of the last hundred years or so, in which the elements of structural theory were developed. One can hope that the new emphasis on applications will result in profound fundamental advances as well. However, it would be a serious mistake to pursue applications to the exclusion of advancement of chemistry itself as a discipline.

Prof. Reinhoudt commented that there is every reason to expect fundamental advances to result from application-driven science, and from efforts that cross the boundaries of what is considered pure chemistry. When the aims are sufficiently bold and lofty, and the target molecules sufficiently complicated, surely pursuing them will lead to new methodologies, new synthetic procedures, new insights into bonding, to the observation of new phenomena, etc.

Prof. Whitesides challenged Prof. Eschenmoser to explain in more detail his views on the origin of creativity in science, and to elaborate on his view of basic versus applied science. Prof. Whitesides expressed serious doubt about the utility of differentiating between "basic" and "applied" research at all, and provided a series of illustrative examples that counter the notion that basic research can lead to fundamental discoveries and applied research cannot. Carnot's discoveries in thermodynamics surely originated in very applied beginnings, but we consider them fundamental science. The basics of information theory resulted from an effort to make a telephone system work well, the background radiation in the universe was discovered as a result of an effort to improve microwave amplifiers, virtually all of polymer chemistry arose from industrial and applied research, genetic engineering is not even applied science, it is applied finance these days, etc. Scientists working on any practical problem always have opportunities to make discoveries and can do extremely exciting applied research. One can also do extremely boring basic research.

2.1.5 *Needed: Discoveries*

In his response, Prof. Eschenmoser resisted the temptation to speculate on the origin of creativity in science and restated his view that chemistry has risen from its modest beginnings to its present state by application-driven research. Occasionally, one has the gut feeling that the applications are a "garstig" affair (a German word that Prof. Eschenmoser refused to translate and which is probably best approximated by "nasty"), but there just is no doubt in his mind that applications are at the very basis of science. Thus, he is not in disagreement with

Prof. Whitesides on this issue.

How does one differentiate good, or strong, and bad, or weak, research? One criterion could be the degree of probability for making a discovery. If this is high, it should not matter whether the topic is basic or applied. Thus, even if a particular applied goal is reached, and the research thus deemed successful from a practical point of view, if there was only low potential for a discovery or none, then from the point of view of fundamental science the research was weak. There is a qualitative difference between a piece of research that increases the sum total of chemical scientific knowledge and one that does not, even if the latter is very successful in its practical application. Prof. Whitesides agreed that there is a difference, and that practical value is a separate issue.

The role of discovery was brought into focus once again when a question from Prof. Sijbesma [Eindhoven University] prompted Prof. Lehn to express the view that although syntheses of supramolecular structures that yield the anticipated result are highly satisfying, those that produce an unanticipated structure are even more valuable, since they teach us something new and represent a true discovery. Prof. Eschenmoser agreed and stated that producing an expected result is in a way trivial, and that discoveries are always more important.

Professor Lehn observed that the current flourishing of supramolecular science is largely a result of timing. Thirty years ago, time would not have been right, while today, it is. All sciences go through their "thousand flowers" periods when imagination flies unbridled and exciting discoveries are on the order of the day, and through periods of consolidation when single-minded attention to nitty-gritty detail is required. The former are much more fun to live in and we are fortunate in that regard. However, the latter are most definitely needed, too.

2.2 FUNDING AND SUPPORT

2.2.1 *Will There Be a Peace Dividend or a Peace Tax for Science?*

The discussion of basic and applied research naturally led to a consideration of public support for chemists' activities. Professor Lehn pointed out that the Ravello meeting was funded by NATO, a military organization, and that one might expect that governments will consider a reallocation of funds that are presumably becoming available at the present time when military spending is being reduced dramatically.

My reading of the view prevalent in the U.S.A. is that no such windfall funds are about to become available. Funds that used to be spent on the military were mostly borrowed from future generations or obtained at the expense of the politically weaker segments of the population. Now, one can anticipate that the scale of such borrowing will be reduced, and social balance perhaps ultimately restored to some degree, but one can hardly expect a flow of previously unavailable cash into science.

To the contrary, there is significant likelihood that public funding of science will be reduced, since politicians will no longer see it as vital to national security. Since we all obviously believe that science has more to offer to the benefit of society now than it ever did in the past, we need to persuade the policy makers that peaceful international competition in non-military arenas, such as those listed earlier by Prof. Whitesides (manufacturing, information, environment), is at least as beneficial as providing national security by military means was in the past. Judging by mankind's past record, we only have a limited period of essentially effortless military security without major threats on the horizon, but one can hope that this period will be long on the scale of human lifespan, and science needs to adjust to the political conditions that will prevail in this period.

2.2.2 *An Uphill Battle for Funding*

Judging by the same past record, scientists are fighting an uphill battle, since traditionally, and sadly enough, science has been supported most ardently at times of war and military tension. If anything, competition for the diminished public funds by a large number of meritorious causes will grow fiercer, and adequate funding for chemistry is by no means guaranteed. It will not be easy to convince the public that scientific support of manufacturing, information and communications, and environment deserves the degree of funding that scientists have become accustomed to in the days of intense military effort. Issues of health are perhaps the most easily grasped by laymen and probably provide the best political argument at present. However, in the long run, environmental issues will clearly become overwhelmingly important.

It is therefore not surprising to me that so much emphasis is placed nowadays on the utilitarian aspects of science, and chemistry in particular. This seems absolutely necessary for the survival of the large-scale scientific endeavor as we know it, and for tapping the wonderful oportunities that we all believe to await us just around the corner.

Professor Lehn listed a common argument in favor of increased spending on science: more knowledge cannot be bad. An example he uses to justify funding for pure science when dealing with politicians and administrators who have no understanding of science at all is a comparison of two slides. One shows the first nuclear magnetic resonance signal obtained by Bloch, the other shows a magnetic resonance image of a part of a human body. Nobody would have ever believed at the time of Bloch's experiment that brain tumor imaging by magnetic resonance might be possible - the jump was just too big. This serves to demonstrate dramatically how practical outcomes that were totally unimaginable in the beginning can result from a seemingly esoteric scientific observation. In general, discoveries in one field that may initially look totally useless or at best secondary may turn out to be of crucial subsequent importance in another field as it evolves. Before monoclonal antibodies were available, it was not easy to imagine how easy medical diagnostics might become. Such examples of mutual enhancement of different areas of science are going to be more frequent in the future, and the present offers wonderful opportunities for science in general.

2.2.3 *Should Additional Instrumentation Centers Be Funded?*

Professor Dalcanale [University of Parma] commented on an aspect of supramolecular science that is particularly closely related to funding. As the frontier moves from molecular recognition to molecular function, analytical tools play an increasingly important role, and become increasingly expensive and exclusive. Their limited availability then becomes a problem. He mentioned a very positive experience that he has had in a collaboration with physicists at the Rutherford Laboratory, and that permitted him to perform some very exotic experiments. Unless chemists develop a similar collaborative spirit and set up centers where expensive and unusual instrumentation is available, perhaps on a Pan-European scale, they may find themselves in a situation where only a few groups can afford to do front-line supramolecular chemistry on molecules tailor-made to function, while the others watch.

The idea was supported by Prof. Lehn. He pointed out that much can be accomplished with instrumentation that is relatively cheap, but for certain types of measurement such resource centers would be very valuable. Having access to the right tools is essential. After all, it was primarily the use of tools that differentiated mankind from other forms of life. Examples he used were electron microscopy, which is now important in his own work, and synchrotron radiation, which permits X-ray structure analyses on extremely small crystals. One does not want to own such machines, so expensive that they must be used all the time, and become their prisoner, but one needs access to them.

Prof. Whitesides also supported the concept of shared facilities, but on a somewhat more sober note. He felt that all too often, we all want to have access to all the instruments rather than just to what is truly needed. Much of what is needed is actually quite inexpensive. He illustrated this in the case of scanning probe microscopy, which, in his view, is one of the most important instrumental advances of the last decade. Much can be done with such simple instruments, and only the truly intractable remaining problems require the use of immensely expensive instrumentation, for which one indeed needs shared facilities. These offer another important benefit as well, as they bring scientists in contact with one another.

I have rather mixed feelings about the funding of research centers myself, at least as it is done in the U.S.A. Obviously, there is no doubt that synchrotrons, neutron sources, and the like absolutely require this type of environment. All too often, however, the various centers that have been set up merely represent a mechanism for siphoning off funds from the research groups of individual investigators, not only to clearly deserving scientists, but also to many who ride the coattails and would never be able to obtain funding on their own in the present highly competitive environment.

2.2.4 *Conference Support Is Available*

Prof. Zwanenburg [University of Nijmegen] returned to the issue of NATO support for science, in his role of a NATO fellow. He was the only chemist on the NATO Advisory Panel for the "Advanced Study Institute" program that considered and approved the proposal for the funding of the Ravello meeting. His physicist and biologist colleagues were rather critical: what is synthesis, what is gnosis, where is the focus? Fortunately, the funding was approved, and he will most enthusiastically report back to the panel that the meeting has been tremendously successful. The organizers responded that it was precisely the breadth of chemical synthesis and its effect on interdisciplinary research that they aimed to emphasize in their proposal.

Most of the NATO-supported meetings are very focussed, and deal with physics, biology, and to a lesser degree, mathematics. This year, about $ 4 million are being spent to organize 60 Advanced Study Institutes, only two of which are in chemistry. Prof. Zwanenburg encouraged the participants, and chemists in general, to be less modest and to send in more applications for NATO ASI's. His suggestion that the Ravello meeting be repeated in three years under the leadership of Professors Chatgilialoglu and Snieckus met with thunderous applause.

2.3 INTERDISCIPLINARY INTERACTIONS

2.3.1 *Interactions across Fields Are the Way of the Future*

On numerous occasions, the discussion turned to the increasingly interdisciplinary nature of modern chemistry in general and supramolecular chemistry in particular. Already in his introductory remarks, Professor Whitesides emphasized that technology transfer between subdisciplines is clearly going to play an increasing role in the future. He exemplified this by the transfer of ideas that are commonplace in inorganic chemistry into innovative assembly of large molecular structures in the hands of Prof. Lehn, by the activities in his own laboratory, where a similar transfer is taking place from material science into organic chemistry, and by the activities of numerous groups that transfer ideas from biology. The transfer from biology is hindered by the existence of a core problem that has not been mastered, namely molecular recognition in aqueous solution. Although some progress has been made, the problem of designing a recognition site for an arbitrary protein surface, with its hydrophobic regions, charges, etc., has proven to be astonishingly intractable.

2.3.2 *Are Chemists Parocchial?*

Prof. Whitesides then remarked that although the breadth of what chemists have to offer is truly breathtaking, outsiders generally view them as some of the most parocchial among scientists. There is so much that a chemist needs to know about his reactions that there usually is no time left for him to learn about other areas. Ask a graduate student in chemistry whether he really understands how a transistor works, what a VCR does, what injection molding involves, etc., and embarrassed silence will result.

Professor Reinhoudt underscored the difficulty that an organic chemist faces when taking the application-driven approach: there is a vast amount of information that needs to be learned first about subjects outside of synthetic chemistry. One needs to learn the physics of devices, the principles of optoelectronics, the language of material engineers, etc., and only after such an investment can one collaborate effectively.

Communication generally seems to be relatively easy with physicists and material scientists, who appear to be very open to the chemists' way of thinking. Unfortunately, it is harder when dealing with biologists, who frequently seem to think that they already know everything that a chemist might be able to contribute. Professor Reinhoudt hypothesized that this may be related to the pecking order among the scientific disciplines, at least as it exists in Holland. At a Dutch university, everyone knows that the physicists are the smartest, then come the chemists, then the biologists, and finally, the medical doctors, even though this is not reflected in the system of financial rewards. Perhaps the physicists are so intelligent that they understand that they need chemists?

Professor Lehn seconded this opinion and confessed that he thought that the pecking order of natural sciences was an originally French invention, with mathematicians coming first, then the physicists, then nobody for a long time, and then chemists and the others. He pointed out that this seems to be changing somewhat now, and that in French schools hands-on science is getting increased emphasis, and even mathematics is taught in less abstract ways.

Prof. Eschenmoser injected that it is easy to understand why a biologist might think that chemistry has nothing to add to what he already knows. After all, to a biologist any self-assembly processes studied by chemists are trivially simple relative to the complex processes of self-assembly in a living cell that he is used to, and the depth of inquiry that is standard for a chemist appears simply irrelevant.

Dr Ripka [Corvas International, Inc.] agreed with Prof. Whitesides concerning the prevailing perception of chemists as excessively parocchial, but felt that it is unjustified. Chemists who are experts in their own area of chemistry also are often quite knowledgeable in other areas of science. Many know a lot of biology and Prof. Whitesides himself is a good example of a chemist who knows much about physics as well. In contrast, it is virtually impossible to find a physicist who knows how to put a molecule together, and some of Prof. Whitesides' own opening comments illustrate this beautifully.

2.3.3 *We Need Popular Books and Articles on Chemistry*

According to Dr. Ripka, the breadth of chemists'knowledge is frequently not recognized, due to insufficient publicity for chemistry in the popular science literature. Few articles express for the general population the rationale behind making new molecules and the excitement associated with it. Compare this with the large number of popular books written by eminent physicists [3], explaining in great detail the elements of very complicated physics, such as particle theory. Even mathematicians [4] do better in this regard than chemists.

Hoffmann and Torrence's [5], Heilbronner and Dunitz's [6], and Djerassi's [7] recent popular books, in which the authors show the relation of many everyday phenomena to chemistry, are exceptions. It is encumbent upon other movers and shakers in chemistry to take the time to write popular explanations of what it is that they are doing. This would be a great service to chemistry as a whole.

In an ensuing exchange between Prof. Whitesides and Dr. Ripka it was agreed that part of the problem also is the vast number of areas outside of chemistry proper in which additional knowledge is needed by chemists, and the limits on suitable mechanisms for acquiring such knowledge. It was pointed out that wide-ranging conferences such as the Ravello meeting are extremely valuable but rare, as most of the meetings chemists go to specialize narrowly in a particular field and would not be useful for a novice who is an expert in a different area.

Professor Lehn remarked that the pictorial and easily visualized nature of many chemical concepts actually makes chemistry quite comprehensible to the general public. He referred to a project involving children's construction sets that introduce the principles of chemistry. Also, the close relation of chemistry to subjects of interest to everyone, such as gastronomy, should provide an easy entry point. He mentioned several successful popular chemistry books available in English and in French [8].

2.4 NOMENCLATURE

2.4.1 *But Do We Need Fancy Words?*

Professor Reinhoudt emphasized the importance of nomenclature. He pointed out that supramolecular chemistry has grown to the present state of vigor from modest beginnings, starting at a time when many traditional chemists viewed "host-guest chemistry", as it was then called, with considerable suspicion as a somewhat quixotic activity. It is perhaps not surprising that many individuals are now coining their own definitions of what is meant by the concepts "supramolecular chemistry" and "supramolecular structure", sometimes motivated by scientific issues, and often by issues related to funding or politics. It is important to clarify the nomenclature in order to facilitate further progress in the field. He provided a word of warning, taken from one of his students' dissertation. She wrote: Frequently, people put more imagination and fantasy into inventing a name for their new molecule than they put into inventing the molecule itself!

Professor Lehn, in contrast, did not see much wrong with that. He pointed out the immense utility of words that may have appeared fancy initially but have served chemists well, such as the boat and the chair forms of cyclohexane. Words play an important role in science, not only because they define things, but also because they suggest things, maybe different things to different people, and that is valuable. In chemistry, definitions are frequently fuzzy, and that is an advantage, since they prompt people to explore peripheral areas. It is often best to attempt to define concepts in a negative way at first: to say what something is not, instead of trying to say what it is. Subsequently, one can come back and sharpen the definitions.

2.4.2 *Self-Assembly and Self-Organization*

Prof. Lehn felt that the definition of supramolecular chemistry is clear and need not be discussed further, but that there are other words whose meaning does need clarification. For instance, when should one talk about self-assembly, and when about self-organization? In his opinion, self-assembly is the more general term and refers to any gathering of molecular entities, the extreme limit being assembly of a gas into condensed phase. He proposed to reserve the term self-organization for that subset of self-assembly processes in which order results. In this regard, there are two types of order. One is order at equilibrium, dictated by thermodynamics, the other is order out of equilibrium, in space and in time. Prof. Lehn recognized that these terms have been used in many different ways and noted the example of a physicist to whom the hydrogen atom was a self-assembled system. If one wished to exclude trivial cases such as this, one could restrict the use of the term to systems that are not at equilibrium.

I pointed out that one can also wonder just how much order is enough to warrant the use of the term self-organization. Is ice self-organized, or merely self-assembled? If ice is self-organized, is a liquid crystal? If so, is an isotropic hydrogen-bonded liquid? Prof. Lehn described some experiments on liquid quinoline that he carried out a long time ago and did not understand at the time. Recently, this non-hydrogen-bonded liquid has been shown to undergo a third-order phase transition in which its internal structure changes. Thus, there is order of a very subtle kind in this isotropic liquid. Clearly, the line between self-assembly and self-organization is fuzzy.

2.4.3 *What Does the "Self" in Self-Assembly Actually Mean?*

Prof. Eschenmoser asked for an expanation of the first half of the word self-assembly or self-organization. What does the self refer to? Prof. Lehn's first response was to recognize the two meanings of "self" in the English language. It can refer to the subject, but it can also mean that a process is spontaneous, automatic. To him, in the present context "self" stands for "auto", i.e., for automatic. Prof. Eschenmoser agreed that this was also the meaning implied in his question, but he wondered about what the "auto" is referring to. To him, it implies "without instructions from the outside", such as changing reaction conditions. However, many of the lectures at Ravello claimed to deal with self-assembly, and yet in the course of the reactions described there were many instructions provided from the outside in between steps that were automatic!

In his opinion, the assembly processes described by Prof. Whitesides, which proceed by themselves once the reagents are mixed, should be called self-assembly or self-organization, unlike those described by Prof. Stoddart, in which there is much chemical manipulation to close rings, etc. If we did not make this distinction, pretty soon all chemical reactions would be called self-assembly, and this would be absurd. Prof. Whitesides, Prof. Reinhoudt, and Prof. Waymouth [Stanford University] agreed. In their view self-assembly implied both spontaneity and the use of more than two components.

2.4.4 *Instructions from the Outside: What Are They?*
Prof. Reinhoudt felt, and Prof. Eschenmoser agreed, that a choice of temperature should not be viewed as instruction from the outside. A system that will self-assemble at room temperature may not do so at another temperature.

I also wondered just how to define an instruction from the outside. Clearly, if a tip of a scanning tunneling microscope pushes atoms along a surface, that is instruction from the outside. Is the choice of a temperature or of a solvent for the self-assembly process not an instruction from the outside? Prof. Heathcock [University of California, Berkeley] added that the structure of the reactants itself is an instruction from the outside. Only in the case of self-replication, when a molecule instructs itself how to produce another copy of itself, as in DNA replication, is the chemist truly out of the loop and is not providing stimuli that cause a certain structure to be formed in each cycle. At this point, I wondered whether living matter will be the only example of self-organization or self-assembly left!

Prof. Lehn agreed that information contained in the reactants is crucial in determining the structure of the self-assembled product. For example, if he had chosen a metal that prefers octahedral instead of tetrahedral coordination for the reactions he presented at Ravello, the products would not have formed. Obviously, one must define the initial conditions: temperature, solvent, pH, reactant structures, etc.

2.4.5 *Self-Assembly and Self-Construction*
It seems to me that once we accept the above, we are back to a situation where the only feature that distinguishes self-assembly or self-organization from any run-of-the-mill synthetic reaction, which also proceeds spontaneously given a suitable set of conditions, will be the requirement that three or more constituent molecules are being assembled. Thus, the old Hantsch synthesis of pyridines from ammonia, formaldehyde, and a ketone, the preparation of sugar mixtures from formaldehyde and base, and all polymerizations will become examples of self-assembly.

If we do not like this, a way out is to require self-assembled species to hold together by weak bonds only, and to use a term such as self-construction for the cases of "covalent self-assembly", as proposed in the introduction (section 1.1.2). This would cause the definition of self-assembly to narrow down to what most chemists, when first asked, would call self-assembly.

However, in Prof. Eschenmoser's view, it is desirable not to narrow down the term self-assembly in this fashion. Indeed, he would view the spontaneous oligomerization of HCN into adenine as a prime example of what self-assembly is. Prof. Whitesides, on the other hand, would prefer not to use the term self-assembly for the oligomerization of HCN, not because the bonds produced are covalent, but because only five molecules are involved. To him, self-assembly might involve a much larger number of subunits, say 10^{14}, although he is willing to accept a smaller number for starters! Prof. Lehn definitely wanted a smaller number as the lower limit. He listed other examples of what could be called self-construction, or covalent self-assembly, similar to the pentamerization of HCN, but was unenthusiastic about calling them self-assembly. These would be perhaps best viewed as ordinary synthetic reactions, and self-assembly could be reserved for targets bound by weaker bonds.

Prof. Eschenmoser defended his position by pointing out that these reactions are in fact quite distinct from the normal synthetic processes, in which each step requires a very specific instruction. He felt that perhaps it will be best to keep the terms covalent and non-covalent self-assembly. I agree that it is useful to distinguish the two cases, but it seems to me more efficient to use the terms self-assembly and self-construction.

2.4.6 *Back to Self-Assembly versus Self-Organization*

Complicating the matter, Prof. Eschenmoser felt that the number of interacting molecules should not be used as a criterion for self-assembly. After all, if two protein molecules aggregate to form a dimer, they may have a large number of interacting groups, and in his view the dimerization represents an intricate example of self-assembly.

However, Prof. Heathcock interjected that one ought to distinguish between self-organization, which the protein dimerization certainly is, and self-assembly requiring a larger number of constituents, which it is not.

Dr. Ripka stated that at this point he is totally confused. He asked where the nomenclature break is in the sequence of polypeptide chain folding, protein dimer formation, tetramer formation, higher assembly formation, ending perhaps in a viral coat formation. At which point in the sequence is the process labeled self-assembly, given that all the steps are based on the same kind of chemical interactions? Prof. Lehn suggested that the first step, chain folding, should not be called self-assembly, since assembly implies the presence of more than a single component. In Prof. Heathcock's view, chain-folding represents a nice example of self-organization, but is distinct from self-assembly.

However, Prof. Hélène [National Museum of Natural History, Paris] objected and pointed out that, contrary to popular perception, most proteins do not fold spontaneously into a unique active conformation, although some do. In fact, there are specific proteins, called chaperons, which make sure in some as yet unknown way that other proteins fold as they should. Left to their own devices, most proteins would fold into a wrong conformation. Thus, if the chaperon proteins represent an instruction from the outside, typical protein folding is not self-organization. I wondered, of course, whether the whole cell should not be viewed as a reaction vessel, in which case the instruction comes from the inside, so the folding is self-organization, after all. Nasty business, these definitions!

2.4.7 *So, What Is Self-Assembly?*

In the end, total agreement among the participants concerning the definition of self-assembly was not reached. Prof. Reinhoudt said that we should not confuse self-assembly with templated synthesis, and should differentiate between covalent and non-covalent bonding (i.e., between self-assembly and self-construction). He also recommended that we should not rely excessively on numbers, particularly not on the difference between two or three constituents. He summarized the situation by reference to an old boss of his at Shell who was in charge of hiring and was not able to describe what makes a prospective employee good, but unerringly recognized one when he talked to him or her. It may be the same with self-assembly. Prof. Whitesides offered a more succinct and rather good-natured summary of his position: I cannot believe that Prof. Eschenmoser is so wrong!

I summarized the outcome of the contentious nomenclature discussion for myself in a way that is not very different from the prejudices I expressed in the introduction: Self-assembly by weak bonds (usually non-covalent) and self-construction by strong bonds (usually covalent) are processes that proceed from several (at least three) or many preferably reasonably complex components in a spontaneous manner once the initial reaction conditions are established. The two types of self-synthesis may but need not introduce well-defined order (regular structure), and both are important in the production of giant chemical entities. Self-organization is the process of spontaneous generation of high order in one or more components by formation of weak or strong bonds.

Although there is no disagreement on what constitutes a supramolecular structure, namely a structure composed from two or more weakly interacting molecules, there clearly will be disagreement as to what constitutes a weak interaction. In my view, a hydrogen bond, however strong, does not tie two molecules into one, while a sufficiently strong coordination bond does. I would apply the term "giant molecules", and not "supramolecular structures", to the beautiful raft designs described by Prof. Lehn and glued together by coordination bonds (e.g., Figure 1). I would view the term modular chemistry as encompassing much of supramolecular chemistry and virtually all of giant molecule chemistry.

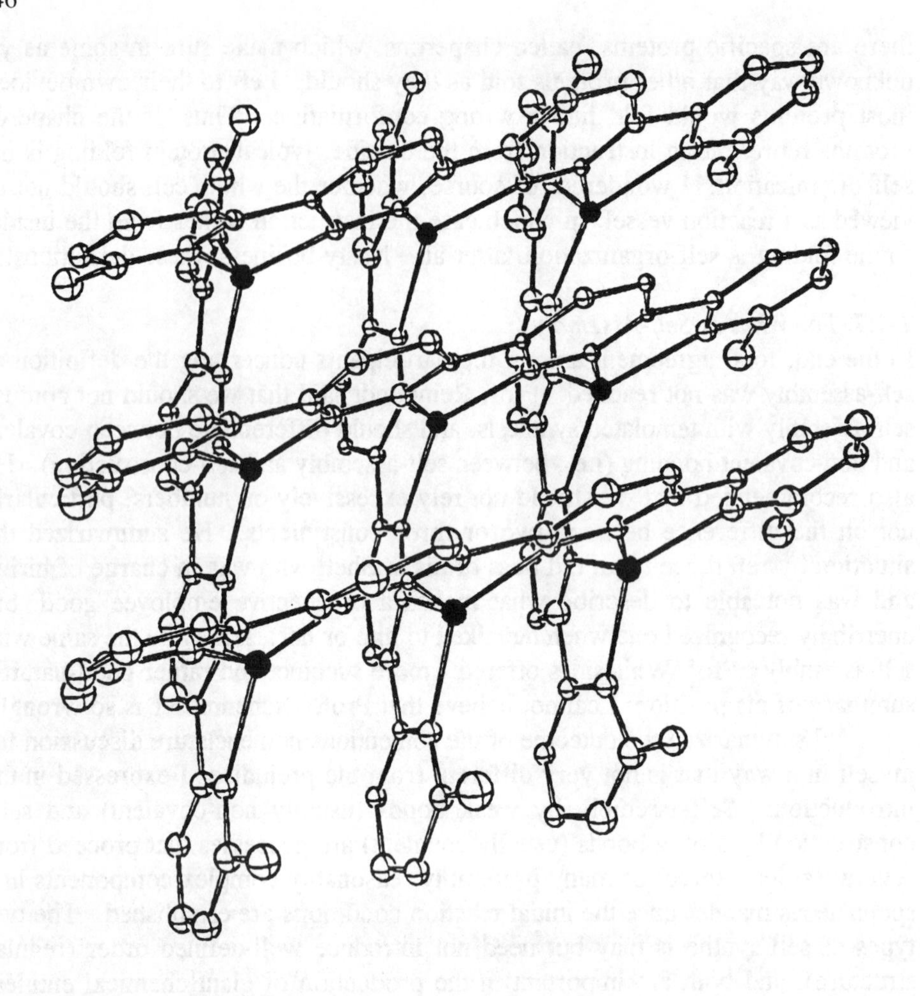

Figure 1. A self-assembled "raft" complex of six molecules of a hexaazaquaterphenyl derivative with nine Ag(I) cations.

2.4.8 *Can Self-Assembly Be Called "Intelligent"?*

Prof. Snieckus [University of Waterloo] brought up another touchy subject. A term that is being increasingly used by those working with molecular recognition is "intelligent". Should we use the expression "intelligent self-assembly", as Prof. Stoddart [University of Birmingham] does? To traditional synthetic chemists, the term "intelligent" used in this context appears redundant. It is also used in a very different context in their subdiscipline. Unfortunately, Prof. Stoddart had already left. Subsequent inquiry revealed that he has no strong feelings regarding the use of the term, which he always places in quotation marks. He uses it primarily to avoid Cram's term "preorganization" and Lehn's term "preprogrammed", and to express his own personality, sometimes simply to get a response.

3. Specific Problems

3.1 ADRESSING INDIVIDUAL MOLECULES

3.1.1 *Molecular Circuitry*

In order to get the discussion of specific problems under way, I returned to an issue that was discussed after one of the morning lectures: given that in our drive towards miniaturization we may eventually be able to build nanoscopic devices in which individual molecules perform functions that are currently performed by the elements of microelectronic or optoelectronic circuits or additional functions that we are presently not able to perform, how shall we address the individual molecules and connect them to the outside world? At times, this has been declared impossible, and much criticism has been leveled at the whole so far non-existent field of molecular electronics as a result.

3.1.2 *Do Individual Molecules Really Need to Be Addressed?*

The problem is actually a bit overstated. First, we shall undoubtedly need redundancy for reliable operation, and will want to address small groups of molecules rather than single individual molecules. Second, not every element of a working circuit will need to be connected to the outside world. If we could hard-wire molecules into intricate circuits, only a small fraction of them would actually need to connect to the outside. Still, the problem is real. Although scanning probe microscopy permits one to obviate the limits posed to far-field scanning by the relatively long wavelength of light that we can afford to use without causing irreversible damage to intricate molecules, it will not permit individual communication with molecules closely packed on a surface. Spreading the molecules apart would defeat the purpose of miniaturization.

3.1.3 *Individual Molecules Already Have Been Addressed!*

Indeed, widely spaced individual molecules already have been addressed with laser light, and the results suggest a way to address much more densely packed molecules individually as well. This is via the use of inhomogeneous static fields, in line with the well-established principles of magnetic resonance imaging and site-selection optical spectroscopy with narrow bandwith laser light. At low temperatures, molecules that are weakly coupled to their environment have extremely sharp spectral lines. It has already been demonstrated that the exact wavelength at which a molecule will absorb or emit is a very sensitive function of local static electric and magnetic fields. In addition to aiming our laser beam or microscope aperture at a particular small spot on the molecular nanocircuit board, and thus addressing a limited number of molecules to start with, we can also tune to a particular wavelength, and thus pick only those molecules in the interrogated area that experience a particular combination of electric and magnetic fields (for a given molecular orientation). These nanoscopically inhomogeneous fields could be provided by situating charged and/or strongly paramagnetic species such as Gd(III) ions (presumably in pairs to lock the direction of the magnetic field in space), along with suitable counterions, at strategic locations within the molecular circuitry. Although this would be difficult to do today, it should be possible in principle, and would probably be no more demanding than the construction of the molecular circuitry itself. Spatial and frequency resolution would then jointly permit the addressing of a small number of molecules at a time, perhaps of just a single one, even if they are packed fairly compactly in three dimensions.

3.2 SYNTHETIC METHODS FOR COMPLEX STRUCTURES

3.2.1 *Additional Foolproof Synthetic Reactions Are Needed*

Professor Whitesides opened this subject by a challenge to the audience: very few known synthetic reactions actually work! I.e., work reliably in a foolproof fashion, in high yield, and in a way that is immune to variation in the structure of the substrates. One of the very few sequences for the coupling of molecules into larger entities that he has found to be fairly reliable is the formation of a chloroacetamide followed by a coupling with a thiolate. It is also possible to react trichloro-1,3,5-triazine with an amine quite reliably. Why are there so few other reactions that are really dependable?

3.2.2 *Are There Foolproof Synthetic Reactions?*

Several synthetic chemists rose to the challenge. Prof. Eschenmoser represented chemists who synthesize natural products when he said that it is a well recognized fact in that field of endeavor that each reaction step, no matter how many times documented in the literature, is a battle until it is accomplished. The circumstances and the structural environment are never the same in two different substrates, and conspire to make each and every reaction intrinsically difficult to accomplish. This should be generally known by all chemists, and nobody who is having difficulties making synthetic reactions work should develop a feeling of inferiority.

With tongue in cheek, Professor Reinhoudt suggested that this leads to the conclusion that natural products synthesis will not teach us anything about useful synthetic methods, and that one of its main *raisons d'être* is therefore spurious. For each individual problem, one needs to fight the battle and find a new solution. Given this, one may as well struggle with the synthesis of a useful molecule and not one that will end up in a bottle on a dusty shelf. Indeed, in his experience, those new and marvellous protecting groups developed for natural product synthesis and touted by Aldrich never work for compounds of interest to him.

Professor Eschenmoser admitted that he may have overstated the problem a bit. He feels that each time a reaction is run under new circumstances one learns more about it, and ultimately the degree of uncertainty about the outcome diminishes. Natural product synthesis therefore is obviously useful. We just do not yet know enough about most reactions.

Of course, a drastic failure of a reaction is not really a failure for chemistry, since one has learnt something new. It is an opportunity to invent a modified or a new reaction. These are very important situations that lead to the invention and progress in chemical synthesis.

Professor Lehn compared a synthetic reaction to a disease. The disease may be well defined, but expresses itself differently in each patient, just like a reaction works a little differently for each molecular substrate. This may not be the answer that Prof. Whitesides would like to hear, but reactions just are not simple. Each substrate will require a different touch, and it is a mark of a great synthetic chemist to have that special touch.

At this point Prof. Snieckus projected a slide with a quote from "La Rotisserie de la reine Pedauque" by Anatole France: "Si les plats que je vous offre sont mal preparés, c'est moins la faute de mon cuisinier que celle de la chimie, qui est encore dans l'enfance". Translated by Prof. Lehn, it read: "If the courses I offer to you are poorly prepared, it is less the fault of my chef than that of chemistry, which is still in its infancy".

3.2.3 *There Are No Foolproof Synthetic Reactions!*

Referring to the short list of reactions that Prof. Whitesides considered foolproof, Prof. Heathcock remembered that in his work several amines could not be acylated easily. He wondered whether any of Prof. Whitesides' amines ever underwent diacylation? Did his α-halo compounds ever reduce instead of substituting by sulfur nucleophiles? It would surely be nice if there did not have to be forty-seven different ways to hydrolyze a dithiane, but there is a need for all of them. Realistically, no reaction is ever going to be totally general, and the short list really is of length zero. There is just too much variety in the possible substrates.

Prof. Whitesides underscored once more that he is actually asking for very little. His aim is not to reproduce exactly some natural structure, but just to hook together pieces that he is trying to assemble for a purpose unrelated to the synthesis itself, and he does not particularly care what the details of the connection look like, just wants reliable connecting reactions. However, he is clearly not going to get what is asking for at the present time, for chemistry is still in its infancy.

3.3 KINETIC VERSUS THERMODYNAMIC CONTROL

3.3.1 *Is Thermodynamic Control Really the Only Way?*

The subject of thermodynamic versus kinetic control was opened by a question posed by Prof. Waymouth. He referred to a statement made in Prof. Whitesides' lecture, concerning a balance between ΔH and $T\Delta S$. If we wish to perform self-assembly strictly under thermodynamic control, all the information has to be placed into the initial units in order to overcome ΔS. The question is, do we really have to rely only on thermodynamics in the synthesis of large supramolecular structures from weakly interacting molecules? Could we take advantage of cooperativity, where we have only a small advantage in ΔH for each assembly unit, but a quite significant total ΔH? How many of the exquisitely organized biological structures are thermodynamically stable? After all, at thermodynamic equilibrium, we would all be dead! Of course, nature does not allow its structures to self-assemble; it has a wonderful machinery for putting them together piece by piece. Could we, too, work with kinetically rather than thermodynamically stable structures based on weak interactions?

To develop the challenge further, I asked how neccessary it is to build with weak bonds. Could we use covalent bonds?

Prof. Lehn pointed out that covalent does not neccessarily imply strong and non-covalent does not neccessarily imply weak. He emphasized the utility of thinking in terms of shallow and deep minima in potential energy hypersurfaces, and considering the effects of temperature. Prof. Eschenmoser emphasized that thermodynamic selectivity control is far simpler than kinetic selectivity control. The fact that we see mostly the former in self-assembly reactions published today is merely an indication that we are only beginners in self-assembly.

3.3.2 *Four Approaches to the Synthesis of Large Structures*
Prof. Whitesides disagreed and felt that thermodynamic control is not all that easy, either, if one wishes to assemble a particular designed structure. He pointed out that there actually are four fundamental strategies for accurately constructing a complex structure. The equilibrium approach relies on the occurrence of repeated association and dissociation steps, which permits a correction of errors, so the system does not get trapped in one of the undesired shallow local minima. The second strategy is allowing mistakes to occur and subsequently editing. This is how much of biology works. The third approach is polymerization, in which a single reaction is repeated, but the resulting structure is not very complex. The fourth strategy is the synthetic organic approach in which mixtures are made and manipulated in a series of separate steps. Prof. Eschenmoser commented that near-equilibrium synthesis represents a special sort of editing.

3.3.3 *Not by Thermodynamics Alone ...*
Prof. Hélène added that biology operates by a combination of thermodynamic and kinetic control and provided an example of a pair of repressors that bind to the lac gene in *Escherichia coli*. One is a protein that binds to a very specific DNA sequence with a high binding constant, 10^{12} M^{-1}. It is an excellent repressor. The other is a mutant that binds to the same sequence with a constant that is 10^2 times higher. From thermodynamics alone, one might expect it to be a super repressor. However, under the same conditions, it is not a repressor at all! Its problem is kinetic: binding to other segments on DNA is also enhanced by a similar factor, and it spends so much time on these other DNA segments that its probability of reaching the best binding segment in the period of time available is virtually nil. At lower temperatures the bacterium division time is much longer and efficient repression by the mutated repressor is then observed [9]. In biological systems, thermodynamic binding is not maximized, it is optimized. There is an important message here for those working with self-organization.

4. Conclusion

Although the discussion panel did not lead to agreement on every issue, particularly not those of nomenclature, the participants appeared quite satisfied and rewarded the panel members by a long applause. The final conclusion of this last session was perhaps best stated by Prof. Jordis [University of Vienna], who said that the two weeks in Ravello have greatly changed his own view of synthetic chemistry and have been extremely valuable. He expressed hope that the thoughts developed at the meeting will not only find their way into the book of proceedings, but that the participants will also use them in their own writing in journals such as ChemTech or the magazines of the various national chemical societies.

5. References

1. Dubois, L.H. and Nuzzo, R.G. (1992) *Annu. Rev. Phys. Chem.* **43**, 437.
2. MacRitchie, F. (1990) *Chemistry at Interfaces*, Academic Press, San Diego, CA.
3. For instance, Wolf, F.A. (1988) *Parallel Universes*, Simon and Schuster, New York, NY; Thorne, K.S. (1994) *Black Holes and Time Warps - Einstein's Outrageous Legacy*, Norton, New York, NY; Feynman, R. (1994) *The Character of Physical Law*, The MIT Press, Cambridge, MA (21st printing); Penrose, R. (1994) *Shadows of the Mind*, Oxford University Press, New York, NY; Lederman, L. (1993) *The God Particle - If the Universe is the Answer, What is the Question?*, Dell, New York, NY; Zee, A. (1989) *An Old Man's Toy - Gravity at Work and Play in Einstein's Universe*, MacMillan, New York, NY; Pagels, H.R. (1985) *Perfect Symmetry*, Simon and Schuster, New York, NY.
4. For example, Salem, L., Testard, F., and Salem, C. (1992) *The Most Beautiful Mathematical Formulas*, Wiley, New York, NY; Gleick, J. (1987) *Chaos - Making a New Science*, Viking Press, New York, NY; Gardner, M. (1986) *Knotted Doughnuts and Other Mathematical Entertainments*, Freeman, New York, NY; Peterson, I. (1988) *The Mathematical Tourist - Snapshots of Modern Mathematics*, Freeman, New York, NY.
5. Hoffmann, R. and Torrence, V. (1993) *Chemistry Imagined - Reflections on Science*, Smithsonian, Washington, DC.
6. Heilbronner, E. and Dunitz, J.D. (1993) *Reflections on Symmetry - in Chemistry and Elsewhere*, Verlag Helvetica Chimica Acta, Basel.
7. Djerassi, C. (1992) *The Pill, Pygmy Chimps, and Degas' Horse*, Harper-Collins, New York, NY.
8. E.g., Atkins, P.W. (1987) *Molecules*, Freeman, New York, NY; McGee, H. (1990) *The Curious Cook*, North Point, San Francisco, CA; Salem, L. (1979) *Molécule la Merveilleuse*, InterEditions, Paris; Lécaille, C. (1991) *Le Monde de la Chimie*, Messidor/La Farandole, Paris.
9. Pfahl, M. (1976) *J.Mol. Biol.* **106**, 857; Hélène, C. and Lancelot, G. (1982) *Progr. Biophys. Molec. Biol.* **39**, 1.

POLYMER SYNTHESIS AND STEREOCHEMISTRY WITH TRANSITION METAL CATALYSTS

R. WAYMOUTH[a], G. W. COATES[a], F. CIARDELLI[b], C. CARLINI[c], A. ALTOMARE[b]

[a]*Department of Chemistry, Stanford University, Stanford, CA 94305, USA*

[b]*Dipartimento di Chimica e Chimica Industriale, Università di Pisa, Via Risorgimento 35, 56100 Pisa, Italy*

[c]*Dipartimento di Chimica Industriale e dei Materiali, Università di Bologna, Viale Risorgimento 4, 40136 Bologna, Italy*

1. Introduction

The discovery of transition metal catalysts for olefin polymerization by Ziegler[1] and Natta[2] in 1955 formed the foundation of today's polyolefin industry, which in 1992 produced over 60 million tons of polymers.[3] Intense research during this new era of polymer synthesis has resulted in a wide range of highly active and selective catalysts for olefin polymerization.[4,5] After about 40 years, commercial Ziegler-Natta catalysts have reached levels of activity and stereospecificity that are unrivaled by any other catalyst system.

The objective of this paper is to provide a fundamental presentation of well established aspects of the synthesis of stereoregular macromolecules and related stereochemical features. Since the discovery of stereoregular polypropylene by Natta and coworkers 40 years ago[6] the number of stereoregular polymers has increased enormously and similar concepts have been extended to natural macromolecules. Therefore, this presentation cannot be really exhaustive and we limit ourselves to polymers of 1-olefins, which were the first studied systems and can be sites of steric isomerism in the main chain (formed during polymerization) and in the side chains (present in the monomer).

All the particular structural properties of macromolecular compounds will be not discussed and the reader is referred to several good reviews which have appeared in the past and recent years.[7] Among the properties related to primary structure of

C. Chatgilialoglu and V. Snieckus (eds.), Chemical Synthesis, 453–473.

macromolecules we shall consider systems where the molecular weight is large enough that the specific property is independent of chain length.

However before dealing with the main object of the present paper it is necessary to recall some basic definitions concerning stereoregular polymers (the product) and stereospecific polymerization (the reaction).

2. Stereoisomerism in Vinyl Polymers

2.1 RELATIVE STEREOCHEMISTRY: TACTICITY

The IUPAC commission on macromolecular nomenclature defines[8] a stereoregular polymer as "a regular polymer whose molecules can be described by only one species of stereorepeating unit", and defines it further as "a configurational repeating unit having defined configuration at all sites of stereoisomerism in the main chain of a polymer molecule". Tactic polymers are considered to be those which exhibit a regular pattern of configurations for at least one type of stereoisomerism sites in the main chain. According to this nomenclature proposal, one can have tactic stereoregular polymers and tactic non-stereoregular polymers, according to whether every type or only one type of stereoisomerism site in the main chain is in a configurationally ordered sequence. Cases in which steric order exists only for stereoisomerism sites in the side chains, the main chain being nontactic (not containing stereoisomerism sites) or atactic are not encompassed within this definition.

This definition refers to ideal structures, but it is conceded that it may be applied also to practical cases where deviations from the ideality are not too large. Despite this, the IUPAC definition is very restrictive. Indeed, stereoregular polymers are not only those which fall under this definition, but also any tactic polymer and any copolymer that contains in the main chain stereoisomerism sites of the same type in a configurationally regular sequence.

The case in which this site (e.g. a double bond or an asymmetric carbon atom) is repeated along the polymer main chain with the same configuration is much more frequent both in natural (natural rubber, natural peptides, poly-β-hydroxybutyrate) and synthetic polymers (isotactic vinyl polymers) than the case in which stereoregularity arises from ordered successions of different configurations. Indeed, up to now, only the ordered alternation of the two possible configurations of asymmetric carbon atoms (syndiotactic vinyl polymers and alternating copolymers of D- and L-amino acids) has been recognized.

In keeping with the established objective of this presentation, macromolecules derived from 1-olefins will be discussed in the main. Within this last frame, both prochiral and chiral 1-olefins will be examined allowing us to consider interrelation between main chain and side chain stereoregularity. It is useful to consider that among the enormously large number of diastereoisomers that can be envisaged for macromolecules formed of head-to-tail units from a prochiral monomer such as polypropylene, attention should be focused on those presenting some kind of short-range order. Two structures stand out immediately: that in which the methyl groups all lie on one side of the chain (*isotactic*),[1,9,10] and that in which they regularly alternate on both sides (*syndiotactic*) (Fig. 1).[11,12] When no order is observed in the positioning of the substituents, the polymer structure is referred to as '*atactic*'.

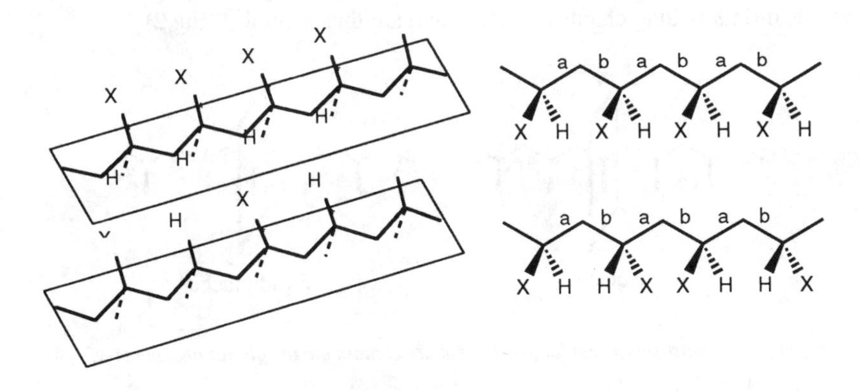

Figure 1.

2.2 ABSOLUTE STEREOCHEMISTRY: CHIRALITY IN VINYL POLYMERS

Analysis of the chirality of a molecule or a macromolecule can be made by looking at its symmetry properties. Macromolecules differ from conventional low molecular weight molecules in that they possess a substantially linear structure. As recently discussed by Farina,[13] small molecules are best described in terms of point symmetry, valid for molecules having definite and "discrete" dimension in all directions. Linear macromolecules are best described in terms of line symmetry, valid for molecules having one infinite dimension. The symmetry criteria for chirality differ slightly for the two types of symmetry classes: in point symmetry, the inversion center and mirror plane must be absent for a molecule to be chiral. In line symmetry, the symmetry elements that are inconsistent with chirality are a mirror plane perpendicular to the chain axis and a mirror glide plane.

Analysis of the symmetry properties of macromolecules can be carried out on the basis of three different models: i) the infinite length chain, ii) the finite length chain with equal end groups, and iii) the finite length chain with different end groups. Point symmetry is most appropriate for the last two models, whereas line symmetry must be used for the first model which implies an infinite dimension. The importance of choosing the appropriate model to describe a molecule was pointed out by Mislow[14] and is illustrated by considering the chirality of polymers as a function of chain length. According to the above concepts, isotactic oligomers can be chiral if the endgroups are different (model iii), but in the limit of high molecular weight, vinyl polymers are best modeled as infinitely long chains (model i), and are thus achiral[13] (Fig.2).

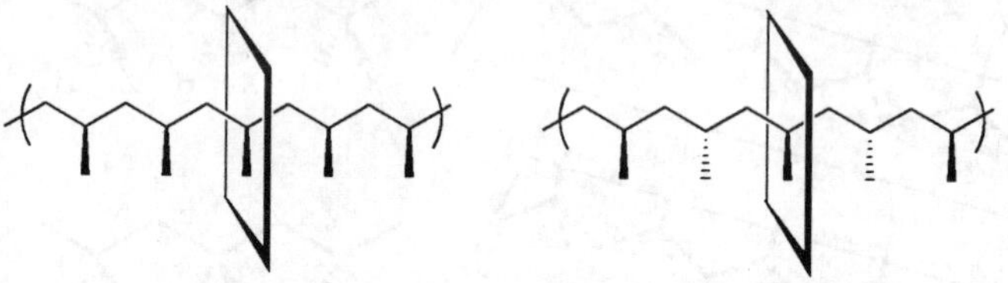

Isotactic Syndiotactic

Figure 2. Most stereoregular vinyl polymers contain mirror planes of symmetry.

If we restrict ourselves to the infinite chain model, the analysis of the derived Fisher projection of an infinite chain indicates that this is chiral when the symmetry plane containing the chain, as well as those perpendicular to the chain and that with translation containing the chain are all lacking. For finite length chains, chirality is granted by the lack of a symmetry plane containing the chain and of the plane perpendicular to the chain in its central point.[13]

All these observations suggest that tactic macromolecules of the above type are substantially, even sometime not formally, achiral and their symmetry properties are better represented by the infinite length chain. Thus in order to prepare macromolecules capable of displaying stable and appreciable chiro-optical properties in all conditions, the symmetry elements rendering the infinite chain model achiral must be eliminated.

The definitions of polymer chirality are also closely connected with the clear description of stereoisomerism in macromolecules. The sites of isomerism in vinyl polymers are comprised of stereogenic centers (rather than asymmetric or psuedo-asymmetric) which can be termed chirotopic or achirotopic depending on whether the polymer is chiral or achiral, according to the above definitions.[13,14]

When propylene or another prochiral 1-olefin is replaced by a chiral 1-olefin such as 3-methyl-1-pentene the stereochemical analysis becomes more complicated. However, when starting with a pure enantiomer the same definition of an isotactic polymer can be valid as for polypropylene, with the additional restriction that all monomeric units have in the side chain an asymmetric carbon atom of a single absolute configuration (Fig. 3).

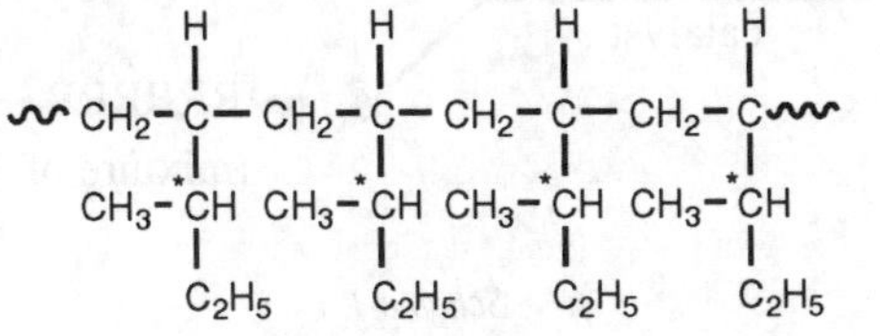

```
    H          H          H          H
    |          |          |          |
~ CH2-C— CH2-C— CH2-C— CH2-C~~
      |          |          |          |
CH3-*CH  CH3-*CH  CH3-*CH  CH3-*CH
      |          |          |          |
    C2H5       C2H5       C2H5       C2H5
```

isotactic poly[(S)-3-methyl-1-pentene]

```
               C2H5                  C2H5
                |                     |
    H      CH3-*CH          H     CH3-*CH
    |           |           |          |
~ CH2-C— CH2-C— CH2-C— CH2-C~~
      |          |           |          |
CH3-*CH          H      CH3-*CH         H
      |                      |
    C2H5                   C2H5
```

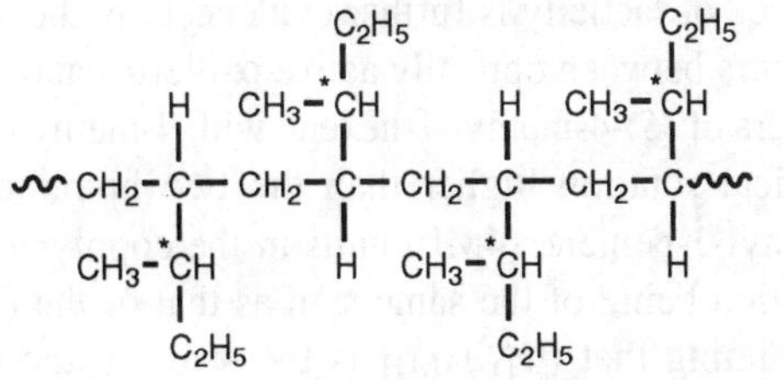

syndiotactic poly[(S)-3-methyl-1-pentene]

Figure 3.

If one starts with a racemic monomer or more generally a mixture of the two enantiomers, different situations can be envisaged. In particular, an isotactic polymer derived from a racemic 1-olefin can give rise to two limiting situations as shown in Scheme 1, corresponding to a random copolymer of the two enantiomers or a mixture of two homopolymers, respectively.[15,16]

Polymers of vinyl monomers having a chiral center may display optical activity, provided the stereogenic center in the side chains has predominantly a single absolute configuration. In these polymers each constitutional unit contains at least two stereogenic centers, one in the main chain and the other in the side chain. The configurational sequence distribution of the former defines the tacticity of the polymer, while the prevalence of a single configuration in the latter determines the enantiomeric purity. Both clearly affect the absolute values of chiroptical properties, which increase with increasing enantiomeric purity and degree of isotacticity.[17,18] Such increase is linear when the side chain stereogenic center is far from the backbone and asymptotic when it is in α or β position, as found for polyolefins.[19,20]

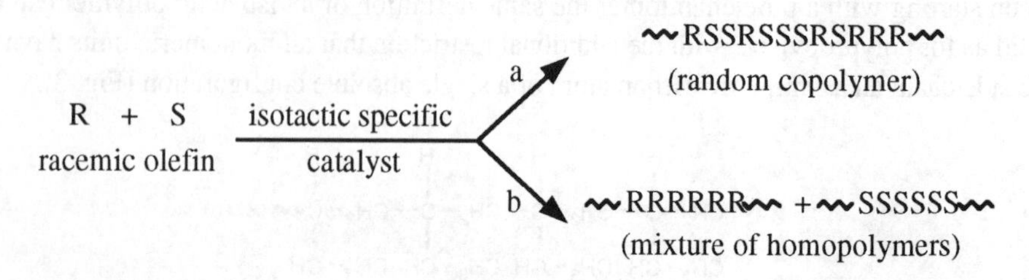

Scheme 1

The cooperative effect anticipated by the asymptotic dependence of chiroptical properties on the degree of tacticity is further evidenced by the study of the chiroptical properties of copolymers between optically active α-olefins and achiral comonomers.[21] Co-isotactic copolymers of (*S*)-4-methyl-1-hexene with 4-methyl-1-pentene show, at any composition, an optical rotation higher than the two homopolymer mixtures, thus indicating that 4-methyl-1-pentene (4MP) units in the copolymer contribute to optical rotation, the contribution being of the same sign as that of the (*S*)-4-methyl-1-hexene (4MH) units. By assuming that $(\Phi_D^{25})_{4MH}$ is the same in the copolymer and in the homopolymer, the values of $(\Phi_D^{25})_{4MP}$ can be derived at each composition by the equation:

$$(\Phi_D^{25}) = N_{4MH}(\Phi_D^{25})_{4MH} + N_{4MP}(\Phi_D^{25})_{4MP} \qquad (1)$$

The value for the most isotactic fraction is +160 against the value +240 calculated for a left handed helical chain of poly(4-methyl-1-pentene).[21] Cooperativity is a critical feature of macromolecular behavior. Studies of the dependence of chirooptical properties on the degree of tacticity in homopolymers and the mole fraction of chiral monomers in copolymers has shed considerable light on the nature of cooperativity in linear macromolecules.

These aspects, as well as the stereochemical structure of the related polymer, are the subject we discuss in the following pages. Many different heterogeneous or supported, as well as homogeneous catalysts can lead to polymers with the above symmetry properties. In the case of heterogeneous catalysts, only hypotheses can be made on the structure of active sites. These aspects will be discussed in better detail in the case of homogeneous catalysts while a more precise relationship between catalyst stereochemistry and polymer stereoregularity is possible with homogeneous metallocene catalysts.

3. Strategies in Stereocontrol

3.1. CHIRAL SITES

Early in 1954, Natta's first experiments with propylene polymerization using complex heterogeneous catalysts yielded products constituted by mixtures of atactic and isotactic polymers. Shortly thereafter, Natta produced polymers consisting primarily of isotactic chains by modifying the composition of the catalyst.[4]

The origin of stereospecificity in Ziegler-Natta catalysis has been an area of intense interest and research since the original discoveries of Natta in the early 50's. To produce isotactic polymers, each olefin insertion must occur on the same enantioface to create stereocenters possessing the same relative configuration, provided that the regio- and stereochemistry of addition to the double bond is always of the same type. To distinguish among two achiral enantiofaces of a simple 1-olefin such as propylene, a chiral environment must be present in the coordination sphere of the metal in order to produce diastereomeric transition states necessary for stereodifferentiation. This chiral environment can be provided by the stereocenters created at the growing polymer chain (chain-end control) or by the coordination geometry of the catalyst site (enantiomorphic site control). One of Natta's early hypotheses was that the origin of the high stereospecificity was due to the regular structure of crystalline transition metal catalyst precursors.[6]

Experimental evidence for the presence of chiral racemic sites on the heterogeneous catalysts proved challenging to obtain, but stimulated some of the most creative and informative experiments in stereospecific polymerization. Among the experimental difficulties in providing a correlation between the absolute stereochemistry of the catalyst sites and the stereospecificity of the reaction is the fact that high molecular weight stereoregular vinyl polymers, despite the presence of stereogenic centers of similar configuration, possess mirror planes of symmetry.

This fact, coupled with the difficulty in resolving racemic sites on a heterogeneous catalyst, stimulated a series of indirect but elegant investigations on the polymerization of chiral monomers.

3.2. POLYMERIZATION OF CHIRAL α-OLEFINS

The polymerization of racemic α-olefins bearing the asymmetric carbon atom at different distances from the double bond gave polymers which could be separated in fractions having optical rotatory power of opposite sign when the asymmetric carbon atom of the monomer was in α- {3-methyl-1-pentene [(*R*)(*S*)3MP] and 3,7-dimethyl-1-octene [(*R*)(*S*)DMO]} or in β-position [(*R*)(*S*)4MH] to the double bond.[22,23] While the chiral discrimination observed decreased from the former to the latter, the enantiomeric prevalence[24-26] was large enough to allow the conclusion that conventional $TiCl_3$ (or $TiCl_4$)/AlR_3 catalysts contain sites capable of discriminating between the two enantiomers of the above racemic α-olefins. In these cases the polymerization of the racemic α-olefin was called "stereoselective".[27] The modest, if any, influence of the growing chain end on chiral discrimination was subsequently confirmed by showing that the average enantiomeric purity of single homopolymer macromolecules was very close to that of random copolymers of the same racemic α-olefin with ethylene.[28]

An application and, at the same time, an additional confirmation of the occurrence of a stereoselective polymerization was obtained by the copolymerization of a racemic α-olefin with a single enantiomer of a chiral α-olefin having a similar structure. Due to the stereoselectivity of the process, the two enantiomers with the same absolute configuration are polymerized on sites having the same chirality giving rise to copolymer macromolecules; the other enantiomer gives homopolymer macromolecules, being polymerized on sites with opposite enantiomeric configuration.[29]

Taking into account the consequent racemic structure of the active sites one could expect to be able to discriminate the two enantiomeric faces of prochiral α-olefins. This was proven by Zambelli et al[30,31] on the basis of NMR studies of polypropylene and ethylene/propylene copolymers, which showed that the chirality of the sites was mainly responsible for the isotactic specific polymerization.

The stereoselective character of the $MgCl_2$ supported Ti catalysts was promptly proved by chromatographic separation into fractions of opposite sign of (*R*)(*S*)4MH polymers obtained in the presence of the $TiCl_4$-ethyl anisate (EA)/$MgCl_2$/$AlEt_3$/EA system.[32] The evaluation of the separation degree showed that for this last catalytic system the stereoselectivity was higher than for $TiCl_3$ "ARA" based non supported catalysts and particularly for the more stereoregular polymer fractions.

The stereoselectivity and then the enantiomorphic structure of the active sites in the supported catalyst has been susequently and simply confirmed[33] by the copolymerization of a racemic α-olefin, (*R*)(*S*)-3,7-dimethyl-1-octene [(*R*)(*S*)DMO] with optically active

(*S*)-3-methyl-1-pentene [(*S*)3MP] in the presence of the $MgCl_2/EtB/PC/AlEt_3/TiCl_4$ catalytic system previously prepared,[34] where the internal base EtB is ethyl benzoate and PC is *p*-cresol. As the active sites discriminate on the basis of chirality rather than chemical structure, a random copolymer of (*S*)DMO with (*S*)3MP was obtained together with the homopolymer of (*R*)DMO as observed for non-supported systems.

Fractionation of the polymer by solvent extraction, optical rotation ($[\alpha]_D^{25}$) and IR absorption (D_{763}/D_{732}) of the fractions indicated that the more soluble polymer fractions consisted, as expected, of copolymer of (S)-monomers, whereas the last fraction was based on poly[(R)DMO)]. In the presence of the above high yield system, activated by $Al(i\text{-}Bu)_3$ and *p*-methyl toluate (MT) as external base, DMO and (S)3MP (3.8/1 molar ratio) gave results very similar to the conventional catalyst.[29] This indicates that the chiral recognition of active centers capable of polymerizing these sterically hindered branched olefins remains substantially the same even in the presence of the support ($MgCl_2$) and the Lewis base.

The strict connection between chiral structure of the active site and polymer stereochemistry, pointed out by the investigations on the heterogeneous Ziegler-Natta catalytic polymerization, has been fully confirmed and extended by the studies concerning polymerization of prochiral α-olefins in the presence of metallocene catalysts.

4. Homogeneous Catalysts

The discoveries by Kaminsky and Sinn that Group 4 metallocenes in the presence of methylaluminoxane were extremely active for olefin polymerization[35,36] caused a considerable resurgence of interest in the application of homogeneous catalysts for olefin polymerization. Surprisingly perhaps, the modification of group IV metallocenes to produce catalysts capable of isospecific polymerization developed more slowly than that of the heterogeneous catalysts by Natta, but has recently seen dramatic success.

Prior to the mid-1980's, catalysts formed using achiral Cp_2MCl_2 precursors were found to produce only atactic polypropylene (which, incidentally cannot be obtained in the pure form directly from heterogeneous catalysts).[37] In 1984, Ewen reported the use of metallocene-based catalysts for the isospecific polymerization of propylene.[38] The polymerization of propylene at -45°C using a Cp_2TiPh_2 (**I**,Fig.4) / MAO catalyst system produced a partially isotactic polymer with an *mmmm* pentad content of 52% (versus 6.25% for a purely atactic polymer). ^{13}C NMR analysis of the polymer revealed the stereochemical errors *mmmr* and *mmrm* in the ratio of 1:1, which is indicative of a stereoblock microstructure (Fig.5). Such a structure is consistent with a chain-end control mechanism,[39] where the stereocenter of the last inserted monomer unit provides stereochemical control during the polymerization.

In the same paper, Ewen reported that the chiral ansa-metallocenes first prepared by Brintzinger[40] were capable of producing isotactic polypropylene (Fig.4).[11] Using a mixture of meso (**II**) and racemic (**IIIa**) ansa-titanocenes, Ewen produced a mixture of atactic and isotactic polymer chains. In 1985, Kaminsky and Brintzinger[41] confirmed Ewen's proposal that the chiral metallocene isomer was responsible for the formation of the isotactic polypropylene by using an isomerically pure sample of a racemic zirconocene analogue (**IIIb**) to produce only isotactic polymer chains.

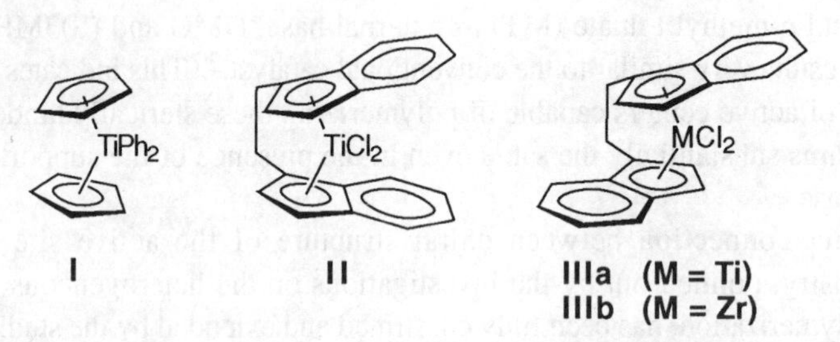

Figure 4. Metallocenes for isospecific olefin polymerization.

The significance of these developments was two-fold. On the one hand, these discoveries first established that it was possible to produce isotactic polypropylene with homogeneous catalysts, thus opening the possibility for the development of a new generation of stereospecific olefin polymerization catalysts.

At the same time, the fact that the homogeneous catalyst precursors are structurally well-defined has provided an extraordinary opportunity to investigate the origin of stereospecificity in olefin polymerization at a level of detail that was difficult if not impossible with the conventional heterogeneous catalysts. For example, ^{13}C NMR analysis of the isotactic polymer produced with **III** revealed the stereochemical errors *mmmr*, *mmrr*, and *mrrm* in the ratios of 2:2:1 (Fig.5). This observation is consistent with an enantiomorphic site control mechanism, where the geometry of the catalyst framework controls the stereochemistry of olefin insertion.[6,30,31] These results established unambiguously a clear experimental correlation between the chirality of the active site, which could be established by x-ray crystallography of the metallocene catalyst precursor, and the isotacticity of the polymer produced.

With the advent of these homogeneous catalysts, extensive investigations followed on the influence of the catalyst geometry on the stereospecificity of olefin polymerization and the origins of stereocontrol. Zambelli employed the chiral metallocene (**IIIa**) to study the origin of enantiofacial selectivity of olefin insertion. He discovered that the insertion of propylene into a M-CH_2CH_2R (R=H, or alkyl) bonds proceeds with a high degree of enantiofacial selectivity, while the insertion into a M-CH_3 bond occurs without selectivity.[42]

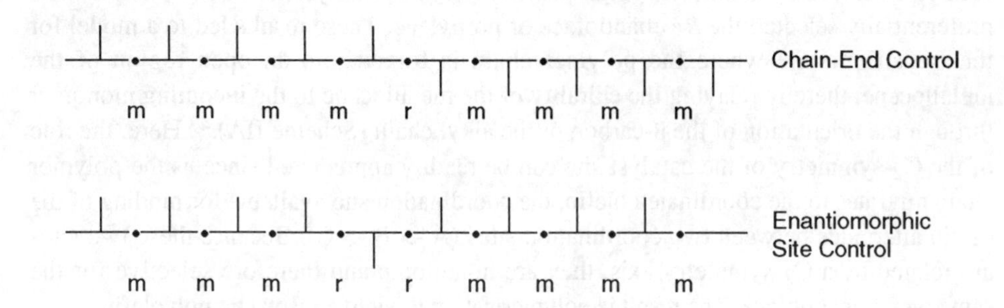

Figure 5. Microstructure of isotactic polypropylene by chain-end control and enantiomorphic site control.

These results suggest that the high degree of stereoselectivity of these catalysts is contingent upon insertion into a M-R bond, where R is an alkyl group consisting of two or more carbon atoms. Conformational modeling studies by Corradini and coworkers suggest that the polymer chain is forced into an open region of the metallocene, thereby relaying the chirality of the metallocene to the incoming monomer through the orientation of the β-carbon of the alkyl chain (Scheme II).[43]

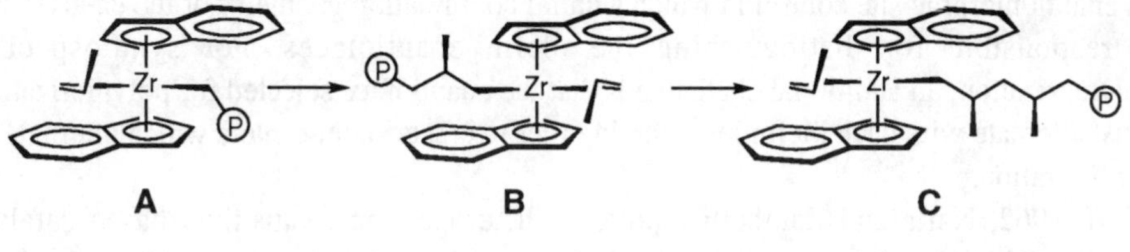

A B C

Scheme II. Origin of isospecificity in olefin polymerization using a chiral, C_2-symmetric metallocene.

One of the key pieces of information was provided by a brilliant experiment by Pino to establish the absolute stereochemistry of propylene polymerization.[44] Brintzinger had demonstrated that the chiral metallocenes could be resolved into their respective enantiomers,[40] and thus for the first time it was possible to correlate the absolute stereochemistry of the catalyst with the prevailing enantioface of the olefin selected for polymerization. Due to the fact that the high molecular weight polymer of propylene is achiral (Fig.3), Pino devised a hydrooligomerization experiment to obtain low molecular weight oligomers whose absolute configuration could be ascertained.

On the basis of this experiment, Pino and coworkers were able to determine that catalysts derived from the (*R*)-ethylenebis(tetrahydroindenyl)zirconium binaptholate preferentially selected the *Re* enantioface of propylene. These results led to a model for the transition state where the polymer chain is forced into an open region of the metallocene, thereby relaying the chirality of the metallocene to the incoming monomer through the orientation of the β-carbon of the alkyl chain (Scheme IIA).[43] Here, the role of the C_2-symmetry of the catalyst site can be readily appreciated since as the polymer chain migrates to the coordinated olefin, the coordination site available for binding of the olefin alternates between two coordination sites (A -> B -> C). Because these two sites are related by a C_2-symmetry axis, they are homotopic and therefore selective for the same olefin enantioface. The result is polymerization to yield an isotactic polyolefin.

4.1. SYNDIOTACTIC POLYMERS

The origin of stereospecificity in the syndiotactic polymerization of olefins has also been an area of considerable interest as syndiotactic polymers, where each stereogenic center alternates in configuration, are also crystalline thermoplastics with properties similar to isotactic polyolefins. As with isospecific polymerization, there are two mechanisms for stereodifferentiation: chain-end control, where the stereogenic centers of the growing polymer chain are responsible for distinguishing the olefin enantiofaces, and an enantiomorphic site control in which a chiral coordination geometry of the catalyst site is responsible for distinguishing the olefin enantiofaces. For syndiospecific polymerization, an additional challenge is that the enantioface selected for polymerization must alternate with each olefin insertion in order to create stereocenters which alternate in configuration.

In 1962, Natta and Zambelli reported a heterogeneous, vanadium-based catalyst mixture which produced partially syndiotactic polypropylene at low polymerization temperatures.[45] The regiochemistry of the insertion was determined to be a secondary insertion of propylene, and a chain-end control mechanism determined the syndiospecificity of monomer insertion. This system suffered from both low activity and low stereospecificity.

Syndiotactic polymers have also been synthesized using metallocene catalysts where the polymer chain end controls the syndiospecificity of olefin insertion. Resconi has shown that $Cp^*_2MCl_2$ (M = Zr, Hf) derived catalysts produce predominantly syndiotactic poly(1-butene) with an approximate 2 kcal/mol preference for syndiotactic versus isotactic dyad formation.[46] At -20°C, $Cp^*_2HfCl_2$ / MAO produces poly(1-butene) with 77% rr triads. The complex $(^iPrCp)_2TiCl_2$ produces either isotactic or syndiotactic polypropylene, depending on the temperature of the polymerization. Using this catalyst, the tacticity of the polymer changed from slightly isotactic at -30°C to atactic at -10°C to slightly syndiotactic at +10°C.[47]

In another significant development, Ewen reported a metallocene catalyst precursor (**IV**) that is highly active for the syndiospecific polymerization of propylene and higher aliphatic α-olefins (Fig.6).[48] The degree of syndiospecificity for this catalyst is extraordinarily high and has provided for the first time a catalyst which is capable of producing highly syndiotactic crystalline polypropylene.

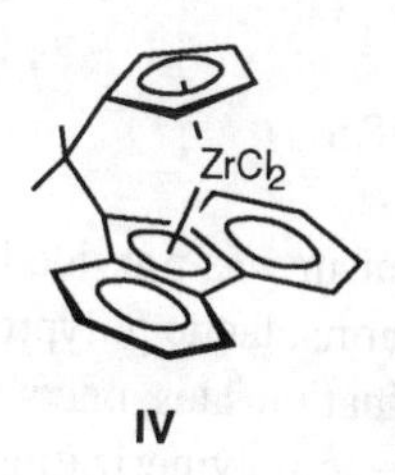

Figure 6. Metallocene for syndiospecific olefin polymerization.

The ^{13}C NMR spectrum for polypropylene produced at 50°C with this catalyst reveals a high degree of stereocontrol (81% *rrrr* pentads). Moreover, analysis of the stereochemical defects (predominantly *rmmr* pentads) were indicative of a site control mechanism. For a site control mechanism to operate in syndiospecific polymerization, the olefin must alternately bind to coordination sites with opposite enantioface selectivity. The model for this polymerization is shown in Scheme III.

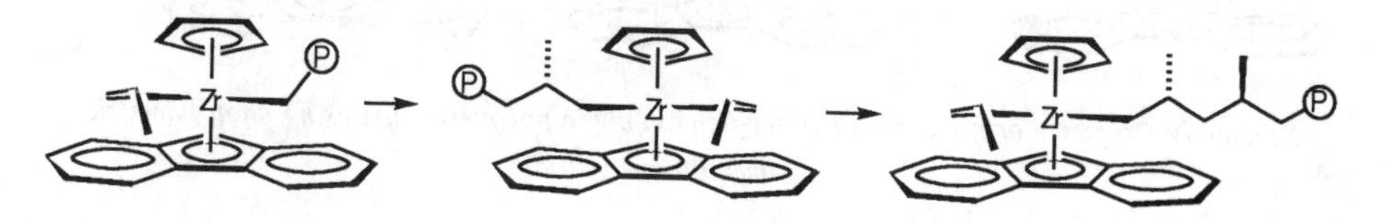

Scheme III. Proposed origin of syndiospecificity in olefin polymerization using a C_S-symmetric metallocene.

This catalyst possesses C_s-symmetry where the two coordination sites bear an enantiotopic relationship. The regularly alternating insertion of olefins at the heterotopic sites of the C_s-symmetric complex results in the formation of syndiotactic polymer (Scheme III). As with the isospecific C_2-symmetric catalysts, occasional misinsertion of the incorrect enantioface is proposed to be the predominant source of defects. Other defects corresponding to chain migration without olefin insertion (*rmrr*) have been detected.[49]

It should be pointed out that, in contrast to the case with the isospecific catalysts, the sense of the enantioface selectivity for olefin insertion has not been determined for the syndiospecific catalysts **IV**. The model shown below is based on that offered by Corradini and coworkers that the orientation of the growing polymer chain is the primary stereodifferentiating element.[50]

4.2. HEMIISOTACTIC POLYMERS

Further support for this mechanism was provided by Ewen in the form of a catalyst which polymerizes propylene to hemiisotactic polypropylene. The metallocene shown in Scheme IV has two different coordination sites, one which is isospecific and one which is aspecific.[51] When used for propylene polymerization, the alternation between iso- and aspecific sites results in a hemiisotactic polymer (Scheme IV). The polymer was readily characterized due to the pioneering work of Farina, who independently prepared this material previously.[52] The rational synthesis of isotactic, syndiotactic, and hemiisotactic polyolefins represents a crowning achievement in the application of transition metal catalysts in stereocontrolled reactions.

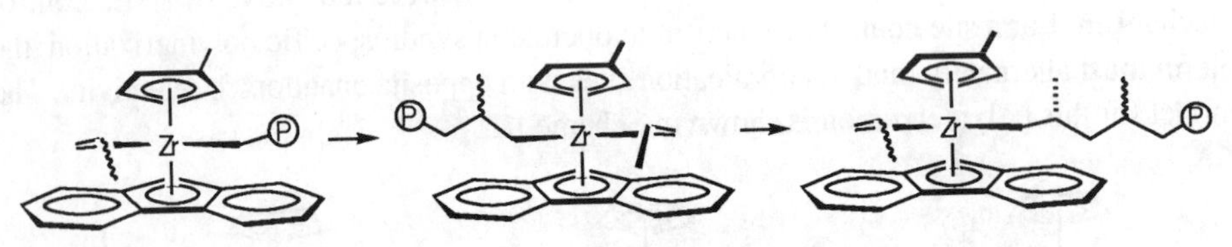

Scheme IV Proposed origin of hemiisospecificity in olefin polymerization using an asymmetric metallocene.

4.3. ENANTIOSELECTIVE SYNTHESIS OF OPTICALLY ACTIVE MACROMOLECULES: A DIRECT CORRELATION BETWEEN CATALYST CHIRALITY AND POLYMER CHIRALITY

Studies with homogeneous catalysts precursors have provided an extraordinary amount of insight into the origin of stereocontrol in Ziegler-Natta polymerization of olefins. These studies have provided more direct evidence for the role of chiral catalyst sites in the stereodifferentiation of olefin enantiofaces. However, as previously discussed, although chiral catalyst sites are quite efficient for stereodifferentiation, the use of enantiomerically pure catalysts will not lead to enantiomerically pure polymers since the high molecular weight polymers are, in general, achiral (Fig. 2).

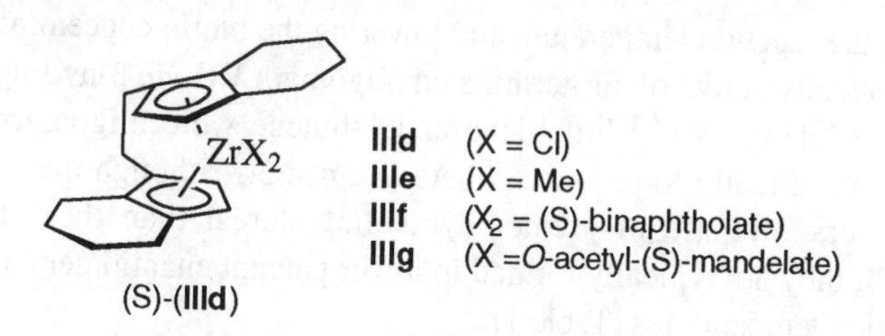

Figure 7 Enantiomerically pure catalyst precursors.

Despite this limitation, the resolved chiral catalysts have been used in enantioselective oligomerization reactions (Fig. 7). Kaminsky attempted the polymerization of propylene with enantiomerically pure (EBTHI)ZrX_2 (**IIId**) / MAO catalysts;[53] as expected, the high molecular weight polymer in solution was optically inactive.[54] Pino used (R)-**IIIf** / MAO to accomplish the asymmetric oligomerization of propylene, 1-pentene, and 4-methyl-1-pentene using hydrogen as a chain transfer agent (Scheme V).[55] In the case of propylene, approximately 90% of the products **VIa** ($n < 47$) had a measurable optical activity. The oligomeric alkane fractions were characterized by polarimetry, and their absolute configurations were used to unambiguously determine the enantiofacial preference of the metallocene catalyst for the first time (as illustrated in Scheme II).[55]

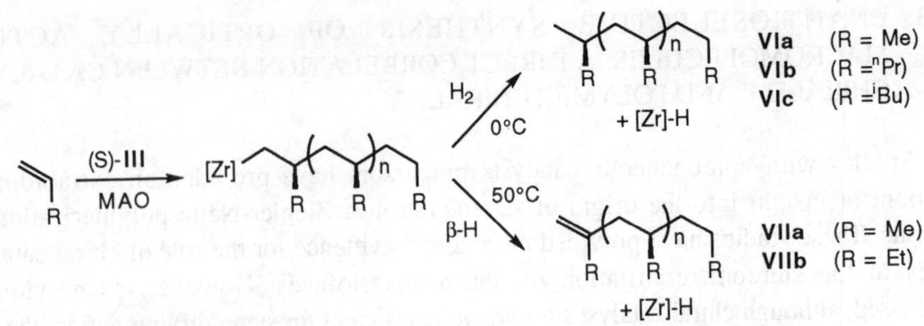

(S)-III
MAO
[Zr]
H_2
0°C
50°C
β-H
VIa (R = Me)
VIb (R = ⁿPr)
VIc (R = ⁱBu)
+ [Zr]-H
VIIa (R = Me)
VIIb (R = Et)
+ [Zr]-H

Scheme V Enantioselective synthesis of optically active α-olefin oligomers.

By raising the reaction temperature and lowering the olefin concentration, Kaminsky synthesized optically active olefin-terminated oligomers **VII** *via* β-hydrogen elimination chain transfer (Scheme V).[56] Propylene and 1-butene were oligomerized using (*S*)-**IIIg**/MAO predominantly to products where $0 < n < 5$. Although these functionalized alkene oligomers **VII** are of greater synthetic interest than the related saturated compounds **VI**, they are typically formed in lower percent enantiomer excess (%ee) due to higher reaction temperatures (Table 1).

TABLE 1. Enantiomer excess of oligomer formation.

Oligomer	Metallocene	Temp. (°C)	ee (%)	Ref.
VIa (n = 1)	(*R*)-**IIIf**	0	100[a]	55
VIb (n = 0)	(*R*)-**IIIf**	25	50[a]	55
VIc (n = 0)	(*R*)-**IIIf**	25	84[a]	55
VIIb (n = 1)	(*S*)-**IIIg**	50	27[b]	56

[a] Determined by comparison of rotation with known compound.
[b] Determined by chiral chromatography.

4.4. CYCLOPOLYMERIZATION

While the availability of enantiomerically pure metallocenes has stimulated interest in the application of these catalysts for enantioselective synthesis,[40,57] this route for obtaining optically active polymers is challenged by the special symmetry limitations of stereoregular vinyl polymers. Waymouth and coworkers have recently discovered a means of overcoming these limitations by taking advantage of a cyclopolymerization reaction to create polymer architectures which are intrinsically chiral. Nonconjugated diolefins can be polymerized in an insertion and cyclization sequence, resulting in a polymer containing rings in the main chain. Whereas vinyl polymers have only two structures of maximum order (isotactic, syndiotactic), cyclopolymers are inherently more complicated (Scheme VI).

Close inspection of Scheme VI reveals that the *trans*-isotactic microstructure contains no mirror planes of symmetry and is thus chiral by consequence of its main-chain stereochemistry.[13] There are two criteria for chirality for this polymer: (1) isotacticity (the same relative stereochemistry of every other stereocenter); and (2) the presence of *trans* rings. The enantiofacial selectivity of the first olefin insertion determines the tacticity of the polymer, and the diastereoselectivity of the cyclization step determines whether *cis* or *trans* rings are formed.

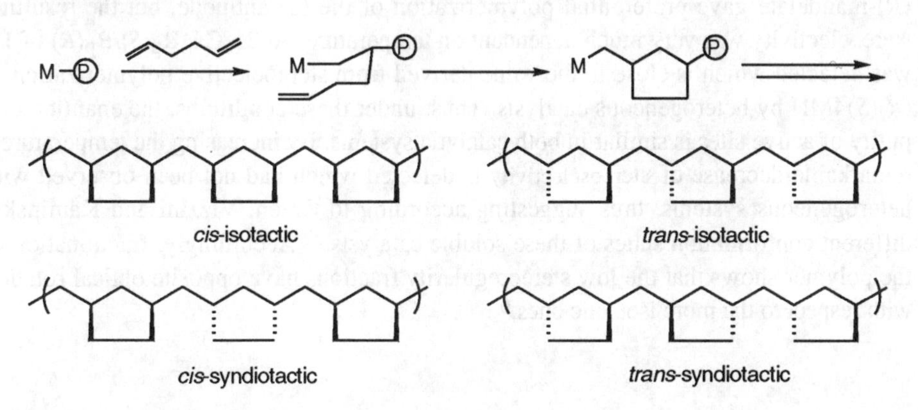

Scheme VI. Four structures of maximum order for poly(1,5-hexadiene).

Using homogeneous Ziegler-Natta catalysts, Waymouth has studied the effect of the catalyst geometry on the diastereoselectivity of ring formation with various α,ω-dienes.[58-60] By employing (*R*)-**IIIf** / MAO, Waymouth found that 1,5-hexadiene could be cyclopolymerized to yield an optically active macromolecule with $[\Phi]^{28}_{405}) = +51.0°$ (Scheme VII).[61] Cyclopolymerization with (*S*)-**IIIf** afforded the enantiomeric polymer, $[\Phi]^{28}_{405} = -51.2°$. Microstructural analysis of the polymer by ^{13}C NMR revealed a trans ring content of 72% and an enantiofacial selectivity for olefin insertion of 91%. The high degree of stereoregularity and predominance of *trans* rings are responsible for the optical activity of the polymer.

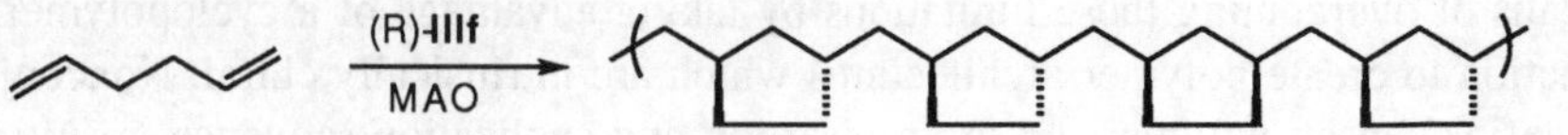

(R)-Hf
MAO

Scheme VII. Synthesis of trans-isotactic poli(1,5-hexadiene)

4.5 POLYMERIZATION OF CHIRAL MONOMERS

These stereorigid metallocene systems appear attractive tool investigating if their proved capacity of discriminating between the prochiral faces of α-olefins (stereospecific polymerization) also implies the capacity to discriminate between two enantiomers (stereoselective polymerization). Indeed polymerization of racemic 4MH[62] in the presence of (*S*)-ethylene-bis (4,5,6,7,-tetra-hydro-1-η^5-indenyl)zirconium di-O-acetyl-(*R*)-mandelate gave preferential polymerization of the (*S*) antipode, but the resulting stereoelectivity was very much dependent on temperature. At 25°C a $R_p(S)/R_p(R)$ of 1.4 was detected which is close to the value derived from stereoelective polymerization of (*R*)(*S*)4MH by heterogeneous catalysts. Thus, under these conditions, the enantiomeric purity of active sites is similar in both catalytic systems. By increasing the temperature a remarkable decrease of stereoselectivity is detected which had not been observed with heterogeneous systems, thus suggesting according to Chien, Vizzini and Kaminsky, different conformation states of these soluble catalysts.[62] Accordingly, fractionation of the polymer shows that the low stereoregularity fractions have opposite optical rotation with respect to the more isotactic ones.[6]

References

1. Ziegler, K., Holzkamp, E., Breil, H., Martin, H. (1955) *Angew. Chem.*, **67**, 426-426.
2. Natta, G., Pino, P., Corradini, P., Danusso, F., Mantica, E., Mazzanti, G., Moraglio, G. (1955) *J. Am. Chem. Soc.*, **77**, 1708-1710.
3. Hunter, D., Young, I. (1993) *Chem. Week* , 26-36.
4. Pino, P., Mülhaupt, R. (1980) *Angew. Chem., Int. Ed.*, **19**, 857-875.
5. Boor, J. (1979) *Ziegler-Natta Catalysts and Polymerizations,* Academic Press, New York.
6. Natta, G., Pino, P., Mazzanti, G., Longi, P. (1957) *Gazz. Chim. Ital.* , **87**, 570-585.
7. See for example "*Comprehensive Polymer Science*" (1989) G. Allen and J.C.C. Bevington (eds), Pergamon Press.
8. Pure & Appl.Chem. (1974), **40**, 479.
9. Natta, G. (1955) *Atti Accad. Naz. Lincei, Mem., Cl. Sci. Fis., Mat. Nat., Sez. 2a*, **4**, 61-71.
10. Natta, G. and Corradini, P. (1955) *Atti Accad. Naz. Lincei, Mem., Cl. Sci. Fis., Mat. Nat., Sez. 2a*, **4**, 73-80.
11. Natta, G. and Corradini, P. (1955) *Rend. Accad. Naz. Lincei*, **19**, 229-239.
12. Natta, G. and Corradini, P. (1955) *J. Polym. Sci.,* **20**, 251-266.
13. Farina, M. (1987) *Top. Stereochem.*, **17**, 1-111.
14. Mislow, K. and Siegel, J. (1984) *J. Am. Chem. Soc.*, **106**, 3319-3328.
15. Pino, P., Oschwald, A., Ciardelli, F., Carlini, C., Chiellini, E. (1975) in J.C.W. Chien (ed), *Coordination Polymerization: A Memorial to Karl Ziegler*, Academic Press, New York, pp. 25-72.
16. Ciardelli, F., Altomare, A., Carlini, C. (1991) *Prog. Polym. Sci.*, **16**, 259-277.
17. Pino, P. (1965) *Adv. Polymer Sci.*, **4**, 236-245.
18. Ciardelli, F., Montagnoli, G., Pini, D., Pieroni, O., Carlini, C., Benedetti, E. (1971) *Makromol. Chem.*, **147**, 53-68.
19. Pino, P., Lorenzi, G.P. (1960) *J. Am. Chem. Soc.*, **82**, 4745-4747.
20. Pino, P., Ciardelli, F., Lorenzi, G.P, Montagnoli, G. (1963) *Makromol. Chem.*, **61**, 207-224.
21. Carlini, C., Ciardelli, F., Pino, P. (1968) *Makromol. Chem.*, **119**, 244-248.
22. Pino, P., Ciardelli, F., Lorenzi, G.P., Natta, G. (1962) *J. Am. Chem. Soc.* **84**, 1487-1488.
23. Pino, P., Montagnoli, G., Ciardelli, F., Benedetti, E. (1966) *Makromol. Chem.*,

93, 158-179.

24. Montagnoli, G., Pini, D., Lucherini, A., Ciardelli, F., Pino, P. (1969) *Macromolecules* , **2**, 684-686.
25. Zambelli, A., Ammendola, P., Sacchi, M.C., Locatelli, P., Zannoni, G. (1983) *Macromolecules* , **16**, 341-348.
26. Sacchi, M.C., Tritto, I., Locatelli, P., Ferro, D.R. (1984) *Makromol. Chem., Rapid Commun.*, **5**, 731-736.
27. Pino, P., Ciardelli, F., Montagnoli, G. (1968) *J. Polym. Sci., Part. C*, **16**, 3265-3277.
28. Ciardelli, F., Locatelli, P., Marchetti, M., Zambelli, A (1974) *Makromol. Chem.*, **175**, 923-933.
29. Ciardelli, F., Carlini, C., Montagnoli, G. (1969) *Macromolecules* , **2**, 296-301.
30. Zambelli, A., Tosi, C. (1974) *Adv. Polym. Sci.*, **15**, 31-60.
31. Zambelli, A., Bajo, G., Rigamonti, E. (1978) *Makromol. Chem.*, **179**, 1249-1259.
32. Pino, P., Fochi, G., Oschwald, A., Piccolo, O., Mulhaupt, R., Giannini, U. (1983) in C.C. Price and E.J. Vandenberg (eds), *Coordination Polymerization*, Plenum Publishing Corporation, New York, pp. 207-223.
33. Carlini, C., Altomare, A., Menconi, F., Ciardelli, F. (1987) *Macromolecules* , **20**, 464-465.
34. Chien, J.C.W., Wu, J.C., Kuo, C.I. (1982) *J. Polym. Sci., Part A1: Polym. Chem.*, **20**, 2445-2460.
35. Sinn, H., Kaminsky, W. (1980) *Adv. Organomet. Chem.*, **18**, 99-149.
36. Sinn, H., Kaminsky, W., Vollmer, H. J., Woldt, R. (1980) *Angew. Chem., Int. Ed.*, **19**, 390-392.
37. Kaminsky, W. (1991) in H. R. Kricheldorf (ed.), *Handbook of Polymer Synthesis*, Marcel Dekker, New York, Vol. 1, pp. 1.
38. Ewen, J.A. (1984) *J. Am. Chem. Soc.*, **106**, 6355-6364.
39. Bovey, F.A., Tiers, G. V. D. (1960) *J. Polym. Sci.*, **44**, 173-182.
40. Wild, F.R.W.P., Zsolnai, L., Huttner, G., Brintzinger, H.H. (1982) *J. Organomet. Chem.*, **232**, 233-247.
41. Kaminsky, W., Külper, K., Brintzinger, H.H., Wild, F.R.W.P. (1985) *Angew. Chem., Int. Ed.*, **24**, 507-508.
42. Longo, P., Grassi, A., Pellecchia, C., Zambelli, A. (1987) *Macromolecules* , **20**, 1015-1018.
43. Corradini, P., Guerra, G. (1991) *Prog. Polym. Sci.*, **16**, 239-257.
44. Pino, P., Cioni, P., Wei, J. (1987) *J. Am. Chem. Soc.*, **109**, 6189-6191.

45. Natta, G., Pasquon, I., Zambelli, A. (1962) *J. Am. Chem Soc.*, **84**, 1488-1490.
46. Resconi, L., Abis, L., Franciscono, G. (1992) *Macromolecules* , **25**, 6814-6817.
47. Erker, G., Fritze, C. (1992) *Angew. Chem., Int. Ed.*, **31**, 199-202.
48. Ewen, J.A., Jones, R.L., Razavi, A., Ferrara, J.D. (1988) *J. Am. Chem. Soc.*, **110**, 6255-6256.
49. Ewen, J.A., Elder, M.J., Jones, R.L., Curtis, S., Cheng, H.N. (1990) in T. Keii and K. Soga (eds), *Catalytic Olefin Polymerization* , Kodansha, Tokyo, pp 439-482.
50. Cavallo, L., Guerra, G., Vacatello, M., Corradini, P. (1991) *Macromolecules* , **24**, 1784-1790.
51. Ewen, J.A., Elder, M.J., Jones, R.L., Haspeslagh, L., Atwood, J.L., Bott, S.G., Robinson, K. (1991) *Makromol. Chem., Macromol. Symp.*, **48/49**, 253-295.
52. Farina, M., Di Silvestro, G., Sozzani, P. (1982) *Macromolecules* , **15**, 1451-1452.
53. Kaminsky, W., Möller-Lindenhof, N. (1990) *Bull. Soc. Chim. Belg.*, **99**, 103-111.
54. Pino, P., Ciardelli, F., Zandomeneghi, M. (1970) *Ann. Rev. Phys. Chem.*, 21, 561-608.
55. Pino, P., Galimberti, M., Prada, P., Consiglio, G. (1990) *Makromol. Chem.*, **191**, 1677-1688.
56. Kaminsky, W., Ahlers, A., Möller-Lindenhof, N. (1989) *Angew. Chem., Int. Ed.*, **28**, 1216-1218.
57. Halterman, R.L. (1992) *Chem. Rev.*, **92**, 965-994.
58. Resconi, L., Waymouth, R.M. (1990) *J. Am. Chem. Soc.* , **112**, 4953-4954.
59. Cavallo, L., Guerra, G., Corradini, P., Resconi, L., Waymouth, R. M. (1993) *Macromolecules* , **26**, 260-267.
60. Coates, G.W., Waymouth, R.M. (1992) *J. Mol. Catal.*, **76**, 189-194.
61. Coates, G.W., Waymouth, R.M. (1993) *J. Am. Chem. Soc.*, **115**, 91-98.
62. Chien, J.C.W., Vizzini, J., Kaminsky, W. (1992) *Makromol. Chem. Rapid Commun.*, **13**, 479-484.

MAKING UNNATURAL PRODUCTS BY NATURAL MEANS

STEVEN J. LANGFORD, J. FRASER STODDART
School of Chemistry
University of Birmingham
Edgbaston, Birmingham B15 2TT, UK

ABSTRACT: The use of the template-directed methodology as a means of achieving self-assembly has led to a large number of mechanically-interlocked catenanes and rotaxanes being constructed from different, but cheap and readily available components. By varying the nature of the components that make up the supramolecular arrays, as well as molecular assemblies, fundamental information has been gleaned about the steric and electronic interactions that stabilize complexes and molecular geometries at a noncovalent level. This account describes some of the molecular assemblies and supramolecular arrays we have produced over the years and our attempts to gain some understanding of the mechanisms of their formation and the dynamic processes the assemblies and arrays exhibit when the molecular compounds or supramolecular complexes are dissolved in solution.

1. Introduction

The concepts of self-assembly,[1] self-organisation,[2] and self-replication,[3] as used so efficiently in the biological world, are now capable of dominating synthetic chemistry. They are ready to be adopted by chemists keen to develop new approaches to chemical synthesis by which highly "intelligent" components are used to construct much larger molecular assemblies and supramolecular arrays. The synthetic targets need not only be highly-ordered molecular and supramolecular structures with well-defined properties, but they must also relate to the pursuit of the understanding at a conceptual level, such that they shed more light upon key biological processes. The synthetic targets of the future will undoubtably assume many different forms and they will require different methodologies if they are to be realised. Realistically however, we need to walk before we can run. We limited ourselves in the first[4] of these two essays to a discussion of the mechanisms behind the template-directed self-assembly of the intriguing mechanically interlocked classes of compounds - the so-called catenanes and rotaxanes. Our objectives in this article are to show how the chemist might gain a full appreciation of the complementarity which must exist between host and guest molecules if they are to serve as efficient synthetic building blocks, and to see how this complementarity might "live on" in self-assembled molecular compounds and supramolecular systems. The breakthrough we made in 1989, when our first wholly-synthetic [2]catenane was self-assembled in 70% yield, has led to the construction of a large number of mechanically interlocked catenanes and rotaxanes, giving us a fundamentally better understanding of the limitations and the opportunities available to us from this methodology in self-assembly in a structural sense. We must take steps to avoid a structural "stalemate" occurring. Furthermore, we need to find out the fundamental consequences of structure upon the properties and functions of new materials in the development of well-defined molecules and collections of molecules.

C. Chatgilialoglu and V. Snieckus (eds.), Chemical Synthesis, 475–510.

One area that is surely set to benefit from research aimed at developing the self-assembly of mechanical systems is molecular electronics. Initially, one can look with considerable interest towards the development of chemical sensors and molecular switches of a quite novel design. Furthermore, the ability to store and manipulate information at a molecular level and on the nanometer-scale is clearly on the horizon. It will open the way for the development of molecular electronics and, ultimately perhaps, the molecular-scale computer. The objective at present is to be able to control the assembly, the form, and the function of synthetic nanometer-scale structures with the same degree of precision that is displayed by nature, so as to make it possible to achieve these ambitious goals. Only now are chemists beginning to learn how to construct molecular assemblies and supramolecular arrays such that information might ultimately be written into them, processed by them, stored in them, and eventually read back out of them.

2. Towards Molecular Machines[5-7]

Rotaxanes provide potentially useful synthetic targets. A major objective is to control their structures in order to create nanoscale devices by influencing somehow the interactions between the ring and the dumbbell components. The mechanically interlocked nature of the complementary molecular components provides, not only an aesthetically-pleasing form, but the resulting molecular structure also has the ability to function through the use of external stimuli such as protons, photons, and electrons. The ability of the supramolecular chemist to take control of the processes available, namely clipping, threading and slipping (***Scheme 1***), to provide entries into nanometer-scale architectures, such as the polyrotaxanes,[8] might hold the secrets to realising the most ambitious goal of all - the molecular-scale computer. Clipping and slipping both require the prefabrication of a dumbbell component by some traditional synthetic approach.

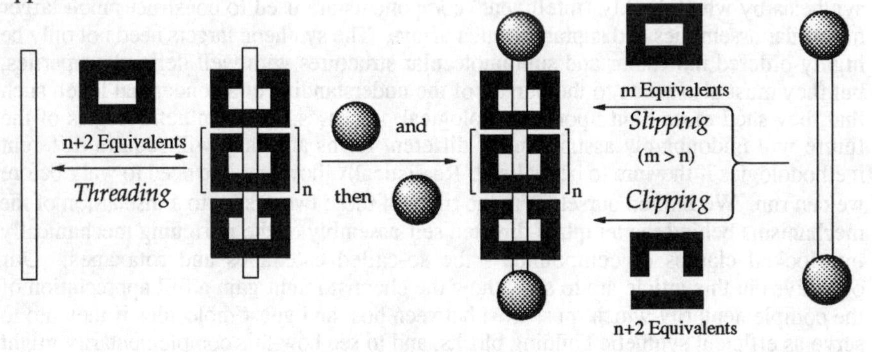

Scheme 1

A number of nanometer-scale molecular assemblies with potential device-like character, based on the [2]rotaxane architecture, have been self-assembled from molecular components.[9-13] All undergo the *clipping* methodology in which the molecular recognition between the π-electron rich dumbbell component and the developing π-electron poor cyclophane is being exploited. Several syntheses have been described[9-13] of dumbbell components comprised of more than one π-electron rich "station" of varying and different "complexation" strengths. In all these cases, the cyclic component (the tetracationic cyclophane) containing the π-electron acceptor units is able to position itself around one of a number of different π-electron donor units on the dumbbell component, which is comprised of a linear polyether chain interrupted by two or three of these π-donor units and terminated in all cases by tetraarylmethane stoppers. The possibility of controlling the translational isomerism in such systems points the way towards simple molecular devices. The simplest case involves the self-assembly of a [2]rotaxane with only one molecular recognition site. Obviously, the complementary macrocycle will reside there or not depending on the nature of the "fit". However, if we add another station and so create two recognition sites, which can be populated by the macrocycle with equal probability or with some predetermined bias, then we have a situation of degenerate or preferred translational isomerism and the makings of a controllable molecular shuttle. The energy barrier which must be overcome to allow inversion and interconversion to take place reflects directly on the relative strengths of recognition between the two components. In the example shown in ***Figure 1***, we have assumed that site B has better recognition for the macrocycle than does site A, but that both sites are more stable than the transition state where the macrocycle occupies a position between the two sites. This simple design becomes the basis for a controllable molecular shuttle.

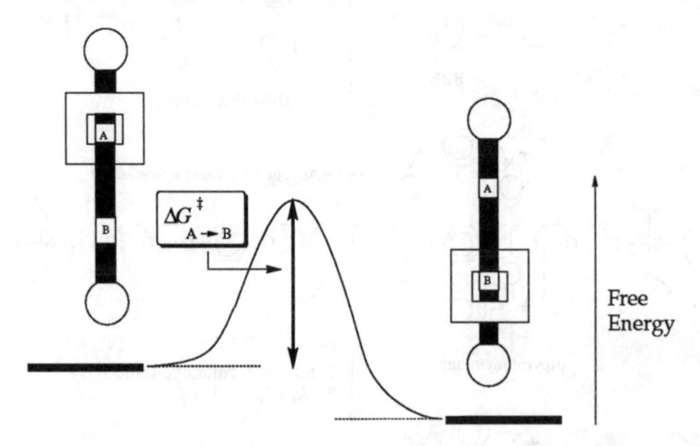

Figure 1. An energy diagram showing the relative stabilities of the translational isomers in a [2]rotaxane containing two different recogniton sites, A and B, on the dumbbell component. We have assumed that site B has much better recognition than site A for the macrocyclic component.

2.1. THE FIRST MOLECULAR SHUTTLE [9]

Our earliest model involved the formation of the [2]rotaxane **2**.4PF_6 in which the two equivalent hydroquinol stations in the dumbbell-shaped compound **1** play host to the tetracationic cyclophane by a template-directed self-assembly process (***Scheme 2***) involving the components BBB and [BBIPYXY][PF_6]$_2$ in the reasonably efficient yield of 32%.[9] Characterisation of the rotaxane **2**.4PF_6 in solution (CD_3COCD_3) by dynamic ^1H NMR spectroscopy reveals that the tetracationic cyclophane component is "shuttling" some 500 times per second between the two degenerate hydroquinol rings in the dumbbell component such that, at any given time, each station is populated equally.

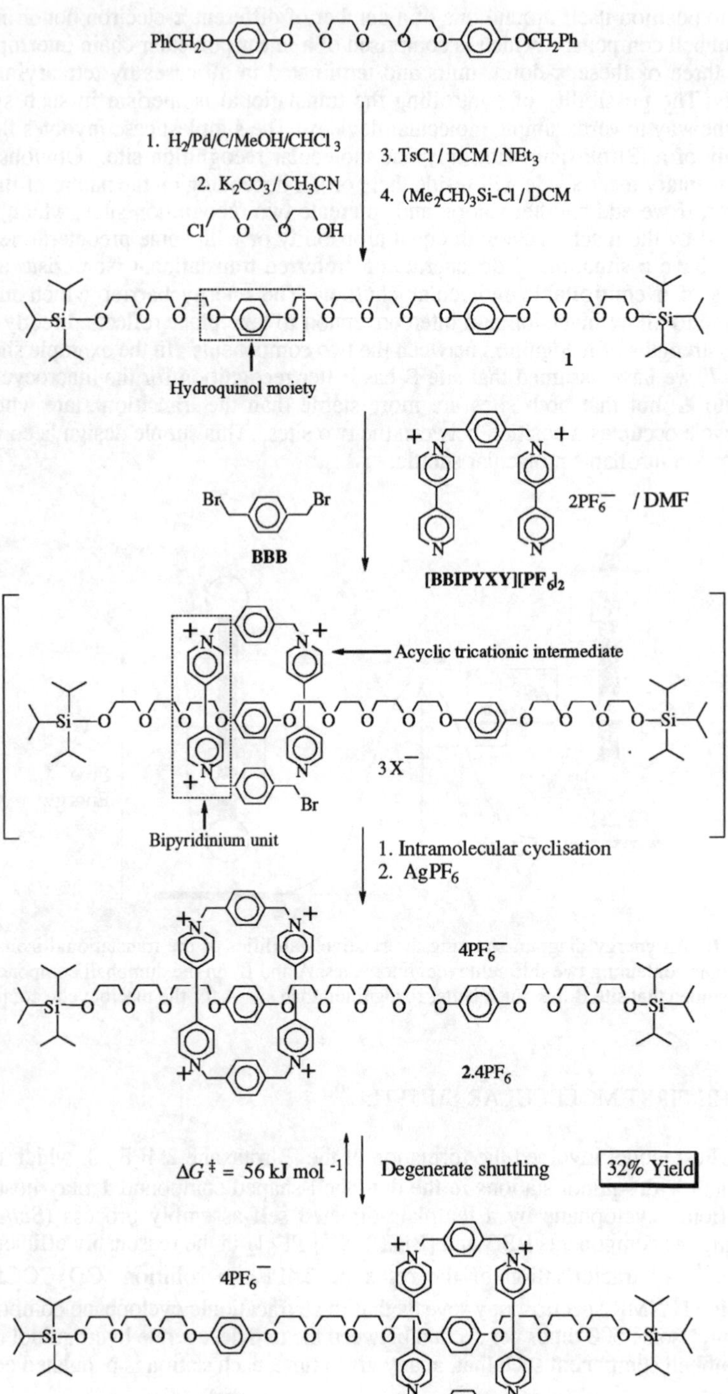

$PhCH_2O$–C6H4–O–…–O–C6H4–OCH_2Ph

1. $H_2/Pd/C/MeOH/CHCl_3$
2. K_2CO_3 / CH_3CN
Cl–O–O–OH

3. TsCl / DCM / NEt_3
4. $(Me_2CH)_3Si\text{-}Cl$ / DCM

Si–O–O–O–O–O–O–O–O–O–O–Si

1

Hydroquinol moiety

Br, Br

BBB

$2PF_6^-$ / DMF

$[BBIPYXY][PF_6]_2$

Acyclic tricationic intermediate

$3X^-$

Bipyridinium unit

1. Intramolecular cyclisation
2. $AgPF_6$

$4PF_6^-$

$2.4PF_6$

$\Delta G^{\ddagger} = 56$ kJ mol $^{-1}$ Degenerate shuttling **32% Yield**

$4PF_6^-$

$2.4PF_6$

Scheme 2

This compound can be viewed as the prototype for more complex systems in which control over the site occupancy by the tetracationic cyclophane is possible. Such systems have the potential to act as binary molecular devices. Encouraged by the emergence of the degenerate molecular shuttle, we needed to draw up a plan by which we might impart some control over the populations of translational isomers.

2.2. TOWARDS CONTROLLING THE SHUTTLING[10-13]

The strategy for the construction of an electrochemically-addressable molecular switch is outlined in ***Figure 2***. Both "stations" A and B should recognise the tetracationic cyclophane but A must recognise it stronger than B and A must have a lower oxidation potential than B. Thus, the tetracationic cyclophane should reside on station A to start with. Perturbation by an external stimulus, *e.g.*, from an electrode, should cause the cyclophane to move from station A to station B. With reference to the degenerate molecular shuttle shown in ***Scheme 2***, it is evident there are two ways to tackle the problem of controlling the translational isomerism. They are to:-

(i) replace one hydroquinol station by a π-electron donor whose affinity for the cyclophane is less than that of a hydroquinol ring. Our previous experience at the time led us to the use of a *p*-xylyl unit;

(ii) replace the hydroquinol station by a π-electron donor with a greater affinity for the tetracationic cyclophane.

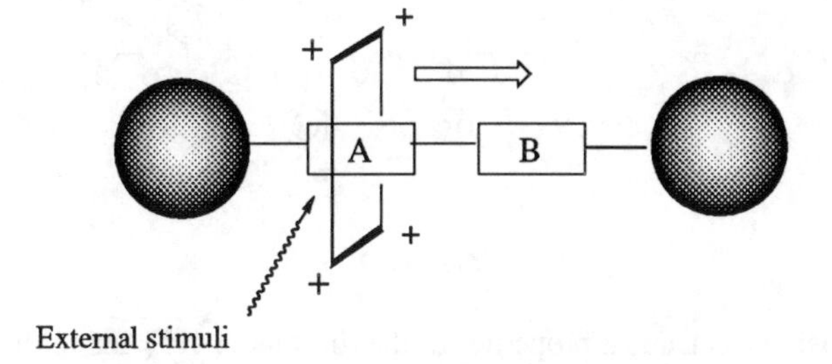

Figure 2. The set of criteria for a controllable molecular shuttle dictates that station A should be a better π-donor than station B and also have a lower oxidation potential.

The self-assembly of the [2]rotaxane **4**.4PF_6 from the dumbbell-shaped compound **3**, containing a unit of lower π-donating ability than a hydroquinol ring, and the components BBB and [BBIPYXY][PF_6]$_2$ was achieved (***Scheme 3***) in 8% yield.[10] The relatively low yield obtained in this particular self-assembly process is probably a reflection of the reduced molecular recognition between the dumbbell component **3** and the forming tetracationic component. Since dispersive interactions between the tetracationic cyclophane and the π-electron rich site account for a sizable fraction of the binding energies involved, we predicted that the cyclophane would occupy preferentially the hydroquinol station. The low temperature ^{1}H NMR spectrum of **4**.4PF_6 in CD_3CN did reveal some control of the site occupancy. However, only 70% of the tetracationic cyclophane populates the hydroquinol ring. This finely balanced situation regarding the populations, and the fact that the hydroquinol ring is neither photochemically nor electrochemically addressable in a mild manner, suggested that a change of plan was necessary.

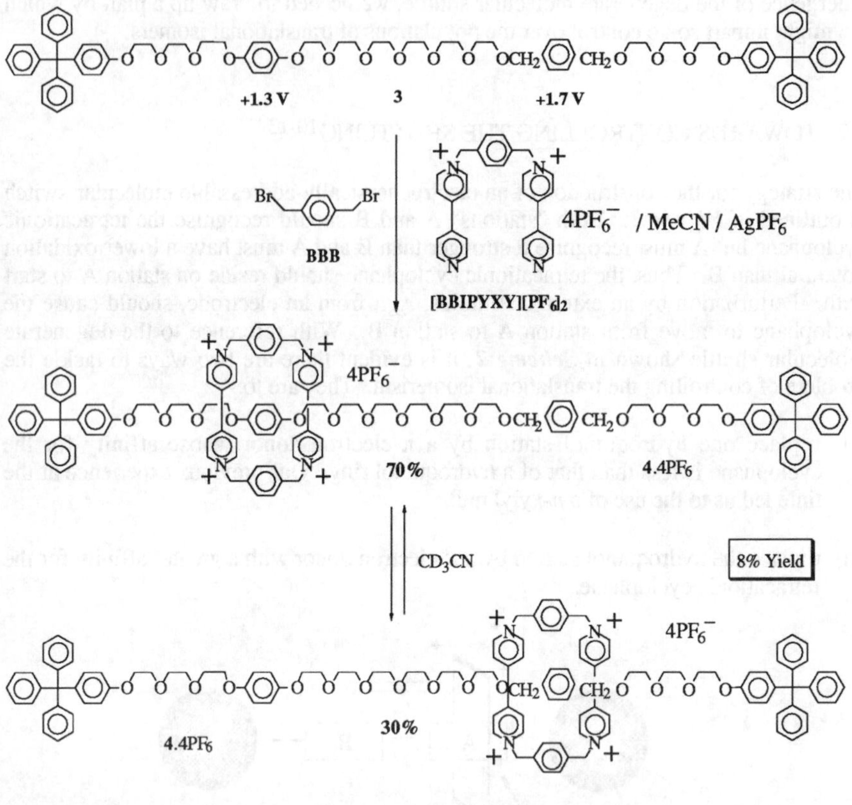

+1.3 V 3 +1.7 V

BBB [BBIPYXY][PF$_6$]$_2$ $4PF_6^-$ / MeCN / $AgPF_6$

$4PF_6^-$ 70% 4.4PF$_6$

8% Yield

CD_3CN

$4PF_6^-$ 4.4PF$_6$ 30%

Scheme 3

The indecisive mechanical properties of the rotaxane $4.4PF_6$ led to the inversion of the design logic. The necessary modification was achieved by replacing the *p*-xylyl group with a 2,3,5-trisubstituted indole, which is more π-electron rich than the hydroquinol ring but has a much lower oxidation potential (*ca.* 1 volt) as well as forming a stable radical cation on electrochemical oxidation.[11] These criteria require that the tetracationic cyclophane should reside on the indole ring in preference to the hydroquinol one until the indole ring is oxidised to the radical cation, at which point the tetracationic cyclophane should be repelled and so should "jump" to the hydroquinol station - the next lowest energy state.

The final step in the construction of the dumbbell-shaped compound **7** involved a Fischer indole synthesis between the protected hydrazine **5** and the ketal **6**, which, on treatment with acid (***Scheme 4***), formed the indole ring system and the dumbbell-shaped compound **7** in 51% yield. Low temperature ^{1}H NMR spectroscopy on the rotaxane $\mathbf{8}.4PF_6$ in CD_3CN showed that in fact the hydroquinol ring was occupied by the tetracationic cyclophane in a ratio of 100:0! This outcome is probably a consequence of steric outweighing electronic factors, leaving the indole to "solvate" the exterior of the tetracationic cyclophane. The quite decisive preference for this rotaxane to yield the "incorrect" translational isomer meant that this system was not suited to our needs.

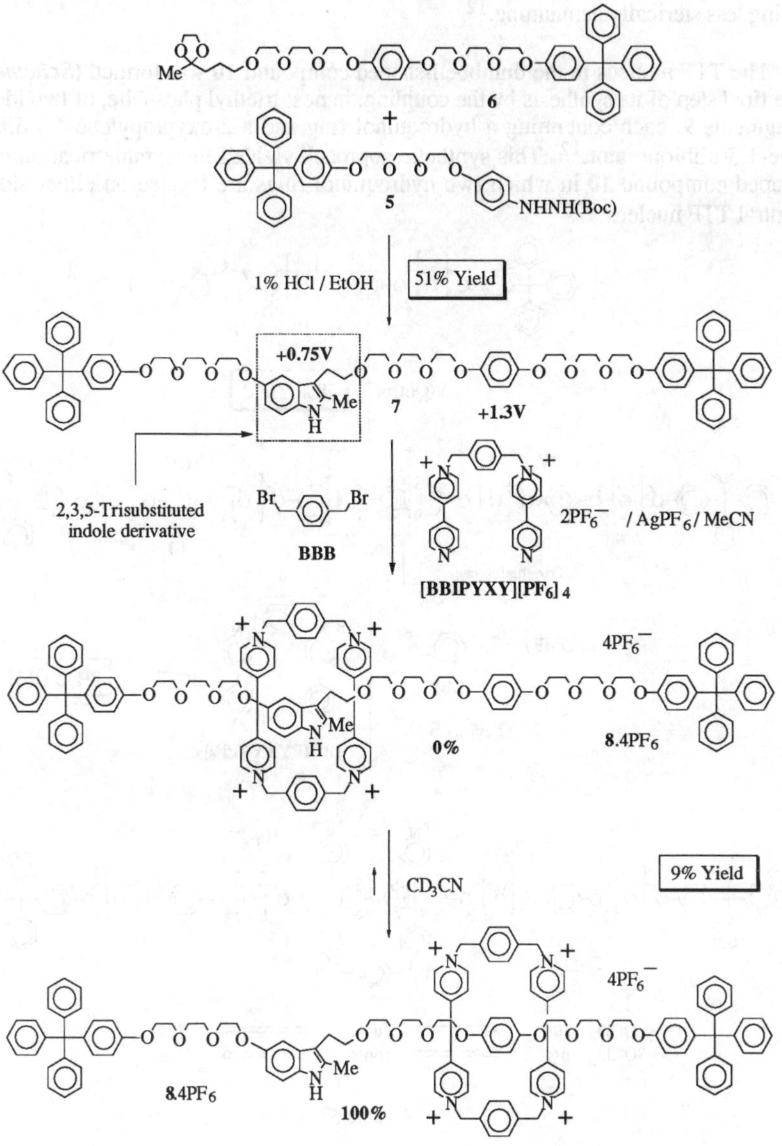

1% HCl / EtOH

51% Yield

+0.75V

+1.3V

2,3,5-Trisubstituted indole derivative

BBB

$2PF_6^-$ / $AgPF_6$ / MeCN

$[BBIPYXY][PF_6]_4$

$4PF_6^-$

0%

8.4PF6

CD_3CN

9% Yield

100%

Scheme 4

The breadth of investigations we conducted in the early research, involving different substrates for the tetracationic cyclophane, left us in the position to be able to choose from a broad selection of π-electron rich "stations". One such substrate, which has a high affinity for the tetracationic cyclophane, is tetrathiafulvene (TTF).[14] The [2]pseudorotaxane nature of the inclusion complex made its introduction into the dumbbell component of a [2]rotaxane an extremely attractive proposition. The targetted TTF derivative (***Scheme 5***) was expected to fulfil the criteria set for the indole group (high affinity for π-electron deficient groups and a low oxidation state), whilst being less sterically demanding.[12]

The TTF nucleus in the dumbbell-shaped compound **10** was formed (***Scheme 5***) in the final step of its synthesis by the coupling, in neat triethyl phosphite, of two identical fragments **9**, each containing a hydroquinol ring and a 2-oxypropylene-4,5-dithio-2-one-1,3-dithione unit.[12] This synthetic approach yielded the symmetrical dumbbell-shaped compound **10** in which two hydroquinol rings are located on either side of a central TTF nucleus.[12]

9

$P(OEt)_3$ 58% Yield

+1.3V +1.3V

+0.7V 10

Tetrathiafulvene portion

1. DMF / BBB

2. NH_4PF_6 / H_2O

$2PF_6^-$ 8% Yield

$[BBIPYXY][PF_6]_2$

$4PF_6^-$

$11.4PF_6$

CD_3COCD_3	50%	0%		50%
CD_3SOCD_3	0%	100%		0%

Scheme 5

The ^{1}H NMR spectrum in CD_3SOCD_3 of the rotaxane **11**.$4PF_6$ revealed the desired complete occupancy of the tetracationic cyclophane around the TTF moiety - a result confirmed by UV/VIS spectroscopy (the charge transfer band is located at 752 nm, compared with the more normal band at 450 nm observed for hydroquinol residues).[12,15] The solvent dependency of the translational isomerism, exhibited by the [2]rotaxane **11**.$4PF_6$ in CD_3SOCD_3 and CD_3COCD_3, was quite an unexpected result (***Scheme 5***). The use of CD_3COCD_3 for low temperature ^{1}H NMR studies revealed this solvent dependent behaviour. The preferred translational isomer in CD_3COCD_3

involves the equal population of the two hydroquinol units by the tetracationic cyclophane which experiences no residence on the TTF unit. This kind of behaviour creates the possibility for the development of a solvent dependent switch. Unfortunately, while an ideal occupancy situation was found to pertain in DMSO, the possibility of exploiting this system for electrochemical studies was not possible because of the lack of reversibility shown by the TTF derivative in this solvent.

Obviously, a total redesign was needed if a significant advance was to be made in this area. This breakthrough was achieved in Miami, where Kaifer and Bissell[13] took the decision to remove the hydroquinol ring from the system. Modelling studies on benzidine and biphenol derivatives showed that they are both included into the cavity of the tetracationic cyclophane and that the benzidine unit is included preferentially.[13] The control in such a system was found to be possible on two fronts - chemical (through protonation of the benzidene nitrogen atoms) and electrochemically (on account of the low oxidation potential of the benzidene unit).

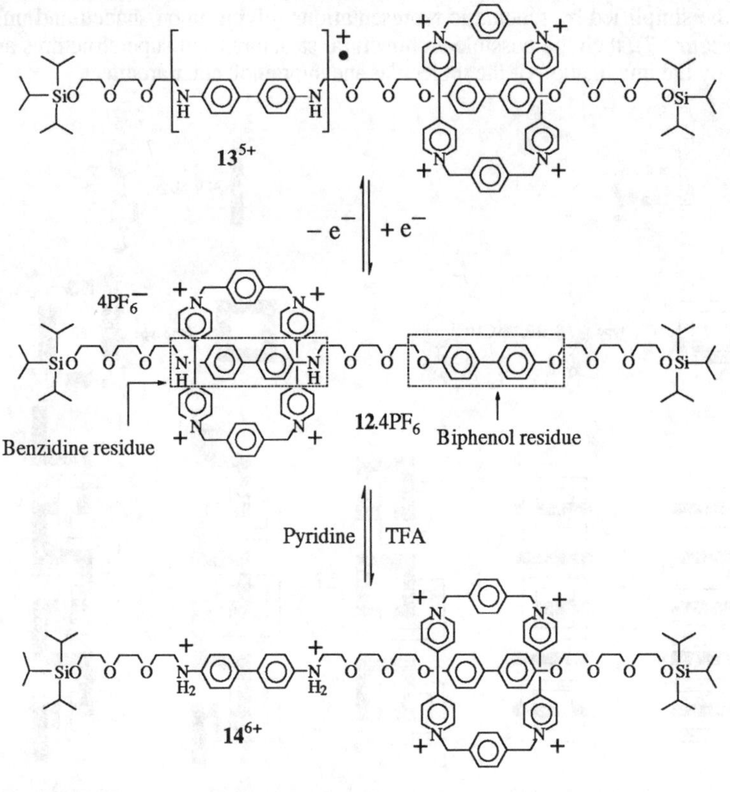

13^{5+}

$-e^-$ $+e^-$

$4PF_6^-$

Benzidine residue

$12.4PF_6$

Biphenol residue

Pyridine TFA

14^{6+}

Scheme 6

The [2]rotaxane $\mathbf{12}.4PF_6$, shown in ***Scheme 6***, was self-assembled in 19% yield from the appropriate dumbbell-shaped compound, BBB, and $[BBIPYXY][PF_6]_2$. Analysis of the rotaxane showed that the anticipated control by external stimuli was possible. The position of the tetracationic cyclophane can be switched from the benzidine station (which corresponds to the major translational isomer with 84%

occupancy) to the biphenol station by either protonation of the basic nitrogen atoms, yielding the [2]rotaxane 14^{6+} (as a result of charge repulsion) or by electrochemical oxidation, yielding the [2]rotaxane 13^{5+} (as a consequence of destroying the dispersive interactions). Although this model represents the state-of-the-art version of a controllable molecular shuttle, there is plenty of room for improvement. While this system is a commendable one, its control relies on the bulk properties of the system. To address this problem, we need to incorporate a sensitiser into the system which will allow the necessary oxidation to be initiated from within the molecule itself. While this goal may prove to be a difficult one to address, the complementary system, wherein the dumbbell incorporates π-electron deficient units, which can be reduced as a result of internal electron transfer, might constitute a more amenable target.

3. Towards Polyrotaxanes and Beyond

When you consider the diversity of host-guest complexes we have studied,[14-26] which can all be simplified by schematic representations relying upon shaded and unshaded parts (***Figure 3***), then the possible architectural structures and superstructures are only limited by the imagination of the molecular and supramolecular architect.

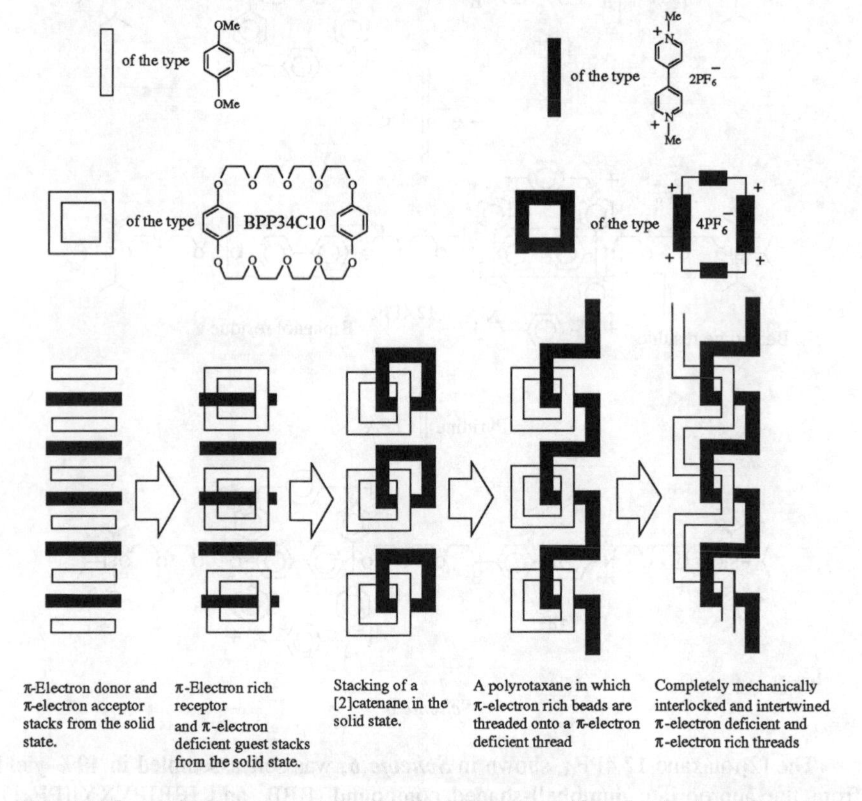
of the type
OMe
OMe
of the type
Me
Me
$2PF_6^-$
of the type
BPP34C10
of the type
$4PF_6^-$
π-Electron donor and π-electron acceptor stacks from the solid state.
π-Electron rich receptor and π-electron deficient guest stacks from the solid state.
Stacking of a [2]catenane in the solid state.
A polyrotaxane in which π-electron rich beads are threaded onto a π-electron deficient thread
Completely mechanically interlocked and intertwined π-electron deficient and π-electron rich threads

Figure 3. Increasingly programmed molecular information. Note that the trend from the molecular complexes, through covalent modification to host-guest systems and catenanes, can be further modified to yield polyrotaxanes and intertwining "double helical-like" systems.

The images of complementary host-guest systems, in which the reversal of recognition roles is possible, conjures up some highly suggestive situations that lead to the marriage of the molecular recognition features of the two host-guest systems, after covalent modification, to afford intricate molecular assemblies based on rotaxanes and catenanes. By increasing the programmed information further, we can look forward (***Figure 3***) to the design of [n]rotaxanes[8] and ultimately, by replacing the interlocking characteristics of the components, by entanglement of some kind, to the self-assembly of polymer-like molecules similar to the double helix of DNA.[27]

3.1. POLYROTAXANES WITH π-ELECTRON DEFICIENT BEADS

An obvious progression along the route from [2]rotaxanes to polyrotaxanes is *via* the pseudorotaxanes that incorporate the repeating [2]rotaxane oligomeric units. Self-assembled pseudorotaxanes have been realised by threading the cyclo-bis(paraquat-*p*-phenylene) onto prefabricated polymeric threads, incorporating π-electron rich hydroquinol rings. The investigation of the interactions of the tetracationic cyclophane within a homologous series of rotaxanes, pseudorotaxanes and pseudo-oligorotaxanes incorporating hydroquinol rings, revealed one possible way forward in this field.[28] The progression started (***Scheme 7***) by synthesising a range of rotaxanes with three, four, and five hydroquinol rings in the dumbbell components and stoppering with adamantoyl groups.[28] While the synthetic yield of the dumbbell-shaped compounds dropped steadily as their size increased, the self-assembly processes of adding the tetracationic cyclophane by clipping actually proceeded more efficiently.[28] Dynamic ^{1}H NMR spectroscopy showed the presence of different translational isomers.

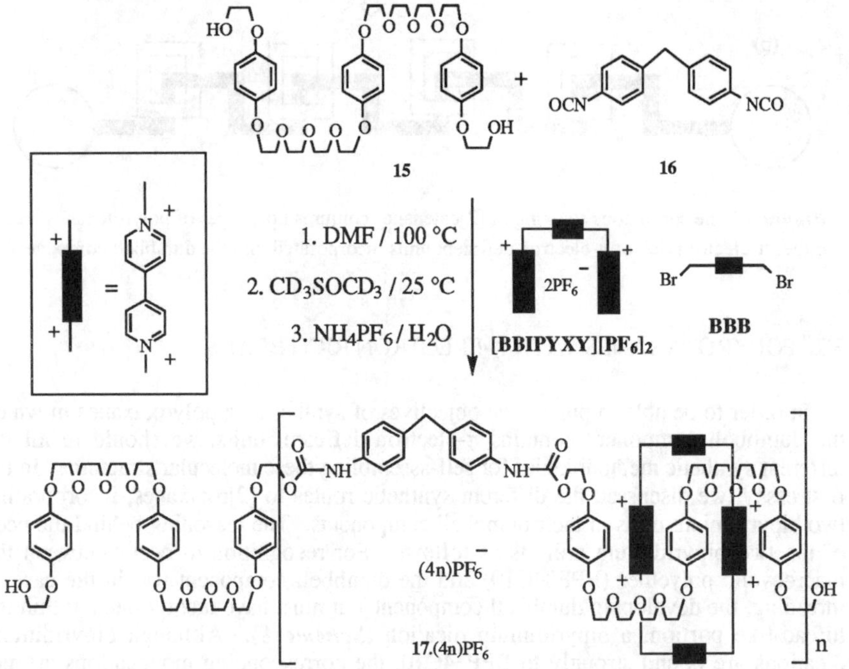

15 + 16

1. DMF / 100 °C
2. CD_3SOCD_3 / 25 °C
3. NH_4PF_6 / H_2O

$[BBIPYXY][PF_6]_2$ BBB

17.(4n)PF_6

Scheme 7

Oligorotaxanes of the order M_n 9000 Daltons (nine repeat units) have been prepared by reacting a polyether chains, symmetrically substituted with hydroquinol rings with methylene bis(4-phenyl-isocyanate).[28] The solubility of the system in typical solvents imposed limitations on characterisation. The intense orange colours did indicate that we had accomplished the objective of self-assembling pseudo-oligorotaxanes.

So far, we have addressed the construction and properties of polyrotaxanes, incorporating one or more tetracationic cyclophanes on π-electron rich dumbbells (***Figure 4a***). Formalistically, we could also extrapolate in our minds from the solid state structures of the [2]catenane to polyrotaxanes which incorporate one or more bis-paraphenylene-34-crown-10 macrocycles onto π-electron deficient dumbbells (***Figure 4b***). We shall now discuss the initial approaches that have been developed for constructing such polyrotaxanes.

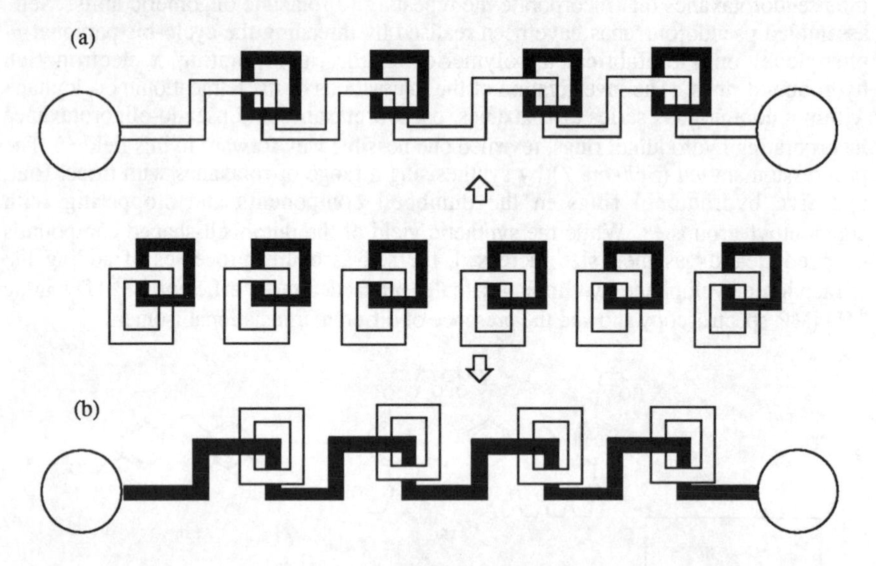

Figure 4. The continuous stacking of [2]catenanes conjures up images of polyrotaxanes with either π-electron rich or π-electron deficient units incorporated into the dumbbell component.

3.2. POLYROTAXANES WITH π-ELECTRON RICH BEADS

In order to be able to pursue the objectives of synthesising polyrotaxanes in which the dumbbell component contains π-electron deficient units, we should recall the different synthetic methodologies for self-assembling these molecular structures. In the first essay, we discussed the different synthetic routes to [2]rotaxanes, incorporating two bipyridinium units in their dumbbell components. The reasoning behind the need of the two bipyridinium units is as follows. For recognition to occur between the macrocyclic polyether (BPP34C10) and the dumbbell component (or, in the case of threading, the developing dumbbell component), it must have incorporated within the thread-like portion, a bipyridinium dication (***Scheme 8***). Although bipyridinium dications are bound strongly to BPP34C10, the corresponding monocations are not complexed. A way to overcome this problem is to link two bipyridine units together through one of their nitrogen atoms by some spacer group and then at each end by some stopper group, linked *via* the other nitrogen atom in each case. The added stability, achieved after complexation of the bead and rod components, allows the self-assembly process to be completed. Indeed, it was found that the [2]rotaxane formed exhibits shuttling of the π-electron rich macrocycle (BPP34C10) between the two π-electron deficient sites at a rate of around 300,000 times a second at room temperature![29] The fact that n+1 bipyridinium sites (n = number of crowns) are required to achieve rotaxane formation *via* threading is a limitation of this approach.

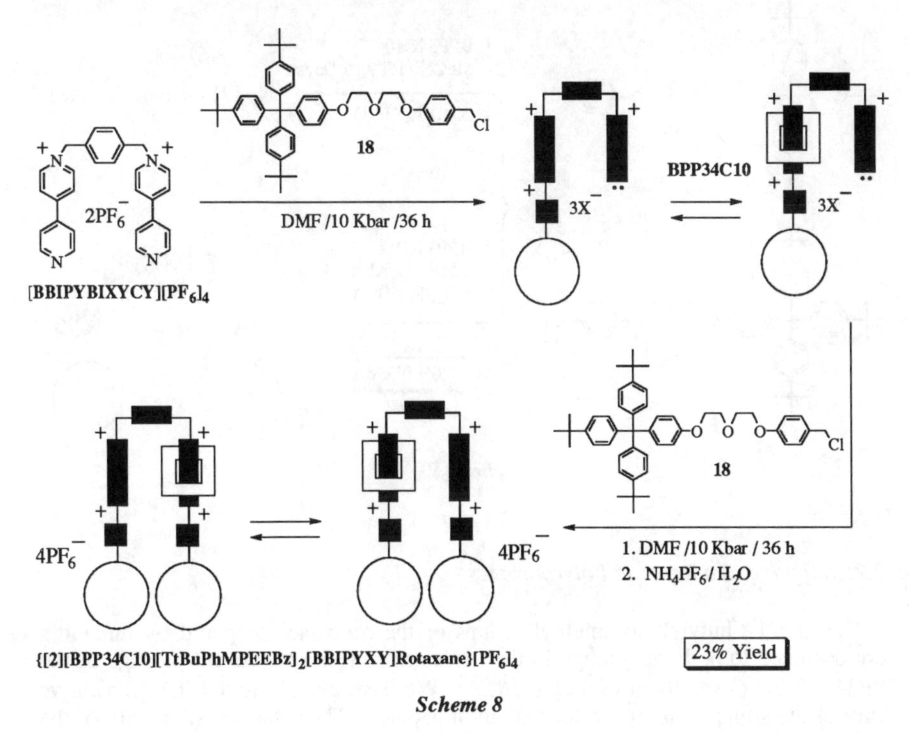

Scheme 8

3.2.1. The Use of Ultra-High Pressure in making Rotaxanes

The preparation of [2]rotaxanes incorporating only one bipyridinium station is possible but special conditions such as the use of ultra-high pressure during the threading process are necessary. This situation is best demonstrated by the following example. When the benzylic chloride **18** and 4,4'-bipyridine BP were mixed together in a 2:1 ratio at ambient temperature in the presence of a 3 molar excess of BPP34C10, no [2]rotaxane was detected, even after prolonged periods.[29] When the same components, in the same molar ratios, were subjected to ultra-high pressure conditions, a 26% yield of the [2]rotaxane was obtained (***Scheme 9***).[29] These conditions - by the Le Chatelier Principle - favour the formation of the [2]rotaxane. The reaction may be influenced further by the increase in reactivity of the second nitrogen atom of the pyridyl-pyridinium monocation under high pressure conditions.[30]

A similar marked increase in reactivity at high pressure is observed in self-assembling systems containing two bipyridinium units. The proportion of the

[2]rotaxane and [3]rotaxane formed during the reaction is dependent on the molar ratio of BPP34C10 used. This result shows that the key steps in threading process - *i.e.*, the formation of a pseudorotaxane intermediate - is influenced strongly by pressure and the relative concentration of the neutral macrocycle and, as such, obeys the Le Chatelier Principle. Continuing along the same vein, we have been able to self-assemble [2]-, [3]-, and even [4]-rotaxanes, albeit in low yields (<3%), based on dumbbell components containing three bipyridinium units.[31,32]

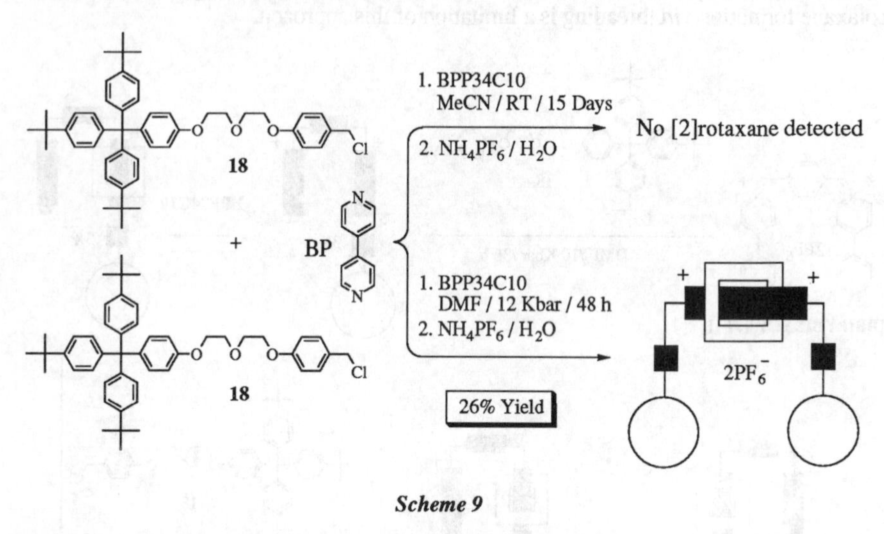

Scheme 9

3.2.2. *Slippage as a Route to Polyrotaxanes*

The tris(4-*t*-butylphenyl)methyl groups of the rotaxanes prepared by threading were designed to be large enough in their roles as stoppers to prevent passage of the BPP34C10 ring over them (***Scheme 10***).[32] We have already noted the preparative utility of the slippage approach for making rotaxanes. The synthetic simplicity of the slippage approach (***Scheme 10***) does much to recommend it as an alternative method for the construction of larger oligorotaxanes and polyrotaxanes (***Table 1***).[33,34] We have applied this methodology to the self-assembly of rotaxanes containing two and three bipyridinium units. Consider the example illustrated in ***Scheme 10***.

Table 1. A comparison between the slippage and threading methods to rotaxane formation for the construction of the [2]- and [3]-rotaxanes. Refer to ***Scheme 10***.

Equivalents of BPP34C10	Conditions	Process	[2]Rotaxane (% Yield)	[3]Rotaxane (% Yield)
1.5[a]	10 Kbar / 36 h	Threading	18	3
4.0[a]	12 Kbar / 72 h	Threading	5	33
4.0[b]	RP[c] / 10 Days	Slippage	31	20
10.0[b]	RP[c] / 10 Days	Slippage	8	55

[a] Refers to the top reaction sequence in Scheme 10. [b] Refers to the equivalents of crown ether used compared with the amount of the dumbbell in the lower reaction described in ***Scheme 10***. [c] Room pressure.

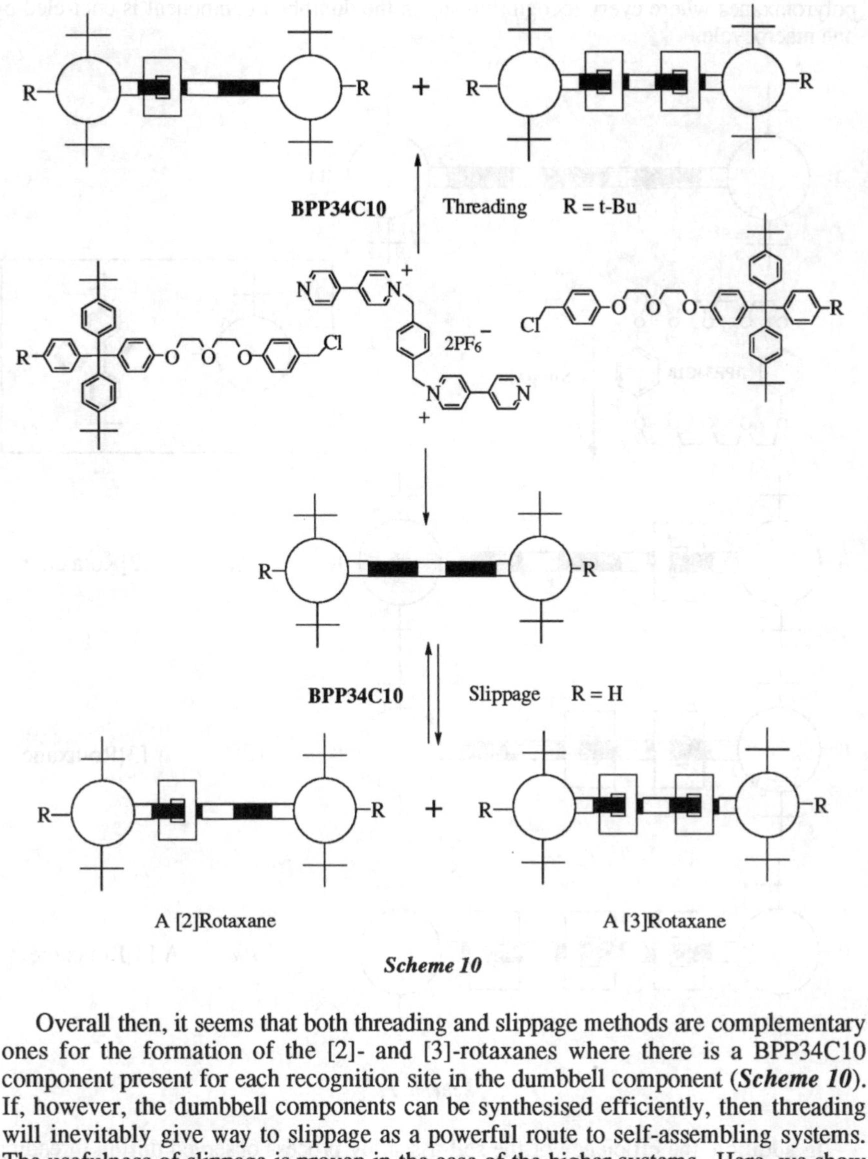

BPP34C10 Threading R = t-Bu

$2PF_6^-$

BPP34C10 Slippage R = H

A [2]Rotaxane

A [3]Rotaxane

Scheme 10

Overall then, it seems that both threading and slippage methods are complementary ones for the formation of the [2]- and [3]-rotaxanes where there is a BPP34C10 component present for each recognition site in the dumbbell component (***Scheme 10***). If, however, the dumbbell components can be synthesised efficiently, then threading will inevitably give way to slippage as a powerful route to self-assembling systems. The usefulness of slippage is proven in the case of the higher systems. Here, we show (***Scheme 11***) a system that includes more than two bipyridinium units. On heating the dumbbell compound with 20 molar equivalents of BPP34C10, we have been able to isolate the [2]rotaxane (2%), the [3]rotaxane (12%), and the [4]rotaxane (19%).[32] The use of threading, in such a case, yields very little of the corresponding rotaxanes.[32]

The self-assembly of [2]-, [3]-, and [4]-rotaxanes demonstrates conclusively that it is possible to construct wholly synthetic systems with potential device-like properties on the nanometre scale by relying upon the molecular recognition between complementary components. Slippage can, in principle, be used for the synthesis of polyrotaxanes where every recognition site in the dumbbell component is encircled by one macrocycle.

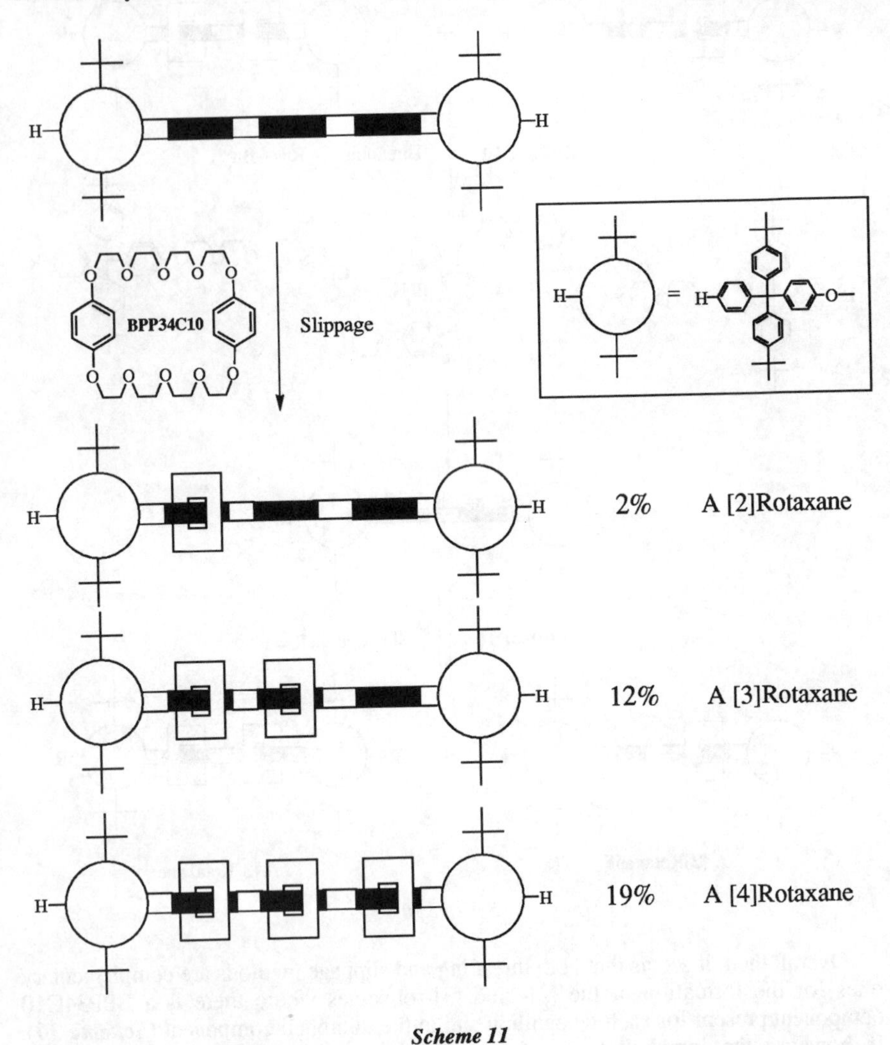

Scheme 11

In addition, the efficiency of the self-assembly process observed during threading under ultra-high pressure is matched by the alternative slippage method which is based solely on size complementarity and the thermodynamic trap which exists between the macrocyclic polyether and the bipyridinium units that allow the preparation of [2]-, [3]- and [4]-rotaxanes to occur at slightly elevated temperatures and ambient pressures with such remarkable yields. Clearly, the method can be extended to the construction of oligorotaxanes and polyrotaxanes. Alternatively, by "undernourishing" the dumbbell component, partial coverage of the polyrotaxane could provide a method of fine tuning the polymer's properties and functions.

The next task we addressed was to modify these [2]rotaxanes by desymmetrising them in order to investigate their potential to be switched either photochemically or electrochemically between different states. Furthermore, the study of other suitable π-electron deficient systems, such as diazapyrenium dications to replace the bipyridinium units, is also being investigated.

4. Light-Driven Rotaxanes Incorporating a Photosynthesiser[35,36] - Towards Photochemical Machines

On the basis of our understanding of the synthesis and dynamics of these mechanically interlocked molecules, one of the priority aims of our research is to extend the design into a system whereby the components are photofunctional. Even at a molecular level, machines need energy to operate. The best source of energy involves light conversion. The concept of a molecular shuttle, which may be addressed photochemically, is an appealing one. A system that could be formed consists of a photosensitiser linked to a π-electron deficient station on which a π-electron rich bead resides (***Figure 5***).

An investigation of systems that use self-assembly and are photoactive has been initiated.[36] This simple system (***Scheme 12***) consists of the unthreading of a pseudorotaxane incorporating the tetracationic cyclophane and 1,5-bis[2-(2-hydroxy-ethoxyethoxy)ethoxy]naphthalene as the thread. In aqueous solution, the unthreading process can be driven by light in the presence of an external photosensitiser, 9-anthracene carboxylic acid, and a sacrificial electron donor, triethanolamine, which prevents the back electron transfer. Since the molecular assembly is based on donor-acceptor type interactions, the reduction of either component - in this case the cyclophane - destroys the π–π stacking interactions, thus destabilising the complex and causing the unthreading. The process can be monitored by absorption and luminesence spectroscopy. The components can be rethreaded by bubbling oxygen through the aqueous solution in the dark.[36]

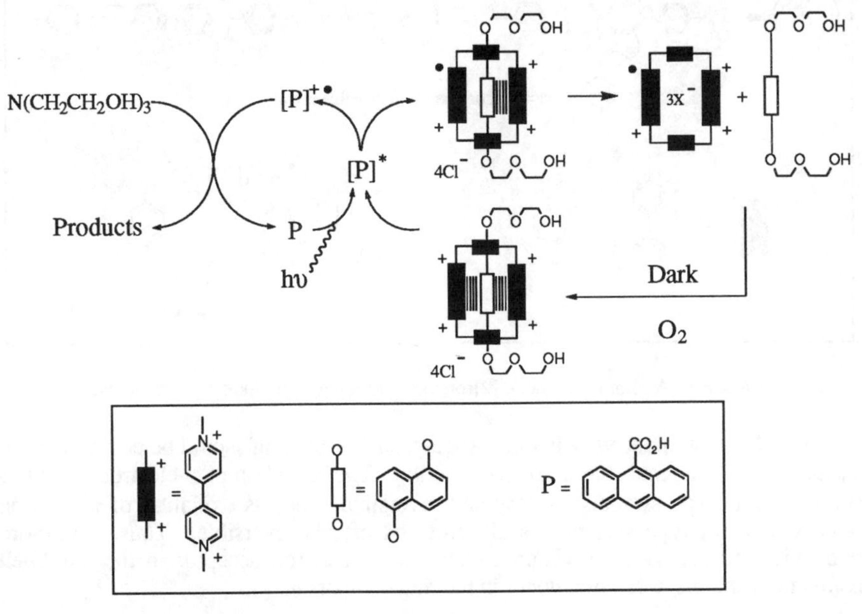

Scheme 12

The strategy described here could be extended to more interesting systems whereby the photosensitiser is incorporated into the dumbbell component of a rotaxane. A particularly appealing system is described in ***Figure 5***. Consider the following situation. Station **A1** is a better receptor than station **A2** and so the macrocyclic donor molecule **D** will reside there. After light excitation, the photosensitiser **P** transfers an electron to the π-electron deficient station **A1**. Since the charge-transfer interaction between the station **A1** and ring is destroyed, the ring **D** leaves the station **A1** to reside at the next station **A2** which is comparable in its stability. The process can be followed by absorption spectroscopy (the reduced π-electron acceptor displays an intense absorption in its visible spectrum) and by emission spectroscopy (the π-electron donor is luminescent, *i.e.*, the fluorescence is quenched by charge-transfer interactions in the pseudorotaxane structure).

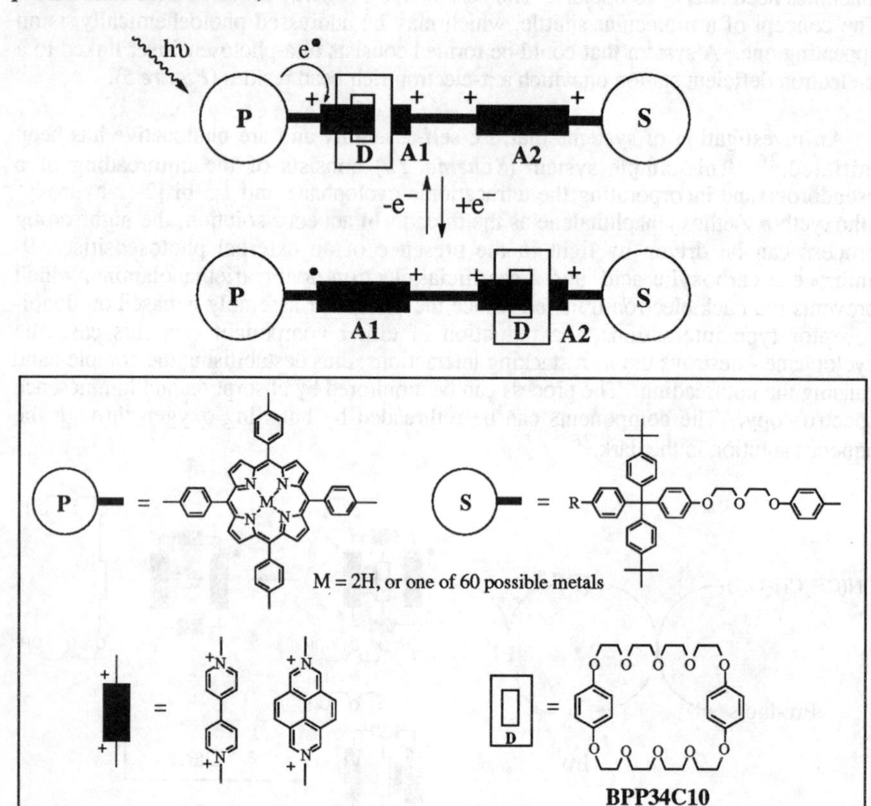

Figure 5. A photoswitchable [2]rotaxane showing some likely components.

After the photoinduced switching process, back switching should be easily effected by allowing O_2 to enter the irradiated solution. The reduction of π-electron acceptors (bipyridinium-type systems) is easy and reversible, whereas oxidation of π-electron donors (aryloxy-type systems) is difficult and often irreversible. Thus, it is more convenient to use systems which contain the π-electron acceptor in the dumbbell component and the π-electron donor in the ring component.

Our original objectives have led us to identify the tetraarylporphyrins as building blocks that could serve the dual purpose of stoppers and of photochemically-active functions in a [2]rotaxane with molecular switching capabilities.[35] Two [2]rotaxanes, composed of a polyether chain intercepted by one or two symmetrically located π-electron rich hydroquinol rings (forming a molecular shuttle) and terminated by either free-base or zinc tetraarylporphyrins (***Figure 6***), have been self-assembled with cyclobis(paraquat-*p*-phenylene) using the clipping procedure. The fact that [2]rotaxanes, with both free-base and metallated porphyrin stoppers, can be self-assembled bodes well for the formation of photochemically-active molecular devices.

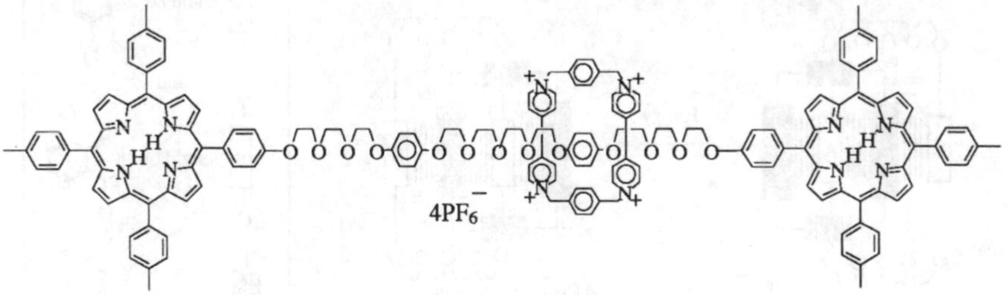

$4PF_6^-$

Figure 6. The identification of tetraaryporphyrins as a photo-active stopper group has led to the synthesis of rotaxanes of this type.

5. Catenanes

The catenanes, as a class of topologically interesting molecules, consist of two or more macrocyclic rings which are mechanically interlocked. Because of this mechanical feature, they have intruiged chemists for years as chemical curiosities.[37] However, it is only recently that there has been a considerable surge of interest towards synthesising these aesthetically-pleasing compounds.[15,38-40] In the past 10 years or so, since the advent of supramolecular chemistry[41] and the consequent development of the chemistry, which relies upon noncovalent bonding interactions to form stable structures, chemists have begun to realise their dreams of making chain-like molecules. While the rotaxanes offer an immediately obvious approach to functioning molecular systems, we shall see that the [n]catenanes provide just as much scope for controlled structural change at a molecular level.

5.1. TRANSLATIONAL ISOMERISM IN [2]CATENANES

The dynamic behaviour of [2]catenanes, such as {[2]-[BPP34C10]-[BBIPYBIXYCY]-catenane}$[PF_6]_4$,[42] is such that both interlocked rings can circumrotate freely through each other at ambient temperature. When we impart some sort of asymmetry into a [2]catenane, either by replacing *one* of the π-electron deficient groups or *one* of the π-electron rich groups with a different one, then there exists the possibility (***Figure 7***) of the catenane existing as a mixture of translational isomers, either of equal populations, or with some predetermined bias for one isomer over the other defined by the strengths of recogniton at the binding sites. The translational isomers differ on account of (i) which unit occupies the cavity inside the other ring component, and (ii) which one takes up a position alongside the other ring component. The kinetic control of this translational isomerism, *i.e.*, the rate of circumrotation of one macrocycle through the other within the [2]catenane, is an important goal in understanding the dynamics of the system.[43] One of the challenges facing us is to learn how to impart control on the dynamic processes of these [2]catenanes through structural change in the components that can be chosen to be photochemically, electrochemically or, simply, chemically active.

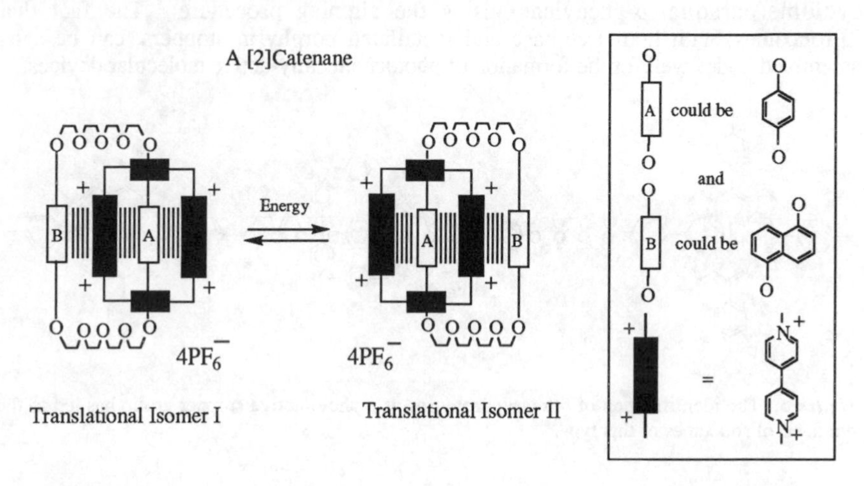

Figure 7. Illustration of the translational isomerism possible by changing the π-electron rich units on the crown. Unit A differs from unit B either sterically or electronically or both.

5.1.1. Changing the Crown Components

The efficiency of the self-assembly process is retained when BPP34C10 is replace by 1,5-dinaphtho-38-crown-10 (1/5DN38C10) with its 1,5-dioxynaphthalene π-donor units (***Scheme 13***). Apparently the naphtho rings are sterically acceptable, while the greater π-donating ability enhances the rate of reaction. The desired [2]catenane[44] was self-assembled in 51% yield from the components, *i.e.*, 1/5DN38C10, BBB, and [BBIPYXY][PF$_6$]$_2$.

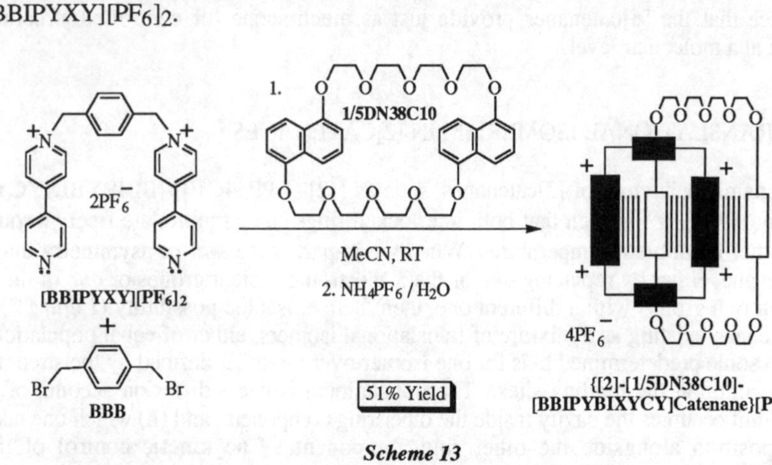

1. 1/5DN38C10

[BBIPYXY][PF$_6$]$_2$

2PF$_6^-$

\+

BBB

MeCN, RT

2. NH$_4$PF$_6$ / H$_2$O

51% Yield

4PF$_6^-$

{[2]-[1/5DN38C10]-[BBIPYBIXYCY]Catenane}[PF$_6$]$_4$

Scheme 13

In principle, the control of translational isomerism may be accomplished by steric[43] and/or electronic means. We chose the macrocyclic polyether 1/5NPP36C10[42,45], containing *one* hydroquinol ring and *one* 1,5-dioxynaphthalene ring system, in order to investigate the influence of steric and electronic factors on the positon of the equilibrium between two translational isomers. It was expected that, if this crown ether could be incorporated into a [2]catenane, then the difference in both the size and the π-donating ability (naphtho > benzo) of the two units would strongly influence the relative populations of the translational isomers in solution.

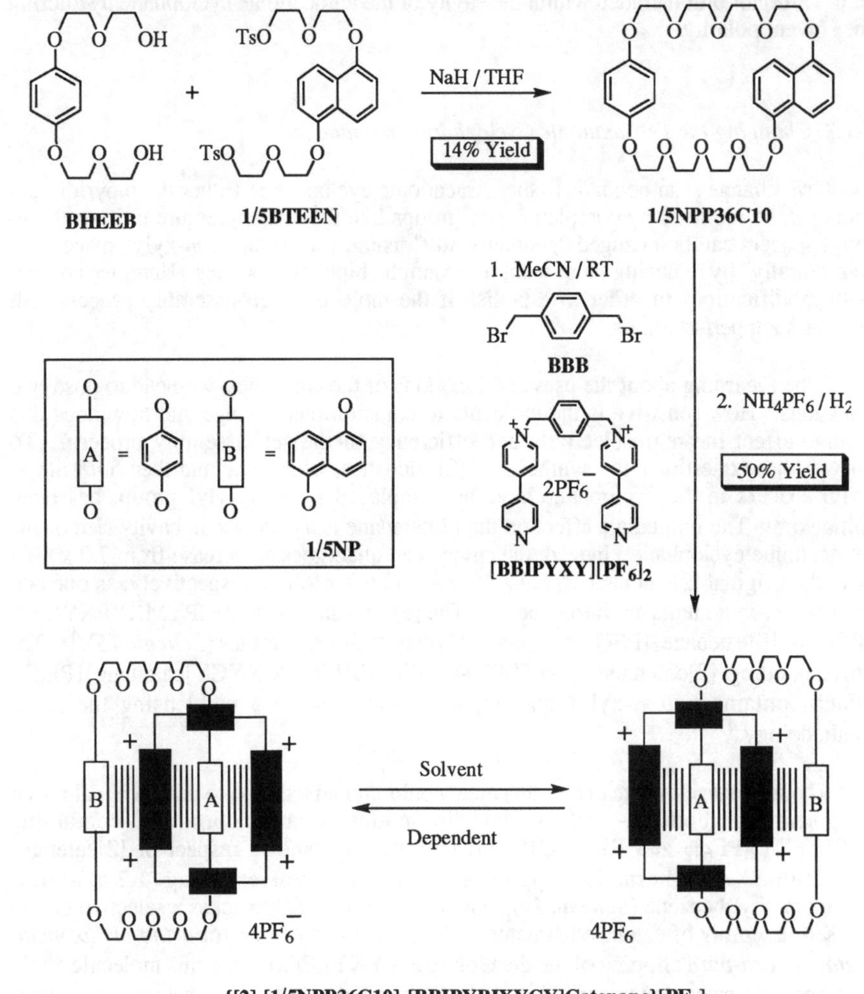

BHEEB + 1/5BTEEN → (NaH / THF; 14% Yield) 1/5NPP36C10

1/5NPP36C10 → (1. MeCN / RT, BBB, $[BBIPYXY][PF_6]_2$; 2. NH_4PF_6 / H_2O; 50% Yield) $\{[2]\text{-}[1/5NPP36C10]\text{-}[BBIPYBIXYCY]Catenane\}[PF_6]_4$ (Solvent Dependent equilibrium)

A = hydroquinol unit; B = 1,5-dioxynaphthalene unit (1/5NP)

Scheme 14

The crown ether 1/5NPP36C10 was prepared (***Scheme 14***) by a convergent synthesis, starting from the precursors, BHEEB and 1/5BTEEN. Under high dilution, these two precursors readily form 1/5NPP36C10 in 14% yield in the presence of a base. The corresponding catenane, {[2]-[1/5NPP36C10][BBIPYBIXYCY]catenane}$[PF_6]_4$, was obtained (***Scheme 14***) in 50% yield using standard self-assembly procedures, starting from BBB, [BBIPYXY][PF_6]$_2$, and 1/5NPP36C10. The self-assembly of {[2]-[1/5NPP36C10][BBIPYBIXYCY]catenane}[PF_6]$_4$ demonstrates that the synthetic strategy which has been applied successfully for the construction of catenanes containing only one kind of π-donating unit can be extended to the self-assembly of systems containing more than one such unit.[46] The relative proportions of translational isomers was found to be highly solvent dependent, such that, the more polar the solvent, the more one translational isomer was favoured over the other. This solvent dependency of translational isomerism raises the possibility of controlling the identity of the π-donor unit included within the cavity of the tetracationic cyclophane by altering the solvent polarity.

5.1.2. Changing the Tetracationic Cyclophane Components

Two changes can be made to the tetracationic cyclophane. Either the bipyridinium units can be replaced by complementary groups like the diazapyrenium unit, or the *p*-xylyl spacers can be changed "geometrically" using, for example, *m*-xylyl spacers, or "structurally" by replacing them with, for example, biphenyl residues. Here, we address both modifications in order to establish if the molecular self-assembly process will survive such perturbations.

When learning about the uses and functions of the catenanes, we need to pose two questions. How sensitive is the molecule to constitutional change and how does this change effect the regioselectivity and efficiency of the self-assembly process? To answer these questions, the synthesis of [2]catenanes, where *one* and then *both* the *p*-xylyl spacers in the cyclophane have been replaced by a *m*-xylyl group, has been initiated.[47] The immediate effect on the [2]catenane is a decrease in cavity size of the tetracationic cyclophane. Indeed, the cavity size undergoes a decrease from 7.0 x 10.3 Å in the original [2]catenane to 6.9 x 10.0 Å and 6.7 x 9.8 Å, respectively, as one and then two *m*-xylyl units are introduced.[47] The [2]catenane, {[2]-[BBIPYMXYPXYCY]-[BPP34C10]catenane}[PF_6]$_4$ was prepared by two different routes (***Scheme 15***).[47] The corresponding [2]catenane, {[2]-[BPP34C10]-[BBIPYBIMXYCY]catenane}[PF_6]$_4$, which contains *two m*-xylyl spacers, was formed in 28% yield using the usual methodology.[47]

The next stage in research in this area would undoubtedly be to add an equal molar proportion of both *m*- and *p*-xylyl dibromides to a reaction pot containing [BBIPYXY][PF_6]$_2$ and BPP34C10 and note the outcome in respect of [2]catenane formation. An alternative approach, however, involves using 1,2,4,5-tetra-(bromomethyl)benzene (***Scheme 16***), which has the potential to act as a selective unit in the self-assembly of catenated structures. There is the potential for *ortho-ortho*, *meta-meta*, or *para-para* clipping of the dication [BBIPYXY][PF_6]$_2$ onto this molecule.[48] In fact, the only catenated material isolated, albeit in low yield (14%), was the *para-para* bis[2]catenane (***Scheme 15***).[48] This result is a demonstration of the kind of regioselectivity which can operate in a self-assembly process. In fact, one might suggest that the presence of an impurity has no effect on the self-assembly process - a feature that characterises self-assembly processes in nature.

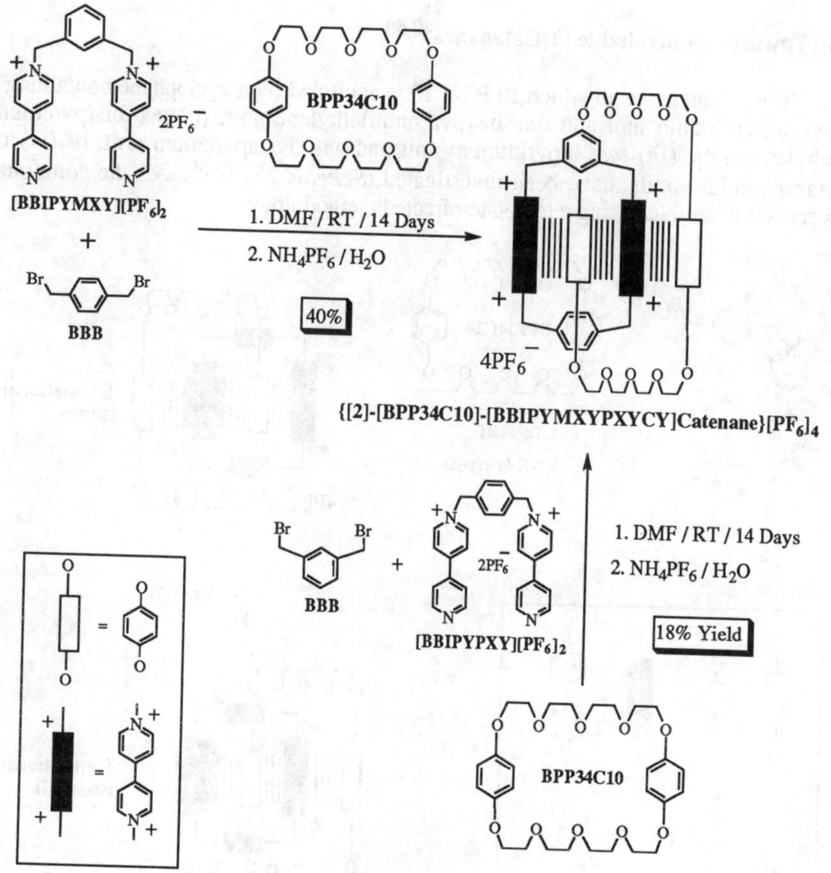

[BBIPYMXY][PF$_6$]$_2$ + BBB → (1. DMF / RT / 14 Days; 2. NH_4PF_6 / H_2O; 40%) BPP34C10 → {[2]-[BPP34C10]-[BBIPYMXYPXYCY]Catenane}[PF$_6$]$_4$

BBB + [BBIPYPXY][PF$_6$]$_2$ + BPP34C10 → (1. DMF / RT / 14 Days; 2. NH_4PF_6 / H_2O; 18% Yield) {[2]-[BPP34C10]-[BBIPYMXYPXYCY]Catenane}[PF$_6$]$_4$

Scheme 15

TBB + [BBIPYXY][PF$_6$]$_2$ + BPP34C10 → (1. DMF / RT / 14 Days; 2. NH_4PF_6 / H_2O; 14% Yield) A Bis[2]catenane ($8PF_6^-$)

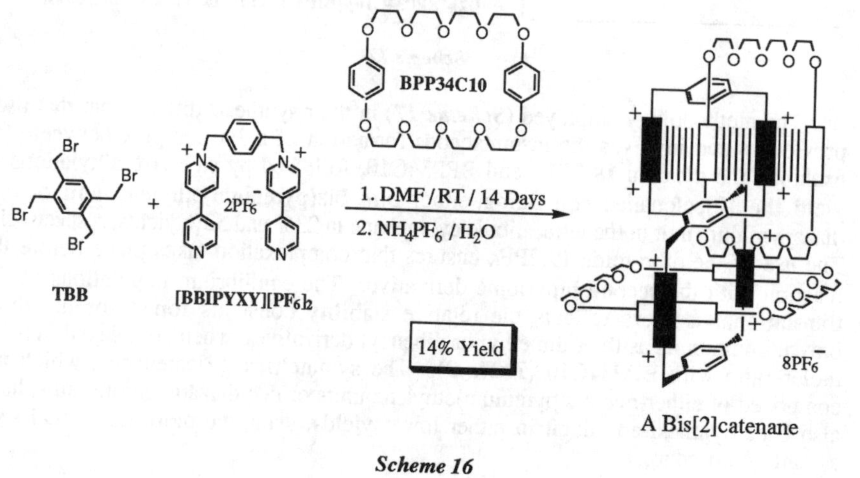

Scheme 16

6. Towards Controllable [2]Catenanes[49,50]

The [2]catenanes in which BPP34C10 is encircled by a cyclophane containing (i) *one* bipyridinium unit and *one* bis(pyrenium)ethylene unit, (ii) *two* bis(pyrenium)-ethylene units, (iii) *one* bipyridinium unit and *one* diazapyrenium unit, or (iv) *two* diazapyrenium units, have been investigated (***Scheme 17***).[49,50] All the compounds were self-assembled using a template-directed methodology.

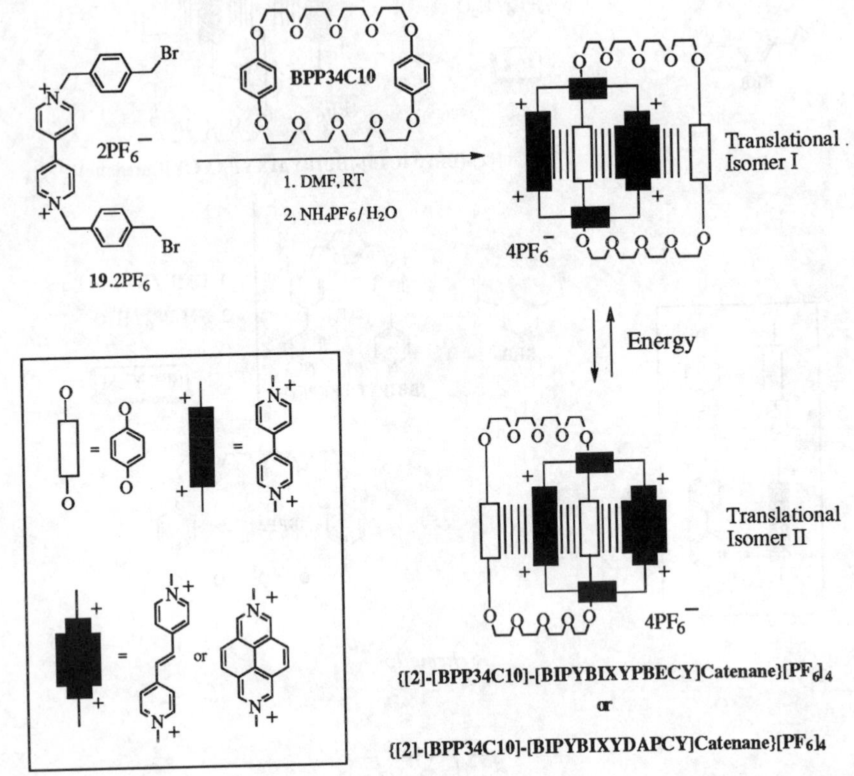

{[2]-[BPP34C10]-[BIPYBIXYPBECY]Catenane}[PF$_6$]$_4$

or

{[2]-[BPP34C10]-[BIPYBIXYDAPCY]Catenane}[PF$_6$]$_4$

Scheme 17

The methodology employed (***Scheme 17***) in their synthesis differs from that used previously and involves the instantaneous formation of a 1:1 complex between, for example, the dication **19**.2PF$_6$ and BPP34C10, followed by stepwise alkylations to yield the [2]catenanes containing either *one* bis(pyrenium)ethylene unit or *one* diazapyrenium unit in the tetracationic cyclophane in 23% and 52% yields, respectively. The use of the dibromide **19**.2PF$_6$ ensures that complexation takes place before the reaction with the second bipyridine derivative. The equilibrium populations of the translational isomers reflects the relative stability constants found for the three bipyridinium units as their dimethyl or dibenzyl derivatives, when complexed in a 1:1 molar ratio with BPP34C10 (***Table 2***). The symmetrical [2]catenanes, which are comprised of either *two* bis(pyrenium)ethylene units or *two* diazapyrenium units, have also been synthesised, albeit in rather lower yields, using the more traditional self-assembly procedure.

Table 2. The difference in population of translational isomers in the [2]catenanes containing unsymmetrical tetracationic cyclophanes at low temperature (<240 K) in CD_3COCD_3. Refer to ***Scheme 17***.

Tetracationic Cyclophane[a]	Percentage Yields	Population (%) of Translational Isomer I	Population (%) of Tranlational Isomer II
$[BIPYBIXYBPECY]^{4+}$	23	8	92
$[BIPYBIXYDAPCY]^{4+}$	52	96	4

[a] The acronyms used in this Table follow Reference 47. The terms BPE and DAP stand for bis(pyridinium)ethylene and diazapyrenium, respectively.

The self-assembly of an optically active [2]catenane involved[51] the incorporation of a chiral hydrobenzoin unit into one of its component rings: one of the *p*-xylyl groups in the tetracationic cyclophane may be substituted by a flexible $CH_2OCH_2CH_2OCH_2CH_2$ chain without impeding catenane formation (***Scheme 18***). However, when both the *p*-xylyl units are replaced, the tetracationic cyclophane loses its ability to complex with π-electron rich aromatic substrates. This achievement has implications for the design of chiral solid-state devices and the construction of asymmetric catalysts.

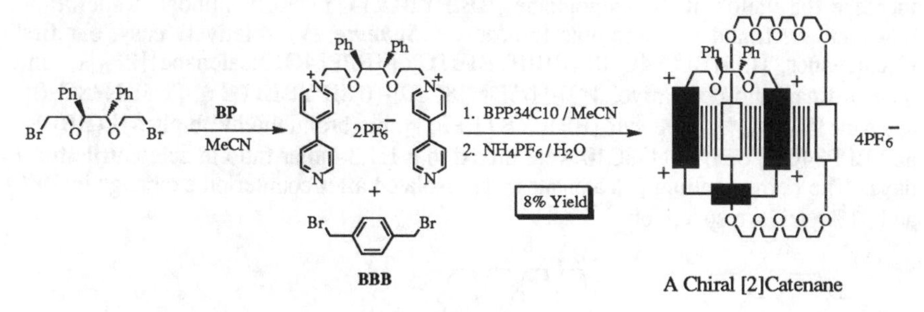

Scheme 18

7. The Self-Assembly of [n]Catenanes

Polycatenanes are themselves interesting compounds in a topological sense. In essence, their architecture resembles chains, where alternating macrocyclic polyethers and tetracationic cyclophanes form a linked combination (***Figure 8***).

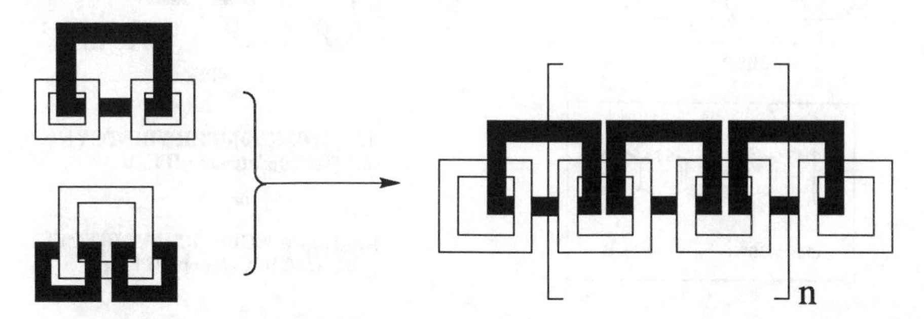

Figure 8. The amalgamation of larger rings to form polycatenanes.

In the macroscopic world, chains conjure up images of strength. It is thought that, on a molecular scale, approaching the nanometer scale, polycatenanes might also be very stable polymers. Obviously, to make polycatenanes, we need to increase the size of both the tetracationic cyclophane and macrocyclic polyether components to allow more than one ring to occupy a position within the cavity of the next ring. [3]Catenanes provide a good starting point towards a template-directed approach to polycatenanes. Extending the methodology, however, was not as straightforward as one might expect. The next section highlights some of the limitations we have encountered in trying to self-assemble [n]catenanes.

7.1. EXPANDING THE CHARGED RING [52]

Early in the quest for [n]catenanes, we targetted a larger cyclophane, such that the two bipyridinium units could each be occupied simultaneously by neutral macrocyclic polyether rings. The tetracation, $[BBIPYBIBTCY]^{4+}$, where the bridging *p*-xylyl spacers units in $[BBIPYBIXYCY]^{4+}$ have been replaced by bitolyl ones, was identified as a good target. Unfortunately, its synthesis was not achievable in the beginning, even with the aid of a template (such as BHEEB), which had been used successfully to increase the yields of the cyclophane $[BBIPYBIXYCY]^{4+}$. In almost desperation, however, we found we were able to produce (***Scheme 19***), relatively easy, our first [3]catenanes, {[3]-[BPP34C10]-[BBIPYBIBTCY]-[BPP34C10]catenane}$[PF_6]_4$, and its 1,5-dinaphtho derivative, {[3]-[1/5DN38C10]-[BBIPYBIBTCY]-[1/5DN38C10]-catenane}$[PF_6]_4$.[52] The salt [BBIPYBT]$[PF_6]_2$, bis(bromomethyl)biphenyl (BBBP), and BPP34C10 or 1/5DN38C10 were mixed in a 1:1:3 molar ratio in acetonitrile for 9 days. The corresponding [3]catenanes were isolated after counterion exchange in 20% and 31% yields, respectively.[52]

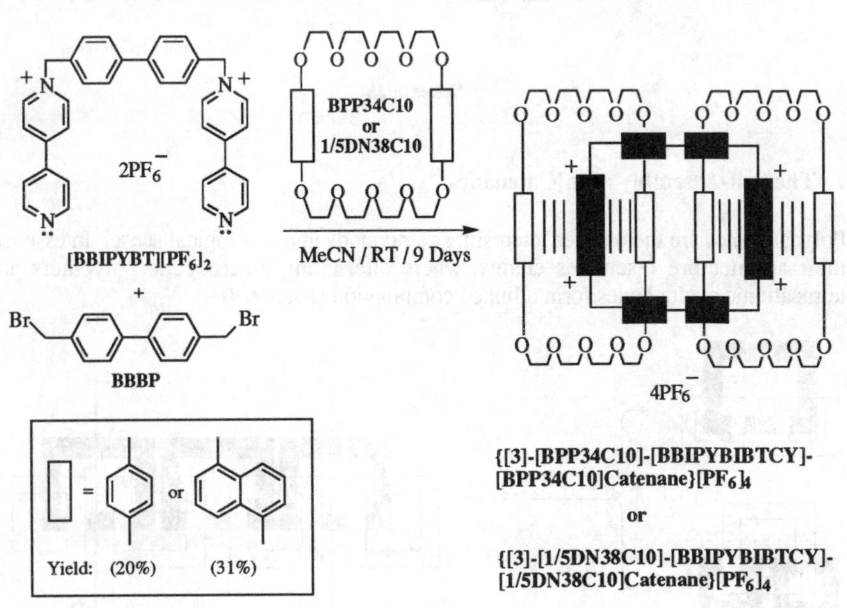

Scheme 19

The cavity of $[BBIPYBIBTCY]^{4+}$ was found to be almost square with inner dimensions of approximately 11 x 10.5 Å, which was fortunately just the right size required for the incorporation of two aromatic units stacked face-to-face. *The take-home message is that it is sometimes easier to make the self-assembled structure than it is to make the molecular components!*

A possible route to the formation of the large cyclophane is to use macrocyclic polyethers incorporating an ester functionality that could be readily hydrolysed in acidic media after catenation, leaving the desired tetracationic cyclophane, $[BBIPYBIBTCY]^{4+}$. These types of crown ethers are being investigated at the moment for their ability to form [3]catenanes. We have yet to demonstrate the viability of the alternative *clipping* approach to rotaxanes and catenanes in which the neutral macrocyclic ring is cyclised around π-electron deficient units in dumbbell or macrocyclic components. Esterification between a diol and a diacid dichloride might permit the clipping of macrocyclic mono- or di-esters around the appropriate dumbbell molecule to afford the desired rotaxanes or catenanes.

7.2. EXPANDING THE NEUTRAL RING[53]

The high dilution, caesium-promoted synthesis of a much larger tetrakisparaphenylene-68-crown-20 in 62% yield from its immediate precursors, the diphenol and the ditosylate, allowed the self-assembly of the [2]catenane, {[2]-[BBIPYBIXYCY]-[TPP68C20]catenane}$[PF_6]_4$, under ambient pressure, and the [3]catenane, {[3]-[BBIPYBIXYCY]-[TPP68C20]-[BBIPYBIXYCY]catenane}$[PF_6]_8$, under ultra-high pressure (***Scheme 20***). The temperature dependence of the ^{1}H NMR spectra show that these catenanes behave like "molecular trains". In the [3]catenane, the "trains" appear to travel together around the "circle line" formed by the four "station" crown. Furthermore, only one translational isomer is observed, *i.e.*, the one in which the two "trains" prefer to keep a free station between them in both directions as they do the loop. Crashes do not occur. The order that is associated with their template-directed synthesis "lives on" in the highly ordered structural behaviour that they exhibit in solution. The reduced preorganisation of the TPP68C20, when compared to BPP34C10, meant, however, that no catenanted materials were produced when [BBIPYBT]$[PF_6]_2$ and BBBP were added to solutions of TPP68C20.

Thus, we have already characterised [3]catenanes from which one could conceive of a route being devised to polycatenanes. In the knowledge that (i) the crown ether TPP68C20, with four hydroquinol rings, can form a [3]catenane with two small tetracationic cyclophanes at ultra-high pressure and (ii) the large tetracationic cyclophane, $[BBIPYBIBTCY]^{4+}$, forms a [3]catenane with two BPP34C10 rings, it seems possible to construct higher order catenanes.

7.3. [4]- AND [5]-CATENANES [54,55]

It is obvious that we can only self-assemble [3]catenanes if we increase the size of both the neutral and tetracationic components. So far, the use of TPP68C20 and the large cyclophane $[BBIPYBIBTCY]^{4+}$ has met with no success. We attribute this lack of catenation to the large dimensions of the two components, and, in the case of TPP68C20, to its flexibility. The macrocyclic polyether TPP51C15 (***Scheme 21***),

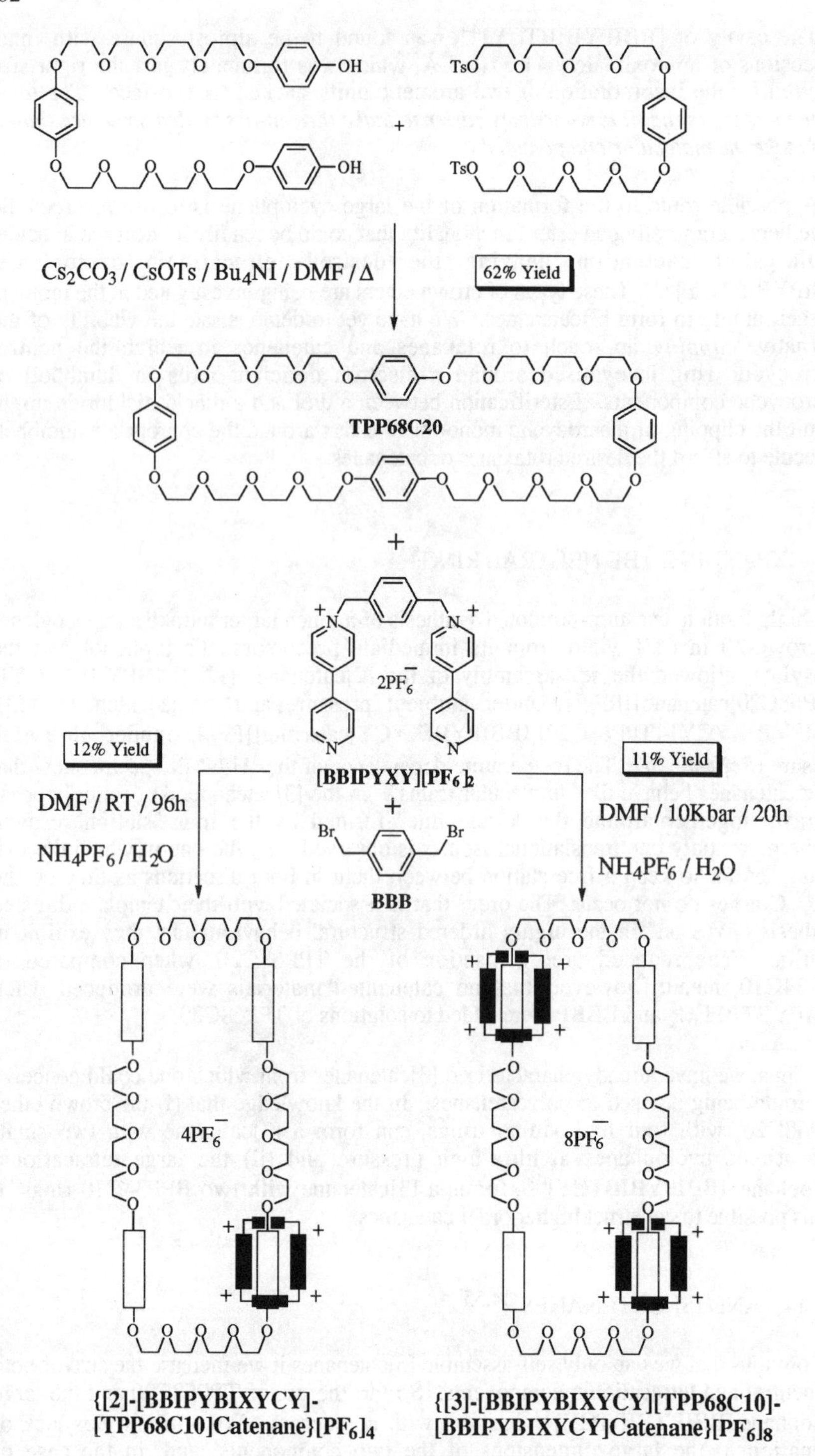
OH
TsO
OH
TsO
Cs_2CO_3 / CsOTs / Bu_4NI / DMF / Δ
62% Yield
TPP68C20
+
$2PF_6^-$
$[BBIPYXY][PF_6]_2$
+
Br
Br
BBB
12% Yield
DMF / RT / 96h
NH_4PF_6 / H_2O
11% Yield
DMF / 10Kbar / 20h
NH_4PF_6 / H_2O
$4PF_6^-$
$8PF_6^-$
{[2]-[BBIPYBIXYCY]-[TPP68C10]Catenane}$[PF_6]_4$
{[3]-[BBIPYBIXYCY][TPP68C10]-[BBIPYBIXYCY]Catenane}$[PF_6]_8$

Scheme 20

which is an intermediate in constitution between the "failed" TPP68C20 and "successful" BPP34C10, should be sufficiently large enough to form catenanes, provided it is not too flexible. In fact, it does work![54]

The crown ether TPP51C15 was formed by the caesium ion promoted reaction in an acceptable yield of 14% from readily available starting materials (***Scheme 21***).[54] The [2]catenane was isolated under ambient pressure in 48% yield while the [3]catenane could only be isolated in 15% yield from an ultra-high pressure reaction. Encouraged by this result, a reaction was carried out using the ingredients to form the large cyclophane at normal pressure. From this reaction we managed to isolate, in low yield, the corresponding [3]catenane, a key intermediate in the formation of the higher [4]- and [5]-catenanes. This result infers that the smaller crown ether contains the necessary molecular recognition information to self-assemble with the usual components to form catenated species.

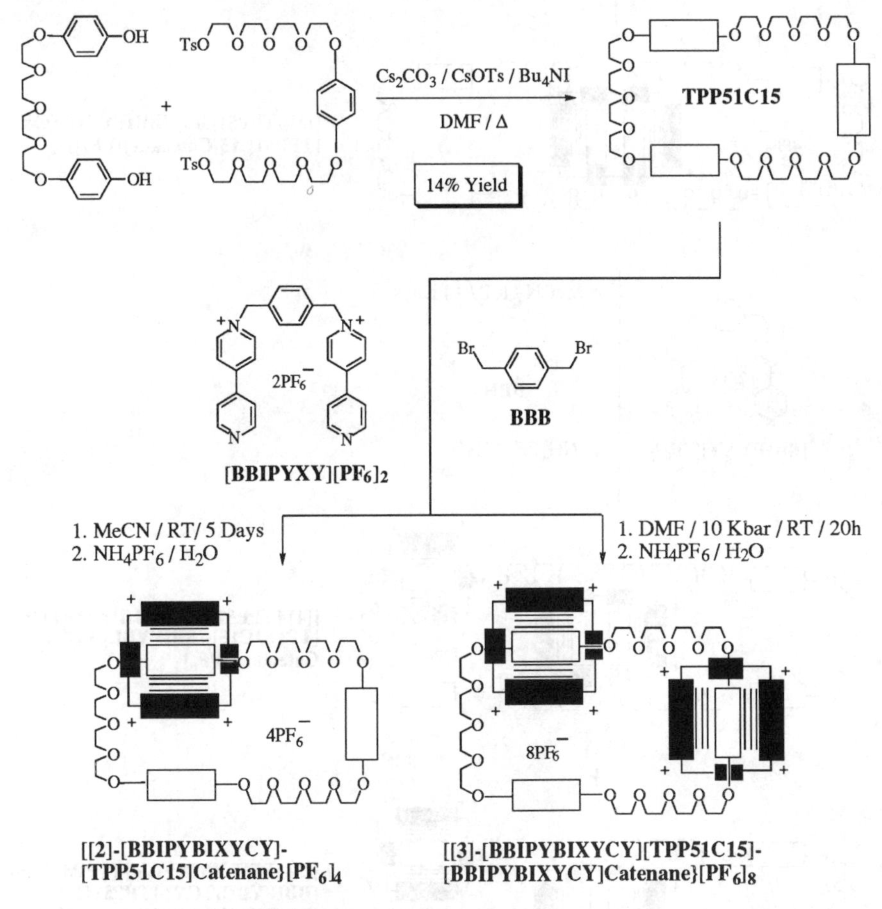

Scheme 21

Subjecting the newly formed [3]catenane to ultra-high pressure in the presence of $[BBIPYBIXYCY][PF_6]_4$ and bis(bromomethyl)benzene yielded the [4]catenane in 22% and a trace of the [5]catenane, which was characterised only by mass spectrometry

(***Scheme 22***).[54] However, when we replaced the hydroquinol rings by 1,5-dioxynaphthalene residues, which are known to bind much more strongly to the bipyridinium dications, our goal of a [5]catenane was achieved.[55] Under the same reaction conditions, and after only fourteen days at ambient pressure, the reaction produced[55] the starting [3]catenane (45%), the intermediate [4]catenane (31%), and the [5]catenane (5%).

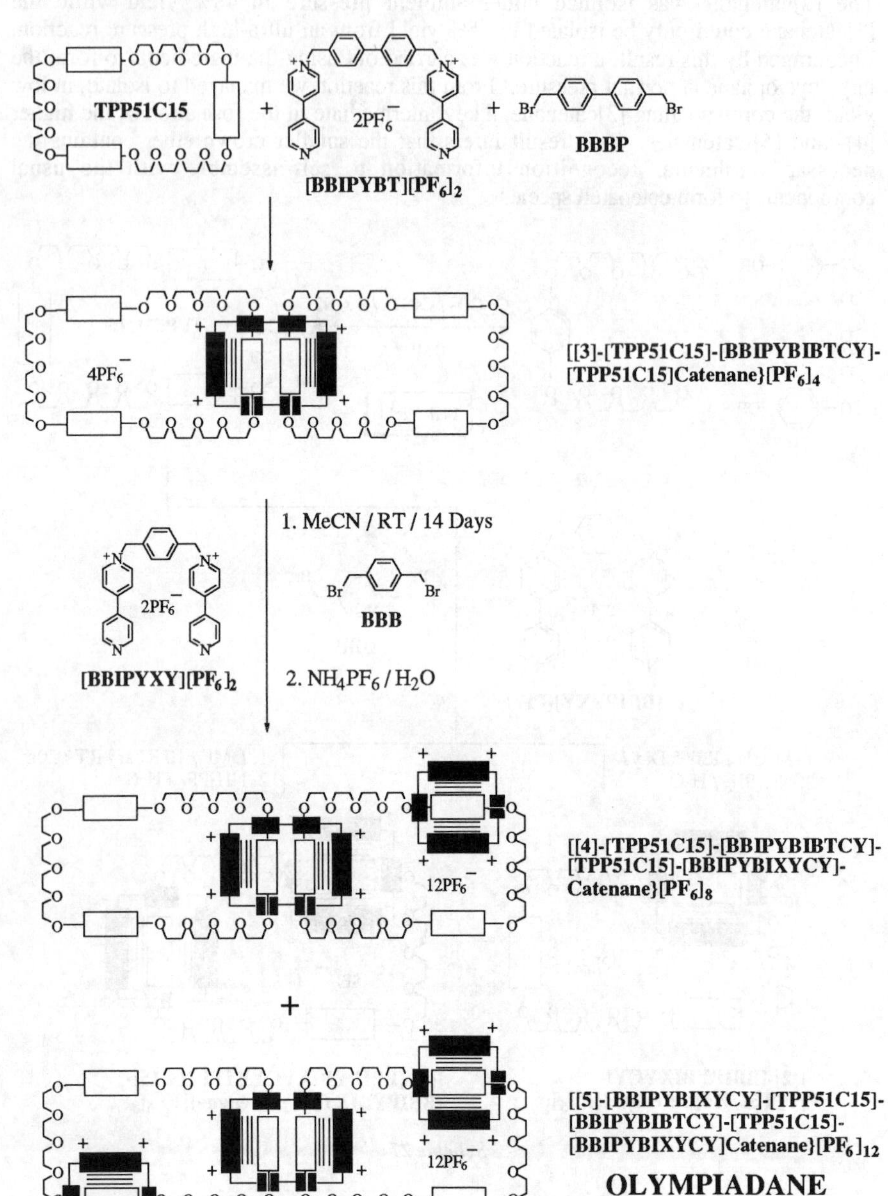

Scheme 22

Hence, by programming sufficient structural and electronic information into the constituent recognition components of both the macrocyclic polyether and tetracationic cyclophanes, the beginnings of a route to polycatenanes are beginning to emerge.

8. Future Directions

A number of different molecular architectures, associated with the possible polymeric structures, are shown in ***Figure 9***. These particular assemblies represent only a very few of the possibilities that exist for self-assembling molecules on the nanoscale in the years to come.

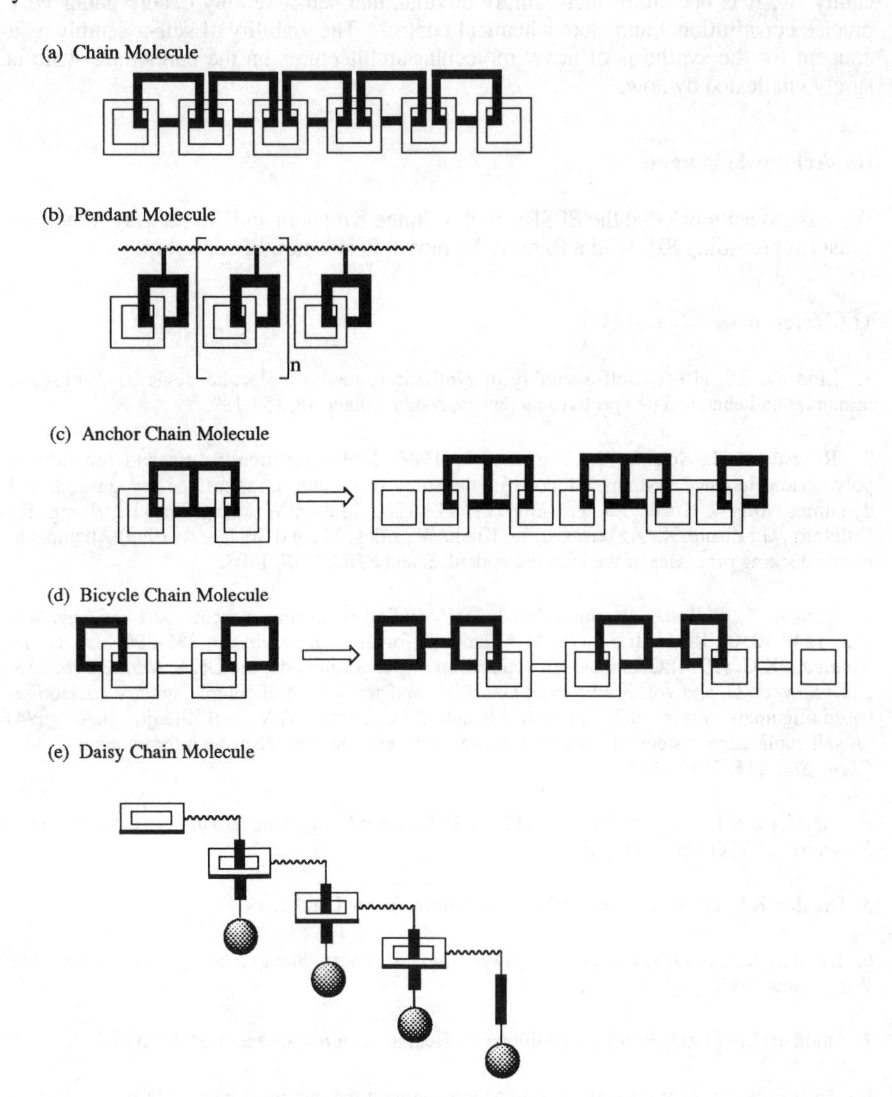

Figure 9. Some of the nanometre-scale targets for self-assembly in the future.

9. Conclusion

It has now been demonstrated beyond any doubt that it is possible to use self-assembly processes to construct molecular assemblies and supramolecular arrays, which are of nanometre-scale, with a high degree of control and precision from molecular components comprised of simple and inexpensive building blocks. Furthermore, it is often easier to make molecular assemblies and supramolecular arrays than it is to make some of the components on their own. The molecular components are "intelligent" in that they hold all of the information necessary for the construction of the precisely assembled structures and superstructures without the need for external reagent or catalysis. It is becoming increasingly obvious that self-assembly occurs under very precise constitutional and stereochemical control. The viability of self-assembly as a concept for the synthesis of novel molecular architectures on the nanometre-scale is surely vindicated by now.

10. Acknowledgments

We wish to acknowledge the EPSRC in the United Kingdom and the Ramsay Memorial Trust for providing SJL with a Ramsay Memorial Postdoctoral Fellowship.

11. References

1. Lindsey, J.S. (1991) Self-assembly in synthetic routes to molecular devices. Biological principles and chemical perspectives : a review, *New J. Chem.* **15**, 153-180.

2. Ringsdorf, H., Scharb, B., Venzmer, J. (1988) Molecular architecture and function of polymeric orientated systems: Models for the study of organisation, surface recognition, and dynamics, *Angew. Chem. Int. Ed. Engl.* **27**, 113-158. Muller, W., Ringsdorf, H., Rump, E., Wildburg, G., Zhang, X., Angermaier, L., Knoll, W., Liley, M., and Spinke, J. (1993) Attempts to mimic docking processes of the immune system, *Science* **262**, 1706-1708.

3. Tjivikva, T., Ballester, P., and Rebek Jr.J. (1990) Self-replicating systems, *J. Am. Chem. Soc.* **112**, 1249-1250. Rebek Jr., J. (1994) A template for life, *Chem. Br.* **30**, 286-290. Li, T. and Nicolaou, K.C. (1994) Chemical self-replication of palindromic duplex DNA, *Nature* **396**, 218-221. Sievers, D. and von Kiedrowski, G. (1994) Self-replication of complementary nucleotide-based oligomers, *Nature* **369**, 221-224. Menger, F.M., Eliseev, A.V., and Khanjin, N.A. (1994) "A self-replicating system": New experimental data and a new mechanistic interpretation, *J. Am. Chem. Soc.* **116**, 3613-3614.

4. Langford, S.J. and Stoddart, J.F. (1994) Self-assembly in chemical synthesis, In *Gnosis to Prognosis*, NATO, Preceeding article.

5. Drexler, K.E. (1990) *Engines of creation*, Fourth Estate, London.

6. Drexler, K.E. (1992) *Nanosystems, molecular machinery, manufacturing and computation*, Wiley, New York.

7. Stoddart, J.F. (1992) Whither and thither molecular machines, *Chem. Aust.* **59**, 576-577.

8. An [n]rotaxane is formed by n components, namely the dumbbell plus (n-1) rings. Thus, a [2]rotaxane consists of a dumbbell component and *one* ring component. A [3]rotaxane consists of one dumbbell and *two* ring components.

9. Anelli, P.L., Spencer, N., and Stoddart, J.F. (1991) A molecular shuttle, *J. Am. Chem. Soc.* **113**, 5131-5133.

10. Ashton, P.R., Bissell, R.A., Spencer, N., Stoddart, J.F., and Tolley, M.S. (1992) Towards controllable molecular shuttles - 1, *Synlett*, 914-918.

11. Ashton, P.R., Bissell, R.A., Gorski, R., Philp, D., Spencer, N., Stoddart, J.F. and Tolley, M.S. (1992) Towards controllable molecular shuttles - 2, *Synlett*, 919-922.

12. Ashton, P.R., Bissell, R.A., Spencer, N., Stoddart, J.F., and Tolley, M.S. (1992) Towards controllable molecular shuttles - 3, *Synlett*, 923-926.

13. Bissell, R.A., Cordova, E., Kaifer, A., and Stoddart, J.F. (1994) A chemically and electrochemically switchable molecular device, *Nature* **369**, 133-137.

14. Philp, D., Slawin, A.M.Z., Spencer, N., Stoddart, J.F., and Williams, D.J. (1991) The complexation of tetrathiafulvalene by cyclobis(paraquat-*p*-phenylene), *J. Chem. Soc., Chem. Commun.* 1584-1586.

15. Anelli, P.L., Ashton, P.R., Ballardini, R., Balzani, V., Delgado, M., Gandolfi, M.T., Goodnow, T.T., Kaifer, A.E., Philp, D., Pietraszkiewicz, M., Prodi, L., Reddington, M.V., Slawin, A.M.Z., Spencer, N., Stoddart, J.F., Vicent, C., and Williams, D.J. (1992) Molecular meccano. 1: [2]Rotaxanes and a [2]catenane made to order, *J. Am. Chem. Soc.* **114**, 193-218.

16. Philp, D. and Stoddart, J.F. (1991) Self-assembly in organic synthesis, *Synlett*, 445-458. Anelli, P.L., Ashton, P.R., Spencer, N., Slawin, A.M.Z., Stoddart, J.F., and Williams, D.J. (1991) Self-assembling [2]pseudorotaxanes, *Angew. Chem. Int. Ed. Engl.* **30**, 1036-1039. Stoddart, J.F. Chirality in drug design and synthesis, C. Brown (Ed.), Academic Press, London, 1990, p. 53-81.

17. Ashton, P.R., Philp, D., Spencer, N., and Stoddart J.F. (1991) The self-assembly of [n]pseudorotaxanes, *J. Chem. Soc., Chem. Commun.* 1677-1679.

18. Ashton, P.R., Philp, D., Reddington, M., Slawin, A.M.Z., Spencer N., Stoddart, J.F., and Williams, D.J. (1991) The self-assembly of complexes with [n]pseudorotaxane structures, *J. Chem. Soc., Chem. Commun.* 1680-1683.

19. Anelli, P.L., Ashton, P.R., Goodnow, Slawin, A.M.Z., Spencer N., Stoddart, J.F., and Williams, D.J. (1991) Self-assembling [2]pseudorotaxanes, *Angew. Chem., Int. Ed. Engl.* **30**, 1036-1039.

20. Allwood, B.L., Spencer, N., Shahriari-Zavareh, H, Stoddart, J.F., and Williams, D.J. (1987) Complexation of paraquat by a bisparaphenylene-34-crown-10 derivative, *J. Chem. Soc., Chem. Commun.* 1064-1066.

21. Brown, C.L., Philp, D., and Stoddart, J.F. (1991) The template directed synthesis of a rigid tetracationic cyclophane receptor, *Synlett*, 462-464.

22. Odell, B., Reddington, M.V., Slawin, A.W.Z., Spencer, N., Stoddart, J.F., and Williams, D.J. (1988) Cyclo(paraquat-*p*-phenylene). A tetracationic multipurpose receptor, *Angew. Chem. Int. Ed. Engl.* **27**, 1547-1550.

23. Goodnow, T.T., Reddington, M.V., Stoddart, J.F., and Kaifer, A.E. (1991) Cyclo(paraquat-*p*-phenylene). A novel synthetic receptor for amino acids with electron-rich aromatic moieties, *J. Am. Chem. Soc.* **113**, 4335-4337.

24. Ashton, P., Odell B., Reddington, M.V., Slawin, A.W.Z., Stoddart, J.F., and Williams, D.J. (1988) Isostructural, alternately-charged receptor stacks, *Angew. Chem. Int. Ed. Engl.* **27**, 1550-1553.

25. Reddington, M.V., Slawin, A.M.Z., Spencer, N., Stoddart, J.F., Vicent, C., and Williams, D.J. (1991) Towards a molecular abacus, *J. Chem. Soc., Chem. Commun.* 630-634.

26. Ashton, P.R., Philp, D., Spencer, N., Stoddart, J.F., and Williams, D.J. (1994) A self-organised layered superstructure of arrayed [2]pseudorotaxanes, *J. Chem. Soc., Chem. Commun.* 181-184.

27. Watson, J.D. and Crick, F.H.C. (1953) Molecular Structure of Nucleic Acids, *Nature* **171**, 737-738.

28. Sun, X., Amabilino, D.B., Parsons, I.W., and Stoddart, J.F. (1993) Self-assembly and shuttling properties of some multisite [2]rotaxanes, *Polymer Reprints*, **34**, 104-105.

29. Ashton, P.R., Philp, D., Spencer, N., and Stoddart, J.F. (1992) A new design strategy for the self-assembly of molecular shuttles, *J. Chem. Soc., Chem. Commun.* 1125-1128.

30. For a discussion of the effects of high pressure on quaternisations at nitrogen, see: Ostazewski, P., Pietraszkiewicz, M., and Salanski, P. (1987) High pressure approach to the synthesis of cryptands and related compounds, *J. Incl. Phenom.* **5**, 553-561.

31. Amabilino, D.B. and Stoddart, J.F. (1993) Self assembly and macromolecular design, *Pure Appl. Chem.* **65**, 2351-2359.

32. Bělohradský, M., Philp, D., Raymo, F.M., and Stoddart, J.F. (1994) The self-assembly of [n]rotaxanes, Proceedings of the International Symposium on Organic Reactivity: Physical and Biological Aspects, Newcastle, 11-16 July 1993, Golding, B.T. (Ed.), Special Publication No 00, Royal Society of Chemistry, Cambridge, In press.

33. Ashton, P.R., Bělohradský, M., Philp, D., and Stoddart, J.F. (1993) Slippage - an alternative method for assembling [2]rotaxanes *J. Chem. Soc., Chem. Commun.*, 1269-1274. Ashton, P.R., Belohradsky, M., Philp, D., and Stoddart, J.F. (1993) The self-assembly of [2]- and [3]-rotaxanes by slippage, *J. Chem. Soc., Chem. Commun.* 1274-1277.

34. Ashton, P.R., Bělohradský, M., Philp, D., Spencer, N., and Stoddart, J.F. (1993) The self-assembly of [2]- and [3]-rotaxanes by slippage, *J. Chem. Soc., Chem. Commun.* 1274-1277.

35. Ashton, P.R., Johnson, M.R., Stoddart, J.F., Tolley, M.S., and Wheeler, J.W. (1992) The template-directed synthesis of porphyrin-stoppered [2]rotaxanes, *J. Chem. Soc., Chem. Commun.* 1128-1131.

36. Ballardini, R., Balzani, V., Gandolfi, M.T., Prodi, L., Venturi, M., Philp, D., Ricketts, H.G., and Stoddart, J.F. (1993) A photochemically-driven molecular machine, *Angew. Chem. Int. Ed. Engl.* **32,** 1301-1303.

37. Frisch, H.L. and Wasserman, E. (1961) Chemical topology, *J. Am. Chem. Soc.* **83**, 3789-3794. Schill, G. (1971) *Catenanes, rotaxanes and knots*, Academic Press, New York. Walba, D.M. (1985) Topological stereochemistry, *Tetrahedron* **41**, 3161-3212. Dietrich-Buchecker, C.O. and Sauvage, J.P. (1987) Interlocking of molecular threads: From the statistical approach to the templated synthesis of catenands, *Chem. Rev.* **87**, 795-810. Philp, D. and Stoddart, J.F. (1991) Self-assembly in organic synthesis, *Synlett* 445-458.

38. Dietrich-Buchecker, C.O., Sauvage, J.P., and Kitzinger, J.P. (1983) Une nouvelle famille de molecules: Les metallo-catenanes, *Tetrahedron Lett.* **24**, 5095-5098. Sauvage, J.P. (1990) Interlacing molecular threads on transition metals: Catenands, catenates, and knots, *Acc. Chem. Res.* **23**, 319-327. Dietrich-Buchecker, C.O. and Sauvage, J.P. (1992) *Supramolecular chemistry*, Kluwer, Amsterdam, 259-277.

39. Amabilino, D.B. and Stoddart, J.F. (1994) Molecules that build themselves, *New Scientist* **141**, 25-29. Ashton, P.R., Goodnow, T.T., Kaifer, A.E., Reddington, M., Slawin, A.M.Z., Spencer N., Stoddart, J.F., Vicent, C., and Williams, D.J. (1989) A [2]catenane made to order, *Angew. Chem., Int. Ed. Engl.* **28**, 1396-1399. Brown, C.L., Philp, D., and Stoddart, J.F. (1991) The self-assembly of a [2]catenane, *Synlett.* 459-461.

40. Hunter, C.A. (1992) Synthesis and structure elucidation of a new [2]catenane, *J. Am. Chem. Soc.* **114**, 5303-5311. Vogtle, F., Meiers, S., and Hoss, R. (1992) One-step synthesis of a fourfold functionalised catenane, *Angew. Chem., Int. Ed. Engl.* **31**, 1619-1622.

41. Lehn, J-M. (1990) Perspectives in supramolecular chemistry, *Angew. Chem. Int. Ed. Engl.* **29**, 1304-1319. Lehn, J.-M. (1988) Supramolecular chemistry - scope and perspective, *Angew. Chem. Int. Ed. Engl.* **27**, 89-112.

42. It will be convienient at this point to describe the acronyms, which are composed of letters and numbers, that identify the neutral and charged compounds displayed throughout this essay. Compounds such as 1,4-dihydroxybenzene and 1,4-dimethoxybenzene are abbreviated to 1/4DHB and 1/4DMB, respectively, and bis-*p*-phenylene-34-crown-10 to BPP34C10. The other acronyms can be deduced from the following rules: B stands for bis when at the beginning, for benzyloxy when in the middle and benzene when at the end of the name. E, H, P, S, and T stand for ethoxy, hydroxy, phenoxy, triisopropylsilyl, and tosyloxy units, respectively. CY, TU, XY, and BIXY represents cyclophane, trioxaundecane, xylylene, and bisxylylene units respectively. Thus 1/5NPP36C10 stands for 1,5-naphthoparaphenylene-36-crown-10.

43. Dietrich-Buchecker, C.O. Sauvage, J.P., and Weiss, J. (1986) Interlocked macrocyclic ligands: A catenand whose rotation of one ring into the other is precluded by bulky substituents, *Tetrahedron Lett.* **27**, 2257-2260. Amabilino, D.B. and Stoddart, J.F. (1993) New approach to controlling catenated structures, *Rec. Trav. Chim. Pays Bas* **112**, 1534-1544.

44. Ashton, P.R., Brown, C.L., Chrystal, E.J.T., Goodnow, T., Kaifer, A.E., Parry, K.P., Philp, D., Slawin, A.M.Z., Spencer, N., Stoddart, J.F., and Williams, D.J. (1991) The self-assembly of a highly ordered [2]catenane, *J. Chem. Soc., Chem. Commun.* 634-639.

45. Ashton, P.R., Blower, M., Philp, D., Spencer, N., Stoddart, J.F., Tolley, M.S., Ballardini, R., Ciano, M., Balzani, V., Gandolfi, M.T., Prodi, L., and McLean, C.H. (1993) The control of translational isomerism in catenated structures, *J. New. Chem.* **17**, 689-695.

46. Blower, M., Stoddart, J.F., and Venner, M.R.W. (1994) University of Birmingham, Unpublished results.

47. Amabilino, D.B., Ashton, P.R., Tolley, M.S., Stoddart, J.F., and Williams, D.J. (1993) Isomeric self-assembling [2]catenanes, *Angew. Chem. Int. Ed. Engl.* **32**, 1297-1301.

48. Ashton, P.R., Reder, A.S., Spencer, N., and Stoddart, J.F. (1993) The self-assembly of a chiral bis[2]catenane, *J. Am. Chem. Soc.* **115**, 5286-5287.

49. Ashton, P.R., Ballardini, R., Balzani, V., Gandolfi, M.T., Marquis, D.J.-F., Pérez-Garcia, L., Prodi, L., Stoddart, J.F., and Venturi, M., (1994) The self-assembly of controllable [2]catenanes, *J. Chem. Soc., Chem. Commun.* 177-180.

50. Langford, S.J., Pérez-Garcia, L., and Stoddart, J.F. (1994) University of Birmingham, Unpublished results.

51. Ashton, P.R., Iriepa, I., Reddington, M.V., Spencer, N., Slawin, A.M.Z., Stoddart, J.F., and Williams, D.J. (1994) An optically-active [2]catenane made to order, *Tetrahedron Lett*. **35**, 4835-4838.

52. Ashton, P.R., Brown, C.L., Chrystal, E.J.T., Goodnow, T.T., Kaifer, A.E., Slawin, A.M.Z., Spencer, N., Stoddart, J.F., and Williams, D.J. (1991) Self-assembling [3]catenanes, *Angew. Chem. Int. Ed. Engl.* **30**, 1039-1042.

53. Ashton, P.R., Brown, C.L, Chrystal, E.J.T., Parry, K.P., Kaifer, A.E., Pietraszkiewicz, M., Spencer, N., and Stoddart, J.F. (1991) Self-assembling [3]catenanes, *Angew. Chem. Int. Ed. Engl.* **30**, 1042-1045.

54. Amabilino, D.B., Ashton, P.R., Reder, A.S., Spencer, N., and Stoddart, J.F. (1994) The two-step self-assembly of a [4]- and [5]-catenane, *Angew. Chem. Int. Ed. Engl.* **33**, 433-437.

55. Amabilino, D.B., Ashton, P.R., Reder, A.S., Spencer, N., and Stoddart, J.F. (1994) Olympiadane, *Angew. Chem. Int. Ed. Engl.* **33**, 1286-1290.

SUPRAMOLECULAR CHEMISTRY AND CHEMICAL SYNTHESIS.

From Molecular Interactions to Self-Assembly

Jean-Marie LEHN
Université Louis Pasteur, Strasbourg and Collège de France, Paris

1. From Molecular to Supramolecular Chemistry

For more than 150 years, since the synthesis of urea by *Friedrich Wöhler* in 1828 [1], molecular chemistry has developed a vast array of highly sophisticated and powerful methods for the construction of ever more complex molecular structures by the making or breaking of covalent bonds between atoms in a controlled and precise fashion.

Organic synthesis grew rapidly and masterfully, leading to a whole series of brilliant achievements, constituting the great syntheses of the last 50 years, where elegance of strategy combined with feats of efficiency and selectivity. There is a long road from *Wöhler's* urea to the synthesis of Vitamin B_{12} by *Robert B. Woodward* [2] and *Albert Eschenmoser* [3], assisted by a hundred or so collaborators!

Molecular chemistry has, thus, established its power over the covalent bond. The time has come to do the same for non-covalent intermolecular forces. Beyond molecular chemistry based on the covalent bond, lies the field of *supramolecular chemistry*, whose goal it is to gain control over the intermolecular bond [4-7]. It is concerned with the next step in increasing complexity beyond the molecule towards the supermolecule and large organized molecular systems, held together by non-covalent interactions.

The field started with the selective binding of alkali metal cations by natural [8, 9] as well as by synthetic macrocyclic and macropolycyclic ligands, the crown ethers [10, 11] and the cryptands [4-6, 12]. The outlook broadened, leading to the emergence of *molecular recognition* [4-6, 12] as a novel domain of chemical research which, by extension to intermolecular interactions and processes in general and by broadly expanding over other areas, grew into the field of supramolecular chemistry. It was defined as ""chemistry beyond the molecule", bearing on the organized entities of higher complexity that result from the association of two or more chemical species held together by intermolecular forces"" [4-7].

A chemical species is defined by its components, by the nature of the bonds that link them together and by the resulting spatial (geometrical, topological) features. The objects of supramolecular chemistry are "supramolecular entities, *supermolecules* possessing features which are as well defined as those of molecules themselves. One

C. Chatgilialoglu and V. Snieckus (eds.), Chemical Synthesis, 511–524.

may say that supermolecules are to molecules and the intermolecular bond what molecules are to atoms and the covalent bond" [5, 6].

Supramolecular species are characterized both by the spatial arrangement of their components, their architecture or superstructure, and by the nature of the intermolecular bonds that hold these components together. They possess well-defined structural, conformational, thermodynamic, kinetic and dynamic properties. Various types of interactions may be distinguished that present different degrees of strength, directionality, dependence on distance and angles, metal ion coordination, electrostatic forces, hydrogen bonding, van der Waals interactions, donor-acceptor interactions, etc. Their strengths range from weak or moderate as in hydrogen bonds, to strong or very strong for metal ion coordination. The former provide associations of stabilities comparable to enzyme-substrate species, whereas the latter makes available, by means of a single metal ion, binding strengths that lie in the domain of antigen-antibody complexes (or higher), where many individual interactions are involved. However intermolecular forces are, in general, weaker than covalent bonds, so that supramolecular species are thermodynamically less stable, kinetically more labile and dynamically more flexible than molecules.

Selective binding of a specific substrate σ to its receptor ρ yields the supermolecule $\rho\sigma$ and involves a molecular recognition process. If, in addition to binding sites, the receptor also bears reactive functions, it may effect a chemical transformation on the bound substrate, thus behaving as a supramolecular reagent or catalyst. A lipophilic, membrane-soluble receptor may act as a carrier effecting the translocation of the bound substrate. Thus, *molecular recognition*, *transformation*, and *translocation* represent the basic functions of supramolecular species. More complex functions may result from the interplay of several binding subunits in a polytopic coreceptor. In association with organized polymolecular assemblies and phases (layers, membranes, vesicles, liquid crystals, etc.), functional supermolecules may lead to the development of *molecular* and *supramolecular devices*. Recent lines of investigation concern *self-processes* (self-assembly, self-organization, replication) and the design of *programmed supramolecular systems*. All these features suppose the presence and operation of *molecular information*, a notion that has become the basic and crucial tenet of supramolecular chemistry.

Supramolecular chemistry may be divided into two broad, partially overlapping areas concerning: (1) *supermolecules*, well-defined, discrete *oligo*molecular species that result from the intermolecular association of a few components (a receptor and its substrate(s)) following a built-in "Aufbau" scheme based on the principles of molecular recognition; (2) *supramolecular assemblies*, *poly*molecular entities that result from the spontaneous association of a large, undefined number of components into a specific phase having more or less well-defined microscopic organization and macroscopic characteristics depending on their nature (such as films, layers, membranes, vesicles, micelles, mesomorphic phases, solid state structures, etc.). It thus covers the rational, coherent approach to molecular associations: from the smallest, the dimer, to the largest, the organized phase, and to their designed manipulation.

The highly selective processes of molecular recognition are, of course, of stereochemical nature. Thus may be defined a *supramolecular stereochemistry* that extends from supermolecules to polymolecular assemblies. Different spatial dispositions of the components of a supermolecule with respect to each other lead to *supramolecular stereoisomers*. Their eventual interconversion will depend on the properties of the interactions that hold them together, i.e. on the variation of the intermolecular interaction energy with distances and angles. There is thus an *intermolecular conformational analysis* like there is an intramolecular one.

2. Supramolecular Chemistry and Chemical Synthesis

The contribution of supramolecular chemistry to chemical synthesis has two main aspects: the production of non-covalent supramolecular species themselves and the use of supramolecular processes for the synthesis of covalent molecular structures. In addition, supramolecular chemistry, of course, requires and depends on molecular synthesis for the construction of its covalent entities: the molecular receptors, catalysts and carriers as well as the components for devices and self-assembling systems.

2.1. SUPRAMOLECULAR SYNTHESIS

Supramolecular synthesis, the generation of supramolecular architectures, consists in the designed association of molecular components directed by the physico-chemical features of intermolecular forces; like molecular synthesis, it requires strategy, planning and control (Figure 1). The conception of the components must, at the outset, already incorporate the supramolecular project.

Supramolecular synthesis may thus be considered to involve two steps: - the synthesis of the instructed molecular components by the formation of strong, kinetically non-labile covalent bonds; - the generation of the supramolecular entities by association of the components in a pre-determined arrangement by means of comparatively weak and kinetically labile interactions following the built-in intermolecular plan [4a]; - the latter may, in addition, follow a given sequence and hierarchy of operations.

The goal is to gain control over intermolecular events in much the same way as molecular synthesis strives to control covalent bond formation. The synthesis of a supramolecular architecture requires similar efforts on the intermolecular level as molecular synthesis on the intramolecular one. In the realm of synthetic chemistry, supramolecular chemistry must claim and engage in the same pursuit of design and control as molecular chemistry, except that it may well be more demanding since it requires, in addition, the correct storage of an intramolecular programme into a covalent framework and thus necessitates expertise at the two levels.

Chemistry has just entered the confines of the supramolecular world and great feats in synthetic power for the elaboration of ever more complex architectures lie ahead, in the same manner as there has been a rich and exciting but long road of tribulations from the synthesis of urea in 1828 to those of vitamin B_{12} [2, 3], palytoxin [13] or calicheamicin [14], to cite just a few major achievements in recent times.

SUPRAMOLECULAR SYNTHESIS

⇨ *Generation of Supramolecular Architectures*

involves two steps

- the synthesis of instructed molecular, covalent, components
- the designed association of these components in a pre-determined arrangement by means of non covalent interactions

implies

- the conception of components containing the supramolecular project through a built-in programme
- the control of intermolecular bonds and events within a supramolecular strategy

requires

- design, planning and expertise at both the intramolecular and the intermolecular levels

Figure 1

2.2. SUPRAMOLECULAR ASSISTANCE TO SYNTHESIS

Supramolecular association may be put into action as an aid to the synthesis of complex covalent species in order to position the components by templating [15] or self-assembly. In this manner, subsequent reactions, deliberately performed on the preassembled species or occurring spontaneously within them, will lead to the generation of the desired architecture. Dismantling of the intermolecular connection(s) may follow, thus liberating the covalently connected structure. This amounts to *supramolecular assistance to synthesis*, that may specifically become *self-replication* if spontaneous reproduction of one of the initial species takes place by binding, positioning and condensation of its parts by itself (Figure 2).

Intermolecular interactions and metal coordination have been used very effectively for the synthesis of novel organic and inorganic entities that would be of difficult access otherwise, leading in particular to the development of a *topological chemistry* [16]. *Inorganic templating* and *self-assembly* provide coordination compounds whose geometries make possible the synthesis of a variety of complex structures, such as cyclic multiporphyrin arrays [15a], multicatenates and catenands [17] and even molecular knots [18].

SUPRAMOLECULAR ASSISTANCE to CHEMICAL SYNTHESIS

⇨ Trough positioning of covalent components by means of non-covalent bonds for subsequent covalent connection

Nature	**Levels** of increasing complexity
ORGANIC	TEMPLATING
INORGANIC	SELF-ASSEMBLY
	SELF-ORGANISATION

With **Reactive Components**: SPONTANEOUS CONDENSATION
REPLICATION SELF-REPLICATION

Figure 2

Organic templating and *self-assembly* may similarly be used to position partners for subsequent reaction, amounting to a template-assisted organic synthesis. Rotaxanes and multicatenanes have been synthesized based on the spontaneous threading of donor-acceptor partners [19]. Similarly, making use of substrate inclusion features of cyclodextrins for inserting linear components through their cavity, rotaxanes and polyrotaxanes have been obtained [20]. The synthesis of such mechanically interlocked superstructures is entirely dependent on the preformation, by self-assembly, of a threaded supramolecular species.

One may expect that, by making use of suitably designed supramolecular features, it will be possible to generate a variety of highly complex architectures that would not be accessible otherwise (or only with low efficiency). Such supramolecular assistance adds a new direction with powerful means to organic synthesis.

2.3. REPLICATION. SELF-REPLICATION

Self-assembled template-directed systems containing reactive groups provide pathways towards systems displaying copying, information transfer and replication. This may lead to spontaneous condensations within the self-assembled entity into the final covalent structure. A special case is the reproduction of the template itself by replication or self-replication [21]. Replicating molecules are of interest both chemically and biologically in view of their relation to the origin of life.

Self-replication takes place when a molecule catalyses its own formation by acting as a template for the constituents whose reaction generates a copy of the template. A problem is the occurrence of product inhibition, when the dimer of the template, formed after the first condensation round, is too stable to be easily dissociated by the incoming components for a new cycle.

Several self-replicating systems have been developed in which the template is generated from two components [21].

The investigation of self-replicating systems will be subject to increasing activity. The information processing that operates in molecular recognition and self-assembly will play a major role, coupled to chemical reactions for connecting the building blocks.

3. Self-assembly. Self-organisation

In line with the key role played by molecular information in the field, the most recent developments of supramolecular chemistry concern the explicit implementation of molecular recognition as a means of controlling the evolution of supramolecular species and devices as they build up from their components and operate through *self-processes*.

Supramolecular chemistry has relied on more or less rigidly organised, synthetically constructed molecular receptors for effecting molecular recognition, catalysis, and transport processes and for producing molecular devices. Beyond *pre*organisation lies the design of systems undergoing *self*-organisation, that is, systems capable of *spontaneously* generating a well-defined (functional) supramolecular architecture by self-assembly from their components under a given set of conditions. It is a *designed assembly* into a discrete supramolecular species, as compared to the spontaneous formation of molecular layers, films, membranes, etc.

The *molecular information* necessary for the process to take place and the algorithm that it follows must be stored in the components and operate through selective molecular interactions. Thus, these systems may be termed *programmed supramolecular systems* that generate organized entities following a defined plan based on molecular recognition events [7]. Several approaches to self-assembling systems have been pursued [6, 7, 19, 22].

3.1. SELF-ASSEMBLY OF ORGANIC SUPRAMOLECULAR STRUCTURES

The self-assembly of organic supramolecular species makes use of interactions such as electrostatic, hydrogen bonding, Van der Waals, stacking or donor-acceptor effects similar to those found in proteins, nucleic acids, liquid crystals, and molecular complexes. The designed use of these forces for the directed self-assembly of a given structure requires knowledge of their strength and of their dependence on distances and angles.

The spontaneous generation of organised structures depends on the design of molecular components capable of self-assembling into supramolecular entities which present the desired architectural and functional features. The nature of the species obtained will depend on the information stored in the components. Thus, the self-assembly process may be directed by molecular recognition between two or more complementary subunits so as to form a given supramolecular architecture. If these molecular units incorporate specific optical, electrical, magnetic, binding, etc. properties, their ordering may induce a range of novel features. Depending on the subunits involved, the association may lead either to supermolecules or to organised assemblies, such as membranes, molecular layers and films, mesophases, polymeric species or solid state lattices.

Mesophases and liquid crystalline polymers of supramolecular nature have been generated from complementary components, amounting to macroscopic expression of molecular recognition (Figure 3) [23, 24].

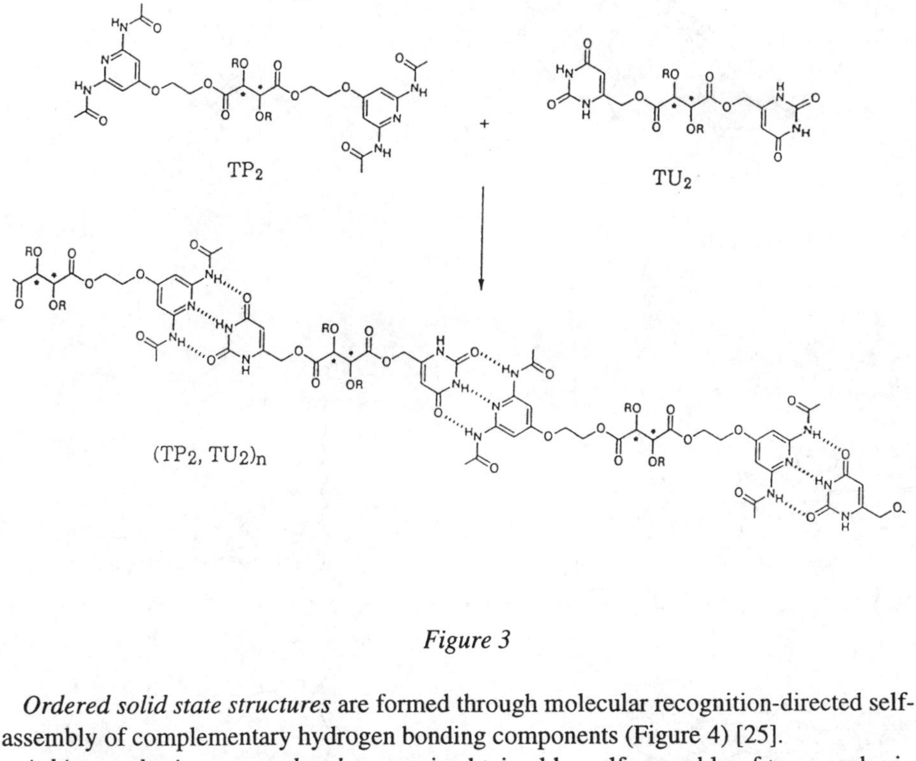

Figure 3

Ordered solid state structures are formed through molecular recognition-directed self-assembly of complementary hydrogen bonding components (Figure 4) [25].

A *bis-porphyrin supramolecular cage* is obtained by self-assembly of two prophyrin components bearing uracil type units that interact through hydrogen bonding with two complementary triaminopyrimidine units (Figure 5) [26].

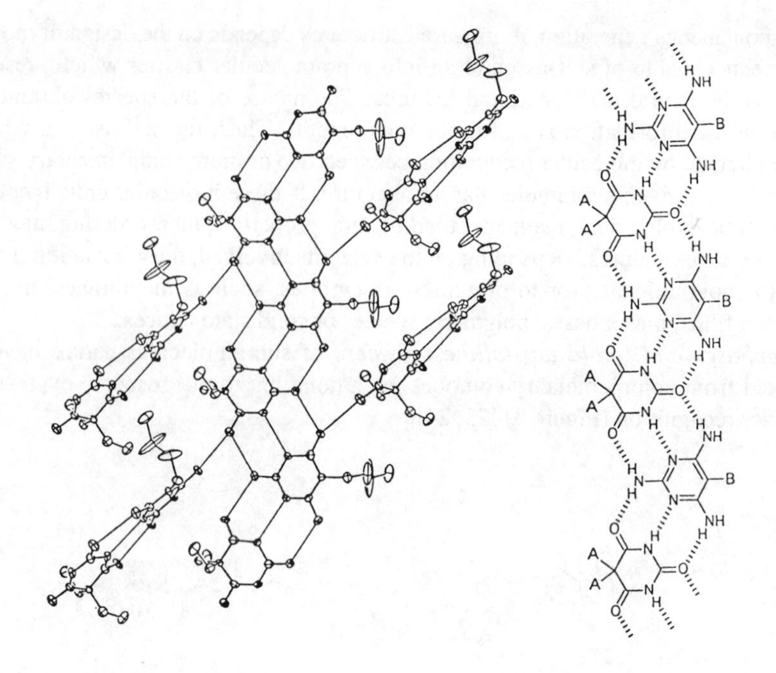

Figure 4

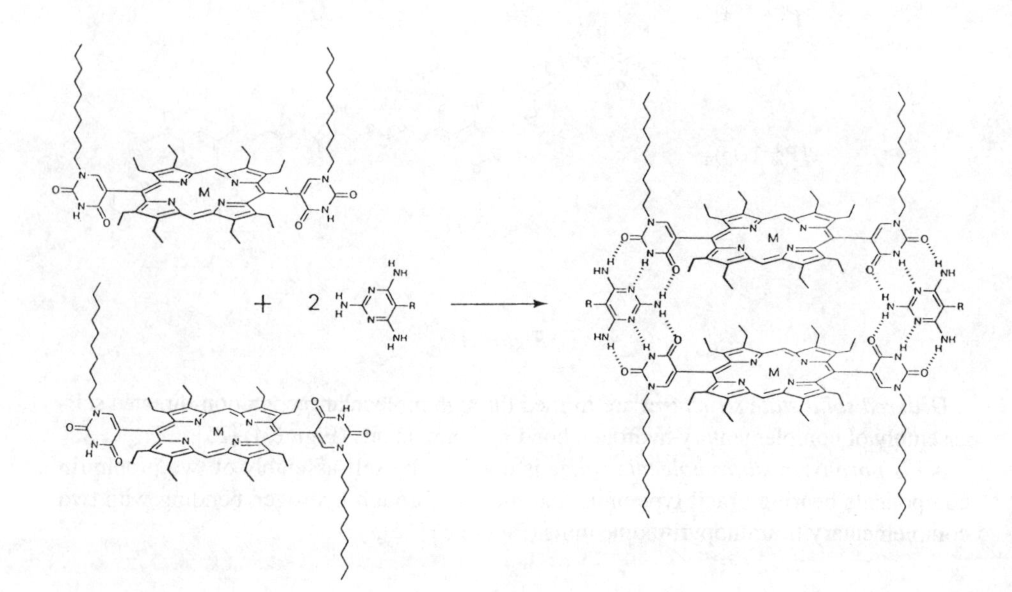

Figure 5

3.2. SELF-ASSEMBLY OF INORGANIC ARCHITECTURES

Inorganic self-assembly and self-organisation involve the spontaneous generation of well-defined supramolecular architectures from organic ligands and metal ions. The latter serve both as cement that holds the ligands together and as a centre orienting them in a given direction. In the process, full use is being made of the structure and coordination features of both types of components, which, in addition, convey redox, photochemical or chemical functionality depending on their nature.

Metal ions have properties of special interest as linkers for self-assembly. They provide a range of coordination geometries, binding strengths and formation and dissociation kinetics, as well as a variety of photophysical and electrochemical properties.

The self-assembly of inorganic structures of several types has been achieved, based on ligand design and on the use of suitable coordination geometries that act as the assembling algorithm.

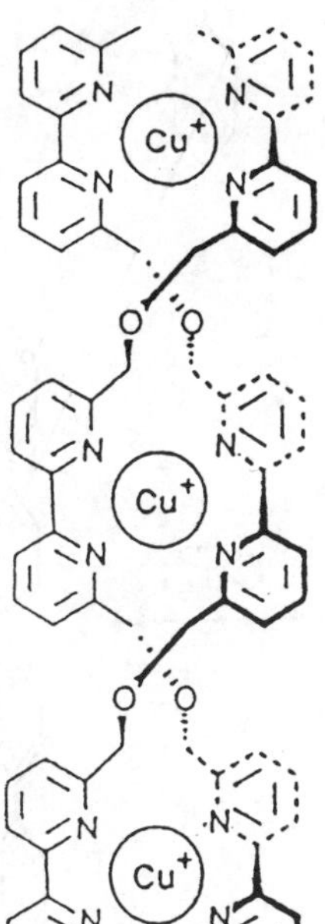

1

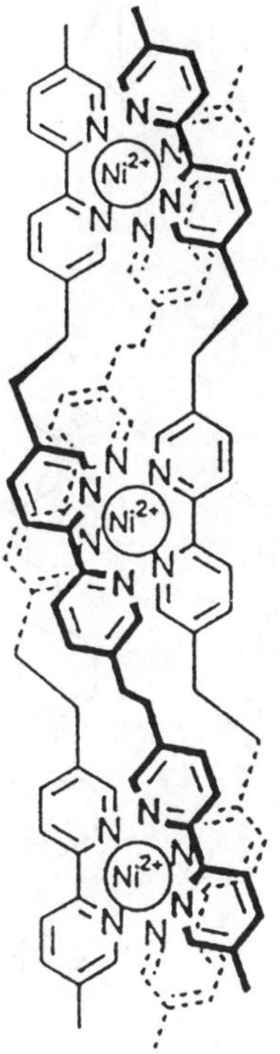

2

Double-stranded and triple-stranded helicates as well as double helical and triple helical metal complexes, are formed by the spontaneous organisation of two or three linear polybipyridine ligands of suitable structure into a double or a triple helix by binding of specific metal ions displaying respectively tetrahedral (Cu^I) and octahedral (Ni^{II}) coordination geometry. These species are illustrated by the trinuclear double helicate **1** [27] and triple helicate **2** [28] (see also [29]).

A *circular complex* **3** is obtained from the assembly of a trinucleating ligand, three 2,2'-bipyridine units and three Cu^I ions [30]. A *capped structure* **4** has also been prepared [31].

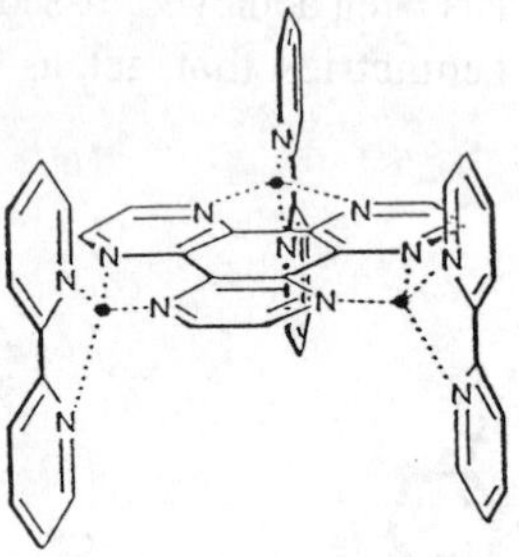

3

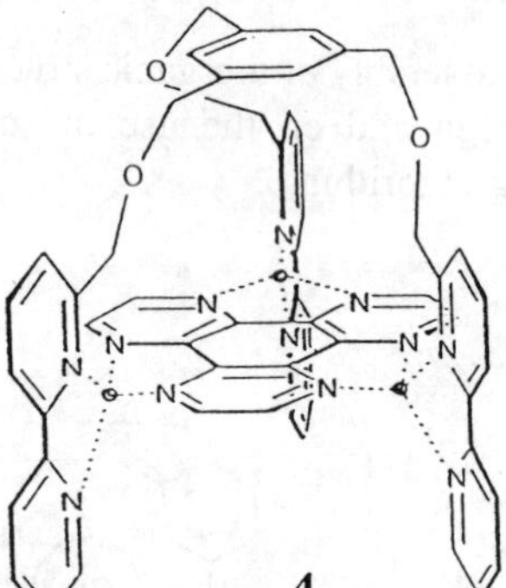

4

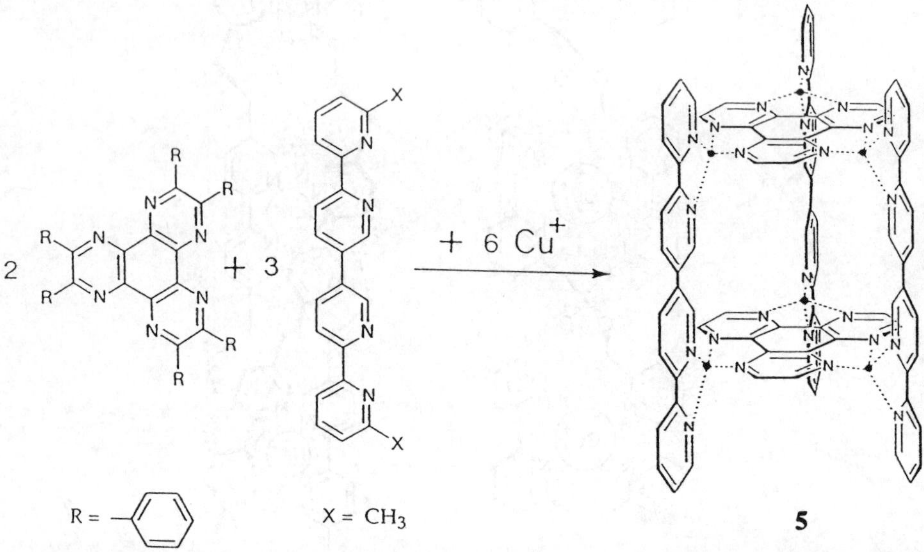

2 + 3 + 6 Cu^+ → **5**

R = C_6H_5 (phenyl)

X = CH_3

C_6H_5 and CH_3 substituents omitted

Figure 6

Multiple component self-assembly leads to the spontaneous generation of a cylindrical superstructure **5** from five ligands of two different types and six Cu^{I} ions [30]. This process represents the remarkable formation of a closed inorganic architecture from several components by the spontaneous and correct assembly, in one stroke, of no less than eleven particles belonging to two types of ligands and one type of metal ion (Figure 6).

Analogous cylindrical architectures of even larger size presenting three (see **6**) and four layered features have been obtained using similar procedures [32].

Grid-like inorganic superstructures, such as the 3 x 3 grid **7**, have been shown to self-assemble from six ligands and nine Ag^{I} ions [33].

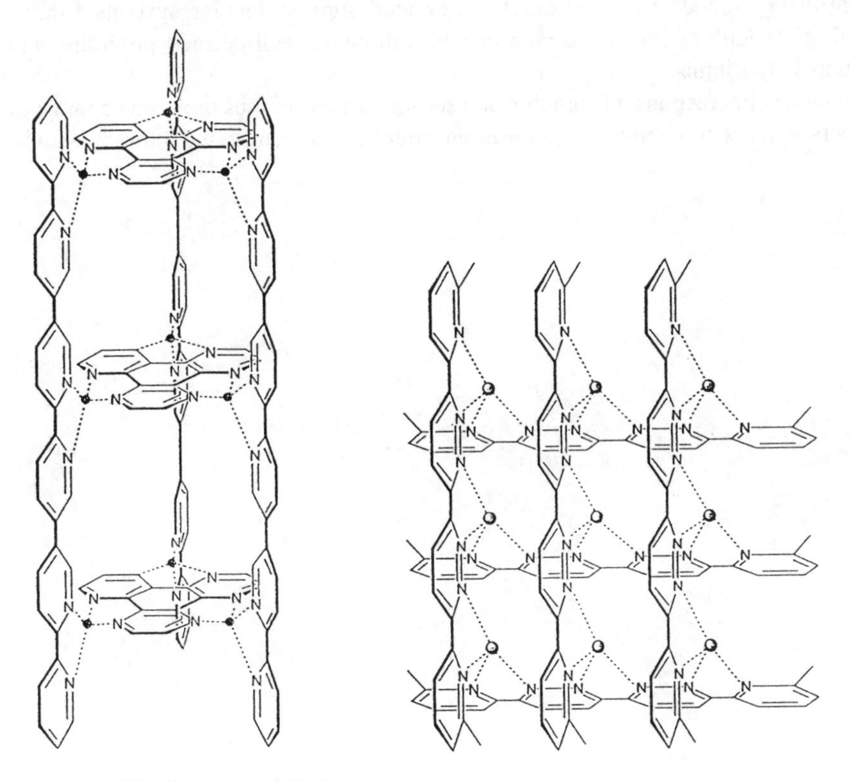

6 **7**

The operation of such instructed supramolecular systems fulfills the three levels of molecular programming and information input (*recognition*, *orientation* and *termination*), that determine the generation of a discrete supramolecular architecture. The steric and binding information contained in the ligand is read out by the metal ions following a given coordination algorithm. These processes represent progressive steps in the control of the self-organization of large and complex supramolecular architectures through molecular programming.

In a study of helicate self-assembly from a mixture of different ligands and different metal ions, it was found that only the "correct" helical complexes are formed through *self-recognition* [34]. In a broader perspective, this points to a change in paradigm from pure compounds to instructed mixtures, that is, from seeking chemical purity to designing programmed systems composed of mixtures of instructed components capable of spontaneously forming well-defined superstructures.

One may venture to predict that this trend will represent a major line of development of chemistry in the years to come: the spontaneous but controlled assemblage of structurally organized and functionally integrated supramolecular systems from a preexisting "soup" of instructed components following well-defined programs and interactional algorithms.

Non-covalent interactions, recognition and self-processes are thus providing a range of novel perspectives to chemical synthesis on both the molecular and supramolecular levels.

References

1 F. Wöhler, *Poggendorfs Ann. Physik*, 1828, *12*, 253.

2 R. B. Woodward, *Pure Appl. Chem.*, 1968, *17*, 519.

3 A. Eschenmoser, *Quat. Rev.*, 1970, *24*, 366; *Chem. Soc. Rev.*, 1976, *5*, 377; *Nova Acta Leopoldina*, 1982, *55*, 5.

4 a) J.-M. Lehn, *Pure Appl. Chem.* , 1978, *50*, 871; b) *Acc. Chem. Res.*, 1978, *11*, 49.

5 J.-M. Lehn, *Science*, 1985, *227*, 849.

6 J.-M. Lehn, *Angew. Chem.*, 1988, *100*, 91; *Angew. Chem. Int. Ed. Engl.*, 1988, *27*, 89.

7 J.-M. Lehn, *Angew. Chem.*, 1990, *102*, 1347; *Angew. Chem. Int. Ed. Engl.*, 1990, *29*, 1304.

8 a) Yu. A. Ovchinnikov, V. T. Ivanov and A. M. Skrob, "*Membrane Active Complexones*", Elsevier, New-York 1974; b) B. C. Pressman, *Annu. Rev. Biochem.*, 1976, *45*, 501.

9 M. Dobler, *Ionophores and their Structures*, J. Wiley, New-York, 1981.

10 C. J. Pedersen, *J. Am. Chem. Soc.*, 1967, *89*, 7017; C. J. Pedersen, *Angew. Chem.* , 1988, *100*, 1053; *Angew. Chem. Int. Ed. Engl.*, 1988, *27*, 1053.

11 G. Gokel, "*Crown Ethers and Cryptands*", Roy. Soc. Chem., Cambridge, 1991.

12 J.-M. Lehn, *Struct. Bonding* , 1973, *16*, 1.

13 R. W. Armstrong, J.-M. Beau, S. H. Cheon, W. J. Christ, H. Fujioka, W.-H. Ham, L. D. Hawkins, H. Jin, S. H. Kang, Y. Kishi, M. J. Martinelli, W. W. McWhorter, Jr., M. Mizuno, M. Nakata, A. E. Stutz, F. X. Talamas, M. Taniguchi, J. A. Tino, K. Ueda, J.-i. Uenishi, J. B. White and M. Yonaga, *J. Am. Chem. Soc.*, 1989, *111*, 7525; *ibid.*, 7530.

14 K. C. Nicolaou, *Angew. Chem.*, 1993, *105*, 1462; *Angew. Chem. Int. Ed. Engl.*, 1993, *32*, 1377.

15 a) S. Anderson, H.K L Anderson and J. K. M. Sanders, *Accounts Chem. Res.*, 1993, *26*, 469; b) R. Hoss and F. Vögtle, *Angew. Chem.*, 1994, *106*, 389; *Angew. Chem. Int. Ed. Engl.*, 1994, *33*, 375.

16 a) D. M. Walba, *Tetrahedron*, 1985, *41*, 3161; b) for a recent overview see: *Topology in Chemistry, New J. Chem.*, Special issue 1993, *17*, 617-763.

17 C. O. Dietrich-Buchecker and J.-P. Sauvage, *Chem. Rev.*, 1987, *87*, 795; *Tetrahedron*, 1990, *46*, 503.

18 a) J.-P. Sauvage, *Acc. Chem. Res.*, 1990, *23*, 319; b) C. Dietrich-Buchecker and J.-P. Sauvage, *New. J. Chem.*, 1992, *16*, 277.

19 D. Philp and J. F. Stoddart, *Synlett*, 1991, 445.

20 a) H. Ogino, *New. J. Chem.*, 1993, *17*, 683; b) A. Harada and M. Kamachi, *J. Chem. Soc., Chem. Commun.*, 1990, 1322; c) A. Harada, J. Li and M. Kamachi, *Nature*, 1992, *356*, 325; d) G. Wenz and B. Keller, *Angew. Chem.*, 1992, *104*, 201; *Angew. Chem. Int. Ed. Engl.*, 1992, *31*, 197.

21 a) L. E. Orgel, *Nature*, 1992, *358*, 203; b) G. von Kiedrowski, J. Helbing, B. Wlotzka, S. Jordan, M. Mathen, T. Achilles, D. Sievers, A. Terfort and B. C. Kahrs, *Nachr. Chem. Tech. Lab.*, 1992, *40*, 578; c) S. Hoffmann, *Angew. Chem.*, 1992, *104*, 1032; *Angew. Chem. Int. Ed. Engl.*, 1992, *31*, 1013; d) J. S. Nowick, Q. Feng, T. Tjivikua, P. Ballester and J. Rebek, Jr., *J. Am. Chem. Soc.*, 1991, *113*, 8831; e) M. Famulok, J. S. Nowick and J. Rebek, Jr., *Acta Chem. Scand.*, 1992, *46*, 315.

22 a) J. S. Lindsey, *New J. Chem.*, 1991, *15*, 153; b) G. M. Whitesides, J. P. Mathias and C. T. Seto, *Science*, 1991, *254*, 1312.

23 a) C. Fouquey, J.-M. Lehn and A.-M. Levelut, *Adv. Mater.*, 1990, *2*, 254; b) T. Gulik-Krzywicki, C. Fouquey and J.-M. Lehn, *Proc. Natl. Acad. Sci. USA*, 1993, *90*, 163.

24 J.-M. Lehn, *Makromol. Chem., Macromol. Symp.*, 1993, *69*, 1.

25 a) J.-M. Lehn, M. Mascal, A. DeCian and J. Fischer, *J. Chem. Soc., Chem. Commun.*, 1990, 479; b) J.-M. Lehn, M. Mascal, A. DeCian and J. Fischer, *J. Chem. Soc. Perkin Trans. 2*, 1992, 461.

26 C. M. Drain, R. Fischer, E. G. Nolen and J.-M. Lehn, *J. Chem. Soc., Chem. Commun.*, 1993, 243.

27 J.-M. Lehn, A. Rigault, J. Siegel, J. Harrowfield, B. Chevrier and D. Moras, *Proc. Natl. Acad. Sci. USA*, 1987, *84*, 2565; J.-M. Lehn and A. Rigault, *Angew. Chem. Int. Ed. Engl.*, 1988, *27*, 1095.

28 R. Krämer, J.-M. Lehn, A. DeCian and J. Fischer, *Angew. Chem. Int. Ed. Engl.*, 1993, *32*, 703.

29 E. C. Constable, *Tetrahedron*, 1992, *48*, 10013.

30 P. Baxter, J.-M. Lehn, A. DeCian and J. Fischer, *Angew. Chem. Int. Ed. Engl.*, 1993, *32*, 69.

31 E. Leize, A. Van Dorsselaer, R. Krämer and J.-M. Lehn, *J. Chem. Soc., Chem. Commun.*, 1993, 990.

32 P. Baxter and J.-M. Lehn, unpublished results.

33 P. N. W. Baxter, J.-M. Lehn, J. Fischer and M.-T. Youinou, *Angewandte Chem. Int. Ed. Engl.*, 1994, *33*, 2284.

34 R. Krämer, J.-M. Lehn and A. Marquis-Rigault, *Proc. Natl., Acad. Sci. USA*, 1993, *90*, 5394.

SYNTHETIC APPROACHES TO FUNCTIONAL POLYMERS

F. CIARDELLI,[a] M. AGLIETTO,[a] O. PIERONI,[a] G. RUGGERI,[a] R. WAYMOUTH,[b] M. KESTI,[b] K. STEIN,[b]

[a]*Dipartimento di Chimica e Chimica Industriale, Università di Pisa, Via Risorgimento 35, 56100 Pisa, Italy*

[b]*Department of Chemistry, Stanford University, Stanford, CA 94305, USA*

1. General Aspects of Synthesis and Structure of Macromolecules

Polymerization reactions are carried out with the purpose to link together small reactive molecules (monomers) through covalent bonds in order to obtain long chain molecules (macromolecules). The monomers are thus incorporated in the chain with moderate modification, depending on the type of polymerization and the nature of the monomer itself. Thus, starting with a certain monomer the local chemical constitution of the related macromolecules is largely predetermined.

On the other hand, the polymerization reaction and the adopted process parameters affect typical macromolecular characteristics such as average molecular weight, molecular weight distribution, stereoregularity, chemical composition and sequence distribution in copolymers. For the synthesis of macromolecules bearing specific functional groups, chemical modification of preformed macromolecules can also be carried out.

This essay describes the main routes to the synthesis of functionalized macromolecules. It will also be shown how the macromolecular structure influences the behavior of the functional groups relative to the monomeric analogs.

An exhaustive description of all possible approaches is not possible within the limits of this contribution. Therefore, selected examples will be described based on work carried out in our laboratories.

2. Functionalization of Polyolefins

Polyolefins can be functionalized either by chemical modification of preformed polymers[1] or by copolymerization of olefins with suitable polar comonomers.[2]

The former method has been largely investigated[3-8] as offering several practical advantages such as: use of commercial polyolefins (in particular ethylene and propylene polymers) as starting materials, large selection of functional groups, retention of the original

C. Chatgilialoglu and V. Snieckus (eds.), Chemical Synthesis, 525–548.

properties of the polyolefins, the ability to functionalize during extrusion[9-10] and surface functionalization.[11-14]

The copolymerization method cannot be applied in a general since the Ziegler-Natta catalysts, which produce high molecular weight and structurally regular polyolefins, are usually sensitive to polar monomers[15-21] (see Section 3).

Herein we describe the method of chemical modification of preformed polymers[22] with reference to work from our laboratories. It was shown that the functionalization of polypropylene with diethylmaleate (DEM) as the unsaturated substrate, in the presence of dicumylperoxide (DCP), is higher when the reaction is conducted under nitrogen, at 200°C in bulk , than at 145°C in *o*-xylene solution.

$$\left(-\overset{|}{\underset{|}{C}}-\right)_n H + \underset{COOEt}{\underset{|}{CH}}=\underset{COOEt}{\underset{|}{CH}} \xrightarrow{DCP} \left(-\overset{|}{\underset{|}{C}}-\right)_n \underset{COOEt}{\underset{|}{CH}}-\underset{COOEt}{\underset{|}{CH_2}}$$

Scheme 1. Functionalization of polyolefins by reaction with dietylmaleate and dicumylperoxide (DCP).

Degrees of functionalization of up to 10-18 % and 20-22 % by wt respectively, were achieved by reacting ethylene-propylene copolymers (EP) and polyethylene (PE) at 200°C in bulk with diethylmaleate in the presence of dicumylperoxide.[23] Apparently, in the case of ethylene-propylene copolymer, the reactive tertiary C-H bonds are less sterically hindered than in polypropylene. In the case of polyethylene, despite the lack of tertiary hydrogens, the higher chain mobility causes a high degree of functionalization which is, however, negatively offset by the crystallinity of polyethylene.

At the end of the functionalization reaction, the recovered polymers are completely soluble in boiling toluene indicating that crosslinking reactions do not take place.[24] In this generally accepted functionalization reaction, an important issue is the difficulty of evaluating accurately polymer structural changes and the molecular structure of the functional group introduced. Free radical reactions, particularly at high temperatures, may follow different pathways involving the reactive group in a variety of ways. While the polar monomer can give rise to a variable length chain, it can also encounter different reaction sites along the polyolefin backbone (primary, secondary and tertiary carbon atoms). The latter sites can be present in the starting polyolefin, as for polypropylene and ethylene/propylene copolymer or formed under the reaction conditions.

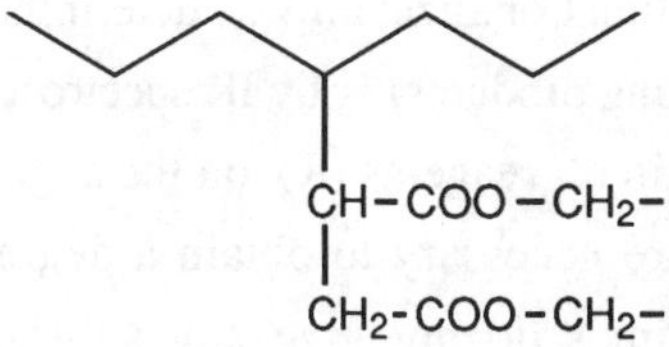

$CH-COO-CH_2-CH_3$

$CH_2-COO-CH_2-CH_3$

Ethyl 3-Carbethoxy-4-n-propylheptanoate **(1a)**.

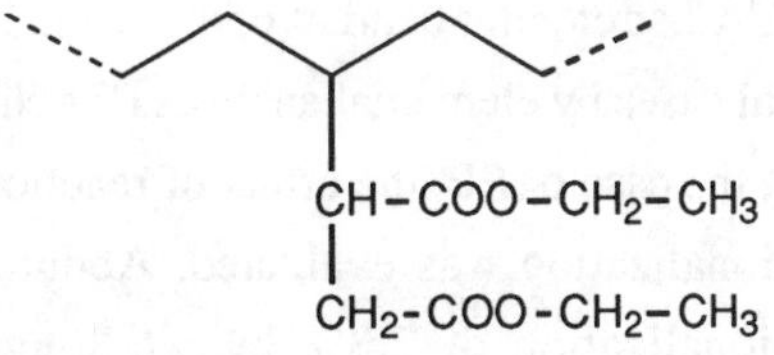

$CH-COO-CH_2-CH_3$

$CH_2-COO-CH_2-CH_3$

Functionalized HDPE **(2)**

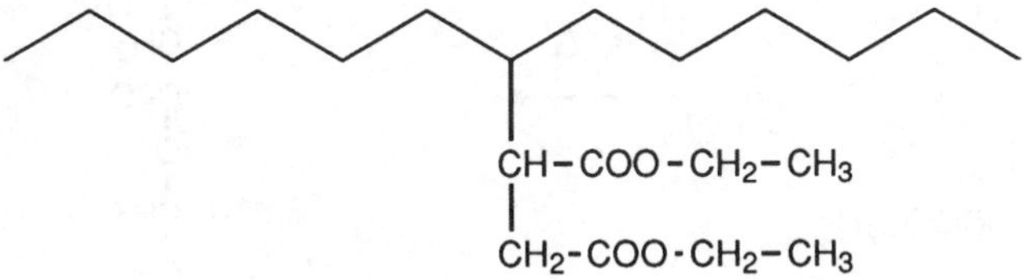

$CH-COO-CH_2-CH_3$

$CH_2-COO-CH_2-CH_3$

Ethyl 3-Carbetoxy-4-n-hexyldecanoate **(1b)**

Scheme 2. Structures of model compounds and of functionalized HDPE

The unequivocal determination of the structure of the functional groups inserted into linear high density polyethylene (HDPE) after reaction at 200°C with diethylmaleate in presence of dicumylperoxide[25] could be carried out by ^{13}C NMR spectroscopy. This was demonstrated by synthesizing two model compounds (**1a** and **1b**) of the functionalized polymer and then comparing the ^{13}C NMR spectra with the analogous spectrum of a sample of functionalized high density polyethylene. The copolymer of structure **2** (Scheme 2) is proposed on the basis of the functionalization mechanism shown in Scheme 1.

The chemical shifts of the functionalized polyethylene and the two model compounds were so close that isolated $CH(COOC_2H_5)(CH_2COOC_2H_5)$ units along the macromolecular chains appear to account very well for the functionalities present in the examined polyethylene sample.

A second functionalization approach was based on the insertion of carbenes into the C-H bond (Scheme 3). The reaction of different polyolefins, such as high density polyethylene (HDPE), atactic polypropylene (aPP), isotactic polypropylene (iPP), ethylene-propylene copolymers (EP) and polyisobutene (iPBu) with diazoesters at 210°C in bulk has been shown to give a product containing the carboxylate functionality attached to the polymer backbone.[26] The presence of the functional groups in the polymer was inferred by IR spectroscopy (carbonyl stretching band at 1740 cm^{-1} and in case of chloroethyldiazoacetate also C-Cl stretching band at 658 cm^{-1}). The degree of functionalization was determined in several cases by elemental analysis (for chlorine containing products) or by IR spectroscopy.

In the case of EP, the effect of reaction time and ratio of reagents (R) on the degree of functionalization was evaluated. About 20 minutes are necessary to obtain a degree of functionalization of 2.3% by wt; longer reaction times do not give any substantial improvement. Results obtained by varying the ratio R indicate that the degree of functionalization remains relatively constant during every reaction.

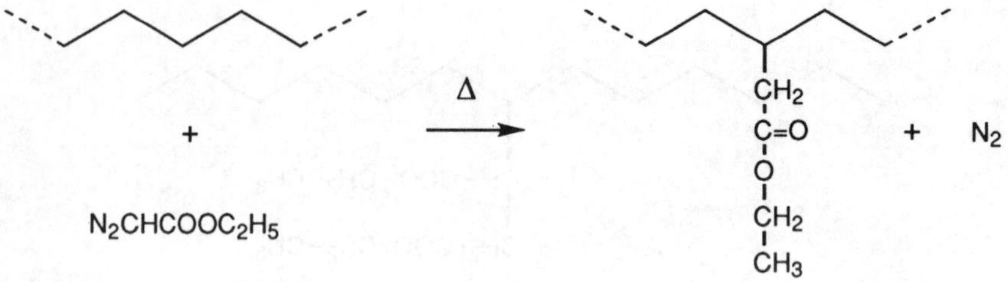

+ $N_2CHCOOC_2H_5$ $\xrightarrow{\Delta}$ CH_2–C=O–O–CH_2–CH_3 + N_2

Scheme 3. Functionalization of HDPE by thermal decomposition of diazoester.

Detailed spectroscopic analysis (NMR, IR) of the functionalized high density polyethylene as compared to low molecular weight structural models allows the unequivocal determination of the structure of the groups attached to the polymer backbone.[27] The ^{13}C NMR spectrum of the functionalized polymer was practically identical with those of ethyl-(3-buthyl)-heptanoate (**4a**), and ethyl -(3-pentyl)-undecanoate (**4b**); the similarities are more evident in the spectrum of **4b** (Scheme 4), in which the influence of the methyl groups on the peaks f and g (Scheme 4) is attenuated by the greater number of methylene groups on both sides inserted between them and the g carbons. The spectral similarities strongly supported that the reaction had proceeded according to Scheme 3, as expected from literature data concerning the reaction of paraffins with carbene derivatives produced *in situ* from ethyl diazoacetate either thermally[27] or by organometallic catalysts.[28]

These results confirm that substantially clean reactions can be performed on polymeric materials even in the melt, provided proper systems and conditions are accurately selected.

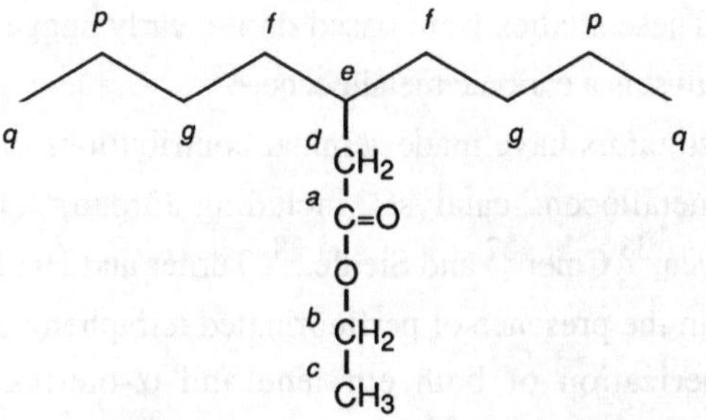

4 a

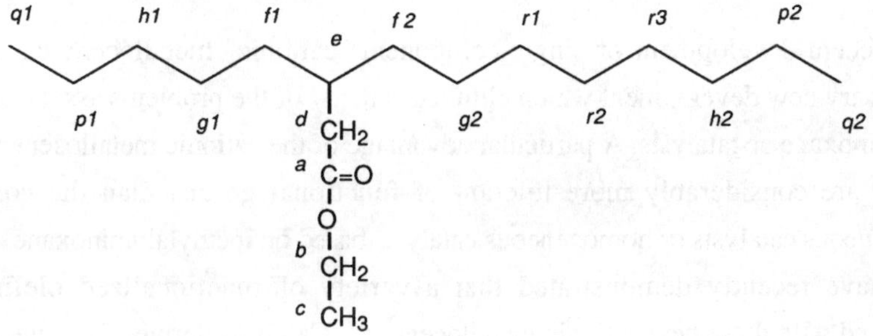

4 b

Scheme 4. Structures of model compounds.

3. Copolymerization of Olefins

3.1. Copolymerization with Functional Monomers

The absence of functionality in hydrocarbon polymers limits applications where good adhesive properties, affinities for dyes, permeability, and compatibility with polar polymers are necessary.[29] One of the limitations of conventional Ziegler-Natta catalysts is their intolerance of functional groups due to the high Lewis acidity of the transition metal component and the presence of Lewis acidic cocatalysts based on alkyl aluminums or aluminoxanes, as was anticipated in Section 2.

The advent of homogeneous Ziegler-Natta catalysts has provided new opportunities for the direct polymerization of functionalized monomers. The basis for the development of more tolerant homogeneous catalysts grew out of mechanistic studies which addressed the nature of the active site for homogeneous catalysts derived from metallocenes and methylaluminoxane. These studies were based on the early suggestion by Shilov that the nature of the active catalyst is a cationic metallocene.[30]

A number of investigators have made seminal contributions in the synthesis of single component cationic metallocene catalysts, including Jordan,[31] Bochmann,[32] Taube,[33] Turner,[34] Marks,[35] Ewen,[36] Chien[37] and Siedle.[38] Turner and Hlatky reported in 1988 that cationic metallocenes in the presence of perfluorinated tetraphenylborate anions are highly active for the polymerization of both ethylene and α-olefins in the absence of an alkylaluminium cocatalyst (Scheme 5).[34].

$$Cp_2ZrMe_2 + [PhNHMe_2]^+[B(C_6F_5)_4]^- \longrightarrow [Cp_2ZrMe]^+[B(C_6F_5)_4]^- + CH_4 + PhNMe_2$$

Scheme 5

The recent development of single-component "cationic" metallocene catalysts is an extraordinary new development which eliminates many of the problems associated with the methyalumoxane cocatalysts. A particular advantage of the cationic metallocene catalysts is that they are considerably more tolerant of functional groups than the conventional hetereogeneous catalysts or homogeneous catalysts based on methylaluminoxane.

We have recently demonstrated that a variety of functionalized olefins can be polymerized with these new cationic metallocenes.[39] Catalysts derived from the reaction of $Cp^*{}_2ZrMe_2$ with $B(C_6F_5)_3$ or $[PhNHMe_2]^+[B(C_6F_5)_4]^-$ are active for the homopolymerization of the functionalized monomers 4-(TMSO)-1,6-heptadiene (**5**) (TMSO = trimethylsiloxy), 1-(TBSO)-4-pentene (**6**) (TBSO = *tert*-butyldimethylsiloxy) and *N,N*-diisopropyl-4-penten-1-amine (**7**) (Scheme 6).

OTMS
(**5**)
$Cp^*{}_2ZrMe_2$
$B(C_6F_5)_3$
OTMS OTMS OTMS
n

R
(**6**) R = OTBS
(**7**) R = $N^iPr)_2$
$Cp^*{}_2ZrMe_2$
$B(C_6F_5)_3$
n
R R R

Scheme 6. Polymerization of functionalized monomers.

5-(*N,N*-Diisopropylamino)-1-pentene was polymerized to poly(5-(*N,N*-diisopropylamino)-1-pentene) with a variety of catalysts derived from the complex Cp'_2ZrMe_2 {$Cp' = Cp^*$, Ind (indenyl), EBI (*rac*-ethylene-1,2-bis(indenyl)), and $Me_2C(Cp)(Flu)$

(isopropyl(cyclopentadienyl-1-fluorenyl)}. The active catalysts for these reactions were derived from the dimethyl metallocenes and the acid, $[PhNHMe_2H][B(C_6F_5)_4]$. These polymerizations are summarized in Table 1.

Polymerization of the tertiary amine monomer in the presence of $[Cp^*{}_2ZrCH_3]^+[X]^-$ (X = $B(C_6F_5)_4$) occurred with the highest level of productivity albeit with the lowest molecular weight of any of the catalyst systems utilized. This catalyst precursor yielded an atactic polymer. A catalyst derived from the chiral metallocene cation $[(EBI)ZrCH_3]^+[X]^-$ yielded an isotactic polymer with the highest molecular weight polymer. The syndiospecific catalyst $[Me_2C(Cp)(Flu)ZrCH_3]^+[X]^-$.afforded a syndiotactic polymer, although this catalysts showed the lowest productivity for this particular monomer. The Lewis acid, $B(C_6F_5)_3$, may be used instead of the $[C_6H_5N(CH_3)_2H][B(C_6F_5)_4]$ to generate a polymerization catalyst. However, these catalysts are less active and produce polymers of lower molecular weight than those using the anilinium salt. This observation is consistent with recent reports by other authors using similar catalyst precursors for the polymerization of ethylene and other α-olefins.[36]

For reference, the polymerization of 1-hexene with several catalysts under similar conditions is also included in Table 1. The productivities for both monomers with the most sterically hindered catalyst precursor, $Cp^*{}_2ZrMe_2$, are almost identical. In contrast, the productivities for the amine monomer using the less sterically hindered precursor, *rac*-(EBI)$ZrMe_2$, are two orders of magnitude lower than those observed with 1-hexene. This is likely a result of coordination of the Lewis basic amine nitrogen to the Lewis acidic metal center. Experiments to address the observed difference in monomer reactivities are currently in progress.

Table 1. Polymerization of Olefins with Cationic Catalysts $LL'ZrMe^+[B(C_6F_5)_4]^-$

Catalyst	Productivity[a] (amine monomer, N(iPr)$_2$)	M_n	Productivity[a] (1-hexene)	$\overline{M_n}$
Ind_2ZrMe_2	1.2	18,500		
rac-$EBIZrMe_2$	4.1	> 35,000	560	3400
$Cp^*{}_2ZrMe_2$	340	1280	325	1052
$Me_2C(Cp)(Flu)ZrMe_2$	5.2	30,100		

[a] g P/ mmol Zr•hr

Collins has reported that the cationic complex $[Cp_2ZrMe(THF)]^+[BPh_4]^-$ is active for the syndiospecific polymerization of methyl methacrylate in the presence of Cp_2ZrMe_2.[40] The reaction is proposed to proceed *via* a group-transfer mechanism. Collins also discovered that the complex $Cp_2Zr(OC(O^tBu)=CMe_2)_2$ rapidly polymerizes methyl methacrylate when activated by $[Et_3NH]^+[BPh_4]^-$. A similar polymerization of methyl methacrylate was reported by Yasuda using the dimer of $Cp^*{}_2SmH$.[41]

3.2. Functionalized End Groups

In the absence of hydrogen, metallocene-based catalyst systems produce well-defined polymers which are olefin- or aluminum-terminated. Mülhaupt has polymerized propylene with a chiral metallocene and MAO under conditions where β-hydrogen elimination was the predominant chain transfer process.[42] In a post-polymerization functionalization, the olefin endgroups of the highly isotactic polypropylene chains were converted to bromo-, epoxy-, anhydride-, ester-, amine-, carboxylic acid-, silane-, borane-, hydroxy-, thiol-terminated polymers as intermediates for the preparation of block copolymers. Using olefin-terminated atactic and isotactic polypropylene formed with MAO-activated Cp_2ZrCl_2 and (EBTHI)$ZrCl_2$ Shiono has synthesized amine-[43] and aluminum-terminated polymers.[44]

We have recently discovered a direct route to the synthesis of aluminum-terminated polyolefins.[45] For the cyclopolymerization of 1,5-hexadiene using MAO-activated $Cp^*{}_2ZrCl_2$ at -25°C, chain transfer to aluminum is the exclusive process (Scheme 7). The aluminum endgroups can be oxidized to give hydroxy-terminated poly(1,5-hexadiene), which can be further used as a macroinitiator for the ring-opening polymerization of ε-caprolactone.[46] Related contributions in the area of metal-terminated polymers have been made by Doi, Soga, and Shiono by using living, vanadium-based systems to prepare hydroxy-, aryl-, vinyl-, aldehyde-, iodide-, and zinc alkyl-terminated polymers.[47,48]

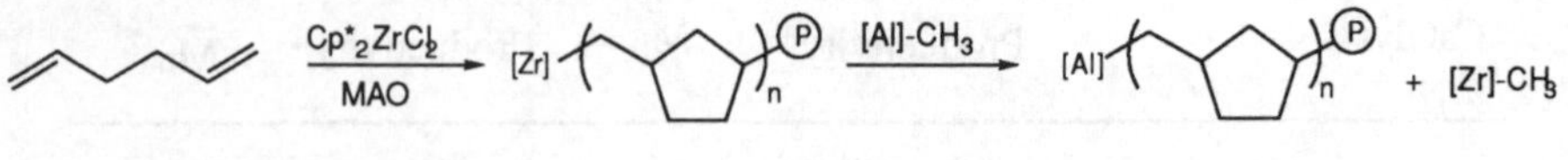

Scheme 7. Synthesis of aluminum-terminated poly (1,5-Hexadiene).

4. Photochromic Polypeptides

Poly-α-amino acids provide a well defined macromolecular matrix which can be chemically modified by attachment of photoresponsive chromophoric groups. This approach may yield a functional polymer in which the light response of the chromophores can be amplified by cooperative effect along the regular macromolecules chain thus providing new photoresponsive materials.

The modification of poly(L-glutamic acid) with p-aminoazobenzene in dimethylformamide (DMF) with dicyclohexylcarbodiimide as a coupling agent yields polypeptides largely contaminated by N-acylureic groups.[49] This side reaction was suppressed by addition of N-hydroxy-benzotriazol (HOBt).[50] These conditions led to the selective attachment of aminoazobenzene to the free carboxyl group of the polypeptide through formation of the corresponding amide (Scheme 8). The same result now has been obtained according to an adapted mixed-anhydrides method, by addition of pivaloyl chloride to poly(L-glutamic acid) in DMF and subsequent reaction with *p*-aminoazobenzene.

The introduction of azobenzene units in the side chains of poly(L-lysine) been achieved by means of various procedures and different azo-reagents.[51] In some experiments, samples of poly(Lysine HCl) or poly(Lysine acetate) were reacted with *p*-phenyl azobenzoic acid in the presence of pivaloyl (trimethylacetyl) chloride (mixed anhydrides method). Although anhydrous conditions must be maintained during the preparation of the anhydride itself, the condensation reaction with the amino compound can be performed in aqueous medium. This is necessary in this case, as poly(Lysine) is insoluble in pure DMF and other organic solvents commonly used in peptide synthesis.

In other experiments, poly(Lysine) was treated with *p*-phenylazobenzoic-*N*- hydroxy-succinimide ester (active esters method) in DMF/water solvent mixture. The reaction seems to be a suitable method to modify poly(Lysine) with organic reagents, and may be carried out both in organic and in aqueous solvents at acidic pH (Scheme 9) Finally, in one experiment, poly(Lysine) was acylated with *p*-phenylazobenzoyl chloride in DMF/water solvent mixture in the presence of excess MgO.

A detailed investigation of photochromism, conformation and CD properties was carried out in recent years on poly(L-glutamic acid) having various contents of azobenzene groups in the side chains.[52,53] In trimethylphosphate (TMP), the azo-polypeptides exhibited the α-helix CD pattern. In water at acid pH, the polymers containing 16 mol % and 21 mol % azobenzene groups adopted the α-helix structure while a β-structure was present in a 36 mol % azo-polypeptide.

In TMP, irradiation at 370 nm produced the *trans* to *cis* isomerization of the azo side chains, but did not induce any variations of the backbone conformation, as shown by the constancy of the CD spectra. In contrast, irradiation in water produced remarkable variations of the CD bands which were completely reversed in the dark.

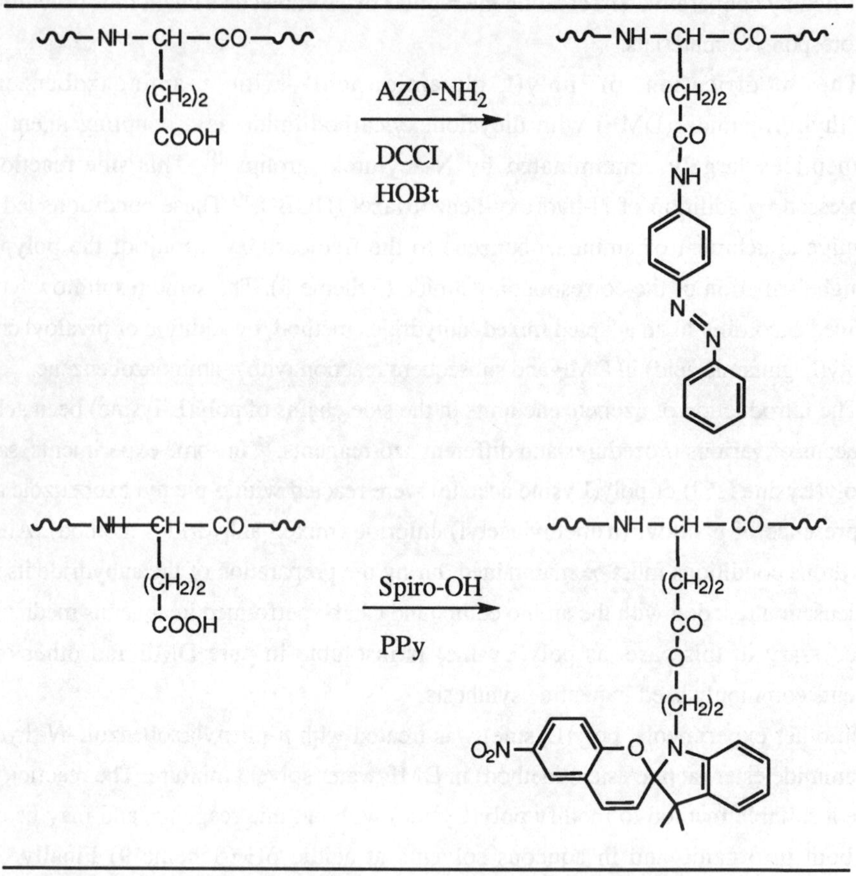

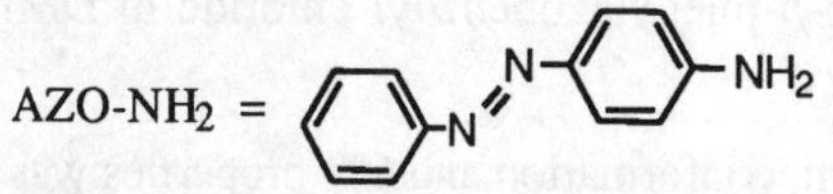

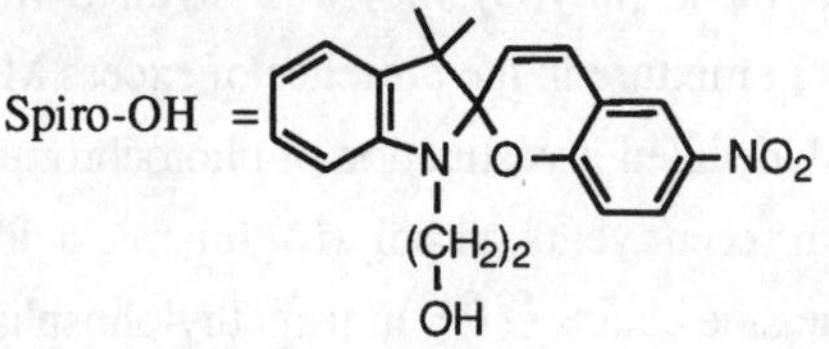

DCCI = Dicyclohexylcarbodiimide
PPy = 4-Pyrrolidino-pyperidine
HOBt = N-hydroxy-benzotriazol

Scheme 8. Modification reactions of poly(L-glutamic acid) with azobenzene and spiropyran reagents

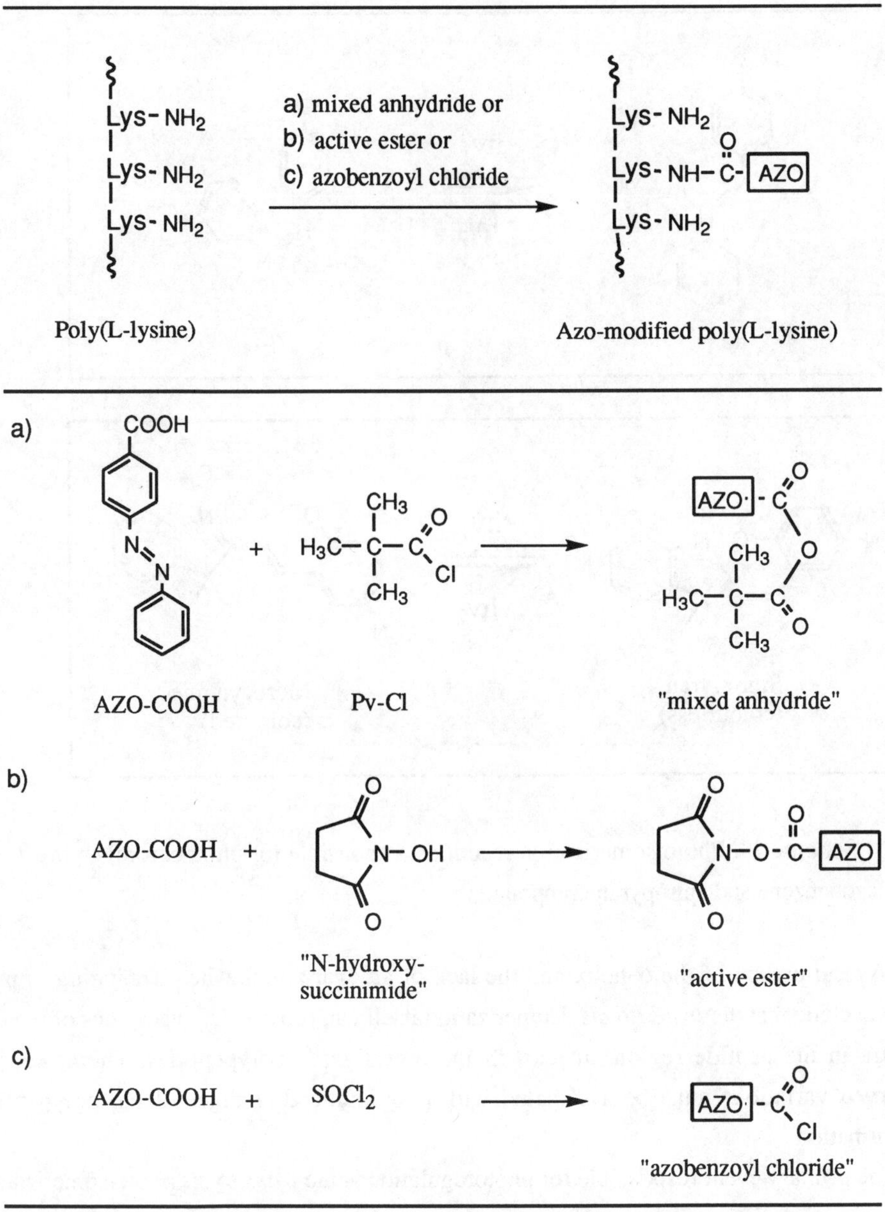

Lys-NH_2
Lys-NH_2
Lys-NH_2

a) mixed anhydride or
b) active ester or
c) azobenzoyl chloride

Lys-NH_2
Lys-NH–C(=O)-AZO
Lys-NH_2

Poly(L-lysine) → Azo-modified poly(L-lysine)

a)

AZO-COOH + Pv-Cl → "mixed anhydride"

COOH, N=N, H_3C–C(CH_3)$_2$–C(=O)Cl, AZO-C(=O)–O–C(=O)–C(CH_3)$_3$

b)

AZO-COOH + "N-hydroxy-succinimide" (N–OH) → "active ester" (N–O–C(=O)–AZO)

c)

AZO-COOH + $SOCl_2$ → "azobenzoyl chloride" (AZO-C(=O)Cl)

Scheme 9. Modification reactions of poly(L-lysine) with an azobenzene reagent

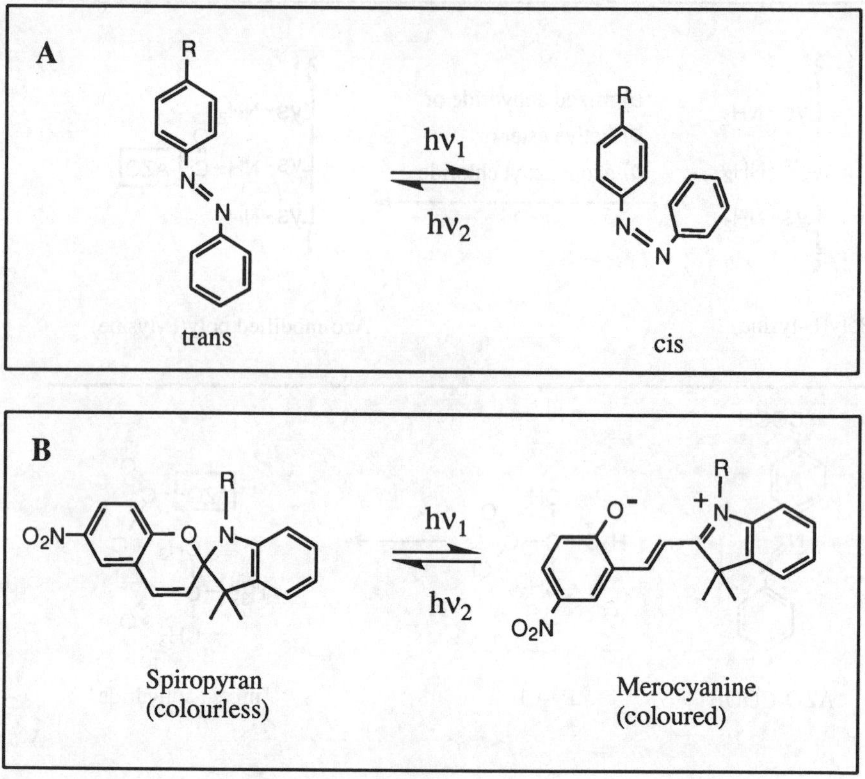

Scheme 10. Photoisomerization reactions responsible for photochromism in azobenzene and spiropyran compounds

The typical pattern of the α-helix and the lack of any variations when irradiating in pure TMP exclude that the *trans* to *cis* isomerization itself can provide any variations of the CD spectra in the peptide region, at least in the investigated polypeptides. Therefore, the observed variations must be associated with photoinduced changes of the polypeptide conformation.

The primary event responsible for photoregulation is the *trans* to *cis* photoisomerization of the azo side chains (Scheme 10A). However, the simple geometry variation during the photoisomerization of the chromophore is not sufficient to induce appreciable conformational changes in the main chain of the macromolecule. The driving force for photoregulation is likely to result from a polarity variation of the environment around the polypeptide backbone, as a consequence of the different polarity and hydrophobicity between the *trans* and the *cis* isomers. Large variations of the CD spectra induced by light are observed in hexafluoro-2-propanol/water/sodium dodecyl sulphate. In this solvent mixture, the dark-adapted samples show the typical CD profile of the β–structure, with a single minimum at 220 nm and an intense couplet in the 330 nm region due to electronic interactions between the azo side chains. Irradiation at 338 nm cancels the side chain CD bands and gives rise to the α-helix CD pattern, thus indicating the occurrence of a photoinduced α- helix to β-structure conformational transition.

Photostimulated aggregation-deaggregation processes have been observed in azobenzene-containing poly(L-glutamic acid).[53] The samples stored in the dark or irradiated at 450 nm (azo groups in *trans* configuration) show slow variations of CD spectra on ageing in aqueous solution. The time-dependence is characterized by the progressive distortion of the α–helix CD pattern, typical of those produced by aggregates of polypeptide chains. Irradiation at 370 nm, with the consequent isomerization of the azo moieties to the *cis* configuration, is accompanied by the complete restoration of the initial CD spectra, indicating dissociation of the aggregates. By dark-adaptation or by irradiating at 450 nm, the spectra revert again to the distorted ones, thus confirming the reversibility of the aggregation-deaggregation process. Recently, examination of a polypeptide having high azobenzene content has confirmed the occurrence of light-induced aggregation changes, together with direct evidence that the polypeptide solubility can be photochemically modulated.[54] The dark-adapted polymer is insoluble in hexafluoro-2-propanol/water solvent mixture, but irradiation of the suspension for a few seconds at 338 nm causes its complete dissolution. The precipitation-dissolution cycles can be repeated without apparent fatigue by irradiation and relaxation in the dark, or by irradiating alternately at 338 and 450 nm.

The polymer precipitation can be due to aggregation between macromolecules through hydrophobic interactions in water and stacking of the azobenzene side chains. These interactions are favoured in the dark-adapted or 450 nm irradiated samples, as *trans* isomers are apolar and essentially planar. Light-induced dissolution, on the other hand, can be due to a deaggregation process, as *cis* isomers have a considerable dipole moment and are not planar, thus inhibiting aggregation and enhancing the polymer solubility.[53,54]

Azo-modified polypeptides were also investigated in connection with their photochromic behaviour caused by the *trans* to *cis* photoisomerization of the azo groups present in the side chains. In methanol/water solvent mixture, the 20% azo-poly(L-lysine) adopts the α-helix conformation. The helix stability was found to be higher when the azo chains are in *cis* than when they are in *trans* configuration. Irradiation at 340 nm (*trans* to *cis* isomerization) and alternately at 450 nm (*cis* to *trans* isomerization) produced reversible variations of the α-helix content. In hexafluoro-2-propanol/water/sodium dodecylsulfate mixture, the 43% azo-poly(L-lysine) adopted a β-structure and promoted the α-helix conformation. The effect was reversed upon irradiation at 450 nm. The photoinduced β to helix change was explained on the basis of the different geometry and hydrophobic character of the *trans* and the *cis* azobenzene units.[51]

Poly(L-glutamic acid) and partially methylated poly(L-glutamate)s with spyropyran side chains were found to exhibit "reverse photochromism" in hexafluoro-2-propanol (HFP).[55,56] The samples kept in the dark were characterized by an intense absorption band in the visible range of the spectrum which was completely erased upon exposure to sunlight (500-550 nm) (Scheme 10B). According to the CD spectra, the macromolecules adopted a random coil conformation in the dark, whereas solutions exposed to light, displayed the typical CD pattern of the α-helix. The back reaction in the dark was accompanied by the progressive decrease of the helix content and recovery of the original disordered conformation. As a consequence of photoinduced conformational changes, large and reversible viscosity variations were obtained.

The introduction of spiropyran groups into the side chains of Poly(L-lysine) was carried out as reported in the Scheme 11.[57] The modified polymers were soluble in HFP where they exhibited adsorption spectra and photochromic behaviour very similar to those described above for poly(L-glutamate)s. In contrast to the photochromic polymers of L-glutamic acid, however, the analogous polymers of L-lysine do not give light induced conformational changes in pure HFP, and they are always random coil, either when the samples are kept in the dark or when they are exposed to light. The different conformational behaviour of spiropyran-modified poly(L-lysine) with respect to spiropyran-containing poly(L-glutamate) is likely to be due to the unmodified lysine side chains protonated by the acid solvent HFP. However, the addition of triethylamine to the HFP solutions does not produce significant variations of the adsorption spectra whereas CD measurements reveal that triethylamine addition induces coil to helix conformational transitions of the macromolecular chains. Most remarkably, the sample stored in the dark and that irradiated show different transition curves. Therefore, at any triethylamine concentration in the range between to the curves, exposure to light and dark conditions produces reversible variations of helix content. Thus, combination of photochromic effects with appropriate triethylamine amounts allows regulation of the extent of the photoresponse. Triethylamine may induce deprotonation of the unmodified amino side chains thus allowing control of polypeptide conformation by isomerization of the photochromic groups.[57]

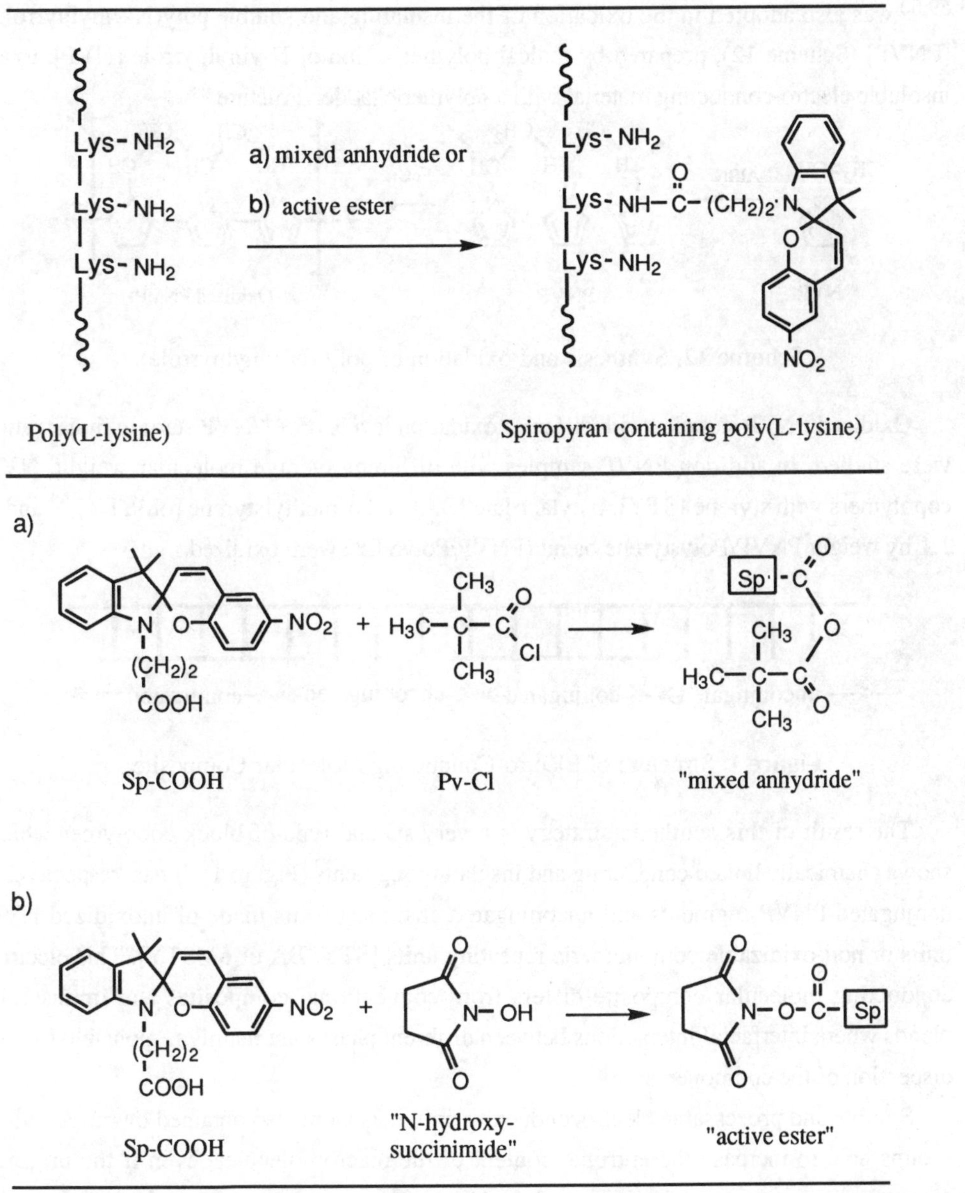

Lys- NH_2

Lys- NH_2

Lys- NH_2

a) mixed anhydride or

b) active ester

Lys- NH_2

Lys- NH– C(=O)– $(CH_2)_2$· N

Lys- NH_2

NO_2

Poly(L-lysine)

Spiropyran containing poly(L-lysine)

a)

NO_2 + H_3C–C(CH_3)(CH_3)–C(=O)Cl

N

$(CH_2)_2$

COOH

Sp-COOH

Pv-Cl

"mixed anhydride"

b)

NO_2 + N– OH

N–O–C(=O)– Sp

$(CH_2)_2$

COOH

Sp-COOH

"N-hydroxy-succinimide"

"active ester"

Scheme 11. Modification reactions of poly (L-lysine) with a spiropyran reagent

5. Electroconducting Polymers

It is well known that the extended coniugation of monomeric residues is the fundamental property which determines the extent of electroconductivity of an organic polymer.[58] The polymer precursor route used to improve the processability of electro-conductive polymers

59,60 was also adopted in the oxidation of the insulating and soluble poly(N-vinylpyrrole) [PNVP] (Scheme 12), prepared by radical polymerization of N-vinylpyrrole [NVP], to an insoluble electro-conducting material with a polymeric ladder structure.[61]

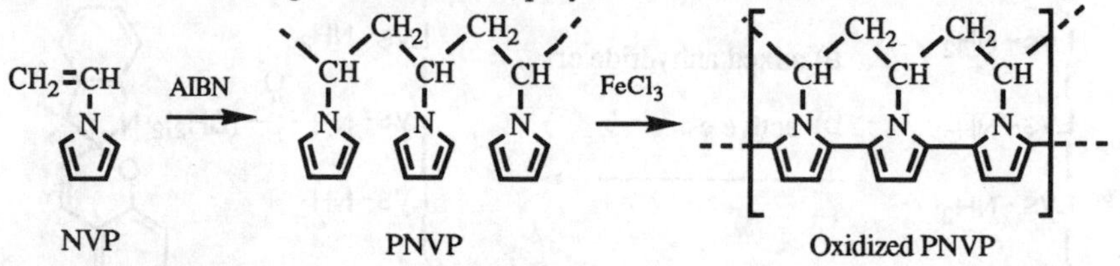

$CH_2{=}CH$ (NVP) $\xrightarrow{AIBN}$ PNVP $\xrightarrow{FeCl_3}$ Oxidized PNVP

Scheme 12. Synthesis and oxidation of poly (N-vinylpyrrole).

Oxidized PNVP samples with different oxidation levels (Fe^{3+}/NVP starting molar ratio) were studied. In addition PNVP samples with different average molecular weight, NVP copolymers with styrene [STY], butylacrilate [BA] and o-methylstyrene [oMSTY],[62] and a 1:1 by weight PNVP/Polystyrene blend (PNVP/PolySTY) were oxidized.

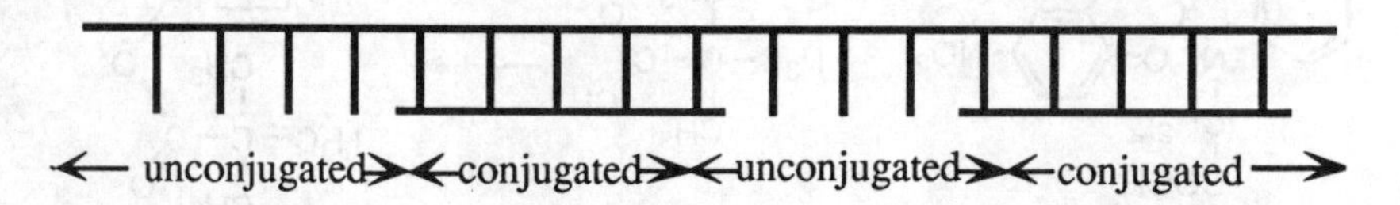

Figure 1. Structure of Electro-Conducting Molecular Composite .

The result of this synthetic strategy is a very special type of block copolymer which shows chemically linked conducting and insulating segments (Figure 1). It has, respectively, conjugated PNVP segments and unconjugated chain segments made of unoxidized NVP units or non-oxidizable comonomeric repeating units [STY, BA or oMSTY]. This electro-conducting molecular composite differs from conventional composites and immiscible blends where interfacial interactions between different phases are mainly responsible for the dispersion of the components.

Soluble and processable electroconducting polymers were also obtained by introducting groups able to increase the entropic content of the macromolecules, even if the original conjugated structure is retained. Thus, 3-decylpyrrole homopolymer, 3-decylpyrrole/pyrrole copolymers and 3-(3,7-dimethyloctyl)pyrrole homopolymer were synthesized from the corresponding monomers by chemical oxidation using various oxidants, reaction solvents and temperatures. The different polymerization conditions, together with the copolymerization of alkyl substituted monomers with pyrrole, result in macromolecules with different structural regularity, average chain length, doping efficiency and distribution of paraffinic side chains.[63] These concepts were used to obtain soluble and tractable electroconductive polymers with pyrrole conjugated chains. N-Vinyl-3-decylpyrrole (NVDP) was prepared starting from pyrrole following the simple route described in Scheme 13. NVDP was either homopolymerised or copolymerised with NVP by free radical initiators to yield a series of polymers (Scheme 14) containing variable amount of pyrrole side chains with n-decyl group. These polymers, schematically represented in Scheme 14, had $\overline{M}n$ in the range 3,200 - 7,200 corresponding to a number of monomeric residues ranging between 18 and 54.

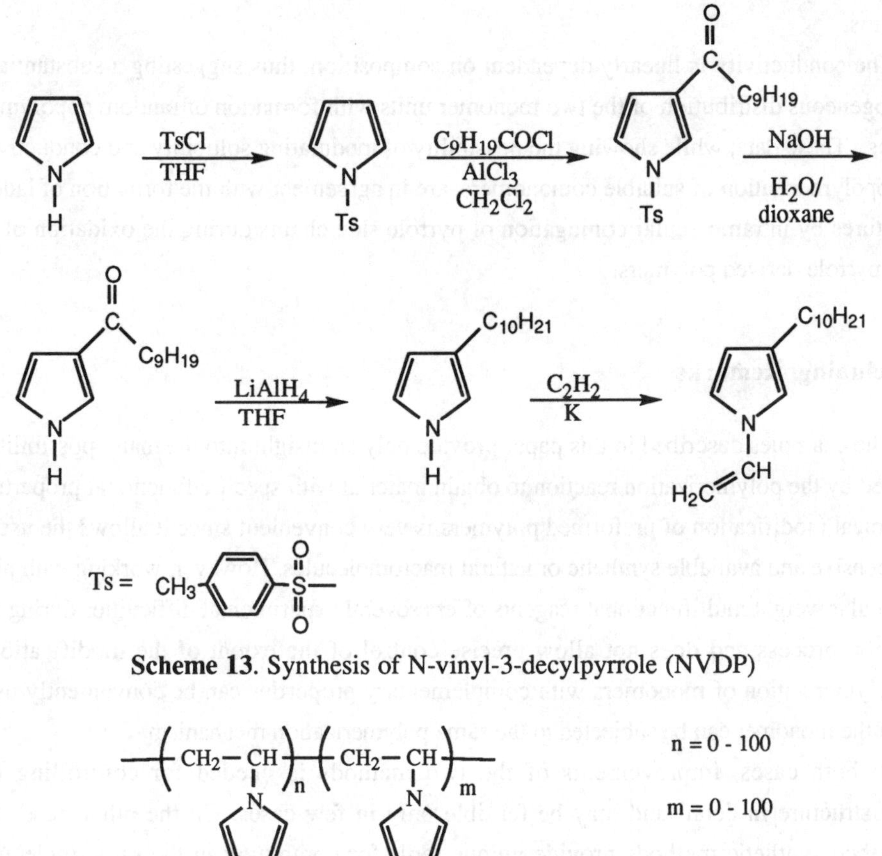

TsCl / THF; $C_9H_{19}COCl$ / $AlCl_3$ / CH_2Cl_2; NaOH / H_2O/ dioxane; C_9H_{19}; LiAlH$_4$ / THF; $C_{10}H_{21}$; C_2H_2 / K; $H_2C{=}CH$; Ts = CH_3-C$_6$H$_4$-SO$_2$-

Scheme 13. Synthesis of N-vinyl-3-decylpyrrole (NVDP)

$-(CH_2-CH)_n-(CH_2-CH)_m-$; n = 0 - 100 ; m = 0 - 100 ; $C_{10}H_{21}$

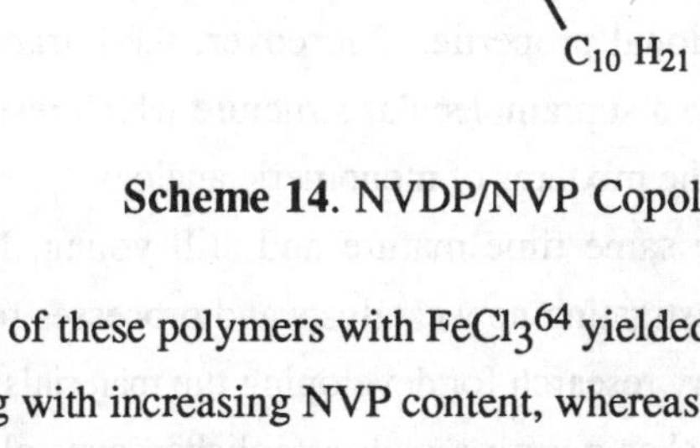

Scheme 14. NVDP/NVP Copolymers Structure

Treatment of these polymers with $FeCl_3$[64] yielded a black material whose conductivity was increasing with increasing NVP content, whereas the solubility in organic solvents was decreasing correspondingly. While oxidized poly(*N*-vinyl-3-decylpyrrole) was completely soluble in methanol, THF, $CHCl_3$ and dioxane at room temperature, oxidized poly(*N*-vinylpyrrole) was completely insoluble in organic solvents.[61] When more then 60% of pyrrole rings are substituted with a decyl group, the corresponding oxidized *N*-vinylpyrrole/*N*-vinyl-3-decylpyrrole copolymer is soluble.

Copolymer conductivity depends not only on the composition, but also on the average polymerization degree. Copolymers with an higher degree of polymerization have a higher probability, after oxidation, of yielding polyconjugated chains with a length >15 pyrrole units, which has been demonstrated as the threshold to achieve conductivity in PNVP systems.

The conductivity is linearly dependent on composition, thus suggesting a substantially homogeneous distribution of the two monomer units with formation of random copolymers chains. These data, while showing the possibility of modulating solubility and conductivity by copolymerisation of suitable comonomers, are in agreement with the formation of ladder structures by intramolecular conjugation of pyrrole side chains during the oxidation of *N*-vinylpyrrole derived polymers.

Concluding Remarks

The examples described in this paper provide only an insight into the many possibilities fered by the polymerization reaction to obtain material with specific functional properties. Chemical modification of preformed polymers is very convenient since it allows the use of inexpensive and available synthetic or natural macromolecules. However, working with high molecular weight multifunctional reagents offers several experimental difficulties during the reaction process and does not allow precise control of the extent of the modifications. Copolymerization of monomers with complementary properties can be conveniently used when the monomer can be subjected to the same polymerization mechanism.

In both cases, improvements of the two methods is needed for controlling the microstructure in detail and may be feasible only in few cases. On the other hand, the described synthetic methods provide unique tools for combining in the same molecular structure several structural and functional properties. Moreover, the intramolecular and intermolecular interactions can provide a supramolecular structure which results in new or amplified properties not observable in the mixtures of monomeric analogs.

Macromolecular science is at the same time mature and still young. New exciting developments can be expected by investigating new catalysts and processes for controlling structure and by a more interdisciplinary research for developing the materials of the future. Indeed the available synthetic approaches permit a well established control of structural properties allowing to design not only elastomers, plastics and fibers, but also light materials to be used in a very large temperature range with mechanical performances comparable to or even better than those of natural and conventional materials. Moreover, the development of functional polymeric materials allowing the modulation also of specialized properties has opened new promising applications possibilities.

The preparation of molecular materials which respond not only to mechanical stress, but also to light, electric or magnetic fields, environmental changes or a combination of these effects is providing new opportunities in important technologies of the future dealing with electronics, optoelectronics, photonics, storage and transfer of information, sensors, thermal and electric conductivity.

References

1. Severini, F. (1978) Graft copolymers from polyolefins, *Chim. Ind.* (Milan), **60**, 743-751.
2. Giannini, U., Bruckner, G., Pellino, E., and Cassata, A. (1968) Polymerization of Nitrogen-containing and Oxygen-containing monomers by Ziegler-Natta catalysts, *J. Polym. Sci. C.*, **22**, 157-175.
3. De Vito, G., Lanzetta, N., Maglio, G., Malinconico, M., Musto, P., and Palumbo, R. (1984) Functionalization of an amorphous ethylene-propylene copolymer by free radical initiated grafting of unsaturated molecules, *J. Polym. Sci., Polym. Chem. Ed.*, **22**, 1335-1347.
4. Pradellok, W., Vogl, O., and Gupta, A. (1981) Functional polymers. XIV. Grafting of 2(2-hydroxy-5-vinylphenyl) 2H-benzotriazole onto polymers with aliphatic groups, *J. Polym. Sci., Polym. Chem. Ed.*, **19**, 3307-3314.
5. Ruggeri, G., Aglietto, M., Petragnani, A., and Ciardelli, F. (1983) Some aspects of polypropylene functionalization by free radical reactions, *Eur. Polym. J.*, **19**, 863-866.
6. Gaylord, N.G. and Mehta, M. (1982) Role of homopolymerization in the peroxide-catalyzed reaction of maleic anhydride and polyethylene in the absence of solvent, *J. Polym., Sci., Polym. Lett. Ed.*, **20**, 481-486.
7. Schlecht, M.F., Pearce, E.M., Kwei, T.K., and Cheung, W. (1986) Radical-promoted functionalization of polyethylene. Controlled incorporation of hydrogen-bonding groups, *ACS Polym. Prep.*, **27**(2), 63-64.
8. Joshi, S.G. and Natu, A.A. (1986) Chlorocarboxylation of polyethylene. I. Preparation and structure of chlorocarboxylated polyethylene, *Angew. Makromol. Chem.*, **140**, 99-112.
9. Lamblà, M., Druz, J., and Satyanarayana, N. (1988) Crosslinking reactions in the molten state: interpolymeric condensation reactions, *Makromol. Chem.*, **189**, 2703-2717.
10. Al-Malaika, S. (1988) Modification of polyolefins in-situ: reactive processing in mixers and extruders, *ACS Polym. Prep.*, **29**(1), 555-556.
11. Morgan, A.W. and Swenson, J.S. (1968) Chemical modification of the solid surface of polymers, *US Patent* 3,376,278 ; (1968) *Chem. Abstr.*, **68**, 105746e.
12. Ranby B., Gao, M.Z., Hult., A., and Zhang, P.Y. (1988) Modification of polymer surfaces by photoinduced graft copolymerization, *ACS Symp. Ser.*, N° 364, 168-186.
13. Lee, K.-W. and McCarthy, T.J. (1987) Surface-selective hydroxylation of polypropylene, *ACS Polym. Prep.*, **28**(2), 250-251.
14. Batich, C. and Yahiaoui, A. (1987) Surface modification. I. Graft polymerization of acrylamide onto low-density polyethylene by Ce^{4+}-induced initiation, *J. Polym. Sci., Polym. Chem. Ed.*, **25**, 3479-3488.
15. Hagman, J.F. and Crary, J.W. (1985) Ethylene-acrylic elastomers in "*Encyclopedia of Polymer Science and Engineering*", J. Wiley, New York, vol. 1, pp. 325-334 and references therein.

16. Klabunde, U., Mulhaupt, R., Herskovitz, T., Janowicz, A.H., Calabrese, J., and Ittel, S.D. (1987) Ethylene-homopolymerization with P, O-chelated nickel catalysts, *J. Polym. Sci., Polym. Chem. Ed.*, **25**, 1989-2003.
17. Ostoja Starzewski, K.A., Witte, J., Reichert, K.H., Vasiliou, G. (1988) Linear and branched polyethylenes by new coordination catalysts, in W. Kaminsky and H. Sinn (eds), *"Transition Metals and Organometallics as Catalysts for Olefin Polymerization"*, Springer-Verlag, Berlin Heidelberg, pp. 349-360.
18. Starkweather, H. (1987) Olefin-carbon monoxide copolymers in *"Encyclopedia of Polymer Science and Engineering"*, J. Wiley, New York, vol.10, pp. 369-373 and references therein.
19. Ciardelli, F., Menconi, F., Altomare, A., and Carlini, C. (1988) Polymerization of olefins by highly active Ziegler-Natta catalysts in presence of vinyl esters as "external" Lewis bases, XVI Congress of Italian Chemical Soc., *Bononiachem '88*, Bologna, October 9-14, Abstracts, p.278
20. Landoll, L.M. and Breslow, D.S. (1987) Polypropylene ionomers, *ACS Polym. Prep.*, **28**(1), 256-257.
21. Purgett, M.D. and Vogl, O. (1988) Functional polymers. XLVIII. Polymerization of ω-alkenoate derivatives, *J. Polym. Sci., Polym. Chem. Ed.*, **26**, 677-700.
22. Natta, G., Beati, E., and Severini, F. (1959) Production of graft copolymers from poly(α-olefin) hydroperoxides, *J. Polym. Sci.*, **34**, 681-698.
23. Benedetti, E., D'Alessio, A., Aglietto, M., Ruggeri, G., Vergamini, P., and Ciardelli, F. (1986) Vibrational analysis of functionalized polyolefins / poly(vivylchloride) blends, *Polym. Eng. Sci.*, **26**, 9-14.
24. Aglietto, M., Bertani, R., Ruggeri, G., and Ciardelli, F. (1992) Radical bulk functionalization of polyethylenes with esters groups, *Makromol. Chem.*, **193**, 179-186.
25. Aglietto, M., Bertani, R., Ruggeri, G., and Segre, A.L. (1990) Functionalization of polyolefins. Determination of the structure of functional groups attached to polyethylene by free radical reactions, *Macromolecules*, **23**, 1928-1933.
26. Aglietto, M., Alterio, R., Bertani, R., Galleschi, F., and Ruggeri, G. (1989) Polyolefins functionalization by carbene insertion for polymer blends, *Polymer*, **30**, 1133-1136.
27. Aglietto, M., Bertani, R., Ruggeri, G., Fiordiponti, N., and Segre, A.L. (1989) Functionalization of polyolefins: structure of functional groups in polyethylene reacted with ethyldiazoacetate, *Macromolecules*, **22**, 1492-1493.
28. Demonceau, A., Noels, A.F., Hubert, A.J., and Theyssié, P. (1984) Transition-metal-catalysed reactions of diazoesters. Insertion into C-H bonds of paraffins catalysed by bulky rhodium (II) carboxylates: enhanced attack on primary C-H bonds, *Bull. Soc. Chim. Belg.*, **93**, 945-948.
29. Padwa, A.R. (1989) Functionally substituted poly(α-olefins), *Prog. Polym. Sci* , **14**, 811-833.
30. Zefirova, A.K. and Shilov, A.E. (1961) Kinetics and mechanism of reaction of alluminum alkyls with titanium halides, *Doklady Akad. Nauk SSSR* , **136**, 509-609; (1961) *Chem. Abstr.*, **55**, 19884f.

31. Jordan, R.F. (1991) Chemistry of cationic dicyclopentadienyl group 4 metal-alkyl complexes, *Adv. Organomet. Chem.*, **32**, 325-387.
32. Bochmann, M. and Lancaster, S.J. (1993) Base-free cationic zirconium benzyl complexes as highly active polymerization catalysts, *Organometallics*, **12**, 633-640.
33. Taube, R. and Krukowka, L. (1988) Complex catalysis. XXX. Cationic alkyl-dicyclopentadienyltitanium complexes as catalysts for ethylene polymerization, *J. Organomet. Chem.*, **347**, C9-C11.
34. Turner, H.W. (1988) Soluble catalysts for polymerization of olefins, *Eur. Pat. Appl.;* 277,004; (1989) *Chem. Abstr.*, **110**, 58290a.
35. Yang, X., Stern, C.L., and Marks, T.J. (1991) "Cation-like" homogeneous olefin polymerization catalysts based upon Zirconocene alkyls and tris(pentafluorophenyl) borane, *J. Am. Chem. Soc.*, **113**, 3623-3625.
36. Ewen, J.A. and Elder, M.J. (1993) Syntheses and models for stereospecific metallocenes, *Makromol. Chem., Macromol. Symp.*, **66**, 179-190.
37. Chien, J.C.W., Tsai, W.-M., and Rausch, M.D. (1991) Isospecific polymerization of propylene catalyzed by rac-ethylenebis(indenyl)methylzirconium "cation", *J. Am. Chem. Soc.*, **113**, 8570-8571.
38. Siedle, A.R.,Lamanna, W.M., Newmark, R.A., Stevens, J., Richardson, D.E., and Ryan, M. (1993) The role of non-coordinating anions in homogeneous olefin polymerization, *Makromol. Chem., Macromol. Symp.*, **66**, 215-224.
39. Kesti, M.R., Coates, G.W., and Waymouth, R.M. (1992) Homogeneous Ziegler-Natta polymerization of functionalized monomers catalyzed by cationic group IV metallocenes, *J. Am. Chem. Soc.*, **114**, 9679-9680.
40. Collins, S. and Ward, D.G. (1992) Group-transfer polymerization using cationic zirconocene compounds, *J. Am. Chem. Soc.*, **114**, 5460-5462.
41. Yasuda, H., Yamamoto, H.,Yokota, K., Miyake, S., and Nakamura, A. (1992) Synthesis of monodispersed high molecular weight polymers and isolation of an organolanthanide (III) intermediate coordinated by a penultimate poly(MMA) unit, *J. Am. Chem. Soc.*, **114**, 4908-4910.
42. Mülhaupt, R., Duschek, T., and Rieger, B. (1991) Functional polypropylene blend compatibilizers, *Makromol. Chem., Macromol. Symp.*, **48/49**, 317-332.
43. Shiono, T., Kurosawa, H., Ishida, O., and Soga, K. (1993) Synthesis of polypropylenes functionalized with secondary amino groups at the chain ends, *Macromolecules*, **26**, 2085-2089.
44. Shiono, T. and Soga, K. (1992) Synthesis of terminally aluminum-functionalized polypropylene, *Macromolecules*, **25**, 3356-3361.
45. Mogstad, A.L. and Waymouth, R.M. (1992) Chain transfer to Aluminum in the homogeneous cyclopolymerization of 1,5-hexadiene, *Macromolecules*, **25**, 2282-2284.
46. Mogstad, A.L. and Waymouth, R.M. (1994) Synthesis of diblock polyolefins and polyester copolymer using zirconium and alluminum catalysts, *Macromolecules*, **27**, 2313-2315.
47. Doi, Y. and Keii, T. (1986) Synthesis of "living" polyolefins with soluble Ziegler-Natta catalysts and application to block copolymerization, *Adv. Polym. Sci.*, **73/74**,

201-248.
48. Shiono, T., Yoshida, K., and Soga, K. (1990) Synthesis of terminally hydroxylated isotactic polypropylene using $Zn(C_2H_5)_2$ and oxygen as chain transfer and quenching reagents, *Makromol. Chem., Rapid Commun.*, **11**, 169-175.
49. Houben, J.L., Fissi, A., Bacciola D., Rosato, N., Pieroni, O., and Ciardelli, F. (1983) Azobenzene containing poly(L-glutamates photochromism and conformation in solution, *Int. J. Biol. Macromol.*, **5**, 94-100.
50. Konig, W. and Geiger, R. (1973) N-hydroxy compounds as catalysts for the aminolysis of activated esters, *Chem. Ber.*, **106**, 3626-3635.
51. Fissi, A., Pieroni, O., and Ciardelli, F. (1987) Photoresponsive polymers: azobenzene-containing poly(L-lysine), *Biopolymers* , **26**, 1993-2007.
52. Ciardelli, F., Pieroni, O., Fissi, A., and Houben, J.L. (1984) Azobenzene-containing polypeptides. Photoregulation of conformation in solution, *Biopolymers*, **23**, 1423-1437.
53. Pieroni, O., Fissi, A., Houben, J.L., and Ciardelli, F. (1985) Photoinduced aggregation changes in photochromic polypeptides, *J. Am. Chem. Soc.*, **107**, 2990-2991.
54. Ciardelli, F., Pieroni, O., and Fissi, A. (1986) Photocontrol of the solubility of azobenzene-containing poly(L-glutamic acid), *J. Chem. Soc., Chem. Comm.*, 264-265.
55. Ciardelli, F., Fabbri, D., Pieroni, O., and Fissi, A. (1989) Photomodulation of polypeptide conformation by sunlight in spiropyran-containing poly(L-glutamic acid), *J. Am. Chem. Soc.*, **111**, 3470-3472.
56. Fissi, A., Pieroni, O., Ciardelli, F., Fabbri, D., Ruggeri, G., and Umezawa, K. (1993) Photoresponsive polypeptides: photochromism and conformation of poly(L-glutamic acid) containing spiropyran units, *Biopolymers*, **33**, 1505-1517.
57. Pieroni, O., Fissi, A., Viegi, A., Fabbri, D., and Ciardelli, F. (1992) Modulation of the chain conformation of spiropyran-containing poly(L-lysine) by the combined action of visible light and solvent, *J. Am. Chem. Soc.*, **114**, 2734-2736.
58. Billingham, N.C. and Calvert, P.D. (1989) Electrically conducting polymers. A polymer science viewpoint, *Adv. Polym. Sci.*, **90**, 1-104.
59. Edwards, J.H., Feast,W.J., and Bott, D.C. (1984) New routes to conjugated polymers: 1. A two step route to polyacetylene, *Polymer*, **25**, 395-398.
60. Gagnon, D.R., Capistran, J. D., Karasz, F. E., Lenz, R.W., and Antoun, S. (1987) Synthesis, doping, and electrical conductivity of high molecular weight poly(p-phenylene vinylene), *Polymer*, **28**, 567-573.
61. Castelvetro, V., Colligiani, A., Ciardelli, F., Ruggeri, G., and Giordano, M. (1990) The oxidation of poly-N-vinylpirrole to electroconducting ladder polymers, *New. Polym. Mater.*, **2**, 93-101.
62. Ciardelli, F., Mori, S., Ruggeri, G., Colligiani, A., Cagnolati, R., Campani, E., Rolla, P.A., and Lucchesi, M. (1992) Structural macromolecular parameters determining electroconductivity of oxidized poly-N-vinylpirrole, *Makromol. Chem., Macromol. Symp.*, **59**, 363-368.
63. Spila, E. (1992) Dr. Degree Thesis, University of Pisa.

64. Ruggeri, G., Spila, E., Puncioni, G., and Ciardelli, F. (1994) Oxidized poly(N-vinyl-3-decylpirrole) and poly (N-vinyl-3-decylpyrrole-*co*-N-vinylpyrrole): soluble electroconducting ladder polymers, *Makromol. Chem., Rapid Commun.*, **15**, 537-542.

TRANSDUCTION OF MOLECULAR INTERACTIONS INTO MACROSCOPIC PROPERTIES

H.C. Visser, R.J.W. Lugtenberg, J.F.J. Engbersen and D.N. Reinhoudt
Department of Organic Chemistry
MESA Research Institute
University of Twente
P.O. Box 217
7500 AE Enschede
The Netherlands

ABSTRACT. Synthetic receptor molecules derived from calix[4]arenes have been used in different technological applications. The use of various functionalized calix[4]arenes in selective membrane transport through supported liquid membranes, selective cation detection with chemically modified field effect transistors, as preorganized donor-π-acceptor systems in non linear optics and in the development of monolayers with receptor headgroups is discussed.

1. Introduction

The spatial organization of functional groups in hydrophobic calixarenes and their analogs offers an excellent possibility to construct molecules with specially designed properties. Based on these compounds we have obtained selective receptors for cations and anions, donor-acceptor functionalized calix[4]arenes with attractive non-linear optical properties, and adsorbates with resorcin[4]arene headgroups which are able to recognize small organic molecules. Technological application of these molecules in materials requires the translation of the molecular properties into macroscopic properties. In the following sections the application of calix[4]arenes in selective membrane transport, selective ion detection, non-linear optics materials, and monolayers of receptor adsorbates on gold will be discussed.

C. Chatgilialoglu and V. Snieckus (eds.), Chemical Synthesis, 549–563.

2. Calix[4]arenes in selective ion transport through supported liquid membranes

Supported liquid membranes (SLMs) containing selective carriers can be used efficiently for separation of aqueous solute species. In SLMs an organic phase is immobilized in the pores of a polymeric support and receptor molecules that selectively form complexes with the solute species are dissolved in the organic layer and serve as carriers for transport. However, the relatively short life-time of SLMs has limited their large scale application. Important parameters for degradation of SLMs are the leaching of the carrier from the membrane phase to the water phases[1] due to the large exchange area and volume ratio of the aqueous phases over the membrane phase, and the progressive wetting of the membrane phase[2], leading to formation of water channels and expulsion of the organic phase by osmotic pressure driven water transport[3]. These problems can be overcome by the use of lipophilic carriers, support, and membrane solvents.

In order to prevent the partitioning of the carrier to the aqueous phase the following approaches have been investigated in our laboratory: (*i*) attachment of long alkyl chains[4,5], (*ii*) attaching the carrier to a polymer[6,7] and (*iii*) constructing carriers by using lipophilic building blocks like calixarenes[8,9].

Besides transport of cations with these modified carriers, also carrier mediated transport of neutral molecules through SLMs has been studied[10,11] and recently also anion[12] and ditopic[13] (combined anion and cation) receptors as carriers for membrane transport have been studied.

As carriers for metal cations especially the lower rim functionalized *p-tert*-butylcalix[4]arenes have been used. The calixarenes **1a**, **1b** with a crown-5 over two opposite aryl moieties are selective for potassium, whereas the tetraamide **2a**, tetraketone **2b** and tetraester **2c** (Figure 1) selectively form complexes with sodium. These neutral receptors are suitable as carriers because of their high lipophilicity and complexation selectivity.

The transport of alkali cations has been studied in single cation and competitive transport. A model has been developed which describes diffusion limited transport in terms of the extraction equilibria (K_{ex}) and the mean diffusion constants (D_m)[8]. This model has also been used to predict the transport selectivity based on the differences in extraction ability

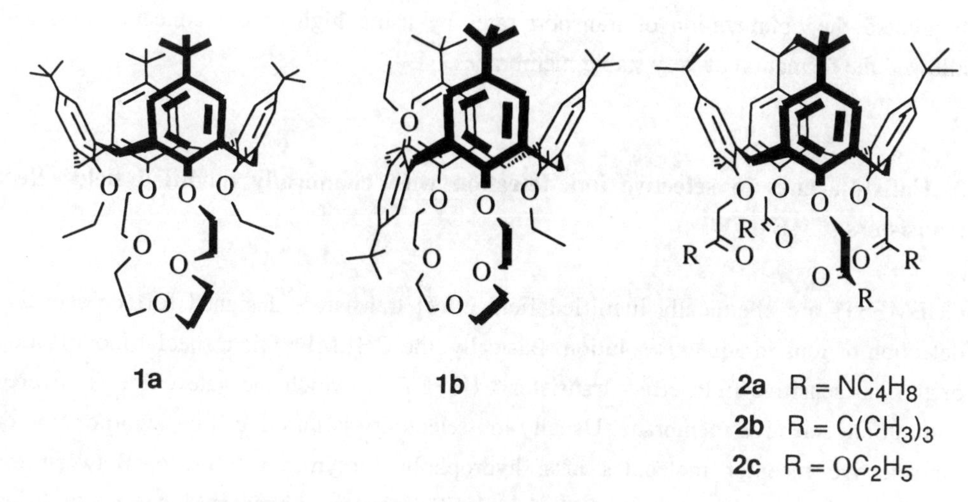

Figure 1. Calix[4]arene carriers in membrane transport.

of the carrier with respect to the different cations[9].

A more thorough investigation of the transport mechanism revealed that the potassium flux with calixcrown carriers is not always limited by diffusion of the complex through the membrane. In addition to the diffusional resistance of the membrane, a phase transfer coefficient has to be inserted into the model, which takes into account slow kinetics of release[7,12]. It was found that the kinetics of release depend very strongly on the conformation of the calixcrowns: the calixarene **1a** in the cone conformation shows pure diffusion limited transport whereas its partial cone analog **1b**, which bind the potassium cation much more strongly, gives much lower transport rates because of slow release[12]. This slow release of K^+ has a negative effect on the transport selectivity for potassium versus sodium. The selectivity can be improved by an increase of the operating temperature and membrane thickness under which condition the transport is determined by the diffusion limited regime[7]. Furthermore, it was found that the rate limiting step of transport could be influenced by variation of the membrane solvent and the co-transported anion. A remarkable effect of the solvent polarity on the kinetics was observed; lowering of the polarity led to an increase of the flux by accelerating the release process[12].

The above calix[4]arene carriers give stable membranes, at least under laboratory conditions. In order to stabilize the SLM even further, the calixarene carriers have been covalently linked to the membrane solvent, rendering them yet more lipophilic and increasing their compatibility with the membrane solvent[12]. This increased compatibility facilitated the optimization of transport rates by using high carrier concentrations and allowed the formation of very stable membranes.

3. Calix[4]arenes in selective ion detection with chemically modified field effect transistors (CHEMFETs)

CHEMFETs are chemically modified field effect transistors designed for the selective detection of ions in aqueous solution. Basically, the CHEMFET is a special modification of the ion sensitive field effect transistor[14] (ISFET) in which the gate oxide is covered with an ion selective membrane. Usually ion selectivity is introduced by incorporation of ion-selective receptor molecules in a hydrophobic polymer membrane. Between the sensing membrane and the gate oxide of the ISFET a buffered hydrogel layer is introduced which serves as inner electrolyte solution and eliminates pH sensitivity and CO_2 interference[15] (Figure 2). CHEMFETs have some important advantages compared to ion-selective membrane electrodes as fast response time, low noise level, small size, and possibilities for mass production. A disadvantage is, however, that due to the small membrane volume plasticizer and electroactive components leach relatively fast from the sensing membrane upon prolonged exposure to the contacting solution. Much research in our group has been focused on the development of a membrane architecture on the sensor device which combines high ion selectivity with long life time. The hydrogel, poly-hydroxyethylmethacrylate (polyHEMA) was covalently attached to the gate oxide surface by photopolymerization of the monomer on the gate oxide which was pretreated with methacryloylpropyltrimethoxysilane. The methacrylated siloxane groups on the surface ensure the covalent binding of the hydrogel. The polyHEMA layer is covered with a polymeric sensing membrane, of which the nature is determined by the desired physical and chemical properties.

The selectivity of the CHEMFET is determined by the selectivity of the ion receptor in the sensing membrane, the partition coefficients of the different ions over the membrane and aqueous phase and the relative concentrations of the ions in the sample solution.
An important part of our sensor research is directed to the design and synthesis of ion receptor molecules which have a high selectivity towards common interfering ions. The

membrane upon prolonged contact of the sensor with the aqueous solution. When these sensors are exposed to a continuous stream of tap water, the durability is less than a week. Therefore, a sensor with a long lifetime requires the chemical attachment of the sensing membrane to the semiconductor. Also the electroactive components in the sensing membrane must be prevented from leaching out, which can also be achieved by covalent attachment. To a sensing membrane suitable for these purposes a number of requirements have to be made: (*i*) The sensing membrane must be an elastomer (glass transition temperature ≤ 0 °C); (*ii*) The membrane must be hydrophobic, but with sufficient conductivity in order to obtain fast response times. For the permselectivity of the analyte ions, ionic sites of opposite charge must be present in the membrane. The presence of charged species in the membrane also lowers the electrical resistance of the membrane. (*iii*) Functionalization of the membrane should be possible to allow covalent binding of the electroactive components to the sensing membrane. (*iv*) In order to make sensor fabrication compatible with IC-technology, linkage of all components to the membrane matrix and attachment of the membranes to the gate oxide surface must occur by photo-polymerization processes.

We have developed sensing membranes based on polysiloxanes which meet these requirements[17]. Good response characteristics were obtained with a copolymer composed of dimethylsiloxane, cyanopropylsiloxane (for modulation of the polarity of the membrane), and methacryloxypropylsiloxane (enabling covalent linkage of receptor and ionic sites) [2]. The synthesis of these membranes is depicted in Scheme 1.

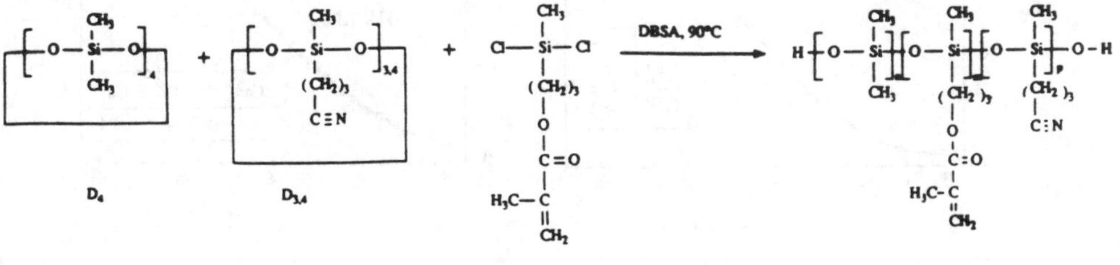

Scheme 1. Synthesis of a photopolymerizable polysiloxane.

The receptor molecules and anionic sites containing photopolymerizable side groups were incorporated in the polysiloxane membrane. Covalent attachment of polysiloxane to the polyHEMA layer was achieved by prior methacrylation of part of the hydroxyethyl groups of polyHEMA. Subsequent photopolymerization yields a stable polymer network in which

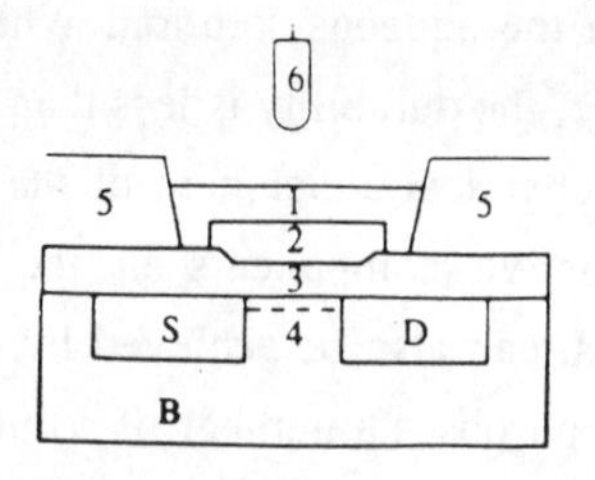

1. hydrophobic membrane
2. poly-HEMA hydrogel
3. gate oxide
4. channel
5. insulating resin
6. reference electrode

S: source
D: drain
B: bulk

Figure 2. Schematic representation of a CHEMFET

use of calix[4]arenes as molecular building blocks have been proven to be particularly versatile in this respect because functionalization of the phenolic oxygens in the cone conformation offers the unique possibility to orient four ligand chains at the same side of the molecule. Moreover, these molecules have a high lipophilicity which gives them a high affinity for the membrane phase.

Using calix[4]arene as building blocks, highly selective receptors have been developed for potassium, sodium, silver, lead, and a number of other cations. For the first test of their performance in CHEMFETs, these receptor molecules were incorporated in plasticized PVC membranes. The excellent response of the CHEMFETs containing the silver and lead receptor, respectively, is given in Figure 3.

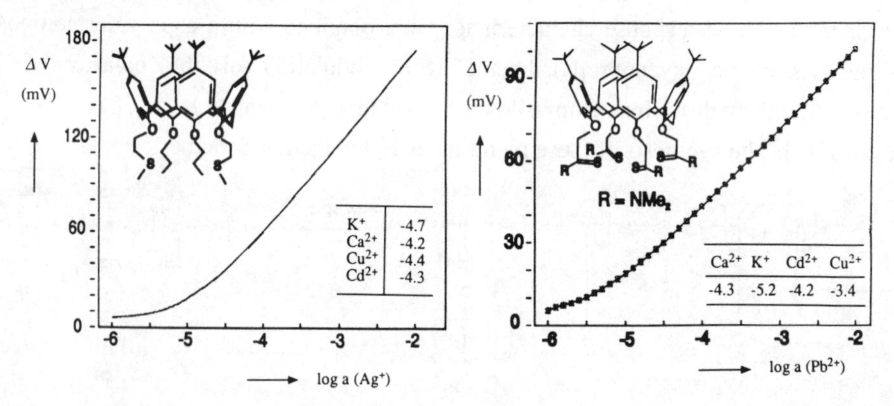

Figure 3. Response of Ag^{+}-selective CHEMFET (a) and Pb^{2+}-selective CHEMFET (b) in the presence of 0.1 M $Ca(NO_3)_2$ and 1 M KCl, respectively. Inset: Selectivity coefficients in the presence of 0.1 M of interfering ion.

Although plasticized PVC membranes with ion selective receptor molecules as sensing membranes give already quite satisfactory results[16] for a limited life time, the adhesion of the membrane to the surface is not stable and plasticizer and ionophore leach out the

all membrane components are covalently anchored. A schematic representation of the architecture of the sensing layers on top of the field effect transistor for a potassium-selective CHEMFET is given in Figure 4.

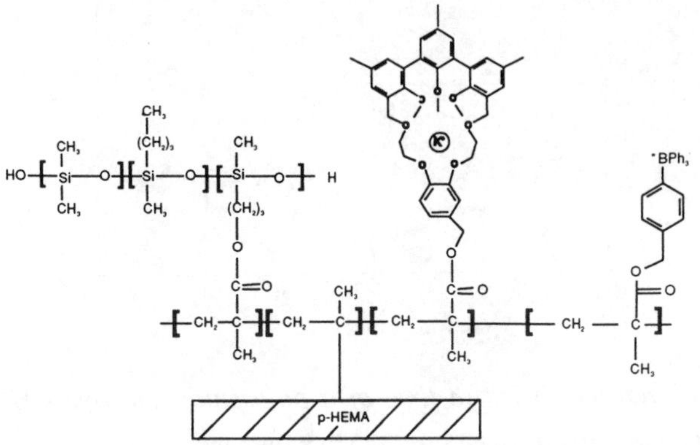

Figure 4. Schematic representation of a polysiloxane sensing layer covalently attached to a polyHEMA layer, with all electroactive components covalently bound in the sensing membrane.

In this schematic representation both ionophore and anionic sites are covalently bound to the membrane matrix. The effects of covalent attachment of the electroactive components in the membrane have been investigated sequentially. In Figure 5 the time-dependent sensitivity is given for potassium-selective CHEMFETs with polysiloxane membranes with potassium tetrakis[3,5-bis(trifluoromethyl)phenyl]borate as anionic sites and in which the ionophore is either free or covalently bound. This figure shows that due to the covalent linkage of the ionophore via the methacrylate group, leaching out from the membrane matrix is prevented and a stability of several months for the sensor in flow cells is obtained. In contrast, membranes with the same ionophore but without covalent attachment lose their activity quickly[18]. The effects of covalent immobilisation of both ionophore and ionic sites in the membrane matrix is given in Figure 6. This figure shows the response of sodium-selective CHEMFETs, based on polysiloxane copolymers with methacrylated sodium ionophore covalently bound by photopolymerisation, and in which the anionic site tetraphenylborate is either free (a) or covalently bound (b) to the membrane matrix.

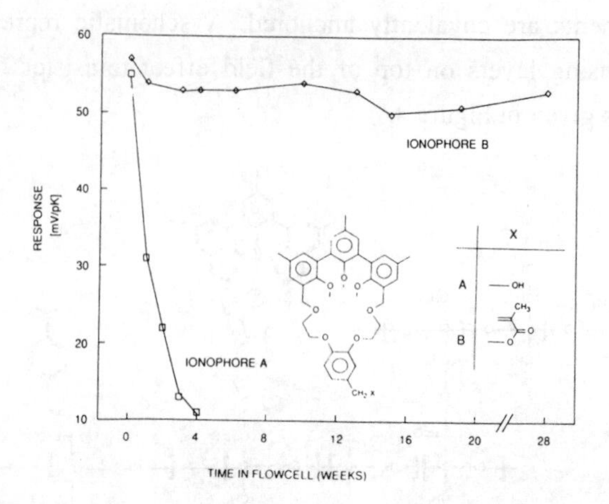

Figure 5. Response characteristics in time of a potassium-selective CHEMFET with the ionophore free and covalently attached to the siloxane membrane.

In the presence of 0.1 M of interfering ions good Nernstian reponse is obtained for activities of sodium of 8.10^{-4} M and higher. The sensor, with all electrical active components covalently bound to the membrane, shows stable response behavior after continuous exposure to 0.1 M NaCl solution for already more than 90 days[19].

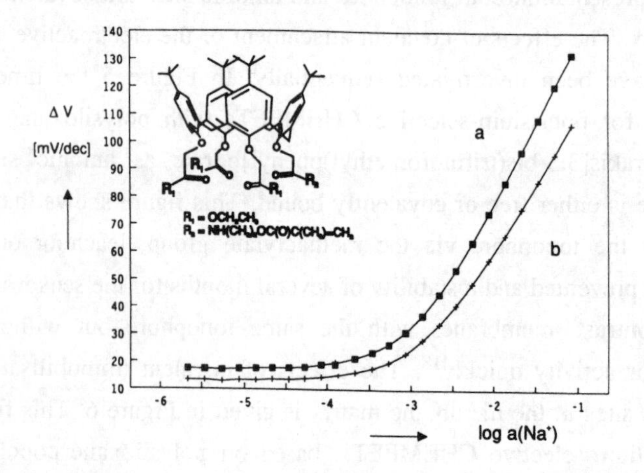

Figure 6. Response of sodium-selective CHEMFETs in the presence of 0.1 M potassium chloride. The ionophore is covalently bound and the tetraphenylborate anions are either free (a) or covalently bound (b) to the membrane matrix.

4. Calix[4]arenes in non-linear optics materials

Non-linear optics is the basis of the strongly emerging photonics technology with important applications in, for example, data processing and data storage. Materials for non-linear optics require a high nonlinear susceptibility ($\chi^{(2)}$), by which optical properties like refractive index and absorbance vary with the intensity of light. Large nonlinear susceptibilities can only be obtained for materials that contain chromophoric units which have a large hyperpolarizability (β), and of which the orientation along their molecular axis is non-centrosymmetrically distributed in the material. Urea crystals, for example, show nonlinear optical behavior because the orientation of the polarizable dipole of the molecules in the crystal is non-centrosymmetric. However, the hyperpolarizability of the urea molecule is too small and the processability of the crystals is too low to allow widespread application. Organic molecules with both electron-donor and electron-acceptor substituents attached to a π-conjugated system (D-π-A systems) are more promising for use in nonlinear optics because of their relatively high β values and their potential ease of processability, especially when incorporated in polymeric materials. Extension of the conjugated system in these D-π-A molecules increases the β, but unfortunately this is accompanied by a long wavelength shift of the absorbance, thereby limiting the applicability for frequency doubling in, e.g. blue laser devices.

Recently, we have successfully exploited the preorganized orientation of the aromatic moieties in calix[4]arene to develop D-π-A molecules (Figure 7) with relatively high dipole moments and β values, without showing the undesired shift of absorption to the blue wavelength region[20]. The β values of the calix[4]arenes given in Table 1 are 2-3 times higher than the monoaromatic analogs. The β value of the tetrakis(nitrostilbene) derivative **6** is comparable with that of *N,N*-dimethylamino-4-nitrostilbene (DANS).

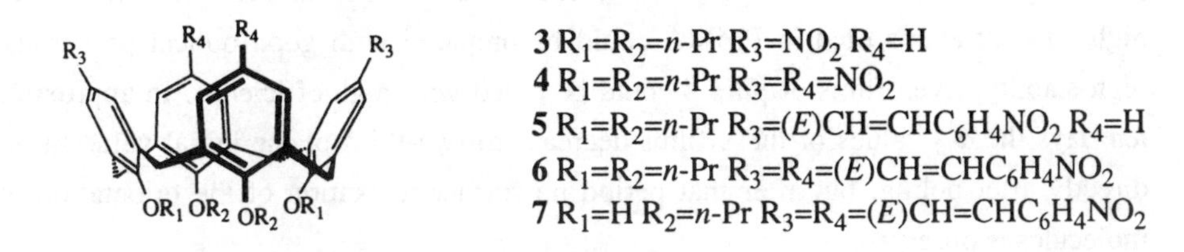

Figure 7. Donor-π-acceptor substituted calix[4]arenes.

Table 1. Dipole moments μ, hyperpolarizability β, and charge transfer band maximum λ_{CT} of donor-acceptor substituted calix[4]arenes[a].

Cmpnd	μ [D]	β [10^{-30}esu]	λ [nm]
3	4.5	16	308
4	13.8	30	291
5	5.2	113	386
6	15.3	280	370
7	13.0	161	356

[a]Measurements were performed in chloroform. The β values (± 20%) were obtained from EFISH measurements with a fundamental wavelength of 1064 nm.

Non-centrosymmetrical orientation of the chromophoric units can be achieved by applying a high electrical DC field over a thin film of the material which orients the dipoles of the molecules towards the direction of the field (poling). The electrical field orients the dipoles of the molecules towards the direction of the field. Relaxation of the poling orientation can be prevented as much as possible by poling at high temperature in a polymer of high glass transition temperature and subsequenly cooling down of the material in the poling field, or by extensive crosslinking of the polymer after poling. Compounds **4, 6,** and **7** have relatively high dipole moments which facilitate poling of these compounds in thin films of the material[21]. In Figure 8 the frequency-doubling coefficients d_{33} are given of various polymeric materials containing different weight percentages of **4**, after corona poling with 200-300 V.μm^{-1} at 110 °C. Films with very high concentrations of **4** in PMMA could be prepared with good optical properties and high stabilty. Even films of pure **4** could be poled with high efficiency. In approximately ten days the d_{33} values of these films decrease to two-third of the initial value measured directly after poling, but after that period no further relaxation of the orientation of the molecules is observed.

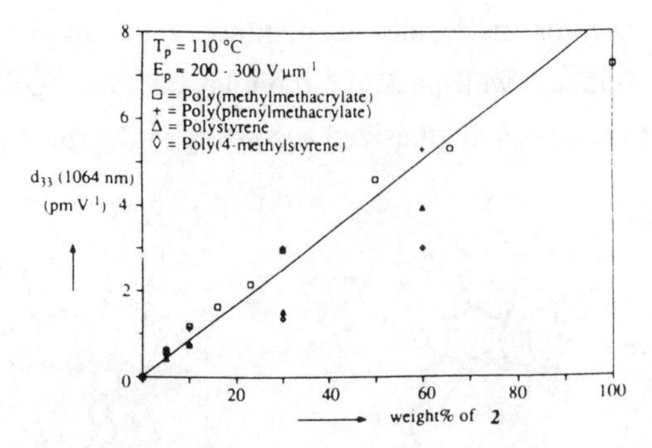

*Figure 8. Frequency doubling coefficient d_{33} as function of the weight% of **4** in different polymeric matrices.*

The frequency doubled signal from the films is now already stable for more than two and a half years. Also for the tetrakis(nitrostilbene) derivative **6**, which has even a higher dipole moment than **4** high d_{33} values have been observed when this compound is poled at 110 °C in PMMA. The frequency doubling coefficient increases linearly with the amount of **6** in the polymer and at 60 weight% of **6** a d_{33}(1064 nm) of 10.8 pm.V^{-1} is measured. This value can be further increased by poling of films of neat **6** at more elevated temperatures. Poling of **6** at 210 °C (which is still 100 °C below its decomposition temperature) gives a d_{33} value of 23 pm.V^{-1} which, after 35% decrease in the first week, remains stable for already more than 1.5 year.

5. Monolayers of resorcin[4]arene receptor adsorbates on gold

An efficient control of molecular interactions and molecular orientation in materials can be obtained by self assembly of molecules in organized mono- and multilayers. Organosulfur compounds with long hydrocarbon chains can form stable monolayers on gold surfaces because of the strong sulfur-gold interaction and the Van der Waals interactions between the hydrocarbon chains[22]. Several self-assembled monolayers of sulfur-adsorbates with functional endgroups (COOH, OH) have been reported. Particularly interesting for sensing purposes are monolayers with cavities protruding to the outside which are able to selective binding of neutral organic molecules. In this category, only one cyclodextrin-based adsorbate is reported, but monolayers of this adsorbate showed many defects, probably due to the inbalance of the cross-sectional area used by the alkyl chains and the cavity headgroup in the monolayer. We have designed and synthesized a number of resorcin[4]arene receptor molecules containing four alkyl sulfide chains which self-assemble into stable, well-packed monolayers on gold substrates[23]. The resorcin[4]arenes **8a-c** have been synthesized according to the route given in Scheme 2.

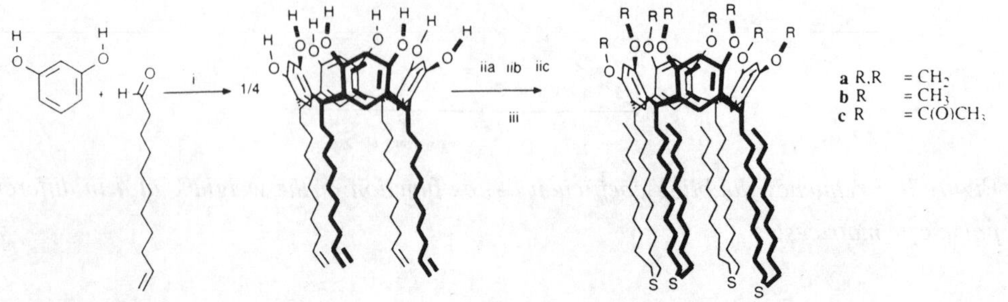

Scheme 2. Synthesis of the resorcin[4]arene adsorbates 8a-c.

The four resorcinol units form a conelike cavity which is supported by four dialkylsulfide chains connected to the methine linkages of the macrocycle. The alkyl chains at the resorcinol units are all oriented parallel to each other and perpendicular to the annulus of the cavity. In compound **8a** the resorcinol oxygens are linked by a methylene bridge, providing a receptor moiety with a rigid cavitand structure. In **8b** the oxygens of the resorcinol unit are methylated, and in **8c** these are acetylated, yielding flexible cavities of different polarity. Immersion of clean gold substrates in ethanol:chloroform (3:1v/v) solutions of the adsorbates at 60 °C for 13 h yielded very stable, well-packed monolayers of **8a-c**. Upon binding of the sulfur atoms to the gold surface the terminal alkyl chains of the sulfide are looped back and the total lateral area of the densily packed underlying alkyl chains (160 Å) is slightly larger than the area occupied by the cavity head group (140 Å). Monolayers formed by adsorption at room temperature are less well organized, probably because the alkyl chain packing is then kinetically, rather than thermodynamically, controlled.

A variety of surface analytical techniques have been used to confirm the well-organized structure of the monolayers as depicted in Figure 9.

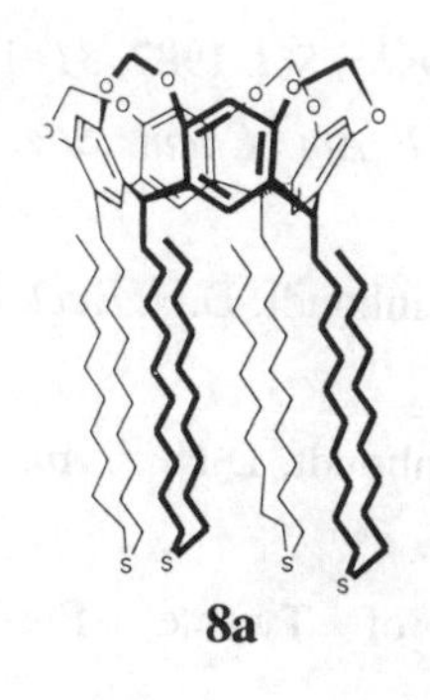

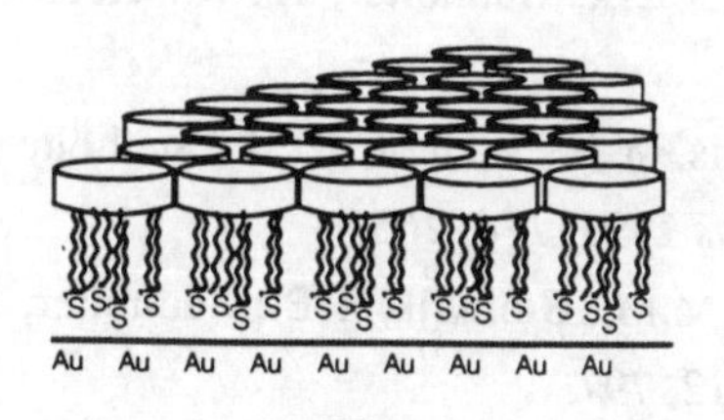

*Figure 9. Schematic representation of monolayers of **8a** on gold.*

X-ray photoelectron spectroscopy (XPS) indicated that the oxygen-rich cavity resides at the outer interface and that the sulfur atoms are located near the gold surface. The thickness of the monolayers, measured by surface plasmon spectroscopy is 22(±2) Å for **8a** and **8c**, and 24(±2) Å for **8b**, which is in accordance with the expected values. Electrochemical measurements show that the heterogenous electron transfer between a gold electrode covered with a monolayer of **8a** and a 1 mM $K_3Fe(CN)_6$, 0.1 M K_2SO_4 solution is effectively blocked, indicating the absence of defects in the monolayer structure.

Monolayers of the cavitand **8a** have been successfully used for the detection of small neutral organic molecules[24]. Therefore, gold electrodes of a quartz microbalance oscillator were covered with monolayers of **8a** and this device was exposed to ppm concentrations of perchloroethylene and other small organic molecules (chloroform, trichloroethylene, and toluene) in synthetic air. Fast, reversible, and relatively large shifts in the fundamental oscillation frequency are observed with remarkable selectivity (> 7-25) for perchloroethylene. In contrast, monolayers of didecyl sulfide exhibit almost no detectable signal. These results indicate that perchloroethylene molecules incorporate into the molecular cavities of the cavitand monolayers. Further studies of these supramolecular assemblies with surface recognition sites are in progress.

REFERENCES

1 Lamb, J.D.; Bruening, R.L.; Izatt, R.M.; Hirashima, Y.; Tse, P.-K.; Christensen, J.J. *J. Membr. Sci.* **1988**, *37*, 13.

2 Takeuchi, K.; Nakano, M. *J. Membr. Sci.* **1987**, *34*, 19.

3 Danesi, P.R.; Reichley-Yinger, L.; Rickert, P.G. *J. Membr. Sci.* **1987**, *31*, 117.

4 Stolwijk, T.B.; Sudhölter, E.J.R.; Reinhoudt, D.N. *J. Am. Chem. Soc.* **1989**, *111*, 6321.

5 Nijenhuis, W.F.; Walhof, J.J.B.; Sudhölter, E.J.R.; Reinhoudt, D.N. *Recl. Trav. Chim. Pays-Bas* **1991**, *110*, 265.

6 Wienk, M.M.; Stolwijk, T.B.; Sudhölter, E.J.R.; Reinhoudt, D.N. *J. Am. Chem. Soc.* **1990**, *112*, 797.

7 Reichwein-Buitenhuis, E.G. *Thesis*, University of Twente, Enschede, The Netherlands, **1994**.

8 Nijenhuis, W.F.; Buitenhuis, E.G.; de Jong, F.; Reinhoudt, D.N. *J. Am. Chem. Soc.* **1991**, *113*, 7963.

9 van Straaten-Nijenhuis, W.F. *Thesis*, University of Twente, Enschede, The Netherlands, **1992**.

10 van Straaten-Nijenhuis, W.F.; van Doorn, A.R.; Reichwein, A.M.; de Jong, F.; Reinhoudt, D.N. *J. Org. Chem.* **1993**, *58*, 2265.

11 van Straaten-Nijenhuis; W.F.; de Jong, F.; Reinhoudt, D.N.; Thummel, R.P.; Bell, T.W.; Liu, J. *J. Membr. Sci.* **1993**, *82*, 277.

12 Visser, H.C. *Thesis*, University of Twente, Enschede, The Netherlands, **1994**.

13 Rudkevich, D.M.; Brzozka, Z.; Palys, M.; Visser, H.C.; Verboom, W.; Reinhoudt, D.N. *Angew. Chem.* **1994**, *in press*.

14 Bergveld, P.; Sibbalt, A. *Comprehensive Analytical Chemistry, Vol XXIII*, Elsevier Amsterdam, 1988.

15 (a) Reinhoudt, D.N.; Sudhölter, E.J.R. *Adv. Mater.* **1990**, *2*, 23.; (b) Sudhölter, E.J.R.; van der Wal, P.D.; Skowronska-Ptasinska, M.; van den Berg, A.; Bergveld, P.; Reinhoudt, D.N. *Anal. Chim. Acta* **1990**, *230*, 67.

16 (a) Brunink, J.A.J.; Haak, J.R.; Bomer, J.G.; McKervey, M.A.; Harris, S.J.; Reinhoudt, D.N. *Anal. Chim. Acta* **1991**, *254*, 75; (b) Cobben, P.L.H.M.; Egberink, R.J.M.; Bomer, J.G.; Bergveld, P.; Verboom, W.; Reinhoudt, D.N. *J. Am. Chem. Soc.* **1992**, *114*, 10573.

17 Reinhoudt, D.N. *Biosensors and Chemical Sensors. Optimizing Performance Through Polymeric Materials (ACS Symposium Series 487)*, Edelman, P.G.; Wang, J., Eds., American Chemical Society, Washington, DC, **1992**, 202.

18 Verkerk U.H.; van den Vlekkert H.H.; Reinhoudt D.N., Engbersen J.F.J.; Honig G.W.N.; Holterman H.A.J. *Sensors and Actuators B* **1993**, *13-14*, 221.

19 (a) Brunink, J.A.J.; Bomer, J.G.; Engbersen, J.F.J.; Verboom, W.; Reinhoudt, D.N. *Sensors and Actuators B* **1993**, *15-16*, 195; (b) Brunink, J.A.J.; Lugtenberg R.J.W.; Brzozka, Z.; Engbersen, J.F.J.; Reinhoudt, D.N. *J. Electroanal. Chem.*, in press.

20 Kelderman, E.; Derhaeg, L.; Heesink, G.J.T.; Verboom, W.; Engbersen, J.F.J.; Van Hulst, N.F.; Persoons, A.; Reinhoudt, D.N. *Angew. Chem. Int. Ed. Engl.* **1992**, *31*, 1075.

21 Kelderman, E.; Heesink, G.J.T.; Derhaeg, L.; Verbiest, T.; Klaase, P.T.A.; Verboom, W.; Engbersen, J.F.J.; Van Hulst, N.F.; Clays, K.;Persoons, A.; Reinhoudt, D.N. *Adv. Mater.***1993**, *5*, 925.

22 (a) Bain, C.D.; Whidesides, G.M. *Adv. Mater.* **1989**, *1*, 506; (b) Dubois, L.H.; Nuzzo, R.G. *Ann. Rev. Phys. Chem.* **1992**, *43*, 437; (c) Lee, T.R.; Laibinis, P.E.; Folkers, J.P.; Whitesides, G.M. *Pure & Appl. Chem.* **1991**, *63*, 821.

23 Thoden van Velzen, E.U.; Engbersen, J.F.J.; Reinhoudt, D.N. *J. Amer. Chem. Soc.* **1994**, *116*, in press.

24 Schierbaum, K.D.; Weiß, T.; Thoden van Velzen, E.U.; Engbersen, J.F.J.; Reinhoudt, D.N.; Gopel, W., to be published.

APPROACHES TO SYNTHESIS BASED ON NON-COVALENT BONDS

George M. Whitesides, Eric E. Simanek, and Christopher B. Gorman
Harvard University
Department of Chemistry, Cambridge, MA 02138

Abstract

Two self-assembling systems — SAMs (self-assembled monolayers) and aggregates based on CA•M (cyanuric acid•melamine lattice) — are not unique in non-covalent synthesis, but they illustrate many of the ideas of non-covalent synthesis, and suggest the ways in which this area differs in its philosophy and objectives from covalent synthesis.

1. Targets for Synthesis

What are the current targets for organic synthesis? The traditional activity of organic chemistry has been to synthesize molecules. The history of organic synthesis is reflected in its target molecules: dyes, polymers, specialty chemicals, natural products and pharmaceuticals have each been a favored subject at some period in the development of synthetic chemistry. A second function of organic synthesis has been to develop technology — often using molecules isolated from nature as stimuli — that is generally useful in synthesis.

The targets of synthetic activity — that is, the areas in which new methods and new strategies are most needed — are defined by the areas of science and technology that are themselves especially active and that rely heavily on molecules and materials. Four areas (among others) seem especially to define the fields that require new syntheses and synthetic technology:

C. Chatgilialoglu and V. Snieckus (eds.), Chemical Synthesis, 565–588.

1.1. MEDICINE

The pharmaceutical industry continues to require a high level of expertise in organic synthesis. While the specific classes of compounds that are required in medicinal chemistry change with time, the strategies used to synthesize them represent an extension of paradigms that are now familiar in organic synthesis. Increasing emphasis and creativity is being placed on the development of low-cost processes, on processes that yield enantiomerically pure compounds, and on environmentally friendly processes. These challenging problems make certain that chemists will continue to be an important part of medicinal chemistry for as long as new drugs are developed.

1.2. BIOLOGY

Biology is now posing a range of new types of challenges to synthesis. Current efforts in the synthesis of molecules relevant to biology are concentrated on the major classes of biomolecules: polypeptides, proteins, nucleic acids and oligosaccharides. The need for efficient methods for the synthesis of biomolecules has been a valuable stimulus to synthetic chemistry. This motivation has resulted in several new methodologies including the incorporation of synthetic methods based on enzymes and fermentation into standard synthetic technology, and in the generation of new methodologies applicable to water-soluble and charged species, and to linear macromolecules such as proteins and polynucleotides. It seems probable that biological synthesis, in the future, will be a hybrid of chemical and biochemical methods, with the particular set of methods chosen depending on the efficiency and appropriateness of each to the problem at hand. Advanced targets for organic synthesis relevant to biology — for example, the synthesis of catalytically functional aggregates, of self-replicating structures, or of simple viruses — are still too complex to attract the attention of other than the most adventurous.

1.3. MATERIALS SCIENCE

Participation in materials science requires a more substantial change in attitude for organic synthesis than does biology. In biology, the targets are molecules with unfamiliar properties (for chemists) in that they have high molecular weights, are water soluble and are often highly charged. They are, nonetheless, still *molecules*. In materials science, the targets of synthesis are often molecules that either form or perform as *aggregates*. In many materials systems, the properties of interest may depend absolutely on aggregation or collective behavior. For example, liquid crystals, organic conductors and polymer matrices for high-performance composites all depend on the behavior of collections of molecules: the characteristics of single molecules are important only to the extent that they contribute to collective behavior. In synthesis directed toward materials, the desired physical properties cannot be dissociated from synthesis, and a more complete understanding of the relationships between synthesis, processing and properties are required.

1.4. ENVIRONMENTALLY FRIENDLY SYNTHESIS

Probably the most serious problem now facing the chemical industry is to develop environmentally compatible technology for synthesis (especially large-scale synthesis). The field of environmentally friendly synthesis is a peculiar one. It is unarguably important, and it presents a range of interesting and engaging challenges. Despite these characteristics, very few academic synthetic chemists are working in it. Why? One reason is that the attention of the community of synthetic chemists has not yet been caught; a second is that there are surprisingly few leads to appropriate processes, and surprisingly few really good new ideas; a third is that economic and regulatory considerations are an integral part of all environmental synthesis, and academic chemists have traditionally been uncomfortable in problems requiring an understanding of economics. Whether the field of environmental synthesis is ultimately attacked by chemistry or by chemical engineering remains to be seen.

2. Covalent and Non-covalent Synthesis

Organic synthesis has been dominated by a single intellectual paradigm — that of "covalent synthesis." In this paradigm, molecules — that is, collections of atoms connected by strong, kinetically stable covalent bonds — are constructed in a series of steps focused on stepwise, efficient formation of covalent bonds. Are there alternatives to this paradigm? Both biology and materials science are replete with instances in which the crucial structural elements involve *non-covalent* bonding: examples include interactions between proteins in aggregates, the interactions that give molecular and liquid crystals their structures and properties, and the interactions that hold together the two strands of DNA itself. In each of these types of structures, covalent interactions are important, but they are not sufficient: it is often possible to modify or even cleave the amino acids in a protein and retain function, but the same polypeptide sequence in native and denatured states have very different function. In short, both covalent and non-covalent interactions are important.

One alternative to covalent synthesis as a paradigm in organic synthesis is its semantic converse — *non*-covalent synthesis. This paradigm is as important in complex systems as covalent synthesis. Instead of energetically strong interactions dictated by bond enthalpies, non-covalent synthesis utilizes weaker, non-covalent interactions that are governed by equilibrium thermodyamics: entropic considerations become as important as enthalpic ones. Non-covalent synthesis will never *replace* covalent synthesis: the components of a complex structure will undoubtedly always be based on covalent bonds. The *assembly* of these components into a functional aggregate can, however, be accomplished by either covalent or non-covalent methods.

What are the advantages of non-covalent synthesis? One way of addressing the question is to ask where non-covalent processes are already established to be important. One answer — and not necessarily the only answer — deals with the situation when very large structures are required, and when the function of these structures must be controlled at room temperature. There is no practical method of making a molecule of the three-dimensional complexity of a protein by covalent synthesis: two-dimensional synthesis based on a reactions with very high yields, followed by non-covalent assembly (folding) is a more practical procedure. The resulting structures are, of course, unstable at even moderately elevated temperatures (one wonders what strategies life might have evolved if it had been necessary to survive environments of 500° C!). Similarly, the synthesis of a macroscopic crystal, one bond at a time, is entirely out of the question. Non-covalent synthesis is a strategy that is suited for preparation of these ensembles, where the yield losses that inevitably accompany covalent synthesis are unacceptable. Non-covalent bonding — aggregation, association, folding, annealing — is an efficient strategy, and one that can, in principle, proceed in 100% yield!

The issue of control is a more complex subject, and often has (at least in biological contexts) to do with transitions between different but well-defined conformational states. Although there are strong arguments for non-covalent structures (and non-covalent synthesis) in systems subject to modulation and control at room temperature, we will not address this subject here other than to note the obvious: only non-covalent structures based on bonds of strengths comparable to kT can be interconverted by thermal processes.

2.1. PRECEDENTS FOR NON-COVALENT SYNTHESIS

There are a wide range of precedents for non-covalent synthesis. Biological structures present a set of strategies based on structures in which hydrogen bonds and hydrophobic interactions are crucial [1]. Molecular crystals provide another very important set (and one in which the basic rules are still not defined!) [2-5]. Liquid crystals [6], black lipid films [7], micelles and liposomes [8], clathrates and co-crystals [9], bubble rafts [10], and phase-separated polymers [11], provide others. Not surprisingly, non-covalent interactions between covalently structured molecules are important and common throughout complex systems; what is missing is a rational process for using these interactions in synthesis.

2.2. THERMODYNAMIC AND KINETIC CONSIDERATIONS IN NON-COVALENT SYNTHESIS

A key idea in non-covalent synthesis is the importation of new ideas of bonding into synthesis. Covalent synthesis is based on the use of strong, kinetically stable networks of bonds to assemble kinetically metastable structures. A wide range of bond-types — coordination bonds, hydrogen bonds, charge transfer interactions, hydrophobic interactions, charge-charge interactions — are available for the construction of new structures, but have been largely ignored as explicit components of *synthetic* strategies (although they have, of course, been the object of extensive interest in physical organic chemistry, molecular recognition, biochemistry, solid-state chemistry and other areas). Other types of considerations are also important: for example, considerations of enthalpy dominate considerations of the energetics of covalent synthesis; in non-covalent synthesis, enthalpy and entropy are both important. When entropy enters considerations of synthetic strategy, ideas such as preorganization become important [12] .

3. Systems

We have focused our work in non-covalent synthesis on two systems: self-assembled, two-dimensional monolayers based on ordered structures of alkanethiolates chemisorbed on gold, and soluble three-dimensional aggregates held together by hydrogen bonds, and based on the lattice formed by the 1:1 aggregate of cyanuric acid and melamine (CA•M).

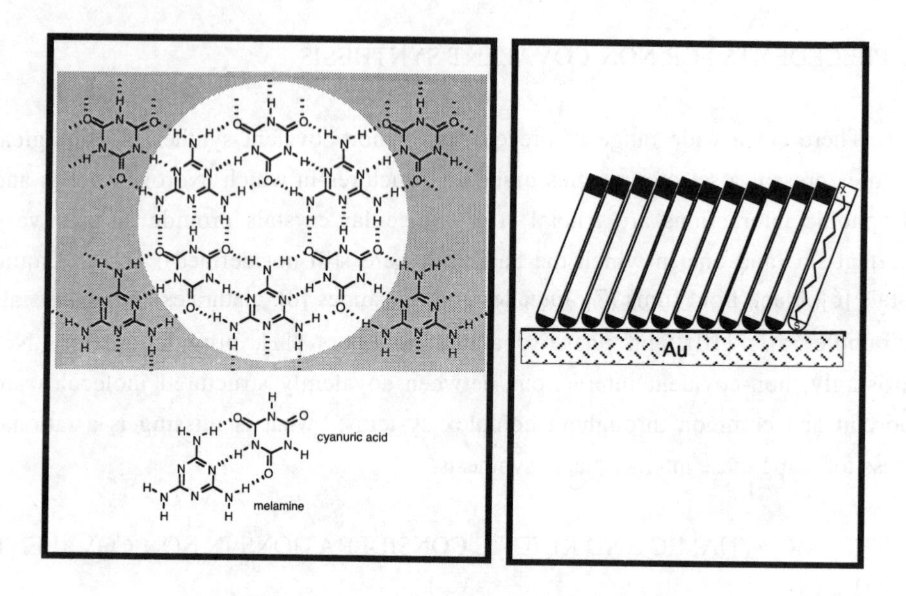

Figure 1. Representations of the CA•M lattice (left) and a SAM (right, X = CH_3, OH, COOH, CN, etc.)

3.1. SELF-ASSEMBLED MONOLAYERS

Self-assembled monolayers (SAMs) of long chain organic compounds on surfaces of metals and metal oxides are increasingly useful in applications which require structurally well-defined substrates. They are easily prepared in a wide variety of structures (thicknesses, degree of order, chain orientation with respect to the surface) and a range of functional groups can be incorporated into them. Several systems of SAMs have been investigated extensively, including those formed from alkanethiolates on gold, silver [13, 14] and copper [13, 15], alkanecarboxylic acids on alumina [16-18], alkanehydroxamic acids on metal oxides [19], alkanephosphonates [20-24] on zirconium and aluminum oxides, alkyltrichlorosilanes or alkoxysilanes on silica [25-28] and Langmuir-Blodgett films on a range of supports [29, 30]. Of these, SAMs of alkanethiolates on gold — a system first used for rational organic surface chemistry by Nuzzo, Allara and coworkers [31-33] — has attracted the most attention. This system is exceptionally easy to work with, and is relatively stable. More importantly, it is very easy to introduce complex functionality onto a surface using it, and it thus offers a high potential for complex synthesis. SAMs are already significantly advanced as systems in materials science, and applications for them are appearing rapidly.

3.2. CYANURIC ACID•MELAMINE LATTICE

A second type of problem that we have examined is the design and synthesis of complex, high-molecular weight aggregates of molecules held together by networks of hydrogen bonds. We have chosen to work with aggregates derived from the lattice of hydrogen bonds formed from the 1:1 aggregate of cyanuric acid and melamine (CA•M) [34] . This system has the advantage that synthesis of the molecules required is relatively straightforward, and that the components have high symmetry. A range of structures based on this system have been prepared. Simply mixing arbitrary derivatives of cyanuric acid and melamine together usually does not generate the desired species containing the cyclic CA_3M_3 "rosette": either the components dissociate in solution, or they precipitate as insoluble disordered aggregates or as linear or crinkled tapes.

Two types of designs have been successful in generating stable aggregates based on the CA•M lattice (Figure 2, left). In one strategy, one set of components (typically the melamines, since they are easier to manipulate synthetically than are the cyanuric acids) is "preorganized" by attachment to a common "hub". This strategy *encourages* formation of the desired aggregates by reducing the change in translational (and perhaps conformational) entropy required to form the aggregate (relative to the corresponding change for the components independently free in solution). The second strategy introduces large substituents into the components to *discourage* formation of tapes and

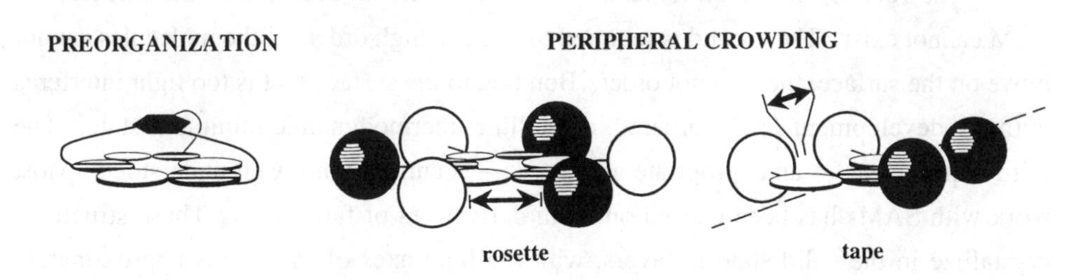

Figure 2. Schematic representations of *preorganization* of three melamine by covalent attachment to a center hub (left) and the formation of a rosette structure due to large groups (depicted here as spheres) by *peripheral crowding* (right). Melamines and cyanuric acid are represented as disks.

other, non-rosette structures by introducing large unfavorable steric interactions into these structures (Figure 2, right). There is, of course, a rich literature in the subject of molecular recognition, with extensive work in systems that exhibit of the phenomena of molecular recognition by Hamilton [34], Rebek [35], Stoddard [36], Lehn [37], Sauvage [38], Zimmerman [39], Breslow [40], Still [41] , Kunitake [42] , and Cram [12, 43], among others.

4. SELF-ASSEMBLED MONOLAYERS

SAMs illustrate a strategy for synthesis based on the idea of reduction in dimensionality. The generic idea underlying SAMs is to use a surface, or some other two-dimensional or pseudo two-dimensional system, as a template and to assemble molecules on it in reasonably predictable geometry using appropriate coordination chemistry to connect the surface with the adsorbed molecules. For this strategy to work, one needs:

- A suitable coupling reaction to connect the adsorbate to the surface.
- Sufficient in-plane mobility for the adsorbate (at some stage in the processes that form the SAM) to allow it to order and to form a highly structured crystalline or quasicrystalline surface phase.
- A geometry for the adsorbate that is compatible with an ordered surface phase.
- The capability to pattern the system in the plane of the surface.

The first of these requirements is obvious: without coupling to the surface, the SAM cannot exist. The second is required to achieve high order. If the molecules cannot move on the surface, they cannot order. Bonding to the surface that is too tight interferes with the development of the ordered, crystalline, thermodynamic minimum state. The third requirement — an appropriate geometry — is currently not well understood. Most work with SAMs has been carried out with derivatives of fatty acids. These structures crystallize in the solid state in layers, with the long axes of the chains approximately parallel. Placing the same molecule in a SAM effectively freezes one plane from the crystal on an appropriate surface. How many other structures that form layers in the crystal can be transferred to a SAM remains to be established; there is virtually nothing known about the SAMs that might be drawn from most classes of organic molecules.

The fourth requirement — for techniques to form patterns in the plane of the SAM — is important for the development of molecule-like structures on surfaces. Consider a circular patch of SAM of a C_{16} hydrocarbon on a surface having a radius of 50 nm (this radius is now achievable using relatively straightforward techniques which we describe below). This patch will contain approximately $4x10^4$ molecules, and have a "molecular weight" of approximately 10^7 D. This size is in the same range as very large polymers, and thus is approaching dimensions familiar to organic synthetic chemists (albeit from the upper range, rather than from the more familiar smaller sizes). Patterning will, we believe, be an essential part of connecting self-assembling and non-covalent synthesis with more conventional methods of synthesis, and also will be essential for many applications requiring structures in the meso range of sizes.

The chemistry and structures of SAMs of alkanethiolates on gold have been extensively studied, and need not be reviewed here [44]. In brief, when fully equilibrated and in their most stable form, these SAMs seem to be two-dimensional quasicrystals, with the sulfur headgroups epitaxial on the gold surface. A 30° tilt of the trans-extended alkane chains brings these chains into van der Waals contact. Functional groups present on the termini of the chains are exposed to the solution. Conformational disorder in the system is concentrated in the terminal regions of the chains.

4.1. CHARACTERIZATION

Characterization of these structures can be accomplished using a number of techniques, with the most useful being XPS [45], polarized infrared external reflectance spectroscopy (PIERS) [46-48], measurement of contact angles, ellipsometry [49, 50] and, increasingly, STM/AFM [51-54]. Computation has also been very useful in understanding the order in these structures [55, 56]

4.2. MESO-SCALE STRUCTURES: MICRO-CONTACT PRINTING (μCP)

We have developed a number of techniques for patterning SAMs in the plane of the monolayer [57-64]. The objective of these methods is to provide procedures for achieving true meso-scale fabrication — that is, fabrication leading to structures with dimensions in the range of 10 nm to 10 μm — using techniques available in synthetic chemical laboratories. The most versatile of these techniques is one based on contact printing (which we call microcontact printing, μCP) [60, 64].

In microcontact printing, a pattern of a SAM (typically that from hexadecanethiol, since it performs well in μCP) is formed by a technique in which an elastomeric stamp is "inked" with the thiol, and then brought into contact with the surface of the gold. Features present on the stamp are transferred to patterns of SAM on the surface with remarkable fidelity: in favorable circumstances, it is possible to produce patterns with feature sizes of 200 nm, and with edge resolution for these features of approximately 50 nm. Once the initial pattern has been produced, the unpatterned areas can be filled in by exposure to a solution of another alkanethiol, an additional pattern can be generated by stamping, or the initial pattern can be used as a mask to protect the gold film from etchants. This technique can, therefore, be used either to generate patterns of SAMs on a continuous gold film, or to generate discontinuous patterns of gold.

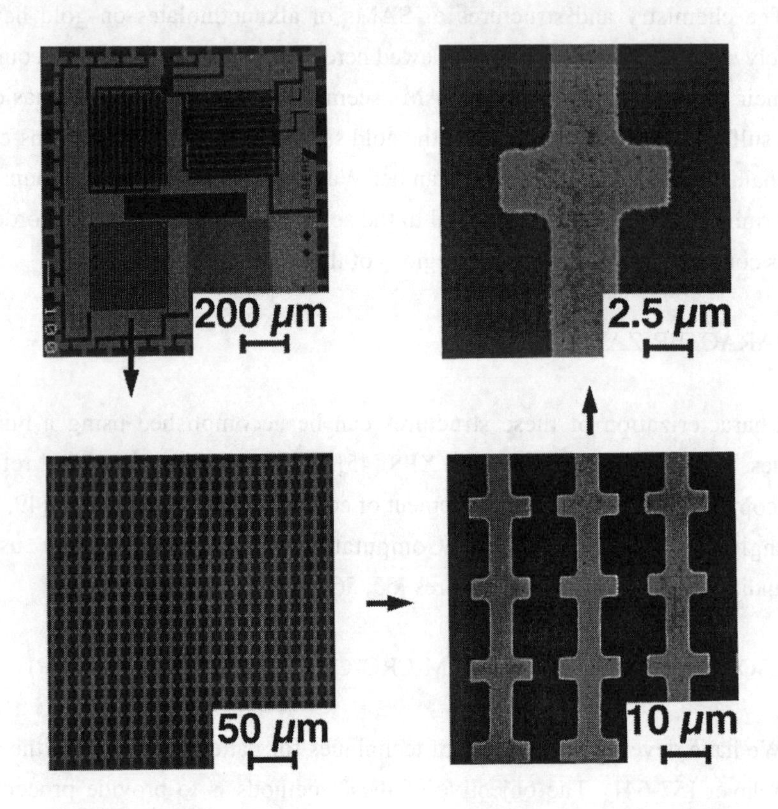

Figure 3. SEM micrographs of a pattern formed by μCP followed by chemical etching to remove gold not protected by the SAM.

The stamp used in μCP is most commonly fabricated by photolithographic procedures. An image is transferred into a film of photoresist, the exposed resist is developed to produce a three-dimensional structure, and this structure is then covered with the prepolymer from which the stamp is to be formed (typically poly(dimethylsiloxane), PDMS). The PDMS is allowed to cure, and then peeled from the master.

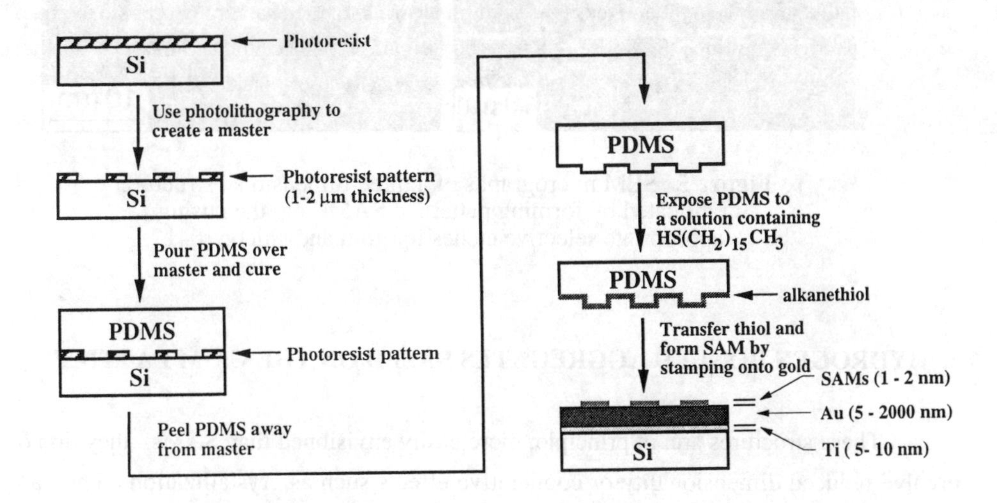

Figure 4. Schematic of the process used for μCP.

Because PDMS is an elastomer, it can generate conformal contact with a surface that is rough. We have not defined its ability to conform to steps, irregular surface topologies and imperfections on the surface of a gold film, but the fidelity of the contact printing does not seem to be limited by this capability. The ability of μCP to generate sharp edges depends on the production of *autophobic* SAMs: in these systems, the extent of reactive spreading of the alkanethiol is limited by the fact that the SAM that it forms is not wet by it, and definition of the edge of the SAM is determined by some combination of the rate of reactive spreading and the rate of formation of an organized, autophobic SAM.

4.3. EXTENSIONS TO QUASI THREE-DIMENSIONAL STRUCTURES.

Patterned gold structures on the surface of silicon/silicon dioxide can be used, in combination with anisotropic etches for silicon, to form three-dimensional structures of the surface of the silicon (Figure 5). Although there are serious limits to the types of structures that can be formed using these procedures, they are unquestionably useful in generating simple surface topologies (grooves, plateaus, ridges, etc.).

Figure 5. SEM micrographs of a three-dimensional structure fabricated by forming patterned SAMs and then using appropriate selective etches for gold and silicon [64].

5. HYDROGEN BONDED AGGREGATES BASED ON THE CA•M LATTICE

These structures are, in principle, more easily envisioned than SAMs: they do not involve reduced dimensionality or cooperative effects such as crystallization. They are effectively the application of the principles of molecular recognition to the assembly of large, three-dimensional structures. In practice, they are remarkably difficult to design, in part because controlling their entropy challenges the depth of understanding in organic chemistry of the second law of thermodynamics.

To achieve three-dimensional, non-covalent aggregates one needs:

- Bonding processes that occur at equilibrium, to permit the structure to reach a thermodynamic minimum
- Interactions that are directional, to simplify the task of design
- Practical synthesis of the components
- Applicable strategies such as "preorganization" to reduce the entropic cost of aggregation.

We have selected systems based on the CA•M lattice as being the simplest that we could identify that seemed to fit these criteria. The system is a very successful one, although it has so far generated aggregates of only intermediate sizes (Figure 6). These systems are easy to synthesize (although not necessarily easy to design!). They provide wide latitude in three-dimensional structure, but stability, characterization and application are at a much earlier stage.

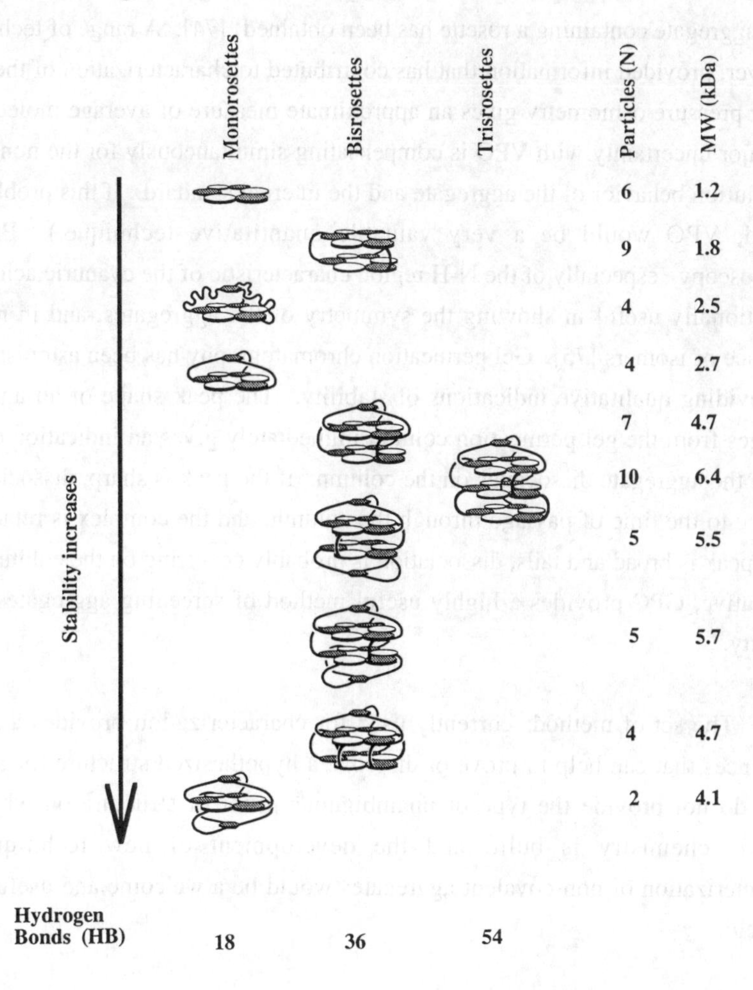

Figure 6. Aggregates based on the CA•M lattice arranged in approximate order of stability and the number of rosettes (mono-, bis-, or tris-) that they incorporate [65-73] . "Particles" refers to the number of separate molecules comprising the aggregate. "MW" is molecular weight in kDa.

5.1. CHARACTERIZATION

Characterization of these hydrogen-bonded aggregates poses some interesting challenges. It has not been possible to use mass spectroscopy with them: under all conditions that we have tried (including electrospray), they dissociate on introduction into the spectrometer. Single-crystal x-ray diffraction has also not been generally useful because it has not been possible to grow crystals (although one relevant crystal structure of an aggregate containing a rosette has been obtained) [74]. A range of techniques have, however, provided information that has contributed to characterization of the complexes. Vapor pressure osmometry gives an approximate measure of average molecular weight. (A major uncertainty with VPO is compensating simultaneously for the non-idealities in the solution behavior of the aggregate and the internal standard. If this problem could be solved, VPO would be a very valuable quantitative technique.) Proton NMR spectroscopy, especially of the N-H region characteristic of the cyanuric acid moieties, is exceptionally useful in showing the symmetry of the aggregates, and in revealing the presence of isomers [75]. Gel permeation chromatography has been astonishingly useful in providing qualitative indications of stability. The peak shape of an aggregate as it emerges from the gel permeation column immediately gives an indication of the rate at which the aggregate dissociates on the column: if the peak is sharp, dissociation is slow relative to the time of passage through the column, and the complex is relatively stable; if the peak is broad and tails, dissociation is probably occurring on the column. Although qualitative, GPC provides a highly useful method of screening aggregates for relative stability.

The set of methods currently used for characterization provides a set of strong inferences that can help to prove or disprove a hypothesized structure for an aggregate. They do not provide the type of unambiguous proof of structure on which much of organic chemistry is built, and the development of new techniques for the characterization of non-covalent aggregates would be a welcome and useful addition to the field.

5.2. LARGE STRUCTURES

The current techniques for the design and preparation of hydrogen-bonded aggregates offer relatively straightforward approaches to structures with molecular weights in the range of 5 - 8 kD; extensions of these techniques should make it possible to prepare aggregates with molecular weights of approximately 10 kD. (The range we have set as a target is substantially larger. Proteins of average size are typically 30 kD; tRNAs are closer to 100 kD.)

6. PROGNOSIS: WHAT CAN NON-COVALENT SYNTHESIS OFFER?

Non-covalent synthesis offers one approach to large, structured collections of molecules (and hence of atoms and functional groups). Covalent synthesis provides the most flexible methodology now available for preparing stable molecules with sizes up to a few hundred atoms and a few thousand molecular weight. Beyond this range, either the flexibility of the synthetic method is compromised in order to achieve the required overall yield (as in polymerization), or yields become very (often impractically) small. Non-covalent structures may provide a compromise between the structural precision but unattractive yields characteristic of covalent synthesis, and the high molecular weights but weak structural specification provided by polymerization. Non-covalent aggregates are, however, generally less stable than covalent molecules. (It is, however, worth noting the obvious: not all covalent molecules are stable; some non-covalent systems *are* very stable — CA•M, which survives for substantial periods of time at 300 °C is an example; high stability is not required in all applications.)

6.1. STRUCTURE

The development of methods for generating large structures by non-covalent synthesis will certainly occur. Aggregates based on the CA•M structure are moving rapidly toward the sizes characteristic of small biopolymers; it should be equally possible to extend aggregates of the types examined by Lehn and coworkers — based on transition metal coordination compounds — into this range of sizes [37]. Whether patterned SAMs should be considered "aggregates" or even the products of non-covalent synthesis is an issue partially of semantics, but these structures are now made in very *large* sizes routinely, and methodology is moving from these macroscopic aggregates (1 cm^2 of a SAM of hexadecanethiolate on gold contains approximately 5×10^{14} molecules of the thiolate, or 3×10^{16} atoms, excluding the gold atoms in the system) to sizes more recognizable as quasimolecular entities (a 100 nm x 100 nm square of SAM — a size that can now be approached experimentally — contains 5×10^4 molecules of alkanethiolate, and has a molecular weight of approximately 10^7).

Non-covalent synthesis is thus approaching molecular aggregates in the range considered "big" from both below and above, and with development will provide a number of new methods of synthesizing (or "fabricating") these types of structures.

6.2. FUNCTION

SAMs are already established as providing a range of functions. They control surface properties such as interfacial free energy [76], wettability [58, 77, 78] and adhesion strength [79-81] ; they influence corrosion [82] and adhesion [83-85]; they can act as resists [59, 60, 86, 87]; they serve as electroactive layers on electrodes [88-94]. Many more functions will undoubted be developed for this class of materials. Hydrogen-bonded aggregates have, as yet, no functions.

7. WHAT LIMITS NON-COVALENT SYNTHESIS?

In the short term, the most serious limitation to the construction of large, *soluble* aggregates by non-covalent synthesis is probably the absence of systems that can be used to build stable aggregates in water. Most work in non-covalent synthesis is presently carried out in the traditional, non-aqueous solvents of organic chemistry. Hydrogen bonding is the bond type most used in assembly (metal-ligand coordination is probably second) of the aggregates. Going to water-soluble systems, and incorporating hydrophobic interactions into the design of the aggregates, would not only make the systems more relevant to biochemistry, but, perhaps more importantly, provide reliable strategies based on incorporation of charged groups to keep the relatively structured and conformationally limited species that are the targets of this kind of work in solution. It is not an accident that the backbone of DNA has a charged phosphate between every base, or that proteins are at their lowest solubility when at their isoelectric point (where they are overall electrically neutral).

In the long term, the most pressing need for non-covalent synthesis and self-assembly of macromolecular assemblies is to find applications for them. The state of these activities depends upon the area. SAMs and some other systems directed toward materials science are finding a number of applications in areas ranging from protective coatings to sensors. Applications for the hydrogen-bonded aggregates are substantially less obvious. Catalysis by design is still unattainable in any field, and it is not likely that enzyme-like activity will emerge soon from these aggregates; applications in materials science are just beginning to be explored.

Acknowledgments. We wish to acknowledge the indispensable contributions to this work of a number of individuals not listed as authors of the paper. In the area of SAMs, Colin Bain, Paul Laibinis, Hans Biebuyck, Nick Abbott and Amit Kumar all contributed vitally important concepts and results. Ralph Nuzzo and Mark Wrighton have been essential collaborators at many stages of the work. In hydrogen-bonded aggregates, Chris Seto and John Mathias laid the foundations, and Jon Zerkowski contributed to understanding them. Financial support came from the Office of Naval Research and ARPA (N00014-93-1-0498), and from the National Science Foundation (CHE-91-22331).

Footnotes and References.

1. Lindsey, J. S. (1991) Self-Assembly in Synthetic Routes to Molecular Devices. Biological Principles and Chemical Perspectives: A Review, *New J. Chem.* **15**, 10988-11002.

2. Zerkowski, J. A., MacDonald, J. C., Seto, C. T., Wierda, D. A., and Whitesides, G. M. (1994) The Design of Organic Structures in the Solid State: Molecular Tapes based on the Network fo Hydrogen Bonds Present in Cyanuric Acid•Melamine Complex, *J. Am. Chem. Soc.* **116**, 2382-2391.

3. Zerkowski, J. A., and Whitesides, G. M. (1994) Steric Control of Secondary, Solid-State Architecture in 1:1 Complexes of Melamines and Barbiturates That Crystallize as Crinkled Tapes, *J. Am. Chem. Soc.* **116**, 4298-4304.

4. Zerkowski, J. A., Mathias, J. P., and Whitesides, G. M. (1994) New Varieties of of Crystalline Architectures Produced by Small Changes in Molecular Structure in Tape Complexes of Melamines and Barbiturates, *J. Am. Chem. Soc.* **116**, 4305-4315.

5. Zerkowski, J. A., and Whitesides, G. M. (submitted) Investigations into the Robustness of Hydrogen-Bonded Crystalline Tapes, *J. Am. Chem. Soc.*

6. Saeva, F. D. (1979) *Liquid Crystals: The Fourth State of Matter*; M. Dekker, New York.

7. Fujiwara, H., and Yonezawa, Y. (1991) Photoelectric Response of a Black Lipid-Membrane containing an Amphiphilic Azobenzene Derivative, *Nature* **351**, 724-726.

8. Menger, F. M. (1991) Groups of Organic-Molecules that Operate Collectively, *Angew. Chem. Int. Ed. Engl.* **30**, 1086-1099.

9. Etter, M. C. (1990) Encoding and Decoding Hydrogen-Bond Patterns of Organic Compounds, *Acc. Chem. Res.* **23**, 120-126.

10. Georges, J. M., Meille, G., Loubet, J. L., and Tolen, A. M. (1986) Bubble Raft Model for Indentation with Adhesion, *Nature* **320**, 342-344.

11. Noshay, A., and McGrath, J. E. (1977) *Block Copolymers: Overview and Critical Survey*; Academic Press, New York.

12. Cram, D. J. (1988) The Design of Molecular Hosts, Guests, and Their Complexes (Nobel Lecture), *Angew. Chem., Int. Ed. Eng.* **27**, 1009-1026.

13. Laibinis, P. E., and Whitesides, G. M. (1992) Omega-Terminated Alkanethiolate Monolayers on Surfaces of Copper, Silver and Gold have Similar Wettabilities, *J. Am. Chem. Soc.* **114**, 1990-1995.

14. Walczak, M. M., Chung, C., Stole, S. M., Widrig, C. A., and Porter, M. D. (1991) Structure and Interfacial Properties of Spontanously Adsorbed n-Alkanethiolate Monolayers on Evaporated Silver Surfaces, *J. Am. Chem. Soc.* **113**, 2370-2378.

15. Laibinis, P. E., Whitesides, G. M., Allara, D. L., Tao, Y.-T., Parikh, A. N., and Nuzzo, R. G. (1991) Comparison of the Structures and Wetting Properties of SAMs of N-Alkanethiols on the Coinage Metals Surfaces, Cu, Ag, Au, *J. Am. Chem. Soc.* **113**, 7152-7167.

16. Allara, D. L., and Nuzzo, R. G. (1985) Spontaneously Organized Molecular Assemblies. 1. Formation, Dynamics, and Physical Properties of n-Alkanoic Acids Adsorbed from Solution on an Oxidized Aluminum Surface, *Langmuir* **1**, 45-52.

17. Allara, D. L., and Nuzzo, R. G. (1985) Spontaneously Organized Molecular Assemblies. 2. Quantitative Infrared Spectroscopic-Determination of Equillibrium Structures of Solution-Adsorbed n-Alkanoic Acids on an Oxidized Aluminum Surface, *Langmuir* **1**, 52-66.

18. Tao, Y.-T. (1993) Structural Comparison of Self-Assembled Monolayers of N-Alkanoic Acids on the Surfaces of Silver, Copper, and Aluminum, *J. Am. Chem. Soc.* **115**, 4350-4358.

19. Folkers, J. P., Buchholz, S., Laibinis, P. E., Gorman, C. B., Whitesides, G. M., and Nuzzo, R. Self-Assembled Monolayers of Long-Chain Hydroxamic Acids on the Native Oxide of Metals, Manuscript in preparation.

20. Putvinski, T. M., Schilling, M. L., Katz, H. E., Chidsey, C. E. D., Mujsce, A. M., and Emerson, A. B. (1990) Self-Assembly of Organic Multilayers with Polar Order using Zirconium Phosphate Bonding between Layers, *Langmuir* **6**, 1567-1571.

21. Schilling, M. L., Katz, H. E., Stein, S. M., Shane, S. F., Wilson, W. L., Buratto, S., Ungashe, S. B., Taylor, G. N., Putvinski, T. M., and Chidsey, C. E. D. (1993) Structural Studies of Zirconium Alkylphosphonate Monolayers and Multilayer Assemblies, *Langmuir* **9**, 2156-2160.

22. Hong, H.-G., Sackett, D. D., and Mallouk, T. E. (1991) Adsorption of Well-Ordered Zirconium Phosphonate Multilayer Films on High Surface Area Silica, *Chem. Mater.* **3**, 521-527.

23. Katz, H. E., Scheller, G., Putvinski, T. M., Schilling, M. L., Wilson, W. L., and Chidsey, C. E. D. (1991) Polar Orientation of Dyes in Robust Multilayers by Zirconium Phosphate-Phosphonate Interlayers, *Science* **254**, 1485-1487.

24. Lee, H., Kepley, L. J., Hong, H.-G., Akhter, S., and Mallouk, T. E. (1988) Adsorption of ordered zirconium phosphonate multilayer films on silicon and gold surfaces, *J. Phys. Chem.* **92**, 2597-2601.

25. Maoz, R., and Sagiv, J. (1984) On the formation and structure of SAMs I. A comparative ATR-wettability study of LB and adsorbed films on flat substrates and glass microbeads, *J. Coll. Interf. Sci.* **100**, 465-496.

26. Gun, J., Iscovici, R., and Sagiv, J. (1984) On the Formation and Structure of SAMs II. A Comparative Study of LB and Adsorbed Films Using Ellipsometry and IR Reflection-Absorption Spectroscopy, *J. Coll. Interf. Sci.* **101**, 201-213.

27. Pomerantz, M., Segmüller, A., Netzer, L., and Sagiv, J. (1985) Coverage of Si Substrates by SAMs and Multilayers as Measured by IR, Wettability, and X-ray Diffraction, *Thin Solid Films* **132**, 153-162.

28. Wasserman, S. R., Whitesides, G. M., Tidswell, I. M., Ocko, B. M., Pershan, P. S., and Axe, J. D. (1989) The Structure of Self-Assembled Monolayers of Alkylsiloxanes on Silicon: A Comparison of Results from Ellipsometry and Low-Angle-X-ray Reflectivity, *J. Am. Chem. Soc.* **111**, 5852-5861.

29. Möbius, D. (1981) Molecular Cooperation in Monolayer Organizates, *Acc. Chem. Res.* **14**, 63-68.

30. Swalen, J. D., Allara, D. L., Andrade, J. D., Chandross, E. A., Garoff, S., Israelachivili, J., McCarthy, T. J., Murray, R., Pease, R. F., Rabolt, J. F., Wynne, K. J., and Yu, H. (1987) Molecular Monolayers and Films, *Langmuir* **3**, 932-950.

31. Nuzzo, R. G., and Allara, D. L. (1983) Adsorption of Bifunctional Organic Disulfides on Gold Surfaces, *J. Am. Chem. Soc.* **105**, 4481-4483.

32. Nuzzo, R. G., Fusco, F. A., and Allara, D. L. (1987) Spontaneously Organized Molecular Assemblies. 3. Preparation and Properties of Solution Adsorbed Monolayers of Organic Disulfides on Gold Surfaces, *J. Am. Chem. Soc.* **109**, 2358-2368.

33. Chidsey, C. E. D., Porter, M. D., and Allara, D. L. (1986) Electrochemical Characterization of N-Alkyl Thiol, Sulfide, and Disulfide Monolayers on Gold, *J. Electrochem. Soc.* **133**, 130.

34. Geib, S. J., Vicent, C., Fan, E., and Hamilton, A. D. (1993) A Self-Assembling Hydrogen-Bonded Helix, *Angew. Chem., Int. Ed. Eng.* **32**, 119-121.

35. Wyler, R., deMendoza, J., and Rebek, J. (1993) A Synthetic Cavity Assembles Through Self-Complimentary Hydrogen Bonds, *Angew. Chem., Int. Ed. Eng.* **32**, 1699-1701.

36. Ballardini, R., Balzani, V., Gandolfi, M. T., Prodi, L., Venturi, M., Philp, D., Ricketts, H. G., and Stoddart, J. F. (1993) A Thermodynamically Driven Molecular Machine, *Angew. Chem., Int. Ed. Eng.* **32**, 1301-1303.

37. Baxter, P., Lehn, J. M., DeCian, A., and Fisher, J. (1993) Self-Assembly, Structure, and Sponteous Resolution of a Trinuclear Triple Helix from an Oligobipyridine Ligand and Ni(II) Ions, *Angew. Chem., Int. Ed. Eng.* **32**, 703-706.

38. Nierengarten, J. F., Dietrich-Buchecker, C. O., and Sauvage, J. P. (1994) Synthesis of Doubly Interlocked 2-Catenanes, *J. Am. Chem. Soc.* **116**, 375-376.

39. Zimmerman, S. C., Saionz, K. W., and Zeng, Z. (1993) Chemically Bonded Stationary Phases that Use Synthetic Hosts Containing Aromatic Binding Clefts: HPLC Analysis of Nitro-Substituted Polycyclic Aromatic Hydrocarbons, *Proc. Natl. Acad. Sci., U.S.A.* **90**, 1190-1193.

40. Breslow, R., and Graft, A. (1993) Geometry of Enolization Using a Bifunctional Cyclodextrin Based Catalyst, *J. Am. Chem. Soc.* **115**, 10988-9.

41. Borchardt, A., and Still, W. C. (1994) Synthetic Receptor Binding Elucidated with an Encoded Combinatorial Library, *J .Am. Chem. Soc.* **116**, 373-4.

42. Kunitake, T. (1992) Synthetic Bilayer Membranes: Molecular Design, Self-Organization, and Application, *Angew. Chem., Int. Ed. Eng.* **31**, 709-726.

43. Helgeson, R. C., Selle, B. J., Goldberg, I., Knobler, C. B., and Cram, D. J. (1993) 18-Membered-Ring Sperands Containing Five Anisyl Groups, *J. Am. Chem. Soc.* **115**, 11506-7.

44. For a review of SAM structure, characterization and physical properties, see: Whitesides, G. M., and Gorman, C. B. (1994) Self-Assembled Monolayers: Models for Organic Surface Chemistry in A. T. Hubbard (ed), *Handbook of Surface Imaging and Visualization*; CRC Press, Boca Raton, Florida.

45. Bain, C. D., and Whitesides, G. M. (1989) Attenuation Lengths of Photoelectrons in Hydrocarbon Films, *J. Phys. Chem.* **93**, 1670-1673.

46. Nuzzo, R. G., Dubois, L. H., and Allara, D. L. (1990) Fundamental Studies of Microscopic Wetting on Organic Surfaces. 1. Formation and Structural Characterization of a Self-Consistent Series of Polyfunctional Organic Monolayers, *J. Am. Chem. Soc.* **112**, 558-569.

47. Nuzzo, R. G., Korenic, E. M., and Dubois, L. H. (1990) Studies of the Temperature-Dependent Phase Behavior of Long Chain N-Alkyl Thiol Monolayers on Gold, *J. Chem. Phys.* **93**, 767-773.

48. Dubois, L. H., and Nuzzo, R. G. (1992) Synthesis, Structure, and Properties of Model Organic-Surfaces, *Annu. Rev. Phys. Chem.* **43**, 437-463.

49. Bain, C. D., and Whitesides, G. M. (1989) Formation of Monolayers by the Coadsorption of Thiols on Gold: Variation in the Length of the Alkyl Chain, *J. Am. Chem. Soc.* **111**, 7164-7175.

50. Bain, C. D., Troughton, E. B., Tao, Y.-T., Evall, J., Whitesides, G. M., and Nuzzo, R. G. (1989) Formation of Monolayer Films by the Spontaneous Assembly of Organic Thiols from Solution onto Gold, *J. Am. Chem. Soc.* **111**, 321-335.

51. Schöenbert, C., Sondag-Huethorst, J. A. M., Jorritsma, J., and Fokkink, L. G. J. What are the "Holes" in Self-Assembled Monolayers of Alkanethiols on gold?, Submitted.

52. Kim, Y.-T., McCarley, R. L., and Bard, A. J. (1993) Observation of N-Octadecanethiol Multilayer Formation from Solution onto Gold, *Langmuir* **9**, 1941-1944.

53. Alves, C. A., Smith, E. L., and Porter, M. D. (1992) Atomic Scale Imaging of Alkanethiolate Monolayers at Gold Surfaces with Atomic Force Microscopy, *J. Am. Chem. Soc.* **114**, 1222-1227.

54. Pan, J., Tao, N., and Lindsay, S. M. (1993) An AFM Study of Self-Assembled Octadecyl Mercaptan Monolayer Adsorbed on Gold(111) Under Potential Control, *Langmuir* **9**, 1556-1560.

55. Hautman, J., and Klein, M. L. (1989) Simulation of a Monolayer of Alkyl Thiol Chains, *J. Chem. Phys.* **91**, 4994-5001.

56. Siepmann, J. I., and McDonald, I. R. (1993) Domain Formation and System-Size Dependence in Simulations of Self-Assembled Monolayers, *Langmuir* **9**, 2351-2355.

57. Abbott, N. L., Folkers, J. P., and Whitesides, G. M. (1992) Manipulation of the Wettability of Surfaces on the 0.1- to 1-Micrometer Scale through Micromachining and Molecular Self-Assembly, *Science* **257**, 1380-1382.

58. Biebuyck, H. A., and Whitesides, G. M. Self-Organization of Organic Liquids on Patterned Self-Assembled Monolayers of Alkanethiolates on Gold, Submitted.

59. Kumar, A., Biebuyck, H. A., Abbott, N. L., and Whitesides, G. M. (1992) The Use of Self-Assembled Monolayers and a Selective Etch to Generate Patterned Gold Features, *J. Am. Chem. Soc.* **114**, 9188-9189.

60. Kumar, A., and Whitesides, G. M. (1993) Features of Gold having Micrometer to Centimeter Dimensions can be formed through a combination of Stamping with an

Elastomeric Stamp and and Alkanethiol "Ink" Followed by Chemical Etching, *Appl. Phys. Lett.* **63**, 2002-2004.

61. Kumar, A., and Whitesides, G. M. (1994) Patterned Condensation Figures as Optical Diffraction Gratings, *Science* **263**, 60-62.

62. López, G. P., Biebuyck, H. A., and Whitesides, G. M. (1993) Scanning Electron Microscopy can Form Images of Patterns in Self-Assembled Monolayers, *Langmuir* **9**, 1513-1516.

63. López, G. P., Biebuyck, H. A., Frisbie, C. D., and Whitesides, G. M. (1993) Imaging of Features on Surfaces by Condensation Figures, *Science* **260**, 647-649.

64. Wilbur, J. L., Kumar, A., Kim, E., and Whitesides, G. M. (in press) Microcontact Printing of Self-Assembled Monolayers: A Flexible New Technique for Microfabrication, *Adv. Mater.*

65. Mathias, J. P., Simanek, E. E., Zerkowski, J. A., Seto, C. T., and Whitesides, G. M. (1994) Structural Preferences for Hydrogen-Bonded Networks in Organic Solution- The Cyclic CA3•M3 'Rosette', *J. Am. Chem. Soc.* **116**, 4316-4325.

66. Mathias, J. P., Simanek, E. E., and Whitesides, G. M. (1994) Self-Assembly Through Hydrogen Bonding: Peripheral Crowding-A New Strategy for the Preparation of Stable Supramolecular Aggregates Based on the Parallel, Connected Rosettes, *J .Am. Chem. Soc* **116**, 4326-4340.

67. Seto, C. T., and Whitesides, G. M. (1993) Molecular Self-Assembly Through Hydrogen-Bonding: Supramolecular Aggregates Based on the Cyanuric Acid•Melamine Lattice, *J. Am. Chem. Soc.* **115**, 905-916.

68. Seto, C. T., and Whitesides, G. M. (1990) Self-Assembly Based on the Cyanuric Acid-Melamine Lattice, *J .Am. Chem. Soc.* **112**, 6409-6411.

69. Mathias, J. P., Seto, C. T., and Simanek, E. E. (1994) Self-Assembly Through Hydrogen Bonding: Preparation and Characterization of Three New Types of Supramolecular Aggregates Based on Parallel, Cyclic CA3•M3 'Rosettes', *J. Am. Chem. Soc.* **116**, 1725-1736.

70. Mathias, J. P., Simanek, E. E., and Seto, C. T. (1993) Self-Assembly Through Hydrogen Bonding: Preparation of a Supramolecular Aggregate Composed of 10 Molecules, *Angew. Chem. Int. Ed. Engl.* **32**, 1766-1769.

71. Seto, C. T., and Whitesides, G. M. (1991) Self-Assembly of a Hydrogen-Bonded 2+3 Supramolecular Complex, *J. Am. Chem. Soc.* **113**, 712-713.

72. Seto, C. T., Mathias, J. P., and Whitesides, G. M. (1993) Molecular Self-Assembly Through Hydrogen-Bonding: Aggregation of Five Molecules to Form a Discrete Supramolecular Structure, *J. Am. Chem. Soc.* **115**, 1321-1329.

73. Seto, C. T., and Whitesides, G. M. (1993) Synthesis, Characterization, and Thermodynamic Analysis of a 1+1 Self-Assembling Structure Based on the Cyanuric Acid-Melamine Lattice, *J. Am. Chem. Soc.* **1993**, 1330-1340.

74. Zerkowski, J. A., Seto, C. T., and Whitesides, G. M. (1992) Solid-State Structures of 'Rosette' and 'Crinkled Tape' Motifs Derived from the Cyanuric Acid•Melamine Lattice, *J. Am. Chem. Soc.* **114**, 5473-5475.

75. Simanek, E. E., Wazeer, M. I. M., Mathias, J. P., and Whitesides, G. M. (submitted) 1H NMR Spectroscopy of the Hydrogen-bonded Imide Groups of Hub(M)3:3CA Provides a Useful Method for the Characterization of these Aggregates, *J. Org. Chem.*

76. Schrader, M. E., and Loeb, G. I. (1992) *Modern Approaches to Wettability*; Plenum Press, New York.

77. Whitesides, G. M., Biebuyck, H. A., Folkers, J. P., and Prime, K. L. (1991) Acid-Base Interactions in Wetting, *J. Adhesion Sci. Technol.* **5**, 57-69.

78. Lee, T. R., Carey, R. I., Biebuyck, H. A., and Whitesides, G. M. (1994) The Wetting of Monolayers Films Exposing Ionizable Acids and Bases, *Langmuir* **10**, 741-749.

79. Chaudhury, M. K., and Owen, M. J. (1993) Correlation between Adhesion Hysteresis and Phase State of Monolayer Films, *J. Phys. Chem.* **97**, 5722-5726.

80. Chaudhury, M. K., and Whitesides, G. M. (1991) Direct Measurement of Interfacial Interactions between Semispherical Lenses and Flat Sheets of PDMS and Their Chemical Derivatives, *Langmuir* **7**, 1013-1025.

81. Chaudhury, M. K., and Owen, M. J. (1993) Adhesion Hysteresis and Friction, *Langmuir* **9**, 29-31.

82. Laibinis, P. E., and Whitesides, G. M. (1992) SAMs of N-Alkanethiolates on Copper are Barrier Films that Protect the Metal Against Oxidation by Air, *J. Am. Chem. Soc.* **114**, 9022-9028.

83. DiMilla, P. A., Folkers, J. P., Biebuyck, H. A., Härter, R., López, G. P., and Whitesides, G. M. (1994) Self-Assembled Monolayers of Alkanethiolates can be Formed on Transparent Films of Gold: Application to Studies of Protein Adsorption and Cell Attachment and Growth, *J. Am. Chem. Soc.* **116**, 2225-2226.

84. Singhvi, R., Kumar, A., Lopez, G. P., Stephanopoulos, G. N., Wang, D. I. C., Whitesides, G. M., and Ingber, D. E. (1994) Engineering Cell Shape and Function, *Science* **264**, 696-698.

85. López, G. P., Albers, M. W., Schreiber, S. L., Carroll, R., Peralta, E., and Whitesides, G. M. (1993) Convenient methods for patterning the adhesion of mammalian-cells to surfaces using SAMs of alkanethiolates on gold, *J. Am. Chem. Soc.* **115**, 5877-5878.

86. Ross, C. B., Sun, L., and Crooks, R. M. (1993) Scanning Probe Lithography. 1. STM Induced Lithography of Self-Assembled N-Alkanethiol Monolayer Resists, *Langmuir* **9**, 632-636.

87. Tiberio, R. C., Craighead, H. G., Lercel, M., Lau, T., Sheen, C. W., and Allara, D. L. (1993) SAM Electron Beam Resist on GaAs, *Appl. Phys. Lett.* **62**, 476-478.

88. Chidsey, C. E. D., Bertozzi, C. R., Putvinski, T. M., and Mujsce, A. M. (1990) Coadsorption of Ferrocene-terminated and Unsubstituted Alkanethiols on Gold: Electroactive SAMs, *J. Am. Chem. Soc.* **112**, 4301-4306.

89. Chidsey, C. E. D., and Loiacono, D. N. (1990) Chemical Functionality in SAMs: Structural and Electrochemical Properties, *Langmuir* **6**, 682-691.

90. Chidsey, C. E. D. (1991) Free Energy and Temperature Dependence of Electron Transfer at the Metal-Electrolyte Interface, *Science* **251**, 919-922.

91. Collinson, M., Bowden, E. F., and Tarlov, M. J. (1992) Voltammetry of Covalently Immobilized Cytochrome c on Self-Assembled Monolayer Electrodes, *Langmuir* **8**, 1247-1250.

92. Finklea, H. O., Avery, S., Lynch, M., and Furtsch, T. (1987) Blocking Oriented Monolayers of Alkyl Mercaptans on Gold Electrodes, *Langmuir* **3**, 409-413.

93. Hickman, J. J., Ofer, D., Zou, C., Wrighton, M. S., Laibinis, P. E., and Whitesides, G. M. (1991) Selective Functionalization of Gold Microstructures with Ferrocenyl Derivatives via Reaction with Thiols or Disulfides: Characterization by Electrochemistry and Auger Electron Spectroscopy, *J. Am. Chem. Soc.* **113**, 1128-1132.

94. Miller, C., Cuendet, P., and Grätzel, M. (1991) Adsorbed Omega-Hydroxy Thiol Monolayers on Gold Electrodes: Evidence for Electron Tunnelling to Redox Species in Solution, *J. Phys. Chem.* **95**, 877-886.

The Functionalization of Saturated Hydrocarbons by Gif Chemistry. Part 1. Use of Superoxide and of Hydrogen Peroxide. Part 2. Use of *t*-Butylhydroperoxide (TBHP).

D.H.R. BARTON
Texas A&M University
Department of Chemistry
College Station, TX 77843-3255
USA

1. Abstract

In Part 1, the Gif systems for the selective functionalization of saturated hydrocarbons based on the reactions of superoxide with Fe^{II} and of hydrogen peroxide with Fe^{III} are described. [1] Both systems are relatively efficient, but not nearly so efficient as the electrochemical system developed in collaboration with Prof. G. Balavoine and Dr. Aurore Gref (Université de Paris-Sud--Orsay, France). [2] All systems afford mainly ketones. This is an unusual selectivity which implies a non-radical mechanism. It has been proven for the Fe^{III}-H_2O_2 system that the activation of the Fe^{III} is independent of the formation of ketone which comes from oxygen. The intermediate for the ketone is a hydroperoxide (derived from oxygen). [3] This intermediate controls the formation of ketone and of secondary alcohol. Addition of a number of trapping reagents such as $BrCCl_3$ diverts the reaction from oxygenation to bromide formation. Although $BrCCl_3$ is indeed a good trap for carbon radicals, the pattern of selectivity across a range of saturated hydrocarbons is completely different for Gif chemistry when compared with normal radical bromination. [4] The chemistry is explained in terms of an Fe^Voxenoid species that inserts itself into secondary carbon-hydrogen bonds (a compromise between bond strength and steric hindrance). [1] This gives an Fe^V intermediate **A** with an iron-carbon bond. This is probably rapidly reduced to the Fe^{III} state by hydrogen peroxide. Then oxygen is inserted into the Fe^{III}-carbon bond. Hydrolysis affords the isolateable intermediate hydroperoxide (intermediate **B**).

In Part 2, a system based on *t*.-butylhydroperoxide (TBHP) is described. This is similar to the above Gif systems, but the kinetic isotope effect is very different and the selectivity for adamantane substitution is different. However, Fe^{III} is activated by TBHP to an Fe^V oxenoid which, after reaction with hydrocarbon, then reacts with oxygen to give hydroperoxide. So the pattern of intermediates **A** and **B** seen in Part 1 is maintained with TBHP. Radical chemistry may be involved in some of the reactions that involve ionic coupling to saturated hydrocarbons.

C. Chatgilialoglu and V. Snieckus (eds.), Chemical Synthesis, 589-599.

2. Part I

2.1 INTRODUCTION

There is universal acceptance that the selective functionalization of saturated hydrocarbons is a research challenge of major proportions. Although paraffins are, as their name implies, usually considered to be inert, Nature has solved the problem of their functionalization using complexes of iron and, to a lesser extent of copper. The iron complexes are either heme based, as in the ubiquitous P_{450} enzymes, or non-heme based, as in the vitally important proline-4-hydroxylase and in the fascinating enzyme, methane monooxygenase. These enzymes contain one Fe^{III}, or two Fe^{III} atoms bridged by an oxo bridge, respectively. Although the protein in these enzymes is responsible for their chemo- and stereo-selectivity, the fundamental activation of the iron should be seen also in appropriate models. Models of P_{450} enzymes have been known for decades and the work of J.T. Groves is particularly noteworthy in this respect. Models of non-heme iron enzymes are much less familiar and, until our publications on Gif type Chemistry, not efficacious. [5]

2.2 USE OF SUPEROXIDE AND OF HYDROGEN PEROXIDE

Our work started with the assumption that when, 3 billion years ago, the blue-green algae started to make oxygen, a unicellular form of life started to concert the oxidation of iron with the oxidation of saturated hydrocarbons. It is agreed that after 1 billion years of life under reductive conditions, the world was full of saturated hydrocarbons, Fe^{II} compounds and metallic iron, as well as, an abundance of hydrogen sulfide, since life was using the reduction of sulfate to sulfide as an energy source.

So our first experiment to test our hypothesis was the oxidation of adamantane in pyridine with some acetic acid in the presence of iron powder, hydrogen sulfide and oxygen. Surprisingly, a major amount of adamantanone was formed and there was only minor formation of the tertiary alcohol. Later, we obtained the same results when we replaced the iron powder with zinc powder and added a catalytic amount of an Fe^{II} salt. This showed the unusual power of the iron species formed to make ketones as principal products.

During systematic studies on adamantane oxidation, we examined the effect of reducing oxygen pressure. We were surprised to find that lowering the oxygen pressure increased the selectivity for adamantanone formation. We defined selectivity as C^2/C^3 where C^2 is total product at the secondary position and C^3 is total product at the tertiary position. Both are normally expressed as mmol. At the secondary position, only ketone and alcohol are formed. At the tertiary position, reduction of oxygen pressure caused formation of *t.*-adamantyl groups attached to the 2- and 4-positions of pyridine, as well as, production of *t.*-adamantanol. When all *sec.* and *tert.* products were taken into consideration, then C^2/C^3 was about 1.0 and was pressure invariant. In contrast, the C^2/C^3 for oxygen radical reactions is about 0.1. By application of

Ockam's razor, we concluded that an intermediate was formed at both *sec.* and *tert.* positions. We postulated that this was formed by insertion of an Fe^{V} oxenoid species into the carbon-hydrogen bond. At the secondary position, the iron-carbon bond was stable, but at the tertiary position, the weaker bond fragmented into radicals which coupled with pyridine in the ordinary way. All the appropriate blank experiments were carried out to show that secondary radicals, had they been formed at the lower oxygen pressures, would have been captured by the solvent pyridine. [6]

Subsequently, potassium superoxide was reacted with Fe^{II} under argon and shown to give an iron species that also gave an adamantane selectivity of about 1.0.

Similarly, Fe^{III} and H_2O_2 showed the same behavior. We can summarize all the results in the following way:

$$Fe^{II} + {}^{\cdot}O_2^- + H^{\oplus} \equiv Fe^{III} + H_2O_2 \longrightarrow Fe^{III}\text{—O—OH}$$

$$^{III}Fe\text{—O—OH} \longrightarrow Fe^{V}\text{=O (the oxenoid species)}$$

$$Fe^{V}\text{=O} + \text{>CH}_2 \longrightarrow Fe^{V}\text{—CH<}$$

The H_2O_2 system is very convenient to use and, since it is homogeneous, it lends itself to C^{13} N.M.R. studies. Using singly labeled cyclohexane, we could show that an intermediate was formed between cyclohexane and cyclohexanone. By comparison with an authentic specimen and by isolation, this was shown to be cyclohexyl hydroperoxide. The reaction, at room temperature, had a half-life of about 2 hours and so was easy to follow by N.M.R.

Further evidence for an intermediate hydroperoxide was found in Zn^{0}-Fe^{II}-superoxide experiments when PPh_3 was added before the formation of the superoxide. This did not change the total amount of oxidation (ketone + alcohol), but did dramatically change the ketone to alcohol ratio in favor of alcohol. Hydroperoxides are, of course, rapidly reduced by PPh_3 to alcohols. Furthermore when trimethyl phosphite is used instead of PPh_3, the products of the reaction are phosphate and ketone. [7] Trimethyl phosphite is a reagent which reduces hydroperoxides at once to alcohols. This new trimethyl phosphite reaction can be understood better when we ask the question how is the hydroperoxide formed?

Using Fe^{III} and H_2O_2, this question can be answered. If we accept that Fe^{III} is oxidized to the Fe^{V} oxenoid, then we can also understand how this is reduced back again to Fe^{III} and O_2 by a second molecule of H_2O_2. Using $^{18}O_2$ labeled oxygen, we can show that the oxygen that appears in the hydroperoxide is indeed formed from oxygen and not from hydrogen peroxide. Does this mean that we have a conventional autoxidation process? The trimethyl phosphite reaction mentioned above precludes this possibility. Hydroperoxyl and alkoxyl radicals are reduced very efficiently by trimethyl phosphite to furnish alkyl radicals. We have indeed confirmed that this is so under Gif conditions. Therefore, the reaction with trimethyl phosphite which forms phosphate (see above) precludes the intermediacy of a hydroperoxyl radical. We propose the mechanism shown in Scheme 1.

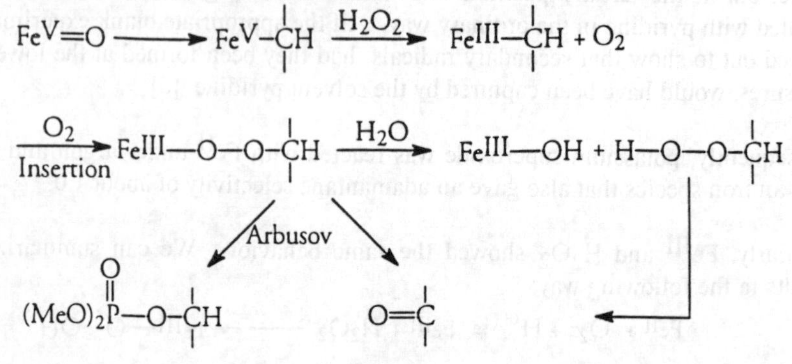

$$Fe^{V}{=}O \longrightarrow Fe^{V}{-}CH \xrightarrow{H_2O_2} Fe^{III}{-}CH + O_2$$

$$\xrightarrow[\text{Insertion}]{O_2} Fe^{III}{-}O{-}O{-}CH \xrightarrow{H_2O} Fe^{III}{-}OH + H{-}O{-}O{-}CH$$

Arbusov

$$(MeO)_2P(=O){-}O{-}CH \qquad O{=}C$$

Scheme 1

The key step here is the insertion of oxygen into the Fe^{III}-carbon bond for which there is precedent in porphyrin chemistry. [3]

Further study of the Fe^{III}-H_2O_2 reaction showed that bromotrichloromethane and its congeners afford bromination instead of oxidation. The selectivity is the same as the oxidation reaction and does not follow a typical radical bromination pattern. [4] Scheme 2 shows a series of hydrocarbons and their relative reactivity normalized per hydrogen for radical bromination and for Gif type Fe^{III}-picolinic acid (PA)-H_2O_2 bromination. In each case, the C-H bond reactivity in cyclohexane is set at 1.0. The two sets of numbers are completely different. Cyclohexane is the most active hydrocarbon in Gif bromination and the least active in radical bromination. The data were

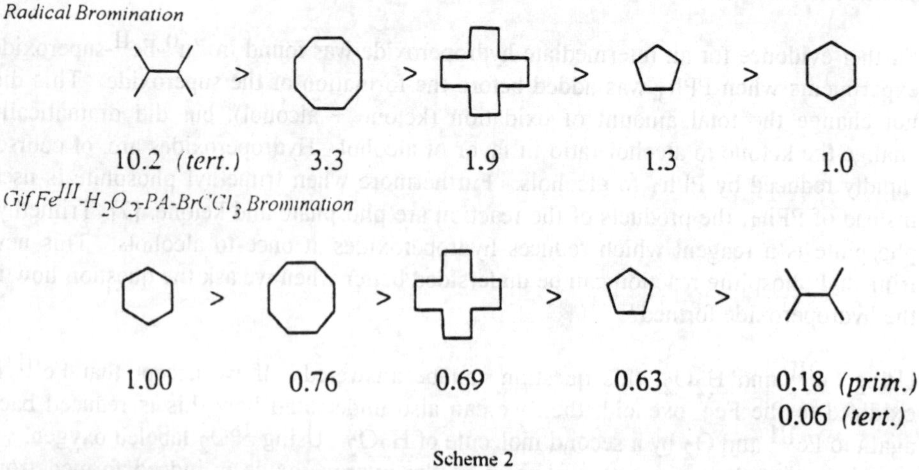

Radical Bromination

10.2 *(tert.)* > 3.3 > 1.9 > 1.3 > 1.0

Gif Fe^{III}-H_2O_2-PA-$BrCCl_3$ Bromination

1.00 > 0.76 > 0.69 > 0.63 > 0.18 *(prim.)* 0.06 *(tert.)*

Scheme 2

accumulated by competing the hydrocarbons mixed together in pairs. This avoids the possibility that radical chain reactions might be different in two separate experiments carried out under nominally the same conditions in different flasks.

The same conclusion could be drawn from a comparison of the relative reactivity of the brominating reagents in radical bromination and in Gif bromination (Scheme 3).

Relative Rates for Radical Bromination (Reagents)

Entry	Halogenating Reagent	cyclohexane (H) *mmol*	cyclohexyl bromide (Br) *mmol*	R-Br / R-H	Total Products *mmol*
1	CBr_4	0.065	0.133	2.05	0.198
2	CBr_2Cl_2	0.108	0.093	0.86	0.201
3	$CBrCl_3$	0.099	0.072	0.73	0.171
4	$(CBrCl_2)_2$	0.124	0.072	0.58	0.196
5	CBr_2F_2	0.120	0.005	0.01	0.125

Relative Rates for Gif Bromination (Reagents)

Entry	Halogenating Reagent	cyclohexanone (O) *mmol*	cyclohexyl bromide (Br) *mmol*	R-Br / R=O	Total Products *mmol*
1	$CBrCl_3$	0.024	0.686	28.6	0.710
2	$(CBrCl_2)_2$	0.031	0.529	17.1	0.560
3	CBr_4	0.070	0.396	5.6	0.466
4	CBr_2Cl_2	0.300	0.418	1.4	0.718
5	CBr_2F_2 [a]	0.352	0.079	0.2	0.431

a. Reaction carried out at 0^oC due to the volatility of CBr_2F_2. Reaction time:48 h.

Scheme 3

For relative rates of radical bromination with the reagents, a competition was set up between the reagents, one by one, and thiophenol. Cyclohexyl radicals were generated by photolysis of the cyclohexyl carboxylic ester of *N*-hydroxy-2-thiopyridone so one measures the ratio cyclohexane (from the thiophenol hydrogen atom transfer) and cyclohexyl bromide from the reagent. For the Gif chemistry, competitive oxidation to cyclohexanone was compared to bromination using the Fe^{III}-PA-H_2O_2-brominating agent. In all cases of Gif bromination, the other product was CO_2 and not the normal Br-X→H-X as seen in radical bromination chemistry. All the results show clearly that Gif bromination and radical bromination are completely different processes.

Another good test for radical bromination is bromination of cyclohexyl bromide. When radical bromination is involved, the *trans*-1,2-dibromide is the major product (Skell-Walling effect). Again with Gif bromination, completely different results are seen (Scheme 4).

Comparison of the Distribution of Dibromocyclohexanes for a radical chain bromination and for a Fe^{III}-PA-H_2O_2 Bromination Reaction

Entry	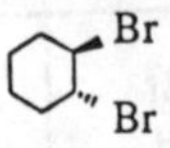 Br Br	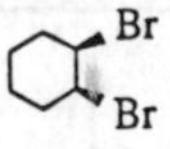 Br Br	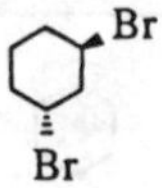 Br Br	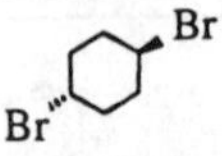 Br Br	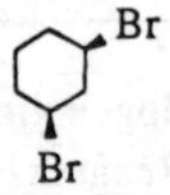 Br Br	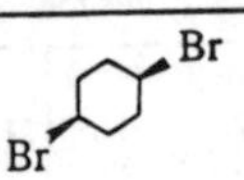 Br Br
1[a]	94.0	3.5[b]	0.9	0.5	0.5	0.5
2[c]	7.7	---[f]	1.9	45.2	27.9	17.3
3[d]	41.0	14.0	0.8	21.0	16.0	7.2
4[e]	33.0	25.0	7.0	15.0	13.0	7.0

a. Values for radical chain bromination The conversion was 20%. *b.* 1,1-dibromocyclohexane and *cis*-1,2-dibromocyclohexane could not be distinguished. *c.* Values obtained in the $GoAgg^{III}$ reaction on cyclohexyl bromide (for reaction conditions see experimental). *d.* Cyclohexyl bromide (1.0 mmol) and $CBrCl_3$ (4.0 mmol) irradiated in a Rayonet photoreactor with four 3000 Å light tubes under Ar at 35-38°C for 22 hs. *e.* Radical reaction initiated by $(PhCOO)_2$ in Py/AcOH. *f.* Not detected

Scheme 4

Our new results for the radical bromination of cyclohexyl bromide are largely in agreement with the literature and completely different from the results of Gif bromination. A major equatorial preference is seen in that much more *cis*-1,3-dibromide (diequatorial) than *trans* (equatorial, axial) is formed, whereas for the 1,4-dibromination, much more *trans*-(diequatorial) than *cis*-(equatorial, axial) dibromide is seen. Insertion of the Fe^{V} oxenoid is subject to steric hindrance and hence equatorial insertion followed by ligand coupling of bromide with retention of configuration would be expected in agreement with the results.

Less extensive data (Scheme 5) are at present available for the oxidation of cycloalkanes to ketones. [8] The results are expressed per C-H bond and cyclohexane is taken as 1.0. Again the hydrocarbons were competed against each other in pairs.

Gif Fe^{III}-H_2O_2 Oxidation

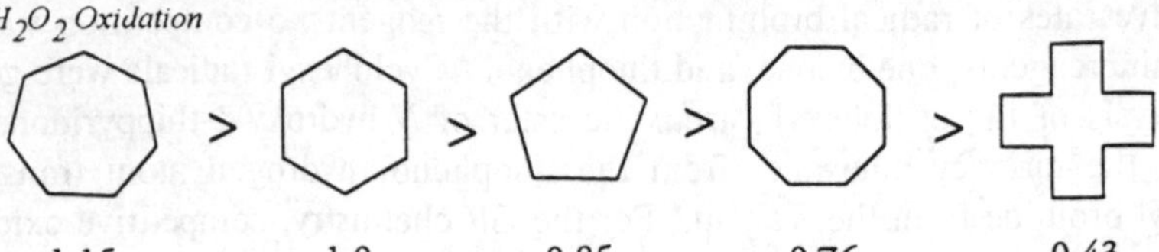

>	>	>	>	
1.15	1.0	0.85	0.76	0.43

Scheme 5

The order is approximately the same as for the Fe^{III}-PA-H_2O_2 bromination, but of course, the ligands are very different.

The oxidation of saturated hydrocarbons at the cathode of an electrochemical cell is (Fe^{II} + superoxide) is a superior procedure from the viewpoint of electronic yield. An extensive study[2] carried out in collaboration with Prof. G. Balavoine, Dr. Aurore Gref and other colleagues at the Université de Paris-Sud gave results partially summarized in Scheme 6.

Hydrocarbon	Electronic Yield %	
	Chemical	Electrochemical
Methylcyclopentane	3.6	20
cis-Decalin	34	50
trans-Decalin	16	42
Adamantane	3	12

Scheme 6

The electronic yield represents the % of the electrons which come from the zinc in the Zn^0-Fe^{II}-O_2 system, or from the cathode which appear in the final product. The electrochemical system is far superior from this point of view and it is also more selective. The electrochemical system uses trifluoroacetic acid, instead of acetic acid. The protonation of the pyridine provides the cations necessary for the conductance of the current.

In any case, the selectivity of the electrochemical system confirms the preferential attack that we have seen on secondary positions during extensive studies on the oxidation of natural products.

The present position for the theory of Gif chemistry is summarized in Scheme 7.

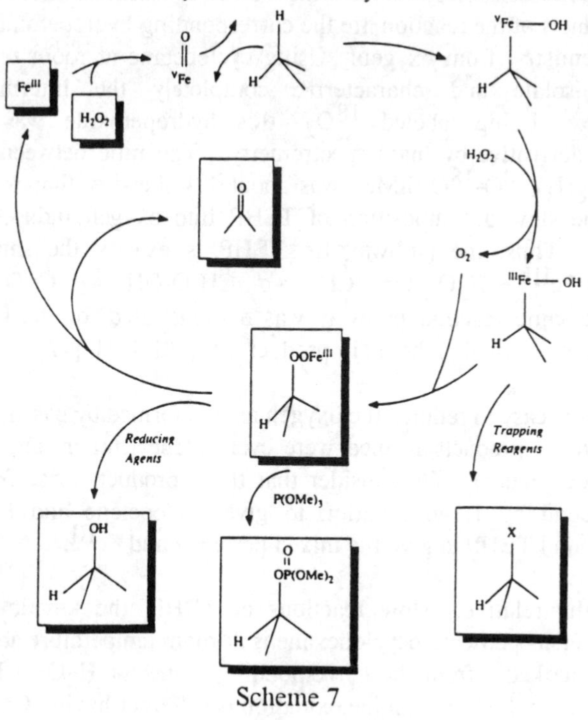

Scheme 7

It is now clear that two intermediates can be discerned in Gif chemistry, **A** and **B**. **B** has been thoroughly characterized as the hydroperoxide which is the precursor of the ketone and of the alcohol. There is evidence (trimethyl phosphite effect) that the hydroperoxide is formed via a form bound to iron. The intermediate **A** is not a carbon radical since it does not react with pyridine even when all the oxygen is removed by a vacuum. Its reactivity towards bromotrichloromethane is very different from standard radical bromination. The postulated iron-carbon bond is still the best hypothesis for **A**.

It is established that the activation of the iron does lead to an attack on the hydrocarbon and that oxygen from gaseous oxygen is introduced in a second step. This contrasts to the theory of P_{450} oxidation where the activation of the iron leads to attack on the hydrocarbon and thence to hydroxylation. Recent work in our laboratory has confirmed experimentally this theory.

3. Part II

3.1 USE OF *t.*-BUTYLHYDROPEROXIDE (TBHP)

The use of TBHP introduces a new dimension to Gif Chemistry. As with Fe^{III}-H_2O_2, the addition of picolinic acid (PA) greatly speeds up the reaction to give mainly ketone. In general, the TBHP reactions are slower than their superoxide or hydrogen peroxide congeners already discussed. Whilst the TBHP is essential for the activation of the Fe^{III}, the products of the reaction are the corresponding hydroperoxide formed, as with the H_2O_2 chemistry, from oxygen. Using cyclooctane at room temperature, it was possible to isolate and characterize completely the intermediate cyclooctyl hydroperoxide. Using labeled $^{18}O_2$, this hydroperoxide was analyzed as its trimethylsilyl derivative by mass spectrometry. The ratio between $C_8H_{15}{}^{16}O$-^{16}O-$SiMe_3$ and $C_8H_{15}{}^{18}O$-^{18}O-$SiMe_3$ was about 1:4 showing that some of the oxygen came from the slow decomposition of TBHP into oxygen induced by Fe^{III} (blank experiments). Thus, the pathway fro TBHP is exactly the same as for Fe^{II} + superoxide or Fe^{III} + H_2O_2 *viz.* $>CH_2 \rightarrow >CH$-O-OH $\rightarrow$ $>C{=}O$ (major) + $>CHOH$ (minor). The same reaction pathway was demonstrated for the solvent acetonitrile although here the alcohol is the main product. [9],[10], [11], [12]

With TBHP, it is easy to remove the oxygen as it is formed by passing a slow current of argon. The main products formed were cyclooctene (minor) and cyclooctyl-*t.*-butyl mixed peroxide (major). We consider that these products arise from the postulated Fe^{V}-carbon bond by 1,2-elimination to give cyclooctene and Fe^{III} or by ligand coupling of bound TBHP to give the mixed peroxide and Fe^{III}.

Apart from the relatively slow reactions of TBHP, the kinetic isotope effect for cyclohexane versus perdeuterocyclohexane is at room temperature about 8 (±0.3). This value differs markedly from the corresponding value for H_2O_2 which is 2.2 (±0.1). Also, the selectivity for adamantane oxidation is different having C^2/C^3 about 0.5.

A study of the oxidation of benzylic methylene groups using TBHP has shown that two mechanisms are available for the formation of ketones. [13] One is oxygen dependent, the other not. The first is analogous to the formation of cyclooctanone from cyclooctane. The second involves the mixed peroxide, analogous to the conversion of cyclooctane into mixed peroxide as described above. The cyclooctyl mixed peroxide is stable under the reaction conditions, but the corresponding benzylic mixed peroxides fragment readily into ketones.
When saturated hydrocarbons are oxidized with Fe^{III}-H_2O_2 or Fe^{III}-PA-H_2O_2 in the presence of chloride or azide ions, the corresponding chlorides or azides are not observed even *in vacuo*. However in TBHP oxidations, the use of $FeCl_3$ and, in particular, the addition of chloride anion gave chloride formation in up to quantitative yield with respect to TBHP. Azide, thiocyanate and other anions also reacted smoothly.

We have interpreted these anionic coupling reactions as ligand coupling on an Fe^V species to which a carbon and a (say) chloride are bonded. [11], [12] This would furnish the observed chloride and Fe^{III}.

Minisci and Fontana [14] have recently offerec an alternative explanation in which an Fe^{II} species (of undefined origin) reacts with TBHP in the Fenton manner to generate a *t.*-butoxide radical, which then reacts with the hydrocarbon to make a carbon radical, which then reacts with the anion attached to Fe^{III} and so makes the derivative and reforms Fe^{II}. Certain of the recorded facts do not agree with this interpretation [15] and further work is needed. However whichever interpretation of the facts is correct, we do have a very efficient method for turning saturated hydrocarbons into their monochloro derivatives.

The formation of dibromides on bromination of cyclohexyl dibromide is a valuable indicator of mechanism (see above). The bromination using bromide ion and TBHP gave, as sole product, the *trans*-1,2-dibromide. Hence, this is formed by a radical mechanism. However, bromination with $BrCCl_3$ gave a different pattern with little formation of the *trans*-1,2-dibromide.

Prof. D.T. Sawyer and his colleagues have made important contributions to Gif chemistry. In addition to optimization studies, they have confirmed that the Fe^{III}-H_2O_2 system produces ketones selectively. [16], [17], [18], [19]

Prof. U. Schuchardt and his colleagues [20], [21] have made a thorough study of the conversion of cyclohexane to cyclohexanone under various Gif systems. The reaction can be optimized to an excellent level of efficiency. We also note the related work by Prof. Patin. [22], [23]

4. References

1. Barton, D.H.R. and Doller, D. (1992) The Selective Functionalization of Saturated Hydrocarbons: Gif Chemistry, *Acc. Chem. Res.* **25**, 504-512.
2. Balavoine, G., Barton, D.H.R., Boivin, J., Gref, A., Le Coupanec, P., Ozbalik, N., Pestana, J.A.X., and Riviére, H. (1988) Functionalisation of Saturated Hydrocarbons. Part X.1. A Comparative Study of Chemical and Electrochemical Processes (Gif and Gif-Orsay Systems) in Pyridine, in Acetone and in Pyridine-Co-Solvent Mixtures, *Tetrahedron* **44**, 1091-1106.
3. Barton, D.H.R., Bévière, S.D., Chavasiri, W., Csuhai, E., Doller, D., and Liu, W.-G. (1992) The Functionalization of Saturated Hydrocarbons. Part 20. Alkyl Hydroperoxides: Reaction Intermediates in the Oxidation of Saturated Hydrocarbons by Gif-Type Reactions and Mechanistic Studies on Their Formation, *J. Am. Chem. Soc.* **114**, 2147-2156.
4. Barton, D.H.R., Csuhai, E., and Doller, D. (1992) The Functionalization of Saturated Hydrocarbons. Part 23. Gif-type Bromination and Chlorination of Saturated Hydrocarbons: A Non-radical Reaction, *Tetrahedron* **48**, 9195-9206.
5. Barton, D.H.R., Gastiger, M.J., and Motherwell, W.B. (1983) A New Procedure for the Oxidation of Saturated Hydrocarbons, *J. Chem. Soc., Chem. Comm.*, 41-43.
6. Barton, D.H.R., Halley, F., Ozbalik, N., Schmitt, M., Young, E., and Balavoine, G. (1989) Functionalization of Saturated Hydrocarbons. 14. Further Studies on the Mechanism of Gif-Type Systems, *J. Am. Chem. Soc.* **111**, 7144-7149.
7. Barton, D.H.R., Bévière, S.D., and Doller, D. (1991) An Unprecedented Chemical Transformation: the Oxidation of Alkanes to Alkyl Dimethyl Phosphates, *Tetrahedron Lett.* **32**, 4671-4674.
8. Barton, D.H.R., Bévière, S.D., Chavasiri, W., Csuhai, E., and Doller, D. (1992) The Functionalization of Saturated Hydrocarbons. Part XXI. The Fe(III)-Catalyzed and the Cu(II)-Catalyzed Oxidation of Saturated Hydrocarbons by Hydrogen Peroxide: A Comparative Study, *Tetra'edron* **48**, 2895-2910.
9. Barton, D.H.R., Bévière, S.D., Chavasiri, W., Doller, D., and Hu, B. (1992) The Fe(III)-Catalyzed Functionalization of Saturated Hydrocarbons by tert-Butyl Hydroperoxide: Mechanistic Studies on the Role of Dioxygen, *Tetrahedron Lett.* **33**, 5473-5476.
10. Barton, D.H.R. and Chavasiri, W. (1994) The Functionalization of Saturated Hydrocarbons. Part 24. The Use of *tert*-Butyl Hydroperoxide: $GoAgg^{IV}$ and $GoAgg^{V}$, *Tetrahedron* **50**, 19-30.
11. Barton, D.H.R., Bévière, S.D., and Chavasiri, W. (1994) The Functionalization of Saturated Hydrocarbons. Part 25. Ionic Substitution Reactions in $GoAgg^{IV}$ Chemistry: The Formation of Carbon-Halogen Bonds, *Tetrahedron* **50**, 31-46.
12. Barton, D.H.R. and Chavasiri, W. (1994) The Functionalization of Saturated Hydrocarbons. Part 26. Ionic Substitutiton Reactions in $GoAgg^{IV}$ Chemistry: The Construction of C-N, C-S and C-C Bonds, *Tetrahedron* **50**, 47-60.
13. Barton, D.H.R. and Wang, T.-L. (1994) The Selective Functionalization of Saturated Hydrocarbons. Part 28. The Acitivation of Benzylic Methylene Groups Under $GoAgg^{IV}$ and $GoAgg^{V}$ Conditions, *Tetrahedron* **50**, 1011-1032.
14. Minisci, F. and Fontana, F. (1994) Mechanism of the Gif-Barton Type Alkane Functionalization by Halide and Pseudohalide Ions, *Tetrahedron Lett.* **35**, 1427-1430.
15. Barton, D.H.R. and Hill, D.R. (1994) Comments on An Article by Francesco Minisci and Francesca Fontana, *Tetrahedron Lett.* **35**, 1431-1434.
16. Sheu, C., Sobkowiak, A., Zhang, L., Ozbalik, N., Barton, D.H.R., and Sawyer, D.T. (1989) Iron-Hydroperoxide Induced Phenylselenization of Hydrocarbons (Fenton Chemistry), *J. Am. Chem. Soc.* **111**, 8030-8032.
17. Sheu, C., Richert, S.A., Cofré, P., Ross, B., Jr., Sobkowiak, A., Sawyer, D.T., and Kanofsky, J.R. (1990) Iron-Induced Activation of Hydrogen Peroxide for the Direct Ketonization of Methylenic Carbon [$c\text{-}C_6H_{12} \rightarrow c\text{-}C_6H_{10}(O)$] and the Dioxygenation of Acetylenes and Arylolefins, *J. Am. Chem. Soc.* **112**, 1936-1942.
18. Sheu, C. and Sawyer, D.T. (1990) Activation of Dioxygen by Bis[(2,6-carboxyl, carboxylato)pyridine]iron (II) for the Bromination (via $BrCCl_3$) and the Monooxygenation (via PhNHNHPh) of Saturated Hydrocarbons: Reaction Mimic for the *Methane Monooxygenase* Proteins, *J. Am. Chem. Soc.* **112**, 8212-8214.
19. Tung, H.-C., Kang, C., and Sawyer, D.T. (1992) Nature of the Reactive Intermediates from the Iron-Induced Activation of Hydrogen Peroxide: Agents for the Ketonization of Methylenic Carbons, the Monooxygenation of Hydrocarbons, and the Dioxygenation of Arylolefins, *J. Am. Chem. Soc.* **114**, 3445-3455.
20. Schuchardt, U., Carvalho, W.A., Spinacé, E.V. (1993) Why is it Interesting to Study Cyclohexane Oxidation?, *Synlett*, 713-718.

21. Schuchardt, U., Krähembühl, C.E.Z., and Carvalho, W.A. (1991) High Efficiency of the GoAgg^{II} system in the Oxidation of Cyclohexane with Hydrogen Peroxide Under an Inert Atmosphere, *New. J. Chem.* **15**, 955-958.
22. Briffaud, T., Larpent, C. and Patin, H. (1990) Catalytic Alkane Activations in Reverse Microemulsions Containing Iron Salts and Hydrogen Peroxide, *J. Chem. Soc., Chem. Commun.*, 1193-1194.
23.. Larpent, C. and Patin, H. (1992) Oxidation of alkanes with hydrogen peroxide catalyzed by iron salts or iron oxide colloids in reverse microemulsions, *J. Mol. Catal.* **72**, 315-329.

Organic Synthesis and the Life Sciences

Duilio Arigoni
Laboratorium fur Organische Chemie
ETH Zentrum - Universitatstrasse 16
CH-8092 Zurich
Switzerland

1. Preamble

The endeavour of the second panel, composed of Hans Bestmann, Albert Eschenmoser, Clayton Heathcock, Claude Hélène, and Don Hilvert was to analyze critically, with the help of the workshop participants, the relationship between organic synthesis and the life sciences. In the guidelines which were distributed, I refrained from including the topics of self-assembly and supramolecular chemistry since these comprised the subject of the third panel discussion. To proceed, I gave a brief introduction, asked for comments from the members of the panel and then invited discussion of the points indicated in the guidelines which, incidentally, as admitted by our co-organizers, had been prepared in full recognition that they may also find use as paper airplanes down the cliffs of Ravello. What follows is an attempt to provide an overview of the major points of the lively discussion which occurred, typical of all the panel discussions. The comments are taken from tape recordings but have been extensively edited and condensed for the purpose of a coherent presentation.

2. Introductory Remarks

During the first panel discussion, Ron Breslow pointed out that organic chemistry, and specifically total synthesis, is rooted only in part in natural sciences. This is certainly true today: most of the compounds that are found in *Chemical Abstracts* are in fact man-made that have little to do with Nature. However, this was not so in the beginning of our science as certified by the very name of the discipline that keeps us busy: it was so defined because it dealt with compounds isolated from organs - that is, from living matter, from cells - be it vertebrates, higher plants or microorganisms.

About thirty years later, when structural theory developed, the need arose to understand chemical reactivity and, to a large extent, most of the practitioners of organic chemistry moved

C. Chatgilialoglu and V. Snieckus (eds.), Chemical Synthesis, 601-619.

away from the original interest and became involved in an aesthetic kind of organic chemistry. This field eventually served as the basis for tackling the problems of biological importance today. In the 1950s, the elucidation of the structure of lanosterol and the suggestion by Woodward and Bloch, and the subsequent proof by Bloch and his co-workers, that lanosterol is a triterpene on the way to cholestrol aroused the interest of chemists who, to that point, had refused to pay any attention to biological processes. There was a chemical message here: it was more appealing to look at lanosterol and cholesterol than to look at the intermediates of the citric acid cycle. Hence, a new branch of organic chemistry was defined as bio-organic chemistry. Parenthetically, this name is somewhat redundant because bio says that it is organic and organic means that it's bio, so it's organic organic or bio bio.

Now, what role has total synthesis played in the development of organic chemistry? Although this has already received discussion at our workshop, may I remind you that, in the beginning, synthesis was attempted mainly to certify that analytical deductions were correct. Synthesis of a compound was the final proof of the correctness of the analytical deductions and Victor Snieckus has provided an early statement by Perkin in illustration of this point (**Scheme 1**).

Turn of the Century Organic Synthesis

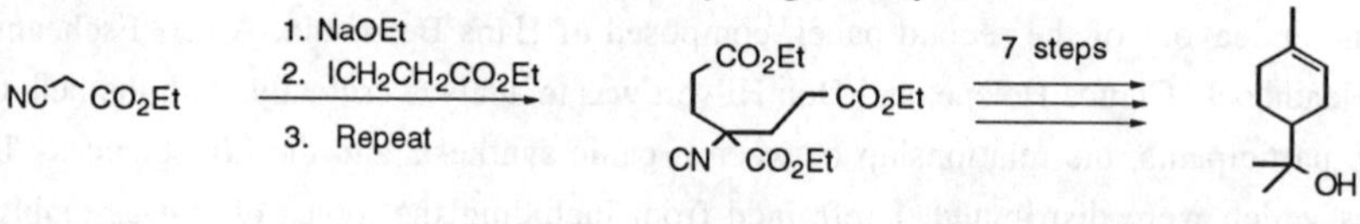

NC⌒CO_2Et —(1. NaOEt; 2. $ICH_2CH_2CO_2Et$; 3. Repeat)→ [CN, CO_2Et, CO_2Et, CO_2Et intermediate] —(7 steps)→ α-terpinol

Perkin. Jr., W. H. *J. Chem. Soc.* **1904**, 654

"This investigation was undertaken with the object of sythesising... terpinol...., not only on account of the interest which always attaches to synthesis of this kind, but also in the hope that a method of synthesis might be devised of such a simple kind that there would be no longer be room for doubt as to the constitution of this important substance."

Perkin, 1904

Scheme 1

Following this remarkable synthesis and statement of Perkin over ninety years ago, chemists proclaimed, whenever they reached a natural compound by total synthesis, that this constituted unequivocal proof for the correctness of the structure under investigation. Personally, I have mixed feelings. I recall the comment of Sir Robert Robinson when he heard that Bob Woodward had completed the synthesis of cholesterol before him. Sir Robert, clearly very frustrated, took the copy of the journal down to Kappa Cornforth and, to paraphrase, said: "allegedly Woodward has now synthesized cholesterol, but I wonder what the value of this synthesis is. It is certainly not a synthesis of industrial interest and it fails to prove what the structure of cholesterol is." One wonders how the very fact that one manages to end up with a few milligrams or even a few micrograms of the final target does indeed prove the structure of that compound. Quite frequently, it happens to be identical to the compound and it is the quest to establish this identity that settles the course of reactions which has been used to reach the target.

Clayton Heathcock raised the issue that total synthesis continues to have analytical value today. I would say definitely so, within a specific topic of compounds. Clayton himself has provided clear-cut evidence for the stereochemistry of the bis-phytyl moiety which represents a important component of the core lipids of archaebacteria. In this case, in view of the complexity of the material, one may raise the question about the unequivocal nature of the proof but I would like to have the answer from the horse's mouth, so to speak, later on.

Synthesis is in an excellent state of health and constitutes a marvellous tool, providing us with substances with which we can test mechanistic aspects of biologically important processes. In the first part of the panel discussion I would like to tackle this role of organic synthesis as a tool. We need not discuss the synthesis of labelled substrates since these only require the modification or extension of existing syntheses. Nevertheless, I remind you that, for the study of biosynthetic pathways, total synthesis is often required to prepare putatative intermediates which display a specific label, a heavy isotope, at a give position. Discussions on previous days has made us all aware of the importance of total synthesis for the obtention of mimics of cofactors or enzymes.

To begin, I would like to challenge my colleagues to make a general, and hopefully an enlightening, statement on the topics to be discussed and then proceed with participation of the workshop audience.

3. Biomimetic Synthesis

Claude Hélène (INSERM, France): In the field of total synthesis of complex natural compounds, one should not forget that, although ultimately one needs to correlate a synthetic product with the natural material, the exercise produces a large number of intermediates which may be useful for other purposes. For example, the recently reported total synthesis of taxol will obviously not be used by pharmaceutical industry for its commercial production; on the other hand, the synthesis has produced many intermediates which may have pharmaceutical, and possibly other, applications such as starting materials for new molecules.

Albert Eschenmoser (ETH, Switzerland): Synthesis and a specific type of comprehension of molecular reactivty are what organic chemists will continue to contribute to the life sciences for a long time to come. We need not debate whether synthesis is important because there cannot be any doubt about that. There is, however, an increasing number of non-chemists (biochemists, biologists) who, more and more, do surprisingly well in doing synthesis themselves. One of our tasks is to develop the "instrument" synthesis and to lead in exploring and utilizing its potential; therefore, we must also continue to do synthesis for the sake of itself.

Duilio Arigoni: I agree with you, **Albert** but I would like to stress that chemists should increasingly ask, in connection with the biological processes, questions that are normally not asked by our more biochemically-oriented colleagues. Some of these colleagues may have the feeling that we are invading their areas of interest. To this I reply by saying, no, we are just complementing what you are doing because we have weapons in our hands which enable us to ask questions which have never been asked by those educated as M.D.s or biochemists. So we must bring back full power to organic chemistry and apply it to a good cause.

Istvan Ujvary (Hungarian Academy of Sciences, Hungary): To pick up on **Professor Eschenmoser's** comment, in my view, the center of scientific, social, economic, and environmental problems is related to material and life sciences. If you consider the material and life sciences as an ellipsoid, then the two focal points of these can be energy and information. Whenever we are exposed to a problem, we require energy and information. To solve the problems posed by society and the environment, especially in the 20th Century, we have to go back to study the structure and organization of living and synthetic matter. Thus we are working with synthesis, materials, catalytic antibodies, and polymers. Whether it be structure-biological activity relationships, or other topics, energy and information are two focal points by which chemistry is being analyzed. Energy and information is very central to chemistry.

Duilio Arigoni: The points that you raising are intrinsically political issues as well. I don't want to minimize the importance of these questions but I personally do not feel competent to take a stance on these issues.

Hans Bestmann (Universität Erlangen-Nürnberg, Germany): From the earlier discussion, it is clear that total synthesis will always be necessary. However, the question arises, as already mentioned by **Professor Eschenmoser** in the first panel discussion, can we leave the pursuit of biological activity aspects soley in the hands of the biologists. The variation of chemical structure to determine biological activity correlations is a very important point for synthetic work. In order to achieve this, of course, you need to develop the knowledge for good bioassays. Synthetic chemists should increasingly enter into areas of biology and the question which needs to be addressed is how we train students to take these directions.

Duilio Arigoni: Now I would like to throw the topic to the audience for comment.

Jim Orr (Memorial University, Canada): I am an organic chemist and have been working in medical schools almost all of my life. I would encourage chemists who want to find problems to sit in on research meetings in their medical schools. These are mostly organized by medical faculty and often the questions that come up have much to do with chemistry and nobody there knows how to answer them properly. I have often found that my knowledge of chemistry has been of help; for example, chemists have clear and penetrating insights into biochemical problems which should be shared. Some of the finer points of biochemistry have been answered by **Duilio Arigoni** in his work on the stereochemistry of enzymatic reactions.

Duilio Arigoni: Thank you, Jim. The long-standing problem is that we may be able to provide answers but we often don't know the questions. On the other side of the barricade, people know the questions but are not able to provide the answers. It is only by fusing the two that we shall make progress.

Ulrich Jordis (Technische Universität Wien, Austria): I would like to address the problem of management in organic synthesis research. Organic synthesis today requires the coordination of a tremendous number of specialists. One needs to be a specialist in pure organic synthesis and have an understanding of mechanisms. However, what is, as I see it, completely impossible is to be at the same time a specialist in NMR, in separation problems, in modelling, in pharmacology, and in medicine. So what I see emerging is a need for a new type of manager of organic synthesis - someone who has the insight into the problems at hand and then directs and coordinates all the specialists which are available.

Duilio Arigoni: You have addressed a very important problem but I am not convinced that we are in need managers. We are in need of people who are able to pick important topics but without resorting to the managerial level. If there is a manager, I fear that there is a real danger that the chemist will degenerate into a technician. The manager will tell you "I need a molecule like this" without telling you why, what he expects, how one would attempt to relate structure to function, and so forth. Perhaps to clarify, are you thinking of management of organic synthesis in industry?

Ulrich Jordis: No, I am thinking of a different type of management. Many of you here are very good examples of managers and coordinators of your groups. Frequently, we are unsatisfied with what we see in the literatuare and we wonder if that is due to poor management in the sense that there has not been positive leadeship, no "Dirigenten," to choose the right problems to pursue. I am aware that this problem also is connected to excellence in teaching.

Duilio Arigoni: You bring back to my memory a discussion that I had with Frank Westheimer on a similar issue. Frank pointed to me that access to scientific libraries is free to everybody. I think it's up to us to take initiative and not to wait for others to teach us what to do. I realize it's a demanding task.

Hans Bestmann: I think you don't need to be a manager, you have to become a chemist with broader knowledge in the natural sciences. In discussions with biologists, I learn that they think that we are the specialists. A big biologist in Germany told me that, for him, chemistry is a service science, nothing else. These times should be over. Chemists should become more and more involved in problems of biology. We do not need managers; we need more scientists who look outside the border of chemistry.

Jean-Marie Lehn (Université Louis Pasteur, France): I would like to comment on **Professor Bestmann's** remark. One may have the impression that synthetic chemists are on the defensive. I think this is a wrong impression and that there is no need for them to be on the defensive. Furthermore, there is nothing bad about being a tool and a service science. Without tools, we cannot do much; after all, mankind began by making tools. To consider synthesis as a tool and a service to solve a problem is not an insult to synthesis. Synthesis is the most basic characteristic of chemistry - it is for making new materials which may be tools for solving particular problems and, as will be discussed in the third panel, for making new compounds which do not exist in Nature. There is a further element which can be illustrated by the following example. When a compound is discovered which regulates a protein by a site which Nature does not use (and this is not an uncommon occurrence), then you have a chemical with a new biological action and this compound is more than a tool because it acts in a way which Nature has not used before.

A final comment on the relationship between synthesis and the life sciences which I hope will not cause the biologists and molecular biologists to jump up: operation is not explanation. Often in biology, one operates, one does things, one doesn't explain them; and one can say that when we begin to understand a biological process, it has become a chemical process. After you have done the "Gedanken" experiments, you have to test and synthesis is for that. I was told that the Chinese character for chemistry is "the study of change." And what is the study of change? It is making new things, it is synthesizing new compounds. So I think one shouldn't be on the defensive. Synthesis is the trunk of chemistry and, at the same time, it spreads over and as a tool, will help to solve problems.

Duilio Arigoni: Thank you, **Jean-Marie**. Your point on operation vs explanation reminds me of a discussion I had with Derek Barton more than twenty-five years. I asked him how he would define the difference between biochemistry and organic chemistry. Without hesitating for a fraction of a second, he replied "it is very simple; if I understand it, then it is organic chemistry."

Albert Eschenmoser: Jean-Marie, I fully agree that we should be satisfied that the significance of synthesis as a tool which helps to solve problems. However, an important question coming up at the interface between chemistry and biology is: who is going to ask the questions which define the problem? One scenario is that this is not done by the chemist. The chemist is a service man. What is required, and I think that is what **Dr. Jordis** meant, are chemists who are able to ask questions which are not narrowly located in synthesis, but who are life scientists having command of a powerful specific tool, namely chemical synthesis. This is what **Professor Bestmann** meant when he said what is required in synthesis are people who see beyond synthesis. Chemists should do the biology that biologists cannot do.

Duilio Arigoni: We have learned to live with UV, with infrared, with MS, with NMR; it's hard for me to see why we should fail to cope with biology.

Jean-Marie Lehn: I would like to add the following remark. Asking a question is often much easier than answering and solving the problem. Now, of course, you have to look further than that. You have to see why you do it; there is a why and not just a how. Nevertheless, the ability to solve and to provide the tool is of great significance. Of course, we are scientists first, chemists second, and organic chemists third. We would like to do everything but we just can't - there is no time for it, we don't have the tools for it. Therefore, we must get together and manage a very complicated system where everybody works together. However, we can at least try to have our own tools strong in hand so that when the problem comes up, we can provide an answer. The problems of biology are definitely asked by biologists at some stage. The biologists are not so much interested in our approaches right now because they are somewhere else. They are working on integration of basic processes, such as physiology. We can talk about physiology but we approach it from a different point of view. In some ways we are acting, while the physiologists are describing. There is a big difference - describing a problem and acting on the problem. The chemist is the actor, the one that acts and not the one who describes.

Claude Hélène: I agree and would like to make a follow up comment. I believe that, whatever you do as an organic chemist, you have to be able to talk and understand the people with whom you are willing to collaborate. So whether you are interested in materials or in living organisms, you have to learn a minimum number of things of the field in which you want to go, if you want to have a positive role - not just taking but also giving. As **Duilio** said, learning enzymology is not complicated; it is probably simpler than learning NMR. It does not mean that you should be a specialist in gene regulation. Learning the main processes of gene regulation can be easily done since, as **Jean-Marie** said, biology is still a very descriptive science and so it is easy to follow the reasoning and the mathematics of biological processes. What has occurred frequently is that organic chemists wanted to solve biological problems without knowing enough biology; they were thinking that they had the right answer to the right problem. In fact, it was wrong because they did not know the right question. So I think that by making an effort to a acquire minimum knowledge of biology, organic chemists can exchange information with biologists and contribute to the life sciences.

Jim Snyder (IRBM, Italy: Most of the comments to now have related to the academic environment. Yet there is a larger world of industrial science which, in many cases, focuses directly on the merger of chemistry and biology to produce solutions to very practical problems. In the pharmaceutical industry, there is a very real reason why chemists should feel defensive and that is that, for over the last 20 years or so, the control of the research environments in the pharmaceutical companies has passed from the hands of the synthetic chemists to biologists, pharmacologists, and doctors. This is because science has moved a few steps down the road. We now have a much better understanding of how biology and biochemistry seem to work in complicated living organisms. As a result, the approaches to solving these problems have been directed by those who understand those mechanisms. Very often, the synthetic chemist provides an extraordinary valuable tool. However, in my opinion, the defensive posture is reasonable not because he or she is not as equipped as a biologist, as an expert in their own field. It is, as has been expressed by several panelists and members of the audience, because the synthetic chemist has yet to learn the language of biology. I would pose the following challenge to the panelists and those who are teachers of the next generation of scientists: how do you generate, how do you produce an expert in synthesis who is critical for the industry (in fact, we wouldn't hire somebody who is not able to put virtually any molecue together) but, at the same time, create in them a sense of curiosity not only about their own field but all the other fields, the language of which they are going to have to learn. This is the major challenge which may allow us to take back the control of direction in the pharmaceutical industry.

Clayton Heathcock (University of California, Berkeley): I am glad that **Duilio**, in his introductory remarks, raised the question whether synthesis can provide structure information because it's widely believed that it doesn't any more due to the availability of X-ray crystallography. I have a little example of a tool that was invented for no other purpose other than for my own amusement. Then it was found that this tool could have some value in solving a structure which was of some significance to biology.

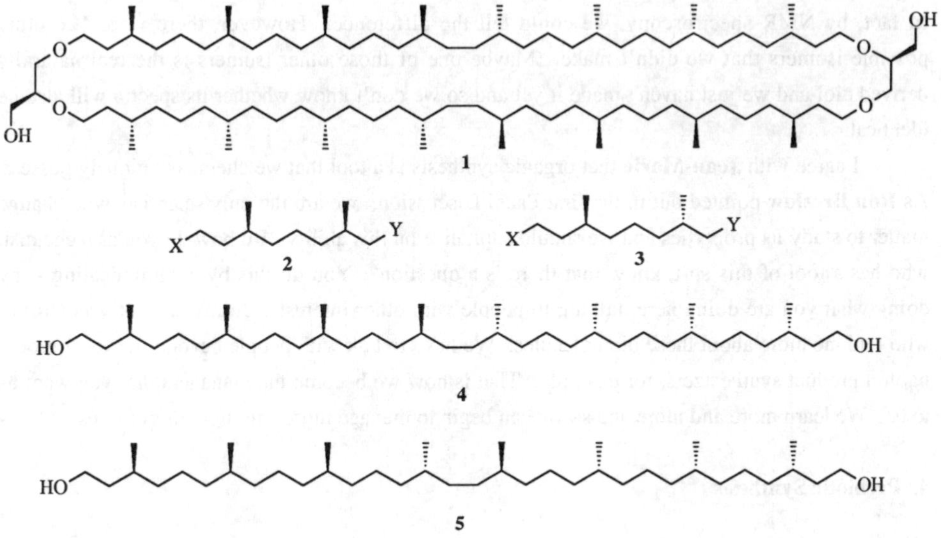

1

2

3

4

5

Scheme 2

Compound **1** (Scheme 2) is an interesting hydrolysis product of a natural material. It is a 74 membered-ring, a tetra-ether, in which there are a lot of methyl branches because it is derived from an isoprenoid structure. This natural product is used by archaebacteria as a sort of a lipid bilayer and has been known for some time. A few years ago, we were doing some work on controlling the stereochemistry of the aldol reaction and had developed a tool to prepare any one of the possible stereoisomers that are represented by structures **2** and **3** in which you have two stereocentres on five carbon repeats. I then looked around to see if there were some interesting questions in biology for which this tool could be applied and found compound **1**. Furthermore, I found that it had been hydrolyzed to the C-40 archaebacterial diol **4** which has 8 stereocentres that come in repeats of five carbons. By the very nature of these substances, both compounds **1** and **4** are waxes, like typical hydrocarbons, and far from crystalline substances! So, although it was easy to determine the constitution of these compounds by mass spectroscopy, nothing was known of the stereochemistry except what schrewd guesses one could make. So we set out to do rather simple synthesis, putting these units together in various combinations, to make some of these isomers. We only made two, compounds **4** and **5** and found that, by our analytical methods, **4** was identical with the substance from the natural product.

Now you could say that here synthesis solved the problem that couldn't be solved by structural methods such as crystallography. On the other hand, the reason we made the second isomer **5** was to be sure that we could tell the difference between two similar isomers of this type. In fact, by NMR spectroscopy, we could tell the difference. However, there were 126 other possible isomers that we didn't make. Maybe one of those other isomers is the real naturally derived diol and we just haven't made it yet and so we don't know whether its spectra will also be identical.

I agree with **Jean-Marie** that organic synthesis is a tool that we chemists uniquely possess. As **Ron Breslow** pointed out in the first Panel Discussion, we are the only scientists who change matter to study its properties and we should capitalize on this ability. Bu how do you as a chemist, who has a tool of this sort, know that there is a question? You do this by communicating - by doing what you are doing here, talking to people with other interests. In my case, it was **Duilio**, who told me more about these biomolecules. We have to talk with people beyond our local group, natural product synthesizers, for example. That is how we become the managers that you want us to be. We learn more and more and so we can begin to manage more and more of our lives.

4. Prebiotic Synthesis

Duilio Arigoni: Speaking of prebiotic synthesis is as blasphemic as speaking of chiral synthesis. There is no such thing as chiral synthesis, only synthesis of chiral compounds. And so there is no such thing as prebiotic synthesis, but there is, I think, synthesis in connection with prebiotic chemistry. Before turning over the matter to **Albert** for comment, let me point out what this includes. There is a real possibility that a number of biologically important compounds were assembled somehow long before life started in ways which have little, if anything, to share with their actual modes of biosynthesis. There are hints in the biological world that this may have been so. If you look at some strange biological routes, you can easily convince yourself that chances are very high that alternative modes of biosynthesis must have been in operation at some time. For instance, in the known mode of biosynthesis of thyamine pyrophosphate, you need a very odd looking C-5 sugar which is assembled in a one-step reaction making use of thyamine pyrophosphate as a cofactor. In other words, thyamine pyrophosphate is catalyzing one of the reactions of its biosynthesis! In a more dramatic example, the biosynthesis of Vitamin B_{12}, no less than eight enzymic methylations are involved which use adenosine methionine, the biosynthesis of which is dependent on the presence of Vitamine B_{12}. With that paradox in mind, I will let **Albert** take over.

Albert Eschenmoser: Most of the discussion about the science of chemistry has to do with the fact that chemistry is so extremely useful for so many practical and technical purposes. So-called prebiotic synthesis has nothing to do with usefulness. It has to do with the task of chemistry to describe the structure and functioning of the living world. It is, in essence, a chapter of what in the future is going to be the chemistry of self-organization, the chapter that refers to the questions of the origin of life. In a way, it is the continuation of the synthesis tradition of natural products chemistry. Asking how Nature biosynthesizes biomolecules eventually leads to asking how did biomolecules originate. When we refer to self-organization today, we mostly mean the formation of structures that are based on non-covalent bonds. In prebiotic chemistry, we are dealing primarily with problems of constitutional self-assembly. In his lecture on combinatorial synthesis, **Paul Bartlett** mentioned the term 'starting material-based synthesis.' Constitutional self-organization of organic matter leading to life was such an example. Let me formulate what the challenge for synthetic bioorganic chemistry in this field of endeavour may have to be. The challenge is to demonstrate experimentally the constitutional self-assembly of a family of autocatalysts which are able to, first: store, retrieve, and mutate combinatorial structural information embodied in constitutional isomerism and second: relate such structural information to auto-catalytic function in such a way that the system can evolve along gradients of increasing efficiency, diversity and control of catalytic function towards partial self-sustainment. You may recognize that what I have said is an attempt to translate what we can extrapolate from biology into terms which are, in principle, amenable to experimental chemical research. The opportunity to meet new chemistry in such an endeavour is enormous.

Claude Hélène: An additional point. When you want to study self-organization, you have to know what are the building blocks you should start with. It is not yet clear what are the molecules that you have to assemble in the prebiotic field. We know now from living organisms what nucleic acids or proteins are made of but whether they started with the monomers, as we know them today, is not clear. Everybody knows the experiment that Miller did 30 years ago in what was considered to be, at that time, the best model of the earth's atmosphere four billion years ago. Then astrophysicists discovered that this was completely wrong even though Miller showed the formation of amino acids and nucleic acid bases. I think we have to be a little careful about the precursors we are talking about.

Albert Eschenmoser: Not to know what we have to look for is not a disadvantage, quite the contrary. It would be good if people would had different opinions about what they should look for and would follow experimentally their own ideas and just report to the scientific community what they observe. It is important to have different ideas about what might be the best candidate of that "family of structures."

Claude Hélène: This is my point exactly: chemists should not limit themselves to the monomers we know today.

Duilio Arigoni: I would like to comment on the time factor alluded to by **Albert**. This kind of research can never reach what we would consider a formal proof. At best, it offers suggestions on how a process may or may not have been followed and, no doubt, there is not only one possible way in which complex molecules may have been assembled in a prebiotic soup. But the very fact that one succeeds in showing that there were such possibilities proves a specific point. You do not show how it was done but you show that it could have been done in this way. And this is all that matters.

Are there other comments on prebiotic formation of biologically important molecules?

Maria do Céu Costa (INETI, Portugal): I have two questions. First, I wonder about the usefulness of research on the understanding of life events if concerted efforts are not made towards the immediate application of results on important goals related to improving the quality of life. By that I mean it is not right, as a priority research, to spend lots of money trying to understand why DNA is assembled if we also have to put in great effort to maintain it protected from the aggressive environment. I may agree with research in prebiotic formation of biologically important molecules if there is always a major program targeting preservation of biologically fundamental mechanisms.

Secondly, I would like to know what this assembly of priviledged free thinkers and experts thinks about the possibility of having government policies which evaluate and direct research projects for better health - which means better life, which means no drugs, which means no old looking and no death - as the main social objective.

Duilio Arigoni: The first issue that you raise is, of course, a political one. I think that we, as a community, lack the competence to decide on such issues. It is very easy to say that quality should be first. A general consensus among competent people would be one way one way to decide. Research on prebiotic science may not impinge immediately on welfare, pollution, and ecology, but it will have to do so in the long run. Understanding is at the basis of what we are doing; applications will follow.

Albert Eschenmoser: If chemists were successful in documenting experimentally how molecules under well-defined conditions can self-organize to a level where the phenomenon of life sets in, such a feat would have a great impact on the way people look at our world. It would be a giant step in that process of demystification of Nature which had started with Wohler's synthesis of urea and has continued, in our time, with the synthesis of complex natural products. There are many biologists today who say: all this is very simple, life started with RNA, and that's it. It is at this point where we chemists should be very critical. In this context, it is our task to demonstrate experimentally what is possible with organic molecules and what is not. The goal is not to be able to propose *how it has* happened, not even to propose *how it could* have happened, it is to demonstrate experimentally *that it can* happen.

Maria do Céu Costa: I think that we could do it better if we have research plans in this very important area. It would be better if the experts could get together and change the situation in research for improving things for human kind.

Albert Eschenmoser: You are right, absolutely, it should be organized. It is organized to some extent in workshops of the kind we are attending. In their case, it may be more difficult to allign ideas because it is a more open subject.

Maria do Céu Costa: My second question relates to the teaching of natural products. We always teach main biosynthetic pathways in classical terms, for example considering coenzyme A as the most important unit in primary and secondary metabolism. I wonder if we should be directing our attention more towards teaching self-assembly and prebiotic chemistry.

Albert Eschenmoser: I think these aspects should not be taught at this moment simply because they are too uncertain. One should teach important fundamental facts and central ideas supported by experiments. What you say is often done, by the way. Students hear a lot about Miller experiments and such things. I think this is dangerous. Earlier, **Duilio** referred to probably the most beautiful case of constitutional self-assembly in which the adenine nucleus is formed. This is science which I believe fulfills the criteria of the type of phenomena which can be taught.

5. Enzymes, Ribozymes, Abzymes, Chemzymes

Duilio Arigoni: We move to the next exciting topic - catalysis. Here we will deal with enzymes that can be modified by chemical synthesis; ribozymes - we see what they do and describe what they do, yet fail to understand how they do it, let alone why they do it; abzymes, about which you heard from **Don Hilvert**; and, last, but not least, chemzymes, which is an odd-looking name, first used by E.J. Corey, for man-made catalysts that mimic the behavior of enzymes. Certainly synthetic chemistry can contribute to these topics. I have asked **Don Hilvert** to give an opening volley on this subject.

Don Hilvert (Scripps Research Institute, U.S.A.): Catalysts allows all living organisms to survive, to multiple, and to reproduce. Through the study of enzymes, ribozymes, abzymes, etc., one can gain insights into fundamental reactivities and to control reactivities and selectivities. The actual application of enzymes in organic synthesis has already seen significant advances. My colleague, Chi-Huey Wong at Scripps, has successfully used aldolases and glucosyl transferases to put together very complex carbohydrates, molecules that are very difficult to synthesize by traditional chemical means. He has provided strategies for making a wide variety of these so that they can be produced and their biological properties can be analyzed. Potentially, therefore, if one could design artificial catalysts - ribozymes, abzymes or chemozymes - to catalyze reactions for which we don't have biological counterparts, we would have a very significant contribution to our science and to the synthetic arsenal that everyone can exploit.

Albert Eschenmoser: Are these good names: abzymes, chemzymes? What is next?

Duilio Arigoni: The only correct one is enzymes and that is from the Greek, enzymos (in the yeast). These were catalysts first isolated from yeast, therefore correctly named enzymes. It is preposterous to have ribozymes, abzymes but they are here.

Don Hilvert: Abzymes and all the rest of these names provide a convenient shorthand. "Catalytic antibodies" is a more accurate, if more cumbersome, term. I say catalytic antibodies.

Gianfranco Cainelli (Universita Degli Studi di Bologna, Italy): I would like to know if the catalyst developed by Noyori, for instance, is a chemzyme or not.

Duilio Arigoni: I would say yes, if the definition is man-made catalysts. Thus the Pfaltz compounds are chemzymes and Noyori's catalysts are chemzymes.

Gianfranco Cainelli: But they are completely artificial. When you say chemzymes, you think the catayst has two meanings, in a sense.

Duilio Arigoni: But they behave catalytically, they do not have to mimic. Number one, I am not responsible for the names; number two, the "zyme" just stands for a catalyst. It is a catalytic entity.

Albert Eschenmoser: What is then an organic catalyst? If the selectively acting organic catalyst is called a chemzyme, what is the rest?

Duilio Arigoni: I can be persuaded to send a cable at the end of this session to E.J. Corey and tell him that his term has been challenged. Who is in favor of abolishing the term chemzyme?

Albert Eschenmoser: As a Swiss, I would have to interfere. Before you vote, you have to carefully discuss and argue about the point. We have not yet done so.

Clayton Heathcock: I am wondering myself about the point that **Albert** is making. If a proton catalyzes an enolization reaction, is that a chemzyme? Is palladium on something made-made, when used as a catalyst, a chemzyme? I think it is a meaningless word. It is a word you invent when you want to make something ordinary seem special.

John Esker (Abbott Laboratories, U.S.A.): I think we all have to be a little careful about how we describe chemical concepts because one aspect that separates biologists, organic chemists, inorganic chemists and other types of scientists, is the language. If you describe a phenomenon that is unique and it requires a unique word, then the language used should be descriptive as well as unambiguous. If you invent a new word to describe a well known phenomenon, you often obscure much of the science that has gone before rather than illuminate it. That is the origin of my own criticism for "chemzyme".

On another point that was brought up earlier: when can biological catalysts be developed and used effecively and when are somewhat less specific man-made catalysts more appropriate? We want a novel solution for every problem rather than redesigning the strategy. We have a tendency to want magic bullets to solve synthetic problems based on a key reaction. In fact, Nature doesn't do this and we shouldn't either. Although Nature has had more time to develop its chemistry, we should also utilize more general and energy efficient approaches. There are situations under which Nature's catalysts are the most efficient and selective but there are many cases, not only in industry, where their properties prevent their use. So we have to appreciate that there are different situations which will require the use of these complementary technologies.

Duilio Arigoni: Absolutely. Let me tell you how I lost a good bottle of wine as a bet to my friend **Albert Eschenmoser.** I told him that I would be willing to bet that **Steve Hanessian** would refuse to use a lipid esterase for carrying out a step in the synthesis for which he could not come up with any other solution. Steve was sitting by and without blinking an eye, said "I wouldn't hesitate for a fraction of second." So it is a matter of efficiency. If you really want to reach your target efficiently and you have an enzyme that performs better than your reagent, use the enzyme.

Hans Bestmann: I agree. We should look to enzymes that might do it better than we can in the synthetic world. We should never be too proud and say, as human being, we are better than Nature. Nature developed for many thousands of years and has achieved optimum qualities. We should be humble towards Nature.

Carl Johnson (Wayne State University, U.S.A.): In view of **Professor Bestmann's** comments, mine may be a little redundant. A good portion of our research is involved with the synthesis of compounds of modest complexity, that is, sugars, disaccharides, trisaccharides, or mimics of these compounds. Our syntheses begin with very simple starting materials like cyclopentadiene, benzene, cycloheptatriene - very modest compounds with no stereochemistry. We have chosen to use enzymes at appropriate points usually to asymmetrize these molecules. The molecules that we end up with often have side-chains that destroy the symmetry of the molecule and then we turn to another very valuable type of catalysis - transition metal catalysis - to modify the enzymatic products. This is the sort of thing that synthetic chemists should think about: use the best tools available to you. The best tools that we have available to do our asymmetrization are clearly enzymes. You could introduce asymmetry by doing a resolution but that is not nearly as effective as choosing to borrow, from Nature, a modest catalyst. But enzymes may not be the best tools in other instances and then you turn to catalysts designed by chemists to do the job. The greatest efficiency in synthesis is achieved when you survey all the tools that are available and apply those that are most appropriate for the task at hand.

Duilio Arigoni: There is consensus that we should not hesitate to use enzymes whenever they prove more efficient than man-made reagents. But there is a second part to this story - we could try to improve on existing enzymes to develop new tools, for example, polymeric materials, that would enable us to perform at a level which is not possible today. So here, too, there is a great challenge.

Carl Johnson: That is certainly an important area. Aside from attempts to mimic enzymes, protein engineering is available to improve enzymes. It may surprise some in the audience to know that their shirts have been washed with detergents that contain enzymes. Many of these are not Nature's enzymes but are enzymes that have been engineered and produced in huge commercial scales. An amino acid here and there has been changed to improve the properties or the stability of these enzymes. Genetic engineering is not only academic, it is in the market place.

Josef Michl (University of Colorado, U.S.A.): Along similar lines, it has been said several times that Nature's way is the best way in synthesis. I wonder whether that is really true under all circumstances because, in spite of the extensive time period that has been available to Nature, the starting materials have been very limited, the reaction conditions have been very restricted, and all the syntheses have been done in very complicated mixtures. When we do synthesis, we isolate what we want and work with that. In spite of the variety of reagents and the variation of conditions, we have been very restricted. So even today, we may not be able to beat Nature in many syntheses of naturally occurring substances. It seem to me that in the future we ought to be able to so.

Don Hilvert: ... given enough time.

6. Combinatorial Chemistry

Duilio Arigoni: We now tackle a very important and current topic which I have asked **Claude Hélène** to address.

Claude Hélène: Because of my own research interest, I will Illustrate combinatorial chemistry in the field of oligonucleotides. The question was raised several years ago: could we design an oligonucleotide that could bind to any chosen molecule? The answer is yes. This is simple to achieve because you can synthesize random sequences of oligonucleotides; at each position, you incorporate all four nucleotides during the synthesis. You can keep the ends of the sequence fixed (for amplification purposes, see below) and have the rest completely variable (all four nucleotides at each position), thus generating a huge mixture of oligonucleotides (4^N molecules if N positions are degenerate). With a selection procedure you can fish out, from a mixture of these oligonucleotides, those which bind to the molecule you are interested in. Then, if you have a procedure to analyze what has been fished out, you can easily determine what is the appropriate ligand for your substrate. In the field of oligonucleotides, this strategy has been helped by the polymerase chain reaction (PCR), an amplification technique which makes use of primers which are complementary to the fixed sequences at the ends of the oligonucleotides and allows you to copy each of the two DNA strands with DNA polymerase. This can be repeated several times and thus at the end of n cycles you obtain 2^n molecules. Thus, using the constant part of the degenerate oligonucleotides, you can easily amplify the selected ones. If you return to the original procedure, you can start again the process of selection and amplification several times. At the end, by using molecular biology techniques, you sequence the oligonucleotides which have been selected. Obviously, you can use this technology to select ligands for proteins whose biological function involves binding to a nucleic acid. However, you can also use it for any kind of protein and one typical example is thrombin, a protein involved in the coagulation cascade and one which has nothing to do with nucleic acids in physiological processes. When the selection procedure was carried out, a series of thrombin-binding oligonucleotides was selected; a consensus sequence was determined to be $G_2T_2G_2TGTG_2T_2G_2$. When you look at the structure of this consensus sequence by NMR spectroscopy, you find a very interesting structure where four guanines form a quartet structure in one plane. The eight guanines of the 14-mer form two adjacent quartet structures. Therefore, you have a very unique structure for this type of oligonucleotides and the variable part is in the loops which connect the two guanine quartets. This is the structure that binds quite strongly to thrombin with an association constant around 1nM. You have selected a nucleic acid which is a strong inhibitor of an enzyme which in its biological function, has nothing to do with nucleic acids.

Another question raised in this area was the following: can one find an oligonucleotide that will selectively recognize a smaller molecule such as Vitamin B_{12}? Again the answer is yes. The structure of the selected RNA is not known in detail because we don't have an X-ray crystal structure and the structure which has been proposed is therefore based on the accessibility of nucleotides to chemical reagents. Using the technology I described above, you can isolate an oligonucleotide that will recognize Vitamin B_{12} with an association constant of the order of 108 mol^{-1}.

A recent report described the preparation of oligonucleotides that discriminate between caffeine and theophylline, compounds which differ by a single methyl group. The change in the association constant is <u>four</u> orders of magnitude. Thus you can select an oligonucleotide which is specific for theophylline with an association constant around 10^6-10^7 mol^{-1}. A few start-up companies in the U.S. (for example Nextar) have developed this strategy and have selected oligonucleotides that bind it to more than a hundred different proteins. In all cases, they found at least one oligonucleotide with an association constant in the nanomolar range with a very specific sequence.

These cases illustrate the power of combinatorial chemistry in the field of oligonucleotides. This strategy can also be applied to oligopeptides and, as **Paul Bartlett** discussed, to reverse peptides. The synthesis can be carried out on beads and the oligomers selected for binding to a particular substrate, e.g., to proteins. Identification of the best ligand is achieved by several techniques, including the use of tagged beads.

7. Conclusions

The field of combinatorial chemistry has been exploding during the last couple of years. The main motivation arises form the pharmaceutical industry which, now, makes use of combinatorial chemistry to screen a huge number of molecules in conjunction with high throughput screening. The technology is no longer restricted to oligomers of nucleotides, aminoacids...It involves the synthesis, e.g. on solid phase, of a large number of molecules carrying a scaffold on which substituents are attached to cover three-dimensional space with as many different functional groups as possible.

One strategy makes use of arrays of molecules that are synthesized, e.g., on a chip or a multi-well reactor. Then the biological entity for which a ligand is to be selected (e.g., a membrane receptor, an enzyme, a structural protein...) is bound to the chip or the multi-well plate. The sites to which strong binding occurs are identified, and, consequently, as the array has been synthesized in a controlled order, the best ligand is identified.

Another strategy makes use of simultaneous synthesis of mixtures of 5 to 100 molecules in a single reactor, keeping the diversity as large as possible. Then a binding screen is used to select the molecule(s) which have the highest affinity within the mixture (by affinity chromatography, gel retardation, separation with magnetic beads, competition with a labeled ligand...). The identification of the selected compound is done with highly sensitive techniques (using e.g., mass spectrometry technologies) or a deconvolution approach is attempted by synthesizing mixtures of molecules with reduced complexity.

The idea of synthesizing mixtures of molecules and trying to identify the one with the expected substrate-binding property might not look attractive to many organic chemists. However this approach raises many interesting questions in chemistry, including new synthetic methods on solid phase, new chemical reactions with quantitative yield applicable to the simultaneous synthesis of arrays of molecules, ... This new approach, i.e. combinatorial chemistry associated with the development of high throughput screening, has led to a lot of excitement on the pharmaceutical industry and to the development of start-up ("chemtech") companies. The future will tell us whether this strategy brings new drugs in the pipeline of therapeutical products, faster than the more conventional medicinal chemistry approach. Meanwhile our increasing knowledge of the three-dimensional structure and of the dynamics of interesting targets for pharmaceutical drugs will provide the basis for the rational design of therapeutically active molecules.

INDEX